AF903018

Handbook of the American Short Story

Handbooks of English and American Studies

Edited by
Martin Middeke, Gabriele Rippl, Hubert Zapf

Advisory Board
Derek Attridge, Elisabeth Bronfen, Ursula K. Heise,
Verena Lobsien, Laura Marcus†, J. Hillis Miller†,
Martin Puchner, Oliver Scheiding

Volume 15

Handbook of the American Short Story

Edited by
Erik Redling and Oliver Scheiding

DE GRUYTER

ISBN 978-3-11-058523-0
e-ISBN (PDF) 978-3-11-058764-7
e-ISBN (EPUB) 978-3-11-058532-2

Library of Congress Control Number: 2021942870

Bibliographic information published by the Deutsche Nationalbibliothek
The Deutsche Nationalbibliothek lists this publication in the Deutsche Nationalbibliografie; detailed bibliographic data are available on the Internet at http://dnb.dnb.de.

© 2022 Walter de Gruyter GmbH, Berlin/Boston
Typesetting: Integra Software Services Pvt. Ltd.
Printing and binding: CPI books GmbH, Leck

www.degruyter.com

MIX
Papier aus verantwortungsvollen Quellen
FSC® C083411
FSC www.fsc.org

Editors' Preface

This De Gruyter handbook series has been designed to offer students and researchers a compact means of orientation in their study of Anglophone literary texts. Each volume – involving a particular historical or theoretical focus – introduces readers to current concepts and methodologies, as well as academic debates by combining theory with text analysis and contextual anchoring. It is this bridging between abstract survey and concrete analysis which is the central aim and defining feature of this series, bringing together general literary history and concrete interpretation, theory and text. At a time when students of English and American literary studies have to deal with an overwhelming amount of highly specialized research literature, as well as cope with the demands of the new BA and MA programs, such a handbook series is indispensable. Nevertheless, this series is not exclusively targeted to the needs of BA and MA students, but also caters to the requirements of scholars who wish to keep up with the current state of various fields within their discipline.

Individual volumes in the De Gruyter Handbook series will typically provide:

- knowledge of relevant literary periods, genres, and historical developments;
- knowledge of representative authors and works of those periods;
- knowledge of cultural and historical contexts;
- knowledge about the adaptation of literary texts through other media;
- knowledge of relevant literary and cultural theories;
- examples of how historical and theoretical information weaves fruitfully into interpretations of literary texts.

Internationally renowned colleagues have agreed to collaborate on this series and take on the editorship of individual volumes. Thanks to the expertise of the volume editors responsible for the concept and structure of their volumes, as well as for the selection of suitable authors, HEAS not only summarizes the current state of knowledge in the field of Anglophone literary and cultural studies, but also offers new insights and recent research results on the most current topics, thus launching new academic debates.

We would like to thank all colleagues collaborating in this project as well as Dr. Ulrike Krauss at De Gruyter without whose unflagging support this series would not have taken off.

Martin Middeke
Gabriele Rippl
Hubert Zapf
January 2022

https://doi.org/10.1515/9783110587647-202

Already Published

VOL 1 Gabriele Rippl (ed.): *Handbook of Intermediality*
VOL 2 Hubert Zapf (ed.): *Handbook of Ecocriticism and Cultural Ecology*
VOL 3 Julia Straub (ed.): *Handbook of Transatlantic North American Studies*
VOL 4 Timo Müller (ed.): *Handbook of the American Novel of the Twentieth and Twenty-First Centuries*
VOL 5 Christoph Reinfandt (ed.): *Handbook of the English Novel of the Twentieth and Twenty-First Centuries*
VOL 6 Ralf Haekel (ed.): *Handbook of British Romanticism*
VOL 7 Christine Gerhardt (ed.): *Handbook of the American Novel of the Nineteenth Century*
VOL 8 Ralf Schneider and Jane Potter (eds,): *Handbook of British Literature and Culture of the First World War*
VOL 9 Martin Middeke and Monika Pietrzak-Franger (eds.): *Handbook of the English Novel, 1830–1900*
VOL 10 Ingo Berensmeyer (ed.): *Handbook of English Renaissance Literature*
VOL 11 Sebastian Domsch, Dan Hassler-Forest and Dirk Vanderbeke (eds.): *Handbook of Comics and Graphic Narratives*
VOL 12 Barbara Schaff (ed.): *Handbook of British Travel Writing*
VOL 13 Stefan Helgesson, Birgit Neumann and Gabriele Rippl (eds.): *Handbook of Anglophone World Literatures*
VOL 14 Philipp Löffler, Clemens Spahr and Jan Stievermann (eds.): *Handbook of American Romanticism*

Forthcoming volumes

Sabine Sielke (ed.): *Handbook of American Poetry*
Katrin Berndt and Alessa Johns (eds.): *Handbook of the British Novel in the Long Eighteenth Century*
Martin Middeke and Martin Riedelsheimer (eds.): *Handbook of Literary Ethics*
Nadine Böhm-Schnitker and Marcus Hartner (eds.): *Handbook of Interdisciplinarity*

Contents

Oliver Scheiding and Erik Redling

0 Introduction: The American Short Story – Past and Present

1 Short Story Studies: The Current State of the Field

Since the onset of the new millennium, the American short story has been praised as the perfect literary form for the twenty-first century. In an age in which brevity rules, the short story appears to meet the demands of mobile devices and limited attention spans. According to some critics, the short story is "the genre of today" (Botha 2016, 201). In its long history, however, the American short story has shown itself to be a highly dynamic and productive narrative form. From early on it embodied the technological and cultural changes that advanced American society from the colonial period to the present. Be it the nineteenth-century expansion of the mass-market newspaper and magazine industry or the late twentieth-century boom of electronic publication outlets, the short story is the art of brevity that makes it an attractive catalyst not only for narrative experimentations, but also for building communities of readers who want to hear and tell stories about themselves and others (↗1 Sketches, Tales, and Stories). While scholars noted that America is certainly not "the land of the short story, but 'the land of the definition of the short story'" (Levy 1993, 1), the current *Handbook of the American Short Story* does not intend to develop a new taxonomy or a typological description of the American short story, but rather reassesses the genre in terms of what has been called "bibliomigrancy" (Venkat Mani 2016, 60). As such, the chapters in this handbook illustrate the emergence, circulation, and distribution of the short story beyond national paradigms or a binary that repeatedly reads it in terms of an apprenticeship for the novel (↗2 Canon Formation). The chapters dedicated to individual authors also document each author's creative contribution to the genre and, by pairing two short prose texts from one author, provide a better understanding of the spectrum of creative experiments performed by each author. Additionally, the volume's paired readings offer both practical and critical analysis of the short form's iteration, its narrative identity in terms of ambiguity, implication, and suggestiveness, as well as the genre's portability in different publishing contexts. Focusing on American short stories from the past to the present and relating them to the changing field of literary production and authorship, the *Handbook of the American Short Story* adopts an approach that "allows the co-ordination of immanent formal analysis of the individual text with the twin diachronic perspective of the history of forms and the evolution of social life" (Jameson 1981, 105). What sets this handbook apart from other accounts is its focus on genre, period traits, and material arrangements exposing the American short story as a cultural text that offers specific social and historical meanings distilled in the

https://doi.org/10.1515/9783110587647-001

short story's aesthetic form and brought to light by a practice of close reading. Thus, this volume highlights the socio-poetic dimensions in the writing, publishing, and reception of the American short story.

Given the genre's communicative power and productivity, why is it that both authors and critics repeatedly feel urged to talk about the future of the short story? Are we dealing with an "endangered species" – as Joyce Carol Oates puts it (Oates 2000, 38–41) – that needs our constant reflections to reassure us that it has a future – or, is it already a "dead form," as the literary magazine *n+1* proposes, "unnaturally perpetuated" by America's "fiction factory"? (Batuman 2006, 63–4). In a recent piece on the American short story in the twentieth century, Charles May speculates that "because American readers are more interested in nonfiction and the occasional novel than the short story, the form is currently in the literary doldrums in the United States" (May 2012, 324). Throughout its history, as Kasia Boddy aptly puts it, the American short story has been praised either as "a highly polished gem" or condemned as "literary fast food" (Boddy 2010, 15). In his astute evaluation of the anglophone short story in the context of today's writing and publishing, Paul March-Russell discusses the genre's major transformations from the pre-web era into the digital age. Given the many online self-publication options and new types of media authorship, he acknowledges the "split identity of the short story," referring to "the ambiguous identification of the short story as both rarefied object and commodified artefact," only to conclude: "the short story remains an indistinct process" (March-Russell 2016, 25). Although this is how the rise-and-fall narrative of the genre often goes, it is remarkable that it is mostly expressed by the lovers of stories highlighting the short story's transience rather than its enduring strength. Others, such as the French critic Claire Larriere, keenly concluded: "if anything, the short story is shorter today than it has been in the past. There are now collections of 'short shorts' and 'sudden fiction' which scarcely existed fifty years ago. The game of hide-and-seek is bound to be more exciting" (Larriere 1998, 198). Larriere's remarks point to the innovative potential of the short form and the resonance it has for readers. A good case in point for the short story's sociability, and its function as a transforming and transfixing encounter between a world in miniature and the reader, is the rediscovery of Lucia Berlin's vividly conversational stories about human environments (see *A Manual for Cleaning Women*, 2015). Public short story reading events, sharing stories both in print and online, and short fiction contests show that the genre is in sync with a new wave of slow and mindful media and its focus "on the fertility of individual and collective imaginations" (Rauch 2018, 10). Given the mobility of the genre as a relational concept that is conditioned by numerous codes (compositional, social, economic, material, temporal) allows us to both consider the short story as a socially symbolic act and discuss the genre's "mediatory function" (Jameson 1981, 105). A genre like the short story not only contains conventions, but also deviates from them. In his incisive study on genre, John Frow argues that genre has no "essence." It is not a container. Genre functions productively and possesses a "historically changing use value" (Frow 2006, 125, 134). Likewise, in a most recent study on genres and functions of the short story in North America, Michael Basseler shows that the short form serves as "an organon of life knowledge" (Basseler 2019, 17; ↗3 Current Approaches). In doing so, he explores how the genre's "complexity of thematic, formal, stylistic, and generic choices" shapes "the reader's 'sense of life' . . . and how it can or cannot be understood" (16). Moreover, this volume's genre approach helps us compare, contrast, and untangle the family resemblances that constitute the American short story. Extending our knowledge of short prose works and their dynamic socio-poetic functions, *The Handbook of the American Short Story* exemplifies the genre's pragmatic and active sense to give us the world via a selection of individual responses highlighting the short story's past, present, and future.

The paired readings and model interpretations in *The Handbook of the American Short Story* illustrate that the short story has always been a demanding form. George Saunders, for instance, admits that it is more challenging to write a good short story than to write a novel, defining the latter as a "a story that hasn't yet discovered a way to be brief" (Saunders 2013a, n. pag.). Robert Stone praises the short story as an art form that explores realities "subjectively experienced from the inside" (Stone 1991, 102). Distinguishing the short story from other short forms like the anecdote or news account, Joyce Carol Oates asserts that "the short story is a prose piece that it is not a mere concentration of events . . . but an intensification of events by the way of events" (Oates 1998, 47–48). Despite the numerous approaches – by both critics and writers – toward explaining the essence of brevity in literature, the short story remains, as Martin Scofield aptly puts it, "the exemplary form for the perception of crisis, crux, turning point" (Scofield 2006, 238). Its narrative economy in terms of time and space records decisive, intimate moments of life that give them a broad social resonance. As such, the short story offers a vibrant field of research (see the new journal *Studies in the American Short Story* 2020).

In this sense, Charles May's prediction (May 2012, 324) is of interest here because he raises an important question: how to write about the American short story in the twenty-first century, especially after the minimalist boom of the 1980s? The most recent surveys and companions tell a progressive history of the American short story which roots in a long-lasting legacy of sketches and tales as it first appears in Washington Irving's influential *The Sketchbook of Geoffrey Crayon, Gent.* (1819–1820; ↗5 Irving). In her study on the "Beginnings of American Literature," Lydia Fash claims that Irving "established the utility of short fiction for conjuring an American people and supporting American literary professionalism" (Fash 2020, 55; see Scheiding and Seidl 2011, 67–80). Scholars deal with the evolution of the form in the nineteenth century and explain how short prose works became an art form in the opening decades of the twentieth century culminating in a neo-realist renaissance of the Eighties and Nighties (Nagel 2015). Departures from this master narrative of the American short story can be found in a number of studies written in the wake of Mark McGurl's *The Program Era* (2009) that saw a sea change in the study of postwar literature. A good case in point is Kasia Boddy's survey of the American Short Story after 1950 in which she focuses on institutional aspects of the form in the context of big magazines, the creative-writing industry in the U.S., and the production of the short story in a literary market depending on specific forms of recognition like literary awards and prizes (2010; see also Delaney and Hunter 2019). By contrast, Paul March-Russell's introduction to the short story (2009) reassesses the genre's stadial history. In doing so, he takes the reader on a transnational journey from the short story's beginnings in Western folk traditions to what he calls in the final chapter "Voyages Out," focusing on the "postcolonial short story." In her 2018 study on the American short story cycle, Jennifer Smith analyzes the expansion of short prose within a specific form of narrative composition. She claims that the "short story cycle has become the primary and best descriptor for the genre, because it most accurately captures the recursiveness central to the genre and privileges the short story as its formative element" (Smith 2018, 4). Along the line of earlier studies, such as James Nagel's study on the contemporary American short story cycle (2001), Smith reads the cycle as a productive format for "identity formations" (7). Since the multiethnic literature of the 1980s and 1990s turned toward multiplicity, polyphony, plurality, and fragmentation, the cycle's composite nature allowed ethnic writers to integrate new elements into the short story and expand it. Recent examples of the cycle's "ethnic resonance" (Nagel 2001) can be found in Junot Díaz's *Drown* (↗33 Díaz) in which he chronicles the world of the Dominican American diaspora unfolding the archipelagic histories of his characters. Díaz and African American writers like Bryan Washington spell out versions of the cycle that become manifest in geography. *Lot,* Washington's 2019 debut collection of stories about Houston, Texas, presents on the first page of the text a stark black-and-white illustration of an urban grid, a drone's-eye view of the "hood" and the arteries of freeways around it. It looks almost like an X-ray, signaling that the ensuing stories have an almost supernatural power to see and know what others cannot know. Washington's stories spell out a regional understanding of the world that challenges major western epistemologies. By 2010, as Smith shows, the trend towards microfiction had changed the narrative outlook of the cycle. She claims that a new genre emerged under the term "novellas-in-flash" or "flash cycles" (Smith 2018, 169–177). These are novellas composed entirely of standalone flash fiction pieces organized into full narrative arcs. Smith not only highlights the cycle's duration as an art form, but also the ways in which the short story form resurges in the digital age. In light of new publishing trends like the flash fiction novel or the novella-in-shorts, her book also raises important questions about the commodification of shortness. If brevity becomes a commercially viable brand (see Amazon's short reads section), the novel becomes a label with weight and significance. By borrowing the label, the short story can share the novel's prestige and convince readers to buy it. Perhaps this is another reason for the booming of the short story cycle as a composite novel over the last decade. The traditional New York publisher Henry Holt and Company released Philip Caputo's collection *Hunter's Moon* (2019) under the label "A Novel in Stories."

Another notable approach to explain the genre's productivity is the 2019 collection of essays published as *The Edinburgh Companion to The Short Story in English*, edited by Paul Delaney and Adrian Hunter. The collection offers an analytical framework built upon five major themes: history, publication, form, place, and identity. In terms of future issues for the study of the short story, the most promising piece is Laura Dietz's essay on "The Short Story and Digital Media" (Dietz 2019, 125–139). Dietz discusses what Ellen McCracken calls "transitional literature" (McCracken 2013, 105). According to McCracken, multiform digital tools produce "electronic texts that mimic the format and appearance of print" and mark "a key stage in the shift between print and digital books underway in the 21st century" (105). Dietz writes that transitional literature adds "few or no features that take advantage of the affordances of screen reading . . . keeping open the possibility of moving, returning, to print without significant alterations" (Dietz 2019, 125). In his story collection *Texts from Jane Eyre* (2015), Daniel Mallory Ortberg condenses famous short stories like Charlotte Perkins Gilman's "The Yellow Wallpaper" into serialized smartphone chats. On its website, *The New Yorker* published Jennifer Egan's short story "Black Box" (2012) in the twitter format over a period of ten days. Egan tells her spy thriller in tweets that play with today's phenomenon of serialization, temporality, and interaction. The story is also a striking example for transmedia storytelling that mediates between scribal practices, online technologies, and print publication. In this sense, Dietz's essay raises important questions about the future of the story's form in the overlapping contexts of print, electronic (self-)publications, and digital fiction networks. In doing so, her analysis draws our attention not only to what the label "short story" means in today's e-age, but also how stories or single-author short story collections climb out on new artistic and commercial limbs like digital-only formats, intersecting legacy publishing with short fiction distributed via digital social media. Recent print market evaluations show, for instance, that digital short-story collections provide an outlet for collections that have no print prospect at all. On the one hand, what becomes necessary are further investigations about new publication strategies and how writing emerges and find its readers today; on the other hand, the "ambitious leaps of genre and world-building" (Machado 2019, xviii), especially in the field of science fiction and fantasy, also concern the very form and consumption of the short story. Given these trends in writing and publishing the short story, we need to consider both the short story's economy and texture to better understand the complex weaving of material and aesthetic "forms that organize texts, bodies and institutions" (Levine 2015, 23; see also Hungerford 2016, 19–39; ↗ 4 Textual Materiality).

2 Current Trends in Writing and Publishing the Short Story

Over the last decade, three factors have shaped the development of the American short story: orality and storytelling, genre writing, and new types of media-based authorships and textualities nurtured by independent publishing models. The new digital technologies have not only sparked the return of several traditional storytelling practices, but also made the short story a global form. Besides the resurgence of serialized literature via social media, there is a major trend towards orality and collaborative forms of consuming stories. While the original twitter format of 140 characters has been widely used to produce flash fiction thus revitalizing the short shorts of the 1920s (Cocchiarale and Emmert 2017), there are storytellers who push their stories beyond this limit and create interactive narrative scenarios that put the reader in charge of the story. British writer Naomie Alderman has created a sports app titled *Zombies, Run!* (2012) that combines genre writing and games. It is an engaging story format that merges the thrill of literary fiction with the choice-based storytelling of games. Moreover, the storytelling website and app *Wattpad* provides readers with thousands of interactive stories with intricate storylines and multiple forking sub-narratives taking the reader into unknown directions. The French company Short Édition has installed short story dispensers all over the world that offer readers in public locations the choice to read stories of one, three, or five minutes. On the company's website, the producers praise the art of storytelling by pointing out that it "goes back more than 17,000 years to the earliest known place, the Lascaux Caves in southwestern France, where ancient stories are told through painted symbols on the walls of the caves. Today in France, Short Édition is keeping this artform alive with their own modern innovation, attesting to the timelessness of literature" ("The Gist of It" 2020). The company's promotional statement highlights the multimodal nature of storytelling that synthesizes visual, oral, and graphic elements (↗24 Barthelme). Returning to the storytelling tradition, *The New York Times Magazine* published during the first Covid-19 pandemic a special issue on short fiction that appeared both online and in print, titled "The Decameron Project" (2020). The editors asked designers and writers to create "lifesaving" stories. In her introduction, Rivka Galchen refers back to Boccaccio and other series of stories in which people overcome "difficulties, dilemmas and war" concluding, "[r]eading stories in difficult times is a way to understand those times, and also a way to persevere through them" (Galchen 2020, n. pag.). Online collections like *The Chronicles of Now* commission authors to write short stories based on headlines. These so-called news stories fuse fictional and non-fictional elements and are available on a weekly basis as both podcasted audio files and online prose, linked to sidebars providing additional background information. Each short story is based on a specific subject. In doing so, the website's editors respond to today's accelerated reporting and publishing, promoting the idea of the short story as a form of slow and sustainable fiction. Their mission statement holds that "we hope these short stories prompt readers to pause and deeply consider a single resonant news story – what it means, why it happened, and why it matters." While the modern short story's depth has been explained in terms of its "poetic language and metaphor" (Patea 2012, 9–10), the carefully curated "packages" of podcasted stories, interviews, background information, and videos offered by *The Chronicles of Now* raise questions relevant to studying the short story, for instance: how the genre works as performance in itself; how it conveys a theory of culture in which aesthetic structures and the presentation of cultural problematics interrelate; how the short story and the practices of text-making are related to online and offline cultures; how we can read the short story as an expressive form alongside its material dimensions and the multiple meanings of such concepts as author, textuality, and media.

In a recent post, the editors asked Native American writer Tommy Orange, author of the bestselling novel *There There* (2018), to reflect upon the reopening of schools and what it means in terms of pandemic fear and uncertainty. The short story was inspired by a headline from *The New Yorker* about the new school year. Unlike Donald Barthelme's highly metafictional short piece of classroom calamities "The School" (1974; ↗24 Barthelme), Orange's short story "Reopening" (2020) is an immersive story told from a school kid's perspective. Young schoolchildren, as he states in an interview, are frequently not interested in political news which is too abstract for them, but rather in imagining how to recreate their own worlds "in which half-human, half-animal mutant zombies, felled by Covid, rise from the dead to get their revenge on the living" (Orange 2020, n. pag.).

Orange's observation on world building hints at the second factor influencing contemporary storytelling, especially in the field of the short story. According to McGurl's analysis of literary production in the era of creative writing education programs, genre fiction has become a new type of entertainment and narrative business. Other than the technomodernism of writers like Thomas Pynchon and John Barth, the low middle-class-modernism of minimalist writers like Raymond Carver (↗26 Carver), and the high cultural pluralism of ethnic writers like Toni Morrison and Philip Roth, today's genre writing is a cross-over of multiple generic traits. It includes many subcategories such as mystery/thriller, horror, romance, western, fantasy, science fiction, and their derivatives like the gothic, dystopia, weird fiction, post-apocalyptic fiction, ghost stories, superhero tales, alternate history, steampunk, slipstream, magic realism, and fairy tales. Critics writing about Kelley Link's collections of fantasy short fiction argue that "her stories are liminal; they lie betwixt and between the spaces within which we ordinarily confine fantastic fiction" (Duncan 2012, 413). In her introduction to *The Best American Science Fiction and Fantasy 2019*, Carmen Maria Machado complains about a "pointless category-enforcing" in short story criticism and "people who refuse to read outside their comfort zones" (Machado 2019, xix). Rather than seeking a too narrow definition of today's storytelling practices, the American short story in the twenty-first century attests to the continuous expansion of the short form. Machado herself refers to Alice Munroe's idea of the story as a multipath house. In one of her introductions, Munro states:

> A story is not like a road to follow [. . .] it's more like a house. You go inside and stay there for a while, wandering back and forth and settling where you like and discovering how the room and corridors relate to each other, how the world outside is altered by being viewed from these windows. And you, the visitor, the reader, are altered as well by being in this enclosed space, whether it is ample and easy or full of crooked turns, or sparsely or opulently furnished. You can go back again and again, and the house, the story, always contains more than you saw the last time. It also has a sturdy sense of itself of being built out of its own necessity, not just to shelter or beguile you. (Munro 1998, xvi–xvii)

Instead of "locking the doors" (Machado 2019, xix), Machado praises the form's productivity and portability across a range of genres and media. As such, even the widespread circulation of speculative fiction over the last decade has not replaced science-fiction with its exclusive focus of possible worlds and technology. Speculative fiction develops further a non-mimetic tradition of storytelling. Extending postmodernism's aleatory aesthetics, it combines fantastic, dystopian and science fiction elements with a very strong focus on the human and the pressing ills of society. Contemporary writers like Machado produce genre-bending speculative fiction collections in which they blend elements of folklore, ghost stories, science fiction, otherworldly scenarios and contemporary media exploring the experience of women. In stories like "Real Women Have Bodies" (2017), Machado creates a liminal fantasy that tells the story of women's erasure as a plague across America has turned women into silent incorporeality. The story fuses ghost and pandemic themes and plays with the reader's cultural imagination. The action takes place in a mall that serves as a capitalist dystopia where both desires and fears are on sale. It is also a multipath story embedding hypertextual links to pop culture and literary fiction with related motifs like madness and the haunted house. However, speculative fiction not only offers the reader a fantastic mix of escapes from reality, but functions as a highly oppositional form criticizing western society's mythology of progress as well as the social norms and values that go along with it.

Ethnic millennials in particular invigorate the former "high cultural pluralism" that McGurl (2009) attributes to African American literature of the 1980s and 1990s. Notable contributions also come from writers of Asian descent such as Yoon Ha Lee (*Conservation of Shadows* 2013) and Ted Chiang (*Exhalation: Stories* 2019) or from the younger generation of colored writers like Sofia Samatar (*Tender:* Stories 2017), N.K. Jemisin (*How Long 'Til Black Future Month?* 2018; ↗35 Jemisin) or Nana Kwame Adjei-Brenyah (*Friday Black* 2018), who, in their stories, integrate elements of fantasy and science fiction creating an Afrofuturist aesthetic for envisioning black history and culture. In the foreword to *Falling in Love with Hominids* (2015), a collection of speculative short fiction, Jamaican-born author Nalo Hopkinson calls the

"Caribbean, a region of the world which has had to be keenly aware of issues of race, class, gender, and privilege. You see the concerns reflected in much of our art and literature" (Hopkinson 2015, 2). In her short story "Shift," she reimagines *The Tempest* mixing oral storytelling, folklore, and fairy tales. In doing so, Hopkinson flips the master-slave perspective and turns Shakespeare's drama into a fantastic tale about race, region, and human relationships. Adjei-Brenyah's *Friday Black* consists of twelve short stories and is considered to be "an unbelievable debut, one that announces a new and necessary American voice" (Orange 2018, n. pag.). Since *Friday Black* tackles numerous present-day issues from consumerism to capitalism and abortion, as well as racism and other possibly uncomfortable topics, Adjei-Brenyah's short "stories are hard to read. They are hard to consume in how present they are, how unflinching. And they are true" (Agbo 2018, n. pag.). He draws from his own life experience as a black person in the United States that becomes manifest in short stories such as "The Finkelstein 5," an account of murdered black children decapitated by a white attacker with a chainsaw pretending to protect his own family against black terror. Adjei-Brenyah's short fiction echoes what Christian Sharp calls "wake work" (2016). In her book *In the Wake: On Blackness and Being*, Sharpe traces the legacy of slavery in four chapters: "The Wake," "The Ship," "The Hold," and "The Weather." Together these chapters metaphorically describe the cycle of black life. The chapter "The Hold," for instance, not only refers to the past and the middle-passage that kept human beings in the hold of the ship, but also mirrors the present situation of black life being now in the hold of an incarceration system, racial profiling, and police violence. For Sharpe, "to perform wake work" means to explore "forms of Black expressive culture [. . .] representing the paradoxes of blackness within and after the legacies of slavery's denial of Black humanity" (Sharp 2016, 14). *Friday Black* demonstrates that "living in the wake means living in the history and present of terror [. . .] as the ground of the everyday black existence" (15). In reference to the killer of Trayvon Martin, Adjei-Brenyah creates the story "Zimmer Land," a theme park for white people to act out brutal fantasies against people of color. Adjei-Brenyah uses the short form to reflect upon what critics call "the *spectacle* of Black death" (Wilderson 2020, 225; original emphasis). In doing so, he gives fresh meaning to earlier stories about Blackness written by Richard Wright, James Baldwin (↗20 Baldwin), and Octavia Butler. His collection also experiments with forms of serialized fiction. *Friday Black* contains a miniseries of three dystopian stories: "The Era," "Zimmer Land," and "Through the Flash" form an Afrofuturist triptych mirroring different postapocalyptic settings. In terms of the short form as narrative tryouts, N.K. Jemisin confesses that "shorts gave me space to experiment with unusual plots and story forms – future tense, epistolic format, black characters [. . .] I started to *enjoy* writing short fiction, for itself and not just as novel practice" (Jemisin 2018, ix; original emphasis; ↗35 Jemisin). Although many of the most recent story collections are precisely edited and sometimes resemble well-made concept albums, single short stories like Kelly Link's "Valley of the Girls" (2015) or

Jenny Zhang's autofictional pieces in *Sour Heart* (2018) are resonant for readers because they emulate Flannery O'Connor's dictum that "a short story should be long in depth and should give us an experience of meaning" (O'Connor 1977, 189; ↗19 O'Connor). Hence, genre writing not only has become a popular and global form of entertainment, it also makes the reader think about how short stories work as "affective forms of reorientation" as Sara Ahmed puts it in her study *The Cultural Politics of Emotion* (Ahmed 2015, 8). Asked in an interview about how he came across the idea of writing "The Semplica Girls Diaries" (2013b) – a story about American suburbia set in the near future – George Saunders (↗32 Saunders) referred to the importance of "feeling" in the writing process. In doing so, the story transforms "two cultures," as Saunders calls it, into a haunting tale in which a dystopian setting becomes reality, using a particular language and ambiguous action. One is a culture "in which the residents harness poor foreign women and hang them in their yards," and the other is grounded in a "complete physical domination in the surgical approach that this culture likes and wants; and that, in turn, says something deep about the lengths to which this culture is willing to optimize its aesthetic landscaping choice, i.e., its 'pleasure'" (Saunders 2012, n. pag.).

Saunder's critical remarks on "aesthetic landscaping" in the context of suburbia sheds light on the resurgence of a new regionalism in American storytelling. The wide-spread emphasis on hybridity, diversity, and transformation has recently become both accompanied and supplemented by reflections on rootedness and persistence. Moreover, a self-reflected, critical regionalism emerges in conflict with the commodified abstractions of transnational, universalist civilization, which threatens to destroy specific cultural ensembles. The conflict between transnational and regionalist tendencies in contemporary western societies has reverberations in contemporary American story writing and the ways in which it is assessed. There is a host of writers who situate their work not only ethnically but also regionally. In her Nevada-set debut collection *Battleborn* (2012), Claire Vaye Watkins chronicles small town America and the lives of ordinary people. Her western stories combine gothic, neo-noir, and autofictional elements to reassess the histories of her characters and the places in which they are wasting their lives. In his rural-noir stories, Donald Ray Pollock stages a gritty realism about a forgotten town in southern Ohio called Knockemstiff, in which nearly all people are "connected by blood through one godforsaken calamity or another" (Pollock 2012, 4). Critical regionalism imagines narratives of community, reviewing the economic, social, and cultural changes that go along with the spread of capitalism and neo-liberalism in rural America. While regionalism in American short fiction frequently engages the spatial and historical complexities of place (see Sherwood Anderson's 1919 short story cycle *Winesburg, Ohio*, for instance; ↗15 Anderson), stories by Ron Rash about Appalachia, Oscar Casares's Mexican border stories, Clairy Vaye Watkins's western gothic, Donald Ray Pollock's rural noir endorse disintegration, community, and mobility as central traits of critical regionalism (Achilles 2020).

The notion that stories serve as affect generators leads us to the third assumption about contemporary American story writing and publishing. Given literary fiction's empathetic power, postcritical scholars promote a socio-material approach towards analyzing literature. Rita Felski's "relational ontology" (Felski 2016, 747) focuses on the affective co-agencies that transform the printed page into a site that entwines countless actors and allies that gain meaning via their relations. Since the publication of Caroline Levine's *Form: Whole, Rhythm, Hierarchy, Network* (2015) there is an ongoing debate on textual affordances or how literary forms invite us to bind together subjects, collectives, and objects. Analyzing the short form in the context of affordances requires us to explore the story as a media object including its internal (paratextual) and external (epitextual) relations. It shows that writing and publishing the short story takes place in a web of relations and co-agencies. Over the past decade, acknowledgements have become a unique paratextual genre positioning the short story within a network of people, places, and institutions. As Emily Callaci notes: "At their best, acknowledgements dismantle the myth of the lone, self-contained genius-at-work, and instead expose the messy interplay of institutional support, finances, intellectual genealogies, and interpersonal chaos that shape how an idea is brought into the world" (Callaci 2020, 127). Acknowledgements illustrate the "messy" entanglements of both stories and authors, their connections and dependences that are often forgotten and made invisible in "making literature" (Hungerford 2016). Mining these paratextual informations (i.e., front cover, epigraphs, quotations, footnotes, etc.), readers become aware of a publication engine and its hidden parts (Moody 2007).

Other paratextual devices like "About the Type" highlight the "scriptality" or "graphic ideologies" that create the story's surface texture. In his sociolinguistic analysis of typography, Jürgen Spitzmüller refers to "scriptality" as "a set of beliefs attributed and expressed by means of graphic phenomena" on the page (Spitzmüller 2012, 255). Jenny Zhang's collection of short stories about New York's Chinese American community uses Zuzana Licko's typeface Filosofia. Licko also co-founded *Emigre*, a magazine for exiles in 1984. The paratextual note draws together Licko's "digital typesetting" (Zhang 2017, n. pag.), her background as a Czech immigrant who "came to the United States in 1968" (n. pag.), and Zhang's world of bifurcated characters. The type design, with its "classical proportions with a strong vertical feeling" (n. pag.), serves as an "immutable mobile" (Latour 1986, 7) that inscribes different agents across space and time. As such, Zhang's story collection uses a specific graphic mode of communication that materializes on the page what it means being Chinese American.

Another powerful and engaging mode of paratextual communication is the cover that picks up and intensifies a book's ambience. Kimberly Glyder, who designed the cover of Carmen Maria Machado's *Her Body and Other Parts*, captures the violence and sensuality of Machado's work. The cover also illustrates the stories' suggestiveness. Glyder designed an "optical illusion" inspired by an "old medical illustration"

(Klemperer 2017, n. pag.). The cover encapsulates the book's key theme that is the disappearance of women's bodies. Talking about the collaboration between designer and author, Machado refers to the cover's visual storytelling with its double mirror effect: "It's either a neck with a jaw but it looks like a corset and the ribbon is either being tied or untied, which is unclear [. . .] There's a lot coming out of it, you can read it in a bunch of different ways and it has so many dimensions" (n. pag.). Likewise, Mark R. Robinson's cover design of *Friday Black* draws upon fantastic elements that Adjei-Brenyah borrows form Ashanti folktales and West African trickster stories. The cover is a visual-graphic mode of communication that matters, combining design and content that boosts the book's iconographic aura. Posted on social media, the cover is part of the author's selfing. In addition, epitexts like the writer's homepage – with its links to titles, texts, events, interviews, book trailers and sellers, publication histories, and newsletters – facilitate the author brand as an ongoing and continually evolving story of the writer's success. This new type of transmedial entrepreneurship embodies creative writing, design, and digital literacy, and often the DIY (do it yourself) ethos of independent publishing. It addresses an expansive domain of taste provided by genre fiction's wide range of short story variants. Therefore, predictions about the short form's future are hard to come by. If anything is certain, it is that the form's flexibility and portability proves grounds for newer short story writers and a continuous unfolding of the art of brevity.

As this overview shows, the American short story has always been characterized by exciting aesthetic innovations and an immense range of topics. This handbook offers students and researchers a compact introduction to the multifaceted genre with a special focus on recent developments due to the rise of new media. Part I provides systematic overviews of relevant contexts ranging from historical-political backgrounds, short story theories developed by writers, print and digital culture, to current theoretical approaches and canon formation. Part II consists of paired readings of representative short stories by eminent authors, charting major steps in the evolution of the American short story from its beginnings as an art form in the early nineteenth century up to the very present and such innovative voices like Lydia Davis, Yiyun Li, and George Saunders (↗31 Davis, 34 Li, 32 Saunders). The handbook examines historically, methodologically, and theoretically the coming together of the enduring narrative practice of compression and concision in American literature. It offers not only original readings relevant to studying the American short story, but also provides ideas for fresh interpretations of and further reflection on this fascinating genre that has profoundly shaped the reception of American culture.

Even though several outstanding handbooks dedicated to the American short story have been published in recent years, a handbook that combines theory with paired readings of short stories and a focus on the rise of new media and new narrative trends is not available at the moment. Edited by Alfred Bendixen and James Nagel, *A Companion to the American Short Story* (2010) provides detailed readings, but does not discuss aspects such as print culture or topics related to current directions in literary criticism. *A Companion to the Amercican Short Story* ends in the 1990s; *A History of the American Short Story: Genres – Developments – Model Interpretations* (2011) edited by Michael Basseler and Ansgar Nünning and James Nagel's *The American Short Story Handbook* (2015) contain original discussions, but do not include chapters on the short story development in the digital age or short fiction published since 2000. Our handbook seeks to fill this gap, offering students and researchers alike a compact introduction to the multifaceted genre with a special emphasis on current developments due to the rise of new media.

The editors wish to thank those who made the publication of this handbook possible: first of all, the people who were willing to share their expertise and knowledge and took part in our project, and those who invested their invaluable energy and proficiency in the process of getting this volume ready for publication, notably Johannes Damaschke. The editors also owe special thanks to Patricia Godsave and her eagle-eyed copy editing as well as her steadfast advice from which both the editors and contributors profited greatly. Finally, we are indebted to Martin Middeke, Gabriele Rippl, and Hubert Zapf, the general editors of the Handbooks of English and American Studies, and the publisher's Content Editor Books, Katja Lehming, who encouraged us in seeing this book to print.

Halle and Mainz, 2021 — Erik Redling and Oliver Scheiding

3 Bibliography

3.1 Works Cited

"About," https://chroniclesnow.com/about. *The Chronicles of Now: Short Fiction Torn from Today's Headlines*. 2020. n. pag. (17 September 2020).

Achilles, Jochen. "Community and Social Disintegration in Anderson's *Winesburgh, Ohio* and Pollock's *Knockemstiff*." *Studies in the American Short Story* 1.2 (2020): 103–129.

Adjei-Brenyah, Nana Kwame. *Black Friday*. New York: Houghton Mifflin, 2018.

Agbo, Elinam. "Who Cares About You? (Family and Life in *Friday Black*)," https://sites.lsa.umich.edu/mqr/2018/12/who-cares-about-you-family-and-life-in-friday-black/. *Michigan Quarterly Review*, 21 December (2018): n. pag. (5 June 2020).

Ahmed, Sara. *The Cultural Politics of Emotion*. 2nd ed. Edinburgh: Edinburgh UP, 2015.

Andersen, Tore Rye. "Staggered Transmissions: Twitter and the Return of Serialized Literature." *Convergence: The International Journal of Research into Media Technologies* 23.1 (2017): 34–48.

Bendixen, Alfred, and James Nagel, eds. *A Companion to the American Short Story*. Chichester: Wiley-Blackwell, 2010.

Basseler, Michael, and Ansgar Nünning, eds. *A History of the American Short Story: Genres – Developments – Model Interpretations*. Trier: WVT, 2011.

Basseler, Michael. *An Organon of Life Knowledge: Genres and Functions of the Short Story in North America*. Bielefeld: Transcript, 2019.

Batuman, Elif. "Short Story & Novel." *n+1*, Issue 4: Reconstruction (Spring 2006): 63–70.
Boddy, Kasia. *American Short Story since 1950*. Edinburgh UP, 2010.
Botha, Marc. "Microfiction." *The Cambridge Companion to the English Short Story*. Ed. Ann-Marie Einhaus. Cambridge and New York: Cambridge UP, 2016. 201–220.
Cocchiarale, Michael, and Scott D. Emmert, eds. *Flash Fiction*. Ipswich: Salem P, 2017.
Callaci, Emily. "On Acknowledgments." *American Historical Review* February (2020): 126–131.
Delaney, Paul, and Adrian Hunter, eds. *The Edinburgh Companion to the Short Story in English*. Edinburgh: Edinburgh UP, 2019.
Dietz, Laura. "The Short Story and Digital Media." *The Edinburgh Companion to the Short Story in English*. Eds. Paul Delaney and Adrian Hunter. Edinburgh: Edinburgh UP, 2019. 125–143.
Duncan, Sydney. "Kelly Link: Introduction." *Journal of the Fantastic in the Arts* 23.3 (2012): 413–414.
Fash, Lydia. *The Sketch, the Tale and the Beginnings of American Literature*. Charlottesville and London: U of Virginia P, 2020.
Felski, Rita. "Comparison and Translation: A Perspective from Actor-Network Theory." *Comparative Literature Studies* 53.4 (2016): 747–765.
Frow, John. *Genre*. London: Routledge, 2006.
Galchen, Rivka. "Introduction," https://www.nytimes.com/interactive/2020/07/07/magazine-/what-is-the-decameron.html. The Decameron Project: New Fiction. *The New York Times Magazine* July 7 (2020), n. pag. (17 September 2020).
Goyet, Florence. *The Classic Short Story, 1870–1925: Theory of a Genre*. Cambridge, UK: Open Book Publishers, 2014.
Hopkinson, Nalo. "Foreword." *Falling in Love with Hominids*. San Francisco: Tachyon Publications, 2015. 1–3.
Hungerford, Amy. *Making Literature Now*. Stanford: Stanford UP, 2016.
James, Nagel. *The Contemporary American Short-Story Cycle: The Ethnic Resonance of Genre*. Baton Rouge: Louisiana State UP, 2001.
Jameson, Fredric. *The Political Unconscious: Narrative as a Socially Symbolic Act*. Ithaca: Cornell UP, 1981.
Jemisin, N. K. "Introduction." *How Long 'Til Black Future Month?* London: Orbit, 2018. vii–xii.
Larriere, Claire. "The Future of the Short Story: A Tentative Approach." *The Tales We Tell: Perspectives on the Short Story*. Eds. Barabara Lounsberry et al. Westport: Greenwood P, 1998. 195–199.
Latour, Bruno. "Visualisation and Cognition: Drawing Things Together." *Knowledge and Society Studies in the Sociology of Culture Past and Present*. Vol. 6. Eds. Elizabeth Long and Hendrika Kuklick. Greenwich, CN: Jai P, 1986. 1–40.
Levine, Caroline. *Forms: Whole, Rhythm, Hierarchy, Network*. Princeton: Princeton UP, 2015.
Levy, Andrew. *The Culture and Commerce of the American Short Story*. Cambridge: Cambridge UP, 1993.
Link, Kelly. "Valley of the Girls." *Get in Trouble*. Edinburgh, London: Canongate, 2015. 131–152.
Klemperer, Liz von. "The Story Behind the Most Haunting Book Cover on the Shelves," https://electricliterature.com/the-story-behind-the-most-haunting-book-cover-on-the-shelves/. *Electric Lit* (Oct. 6, 2017): n. pag. (17 September 2020).
Machado, Carmen Maria. *Her Body and Other Parts: Stories*. Minneapolis: Graywolf P, 2017.
Machado, Carmen Maria. "Introduction." *The Best American Science Fiction and Fantasy* 2019. Ed. Carman Maria Machado. Boston and New York: Mariner Books, 2019. xvii–xx.
McCracken, Ellen. "Expanding Genette's Epitext/Peritext Model for Transitional Electronic Literature: Centrifugal and Centripetal Vectors on Kindles and iPads." *Narrative* 21 (2013): 105–124.

Mani, Venkat B. "Weltliteratur als *bibliomigrancy*: Auf Emine Sevgi Özdamars 'Sprachzügen.'" *Text und Kritik: Zeitschrift für Literatur* 211 (2016): 59–70.
March-Russell, Paul. "Writing and Publishing the Short Story." *The Cambridge Companion to the English Short Story*. Ed. Ann-Marie Einhaus. Cambridge and New York: Cambridge UP, 2016. 15–27.
March-Russell, Paul. *The Short Story: An Introduction*. Edinburgh: Edinburgh UP, 2009.
May, Charles. "The American Short Story in the Twenty-First Century." *Short Story Theories: A Twenty-First-Century Approach*. Ed. Viorica Patea. Amsterdam: Rodopi, 2021. 299–324.
Moody, Nickianne, ed. *Judging a Book by Its Cover: Fans, Publishers, Designers, and the Marketing of Fiction*. London: Routledge, 2007.
Munro, Alice. "Introduction." *Selected Stories*. Toronto: Penguin, 1998. i–xvii.
Nagel, James. *The Contemporary American Short-Story Cycle: The Ethnic Resonance of Genre*. Baton Rouge: Louisiana State UP, 2001.
Nagel, James. *The American Short Story Handbook*. Chichester: Wiley-Blackwell, 2015.
Oates, Joyce C. "An Endangered Species." *New York Review of Books* 29 June (2000): 38–41.
Oates, Joyce Carol. "The Origins and Art of the Short Story." *The Tales We Tell: Perspectives on the Short Story*. Eds. Barbara Lounsberry et al. Westport: Greenwood P, 1998. 47–52.
O'Connor, Flannery. "Writing Short Stories" (1969). Eds. Alfred Weber and Walter F. Greiner. *Short Story Theorien (1573–1973): Eine Sammlung und Bibliographie englischer und amerikanischer Quellen*. Kronberg: Athenäum Verlag, 1977. 188–190.
Orange, Tommy. "*Friday Black* Paints a Dark Portrait of Race in America," https://www.nytimes.com/2018/10/23/books/review/-nana-kwame-adjei-brenyah-friday-black.html. *The New York Times*, 23 Oct. (2018): n. pag. (5 June 2020).
Orange, Tommy. "Reopening," https://chroniclesnow.com/stories/reopening. Stories About Now No. 91. *The Chronicles of Now: Short Fiction Torn from Today's Headlines*. 2020. n. pag (17 September 2020).
Patea, Viorica. "The Short Story: An Overview of the History and Evolution of the Genre." *Short Story Theories: A Twenty-First-Century Approach*. Ed. Viorica Patea. Amsterdam: Rodopi, 2012. 1–24.
Pollock, Donald Ray. *The Devil All the Time*. New York: Anchor Books, 2012.
Rauch, Jennifer. *Slow Media: Why "Slow" is Satisfying, Sustainable, and Smart*. Oxford and New York: Oxford UP, 2018.
Samatar, Sofia. *Tender: Stories*. Easthampton, MA.: Small Beer P, 2017.
Saunders, George. "This Week in Fiction: George Saunders. By Deborah Treisman," https://www.newyorker.com/books/page-turner/this-week-in-fiction-george-saunders. *The New Yorker* October 4 (2012): n. pag. (18 September 2020).
Saunders, George. "George Saunders and Andy Ward: Author-Editor Interview," https://slate.com/culture/2013/01/tenth-of-december-author-george-saunders-in-conversation-with-his-random-house-editor-andy-ward.html. *The Slate Book Review* 7 January (2013a): n. pag. (11 September 2020).
Saunders, George. *Tenth of December: Stories*. New York: Random House, 2013b.
Scheiding, Oliver, and Martin Seidl. "Early American Short Narratives: The Art of Story-Telling Prior to Washington Irving." *Handbook of the American Short Story: Genres – Classics – Model Interpretations*. Eds. Michael Basseler and Ansgar Nünning. Trier: WVT, 2011. 67–80.
Scofield, Martin. *The Cambridge Introduction to the American Short Story*. Cambridge: Cambridge UP, 2006.
Sharpe, Christina. *In the Wake: On Blackness and Being*. Durham and London: Duke UP, 2016.
Smith, Jennifer J. *The American Short Story Cycle*. Edinburgh: Edinburgh UP, 2018.

Spitzmüller, Jürgen. "Floating Ideologies: Metamorphoses of Graphic Germanness." *Orthography as Social Action: Scripts, Spelling, Identity and Power*. Eds. Alexandra Jaffe, Jannis Androutsopoulos, Mark Sebba, Sally Johnson. Berlin and Boston: De Gruyter, 2012. 255–288.
Stone, Robert. "Interview with Allan Vondra." *Mississippi Review* 20.1/2 (1991): 98–109.
"The Decameron Project: New Fiction," https://www.nytimes.com/interactive/2020/07/07/-magazine/decameron-project-short-story-collection.html. *The New York Times Magazine* July 12 (2020): n. pag. (12 September 2020).
"The Gist of It." 2020. Short Édition. n. pag Web. 12 September 2020.
Wilderson, Frank B. *Afropessimism*. New York: Liveright, 2020.
Zhang, Jenny. *Sour Heart*. London and New York: Bloomsbury, 2017.

3.2 Further Readings

Achilles, Jochen, and Ina Bergmann, eds. *Liminality and the Short Story: Boundary Crossings in American, Canadian, and British Writing*. New York and London: Routledge, 2015.
Brosch, Renate. *The Short Story: Textsorte und Leseerfahrung*. Trier: WVT, 2007.
Collins, Michael, and Gavin Jones, eds. *Cambridge Companion to the American Short Story*. Cambridge: Cambridge UP. Forthcoming.
Engler, Bernd, and Oliver Scheiding, eds. *Re-Visioning the Past: Historical Self-Reflexivity in American Short Fiction*. Trier: WVT, 1998.
Henderson, Eric, and Geoff Hancock, eds. *Short Fiction & Critical Contexts: A Compact Reader*. Oxford: Oxford UP, 2010.
Lohafer, Susan, and Jo Ellyn Clarey. *Short Story Theory at a Crossroads*. Baton Rouge and London: Louisian State UP, 1989.
Scheiding, Oliver, and Martin Seidl, eds. *Worlding America: A Transnational Anthology of Short Narratives before 1800*. Stanford: Stanford UP, 2015.
Scheiding, Oliver. "The American Short Story." *Handbook of Transatlantic North American Studies*. Ed. Julia Straub. Vol. 3. *Handbooks of English and American Studies*. Berlin/Boston: De Gruyter, 2016. 234–250.
Winther, Per, Jabok Lothe, and Hans H. Skei, eds. *The Art of Brevity: Excursions in Short Fiction Theory and Analysis*. Columbia: U of South Carolina P, 2004.

Part I: **Systematic Questions**

Lydia G. Fash

1 Of Sketches, Tales, and Stories: Theoretical Reflections on the Genre of the Short Story

Abstract: Before the "short story" existed, U.S. writers thought about and composed sketches and tales. These two forms, one more visual and less plot-driven and the other often mythic and fantastical, helped describe a United States that was growing quickly in size and cultural ambition. Sketches supplied calm reflection and travel descriptions, and romantic tales told of the wonderful history of the self-consciously young nation. Sketches and tales remained wildly popular until the massive trauma of the Civil War shifted short fiction towards realist and naturalist depictions – and eventually towards a new genre called the short story. By the final quarter of the nineteenth century, a "short story" meant not a narrative which happened to be modest in length but a genre term with recognizable characteristics including tightly controlled atmosphere and action and, for the western humorists, much droll absurdity. Although the formal unities of limiting the story time to a short window (often a single day) and striving for a single mood still influence the genre today, the twentieth century – and its two great world wars – moved authors towards the modernist and postmodern belief that no rules could adequately explain the world or literature. Instead, authors sought their own versions of truth through spare prose, twist endings, and moments of revelation. While the elements of the contemporary short story are thus variable, writers still seek the form out as a way to forge an uninterrupted connection with readers who can sit and read an entire short story despite living in a busy and fragmented world. This chapter traces the development of this history, highlighting the words of notable U.S. short story writers who have reflected on what the genre means and how it works.

Keywords: Sketch, tale, short story, unity of effect, style

1 Beginnings: Sketches, Tales, and the First Half of the Nineteenth Century

Although there is short fiction published in the United States before Washington Irving's *The Sketch Book of Geoffrey Crayon, Gent.* (1819–1820; ↗5 Irving), that collection and its massive success vaults U.S. short fiction to the fore and gives a powerful model for the two forms – the sketch and the tale – that come to dominate the nineteenth-century American short fiction scene. Published in seven parts in New York and in a slightly expanded two-volume form in London, Irving's *Sketch Book* is the first transatlantic bestseller written by a native-born Euro-American. Yet Irving's

https://doi.org/10.1515/9783110587647-002

name does not appear on the title page. Instead, Irving uses two pseudonymous narrators – Geoffrey Crayon, an urban traveler, and Diedrich Knickerbocker, an absurd historian of Old Dutch New York – to tell, respectively, sketches and tales. Through Geoffrey Crayon, Irving articulates the sketch form as a genre produced by a genial, educated outsider who graciously shares his observations with a reader. Indeed, Crayon introduces himself as "always fond of visiting new scenes, and observing strange characters and manners" (Irving 1822a, 1). This inclination to roving and looking sends Crayon, "pencil in hand" to "get up" one of those "portfolios filled with sketches" for his "friends" (6). Sketches, which would become the dominant genre of travel writers, are about conjuring a scene and a feeling, not about developing a plot. Like a quick drawing, verbal sketches are glimpses. They are incomplete. They are fragments. And they derive authority from that incompleteness: less polish conveys authenticity and essence over artifice. As the visual metaphor of the genre name implies, these sketches also hinge on observation – Crayon likes "observing strange characters and manners" – and thus project a form of proto-objectivity (our own notion of journalistic and other types of objectivity being a later nineteenth-century development) and thereby truth (1). Yes, Crayon did judge places, but he did so by looking carefully and responding generously, so much so that British critics not only accepted but praised Crayon when he chastised the "prejudiced accounts" of English writers reporting on America (96).

In contrast to the urbane, gentlemanly, and measured Crayon, the blustery Diedrich Knickerbocker relates two tales in *The Sketch-Book.* Now the best-remembered parts of Irving's collection, "The Legend of Sleepy Hollow" and "Rip Van Winkle," dramatize the mystical world of Old Dutch New York. Associated as it is with the American folklore, the tale is a vehicle for locally specific stories. As the introduction to "Rip Van Winkle" tells us, Knickerbocker searches for the stories of the "old burghers, and still more, their wives," understanding these people as books for his historical research (53). While Knickerbocker claims to relate history, then, he has no interest in objective truth. Rather he celebrates how the supernatural comes to exist in a community and becomes real to the townsfolk. In "Sleepy Hollow," Ichabod Crane is run out of town when a rival suitor for Katrina Van Tassel dresses as a headless horseman and chases him through the night. While the specter that Ichabod sees has a logical explanation, "the old country wives [. . .] who are the best judges of these matters, maintain to this day, that Ichabod [is] spirited away by supernatural means" (354). In "Rip Van Winkle," the supernatural is very much alive when Rip meets the ghost of Henry Hudson, drinks from a magic keg, falls asleep in the Catskills, and awakes twenty years later to return to a town transformed by the American Revolution. He ends in the role of town storyteller, recounting and embellishing his own account of that magical night. Through both these famed stories, Irving says that tales are about regional identity and embedded in a community, both through characters who, like Ichabod and Rip, tell stories, and in the frequent oral framing of tales (here the origin of the tales in the stories of Old Dutch New Yorkers). Unlike the sketch, tales highlight change and conflict through plots that advance temporally. Although Crayon and Knickerbocker, and thus their sketches and tales, were wildly different, Irving's collection provided a single powerful rebuke to Sydney Smith's 1820 taunting question in the *Edinburgh Review*: "Who reads an American book?" Because of *The Sketch-Book*, the answer was many (Smith 1820, 79).

It followed, therefore, that a number of American authors immediately tried their hand at sketch collections, including Henry Wadsworth Longfellow (1807–1882), the poet, who started his publishing career with *Outre-Mer; A Pilgrimage Beyond the Sea* (1833), a series of sketches about France, Spain, and Italy that recorded in part how "the idler and the man of leisure" very much in the mold of Geoffrey Crayon might "interchange the din of the city, the movement of the crowd, and the gossip of society, with the silence of the hamlet, the quiet seclusion of the grove, and the gossip of a woodland brook" (Longfellow 1859, 50). So too Nathaniel P. Willis's *Pencillings By the Way* (1835) and William Wells Brown's *Sketches of Places and People Abroad* (1855 in the U.S.), the first sketch-collection by an African American, offered accounts of learned gentlemen visiting new places, escaping the humdrum of home life (also, in Brown's case, the dangers of the Fugitive Slave Law), and serving as genial tour guides for their readers. If less commonly, women also wrote travel sketches, including Margaret Fuller's *Summer on the Lakes* (1844) in which Fuller critiques some of Irving's later writing about the American West for having a "second hand air" (Fuller 1844, 33). Irving, Fuller charges, "would have made the best of pictures, if he could have used his own eyes for studies and sketches; as it is, his success is wonderful, but inadequate" (33). In short, Fuller articulates the expectation that sketches, which are supposed to capture the essence of a place, are drawn from life. In providing such snapshots of other places, the sketch was culturally useful at a time when infrastructure made travelling difficult and expensive but literacy (and concomitant curiosity) was rising. Readers voyaged through the sketches in their newspapers, magazines, and volumes to places as far as Europe and as near as the next town over.

It is important to emphasize that not everything in a sketch was necessarily totally factual. Still, the sketch was often understood as a slightly looser nineteenth-century version of what we might now call creative nonfiction. In *A New Home – Who'll Follow* (1839), Caroline Kirkland declares in her account of midwestern homesteading: "I claim for these straggling and cloudy crayon sketches of life and manners in the remoter parts of Michigan the merit of general truth of outline" (Kirkland 1839, v). This strong association with truth and authenticity is how the sketch form underpins longer works like the slave narrative and abolitionist novels whose efficacy was predicated on their perceived facticity. Hence, Harriet Beecher Stowe calls *Uncle Tom's Cabin* (1852), the novel that Abraham Lincoln later supposedly credited with starting the Civil War, a series of "sketches," and Harriet Jacobs, an African American woman who hid for seven years in a tiny attic to escape the sexual predations of her white enslaver, refers to the "strictly true"

Incidents in the Life of the Slave Girl as "a sketch of [her] life" (Stowe 1852, vii; Jacobs 1861, 5, 6). The sketch was so closely associated with authenticity that it conferred validity to many accounts critical in the long battle against slavery.

Even sketches which did not present themselves as nonfiction benefited from the genre's association with authenticity. In many of the fictional character sketches written by women authors, for example, the truth was a valid expression of the essence of a type. Such sketches might be best exemplified by a series (and later collection) called "Sketches of American Character" written by Sarah Josepha Hale (1788–1879), who started the first successful U.S. magazine targeting a women readership and then went on to edit the single most successful periodical in nineteenth-century America, *Godey's Lady Book*, for four decades. In the first volume of her *Ladies' Magazine* (1828), Hale used her "Sketches of American Character" to delineate what exactly was particular and special about American citizens: "It is the *free* expression of that spirit, which, when irradiated by liberty, and instructed by knowledge, is all but divine, that gives to Americans their peculiar characteristics. To exhibit some of these traits, originated by our free institutions, in their manifold and minute effects on the minds, manners, and habits of the citizens of our republic, is the design of these Sketches" (Hale 1830, 8; original emphasis). In positing a sketch character as a representative of a larger class of people, Hale attempted to define what it meant to be American – and to include white women as central to the category. Similarly, when Catharine Sedgwick wrote a sketch entitled "Old Maids" (1833), she relied mostly on one character, Agnes Gray, to articulate the redeeming qualities of all those women who, like Sedgwick herself, remained unwed. While editors paid Sedgwick handsomely for her sketches and tales, the best paid nineteenth-century sketch writer would have been Fanny Fern (Sarah Payson Willis, 1811–1872) who became the first woman columnist in the U.S. in 1851 and shortly thereafter the highest paid columnist in the nation. These mentions of payment are important because they illustrate how, in a young literary market unprotected by stringent copyright laws, short fiction could be lucrative and was essential to developing U.S. literary talent.

Within that group of U.S. talent is, of course, Nathaniel Hawthorne (1804–1864; ↗6 Hawthorne) and Edgar Allan Poe (1809–1849; ↗7 Poe), the two most famous tale writers of the first half of the nineteenth century. While each is famous in his own right, Hawthorne and Poe come together in what many modern scholars understand as the birth of short story theory, an 1842 review Poe wrote of Hawthorne's *Twice-Told Tales*. Praising Hawthorne's originality, Poe offers a definition of "the tale proper," which "affords unquestionably the fairest field for the exercise of the loftiest talent, which can be afforded by the wide domains of mere prose" (Poe 1842, 298). Read in a single sitting of "a half-hour to one or two hours," tales, unlike novels, can control the effect a text has on a reader (298). More than that, Poe believed that fully controlling this "unity of effect" or "unity of impression" is what made a tale-writer great (298). In practice, for Poe, this means that an author determines the impression that he or she intended to create before composing and then assured every moment of the story supports the conjuring of this impression:

> In the whole composition there should be no word written, of which the tendency, direct or indirect, is not to the one pre-established design. And by such means, with such care and skill, a picture is at length painted which leaves in the mind of him who contemplates it with a kindred art, a sense of the fullest satisfaction. The idea of the tale has been presented unblemished, because undisturbed; and this is an end unattainable by the novel. (299)

The aesthetics of the narration are congruent with the themes of Poe's most famous and successful stories, including "Ligeia," in which the narrator's first wife chillingly revivifies in the body of his second wife, and "The Murders in the Rue Morgue," the first modern detective story in which C. Auguste Dupin's ratiocination solves a baffling whodunnit. From the horrifyingly Gothic to the highly rational, Poe's aesthetic decisions conjure strong feelings in the reader. Hawthorne's psychological probing has the same effect as Poe himself recognized: in such stories as "The Minster's Black Veil" in which a Reverend Hooper covers his face with a piece of black crepe which his congregants understand to symbolize some secret sin (Poe speculated it was sexual) but which Hooper refuses to confirm or deny, readers are left wondering how to interpret Hooper's actions and the secret sin that Hooper charges exists in each and every one of us. Like Hooper's congregants, we are unsettled.

It is worth pausing for a moment here to acknowledge that the sketch and the tale could and did overlap before the Civil War. One can find sketches with tales embedded in them and tales that have sketch-like qualities. Yet readers and authors mostly understood these two forms as distinct and complementary, hence the frequent collection title or subtitle of "tales and sketches": Catharine Maria Sedgwick had two different collections entitled *Tales and Sketches* (1835 and 1844), and William Leete Stone, Chandler Robbins Gilman, Benjamin Smith, Benjamin Drake, Nathaniel Greene, James Lawson, and a number of anonymous authors also penned volumes with *Tales and Sketches* in their titles or subtitles between 1819 and 1844. Whether published together or separately, sketches and tales were the dominant short fiction genres until the Civil War. They were integral in establishing an aesthetic program for the later short story and in supporting the growth of U.S. literary culture.

2 To the Short Story: The Second Half of the Nineteenth Century

While more venues for short fiction came into being in the 1830s and 1840s, these venues – magazines, newspapers, and the like – generally became more financially stable postbellum. As a result, more short fiction writers were able to earn more money. Periodicals which were part of larger publishing syndicates could and would run the same story simultaneously, increasing the reach of a piece. And the short fiction readership, what with immigration and rising literacy, continued to grow. Short fiction thus became even more culturally visible and important. And the form it took was the modern short story. The *Oxford English Dictionary* cites 1877 as the first occurrence of the term "short story," and indeed it is the last quarter of the century where the phrase becomes a genre term rather than just a descriptive adjective ("short") paired with a noun ("story").[1] In academic circles, Brander Matthews (1852–1929), a scholar and author whose literary writing is now all but forgotten, was significant in starting short story criticism as a field of study with his 1885 *Lippincott's* article. Matthews believed the short story to be a circumscribed genre that "fulfils the three false unities of the French classic drama: it shows one action in one place on one day" (Matthews 1885, 371). Matthews continues, "a short-story deals with a single character, a single event, a single emotion, or the series of emotions called forth by a single situation," thus exemplifying the "single effect, complete and self-contained" that Poe theorized (371). As the term "short story" came into wide use, in other words, authors and scholars were defining the form through the terms Poe had used to laud Hawthorne's tales. The short story told of a circumscribed event and strove to evoke a certain feeling, a unity of effect. For some, like Mark Twain (↗9 Twain), the short story had a strong oral component, reminiscent of Irving's best tales about Old Dutch New York. And the ascendancy of literary Realism, with its emphasis on the observed and objective, meant that the late nineteenth-century short story also shared much with the antebellum sketch. In the Local Color movement, a subsection of Realism that sought to record the uniqueness of particular non-urban places and people, the short story even recalled how the sketch became a mode of recording far-away places for urbanites back home. And, at the end of the century, social fiction – short stories explicitly intended to advance a social cause – appeared harkening back to the abolitionist work of the sketch. Early twentieth-century critics often ignored these continuities between early and later nineteenth-century short fiction. When they did not dismiss them, they frequently explained them by singling out Poe and Hawthorne as categorically different from their contemporaries. Instead of writing pedantic allegory, this train of thought goes, Poe and Hawthorne wrote *modern* short stories full of psychological drama and productive ambiguity. Thus, the modern short story was logically linked to Poe's and Hawthorne's work.

1 As literary critic Douglass Tallack points out, John Winthrop's speech "A Short Story of the Rise, Reign, and Ruine of the Antinomians Familists, and Libertines that Infected the Churches of Massachusetts Bay" (1644) uses the term, but despite Tallack's enthusiasm, we can be sure that Winthrop did not mean what we mean by the term "short story."

The author Ambrose Bierce (1842–c.1913) was one of those who followed in Poe's theoretical footsteps while seeing a marked break between the early and late nineteenth century. As Bierce wrote his satirical *Devil's Dictionary,* a novel is:

> A short story padded. A species of composition bearing the same relation to literature that the panorama bears to art. As it is too long to be read at a sitting the impressions made by its successive parts are successively effaced, as in the panorama. Unity, totality of effect, is impossible; for besides the few pages last read all that is carried in mind is the mere plot of what has gone before. [. . .] The art of writing novels, such as it was, is long dead everywhere except in Russia, where it is new. Peace to its ashes – some of which have a large sale. (Bierce 1911, 231)

For Bierce, the short story should be read in a single sitting and should be characterized by a unity of effect. More than that, though, Bierce believed that the short story was more modern than the novel. The short story was a sleek vehicle for articulating the psychological and emotional challenges of a new era. "An Occurrence at Owl Creek Bridge" (1890), his most famous story, articulates this idea through its plot. A Confederate sympathizer is set to be hanged from the Owl Creek Bridge and imagines, we see his escape – jumping off the bridge, swimming down the river, running through a forest, and opening the gate to his plantation – before the story rudely ends with the noose breaking the man's neck. It becomes apparent that the escape never actually took place but rather was imagined by the condemned in the time between when he was pushed from the bridge and when the hanging rope snapped tight. A Union Army veteran, Bierce himself saw little glory in war and found the short story effective for exploring the delusion, devastation, and trauma caused by fighting. So too Stephen Crane's short stories used brevity to explore the psychological implications of intense events and human vulnerability to social and natural forces, including some thirty-six odd exhausting hours in the ocean after a shipwreck in "The Open Boat." For Crane (1871–1900) and Bierce, the modern short story was a masculine one closely binding the honest Realist representation of the here and now and the Modernist fascination with consciousness and perception.

Along with Crane, Frank Norris (1870–1902) and Jack London (1876–1916; (↗13 London), writing similarly at the *fin-de-siècle*, are often identified as Naturalists, which many critics see as a subsection or flavor of Realism. Essentially, Naturalism used a Realist objectivity to expose social ills and reveal the cruelty of nature and society. London, who lived and wrote longest of the three, thought the brutality of life communicated something centrally awful about the world: "The great short stories in the world's literary treasure-house seem all to depend upon the tragic and terrible for their strength and greatness. Not half of them deal with love at all; and when they do, they derive their greatness, not from the love itself, but from the tragic and terrible with which the love is involved" (London 1999, 209). London believed so strongly in the worth of such "tragic and terrible" stories that he complained about editorial squeamishness over publishing them. "Editors did not like to publish [Poe's] stories nor people to read them, yet they were read universally and discussed and remembered, and went the round of the foreign newspapers," London noted (209). People think they do not like to be scared, to be bothered, but in reality those are the stories that most seize a reader – even if they have a hard time finding the practical means of reaching the reader. The popularity of London's most famous story, "To Build a Fire" (1908), which records the harrowing attempts of a man to kindle a fire in the frozen Yukon backcountry before finally freezing to death, supports his claim as does, in a different sense, the long tradition of short stories about ghosts and other supernatural phenomena, including spine-tingling examples from Washington Irving, Nathaniel Hawthorne, Henry James (↗12 James), Ambrose Bierce, Mary E. Wilkins Freeman, and, of course, Edgar Poe.

Unlike Jack London, Mark Twain (1835–1910) was not at all interested in finding a single tight moment to elucidate in the short story. Rather, Twain focused on what he called the humorous story, stressing that the laughs lie entirely in "the *manner* of the telling" rather than in the subject matter (Twain 1997a, 150; original emphasis). The "narrative should flow as flows the brook down through the hills and the leafy woodlands, its course changed by every bowlder," Twain wrote (Twain 1929, 237). If the idea was good, the story could meander and digress, reflecting oral taletelling practices, rather than tight plotting. Twain's "The Celebrated Jumping Frog of Calaveras County," then, consists of a man looking for a Leonidas Smiley but derailed by Simon Ward's stories about a *Jim* Smiley. While the narrator's ostensible purpose is to find Leonidas Smiley, Twain's purpose is clearly to let the reader hear Ward's dialect tale about a celebrated jumping frog. Indeed, the humor is created by Twain's masterful control of Ward's speech; a humorous tale, Twain wrote, must be "told gravely; the teller does his best to conceal the fact that he even dimly suspects that there is anything funny about it" (Twain 1997a, 150). While his long-winded narrators might appear to contradict such a fact, Twain's best short stories hinge on masterful pacing and calculated comedic timing. They also rely on careful diction; as Twain would remind us, "the difference between the *almost right* word and the *right* word is really a large matter – 'tis the difference between the lightning-bug and the lightning" (Twain 1997b, 137; original emphasis).

As writers of western humor, Twain and his quondam friend Bret Harte (1836–1902) are frequently linked. In his 1899 essay "The Rise of the 'Short Story,'" Harte argued that the short story only finally separated itself from English literary forms when "*Humour* – of a quality and as distinct and original as the country and civilization in which it was developed" became incorporated into the "concise and condense, yet suggestive" genre (Harte 1899, 3). For Harte, humor did not always or even usually mean conjuring side-splitting laughter. Rather humor was in the irony, cynicism, and unnatural juxtapositions created by a refined, restrained, and sympathetic narrator describing unsavory characters, who speak in dialect and often dress in dirty clothes. For example, in one of Harte's most famous stories, "The Outcasts of Poker Flat," the narrator describes two exiled sex workers in the following way: "I regret to say that some of these [exiled persons] were ladies. It is but due to the sex, however, to state their impropriety was professional, and it was only in such easily established standards of evil that Poker Flat ventured to sit in judgement" (Harte 1916, 30). The amusement comes from our sense that the professionalism of the women's sexual adventures makes things worse, not better, and that the "secret committee" of Poker Flat, as Harte reveals, was deeply hypocritical in how it determined whom to cast out (30). As the rest of the story bears out, lines of good and evil are not so clearly drawn. Such "treatment of characteristic American life, with absolute knowledge of its peculiarities and sympathy with its methods; with no fastidious ignoring of its habitual expression, or the inchoate poetry that may be found even hidden in its slang; with no moral determination except that which may be the legitimate outcome of the story itself; with no more elimination than may be necessary for the artistic conception, and never from the fear of the 'fetish' of conventionalism" is "the secret of the American short story" (Harte 1899, 8). And, Harte emphasizes, "the germ of American literature to come" (8).

By this point, then, the short story was broadly seen as a means to describe the parts and people of the United States: it was integral to the Realism (and Naturalism) that characterized much postbellum U.S. literary writing and to the Local Color movement that served as a (usually less prestigious) subsection of Realism. Writing about parts of New England and of the South, Mary E. Wilkins Freeman (1852–1930), Sarah Orne Jewett (1849–1909), Rose Terry Cooke (1827–1892), Constance Fenimore Woolson (1840–1894), Grace King (1851–1932), and Kate Chopin (1850–1904; ↗11 Chopin) were among the later nineteenth-century women short story writers regarded as Local Colorists. While their stories did record the particular conventions and cultures of various regions, they also questioned social conventions and frequently advanced a feminist agenda. Freeman's "The Revolt of Mother," for example, depicted a New England wife refusing her husband's authority by claiming his just-built barn as a new house for her and her family. Relatedly, Chopin's work, like "The Storm" which shows a woman's satisfying adulterous encounter, daringly explored the sexual desires and frustrations of women in Louisiana. With their high emotions and taut, careful construction, these stories offer an update on Poe's unity of effect, something Chopin herself seemed to experience when writing: "Story-writing – at least with me – is the spontaneous expression of impressions gathered goodness knows where. To seek the source, the impulse of a story is like tearing a flower to pieces for wantonness" (Chopin 1996, 107). In other words, even as they were celebrated for their accuracy in regional depiction, these stories were as much if not more about writing certain feelings into the past, present, and future. These women wrote these stories to make it acceptable for real women to feel power to act in the real world, first vicariously and then perhaps actually. Freeman herself recognized the improbability of "The Revolt of Mother": the story "is not true. [. . .] There never was in New England a woman like Mother. If there had been she certainly would have lacked the nerve. She would also have lacked the imagination. New England women of that period coincided with their husbands in thinking that the sources of wealth should be better housed than the consumers" (Freeman 1917, 75). The story projected an alternate hopeful history hidden in the corners of the United States.

Given the historical moment in which they wrote, it is unsurprising that various nineteenth-century women writers used the short story to consider the limits in which women authors worked: Mary E. Wilkins Freeman's "The Poetess," Kate Chopin's "Elizabeth Stock's One Story," and Constance Fenimore Woolson's "Miss Grief" are among the most notable. In fact, many critics have read Woolson's depiction of a successful but callous male writer in "Miss Grief" – one who cannot quite accept the brilliance of a female author acquaintance – as a fictionalized version of Woolson's friend Henry James (↗12 James). This male writer serves as a mouthpiece for a patriarchal literary structure that inhibits and excludes the voices of women. When he concludes, then, that "she, with the greater power, failed – I, with the less, succeeded," the story intends us to see the full injustice of that situation, even when the male writer himself does not (Woolson 2016, 185). Henry James's famous praise for Jewett's masterwork, *The Country of the Pointed Firs* (a series of linked short stories about the Maine coast, 1896) gestures to Woolson's plot. Although James no doubt meant well when he said *The Country of the Pointed Firs* was a "beautiful little quantum of achievement" and praised Jewett's "elegance of humility or fine flame of modesty," the backhanded compliments reflect the cultural expectation and regard for short fiction written by women (James 1915, 30). James himself wrote at least 112 pieces of short fiction during his lifetime and remarked, "To write a series of good little tales I deem ample work for a lifetime" (James 2015, 50). Yet Local Color remained a diminutive genre, thought to be appropriate for women and for writers of color.

Charles Chesnutt (1858–1932; ↗10 Chesnutt), a Local Colorist and the most prominent African American short story writer at the end of the nineteenth century, is a case in point. Collected in *The Conjure Woman* (1899), Chesnutt's stories are told by Uncle Julius, an older formerly enslaved African American with a thick accent, to a white couple from Ohio who have bought a rundown Southern plantation after the Civil War. Countering the then-recognized genre of "plantation literature" which, in the hands of Thomas Nelson Page and others, glorified the antebellum past with racist stereotypes and nostalgia, Chesnutt's stories illustrated the humanity and the savvy of African American characters who exercised some control over their lives despite being classified as chattel. For example, in "Po' Sandy," Julius tells of the enslaved Sandy who was turned into a tree by his lover so as to avoid being sold away from the plantation; the tree is then cut down, turned into lumber, and built into a schoolhouse, which the unhappy Sandy haunts. Julius's warnings about Sandy's ghost convince the narrator's wife to give up the plan to dismantle the schoolhouse and repurpose the wood for a kitchen. Days later, Sandy requests use of the building for church meetings – and conjures a donation from the narrator to support them. As Chesnutt used it, the genre was a powerful form for "goophered, cunju'd, bewitch[ed]" language, a sort of linguistic trickery in which the less powerful can use to manipulate or insult the more powerful – what modern literary critic Henry Louis Gates, Jr. calls "signifyin(g)'" (Gates 1983, 685). And the aesthetics of brevity were key to that aim: the compression of the short story enabled an introduction and development of an adequately complicated plot, while still assuring that the dénouement arrived quickly enough to have full effect, as when we discover that, in Chesnutt's words, "in every instance Julius had an axe to grind, for himself or his church, or some member of his family, or a white friend" (Chesnutt 1998b, 102). Julius's cleverness is meant to amuse only to achieve a larger purpose. As Chesnutt wrote, "the object of my writings would be not so much the elevation of the colored people as the elevation of the whites, – for I consider the unjust spirit of caste which is so insidious as to pervade a whole nation, and so powerful as to subject a whole race and all connected with it to score and social ostracism – I consider this a barrier to the moral progress of the American people" (Chesnutt 1998a, 87).

Frances Ellen Watkins Harper (1825–1911) shared with Chesnutt a deep concern with social justice and a belief that short fiction was an appropriate genre to fight for it. "The Two Offers," which ran in *The Anglo-African Magazine* in 1859 and is widely accepted as the first published short story by an African American, ends with comments on how writing can be a force for good.[2] In a sentimental plot in which a woman marries a drunkard and consequently dies from neglect and misuse, the woman's cousin, chastened by her former playmate's death, resolves to "kindle the fires of her genius on the altars of universal love and truth" (Harper 1990, 114). Thus, "in her the down-trodden slave found an earnest advocate; the flying fugitive remembered her kindness as he stepped cautiously through our Republic, to gain his freedom in a monarchial land, having broken the chains on which the rust of centuries had gathered" (114). While the story seems to be about white characters, then, Harper concludes with a call for political activism and a suggestion that the short story and the short story writer are critical to abolition and civil rights. At the end of the century, Charlotte Perkins Gilman (1860–1935), whose famous long short story (also sometimes called a novella) "The Yellow Wallpaper" details a woman's decent into insanity after being treated with the "rest cure," similarly endorsed social fiction: "In my judgement it is a pretty poor thing to write, to talk, without a purpose" (Gilman 1997, 111; ↗4 Textual Materiality). Sui Sin Far (Edith Maude Eaton, 1865–1914), the first Asian American to publish a short story in the U.S.; Abraham Cahan (1860–1951), a Russian-Jewish socialist; and Zitkala-Ša (see Bonnin Gertrude Simmons, 1876–1938; ↗14 Ša), a Sioux writer, are among the later nineteenth-century writers who told sensitive stories about their cultures for wider, often white, audiences to combat prejudice and injustice. They found in the short story a genre helpful for

2 Frederick Douglass does publish a novella entitled "The Heroic Slave" some seven years earlier.

bringing readers into unfamiliar worlds and inspiring those readers to act and feel for socially disadvantaged others. Brief and easy to place in a periodical, short stories were ideal for reaching a wide American audience.

3 No More Rules: The Twentieth Century and Today

Strong traditionalist strains remained in many a neatly plotted twentieth-century story. Indeed Eudora Welty (1909–2001) herself was deeply influenced by Poe's "unity of effect," writing of the short story: "It is when the plot, whatever it is, is nearest to becoming the same thing on the outside as it is deep inside, that it is purest. When it is identifiable in every motion and progression of its own with the motions and progressions of the story's feeling and its intensity, then this is plot put to its highest use" (Welty 2002, 14). Yet, it is indisputable that the two massive wars drastically changed the perception of reality. As Sherwood Anderson (1876–1941; ↗15 Anderson), a great short story writer himself, put it in a letter: "And then came the Great War. [. . .] It was terrific in its physical aspects – bodies mangled the young manhood of England, Germany and France blown away or bled white [. . .]. But something else got blow up in that war too. [. . .] It may be that what happened in the Great War, because of it, was a truer estimate of life. The young men who went into it got, must have got, a profound sense of life's cheapness" (Anderson 1992, 65). This trauma forced a new type of narrative that rejected the plots, dialect, and language of nineteenth-century short stories. Anderson, for one, "was certain none of [the Americans he had seen] lived, felt or talked as the average American novel made them feel and talk, and as for the plot short stories of the magazines – those bastard children of De Maupassant, Poe and O. Henry – it was certain there were no plot short stories ever lived in any life [he] had known anything about" (Anderson 2005, 362). By questioning the tenets of Realism and the conceits of Romanticism, Anderson shifted the short story towards more meditative, ambiguous, and subjective pieces, as exemplified in his short story cycle *Winesburg, Ohio.* He thereby influenced later heavy-hitters including William Faulkner (1897–1962; ↗18 Faulkner) and Ernest Hemingway (1899–1961; ↗16 Hemingway). By the third and fourth decades of the twentieth century, as John Steinbeck (1902–1968) put it in a letter to his former professor of creative writing, Edith Mirrieless, the genre felt wide open:

> The basic rule you gave us was simple and heartbreaking. A story to be effective had to convey something from writer to reader, and the power of its offering was the measure of its excellence. Outside of that, you said, there were no rules. A story could be about anything and could use any means and any technique at all – so long as it was effective. As a subhead to this rule, you maintained that it seemed to be necessary for the writer to know what he wanted to say, in short, what he was talking about. As an exercise we were to try reducing the meat of

> a story to one sentence, for only then could we know it well enough to enlarge it to three or six or ten thousand words. (Steinbeck 1989, 136)

Steinbeck's exercise of condensing a story to a single sentence by stripping away all but the essence highlights a concern of many early- and mid-century writers: finding a sleek, uncluttered style that concentrated meaning into fewer words. In the last decade of the nineteenth century and the first two decades of the twentieth, Willa Cather (1905–1947) wrote short stories beautiful in their stream-lined clarity. Deriding the novel for being "over-furnished," Cather associated the short story with a spareness, and urged long-form writers to embrace that same aesthetic (Cather 1922, 5). Many others expressed similar sentiments. Stephen Crane declared that his "chiefest desire was to write plainly and unmistakably, so that all men (and some women) might read and understand" (Crane 1989, 95). F. Scott Fitzgerald (1896–1940), who believed that a writer's goal is to articulate "the *profound* essence of what happened," similarly eschewed overburdened prose (qtd. in Kuehl 1991, 149; original emphasis). But Ernest Hemingway's mid-century writing most famously exemplifies this paring down. As Hemingway commented, "If a writer of prose knows enough about what he is writing about he may omit things that he knows and the reader, if the writer is writing truly enough, will have a feeling of those things as strongly as though the writer has stated them. The dignity of movement of an ice-berg is due to only one-eighth of it being above water" (Hemingway 1960, 154). The spare prose which could, at its most extreme, give a sense of telegraphed language appears in Hemingway's novels, but it is perhaps in his short stories where the aesthetic is best married to the form. In his "A Very Short Story," Hemingway tells quickly of a man being wounded, convalescing, falling in love with his nurse, returning to the front, celebrating the war end, returning States-side to get a job, having his fiancée break up with him, and then "contract[ing] gonorrhea from a sales girl in a loop department store while riding in a taxicab through Lincoln Park" (Hemingway 1970, 85). Hemingway packs a huge amount of information into this short story by stripping away everything but the most pointed and relevant details. Such bareness, Hemingway thought, made the stories last for the reader: "The stories where you leave it all in do not re-read like the ones where you leave it out. They understand easier, but when you have read them once or twice you can't re-read them" (Hemingway 1989, 140). The mystery of a (modernist) short story is the seventh-eighths of the iceberg that floats darkly underwater.

Such a fixation on style could not displace the importance of plot. Before two world wars, O. Henry (William Sidney Porter, 1862–1910) perfected the surprise ending in a plethora of stories published in the turn of the twentieth century. In "The Gift of the Magi," Jim buys his wife a set of beautiful hair combs for Christmas, and Della buys Jim, her husband, a beautiful watch chain for his cherished heirloom watch. Only at the end of the story do we learn that Jim has pawned his watch to buy the combs, and Della has sold her hair to buy the watch chain, rendering both gifts useless. The surprise ending was perhaps first made notable by Thomas Bailey Aldrich's successful story, "Marjorie Daw" (1873), in which a friend invents tales of a young woman to enliven his letters to a bedridden friend, only to have to confess to the ruse when the recovered friend arrives, besotted and determined to woo the lady. Even in O. Henry's day, critics like Sherwood Anderson found such surprising endings trite or contrived. Readers, however, loved them, and *The O. Henry Award* remains a prestigious prize annually granted to the best English-language short stories published in the U.S. or Canada, including Flannery O'Connor's story "The Life You Save May Be Your Own" (↗19 O'Connor).

While O'Connor's stories never had the sentimentality of O. Henry's, they did usually have a twist towards the end. As she revealed in a 1962 lecture, O'Connor (1925–1964) believed that "in the South the general conception of the man is still, in the main, theological" rather than psychological (qtd. in Dowell 1965, 236). A devout Catholic herself, O'Connor organized her short stories around a "moment of grace" in which the main character realizes the significance of their actions and their worldview snaps into focus – a focus that is defined by spiritual awakening (236). In "A Good Man is Hard to Find," one of O'Connor's most famous stories, the nagging Grandmother says to the Misfit, a violent criminal, right before the Misfit fires three bullets into the Grandmother: "Why you're one of my babies. You're one of my own children!" (O'Connor 1979, 132). The point is that the Grandmother after having watched her family be murdered, realizes that she herself has not been living in a Christian manner (but rather worrying selfishly about herself) and is thus aligned with the evil of the criminal who similarly has rejected Jesus. For O'Connor this spiritual struggle – the contest between grace and evil – was the most meaningful drama in life, and thus the proper subject for fiction. O'Connor remarked that "all good stories are about conversion, about a character's changing" and that all her stories "are about the action of grace on a character who is not willing to support it" (qtd. in Kinney, 75, 76). Stories, in their accessibility, were the mechanism by which O'Connor could depict such conversions and help her readers feel that spiritual enlightenment. While O'Connor's worldview and terminology ("moment of grace") was deeply Catholic, the idea of an organizing a short story around a moment of realization – an epiphany – takes broader root in the post-WWII period when religiosity saw a marked decline and a transcendent discovery that converted the physical world into spiritually resonant meaning held great appeal. Hence, one finds a related idea in Isaac Bashevis Singer's "The Son from America" (1973 in English) wherein Samuel, a man who has found monetary success in the United States, realizes, at the end of the story, that his money and new ideas cannot and should not change the culture and contentment his parents have in the Old Country. Singer, who lived in Brooklyn and wrote in Yiddish after fleeing the growing Nazi threat in Europe, mined Jewish symbols and Yiddish culture for his stories, often building to a revelation quite similar to O'Connor's epiphanies.[3]

Even those who were not overtly religious sought some morality in the act of writing. The postmodern Raymond Carver (1938–2008; ↗26 Carver) endorsed the poet Ezra Pound's maxim: "Fundamental accuracy of statement is the ONE sole morality of writing" (qtd. in Carver 1992, 94). Carver's point was that there need be no neat morality to the plot, but rather a sense of truth and fearlessness in the telling. Kay Boyle (1902–1992) similarly articulated a morality to truth: "It is a correction of reality that my stories have attempted. Good or bad, they are affirmation of Camus' belief that man, once he has recognized and acknowledged his own future, may surmount it, and thus give to the apparent absurdity of life a new dimension and dignity" (Boyle 1992, 121). It is to this sense of true, and thus moral, writing that John Cheever (1912–1982) and John Updike (1932–2009), those mainstays of the mid-century *New Yorker* fiction section, subscribed in their portraits of cosmopolitan manners and suburban ennui. Updike set himself the task "to transcribe middleness with all its grits, bumps, and anonymities, in its fullness of satisfaction and mystery" (qtd. in Miller 2006, 15). Although he was often criticized for not tackling grander topics and themes, he believed in the profound interest and worth of the quotidian and even the mundane. As Updike queried: "is it possible [to transcribe middleness] or, in view of the suffering that violently colors the periphery and that at all moments threatens to move into the center, worth doing? Possibly not; but the horse-chestnut trees, the telephone poles, the porches, the green hedges recede into a calm point that in my subjective geography is still the center of the world" (15). Cheever had a similar orientation towards fiction. In his "Five-Forty-Eight," which was published in the April 10, 1954 issue of the *New Yorker*, Blake, a selfish young man who fires his secretary after he has a one-night stand with her, feels relief, not realization, when she, after leading him at gun point into an alley, abandons him with his face in the dirt. Unlike Flannery O'Connor's stories, Cheever's stories offer a portrait rather than an epiphany. Indeed, Blake does not truly understand his cruelty, his unhappiness, or his shallow existence. That lack of realization was, for Cheever, a form of truth. In a 1959 essay, Cheever admitted to enjoying the notion that life and story were centrally similar: "I know almost no pleasure greater than having a piece of fiction draw together incidents as disparate as a dance in Minneapolis and a backgammon game in the mountains so that they relate to one another and confirm that feeling that life itself is creative process, that one thing is put

3 Although Jewish rather than Catholic, Singer joined O'Connor in bringing a deeply religious outlook to the mid-century short story. Singer believed that Jewish symbols helped him explore the position of evil in the world and the condition of living with that evil: "by using Satan or a demon as a symbol, one can compress a great many things [. . .] Devils symbolize the world for me, and by that I mean human beings and human behavior [. . .] I really love this system and I am always finding new symbols and new stories" (Singer 1990, 85).

purposefully upon another, that what is lost in one encounter is replenished in the next and that we possess some power to make sense of what takes place" (Cheever 1989, 108). While the boredom and dissatisfaction of Cheever's characters feel are linked to the lack of neat resolution, Cheever's act of writing demonstrates the narrative sense of – and thus a moral order to – the seemingly random world around us. If a number of twentieth-century writers found in the short story a sort of morality, many also felt that the genre was, above all, about making a connection with a reader. In fact, Shirley Jackson (↗21 Jackson) felt that interesting and leading a reader must be the driving force in writing a story:

> A short story – in common, of course, with all other written work – is of no use whatsoever without a reader. The act of writing is after all a two-way process, one of communication rather than complacent self-expression. The function of the story, most simply, is to be read. A good story must engage its reader, persuade him that he wants to belong in the story of as long as it lasts, so that he will participate eagerly in the act of communication which the author has initiated. There are no set rules, really, for the writing of stories, and no single point at which a story stops being a story and becomes something else, but surely a story which does not so engage the reader must be considered a failure. (Jackson 1993, 113)

And yet, that is not to say that the short story prides itself on being comfortable for readers. Jackson's own very famous story, "The Lottery," which depicted a brutal ritual stoning in a small New England town, catalyzed a wave of canceled subscriptions when it was published in *The New Yorker* in 1948. Morality is in finding and connecting with – even if that means shocking – the reader, not in staid or sentimental topics. The short story is a conversation. In Eudora Welty's words: "Reader and writer, we wish each other well. Don't we want and don't we understand the same thing? A story of beauty and passion, some fresh approximation of human truth?" (Welty 2002, 28).

Short stories are arguably the most appropriate genre for depicting alternative and sometimes uncomfortable ways of being and thinking. Indeed Frank O'Connor, the twentieth-century literary critic, saw the short story as offering a particular opportunity for telling the stories of those who are marginalized. In novels, the characters are surrounded by copious others, by layered social contexts, by history and busy-ness. But in the short story, the character comes into sharp focus on a bare stage. For example, William Faulkner's short story "Wash" (1933) zooms in on the events (Wash Jones's murder of Thomas Sutpen) which would become the events of chapter 7 of *Absalom! Absalom!* (1936), a novel in which Wash Jones is just a minor character. Despite the much greater size of the novel, Faulkner excised discussion of Wash's life and motives. The poor jobless squatter comes to life only in the short story because the genre focuses in on a character's life and makes it matter fully and completely. Short stories privilege the individual even if that individual is unsavory, unsympathetic, or stigmatized. Hence Jean Toomer's *Cane* (1923), a pivotal work of the Harlem Renaissance, could explore, in twelve short stories (linked with poetry and song), the significance of race and gender in the South. Richard Wright's

"The Man Who Lived Underground" (1945) and James Baldwin's "Going to Meet the Man" (1965; ↗20 Baldwin) both explore the racist social practices of the United States and the dehumanization of African Americans that results. Short stories have also been useful to exploring the experiences of Native Americans, Latinx and other racial and ethnic groups as well as religious and gender minorities as in the work of Sandra Cisneros (↗29 Cisneros), Louise Erdrich (↗30 Erdrich), Denise Chávez, Sherman Alexie, Maxine Hong Kingston, Jamaica Kincaid, Saul Bellow, and Z. Z. Packer. In an interview, Puerto Rican short-story writer Judith Ortiz Cofer (1952–2016) remarked: "I feel every time I write a story where a woman is strong or a woman is victimized, that I am making a statement about being a woman. And every time I write a story where Puerto Ricans live their hard lives in the United States, I am saying, look, this is what is happening to all of us" (Ortiz Cofer and Acosta-Belen 1993, 86). Like the work of other writers, Ortiz Cofer's stories make those who are often invisible, visible.

Joyce Carol Oates (b. 1938) once reflected, "As to the nature of short fiction? There is no nature to it, but only natures. Different natures. Just as we all have different personalities, so the dreams of our personalities will be different" (Oates 1994, 127). American authors have, since the very beginning, found short fiction to be a flexible and elastic form, adoptable to the cultural needs and moods of the moment. Developing from a bifurcation between the fantastic tale and the staid sketch, the short story was able to index the Realism and Naturalism of the late nineteenth century and the Modernism and Postmodernism of the twentieth. Perhaps in part for this reason, many authors, critics, and readers have found the short story to be the most American of genres. Brander Matthews, who wrote the first major work of literary criticism on the "short story," contended that "Almost as soon as America began to have any literature at all it had good Short-stories. [. . . F]or fifty years the American Short-story has had a supremacy which any competent critic could not but acknowledge" (Matthews 1885, 371). His argument was widely accepted then and now, including by W. Somerset Maugham, the English novelist, playwright, and short story author, who once acknowledged, "English writers on the whole have not taken kindly to the art of the short story" (qtd. in Korte 2003, 8). For Anglophone writers, the United States is still the country where short fiction is best supported, as measured by the number and prestige of publishing outlets and short fiction awards. The short story is an American tradition and a revered and useful genre. As Amiri Baraka (1934–2014) put it, "The short story should be a sacred form – since it's the most common way we tell our lives and everybody else's. That's why, in my opinion, the most effective kind of story is short indeed, very short and pointed. Short enough and pointed enough to make your teeth curl" (Baraka 2005, 3).

4 Bibliography

4.1 Works Cited

Acosta-Bélen, Edna, and Judith Ortiz Cofer. "A MELUS Interview: Judith Ortiz Cofer." *MELUS* 18, no. 3 (1993): 83–97.

Anderson, Sherwood. *A Story Teller's Story: The Tale of an American Writer's Journey Through His Own Imaginative World and Through the World of Facts, with Many of His Experiences and Impressions Among Other Writers–Told in Many Notes–in Four Books–and an Epilogue*. Ann Arbor, MI: U of Michigan P, 2005.

Anderson, Sherwood. "Four Letters (1916–1939)." *Sherwood Anderson: A Study of the Short Fiction*. Ed. Robert Allen Papinchak. Twayne's Studies in Short Fiction; No. 33. New York: Twayne, 1992. 64–73.

Baraka, Amiri. "Short Story and Poetry." *Writers on Writing: The Art of the Short Story*. Ed. Maurice Angus Lee. Contributions to the Study of World Literature; No. 128. Westport, CT: Praeger, 2005. 3–6.

Bierce, Ambrose. *The Collected Works of Ambrose Bierce: The Devil's Dictionary*. Vol. 7. 12 vols. New York: Neale Publishing Company, 1911.

Boyle, Kay. "About These Stories." *Kay Boyle: A Study of the Short Fiction*. Ed. Elizabeth S. Bell. Twayne's Studies in Short Fiction; No. 34. New York: Twayne, 1992. 121.

Carver, Raymond. "On Writing." *Raymond Carver: A Study of the Short Fiction*. Ed. Ewing Campbell. Twayne's Studies in Short Fiction; No. 31. New York: Twayne, 1992. 93–97.

Cather, Willa. "The Novel Démeublé." *The New Republic*, 5–6, April 12, 1922.

Cheever, John. "What Happened." *John Cheever: A Study of the Short Fiction*. Ed. James Eugene O'Hara. Twayne's Studies in Short Fiction; No. 9. Boston: Twayne Publishers, 1989. 106–108.

Chesnutt, Charles W. "Journal Entry: May 29, 1880." *Charles W. Chesnutt: A Study of the Short Fiction*. Ed. Henry B. Wonham. Twayne's Studies in Short Fiction; No. 72. New York: Twayne, 1998a. 87–88.

Chesnutt, Charles W. "Post-Bellum–Pre-Harlem." *Charles W. Chesnutt: A Study of the Short Fiction*. Ed. Henry B. Wonham. Twayne's Studies in Short Fiction; No. 72. New York: Twayne, 1998b. 102–107.

Chopin, Kate. "On Certain Brisk, Bright Days." *Kate Chopin: A Study of the Short Fiction*. Ed. Bernard Koloski. Twayne's Studies in Short Fiction; No. 65. New York: Twayne, 1996. 107.

Crane, Stephen. "Selected Letters." *Stephen Crane: A Study of the Short Fiction*. Ed. Chester L. Wolford. Twayne's Studies in Short Fiction; No. 7. Boston: Twayne, 1989. 93–109.

Dowell, Bob. "The Moment of Grace in the Fiction of Flannery O'Connor." *College English* 27.3 (1965): 235–239.

Freeman, Mary E. Wilkins. "Mary E. Wilkins Freeman, an Autobiography." *The Saturday Evening Post*, December 18, 1917.

Freeman, Mary E. Wilkins. "Author's Preface to the Edinburgh Edition of *A Humble Romance*." *Mary Wilkins Freeman: A Study of the Short Fiction*. Ed. Mary R. Reichardt. Twayne's Studies in Short Fiction; No. 69. New York: Twayne, 1997. 119.

Fuller, Margaret. *Summer on the Lakes, in 1843*. New York: Little and Brown, 1844.

Gates, Jr., Henry Louis. "The 'Blackness of Blackness': A Critique of the Sign and the Signifying Monkey." *Critical Inquiry* 9.4 (June 1983): 685–723.

Gilman, Charlotte Perkins. "From The Living of Charlotte Perkins Gilman." *Charlotte Perkins Gilman: A Study of the Short Fiction*. Ed. Denise D. Knight. Twayne's Studies in Short Fiction; No. 68. New York: Twayne, 1997. 109–113.

Hale, Sarah Josepha Buell. *Sketches of American Character*. Boston: Putman & Hunt, and Carter & Hendee, 1830.
Harper, Frances Ellen Watkins. *A Brighter Coming Day: A Frances Ellen Watkins Harper Reader*. New York: Feminist P at CUNY, 1990.
Harte, Bret. *The Luck of Roaring Camp, The Outcasts of Poker Flat, Tennessee's Partner*. San Francisco: Book Club of California, 1916.
Harte, Bret. "The Rise of the 'Short Story.'" *The Cornhill Magazine*. Ed. William Makepeace Thackeray. London: Smith, Elder, & Co., 1899. 1–8.
Hemingway, Ernest. *Death in the Afternoon*. New York: Scribner, 1960.
Hemingway, Ernest. *In Our Time: Stories*. New York: Scribner, 1970.
Hemingway, Ernest. "Preface to *The First Forty-Nine*." *Ernest Hemingway: A Study of the Short Fiction*. Ed. Joseph M. Flora. Twayne's Studies in Short Fiction; No. 11. Boston: Twayne, 1989. 127–144.
Irving, Washington. *The Sketch Book of Geoffrey Crayon, Gent*. New Edition. Vol. 1. 2 vols. London: John Murray, 1822a.
Irving, Washington. *The Sketch Book of Geoffrey Crayon, Gent*. New Edition. Vol. 2. 2 vols. London: John Murray, 1822b.
Jackson, Shirley. "The Unloved Reader." *Shirley Jackson: A Study of the Short Fiction*. Ed. Joan Wylie Hall. Twayne's Studies in Short Fiction; No. 42. New York: Twayne, 1993. 113–115.
Jacobs, Harriet Ann. *Incidents in the Life of a Slave Girl*. Republished by the Negro History P, Detroit Michigan. Boston: Pub. for the author, 1861.
James, Henry. "From James's *Prefaces*." *Henry James: A Study of the Short Fiction*. Ed. Richard A. Hocks. Twayne's Studies in Short Fiction; No. 17. Boston: Twayne, 1990. 125–128.
James, Henry. "Mr. and Mrs. James T. Fields." *The Atlantic Monthly*, July 1915. 21–31.
James, Henry. *The Letters of Henry James*. Portland, Oregon: The Floating P, 2015.
Kinney, Arthur F. "Flannery O'Connor and the Fiction of Grace." *The Massachusetts Review* 27.1 (1986): 71–96.
Kirkland, Caroline Matilda. *A New Home – Who'll Follow?: Or, Glimpses of Western Life*. New York: C. S. Francis, 1839.
Korte, Barbara. *The Short Story in Britain: A Historical Sketch and Anthology*. Tübingen, Germany: Francke, 2003.
Kuehl, John. "Scott Fitzgerald's Critical Opinions." *F. Scott Fitzgerald: A Study of the Short Fiction*. Ed. John Kuehl Twayne's Studies in Short Fiction; No. 22. Boston: Twayne, 1991. 142–159.
London, Jack. "From 'The Terrible and Tragic in Fiction.'" *Jack London: A Study of the Short Fiction*. Ed. Jeanne Campbell Reesman. Twayne's Studies in Short Fiction; No. 75. New York: Twayne, 1999. 206–210.
Longfellow, Henry Wadsworth. *Outre-Mer: A Pilgrimage Beyond the Sea*. Boston: Ticknor and Fields, 1859.
Matthews, Brander. "'The Philosophy of the Short-Story.'" *Lippincott's Magazine of Popular Literature and Science* (1885): 366–374.
Miller, D. Quentin. "Updike, Middles, and the Spell of 'Subjective Geography.'" *The Cambridge Companion to John Updike*. Ed. Stacey Olster. New York: Cambridge UP, 2006. 15–28.
Oates, Joyce Carol. "The Nature of Short Fiction; or, The Nature of My Short Fiction." *Joyce Carol Oates: A Study of the Short Fiction*. Ed. Greg Johnson. Twayne's Studies in Short Fiction; No. 57. New York: Twayne, 1994. 122–128.
O'Connor, Flannery. *The Complete Stories*. New York: Farrar, Straus & Giroux, 1979.
Poe, Edgar A. "Review: Twice-Told Tales by Nathaniel Hawthorne." *Graham's Lady's and Gentleman's Magazine*, May 1842. 298–300.

Singer, Isaac Bashevis. "Interview (1963)." *Isaac Bashevis Singer: A Study of the Short Fiction.* Ed. Edward Alexander. Twayne's Studies in Short Fiction; No. 18. New York: Twayne, 1990. 84–86.

Smith, Sydney. "ART. III. Statistical Annals of the United States of America." *Edinburgh Review, or Critical Journal* 33.65 (January 1820): 69–80.

Steinbeck, John. "Letter to Edith Mirrielees." *John Steinbeck: A Study of the Short Fiction.* Ed. R. S. Hughes. Twayne's Studies in Short Fiction; No. 5. New York: Twayne, 1989. 136.

Stowe, Harriet Beecher. *Uncle Tom's Cabin: Or, Life Among the Lowly.* Vol. 1. 2 vols. Cleveland, OH: J. P. Jewett, 1852.

Tallack, Douglas. *The Nineteenth-Century American Short Story: Language, Form, and Ideology.* New York: Routledge, 1993.

Twain, Mark. "How to Tell a Story." *Mark Twain: A Study of the Short Fiction.* Ed. Tom Quirk. Twayne's Studies in Short Fiction; No. 66. New York: Twayne, 1997a. 150–154.

Twain, Mark. "Reply to the Editor of 'The Art of Authorship.'" *Mark Twain: A Study of the Short Fiction.* Ed. Tom Quirk. Twayne's Studies in Short Fiction; No. 66. New York: Twayne, 1997b. 136–137.

Twain, Mark. *The Writings of Mark Twain: Autobiography.* Ed. Albert Bigelow Paine. Stormfield ed. Vol. 36. 37 vols. New York: Harper & Brothers, 1929.

Welty, Eudora. *On Writing.* Modern Library edition. New York: Modern Library, 2002.

Wharton, Edith. "Telling a Short Story." *Edith Wharton: A Study of the Short Fiction.* Ed. Barbara A. White. Twayne's Studies in Short Fiction; No. 30. New York: Twayne, 1991. 128–138.

Williams, Herbert P. "Interview." *Stephen Crane: A Study of the Short Fiction.* E. Chester L. Wolford. Twayne's Studies in Short Fiction; No. 7. New York: Twayne, 1989. 90–92.

Woolson, Constance Fenimore. *Miss Grief and Other Stories.* E. Anne Boyd Rioux. New York: Norton, 2016.

4.2 Further Reading

Fash, Lydia. *The Sketch, the Tale, and the Beginnings of American Literature.* U of Virginia P, 2020.

Hamilton, Kristie. *America's Sketchbook: The Cultural Life of a Nineteenth-Century Genre.* Athens: Ohio UP, 1998.

Pasco, Allan H. "On Defining Short Stories." *The New Short Story Theories.* Ed Charles E. May. Athens: Ohio UP, 1994. 114–130.

Pratt, Mary Louise. "The Short Story: The Long and the Short of It." *The New Short Story Theories.* Ed Charles E. May. Athens: Ohio UP, 1994. 91–113.

Stevick, Philip. *The American Short Story, 1900–1945: A Critical History.* Boston: Twayne, 1984.

Scofield, Martin. *The Cambridge Introduction to the American Short Story.* New York: Cambridge UP, 2006.

Julia Straub

2 Canon Formation and the American Short Story

Abstract: The development of the short story took place as part of the progress towards 'American literature' as an organized body of works in the nineteenth century. This chapter unravels the intersections between the short story as an emergent genre and the manifestation of an American national literature. It examines various institutions, agents, and discourses related to literature in particular consideration of the contested concept of the literary canon.

Keywords: Literary canon, national literature, literary history, anthologies, periodicals

1 The Canon and National Literature

The beginnings of the short story in America, situated roughly in the first third of the nineteenth century, were framed by notorious prophecies of doom (↗1 Sketch, Tales, and Stories). William E. Channing, in an 1823 lecture on "The Means of a National Literature" hesitantly asked whether "[Americans] possess, indeed, what may be called a national literature?," quickly conceding that "[t]he few standard works which we have produced, and which promise to live, can hardly, by any courtesy, be denominated a national literature" (Channing 1830, 13). Sydney Smith in the *Edinburgh Review* of December 1818 claimed, darkly, that Americans did not have the privilege of a national literature at all: "Literature the Americans have none – no native literature, we mean. It is all imported" (Smith 1818, 132). There are countless reiterations of the theme of belatedness and inadequacy by Americans and foreigners (such as Smith, who was British) alike. William Cullen Bryant, in an essay for the *North American Review* issue of July 1818, bemoaned that

> [we] make but a contemptible figure in the eyes of the world, and set ourselves up as objects of pity to our posterity, when we affect to rank the poets of our own country with those mighty masters of song who have flourished in Greece, Italy and Britain. (Bryant 1818, 198)

And earlier on, in 1803, Samuel Miller in *A Brief Retrospect of the Eighteenth Century*, had detected another reason why America was an unfavourable soil for literary industry. In the second volume he wrote that

> [i]n the United States, the rewards of literature are small and uncertain. The people cannot afford to remunerate eminent talents or great acquirements. Booksellers, the great patrons of learning in modern times, are too poor to foster and reward the efforts of genius.
> (Miller 2001, 406–407)

https://doi.org/10.1515/9783110587647-003

These four examples all point towards a perception of deficiency: Channing, Smith, and Bryant complain about the absence of a national literature belonging to the United States and worthy of comparisons with the ancient masters. In this debate, Miller takes a somewhat more apologetic stance, pointing to the relative poverty of Americans, a people too busy building a country with no time to promote literary talent. Miller also brings up the pragmatic, material dimension of any national literature: without a publishing industry and the efforts of committed individuals, no national literature can arise. 'National literature' in this context could easily be replaced with the term 'canon.' Like the canon, national literatures fulfil an important function. As Stuart Hall stated, nations "are composed not only of cultural institutions, but of symbols and representations" (Hall 1994, 292–293). In order for citizens to identify with a nation, the nation needs to produce its own stories and images that will persist in the collective imagination. Sarah Corse uses the phrase "reflection theory" when describing the relationship between 'national literature' and the canon. "Reflection theory" rests on the assumption commonly held in the nineteenth and parts of the twentieth centuries that the nation comes first, and that the canon is derived from it (Corse 1991, 13), a mirror of the distinctive features that set the nation apart from others. Following such an understanding, literary texts contain some singular meaning and do not stand in interaction with social realities, as the idea of literary works as cultural texts would imply (14). Another point of criticism is that such 'reflective' approaches also fail to see the processes of which canon formation consists (15).

In this chapter, 'canon' is understood as a set of literary, secular texts that a certain group deems valuable and worthy of preservation. This is a definition in line with Aleida Assmann's understanding of canonization as a process by which "[w]hatever has made it into the active cultural memory has passed rigorous processes of selection, which secure for certain artifacts a lasting place in the cultural working memory of a society" (Assmann 2010, 100). In the nineteenth century, this society was the nation. Such a simple equation between the canon and national literature is informed by what in canon studies is referred to as a normative understanding of the canon: the assumption that a text deemed canonical possesses some inherent qualities that make it valuable and a compulsory read (see Freise 2013, 50–51). A normative approach may seem quaint given the pervasive revisionary work on the canon starting in the twentieth century (and to which we will turn below), but it has proven to be persistent. Any account of a 'great books' tradition reflects the belief that there is something like a list of particularly worthy 'masterpieces': Harold Bloom's *The Western Canon* (1994) serves as an academic example; the unbroken popularity of 'top ten' lists of novels or writers reflects an abiding need to manage a vast body of literature.

Over the last decades, much revisionary work has been done on the literary canon, highlighting the need for its plural form rather than for the one binding list that defines what is 'good' literature for everyone to read. The so-called canon wars in the 1980s triggered an enthusiastic revision of the old, conservative canon – composed of white Anglo-Saxon Protestant (WASP) authors and their works – inviting a multiplicity of voices. The canon opened up towards female authors and ensuing calls for ethnic and religious diversity cast American literature into a new, fluid shape with open boundaries, allowing space for e.g., Native American, African American, Asian American, or Chicano writers. John Marx identified the three processes of repudiation, revision, and definition of the canon, particularly in post-colonial discourses, which exposed the inequalities in both political and cultural power upon which Western imperial supremacy could thrive. In fact, much of the canon revision that affected U.S.-American literature gained impetus from post-colonial theory, but also from the Civil Rights Movement and feminism of the 1960s and 1970s. All three aimed to uncover the disenfranchisement of minorities and to lay bare the social, ethnic, and educational differences inscribed into Western societies. Regardless of whether we are looking at feminist key works such as Ellen Moers' *Literary Women* (1976) or Elaine Showalter's *A Literature of Their Own* (1977; both concerned with British women writers but triggering reverberations also with regard to the American canon) or the work of African American revisionists such as Henry Louis Gates, Jr.: the goal was to unearth forgotten or repressed texts; to shift perspectives away from male-centered or Euro-centric readings in an overall attempt to decenter models of power and knowledge (Rippl and Straub 2013, 111).

Academic discussions of the term 'canon' have multiplied definitions and challenged underlying assumptions of value and authority. Multifarious as approaches may be and regardless of the vehemence with which the canon has been critiqued, there seems to be some consensus on the canon as being the result of complex processes of selection and interpretation. These processes are historically variable and involve multiple institutions (e.g., universities, schools) and interest groups (e.g., political activists, associations). They combine both literary factors (e.g., formal or aesthetic features of texts) and extra-literary factors (social and cultural conditions at the time of production) (Rippl and Winko 2013, 2). Any approach that seeks to endow the canon with the normative power to guide a community or to convey standards of knowledge or interpretation seems dated in today's plurimedial world where the print canon does not adequately reflect the diversity of knowledge and cultural discourses disseminated through various ways (2). As a consequence, over the last few decades approaches to the canon have been much more descriptive and context-sensitive as they build up on sociological or political models of the literary canon that highlight underlying dynamics of power politics (Beilein 2013, 66). For example, John Guillory in *Cultural Capital: The Problem of Literary Canon Formation* (1993) argued that English replaced Latin as the language of public administration and education in eighteenth-century Britain and that this change of language was the driving force behind the formation of its canon. For Guillory, the school was the cradle of the modern English canon (Guillory 1993, 56) and not so much the ideals of elegance, unity or representativeness inherent in the texts themselves (which would of course be the qualities that neo-classical authors in the seventeenth and eighteenth centuries strived for).

Following such a relational or contextual approach, canon formation takes place in subtle forms and through multiple, often covert, channels. It does not require grand gestures of self-declaration the kinds of which Smith, Bryant, and Channing were calling out for. Rather, it makes sense to factor into our understanding of canon making processes what Simone Winko has described as "invisible hand" phenomena: acts or decisions within processes of canon formation that are carried out by many individuals and institutions on many different levels and that are based on acts of selection and judgment (Winko 2002, 13). An editor's acceptance of a short story for publication; an English Department's reading list or a book vendor's decision to stock up on a new publication in bulks and to put it on display would be examples. By deconstructing the notion of inherent value, such an invisible hand model links back to Barbara Herrnstein Smith's seminal work on *Contingencies of Value* (first published in 1988). Smith spoke of the "visibility" of literary texts, which depends on the cultural climate of a certain time (Smith 1991, 10). She thereby paved the way for our understanding of 'careers' of literary texts, with texts dropping off the radar of public attention at a certain point in time, while claiming the limelight at another. In other words: Smith put to the fore the contingency of value as something which traditional and essentialist approaches had remained oblivious to. "[W]ith respect to value, everything is always in motion with respect to everything else" (15) she argued, thereby undermining value as something stable and detached from the swerves of time and whims of public opinion. Similarly, Jane Tompkins's 1985 monograph *Sensational Designs: The Cultural Work of American Fiction* questioned the myths of a meaningful evolution of literary history. Why of all writers was a figure such as Nathaniel Hawthorne perceived as the American response to Charles Dickens or William M. Thackeray? In the end, most of what might appear to be talent, or in Samuel Miller's words "genius" (Miller 2001, 407), comes down to luck: "There is never a case in which circumstances do not affect the way people read and hence what they read – the text itself" (Tompkins 1985, 10) and these circumstances very often were defined by private and professional networks, and the social background of writers (8).

2 The Short Story and Literature in the Nineteenth Century

For students and scholars, it is tempting to accept the expression 'nineteenth-century American literature,' just like any period in American literary history, as a set package. Any textbook mentions Realism and Naturalism as central aesthetic paradigms, Transcendentalism as a philosophical current, historical sub-divisions such as the

Reconstruction era or the Gilded Age, periodical publishing as a major print culture factor, and, of course, a whole list of towering figures: Mark Twain (↗9 Twain), Herman Melville (↗8 Melville), Harriet Beecher Stowe, E. A. Poe (↗7 Poe), Kate Chopin (↗11 Chopin), Nathaniel Hawthorne (↗6 Hawthorne), Ralph Waldo Emerson, to name but a few. What sometimes goes unnoticed is that just like the short story, the notion of an American literature was work-in-progress in the nineteenth century. The purposeful categories used to divide and sub-divide, to make sense of seemingly disconnected events in terms of a narrative of American literature – they are all products mainly of the twentieth century.

The short story plays a crucial part in this process of canon formation. First, its emancipation into what has been called "America's most identifiable homegrown literary genre" (Patea 2012, 16) falls into this period, and some of its most central authors were practitioners of this genre. Second, the short story was to a certain extent embroiled in a number of social and political changes. As James Nagel has argued, the short story has, from its earliest days onwards, been a place of experimentation for writers of various backgrounds trying to find answers to the question of "what values and standards are to be associated with [an American citizen]" (Nagel 2015, 4; ↗0 Introduction). According to Alfred Bendixen, an author such as Washington Irving (↗5 Irving) used the short story to "engage [. . .] with issues of national identity, with the changing demands of a democratic society, with the possibilities entailed in a society marked by multiplicity and fluidity" (Bendixen 2010, 5). African American authors, women writers, voices from the margins – the short story offered a viable gateway towards becoming a published author when other doors remained closed in the world of publishing. Frances Ellen Watkins Harper's short story "The Two Offers," for example, was published in *The Anglo-African Magazine* in 1859 and counts as the first short story to be published by an African American woman in the United States (Nagel 2015, 189; ↗1 Sketches, Tales, and Stories). From today's perspective it would figure as an account of intersectionality, wrapping issues of racial discrimination up in a critique of sexism in society at large. By all means, the short story did important cultural work at a time of political and social reform, and here lies one of its links to notions of a national literature or canon.

In addition to offering platforms for socio-political debate, the short story has always been rather versatile with regard to aesthetic approaches and owed much of its success to its crossing of boundaries (see Achilles and Bergmann 2015). In the nineteenth century, it catered to the needs of both realist and naturalist writers; it offered space for regional authors, allowed the supernatural imagination to unfold, and lent itself to the comic mode. As a genre it was flexible enough to allow for the co-existence of both Romanticism and Realism in the nineteenth century (Patea 2012, 2). Thus, itself a highly adaptable literary form, it had the potential of reaching a general readership while offering itself to the purposes of marginalized writers.

Most recently, Nancy Glazener, in a study tellingly titled *Literature in the Making*, examined U.S.-American literary culture and its emergent institutions. Literature began to stand for imaginative writing with an aesthetic appeal, rather than, as had been the case up until the end of the eighteenth century, for the general body of published writing:

> During the first half of the 19th century, the kinds of writings widely understood to count as literature [. . .] narrowed, so that literature came to be associated with a particular set of imaginative genres rather than the tapestry of learned culture it had designated before.
>
> (Glazener 2016, 4)

A similar point was made by Trevor Ross, who showed how "literature" in a modern sense emerged via the intermediary stages of "poesy" and "belles lettres" in eighteenth-century Britain (Ross 1996, 404). It is in the nineteenth century only that a "transgeneric canon," a general canon of literature that covers more than just e.g., poetry, developed (Glazener 2016, 44). This "zest for generic regulation" (186) also affected the short story in the course of the century. While first seen as an "utterly new literary phenomenon" (186), it underwent continued formalization and theorization later on, resulting in first academic studies such as Brander Matthews's *The Philosophy of the Short Story* (1902).

This morphing of literature into a shape that is recognizable to us today required structuring and sculpting. Taxonomies of genres, charts of literary periods, authors' biographies, publishing histories – these are all tools meant to tell the grand narratives of literature that scholarship has traditionally relied upon, and to thereby create and maintain a canon. Discussing the situation in eighteenth-century Britain, Jan Gorak, in *The Making of the Modern Canon*, claimed that

> [. . .] eighteenth-century authors begin to push the idea of *canon* in its modern direction when they develop a new self-consciousness about the value of national authors, whose works they think deserve a new 'scientific criticism' in possession of its own independent canons.
>
> (Gorak 1991, 47)

This "scientific criticism" would take some time to develop in the United States. Kermit Vanderbilt gives an impression of the complicating factors that made the U.S.-American situation special:

> The staple issues for the literary historiographer, as for the literary historian, lie in the utterly basic questions: How do we define and explore the obvious terms American, literary, and history? The *American* aspects include the conception of new-world literary devices and cultural themes (including ideas, subjects, myths, indigenous and archetypal characters) leading to a national ethos [. . .]. The literary aspects involve, after a framing definition of literature itself, an explanation of distinctive periods, movements, and schools; a definition of genres, subgenres, and modes; the structure of themes; explication of substance into appropriate form; and critical judgments that ultimately establish a 'canon' of American literary works.
>
> (Vanderbilt 1986, 93; original emphasis)

Thus, canon formation in the nineteenth-century United States meant not only a honing of concepts of literature and its tools and categories, but it was also fraught with the more existential question of what qualities define American writing in relation to both the indigenous presence and 'Old World' traditions. While the scope of this article does not allow for an in-depth discussion of these complexities, Vanderbilt reminds us that even though calls for an American canon were loud, its basic assumptions of cultural and political identity were unstable. In eighteenth-century Britain, one of the challenges for the formation of a canon had been the conflict between the old and the modern (a conflict in aesthetic and moral thought starting in the seventeenth century, also known as the *Querelle des Anciens et Modernes*). In the United States the classical educational ideal was of course engrained into the syllabi of the colleges and permeated learned culture (see Winterer 2007), but the classical legacy did not press quite as hard on the debates as the definition of Americanness itself.

3 Canon Formation and the Short Story in the Nineteenth Century

Literary Magazines: The connection between the short story and periodical publishing, which includes newspapers, magazines, and other miscellany publications, but will here refer specifically to the literary magazine, is close and well-known (↗4 Textual Materiality). Both Hawthorne and Poe, for example, had trouble finding publishers for collections of their short stories; in a vein typical for the fate of other writers at their time, their short stories were scattered across various publications, magazines, or so-called gift-books (see Bendixen 2010, 8). The limited amount of space and the fast-paced serial publication rhythm of historical magazines made them ideal vessels for the short story as a literary form consecrated to brevity and succinctness. Credits here go in particular to Poe, who "exalted the short story for the traits it shared with the magazine itself (i.e., disposability, space conservation)" (Levy 1993, 11). In other words, there was a perfect match between the structural requirements of magazine publishing and the formal features of short stories.

Literary magazines existed in the United States from the middle of the eighteenth century onwards, with Andrew Bradford's *American Magazine* and Benjamin Franklin's *General Magazine, and Historical Chronicle* being the first two examples from the 1740s (on periodical publishing in the eighteenth and nineteenth centuries see Richardson 1931, Tebbel and Zuckerman 1991; more recent interventions are Lanzendörfer 2013, Okker 2003, Straub 2017). Despite their editors' ambitions, most magazines from this period were short-lived and unsuccessful (see Gardner 2012) and only few titles developed an innovative and memorable profile like e.g., the *Massachusetts Magazine*, the *Columbian Magazine*, or the *Port-Folio*, which turns them into valuable sources today. But with the advent of the nineteenth century at the latest, periodicals were both a commercial and a cultural force to be reckoned with. Magazines became increasingly literary in their interests, developing into much sought after platforms for the publication of original poetry but also short prose fiction – what would eventually become short stories. Titles such as *The North American Review*, which published Washington Irving's *The Sketch Book of Geoffrey Crayon, Gent* in 1815, *Graham's Magazine*, *Atlantic Monthly*, *The New England Magazine*, or *Godey's Lady's Book* became well-known for their publication of short stories by Poe, Hawthorne, or influential reviews of short stories. Poe's influential review of Hawthorne's *Twice-Told Tales*, for example, first appeared in two 1842 issues of *Graham's Magazine*; he himself was editor of *The Southern Literary Messenger* from 1835–1837.

However, the almost symbiotic relation between this publishing format and the short story also matters with regard to canon formation. Recent studies have re-examined the importance of precursory forms of short prose writing from the eighteenth century for the development of the short story. As Oliver Scheiding (2015) has shown, short narratives – e.g., sentimental stories, tales, sketches, moral anecdotes, captivity narratives, and fragments – had been essential elements in periodical publishing. Often anonymous or lifted from unidentified sources in reprint, these texts are somewhat unruly since they lack personalized authorship. Their provenance can thus not always be defined. These early forms, often randomly spread across magazine issues, were occasionally accommodated in fluctuating sections titled e.g., "Tales, & Co." (*American Museum*, November 1789). While notorious for their reprinting practices and unscrupulous attitude towards harnessing filling material for their publications (see Cohen 2012, McGill 2007), magazine editors did significant work on the canon by carving out a space for short stories, creating rubrics, and thus forging a terminology. They brought short prose writing to the attention of an ever-increasing readership and to a certain extent helped to establish short narratives as regular features of the literary industry. As James Nagel argued, the short story and its precursory forms, often from oral traditions, represented a "blending of narrative traditions from a broad spectrum of sources and languages" (Nagel 2015, 5). Magazines provided showcases of diverse facets of cultural identity in terms of multilingualism, transatlantic, and global circulation processes (e.g., on a worlding approach to American literature see Scheiding and Seidl 2015). Thus, the process of accommodating short prose pieces and later on proper short stories as part of an ongoing cultural and literary discourse enabled by these magazines opened readerships and caused awareness of this new form of literature.

Literary Histories and Anthologies: There have always been strong commercial investments in the magazine, but also the anthology as a canon-making device (see Glazener 2016, 44). Anthologies are remarkable tools of canon formation because of their inevitably rather unabashed exclusionary strategies. As Barbara Benedict states, anthologies "are characterized as volumes that contain material selected self-consciously for consistency and quality" (Benedict 2003, 231). Anthologies are by definition meant to narrow down a field of interest to what is perceived as most relevant or precious; coming from the Greek, the word means "a collection of flowers" (*OED*). In a cultural environment where the literary production is rich, anthologies are meant to curb the overabundance. Still today – and despite the continuous revisionary work on the canon – anthologies persist as popular devices, in the classroom in particular. Often endowed with a mass-appeal, they can be seen as the "staple of literary mediation and cultural memory" (Korte 2000, 5). No matter whether we look at the *Heath Anthology of American Literature* or the *Norton Anthology of American Literature*, both of which (and especially the *Heath*) have undergone scrupulous revision over the past decades and are indebted to a self-critical scrutiny of their own agenda, eschewing the "monocultural curricula of the 1950s" (Lauter 1994, 181): the short story figures prominently in these works that suggest that the reader buys not only a book, but also the illusion of 'owning' American literature. Hardly any other publishing format sells the idea of a canon as well and efficiently as these major anthologies.

What counts as the first anthology published in America was a collection of shorter prose pieces entitled *Select Essays, with some few miscellaneous copies of verse drawn by Ingenious hands*. It came out in 1714 and only one copy survived (see Boys 1945, 129). Late eighteenth-century anthologies by Elihu Hubbard Smith and Mathew Carey focused on poetry only. However, the market for anthologies became busy relatively quickly in the nineteenth century due to the likes of Rufus Griswold, whose tireless editing work led to the publication of *Prose Writers of America* in 1847. It included short stories previously published elsewhere such as "The Fall of the House of Usher" by Poe or Irving's "Rip van Winkle," which still today appear in anthologies. But there were also many excerpts from longer prose pieces (e.g., "Niagara" from Margaret Fuller's *Summer on the Lakes*) or other kinds of prose writing such as essays (e.g., Benjamin Franklin's "The Way to Wealth"). Included women writers were, next to Fuller, Lydia M. Child, and Caroline M. Kirkland. Lynda Prescott has identified a "standard route" of how the short story often found its way into the canon and which

> would probably begin with the short story being published in a magazine, then, if the writer is sufficiently successful, it might be gathered with other stories in a single-authored collection; from here an anthologist might single it out for reprinting amongst a number of stories by other writers. (Prescott 2016, 565)

The transmission of short stories and their persistence in the literary canon required an ongoing circulation and re-appropriation in different media and formats: from the magazine to the single-author collection, into 'best of' anthologies.

Authors had to struggle for recognition. Claudio Guillén argued that it was in the late eighteenth century that literature took concrete shape through the lives of individual author figures. Authors' biographies served the interpretation of their works (Guillén 1971, 5–6). While Britain saw its first literary celebrity with Lord George Gordon Byron, the United States revered Nathaniel Hawthorne the most (see Tompkins 1985, 25), but Poe also appeared in the public eye, his image wavering between that of an eccentric outsider and that of a misfit (Tomc 2012, 127). Works such as Samuel Johnson's *Lives of the Poets* (1779–1781) and James Boswell's *Life of Johnson* (1791) had certainly paved the way for such a personalized, biography-based approach to literature. The author as agent in a drama played out in literature – this idea affected literary historiography even more. Moses Coit Tyler's *A History of American Literature* was published in 1878, covered the Puritan period up to the end of the Revolution, and is often seen as the first cohesive account of American literature. Fred Lewis Pattee's *A History of American Literature since 1870* (1896) and Stedman and Hutchinson's eleven-volume *A Library of American Literature from the Earliest Settlement to the Present Time* (1891) were further attempts at narrating a smooth flow of literature unfolding across time with writers akin to characters. These early literary histories, Nina Baym argued, helped define American literature as a subject to be taught at schools and universities throughout most of the twentieth century:

> The story that they told, put simply, is this: in the face of numerous obstacles, a bona fide literary culture had developed in America before the Civil War. Before about 1830 there was no such culture in America; after 1830 there was. (Baym 1989, 459–460)

Pattee devoted an entire chapter to the "Triumph of the Short Story" in his study. He found praise for Poe and Hawthorne but bemoaned that "the short story of the mid-century, even in its best specimens, was an imperfect thing" (Pattee 1915, 357). Only towards the end of the century did the form reach maturity. Pattee's canon of short story writers lists "Aldrich, Stockton, Bunner, Garland, Allen, Bierce, Grace King, Mrs. Chopin, Stephen Crane, and Frank Norris," who, "from the standpoint of mere art at least, come near to perfection" (359). Interestingly, apart from Chopin, Crane, and Norris, none of the mentioned writers would be considered as household names for a more general readership today.

4 Conclusion

This chapter focused on processes of canon formation that took place concomitantly with the rise of the short story as a genre in the nineteenth century. The short story continued to thrive throughout the twentieth and into the twenty-first centuries due to the acts of canonization described above (among others): the consolidation of the form in popular media, circulation among a relatively wide readership, reprinting in multiple sources across lengthy periods of time in various kinds of publishing formats, reception histories in the form of reviews, and inclusion in scholarly works such as literary histories. But there are of course processes and means which shape today's canon, many of which allowing or even requiring a considerable level of interactivity (e.g., lay criticism on amazon.com or goodreads.com). Does the internet phenomenon of "going viral" count as a canon-making, invisible hand phenomenon? If so, *The New Yorker*'s publishing of a short story titled "Cat Person" by Kristen Roupenian in 2017 would serve as a good example of the cultural work done by the short story and the involved gate keepers (↗4 Textual Materiality). Based on a date gone awry, the story drew wide circles as part of the #MeToo debate and brought immediate fame upon its hitherto little-known author.

In 2013, Alice Munro, a Canadian author of short stories, received the Nobel Prize for literature, the epitome of all canonization processes. Not only did this prize honor Munro as an outstanding author, but it was also an ennobling gesture, or what Munro herself termed a "wonderful thing" (Allardice 2013), for the genre of the short story. To be sure, this was also a sorely needed act of reverence, seeing that, as Lydia Prescott put it, "the market for single-authored short story collections is uncertain" (Prescott 2016, 565), i.e., mono-authored books of short stories tend to not sell well. The circulation pathways for short stories today, and thus also their admittance into processes of canon formation, still resembles that of the nineteenth century. *The New Yorker* magazine, the short story writer's "Mecca" (Levy 1993, 124), began publishing "identifiably serious short fiction" in 1928, and in 1940 its own anthology, *Short Stories from the New Yorker*, came out (Yagoda 2000, 151). *The New Yorker* is a canon-making institution *sui generis* in today's unstable media landscape because of its ongoing (re-)anthologizing of its own short stories and its branding of its own type of short story, associated e.g., with John Cheever and John Updike. Its short stories are characterized by

> carefully wrought, many-comma'd prose; its long passages of physical description, the precision and the sobriety of which created a kind of negative emotional space, a suggestion of feeling without the naming of it; its well-educated white characters, who could be found experiencing the melancholies of affluence, the doldrums of suburban marriage, or the thrill or the desolation of adultery; and, above all, its signature style of ending, which was either elegantly oblique or frustratingly coy, depending on your taste. (Franzen 2015)

Furthermore, *The New Yorker* often leveraged careers and helped 'canonize' previously little-known writers (consider the example of Bobbie Ann Mason's "Shilo," first published in *The New Yorker* in 1982 before being hailed as the "best the 1980s short story had to offer" [Levy 1993, 117]).

In general, discussions of the canon today need to take into account its aforementioned diversification, reflecting the hyphenization of American literature, and a general tendency in current literary scholarship towards global perspectives in the wake of what is sometimes called a transnational turn (e.g., Jay 2010; Fluck et al. 2011). Add to that the challenges and possibilities provided by the new, i.e., digital, media and their impact on literature, genres, and publishing practices, and it becomes obvious that more than ever terminologies and categories are in flux.

5 Bibliography

5.1 Works Cited

Achilles, Jochen, and Ina Bergmann, eds. *Liminality and the Short Story: Boundary Crossings in American, Canadian, and British Writing*. New York: Routledge, 2015.

Allardice, Lisa. "Interview: Nobel prizewinner Alice Munro: 'It's a wonderful thing for the short story,'" https://www.theguardian.com/books/2013/dec/06/alice-munro-interview-nobel-prize-short-story-literature. *The Guardian* 6 December 2013. (20 March 2020).

Anon. Eds. *Select essays, with some few miscellaneous copies of verse drawn by ingenious hands*. Boston: n.p., 1714.

Assmann, Aleida. "Canon and Archive." *A Companion to Cultural Memory Studies*. Eds. Astrid Erll and Ansgar Nünning. Berlin: De Gruyter, 2010. 97–107.

Baym, Nina. "Early Histories of American Literature: A Chapter in the Institution of New England." *American Literary History* 1.3 (1989): 459–488.

Beilein, Matthias. "Literatursoziologische, politische und geschichtstheoretische Kanonmodelle." *Handbuch Kanon und Wertung*. Eds. Gabriele Rippl and Simone Winko. Stuttgart: Metzler, 2013. 66–76.

Bendixen, Alfred. "The Emergence and Development of the American Short Story." *A Companion to the American Short Story*. Eds. Alfred Bendixen and James Nagel. Malden, MA: Blackwell, 2010. 3–19.

Benedict, Barbara. "The Paradox of the Anthology: Collecting and Différence in Eighteenth-Century Britain." *New Literary History* 34 (2003): 231–256.

Bloom, Harold. *The Western Canon: The Books and School of the Ages*. New York: Harcourt Brace, 1994.

Boys, Richard C. "The Beginnings of the American Poetical Miscellany, 1714–1800." *American Literature* 17.2 (1945): 127–139.

Bryant, William C. "Art IX – An Essay on American Poetry." *North American Review and Miscellaneous Journal* 7.20 (1818): 198–212.

Channing, William E. *The Importance and Means of a National Literature*. London: Edward Rainford, 1830.

Cohen, Lara L. *The Fabrication of American Literature: Fraudulence and Antebellum Print Culture*. Philadelphia: U of Pennsylvania P, 2012.

Corse, Sarah M. *Nationalism and Literature: The Politics of Culture in Canada and the United States*. Cambridge: Cambridge UP, 1991.

Fluck, Winfried, Donald E. Pease, and John Carlos Rowe, eds. *Re-Framing the Transnational Turn in American Studies*. Lebanon: Dartmouth College P, 2011.

Franzen, Jonathan. "The Birth of 'The New Yorker Story,'" https://www.newyorker.com/-books/page-turner/the-birth-of-the-new-yorker-story. *The New Yorker Magazine* 27 October 2015. (16 January 2019).

Freise, Matthias. "Textbezogene Modelle: Ästhetische Qualität als Maßstab der Kanonbildung." *Handbuch Kanon und Wertung*. Eds. Gabriele Rippl and Simone Winko. Stuttgart: Metzler, 2013. 50–58.

Gardner, Jared. *The Rise and Fall of American Magazine Culture*. Champaign: U of Illinois P, 2012.

Glazener, Nancy. *Literature in the Making: A History of U.S. Literary Culture in the Long Nineteenth Century*. Oxford: Oxford UP, 2016.

Griswold, Rufus W., ed. *The Prose Writers of America with a Survey of the History, Condition, and Prospects of American Literature*. Philadelphia: Carey and Hart, 1847.

Gorak, Jan. *The Making of the Modern Canon: Genesis and Crisis of a Literary Idea*. London: Athlone, 1991.
Guillén, Claudio. *Literature as System: Essays Toward the Theory of Literary History*. Princeton: Princeton UP, 1971.
Guillory, John. *Cultural Capital: The Problem of Literary Canon Formation*. Chicago: Chicago UP, 1993.
Hall, Stuart. "The Question of Cultural Identity." *Modernity and Its Futures*. Eds. Stuart Hall, David Held, and Tony McGrew. Cambridge: Polity P, 1994. 273–316.
Jay, Paul. *Global Matters: The Transnational Turn in Literary Studies*. Ithaca: Cornell UP, 2010.
Khazan, Olga. "A Viral Short Story for the #MeToo Moment," https://www.theatlantic.com/technology/archive/2017/12/a-viral-short-story-for-the-metoo-moment/548009/. *The Atlantic* 11 December 2017. (05 February 2021).
Korte, Barbara. "Flowers for the Picking: Anthologies of Poetry in (British) Literary and Cultural Studies." *Anthologies of British Poetry: Critical Perspectives from Literary and Cultural Studies*. Eds. Barbara Korte, Ralf Schneider, and Stefanie Lethbridge. Amsterdam: Rodopi, 2000. 1–32.
Lanzendörfer, Tim. *The Professionalization of the American Magazine: Periodicals, Biography, and Nationalism in the Early Republic*. Paderborn: Schöningh, 2013.
Lauter, Paul. "The Heath Anthology and Cultural Boundaries." *English Studies/Cultural Studies: Institutionalizing Dissent*. Urbana: U of Illinois P, 1994. 180–190.
Levy, Andrew. *The Culture and Commerce of the American Short Story*. Cambridge: Cambridge UP, 1993.
Marx, John. "Postcolonial Literature and the Western Literary Canon." *The Cambridge Companion to Postcolonial Literary Studies*. Ed. Neil Lazarus. Cambridge: Cambridge UP, 2004. 83–96.
McGill, Meredith. *American Literature and the Culture of Reprinting, 1834–1853*. Philadelphia: U of Pennsylvania P, 2007.
Miller, Samuel. "Nations Lately Become Literary." *A Brief Retrospect of the Eighteenth Century*. 1803. Vol. 2. Bristol: Thoemmes, 2001. 302–410.
Moers, Ellen. *Literary Women: The Great Writers*. 1976. Oxford: Oxford UP, 1977.
Nagel, James. *The American Short Story Handbook*. Malden, MA: Wiley Blackwell, 2015.
Okker, Patricia. *Social Stories: The Magazine Novel in Nineteenth-Century America*. Charlottesville: U of Virginia P, 2003.
Patea, Viorica. "The Short Story: An Overview of the History and Evolution of the Genre." *Short Story Theories: A Twenty-First-Century Approach*. Ed. Viorica Patea. Amsterdam: Rodopi, 2012. 1–24.
Pattee, Fred Lewis. *A History of American Literature since 1870*. 1896. New York: D. Appleton-Century Company, 1915.
Prescott, Linda. "The Short Story Anthology: Shaping the Canon." *The Cambridge History of the English Short Story*. Ed. Dominic Head. Cambridge: Cambridge UP, 2016. 564–580.
Price, Kenneth, and Susan Belasco Smith, eds. *Periodical Literature in Nineteenth-Century America*. Charlottesville: U of Virginia P, 1995.
Richardson, Lyon. *A History of Early American Magazines, 1741–1789*. New York: T. Nelson and Sons, 1931.
Rippl, Gabriele, and Simone Winko. "Einleitung." *Handbuch Kanon und Wertung*. Eds. Gabriele Rippl and Simone Winko. Stuttgart: Metzler, 2013. 1–5.
Rippl, Gabriele, and Julia Straub. "Kanon und Macht (Gender, Race, Postcolonialism)." *Handbuch Kanon und Wertung*. Eds. Gabriele Rippl and Simone Winko. Stuttgart: Metzler, 2013. 110–119.

Ross, Trevor. "The Emergence of 'Literature': Making and Reading the English Canon in the Eighteenth Century." *ELH* 63 (1996): 397–422.
Roupenian, Kristen. "Cat Person," https://www.newyorker.com/magazine/2017/12/11/cat-person. *The New Yorker Magazine* 11 December 2017. (05 February 2021).
Scheiding, Oliver. "'Small Tales': Brevity and Liminality in Early American Magazines." *Liminality and the Short Story: Boundary Crossings in American, Canadian, and British Writing*. Eds. Jochen Achilles and Ina Bergmann. New York: Routledge, 2015. 121–133.
Scheiding, Oliver, and Martin Seidl, eds. *Worlding America: A Transnational Anthology of Short Narratives before 1800*. Stanford: Stanford UP, 2015. 1–17.
Showalter, Elaine. *A Literature of Their Own: From Charlotte Brontë to Doris Lessing*. London: Virago, 2009.
Smith, Barbara H. *Contingencies of Value: Alternative Perspectives for Critical Theory*. Cambridge: Harvard UP, 1991.
Smith, Sydney. "Travels in Canada and the United States, in 1816 and 1817, By Lieutenant Francis Hall." *Edinburgh Review, or Critical Journal* 31.61 (1818): 132–150.
Stedman, Edmund C., and Ellen Mackay Hutchinson, eds. *A Library of American Literature from the Earliest Settlement to the Present Time*. 11 vols. New York: Charles L. Webster, 1889–1891.
Straub, Julia. *The Rise of New Media, 1750–1850: Transatlantic Discourse and American Memory*. New York: Palgrave, 2017.
Tebbel, John, and Mary E. Zuckerman. *The Magazine in America, 1741–1990*. New York: Oxford UP, 1991.
Tomc, Sandra. *Industry and the Creative Mind: The Eccentric Writer in American Literature and Entertainment, 1790–1860*. Ann Arbor: U of Michigan P, 2012.
Tompkins, Jane. *Sensational Designs: The Cultural Work of American Fiction, 1790–1860*. New York: Oxford UP, 1985.
Tyler, Moses C. *A History of American Literature, 1607–1667*. New York: G. P. Putnam, 1878.
Vanderbilt, Kermit. *American Literature and the Academy: The Roots, Growth, and Maturity of a Profession*. Philadelphia: U of Pennsylvania P, 1986.
Winko, Simone. "Literatur-Kanon und 'invisible hand'-Phänomen." *Literarische Kanonbildung*. Ed. Heinz L. Arnold. Special issue of Text + Kritik. Munich: Boorberg Verlag, 2002. 9–24.
Winterer, Caroline. *The Mirror of Antiquity: American Women and the Classical Tradition, 1750–1900*. Ithaca: Cornell UP, 2007.
Yagoda, Ben. *About Town: 'The New Yorker' and the World It Made*. Cambridge, MA: Da Capo P, 2000.

5.2 Further Reading

Kramnick, Jonathan. *Making the English Canon: Print-Capitalism and the Cultural Past, 1700–1770*. Cambridge: Cambridge UP, 1999.
Lockard, Joe, and Jillian Sandell. "National Narratives and the Politics of Inclusion: Historicizing American Literature Anthologies." *Pedagogy* 8.2 (2008): 227–254.
Olsson, Anders. *Managing Diversity: The Anthologization of 'American Literature.'* Uppsala: Uppsala U Library, 2000.
Palumbo-Liu, John. *The Ethnic Canon: Histories, Institutions, and Interventions*. Minneapolis: U of Minnesota P, 1995.

Patey, Douglas L. “The Eighteenth Century Invents the Canon.” *Modern Language Studies* 18 (1988): 17–37.
Stokes, Claudia. *Writers in Retrospect: The Rise of American Literary History, 1875–1910*. Chapel Hill: U of North Carolina P, 2006.
Straub, Julia. “Early American Literature: Canon Theory in a Transatlantic Context.” *Comparative American Studies* 9.2 (2011): 104–116.

Michael Basseler

3 Current Approaches to the American Short Story

Abstract: The aim of this chapter is to sketch out a broad variety of recent approaches to the American short story. Although short story criticism may not be among the most innovative fields of research in 21st-century literary studies, there have actually been a number of interesting and original critical interventions off the beaten tracks of earlier approaches to the genre. Taking a closer look at the handbooks, essay collections, and monographs on the American short story since 2000, the chapter identifies some of the most prevalent themes, innovative approaches and forms of social and historical contextualization, providing students of the American short story with some guidance and suggestions for further research.

Keywords: Short story theory, literary history of American short story, contextualization, transnational and transgeneric approaches, thematic definitions, digital analysis, form

1 Introduction

The task of writing a chapter about current approaches to the American short story two decades into the twenty-first century feels a bit like being asked to write an obituary (↗0 Introduction). Various critics and writers have lamented the deplorable condition of one of literary America's once most thriving genres; we may think of Stephen King's 2007 essay "What Ails the Short Story?" in the *New York Times*, tellingly illustrated with an image of a bleeding volume of American Short Stories. Short story criticism, too, has had its days long ago, in a distant time when literary studies used to be a philological island largely unaffected by cultural studies and interdisciplinarity, and when other media were no serious contender to the preeminence of literature in English departments. It already says a lot that four of the most comprehensive volumes on short story theory to date – Clare Hanson's *Re-Reading the Short Story* (1989), Susan Lohafer and Jo Ellyn Clarey's *Short Story Theory at a Crossroads* (1989), and Charles May's *Short Story Theories* (1977) and *New Short Story Theories* (1994) – were all published within a relatively short time span some thirty years ago (see Basseler 2011, 41–64).

Since then, one mostly searches in vain for theoretical and methodological innovations that breathe new life into short story criticism. As Melissa Bostrom contends, contemporary short story criticism has become a highly conservative field, "the final vestige of New Criticism in twenty-first century literary studies," a domain

https://doi.org/10.1515/9783110587647-004

that has "managed to seal itself in a time capsule as the genre it celebrates becomes less celebrated, less visible, less valuable in American literary life" (Bostrom 2007, 8). And Sarah Copland, in a recent essay published as a response to a special issue on short story theory in *Narrative*, laments the mutual lack of interest for each other's work that short story scholars and other literary scholars, especially narrative theorists, have shown over the last decades. While she asks the former why short story theory has "not availed itself of the tools and methodologies of narrative theory in general in a more sustained, systemic, and widespread way" (Copland 2014, 132), the latter have to answer whether and how short stories constitute a particular kind of narrative and therefore require a specific approach that is fine-tuned to the short story's generic particularities. To put it bluntly, then, there does not seem to be much in the way of 'current approaches' to the short story that would be worth the paper on which they are printed.

Rather than stopping at this sobering diagnosis, however, the aim of this chapter is to provide a somewhat less downbeat overview of more recent approaches to the American short story. The chapter will demonstrate that despite the general doldrums that short story criticism arguably still finds itself in at the beginning of the twenty-first century, there have actually been a number of interesting and original critical interventions, especially over the past ten years or so, and that there might be reason to believe in a future for (American) short story criticism off the beaten tracks. Taking a closer look at the handbooks, essay collections, and monographs on the American short story since 2000, the aim is to identify some of the most prevalent themes, innovative current approaches and forms of social and historical contextualization, and thereby to give students of the American short story some guidance and suggestions with regard to further research.

2 Recurring Themes, Emergent Approaches, New Forms of Contextualization

The formalist legacy of the first big wave of short story theory that arguably started with Norman Friedman's essay "What Makes a Short Story Short?" (1958) and lasted for some three decades continues to shape the field until today. While critical debates often still persistently revolve around Friedman's seemingly simple question, there has also been a general tendency to circumvent it altogether, as it too often has led to little productive discussions about the short story's essence in distinction to other literary genres, especially the novel. For example, as Blanche Gelfant, editor of the *Columbia Companion to the Twentieth-Century American Short Story*, remarks, the aim of that volume is to offer "practical criticism rather than theory," and to suggest "ways of reading for understanding and pleasure" (Gelfant 2000b, 2). She thereby deliberately bypasses the "vexed questions" (2) about the essence of the form that have governed short story criticism in the twentieth century. Apart from a hundred very short essays on individual short story writers from Alice Adams and Sherwood Anderson to Hisaye Yamamoto and Anzia Yezierska, the volume features eleven thematic essays. The foci of these essays do not display any systematic approach but range from formal ("The American Short Story Cycle") to ethnic and other identity-related issues (e.g., "The African American Short Story," "The Chicano-Latino Short Story," "Lesbian and Gay Short Stories") to some very specific content-based classifications (e.g., "The Ecological Short Story," "American Short Stories of the Holocaust").

In his chapter on the African American short story, for instance, David Lionel Smith (Smith 2000, 25–33) explains the relatively low popularity of the short story genre among African American writers, who historically turned to autobiographical writing (slave narratives) and other non-fictional forms in their efforts to confront a white readership with the ills of slavery and segregation. Only in the Reconstruction era did writers like Charles Chesnutt (↗10 Chesnutt) begin to employ the short story form for their, now also formally more experimental, explorations of race relations in the United States. At the first heyday of African American letters during the Harlem Renaissance, writers like Jean Toomer, Langston Hughes, and Zora Neale Hurston (↗17 Hurston), and later also Richard Wright, paved the way for later generations of African American writers for whom the short story provided an important arrow in their generic quiver even if other forms, especially drama, poetry, and then the novel outplayed the short story in their importance for African American writers' struggles for recognition and cultural self-expression, especially during the Black Arts Movement.

While Smith concludes his survey with the claim that the short story remains a minor genre for African American writers and that there is "no distinctively African American style or tradition of the short story" (33), such a view has been challenged recently. Focusing on Richard Wright's stories, for instance, Glenda R. Carpio works out the "intercultural and intertextual liminalities" (Carpio 2015, 222) of the African American short story, thus valorizing the role that African American writers have played in the development of the genre more broadly. Juxtaposing the major works of contemporaries Jean Toomer and Sherwood Anderson (*Cane* and *Winesburg, Ohio*, respectively; ↗15 Anderson), Mark Whalan (2007) has persuasively argued that the short story cycle provided the ideal vehicle for their seismographic accounts of socio-cultural changes, and especially of race and gender-related issues during the modernist period. Race also serves as the main analytical category for James Nagel's *Race and Culture in New Orleans Stories* (2014). Nagel offers close and contextual readings of story cycles by George Washington Cable, Grace King, Alice Dunbar-Nelson, and Kate Chopin (↗11 Chopin), all of which pertain to the local color tradition of the New Orleans region. Delineating the co-emergence of local color writing and the short story in the nineteenth century, he shows how the multiethnic culture of the region, with all its social tensions, conflicts, and legal regulations, "provided a perfect setting for the Local Colorists of the area" (Nagel 2014, 8). All of this resonated particularly strongly in the short story (cycle), especially in its settings and precise descriptions of space, its common folk characters and their dialect, the specific regional customs and traditions portrayed, as well as the moral dilemmas these stories presented.

Notions of the short story as a "minor literature" (Deleuze and Guattari 2003) that allows writers to subversively challenge hegemonic power structures and forms of life are fairly widespread among current approaches to the genre. Amy Ling (2000) depicts the "Asian American short story" as a thriving genre that, because of its flexibility and accessibility, provides the reader with rich insights into migration experiences and identity formation processes within the various communities and groups subsumed under the broad label "Asian American." In his chapter on "The Chicano-Latino Short Story" in the same volume, Ramón Saldívar goes even further by arguing that the "collective histories spoken by the characters of Chicano-Latino authors are like morality tales that pluralize the meaning of American political and social life, that violate the taboos erected by uncritical classist, racist, and sexist ruling orders, that politicize the word and proclaim its transforming potential" (Saldívar 2000, 48). In his view, the short story (by authors like Américo Paredes, Sandra Cisneros ↗29 Cisneros, and Dagoberto Gilb) provides the ideal form for this pluralization and democratization of voices in the service of an ongoing social transformation in the United States at least since the late twentieth century. Similarly, short stories by Native American writers have frequently been read as satirical attacks on racial stereotypes about indigenous people. Particularly the work of recent writers such as Sherman Alexie and Thomas King has been instructive in that regard. Yet Native American writers have also significantly contributed to the generic formation and development of the short story cycle, of which Louise Erdrich's *Love Medicine* (1984; ↗30 Erdrich) is perhaps the prime example. As James Ruppert therefore argues, short stories and short story cycles by Native American writers have often "reinforced a belief in the continuance and continuity of Native cultures and values" (Ruppert 2000, 67), rather than dwelling on historical oppression.

A relatively recent and still quite small branch of research has begun to tackle the almost invisible abundance of American short stories written in languages other than English. As part of his research on multilingual American literature (see Sollors and Shell 2000a), Werner Sollors has sketched a history of non-English American short stories: "American literature in Yiddish, Polish, Swedish, Welsh, Norwegian, Portuguese, Spanish, Chinese, and German – the list goes on and on – offers fascinating insights into American ethnic diversity in formally accomplished and thematically provocative works" (Sollors 2000b, 72). A multilingual, transnational, and transcultural approach to the American Short Story is still in its fledgling state and thus offers great potential for future research, especially in our days of increasing digitalization and accessibility of archival material. Using the Jewish American writer Abraham Cahan as one of his examples, Sollors shows how an analysis of Cahan's most famous story about "Yekl" (or, "Yankel") in the Yiddish original and English translation reveals significant differences between versions when it comes to strategies of cultural mediation as well as social criticism: "Cahan's Yiddish-language tales include omniscient socialist narrators instead of English-language happy endings; more social criticism, more profanities in characters' dialogues, and more sexual frankness than those 'same' tales would contain in English" (74). Although it is merely a first sketchy outline to set the ball rolling, Sollors' essay makes a strong case for the vast hidden treasures of non-English American short fiction that still wait to be retrieved.

In the following, I will further summarize what are arguably the most sustained efforts in reconceptualizing the American short story and its sub-genres. While this list is, of course, by no means comprehensive, there are also various overlaps between the current approaches:

Short Story Cycles: Among the most flourishing current approaches to the American short story are those that look at how individual stories are collated, interconnected, and integrated into longer narrative ensembles. While the terms to describe influential literary works from Sherwood Anderson's *Winesburg, Ohio* (1919) to Jennifer Egan's *A Visit from the Goon Squad* (2010) vary from short story cycle to short story sequence to composite novel, the main focus is always on the "progressive unfolding and cumulative effects" (Kennedy 1995, viii) of the juxtaposition of single stories in a bounded whole (see Nagel 2000, 2001; Birkenstein 2010; Smith 2015, 2018). While early works on the short story cycle highlighted the community aspect resulting from inter-story connections, for James Nagel the short-story cycle has traditionally addressed issues related to aspects of ethnic identity: "The short story cycle in twentieth-century American literature is decidedly a multiethnic tradition, perhaps because the brief narrative has its origins in the oral tradition and descends through cultures in every part of the world, uniting them in a legacy of universal storytelling" (Nagel 2000, 10). In his parsing, while the novel favors the model of a central character-protagonist and therefore also a mono-perspectival narrative situation (ideologically aligned with individualism), the short story cycle typically presents stories from a multitude of perspectives that constitute a "community of tellers" (10) rather than a singular point of view:

> In a contemporary world characterized by progressive fragmentation and alienation, an episodic mode better reflects the psychic nature of modern life than would the extended flow of experience represented in long fiction. It would seem also that the changing nature of the literary market, with publishing houses eager for blockbuster novels, makes it easier for writers to establish themselves by writing stories for magazines and assembling them later to form cycles. (12)

The short story cycle or sequence, for Nagel, "offers not only a rich literary legacy but also a vital technique for the exploration and depiction of the complex interactions of gender, ethnicity, and individual identity" (14). In the most recent book-length study on the American short story cycle, Jennifer Smith (2018) opts for a broader understanding of the genre that includes a variety of sub-forms such as sketchbooks, composite novels, and story collections. Arguing that the cycles "reflect in form and content a central narrative problem of modern literature: how to articulate subjectivity" (5), Smith delineates what for her are the genre's main characteristics: the rejection of progression and solution, episodic storytelling, non-chronological temporality, and multiplicity of point of view. Taken together, these narrative techniques give shape to the cycle's preference for what she calls "provisional identities" (7).

Cross- and Transgeneric Approaches: Cross-generic approaches to the (American) short story focus on hitherto untended entanglements between the short story and other literary forms. While more traditional cross-generic approaches used to be interested in textual (i.e., formal, stylistic, thematic) interrelations, more recent works are more sensitive to questions of media history and literary markets (see Levy 1993). A very original example for such a transgeneric approach is Michael Collins's *The Drama of the American Short Story 1800–1865* (2016), in which Collins rehistoricizes the development of the American short story in light of broader cultural contexts usually not directly associated with the short story:

> in order to fully understand how the short story operated in early to mid-nineteenth century American society, one must consider three elements in conjunction: the printed context in which the work appeared, the performance and expressive culture surrounding that work, and the role played by the story within a transnational framework of memory, exchange, and cultural borrowing. (Collins 2016, 3)

Collins is interested in how the New York theater scene around 1800 affected a group of journalists and literati including the young Washington Irving (↗5 Irving). He demonstrates how theater – in this specific case, the Park Theatre that opened in 1798 – became a site of demarcation dispute in which not only questions of an emerging national literature and culture vis à vis the British influence were discussed, but which also prepared the field for the birth and development of the American short story in the subsequent decades (6–7): "It is no coincidence that it is in this context that Irving develops the short story as a highly performative style of textual narrative to comment upon an era in which cultural production was so very politicized" (12). Hence, theatricality and performance as a "political strategy" (16) link the short story to drama and theater: "the world of print and the world of the stage are brought productively into dialogue in the short story through the various ways the genre and the media in which it first appeared dramatized abundance, choice, and aesthetic play" (20). The book delineates a story of origin that firmly places the emergence of the short story in a nineteenth-century American social-cultural context in which theatricality, performance and performing subjects played a prominent role in the negotiation of, for example, the relation between public/private spheres, national and transnational positionalities as well as class-based social stratification processes (on theatricality, see also Lepaludier 2008). Other recent cross- or transgeneric approaches have discussed the short story's affinities with cinema (see Collinge-Germain 2012), photography and visual arts (see, e.g., Collinge-Germain 2011; see also Scofield 2006, 109–112). In contrast to more conventional approaches to the (American) short story, recent cross-generic discussions of the American short story tend to emphasize the fluidity and entanglements of, rather than erect solid boundaries between, genres.

Transnational Approaches: In the wake of a 'transnational turn' in American studies since the early 2000s, some scholars have already begun to reassess and rewrite the history of the American short story, a history that used to be limited by a "national fallacy" (Scheiding and Seidl 2015, 9), neglecting the genre's global interconnections, textual circulation, manifold cultural exchanges and influences as well as complex networks of publication. The already mentioned pioneering work of Werner Sollors (Sollors and Shell 2000a) is a case in point, as would be the essays by Christine Consentino (2000) and Rolf Lundén (2010) that explore the influence of the American short story on German and, respectively, Swedish literature. As the editor of a special issue of the *Journal of the Short Story in English* on "The Transatlantic Short Story" (2013) argues, the genre "calls for a reexamination of its location outside the confines of a fixed identification with the areas where it developed" (Ventura 2015, 2). Discussing stories by Henry James (↗12 James), Tennessee Williams, Gail Jones and other writers, the contributions in this issue emphasize deterritorialization rather than territorialization and fixed locales, "extending the temporal, spatial and representational limits" (2) of these narratives.

Such a transnational widening of our understanding of the American short story has arguably been most effective and compelling, thus far, when it has tackled the historical origins of the genre (see Hochbruck et al. 2008). In their anthology *Worlding America*, Oliver Scheiding and Martin Seidl persuasively challenge "the tendency to reduce earlier short narratives to forerunners of a nineteenth-century form" (2015, 2). Instead, they champion a perspective on short narratives before 1800 that contributes to a "reevaluation of American literature as a network of transnational literary exchanges" (1; see also Scheiding and Seidl 2011). The narratives gathered in their volume, spanning a period from the early seventeenth to the early nineteenth century, along with the approach of 'worlding America' outlined in their introduction, suggest a truly innovative revaluation of short narrative forms prior to Washington Irving. The volume challenges established conceptions of genre and authorship as well as writing and publication practices by reconstructing "literary histories of transmission and circulation in the Americas that have so far received little critical attention and have been disregarded as not fitting modern criteria of aesthetic value or the concept of originality" (10). In a similar tone, Michael Collins, in his contribution to the *Edinburgh Companion to the Short Story in English*, detects an "overemphasis on Romantic nationalism" (Collins 2019, 11) in twentieth-century short story scholarship. Problematizing especially the widespread notion of the short story as an 'American' literary form, Collins reminds us that the short story in the nineteenth century was rather shaped by the "exchange and diffusion" (12) of a plethora of different sources and geographical origins: "the short story developed out of transatlantic, transnational or global networks that signified differently according to local particularities" (14). His analysis of two early stories, Benjamin Franklin's "The Speech of Miss Polly Baker" and James Hogg's "The Story of Two Highlanders," demonstrates how these narratives performed a distinct national identity or consciousness, yet were at the same time embedded in far more complex transnational political, juridical, and cultural networks as well as struggles over meaning.

Epistemological and Cognitive Approaches: Short story theory has long seen approaches that aim to identify the genre with a certain epistemological quality and/ or certain cognitive mechanisms resulting from the brevity, narrative compression as well as certain thematic preferences of many short stories (see Basseler 2011). In one of his later essays, Charles May, who has championed such a perspective for several decades, summarizes what for him constitutes 'the nature of knowledge' in short fiction:

> The very shortness of the short story, as well as the necessary artistic devices demanded by this shortness, force it to focus not on the whole of experience (whatever that is) in all its perceptual and conceptual categorization, but rather on a single experience lifted out of the everyday flow of human actuality and active striving, an experience that is lifted out precisely because it is not a slice of that reality, but rather a moment in which 'reality' itself is challenged. (May 2004, 24; see also May 1994b)

Taking their cue from such observations about the short story's ability to probe the limits of knowledge, and drawing on such diverse disciplines as psycholinguistics, cognitive psychology, and cognitive narratology, various scholars have attempted to describe what happens when we read a (short) story, how readers 'chunk' a text into meaningful units (see Lohafer 1994), or how they (unconsciously) use their world knowledge to make sense of a story (see Basseler 2011, 56). Renate Brosch's model of "projective reading" (Brosch 2007) conceptualizes the short story as a genre that, in comparison with longer narrative forms, tends to invite a higher level of readerly participation in the meaning-making process. Provoking such projective readings, short stories are capable of revising and expanding the cognitive predispositions of the reader, thwarting his or her habitual reactions, and shaping future reading experiences (103). Not unlike Brosch, who draws on visual theory to describe the short story's capacity of producing lasting images in the reader's memory, Michael Collins recently claimed that "the short story *gestures* outwards from the text towards a wider culture; it puts out feelers and its images latch on to other media," thus ascribing to the genre an "especial porousness" (Collins 2016, 22–23). The question about the American short story's cultural knowledge is at the center of my own book *An Organon of Life Knowledge* (2019). Following Ottmar Ette's programmatic notion of literature as a source of knowledge that is both complementary and corrective to the "bio- and natural-scientific concept" (Ette 2010, 983) of life that has become a dominant in our present societies, I discuss how literary 'life knowledge' is generically formed in the American short story of the late nineteenth to early twenty-first century. While the first sections of *Organon* unfold a theory of the generic specificities and constructedness of such knowledge, its last part demonstrates how the American short story, in a variety of sub-genres (e.g., stories of initiation, stories of 'unlived' life, stories of remembered life), engages in the production of cultural life knowledge and shapes the "stories Americans live by" (McAdams 2006).

Short Story Criticism and Recent Theorizing in Cultural Studies: Though it is somewhat surprising that the bulk of short story criticism, especially if compared to the study of the novel, has only reluctantly engaged in interdisciplinary discussions based on recent concepts and theories in literary and cultural studies (see Copland 2014), there are some recent examples of how such an interdisciplinary approach can effectively revitalize short story scholarship. Inspired by affect theory and especially Sara Ahmed's notion of 'affective stickiness,' for instance, Paul Ardoin and Fiona McWilliam have described the peculiar stickiness of the short story and short story cycle (Ardoin and McWilliam 2016). Understanding stickiness as "*an effect of the histories of contact between bodies, objects, and signs*" (Ahmed in Ardoin and McWilliam 2016, n.p.; original emphasis), Ardoin and McWilliam approach the genre through its affective value, that is, the impressions it leaves upon readers. Likewise, Rachel Lister (2016, n.p.) explores how the short story cycle offers a generic vehicle that can challenge and debunk hegemonic, normative notions of happiness, notions that are usually associated with a certain telos or goal pursued by "affective communities." Instead, Lister reckons, the short story cycle presents alternative concepts of happiness that are realized only temporarily in short, fleeting moments. Looking at the works of Katherine Anne Porter, Eudora Welty, and Alice Munro, she examines how the female protagonists of these cycles experience such moments of happiness beyond social norms and gender roles: "These moments enable the heroines to reclaim abandoned or suppressed feelings and sensations and to comprehend what they did not initially understand" (n.p.).

Other important impulses have come from scholars working on the intersections between feminist theory, queer theory, and literature. Focusing on queer short stories, Corber (2000; see also Eeckhout 2015) outlines the cultural work that stories published in gay and lesbian magazines performed for the political mobilization of the community in the liberation movement fueled by the Stonewall riots in 1969. While homosexual themes found their generic expression mostly in coming-out stories, Corber observes a certain discrepancy between their progressive content and conventional form: "Despite their explicit treatment of homosexuality, lesbian and gay short stories are fairly traditional in their structure and style" (Corber 2000, 59). Moreover, whereas in gay stories a white, middle-class perspective tended to marginalize other forms of gay culture, lesbian short stories by contrast often address the multiple oppression, or intersectionality, of homosexual subjects and communities (60). In her analysis of stories from the 1980s to the beginning of the new millennium that center around mother-daughter relationships, Melissa Bostrom contests the notion that the short story constitutes a 'revolutionary' genre capable of challenging social norms. To the contrary, she finds the short story to be a "deeply conservative" genre confining "female sexuality through rigid economies between mothers and daughters" (Bostrom 2007, 6) as well as through racial restrictions. The notion of economy is crucial in her approach, as Bostrom interprets the short story's conservatism as a result of the waning possibilities to publish short fiction, especially for up-and-coming younger authors. At a time when *The New Yorker* remains the last significant outlet through which short story writers can reach a larger audience, this situation "produces the kinds of stories that are limited by cultural anxieties about female sexuality and race, because it targets those readers that are most likely to be racially and socioeconomically white" (153).

One of the most advanced recent conceptual approaches to the study of the (American) short story was propounded by the contributions in the volume *Liminality and the Short Story* (Achilles and Bergmann 2015a). In their introduction, the editors describe the short story as "the liminal genre *par excellence*" (Achilles and Bergmann 2015b, 4), a quality that resides both in the typical content of many short stories (e.g., in stories of initiation) and in their poetics as well as the short story's generic positionality between other genres such as the essay, sketch, and novel. Drawing on concepts and theories of liminality from Arnold van Gennep through Victor Turner, Michel Foucault, and Edward Soja, the volume probes into the value and uses of liminality for the study of the short story for generating new methods of approach, conditions of publication, contexts of writing as well as representative topics:

> Inscribing liminality in its aesthetic structures, the short story turns into a forum of crisis and change. In American short-story writing two techniques of such inscription are particularly prominent: firstly, performative dynamics resulting from transgressions of semantic fields, and secondly, condensations of different normative codes in individual characters or objects. (8)

The contributions in the volume explore different kinds of liminality in the short story, from intra- and intertextual aspects (e.g., liminal or 'heterotopic' settings, ambivalent characters, staging of social drama) to aspects of reader-response and cognitive particularities to matters of circulation and publication, particularly from a historical perspective. In a similar vein, Collins (2016, 22–23) also emphasizes the "inbetweenedness" of the nineteenth-century short story with regard to its place between romance/romanticism and realism, between "histrionics and textuality that defined the early to mid-nineteenth-century transatlantic literary world" (23) as well as between conflicting orders of national and symbolic meaning.

Thematic Approaches: Next to the transnational, cross-generic, epistemological, cognitive, and other conceptual or theory-driven approaches sketched above, there is also a wide array of critical works that deal with emergent thematic constellations in the American short story, constellations that are often also tied to broader societal and political debates. These thematic constellations include such diverse subjects as the short story and 9/11 (Böger 2011), obesity and the short story (Kelly 2001), and food and the short story; Birkenstein (2015) reads, for instance, the significance of food in the short story as an expression of immigrant experience in the United States. Despite these individual contributions, however, it is striking that the short story as a genre as well as short story criticism still play a rather marginalized role in the wider critical debates in which they are embedded. This becomes obvious, for example, in the field of ecocriticism, where research tends to focus almost exclusively on the novel. As Glen Love has argued, ecological issues in the short story are centered around three ideas, "a sense of a degraded environment, a sharpened awareness of geographic place, and an examination of animal lives" (Love 2000, 51). The presence of ecological matters in contemporary short fiction for Love stands in a longer tradition that ranges from Romanticism (and particularly Leo Marx's interpretation of the pastoral tradition) through the nuclear threats of the Cold War era through more recent eco-science fiction by authors like Ursula LeGuin. Love suggests the "ecological" not only as an analytical category for reading contemporary short fiction directly addressing issues of climate change, but also as a critical lens for re-readings of canonized works by writers such as Jack London (↗13 London), Ernest Hemingway (↗16 Hemingway), Eudora Welty, John Steinbeck, and Flannery O'Connor (↗19 O'Connor).

New Approaches in Contextualization: Whereas for many decades short story criticism used to be dominated by formal and structuralist approaches, there have recently been a number of studies that aim to shed new light on the genre by offering various perspectives for a (re-)contextualization of the American short story. For example, the essays in the volume *Re-Visioning the Past: Historical Self-Reflexivity in American Short Fiction* (Engler and Scheiding 1998) examine how short stories from the early nineteenth to the late twentieth century have grappled with positivistic notions of history and the forms as well as truth status of historiography, thus reading the genre alongside broader philosophical debates. In a very recent study, Lesleigh Cushing Stahlberg and Peter Hawkins start from the assumption that the short story as a genre is marked by "the peculiar ability [. . .] to convey religious experience" (Stahlberg and Hawkins 2018, x), tracing the Bible's role in the history of American literary production as well as intertextual references between the Bible and a number of short stories by writers from Flannery O'Connor (↗19 O'Connor) to John Updike and Tobias Wolff (in the Christian tradition), Bernard Malamud and Nathan Englander (in the Jewish tradition). While the decision to focus on short fiction (instead of the novel or poetry) might be new, the volume does not offer much in the way of a coherent, new approach to the short story that would go significantly beyond the identification of intertextual references in these authors' works. Sam Reese's *The Short Story in Mid-Century America* (2017) offers another kind of recontextualization. It examines the short story as a countercultural form in the mid-twentieth century, a form through which writers like Mary McCarthy, Tennessee Williams, Paul Bowles, and Eudora Welty were able to run "against the grain of contemporary expectations of literary and cultural narratives" (Reese 2017, 2). While the concept of "countercultural fiction" (4) serves as the study's main concept, Reese explores why the short story – a literary form he associates with formal containment rather than openness – nevertheless provided an ideal outlet for the articulation of alternative cultural perspectives. Dealing with roughly the same period, Kasia Boddy's (2010) account of the development of the American short story since the 1950s is also very attentive to its various contexts. Boddy is particularly sensitive to the institutional frameworks in which American short stories were produced, published, and read in the mid- to late-twentieth century. Her book delineates a genre history in which each decade witnessed the dominance of a certain 'school' – the *New Yorker* story in the 1950s, experimental fiction in the 1960s and 1970s, minimalism and neo-realism in the 1980s –, carefully working out how the genre has been tightly interwoven with the institution of the university, e.g. through MFA programs like the Iowa Writer's Workshop.

Short-Short Stories: While formal discussions of the short story have always centered on the question of length, the growing popularity of very short stories – also frequently referred to as 'flash fiction', 'short-shorts' or 'microfiction' – has rekindled scholarly interest in how brevity determines the genre's storytelling devices as well as its effects on readers. And while there are certainly numerous examples of very short stories in the history of the genre (think, for instance, of Hemingway's "A Very Short Story"), the past few decades have witnessed a remarkable trend toward extreme brevity, e.g. in the works of writers like Lydia Davis (↗31 Davis) and Dave Eggers. Anthologies of very short stories such as *Short Shorts* (Howe and Howe 1982), *Sudden Fiction* (Shapard and Thomas 1983), and *Flash Fiction* (Thomas et al. 1992) corroborate this notion, as they all proceed on the conviction that the stories they present are categorically different from the more traditional short story. In a recent essay, William Nelles has suggested the term 'microfiction' to account for this new type of short fiction, regarding it as a different species operating on a "different set of narrative principles" (Nelles 2012, 91). Others have interpreted the emergence of very short narrative forms in light of technological innovations and a changing media ecology, particularly the advent of the internet (Chantler 2008). Despite various attempts at coming to terms with the formal-aesthetic, socio-cultural, and epistemological specificities of short-short fiction (see Basseler 2019), monographs and other systematic, book-length studies of very short narrative forms still remain to be written.

Digital Approaches to the American Short Story: It is surprising, perhaps, that the increasing importance of research from the field of digital humanities (DH) has thus far not led to any systematic approach to the ways we 'read' short stories. Within literary studies, the most prominent interventions in DH scholarship were developed on more prestigious genres such as the novel, as in Franco Moretti's methodological suggestion of 'distant reading' that may help the discipline to develop a more comprehensive understanding of the literary field in a given period. In a recent essay, Stephanie Siewert and Nils Reiter (2018) have proposed using quantitative methods from computational linguistics to analyze certain shifts in content and style in nineteenth-century American short stories. They suggest that the increasing commercialization of the genre and its publication outlets (i.e., magazines) is directly reflected in a number of quantitative textual parameters, such as sentence length or vocabulary richness. Particularly the use of computerized, automated literary pattern recognition, the authors argue, "can be tied back to historical and socio-economic conditions" (Siewert and Reiter 2018, 226). Moreover, similar to Moretti's notion of distant reading (2005), Siewert and Reiter contend that quantitative methods "will bring us closer to a perspective on short fiction of the nineteenth century that moves beyond the canon" (Siewert and Reiter 2018, 226). Whether this will automatically lead to "many fascinating questions for future work in the field" (2018, 226) remains to be seen. Certainly, the new accessibility of a large body of short fiction through databases and digital literary archives such Early American Imprints (Readex) and Proquest (see Scheiding and Seidl 2011, 69; see also Scheiding 2012) promises immense potential for a revaluation and expansion of the history of the American short story.

3 Conclusion and Suggestions for Research

The variety of the current approaches sketched in this chapter render any predictions about the future of scholarship on the American short story inherently difficult. It seems relatively safe to claim, however, that there is a certain momentum and potential for overcoming the stale conservatism as well as the methodological and theoretical isolation which has marked short story criticism for some decades now. To conclude, I will end this survey with some tentative remarks on what seem to be the most productive assumptions and vital trajectories of current and future research.

First, a critical revaluation of the short story's relation to notions of a distinctive American national literature and culture (see Collins 2016, 27–28) promises to yield important insights into, and revisions of, the canon (↗2 Canon Formation) of what we call "the American short story." Instead of repeating the widespread view that the modern short story is a decidedly American literary form (articulated influentially by scholars from Frank O'Connor to Charles May), current approaches tend to question this exceptionalist narrative by pointing to the various transnational entanglements of the genre. At the same time, however, debates in the field of American studies around a new, 'critical' regionalism (Lösch, Paul and Zwingenberger 2016) render the short story an ideal case study to explore how notions of spatial and cultural belonging are played out in the realm of literature. A form traditionally associated with regionalism and local color writing, the American short story has always shaped such notions, and the tension between regionalist and transnational tendencies becomes particularly visible in the genre's history. In this regard, a multilinguistic approach to the 'American' short story as suggested by Sollors (2000), as well the insistence on its transnational networks of circulation (Scheiding and Seidl 2015; Collins 2019) seem to be particularly fruitful.

Second, the formalist legacy of short story criticism might be fruitfully merged with, and eventually transformed into, critical approaches interested in the ways in which aesthetic forms overlap or collide with social forms, thus both affording certain possibilities *and* posing certain constraints (see Levine 2015). A "renewed attention to *form*" (Collins 2016, 24; original emphasis) that does not collapse on the old formalism of mid-twentieth century short story scholarship would arguably allow us to come to terms with the per*form*ative work that short stories served to fulfill in certain historical moments (as in Corber's reading of 'queer' short stories, see above). In his widely noticed *Reality Hunger*, David Shields (2010) has argued that narrative fragmentation and (extreme) brevity are closely entwined with the mentality and media ecology of the contemporary age. Similarly, Ansgar and Vera Nünning have conceptualized various "discontinuous and fragmentary stories and forms of storytelling" (Nünning 2016, 37) under the label of "broken narratives," arguing that these forms have become symptomatic for the increasing disruptions and discontinuities of life, particularly in a North American context. Along these lines, future approaches to the American short story should arguably pay close attention to how its past and present form(s) and subgenres, including short story cycles as well as flash fiction, are expressive of broader sociocultural arrangements and developments. "Literary form does not operate outside of the social but works among many organizing principles, all circulating in a world jam-packed with other arrangements," writes Levine (2015, 7). Trying to understand how the American short story operated and continues to work among other social forms and arrangements is perhaps one of the most promising questions for future short story scholarship.

Third, this emphasis on form and its broader functions and social entanglements might signal a shift away from the vexing problem of what short stories *are* to more productive and interdisciplinarily relevant questions about what they *do*, i.e., what kind of cognitive and cultural work they perform within specific sociocultural constellations. While several of the recent works sketched above already helped pave the way for such a 'performative turn' in short story criticism, more research is needed that adopts a 'neo-formalist' perspective in order to overcome the old formalism that had led short story theory to a crossroads at the end of the last century (see Lohafer and Clarey 1989). How the various forms and genres of short fiction in North America have been interlaced with broader political, social, and cultural arrangements, and how they have thereby shaped the "stories Americans live by" (McAdams 2006), are perhaps some of the most intriguing questions for current and future approaches to the American short story.

4 Bibliography

4.1 Works Cited

Achilles, Jochen and Ina Bergman, eds. *Liminality and the Short Story*. London and New York: Routledge, 2015a.

Achilles, Jochen and Ina Bergman. "'Betwixt and Between': Boundary Crossings in American, Canadian, and British Short Fiction." In: Achilles/Bergmann 2015b, 3–34.

Ardoin, Paul and Fiona McWilliam. "Introduction: On the Stickiness of the Short Story and the Cycle," https://journals.openedition.org/jsse/1677. *Affect and the Short Story and Cycle*. Eds. Paul Ardoin and Fiona McWilliam. *Journal of the Short Story in English* 66 (2016). (20 March 2020).

Basseler, Michael. "Theories and Typologies of the Short Story." *A History of the American Short Story. Genres – Developments – Model Interpretations*. Eds. Michael Basseler and Ansgar Nünning. Trier: WVT, 2011a. 41–64.

Basseler, Michael. *An Organon of Life Knowledge: Genres and Functions of the Short Story in North America*. Bielefeld: transcript, 2019.

Birkenstein, Jeff. "'Should I Stay Or Should I Go': American Restlessness and the Short-Story Cycle." *A Companion to the American Short Story*. Eds. Alfred Bendixen and James Nagel. Malden, MA/Oxford: Wiley Blackwell, 2010. 482–501.

Birkenstein, Jeff. "How Significant Food Can Make a Short Story into a Meal: The Hyphenated Immigrant Experience in Contemporary American Short Fiction." In: Achilles/Bergmann 2015a. 198–209.

Böger, Astrid. "New Directions in the American Short Story After 9/11: Don DeLillo's 'Baader-Meinhof' and Nicole Krauss's 'Future Emergencies." *A History of the American Short Story: Genres – Developments – Model Interpretations*. Eds. Michael Basseler and Ansgar Nünning. Trier: WVT, 2011. 413–426.

Boddy, Kasia. *The American Short Story since 1950*. Edinburgh: Edinburgh UP, 2010.

Bostrom, Melissa. *Sex, Race, and Family in Contemporary American Short Stories*. New York: Palgrave Macmillan, 2007.

Brosch, Renate. *Short Story: Textsorte und Leseerfahrung*. Trier: WVT, 2007.

Carpio, Glenda R. "Liminal Subjects, Mixed Genre: Richard Wright and the African American Short Story." *Liminality and the Short Story: Boundary Crossings in American, Canadian, and British Writing*. New York/London: Routledge, 213–224.

Chantler, Ashley. "Notes Towards the Definition of the Short-Short Story." *The Short Story*. Ed. Ailsa Cox. Newcastle upon Tyne: Cambridge Scholars, 2008. 38–52.

Collinge-Germain, Linda, ed. "Special Issue: The Image and the Short Story in English," https://journals.openedition.org/jsse/1124?lang=en. *Journal of the Short Story in English* 56 (2011). (25 January 2021).

Collinge-Germain, Linda, ed. "Special Issue: The Short Story and Cinema," https://journals.openedition.org/jsse/1272. *Journal of the Short Story in English* 59 (2012). (25 January 2021).

Collins, Michael J. *The Drama of the American Short Story, 1800–1865*. Ann Arbor: U of Michigan P, 2016.

Collins, Michael J. "Transnationalism and the Transatlantic Short Story." *The Edinburgh Companion to the Short Story in English*. Eds. Paul Delaney and Adrian Hunter. Edinburgh: Edinburgh UP, 2019. 9–23.

Copland, Sarah. "To Be Continued: The Story of Short Story Theory and Other Narrative Theory." *Narrative* 22.1 (2014): 132–149.

Corber, Robert J. “Lesbian and Gay Short Stories.” *The Columbia Companion to the Twentieth-Century American Short Story*. Ed. Blanche H. Gelfant. New York: Columbia UP, 2000. 56–63.

Cosentino, Christine. “Ingo Schulzes ‘Simple Storys’ und die Tradition der amerikanischen Short Story.” *Germanic Notes and Reviews* 31.2 (2000): 134–138.

Deleuze, Gilles, and Félix Guattari. *Kafka. Toward a Minor Literature*. 1986. Trans. Dana Polan. Minneapolis and London: U of Minnesota P, 2003.

Eeckhout, Bart. “Theorizing the Queer Modernist Short Story: From Failures of Sociality to Negative Affect.” *Journal of the Short Story in English* 64 (2015): 121–136.

Engler, Bernd, and Oliver Scheiding, eds. *Re-Visioning the Past: Historical Self-Reflexivity in American Short Fiction*. Trier: Wissenschaftlicher Verlag Trier, 1998.

Ette, Ottmar. “Literature as Knowledge for Living, Literary Studies as Science for Living.” *PMLA* 125.4 (2010): 977–993.

Gelfant, Blanche H., ed. *The Columbia Companion to the Twentieth-Century American Short Story*. New York: Columbia UP, 2000a.

Gelfant, Blanche H. “Introduction.” *The Columbia Companion to the Twentieth-Century American Short Story*. Ed. Blanche H. Gelfant. New York: Columbia UP, 2000b. 1–5.

Hanson, Clare, ed. *Re-Reading the Short Story*. New York: St. Martin’s P, 1989.

Hochbruck, Wolfgang, Aynur Erdogan, and Philipp Fidler, eds. *Origins of the American Short Story*. Los Gatos: Slack Water P, 2008.

Howe, Irving, and Illana Howe, eds. *Short Shorts: Anthology of the Shortest Short Stories*. Boston: David R. Godine P, 1982.

Kelly, Lionel. “American Fat: Obesity and the Short Story.” *Yearbook of English Studies* 31 (2001): 218–229.

Kennedy, J. Gerald. *Modern American Short Story Sequences: Composite Fictions and Fictive Communities*. Cambridge: Cambridge UP, 1995.

Lepaludier, Laurent. “Theatricality in the Short Story: Staging the Word?” *Journal of the Short Story in English* 51 (2008): 17–28.

Levine, Caroline. *Forms. Whole, Rhythm, Hierarchy, Network*. Princeton and Oxford: Princeton UP, 2015.

Levy, Andrew. *The Culture and Commerce of the American Short Story*. Cambridge: Cambridge UP, 1993.

Ling, Amy. “The Asian American Short Story.” *Columbia Companion to the Twentieth-Century American Short Story*. Ed. Blanche H. Gelfant. New York: Columbia UP, 2000b. 34–41.

Lister, Rachel. “‘Preposterous Adventures’: Affective Encounters in the Short Story Cycle,” https://journals.openedition.org/jsse/1683. *Journal of the Short Story in English* 66 (2016). (20 March 2020).

Lohafer, Susan. “A Cognitive Approach to Storyness.” *The New Short Story Theories*. Ed. Charles E. May. Athens: Ohio State UP, 1994. 301–311.

Lohafer, Susan, and Jo Ellyn Clarey, eds. *Short Story Theory at a Crossroads*. Baton Rouge and London: Louisiana State UP, 1989.

Love, Glen A. “The Ecological Short Story.” *The Columbia Companion to the Twentieth-Century American Short Story*. Ed. Blanche H. Gelfant. New York: Columbia UP, 2000. 50–55.

Lösch, Klaus, Heike Paul, and Meike Zwingenberger, eds. *Critical Regionalism*. Heidelberg: Winter, 2016.

Lundén, Rolf. “Direct and Indirect Influence: The Impact of the American Short Story on Swedish Literature.” *American Studies in Scandinavia* 42.2 (2010): 21–36.

May, Charles E., ed. *Short Story Theories*. Athens: Ohio UP, 1976.

May, Charles E., ed. *The New Short Story Theories*. Ohio: Ohio UP, 1994a.

May, Charles E. "The Nature of Knowledge in Short Fiction." *The New Short Story Theories*. Ed. Charles E. May. Athens: Ohio UP, 1994b. 131–146.
May, Charles E. "Why Short Stories are Essential and Why They are Seldom Read." *The Art of Brevity: Excursions in Short Fiction Theory and Analysis*. Eds. Per Winther, Jakob Lothe, and Hans K. Skei. Columbia: U of South Carolina P, 2004. 14–25.
McAdams, Dan P. *The Redemptive Self: Stories Americans Live By*. Oxford and New York: Oxford UP, 2006.
Moretti, Franco. *Graphs, Maps, Trees. Abstract Models for a Literary Theory*. London: Verso, 2005.
Nagel, James. "The American Short Story Cycle." *The Columbia Companion to the Twentieth-Century American Short Story*. Ed. Blanche H. Gelfant. New York: Columbia UP, 2000. 9–15.
Nagel, James. *The Contemporary American Short-Story Cycle: The Ethnic Resonance of Genre*. Baton Rouge: Louisiana UP, 2001.
Nagel, James. *Race and Culture in New Orleans Stories: Kate Chopin, Grace King, Alice Dunbar-Nelson, and George Washington Cable*. Tuscaloosa: U of Alabama P, 2014.
Nagel, James. *The American Short Story Handbook*. Malden, MA/Oxford: Wiley Blackwell, 2015.
Nelles, William. "Microfiction: What Makes a Very Short Story Very Short?" *Narrative* 20.1 (2012): 87–104.
Nünning, Ansgar, and Vera Nünning. "Conceptualizing 'Broken Narratives' from a Narratological Perspective: Domains, Concepts, Features, Functions, and Suggestions for Research." *Narrative im Bruch: Theoretische Positionen und Anwendungen*. Eds. Anna Babka et al. Göttingen and Vienna: Vienna UP, 2016. 37–86.
Reese, Sam V. H. *The Short Story in Midcentury America: Countercultural Form in the Work of Bowles, McCarthy, Welty, and Williams*. Baton Rouge: Louisiana State UP, 2017.
Ruppert, James. "The Native American Short Story." *The Columbia Companion to the Twentieth-Century American Short Story*. Ed. Blanche H. Gelfant. New York: Columbia UP, 2000. 64–71.
Saldívar, Ramón. "The Chicano-Latino Short Story." *Columbia Companion to the Twentieth-Century American Short Story*. Ed. Blanche H. Gelfant. New York: Columbia UP, 2000b. 42–49.
Scheiding, Oliver. "Migrant Fictions: The Early Story in North American Magazines." *Mobility in Literature and Culture*. Eds. Ingo Berensmeyer, Herbert Grabes, and Christoph Ehland. Tübingen: Narr, 2012. 197–218.
Scheiding, Oliver, and Martin Seidl. "Early American Short Narratives: The Art of Story-Telling Prior to Washington Irving." *A History of the American Short Story: Genres–Developments–Model Interpretations*. Eds. Michael Basseler and Ansgar Nünning. Trier: WVT, 2011. 67–80.
Scheiding, Oliver, and Martin Seidl. "Introduction." *Worlding America: A Transnational Anthology of Short Narratives Before 1800*. Eds. Oliver Scheiding and Martin Seidl. Stanford: Stanford UP, 2015.
Shapard, Robert, and James Thomas. *Sudden Fiction: American Short-Short Stories*. Gibs Smith P, 1983. Standford, CA: Stanford UP, 2015. 1–18.
Scofield, Martin. *The Cambridge Introduction to the American Short Story*. Cambridge: Cambridge UP, 2006.
Shields, David. *Reality Hunger*. London: Penguin, 2010.
Siewert, Stephanie, and Nils Reiter. "The Explorative Value of Computational Methods: Rereading the American Short Story." *Amerikastudien/American Studies* 63.2 (2018): 199–230.
Smith, David Lionel. "The African American Short Story." *Columbia Companion to the Twentieth-Century American Short Story*. Ed. Blanche H. Gelfant. New York: Columbia UP. 25–33.
Smith, Jennifer J. "The Short Story Cycle in American Fiction." *Critical Insights: American Short Story*. Eds. Michael Cocchiarale and Scott D. Emmert. Amenia, NY: Grey House, 2015. 3–19.
Smith, Jennifer J. *The American Short Story Cycle*. Edinburgh: Edinburgh UP, 2018.

Sollors, Werner and the Longfellow Institute. "Non-English American Short Stories." *The Columbia Companion to the Twentieth-Century American Short Story*. Ed. Blanche H. Gelfant. New York: Columbia UP, 2000. 72–80.
Sollors, Werner, and Marc Shell, eds. *The Multilingual Anthology of American Literature*. New York: New York UP, 2000.
Stahlberg, Lesleigh Cushing, and Peter S. Hawkins, eds. *The Bible in the American Short Story*. London: Bloomsbury, 2018.
Thomas, James et al. *Flash Ficiton: 72 Very Short Stories*. New York: W.W. Norton & Company, 1992.
Ventura, Héliane. "Introduction: The Transatlantic Short Story," https://journals.openedition.org/jsse/1368. *Journal of the Short Story in English* 61 (2015). (20 March 2010).
Whalan, Mark. *Race, Manhood, and Modernism in America*. Knoxville: U of Tennessee P, 2007.

4.2 Further Reading

Basseler, Michael, and Ansgar Nünning, eds. *A History of the American Short Story. Genres – Developments – Model Interpretations*. Trier: WVT, 2011.
Bendixen, Alfred, and James Nagel, eds. *A Companion to the American Short Story*. Malden, MA/Oxford: Wiley Blackwell, 2010.
Cocchiarale, Michael, and Scott D. Emmert, eds. *Critical Insights: American Short Story*. Amenia, NY: Grey House, 2015.
Ibánez, José R., José Francisco Fernández, and Carmen M. Bretones, eds. *Contemporary Debates on the Short Story*. Bern: Peter Lang, 2007.
Iftekharuddin, Farhat et al., eds. *Postmodern Approaches to the Short Story*. Westport: Praeger, 2003.
Lohafer, Susan. *Reading for Storyness: Preclosure Theory, Empirical Poetics and Culture in the Short Story*. Baltimore: Johns Hopkins UP, 2003.
Patea, Viorica, ed. *Short Story Theories: A Twenty-First Century Perspective*. Amsterdam: Rodopi, 2012.

Oliver Scheiding

4 Textual Materiality, Magazine Culture, and the American Short Story

Abstract: This chapter explores the American short story in the context of material culture studies, print culture, and magazines studies. Since the development of the short story has been frequently read in close relationship to the expanding magazine business in the nineteenth and twentieth centuries, the bourgeoning field of magazine studies offers new insights into both the economy and the aesthetics of the genre. It also invites us to reconsider the short form's material textuality within the multimodal environment of magazine publications. Since the digital age has not made print disappear, the chapter's material approach to examining the short story not only rereads the genre in publishing terms, but also highlights the form's innovative dynamics and how it adapts to the conditions of both online and offline demands.

Keywords: Short story, storytelling, materiality, multimodality, magazines, print culture

1 Rethinking the American Short Story in Material Terms

When talking about the American short story, we frequently regard formal and thematic innovations stressing the literary artifice in the short fiction of writers like Flannery O'Connor (↗19 O'Connor), Raymond Carver (↗26 Carver), Lorrie Moore, or George Saunders (↗32 Saunders). However, we forget that – since the nineteenth century – the American short story is one of the few literary forms whose modern contributions to American literature depended on major advancements in media technology that made it possible to expand the periodical press, to accelerate the distribution of literature, and to establish a book publishing industry (see Levy 1993; Chan 2007; March-Russell 2016; Norrick-Rühl 2018). The increasing impact of the telephone and the telegraph on nineteenth-century America gave rise to the newspaper syndicate, so that "scores of magazines and newspapers could run the same story on the same day" (Nagel 215, 26) turning the short story and serialized fiction into the preferred form of writing among authors, editors, and readers. In the twentieth century, mainstream literary magazines like *The New Yorker* or *The Paris Review* helped flourish the short-story form and kept writers alive (Boddy 2010; Finkelstein 2015). Presently, the creative-writing education industry, the culture of

https://doi.org/10.1515/9783110587647-005

independent literary magazines, and e-publishing platforms nurture the short story and an entire generation of newer writers (see McGurl 2009; Hamilton 2010; Glass 2016; Dietz 2019; see also Jemisin 2018, vii–xii; ↗35 Jemisin).

In his seminal book *The Development of the American Short Story*, published in 1923, Fred Lewis Pattee calls this development the "journalization of the short story" (Pattee 1923, 337–356). Pattee describes the institutionalization of the short story in the context of magazines and newspapers in the 1890s (↗2 Canon Formation). At the dawn of modernism, commercial pressures prompted modernist formal innovation in popular magazines. The American short story under the new conditions of an overabundant magazine production not only helped establish schools of writers (Pattee mentions Jack London ↗13 London, Stephen Crane, and O. Henry), but also invites us to consider the short story as a field to map out exchanges not just among authors, but also among editors, publishers, reviewers, readers, and the physical text, with its advertisements, illustrations, and editorial gimmicks. This chapter suggests what makes the modern American short story is both its economy and aesthetics. It proposes to reconsider the short story as an expressive form alongside its material dimensions and multiple meanings in the context of (contemporary) print culture.

In a recent book titled *On Company Time: American Modernism in the Big Magazines*, Donal Harris explores the symbiosis between mass magazines that kept "commercial success and enormous readerships at the center of their business model" (Harris 2016, 7) and several major American modernist writers like Willa Cather, F. Scott Fitzgerald, and Ernest Hemingway (↗16 Hemingway). Harris's book brings modernist studies to bear on commercial magazines and vice versa. It demonstrates how the organizational and aesthetic developments in specific magazine offices feed forward into the literary writing of key figures in American literary modernism. Hemingway, for instance, contributed 25 columns and six short stories to *Esquire*, from its first issue in October 1933 to February 1939, when the magazine had established itself as the decade's "periodical embodiment of fashionable masculine culture" (175). *Esquire* gave Hemingway a "reliable access to one of the largest, most sustained readerships of any American author in the twentieth century" (176). Hemingway went on to sell *The Old Man and the Sea* to *Life* magazine, where it appeared one week before it became available as a stand-alone novella. In terms of the short story as a field, it appears that Hemingway is not modern for his terse language, but he is modern in the way the magazines are, especially in terms of the physical text and its design, bringing together prose, graphics, illustration, photography, and typography for purposes of instruction, information, or persuasion.

As distinct material objects, magazines can be classified into what sociologists call semiotic-imaginary artifacts, or "things produced primarily with the intention of transporting signs, images and imaginations calculated to affect people" (Reckwitz 2016, 124). These include texts, images, or acoustic signals. Such multimodal print objects "aim to arouse feelings of identification or to change people's lives" (124) and their visual components produce "affects of fascination for certain forms of subjectivity or compassion for discriminated social groups" (124). Magazines in general offer different points of view on a wide range of topics. Both their themes and their design function as bearers of positive or negative affects. Sometimes the emphasis on design appears to be stronger than on content. Next to their unique design, however, most magazines offer strong and compelling writing. A good case in point is the publication of Hemingway's Spanish Civil War short story "The Butterfly and the Tank" in the December 1938 issue of *Esquire*.

Fig. 1: Ernest Hemingway, "The Butterfly and the Tank," *Esquire* 10.6 (Dec. 1938), 50–51.

The spread (see Fig. 1) demonstrates how *Esquire* uses a specific multimodal page arrangement not to create two pages separated but two pages that work together to create one unit. The ad's imagery and words on page 50 about "the finest fabrics" mirror Hemingway's "superb" skill as a tailor of words and how he weaves the textures of short stories. As depicted in the ad's image, feeling the fabric of the man's jacket has a similar affect upon the reader when holding the magazine in his "hands." The reader experiences the "truly fine materials" and "smarter style" of its physical pages. *Esquire*'s "careful tailoring" of visual and textual elements establishes the magazine's "real character." The magazine's "body" creates a synergy of visual and textual modes intertwining editorial content and art direction, and, in doing so, adapts the series of Hemingway's three Spanish Civil War short stories ("The Denunciation," Nov. 1938; "Night Before Battle," Feb. 1939) to the corporate style of a mass-audience mainstream magazine.

Hemingway's relationship to print culture signals in many ways the transition from an experimental style in modernist writing initiated by independent little magazines of the 1920s (see Ernst 2018) into the mainstream literary marketplace of the 1940s. Earlier, Edgar Allan Poe, editor and promoter of *Graham's Magazine*, described himself as "essentially a Magazinist" (Poe 2008, 470; see Phillips 2019, 479–498; ↗7 Poe), struggling to win his magazine a foothold in the lucrative U.S.-periodical market. After numerous failures to establish himself as a novelist, Herman Melville likewise launched his magazinist career (↗8 Melville) and contributed a number of short stories to leading mid-nineteenth-century journals like *Harper's Monthly Magazine* and *Putnam's Monthly Magazine of American Literature, Science, and Art*, declaring the magazine "as essentially an organ of American thought as possible" (Melville 1993, 627). Through their short stories and magazinist writings, both writers shaped "print modernity" (McGettigan 2017, 2). Moreover, Poe explored how print production and typographic design influenced the reader's attention. His major contributions to the American short story, such as "the unity of effect or impression" (Poe 2006, 532) and its "pre-established design" (533), are probably as entrepreneurial as they were aesthetic.

The examples also make us reflect upon the fact how – throughout the genre's history – the short story readily lends itself to the latest media. Highlighting Hugh Howey's 2011 short story "Wool" released on Kindle Direct Publishing to analyze the evolution of literary forms in today's digital age, Mark McGurl reasons that "fiction in the Age of Amazon *is* genre fiction, a highly gendered and age-differentiated genre system complexly structured by the poles of epic and romance and their characteristic modes of wish fulfillment" (McGurl 2016, 460; original emphasis). As valid as this observation may be, it does not recognize that technological advances have also created new categories of short-form content and narrative experimentations – just think of twitterature, nanofiction or crowd-sourced narratives. In addition, short story dispensers in train stations, hospitals, airports, and malls make us choose what we want to read in one to five minutes, thus adapting literature to the modern, on-the-go world (see Dietz 2014; see also Cocchiarale 2015, 2017; Long 2016; ↗0 Introduction). Against this backdrop of today's e-age, Sam Baker calls the short story the "perfect literary form of the 21st century" and concludes that "it is technology that has cemented the short story's popularity this century. [. . .], the short story finds itself the perfect fit for our attention spans and mobile devices" (Baker 2014, n. pag). Given the close relationship between the short story's modern textuality and an emerging media culture, it is an opportune moment for a material history of the genre's form and its contributions to American literature to understand the changes in publication, the capitalization of fiction, and the relationships between authors and their readers in a digital culture in which brevity rules.

The following close readings examine the American short story and its (textual) materiality in connection with (modern) print artifacts. While recent directions in reading the short story focus on the textual conditions in the field of Book History (McGann 1991) frequently highlighting "the impact of digital media on the future of short story publishing" (Delaney and Hunter 2019, 4), this chapter suggests a material cultural study approach to reassessing the "irresistible rise of the short story" (Baker 2014, n. pag.) from the nineteenth century to the present. It offers paired readings of two different American short stories and their material conditions. First, I re-examine Charlotte Perkin Gilman's canonical short story "The Yellow Wall-Paper" (1892) in the context of the material turn in literary studies. In doing so, I will engage the materiality of Gilman's magazinist writing to understand its multiple lives and meanings. While Gilman's later publications on housing and architecture made her known as a "committed material feminist" (see Farquharson 2009, 4), the second reading proposes to consider the contemporary American short story in the context of indie magazine publications. I will examine the indie magazine *Zoetrope: All-Story* and the publication of Tommy Orange's Native American short story "Freyr" that appeared in the magazine's 2018/2019 winter issue. This reading looks at the indie magazine as a zone of transaction not only inspiring innovation and experimentation but also one of collaboration between author agency and the magazine's curated design. Finally, I will discuss the variety of the twenty-first-century American short story as a result of new material textualities, changing market demands, and the short-form innovations that go along with it.

2 Close Readings

For decades "The Yellow Wall-Paper" has been read as a canonical text demonstrating that the female protagonist's fictional diary serves as a congenial work questioning the constraints of patriarchal norms and expectations. Commodifying the short story in many ways – from feminism to narrative medicine – Gilman's narrative about mental disorder evolved as an object with its own legendary biography. Presently, Gilman's short story gains new life by being reassessed in terms of material culture studies that considers the question of how fiction enables objects to come alive *in* rather than *around* us. Critics claim that Gilman's wallpaper story challenges domestic environments in which people and things not only cohabitate, but where objects help produce an ecological balance between the demands for domestic order and the characters' emotional state (Suess 2003; Wolter 2009). Contrary to a harmonious domestic animism with its therapeutic artifacts providing intimacy – see, for instance, Catherine Beecher and Harriet Beecher Stowe's domestic advice book (1869) – Gilman's short story portrays tyrannical things. According to some critics, Gilman's protagonist experiences the home as a form of total objecthood and finally resigns to becoming just one more thing inside the home's material continuum of objects. I would like to reassess such readings and the ways in which they address things in American literature.

Although the material turn in literary studies has produced a wide array of different approaches, critics share a common critical effort to disentangle the complicated relationship between American fiction and material culture. Present work refers to the "complex interplay of materiality and signification" (Berliner 2010, 81) and proposes a "textual materialism," as Bill Brown (2010, 26) does, that goes beyond a mere study of books as objects. Such approaches share with present material culture studies "an interest in determining how subjects are formed and transformed by the material world" (26). Likewise, in his introduction to "materiality," anthropologist Daniel Miller suggests that what is needed is "to show how the things that people make, make people" (Miller 2005, 38).

Recent studies also gain momentum from Bruno Latour's *Reassembling the Social* (2005), where he argues that unlike empirical data, literature provides a freer environment for exploring material life. Latour claims a radical symmetry between non-human and human actors and studies their complex associations in specific social arrangements or constellations. Latour also writes about the chains of events behind things that make both people and things become co-agents. For instance, in writing about the so-called Berlin key (Latour 2010, 10–21), a "forced locking key" designed to force people to close and lock the main entrance door leading into a common yard or tenement block, Latour reconstructs a collective history of technology, architecture, police laws, urbanization, and city dwellers in late nineteenth-century Berlin that evolves from a network of actors operating on the levels of origination (lock-smith industry), organization (urban housing, laws, regulations), and interaction (the key and its users). The Berlin key is thus not simply a thing, but rather it unfolds a ramified and conflicting network of both non-human and human actors. To follow these complex associations between human beings and objects does not simply result in studying the social life of things; to follow them means to situate artistic, technological and everyday practices within historical perspectives and to investigate the interplay between material and social practices that occur in these very specific arrangements of human and non-human relations.

In this context, I argue that Gilman's short story "The Yellow Wall-Paper" conflates the material and the social worlds, linking environment, bodily sensations, and writing to create a specific effect she calls "progressive insanity" (Gilman 1998, 87). Acknowledging the non-human – the "inanimate things" (Gilman 1994a, 43) in the story – "The Yellow Wall-Paper" suggests alterations in the social organization of modern life. If we consider what reactions emerged after its publications, we recognize the story's materiality effect. Since its first publication in the pages of the *New England Magazine* in 1892, readers not only appreciated the story's chilling qualities as a Gothic tale, but also realized its subversive quality and radical social meaning. The wallpaper is, therefore, "more than a metaphor," as Babette Tischleder puts it in her astute book *The Literary Life of Things: Case Studies in American Fiction* (Tischleder 2014, 110). The wallpaper is a willful actor within an artistic process that enmeshes audiences in a network of connections and references ranging from psychology, biology, society, industrialism, and the arts. Creating a performative context for both human beings and things, "The Yellow Wall-Paper" addresses where and how materials obstruct, disrupt, or interfere with social forces, allowing repressed, messy, or unstable substances and impure formations to surface.

Jennifer Lunden (2011), in her fresh rereading of neurasthenia in the context of the turn of the century arsenic craze, discusses the wallpaper industry and its poisoning products polluting the domestic sphere. The emotional and psychic disorder it causes in Gilman's narrator discloses the trajectory of a "multi-system illness" that implicates – according to Lunden's rereading – "biology, psychology, and society as mutually interactive contributors to poor health, challenging traditional

Cartesian notions of the mind-body divide. In order for neurasthenia and its contemporary corollaries to be thoroughly understood, they must be viewed through a biopsychosocial lens" (Lunden 2011, 174).

In extending this line of thought, my reading stresses the socio-material lens illuminating the intersections among the artistic, social, and material forces that shape Gilman's critique of Victorian domestic environmentalism. However, restoring Gilman's "The Yellow Wall-Paper" to material culture also means to read the wallpaper and the ensuing mental disorder within an artistic framework from which emerges a story unimaginable from the vantage point of the Cartesian master minds in the story, i.e., the medical professionals, the narrator's husband John, her brother, and the psychotherapist Weir Mitchell who became known for his "rest treatment" to cure states of mental disorder. The narrator's intention to "write about the house" (Gilman 1994a, 40) as stated in the autograph manuscript, creates a metafictional writerly text that reveals the knowing self's partiality referred to by the central theme of mental illness. The story's growing complicity between the material world of the "horrid" (42) room and Gilman's narrator suggests that if the self is never whole, it can join with another, to see together, without claiming to be another. The wallpaper's material substance – the female figure shining through its patterned surface – pushes this seeing together as it becomes translated into the story's unfolding process of the narrator's diary writing and as it is signaled by the story's opening letter "I," a homonym (I/eye; see Fig. 3) that stresses the wallpaper's sensual and affective dimension. Gilman's story is not concerned, however, with rendering an anthropomorphic view of things simply triggered by the incremental changes of the narrator's mental state, but rather acknowledges the material to detect a maze of dirty secrets.

Let us act with the material and follow the wallpaper. The three illustrations by Joseph Henry Hatfield that accompany the story's first printed magazine version reveal a kind of invitation. The opening illustration shows a room with a barred view and portrays a Victorian female figure immersed in the act of writing – an act that both violates the rest cure's restrictive handling of creative activities and anticipates the narrator's writing as a rebellious act against standard medical knowledge (see Fig. 2).

This first illustration and the ending of the first diary entry echo each other; the narrator stops writing and addresses her audience: "There comes John, and I must put this away – he hates to have me write a word" (Gilman 1994a, 41). Text and visual icons offer observations of a second order revealing the blind spots in a domestic drama that consists of highly selective perceptions. The illustration is followed by a pictorial capital "I" (see Fig. 3) which appears amidst two triangles that form a large square, mirroring the "sickroom" the first-person narrator occupies.

The bottom triangle is blank, but the top one contains a design that may refer to the arabesque wallpaper design that puzzles the narrator. Of course, Hatfield used similar ornamental letters for other imprints, but the decorative contrast between the two triangles (plain vs. patterned) and the decorative vine motif climbing to the

Fig. 2: Joseph Henry Hatfield, illustration from *The New England Magazine*, n.s. 5 (January 1892): 647; copyright in the public domain.

top of the letter, recall late nineteenth-century changes in interior decoration, especially the use of wallpaper.

The narrator's attic bedroom had apparently been wallpapered in the decades when arabesque designs and French figural patterns were still in vogue. Hence, the wallpaper in her room is slightly old-fashioned for the last decade of the nineteenth century. In addition to contemporary reactions against overelaborate interiors, domestic environmentalism sparked a controversy about truth and falsity in wallpaper design reminiscent of William Dean Howells's campaign for realism in fiction. Referring to Catherine Lynn's study *Wallpaper in America* (1980), Jürgen Wolter claims that critics began "to find fault with the eye-fooling concoctions in the traditional wallpapers of the French design, with their fakery that would deceive even a practiced eye" (Wolter 2009, 197). Critics would increasingly demand honesty and simplicity.

The cessation of interest in patterned wallpaper at the end of the century was certainly a rejection of the previously popular French designs, considered by many to be psychologically pathogenic. Even more generally discussed was that medical research had discovered that many dyes used in wallpaper contained poisonous chemicals. Chrome yellow and flake white, for example, contained lead; King's

Fig. 3: Joseph Henry Hatfield, illustration from *The New England Magazine*, n.s. 5 (January 1892): 647; copyright in the public domain.

yellow and many of the green pigments used in wallpapers and fabrics contained arsenic. Gilman saw a connection between the exposure of the human body to chemical substances and physical and psychic illnesses. Only two days after finishing the draft of "The Yellow Wall-Paper" she wrote in her diary: "*Miserable* night – over dose of acid phosphate gives *terror*" (Gilman 1994b, 418; original emphasis). In Gilman's story, John prescribes "phosphates or phosphites" (Gilman 1994a, 39) to make his wife "well again" (39), causing us to question what "well" really means here. Does John intend to poison her?

The narrator's husband, as a physician of "high standing" (39) would certainly be familiar with this current debate about the toxic effect of certain kinds of wallpaper, and indeed, John rarely enters the room. Earlier critics even suspect that he spends the nights in town with another woman, a theory supported by the narrator who states at one point that John "asked me all sorts of questions [. . .] and pretended to be very loving and kind. As if I couldn't see through him!" (51). Why would John confine his ill wife to a room that is not only unpleasantly old-fashioned, but toxic? And why – since he obviously knows about his wife's physical struggle with the wallpaper – does he object to its removal, especially when it was a well-known fact that it could cause death? A reading of the story in the context of material culture raises questions on whether John intends to kill his wife in such a slow and insidious manner or fails in an attempt to help her regain her health. Some critics argue that murder in the context of Gilman's crime fiction "is ultimately a collective act and the unravelling of the mystery is concerned not so much with the solution of the crime but the crime of patriarchy" (Beer 1998, 60).

Such a plot or papered "subpattern" (Gilman 1994a, 43) can be brought in line with the color yellow in the story. Critics frequently argue that Gilman decided to use the yellow wallpaper to symbolize Victorian patriarchy. In her essay "The Philosophy of Beauty Applied to House Interiors" (1893), decorative artist Candace Wheeler, one of Gilman's contemporaries, wrote that "color is the most powerful mental influence of the home" (Wheeler 1893, 7) and that the whole period became known as "The Yellow Decade" (Wolter 2009, 200). Some critics even suggest that Gilman refers to her great aunt Harriet Beecher Stowe who in her *House and Home Papers* (1865) celebrates domesticity and has the model wife wish for a wallpaper that produces warm yellowish reflections (see Gaskill 2018).

By contrast, material co-agency in Gilman's story dirties everyone involved. Gilman's "unclean" yellow wallpaper with its "sickly sulphur tint" (Gilman 1994a, 41) mimics the contemporary discourse of the yellow decade that ranges from aestheticism and homosexuality to nativism and disease. The obtrusive and recalcitrant yellow wallpaper might suggest the decadence and decay of the conventional male-dominated marriage and the socio-material environment represented by John whose clothes are also ruined by "yellow smooches" (48). The yellow smell, then, is that of the rotting wallpaper and of the society it materializes. Yellow was also the color of the feminist movement. Women used yellow roses, yellow ribbons, and yellow umbrellas to symbolize their suffrage cause, so the yellow wallpaper also connotes the narrator's furtive rebellion. At the beginning of the mental journey out of her husband's world, she repeatedly emphasizes John's authority. Her protest is still suppressed in her subconscious, and the yellow wallpaper is still repellent to her. Gradually, however, the rebellion and the yellow wallpaper become part of her being, culminating in a conflation of expressive objects in the home and her confessions about "my wall paper" (48).

One of the first readerly reactions the story prompted was a short, anonymous review entitled "Perilous Stuff" that appeared in *The Boston Evening Telegraph* in 1892. This note to the editor states that

> In a well-known magazine has recently appeared a story entitled "The Yellow Wallpaper." It is a sad story of a young wife passing through the gradations from slight mental derangement to raving lunacy. It is graphically told in a somewhat sensational style which makes it difficult to stay aside, after the first glance, til it is finished, holding the reader in morbid fascination to the end. It certainly seems open to serious question if such literature should be permitted in print.
>
> (qtd. in Gilman 1998, 88)

Admittedly, such contemporary responses can be read as warnings against female penmanship and the dangers of an overheated imagination that were found on both sides of the gender line in late nineteenth-century America; but the note also exposes a "fascination" with the materiality of the story's graphic, visual, and thematic dimensions. The critic's reference to "glance" illustrates the story's invitation for a seeing together and becoming involved in the wallpaper's agency, similar to the color yellow itself.

The messiness of a yellow world with its sensual, intriguing, and mysterious stuff has produced a proliferating critical industry mirroring cross-disciplinary trends of investigation. Gilman's case study exhibits, as one critic recently put it, "the adaptability of human brains to physical conditions, and it leaves us wondering whether woman's freedom was lost with the Garden, as the story goes, or whether it creeps behind the bars of socialization within each of our minds" (Scott 2009, 202). Other critics read the wallpaper as a symbol of an arabesque prison for women. As such it belongs to an American orientalism as practiced by Edgar Allan Poe (↗7 Poe). Marty Roth concludes: "As a haunted house tale, Gilman's 'The Yellow-Wallpaper' offers a unique terrain on which to examine turn-of-the-century American fiction as a place where imperialist and feminist anxiety meet, where the abjection of the woman and the abjection of the Eastern other collide and negotiate a release of terrifying energy" (Roth 2001, 161). Another critic historicizes Gilman's story in the context of the turn of the century sensationalism promulgated by the rise of the yellow press (see Sari Edelstein 2011).

Extending these arguments, I will conclude by shedding some light on the story's toxic ending. There are innumerable accounts about the story's moment of mental and physical self-destruction in the closing scene where John faints and his wife "creep[s] over him!" (Gilman 1994a, 53). This is certainly a very ambiguous moment of liberation, as the narrator's exclamation "I've got out at last" (53) seems possible only through deliberate self-destruction. But let me highlight another invitation the story offers at its very ending. In Gilman's autograph version, the wife simply creeps over the fainted body of her husband. In the printed magazine version she says: "I had to creep over him every time!" (Gilman 1892, 656), a temporal extension that refers back to the narrator's deranged crawling amidst the walls of her room. The editors who made this and other changes acted according to generic conventions because the literary market obviously preferred grotesque stories of mad horror. These alterations were out of Gilman's (and perhaps the editors') control.

Hatfield's tailpiece illustration (see Fig. 4) shows the scene playing out quite differently: as the narrator creeps over John's fainted body, she deliberately and forcefully pushes her husband's face into the floor's toxic dirt covered by shreds of the wallpaper she had torn down in her repeated ramblings. One might argue that perhaps Hatfield, as one of the story's first readers, sensed and ironically commented on – using his pen and ink – the submerged plot of a toxic substance and John's insidious plan to kill his wife. It also shows that the wallpaper exemplifies a complex network of material co-agency that has many ties to the world.

Fig. 4: Joseph Henry Hatfield, illustration from *The New England Magazine*, n.s. 5 (January 1892): 656; copyright in public domain.

The "Yellow Wall-Paper" emphasizes the literary life of things because "to creep" over the story "every time" produces trans-temporal affinities and connections. The wallpaper therefore not only symbolizes patriarchy but highlights the literary artifact as a non-human actor. As Rita Felski notes, this notion of the non-human actor in literary studies "assumes no particular measure of scale, size, or complexity. It can include not only individual novels or films but also fictional characters, plot devices, literary styles, filming techniques, and other formal devices that travel beyond the boundaries of their home texts to attract allies, generate attachments, trigger translations, and inspire copies, spin-offs, and clones" (Felski 2015, 168–169). If we accept Sari Edelstein's claim that "Gilman had to succumb to 'The Yellow Wall-Paper' before she could align her aesthetics with her politics" (Edelstein 2011, 195), this could only be achieved by a host of co-actors and the literary artifact's sociability in drawing things together.

Following the material approach, the second example moves beyond canonical texts and explores contemporary short fiction and how it is designed, curated, and printed within the pages of independent literary magazines. While most current scholarship in the field of publishing the short story discusses either institutional or economic themes – ranging from the modernist little magazines and anthologies (↗3 Current Approaches) to the short story and digital media (see Bulson 2017; Bazin 2019; Einhaus 2016; Delaney and Hunter 2019) – the following analysis proposes to read the American short story together with today's independent magazine publishing that is currently leading a print renaissance bucking the digital trend with stylishly printed literary magazines. In doing so, I call into question an opinion saying that the contemporary short story is a "dead form, unnaturally perpetuated"

(Batuman 2006, 63). Critics increasingly consider the short story an exhausted genre. As such, the short story continues "to be propagated in America by a purely formal apparatus" (63) – i.e., the big mainstream magazines, which, if they print fiction at all, "sandwich one short story per issue between features and reviews" (64). Workshop-based creative writing programs and their attendant literary journals do the rest. Elif Batuman judges that "today's short stories all seem to bear an invisible check mark, the ghastly imprimatur of the fiction factory" (64). Such negative evaluation of the short story's market value neglects the resurrection of independent literary magazines in the digital age (see Masurier 2020). Indie magazines like *Zoetrope*, *American Chordata*, Dave Egger's *McSweeney's Quarterly*, and the science fiction lit mag *Visions* re-imagine our experience of serialized print and reading short fiction (see Scheiding 2022). Since their print run is small, most do not carry advertising. They are visually and materially seductive, and their arrangements of content, photography and paper stocks convey a different view of the world. Rather than seeing them as fossils of the pre-web era, I would argue that the wave of these new magazines is the result of the possibilities offered by the new technology that was supposed to kill print culture. They sell and distribute online, they crowdfund, and they invent their own business models frequently creating a whole cultural infrastructure around these emerging niche magazines (for instance, Pop-up magazines as public storytelling events). They curate storytelling, often around a single theme and offer their readers narrative journeys combining ideas, pictures, and activities (Lupton 2017). In her essay "Independent Magazines and the Rejuvenation of Print," Australian media scholar Megan Le Masurier claims that the return to the printed magazine is not a "retro-abreaction" (Le Masurier 2012, 394) but an informed response to what the magazine as a printed object can offer that digital screens cannot. In this sense, she refers to the core elements of the printed magazine pushed to the foreground by the indie magazines producers: "the quality and tactility of the paper, format (size, shape, binding), the integration of words, images, space as an expression of graphic design specific to print, and the life and function of the magazine as a material object" (394). Scholars of print culture suggest that the material object of the magazine not only involves textual competence but also material competence – an ability to read the semiotics of the concrete forms that embody, shape, and condition the meanings of texts (see Moylan and Stiles 1996).

Zoetrope: All-Story was originally founded by Francis Ford Coppola in 1997. The magazine's launch was inspired by Alfred Hitchcock who saw film and the short story in close relationship to each other. Harking back to Poe's idea of a "unity of impression" (Poe 2006, 532), he appraised film and short story as a medium to be consumed in one sitting (see Coppola 2000, vii–xi). As the magazine's subtitle signals, the San-Francisco based quarterly echoes *All-Story*, one of the earliest pulp magazines that was published in the opening decades of the twentieth century. *Zoetrope* amalgamates different legacies of American print culture and, like its predecessor, it publishes a wide variety of new short fiction. *Zoetrope* has become known as a publication for emerging and established authors. Each issue includes a classic reprint. Alongside previously unpublished fiction and one-act plays, the classic reprint illustrates a piece of short fiction that has been adapted to film or inspired a movie. The winter issue contains "The Discipline of D.E." (aka "do easy"), a story form William Burroughs' collection *Exterminator!* (1973) turned into a short film by Gus Van Sant in 1977. What distinguished the earlier pulp-magazines was their engaging well-known American artists doing either the cover design or featuring interior illustrations depicting elements of the story. Sometimes covers would be designed first and authors would then be shown the cover art and asked to write a story to match. *Zoetrope* follows this practice by inviting a guest designer to construct the quarterly's issue. Since Helmut Newton was invited to design the magazine in 1998, famous artists, musicians, actors, and film directors have contributed to the magazine's visual aesthetic. The guest designers' selection is based on the policy rule suggesting that the commissioned person has never designed for print before. Moreover, the magazine's Greek title refers to the changing and turning of life as well as continuous motion. Invented in the early nineteenth century, the zoetrope became known as an optical toy. It produces an illusion of motion from a rapid succession of static images. The guest designer concept highlights the magazine's commitment to change, reinventing itself from issue to issue.

Having been awarded the 2018 National Magazine Award for Fiction, *Zoetrope* has also gained cultural capital in the print market. It serves not only as an archive for short fiction (or the "world greatest story department," as Coppola calls it; 2000, x), but – given its experimental feel – differs from mainstream imprints like *The New Yorker* or *The Atlantic Monthly*. Besides, its format mirrors an earlier literary tradition that comes from commissions for periodicals, like the WPA publications in the 1930s, when writers needed assignments to make a living. Unlike traditional publishers that specialize in high volume commercial fiction frequently squeezing work into ill-suited existing genres, *Zoetrope* markets the short story within an inspiring and multimodal magazine design.

The winter volume 2018/19 was guest-edited by David Lynch. The issue combines his surreal and grotesque artwork with contemporary writing. The magazine's front section opens with Tommy Orange's short story "Freyr" (see Fig. 6). His best-selling debut novel *There There* turned Orange into one of the most talked-about authors in 2018. Critics praised him as a new and leading voice in contemporary American fiction. Orange is the son of a white mother and Native father. Like most of his book's characters, he is an "urban Indian," born and raised in Oakland, California. Published by Alfred Knopf and marketed as "A novel" on the cover, his first book reads more as a short-story cycle (↗3 Current Approaches) fusing multi-generational stories about Native Americans living in the area of Oakland, than a typical novel. Since the book came out in June of 2018, one of its chapters titled "Thomas Frank" (Orange 2018c, 208–225) was published earlier as a short story in *The New Yorker* (see Fig. 5). The magazine's fiction editor, Deborah Treisman, decided

Fig. 5: Tommy Orange, "The State," *The New Yorker*, March 26, 2018; Henning Wagenbreth, The New Yorker © Conde Nast.

Fig. 6: Tommy Orange, "Freyr," *Zoetrope: All-Story* 22.4. (Winter 2018/19): 12–13; reprinted with permission.

to print the piece as a teaser supporting the novel's sale. The chapter's new title "The State" (Orange 2018b) and the illustration that highlights some of the story's events market Orange's work under the challenges of living with an ethnic identity, especially those that keep the protagonist confined in the messy "state" of alcoholism. In this sense, *The New Yorker*'s document and product design meet stereotypical expectations about Native Americans' lives. The allegedly fictional self-portrait is framed by the illustration's cliché ethnic colors, a commissioned work drawn by the German graphic artist Henning Wagenbreth (see Orange 2018b).

For the winter issue, *Zoetrope* prints Orange's first original short story "Freyr." It is a grotesque and Kafkaesque story about a "guy with the hand that came out of his chest" (Orange 2018/19, 14). Given what Orange calls the experience of being "ambiguously nonwhite" (Orange 2018c, 28) the story's central theme deals with "our clumsiness regarding all things abnormal" (Orange 2018/19, 14). In the story it says: "Normal is vanilla, nonspecific, flavorless, colorless, your basic blah nothing default white guy, for example – the stick they have everyone measure himself by" (14). "They" are the "coworkers" of an unspecified Oakland "company." Both the unnamed first-person narrator and Frank, the guy with three hands, serve as office clerks. As Native Americans both are avoided by their co-workers as being either weird or ugly; their otherness however makes people gaze at them. The narrator reasons: "My kind of ugly, it's the sort people think it's rude to look at, but they can't look away, either" (14). Frank tries desperately to hide his secret under thick layers of clothing. His secret becomes public during a summer company picnic, then everyone in an "awful moment" (14) of Frank's nudeness – he is pulling of his sweaters – suddenly sees the hand coming out of his chest. Subsequently, both characters develop an intimate friendship, but Frank, falling into a depression, finally commits suicide. At his funeral, the narrator meets Frank's sister and learns more secrets about the family's past. The parents were members of an "Icelandic community" (18) in Utah and Frank was named after the old Norse god of virility and prosperity, Freyr. Another secret concerns his friend's gene mutations causing his bodily overgrowth and the siblings' survival after their parent's death in a car crash. At the end of the story, Freyr's sister sings one of the ancient secret songs used for ceremonial purposes.

David Lynch's framing artwork explores the short story's obsession with embodiment and boundary work and how it extends our understanding of vulnerability, disgust, horror, and uncertainty. The images used for the story's layout derive from Lynch's photo series "Distorted Nudes." In 1999, David Lynch scanned pictures from Uwe Scheid's book called *1,000 Nudes* displaying women in boudoirs and brothels in the 1920s. For his own photo series, Lynch digitally manipulated these erotic photographs into strange shapes, some of which resemble Francis Bacon's paintings cutting off heads and limbs with Photoshop and stretching the human body's shapes. Lynch morphs his images and the classical typesetting of Orange's short story against a black backdrop (see Fig. 7).

Fig. 7: Tommy Orange, “Freyr,” *Zoetrope: All-Story* 22.4. (Winter 18/19): 19; reprinted with permission.

The kaleidoscopic color scheme mirrors the two-page title spread creating a semantic-semiotic environment for projective readings signaled by the sister's affective song that appeals to "your ears, stomach, and heart [. . .]. You went up and down [. . .], oscillating between dread and hope, grief and joy, ever moving beyond [. . .]" (Orange 2018/19, 18).

The story's suggestive looking through the veils of secrets and the drowning in their meanings corresponds with Lynch's use of the color black. Lynch has his own theory of black. For him it is "depth." It makes "you start seeing what you're afraid of. You start seeing what you love, and it becomes like a dream" (Lynch 1997, 20). The dreamlike and fantastic atmosphere connected with the story is expressed by the fragmented torso on the title-spread and the tongue that sticks out of it blasting the word "sleep" (see Fig. 6). The surreal visual setting that surrounds the story is doubled by the short story's intertextual reference to Kafka's highly ambivalent and dreamy novella "The Judgement" (1913).

Zoetrope's densely knitted visual and textual arrangement responds to the abject, where identity, order or systems are thrown into disarray. The abject can also be social situations and activities that are polluting, betrayals of what is considered 'right,' and social positions that are considered defiled, stigmatized or associated with what destabilizes our sense of "normality" (18), as Fryer's sister puts it. For both, Lynch and Orange, it is the abject body. In the story, the narrator confesses: "Ugliness *came out* of my face. Protruded. Think of a Native American Bill Murray, with Steve Buscemi eyes" (Orange 2018/2019, 14; original emphasis). The page as a visual artefact and the magazine as a material object afford the possibility of voicing both the significance and incomprehensibility of disorderly bodies. It also highlights the body's multiplicity. Given society's notions of ideal beauty and the normative body, Freyr's transgressive (Native American) corporeality is under constant repair. Born without thumbs, he is put under society's scalpel of adjustment. Fryer's sister reveals that "they used parts of his – the other hand [. . .] to build two working thumbs" (16). Both artist and writer emphasize, however, the illusion of wholeness and boundedness. In this sense, *Zoetrope*'s experimental setting expands the author's role. The magazine serves as a zone of transaction where editors, readers, artists, and writers collaborate to guide reception of a work. *Zoetrope* is thus a multimodal medium to communicate both "the multi-storied landscape of this country," as Orange (2018b, 81) writes, and the constitutive nature of its (racial) differences.

3 The Short Story as Literary Form for the Twenty-First Century

The short story "Cat Person" was published by *The New Yorker* in the December issue of 2017. When the magazine dropped it on the internet, the story – written by the previously unknown writer Kristen Roupenian – went viral within days and became the most widely read short story released by *The New Yorker* (↗2 Canon Formation). It struck a chord with thousands of readers. The short story received almost two million pageviews and made a splash in social media communities around the globe. Fiction – especially the short story – is not on the list of things that usually go viral, but this bad-date story has fired up readers at exponential rates. Despite numerous complains about the short story's existence in "the margins, the secret shelves between the secret sections in the bookstore," as Michael Chabon puts it (Chabon 2009, 14), Roupenian's short story demonstrates that the genre is neither commercially declining nor dead. Such predictions become irrelevant as the short story adapts to the paperless and borderless media of today's e-age.

Roupenian's story is about a 20-year-old woman named Margot who meets a 34-year-old man, Robert, at an "artsy movie theater" (Roupenian 2017, n. pag.) where Margot works at the concession stand. After a few weeks of friendly banter, Robert gets her phone number. They proceed to text for a few more weeks and then go out on a date that leads to a very unsatisfying sexual encounter, after which Margot breaks off the budding relationship via text-message. A month later they run into each other at a bar, and Robert hurls misogynistic insults at her, also via text, calling her a "whore" (n. pag.). Roupenian's short story kindled a firestorm on social media since many readers interpreted it as a news story or a non-fiction personal essay. Molly Roberts, writing for the *Washington Post*, addressed the issue of fiction and online communities and concluded: "A New Yorker story went viral because, for one of the first times, something in the magazine seemed to capture the experience not of print-oriented, older intellectuals but of millennials. As some Twitter users said, young people rarely see the phenomenon of modern dating – meeting online, talking through text messages, moving to in-person encounters shadowed with the expectation of sex – taken seriously" (Roberts 2017, n. pag.).

"Cat Person" certainly hit a timely nerve and can be tied to the MeToo movement, which may explain why the story gained traction and was found "relatable" by many readers. Despite the fact that the story channeled the subconscious collective, Roupenian's story also speaks to the genre of the short story in an electronic age of unstable borders. As the reader responses indicate, these unstable borders concern the role of literary forms and the skill of writing fiction in a journalistic landscape of breaking news and non-fiction readers. It also makes us reflect upon how – throughout the genre's history (↗2 Canon Formation) – the short story readily lends itself to the latest media. Roupenian's piece also demonstrates that storytelling is a "public art." In *The Lonely Voice* (1962), Frank O'Connor's study of the short form, he asserts that "the short story, like the novel, is a modern art form [. . .] it represents our own attitude to life" (O'Connor 1976, 83). O'Connor suggests that the short story is not about heroes but an expression about "a submerged population group," for example, "Chekhov's doctors and teachers" or "Sherwood Anderson's provincials" (86). Roupenian's short story is, therefore, not a story about an agenda, but about characters more complex and flawed than at any other age. As such, the story gives a generational insight into the reality of rampant misogyny and the ironies of heterosexual relationships (cf. Ernest Hemingway's "Hill like White Elephants" or David Forster Wallace's "Good People"). Like Charlotte Perkins Gilman's "The Yellow Wall-Paper," "Cat Person" may be read as a new allegory for patriarchal abuses. However, Gilman's story became a sacred feminist text not only because it related a harsh reality, but because the short story illustrates her consummate skill in writing and storytelling as a public art distributed via mass magazines. Other stories that come to mind are Shirley Jackson's "The Lottery" (1948; ↗21 Jackson) and Lorrie Moore's "You're Ugly, Too" (1989), both published in *The New Yorker*'s pages.

Although *The New Yorker* has almost no rivals among periodicals publishing short stories in the United States and abroad, it is an exceedingly rare event that a single short story attracts any attention at all. Perhaps it is too early to make predictions about Roupenian's persistence as a writer of fiction, but in the history of the genre, there are storytellers who revitalize a genre that – since the invention of television or digital media – has been continually pronounced both moribund and recovering from the moribund (↗24 Barthelme and ↗26 Carver). It becomes also evident that the connection between the reader and the short story is both a function of aesthetics and economy.

A good case in point is the current wave of micro-fictions frequently published via Twitter or other e-platforms. Contemporary writers like Jennifer Egan ("Black Box"), Teju Cole ("Seven Short Stories about Drones"), and John Wideman ("Two Stories") experiment with novel ways to distribute their new work and try out alternative practices for sharing it with their readers. In an unusual move for an established author, Wideman teamed up with the self-publishing and print-on-demand service Lulu.com to release his collection of micro stories *Briefs: Stories for the Palm of the Mind* (2010). The new book went on sale exclusively through Lulu.com, accompanied by a series of live readings from the book, which were held in New York and Los Angeles. Wideman also made several of his backlist titles available for sale through Lulu.com. While his previous books were published almost exclusively by Houghton Mifflin, self-publishing expresses Wideman's desire for more control over the publishing process and the reach of his narratives. The story "Long Story," for instance, echoes Wideman's critique of what he calls mainstream publishing's "blockbuster syndrome" – the tendency for large trade publishing houses to focus their resources on books with bestseller potential. Wideman laments that the trade "has gotten out of hand. Unless you become a blockbuster, your book disappears quickly. It becomes not only publish or perish, but sell or perish" (Graff 2010, n. pag.).

Wideman's collection of micro stories challenges established norms in literature and establishes norms of its own. Its texture is constituted within a relational field of interwoven stories. The fragments germinate by interaction and come alive by being read and commented upon, as happens in informal blog entries, reading clubs, or online discussion groups. Wideman's micro stories evolve in the interaction with a readerly environment. This is probably one of the reasons why – with the publication of *Briefs* – Wideman ventured out into the blogosphere since readers need stories and stories need readers to bring out the potential of both the stories and the readers. "Small stories can offer," states Wideman, "quick exit and re-entry into the immensity surrounding them" (Wideman 2010b, n. pag). *Briefs* performs stories that move through the world. As the examples in this chapter have shown, the modern American short story is a coming together of people and things in changing media environments. The short-story form offers the reader informal encounters of human becomings, rather than human beings (see Basseler 2019).

The observations made here encourage rethinking the American short story in publishing terms (see Hungerford 2016). In a recent piece on the American short story in the twenty-first century, Charles May speculates that "because American readers are more interested in nonfiction and the occasional novel than the short story, the form is currently in the literary doldrums in the United States" (May 2012, 324). May's statement is of interest because he raises an important question: how should we write about the American short story in the twenty-first century, especially after the minimalist boom of the 1980s? The most recent surveys and companions tell a progressive story that deals with the evolution of the form the nineteenth into the twentieth century culminating in neo-realist renaissance of the Eighties and Nineties. Departures from this master narrative of the American short story can be found in a number of studies written in the wake of Mark McGurl's groundbreaking book *The Program Era* (2009) that saw a sea change in the study of postwar literature. A good case in point is Kasia Boddy's slim volume (2010) on the American short story after 1950 in which she focuses on institutional aspects of the form in the context of big magazines, MFA programs, and the production of the short story in a literary market depending on specific forms of recognition like literary awards and prizes. Other studies, such as Paul-March Russell's introduction to the short story (2009), are more conservative in reassessing the genre as a literary art-form. He travels along the genre's stadial history from its beginnings in folk traditions to what he calls in the final chapter "Voyages Out" focusing on the "postcolonial short story." In her 2018 study on the American short story cycle, Jennifer J. Smith analyzes the expansion of short prose within a specific form of narrative composition. She claims that the "short story cycle has become the primary and best descriptor for the genre, because it most accurately captures the recursiveness central to the genre and privileges the short story as its formative element" (Smith 2018, 4). Along the line of earlier studies, like James Nagel's study on the contemporary American short story cycle, she reads the cycle as a productive format for "new identity formations" (7). Since the multiethnic literature of the 1980s and 1990s turned toward multiplicity, polyphony, plurality, and fragmentation, the cycle's composite nature allowed ethnic writers to integrate new elements into the short story and expand it. Smith's concluding chapter deals with an art form that has become known over the last decade as sudden fiction, quick fiction, fast fiction, and flash fiction referring to stories around and under 1,000 words achieving a particular effect on its audience based on the length of engagement (see *The New Yorker*'s summer online flash fiction series 2017–2019). By 2010 a new genre emerged under the term "novella in shorts" or "novella-in-flash" (170). These are novellas composed entirely of standalone flash fiction pieces organized into full narrative arcs. Smith not only highlights the cycle's duration as an art form, but also the ways in which the short-story form resurges in the digital age.

Another notable approach to explaining the genre's productivity is the 2019 collection of essays published as *The Edinburgh Companion to The Short Story in English*, edited by Paul Delaney and Adrian Hunter. The collection offers an analytical framework built upon five major themes: history, publication, form, place, and identity. In terms of future issues for the study of the short story, the most rewarding piece is Laura Dietz's essay on "The Short Story and Digital Media" (Dietz 2019, 125–139). The essay raises important questions about the future of the story as a portable genre that moves in-between electronic publications, digital fiction networks, and traditional imprints. In doing so, her analysis draws our attention not only to what the label "short story" means in today's digital age, but also to the ways in which traditional publishers and authors alike launch new artistic and commercial endeavors. While digital-only formats seem to push single-author short story collections to the "margins of the print book market" (129), recent publishing evaluations also show that digital short-story collections provide an outlet for collections that have no print prospect at all. On the one hand, this raises questions about selectivity, authorship, the production, distribution, the consumption of short stories, and the affordances of reading; on the other hand, it also concerns the very form of the short story. If brevity becomes a commercially viable brand – see the many labels of the short fiction form on Amazon – the novel becomes a label with weight and significance. New York publisher Henry Holt and Company released Philip Caputo's cycle of short stories *Hunter's Moon* (2019) under the label "A Novel in Stories," for instance. By borrowing the label, the short story can share the novel's prestige and convince readers to buy it. Perhaps this is another reason for the booming of the short story cycle as a composite novel over the last decade. In the light of the short story's progress in global and publishing terms, this chapter has shown that we need studies focusing on both its economy and formal innovations (see Seybold and Chihara 2019). Highlighting the short story's "portability" (Dietz 2019, 137), these studies would involve a different kind of literary analysis that pays close attention to the complex material and aesthetic forms that "organize texts, bodies and institutions" (Levine 2015, 23).

4 Bibliography

4.1 Works Cited

Baker, Sam. "The Irresistible Rise of the Short Story," https://www.telegraph.co.uk-/culture/books/10831961/The-irresistible-rise-of-the-short-story.html *Telegraph* (18 May 2014), n. pag. (23 March 2020).

Basseler, Michael. *An Organon of Life Knowledge: Genres and Functions of the Short Story in North America*. Bielefeld: Transcript, 2019.

Batuman, Elif. "Short Story & Novel." *n+1*, Issue 4: Reconstruction (Spring 2006): 63–70.

Bazin, Victoria. *Modernism Edited: Marianne Moore and the Dial*. Edinburgh: Edinburgh UP, 2019.

Beecher, Catherine E., and Harriet Beecher Stowe. *The American Women's Home*. New York: J. B. Ford, 1869.

Beer, Janet. "Charlotte Perkins Gilman and Women's Health: The Long Limitation." *A Very Different Story: Studies on the Fiction of Charlotte Perkins Gilman*. Eds. Val Gough and Jill Rudd. Liverpool: Liverpool UP, 1998. 54–67.

Berliner, Jonathan. "Written in the Birch Bark: The Linguistic-Material Worldmaking of Simon Pokagon." *PMLA* 125.1 (2010): 73–91.

Boddy, Kasia. *The American Short Story Since 1950*. Edinburgh: Edinburgh UP, 2010.

Brown, Bill. "Introduction: Textual Materialism." *PMLA* 125.1 (2010): 24–28.

Bulson, Eric. *Little Magazine, World Form*. New York: Columbia UP, 2017.

Chabon, Michael. *Maps and Legends: Reading and Writing Along the Borderlands*. New York: Harper Perennial, 2009.

Chan, Winnie. *The Economy of the Short Story in British Periodicals of the 1890s*. New York: Routledge, 2007.

Cocchiarale, Michael, and Scott D. Emmert, eds. *Critical Insights: Flash Fiction*. Ipswich: Salem P, 2017.

Cocchiarale, Michael, ed. *Critical Insights: American Short Story*. Ipswich: Salem P, 2015.

Coppola, Francis Ford. "Introduction." *Francis Ford Coppola's Zoetrope: All-Story*. Orlando, FL: Harcourt, 2000. vii–xi.

Delaney, Paul, and Adrian Hunter, eds. *The Edinburgh Companion to the Short Story in English*. Edinburgh: Edinburgh UP, 2019.

Dietz, Laura. "Online versus Print: The Reputation of Literary Fiction Magazines." *Short Fiction in Theory & Practice* 4.1 (2014): 7–21.

Dietz, Laura. "Short Story and Digital Media." *The Edinburgh Companion to the Short Story in English*. Eds. Paul Delaney and Adrian Hunter. Edinburgh: Edinburgh UP, 2019. 125–143.

Edelstein, Sari. "The Yellow Newspaper: Charlotte Perkins Gilman and Sensational Journalism." *Charlotte Perkins Gilman: New Texts, New Contexts*. Eds. Jennifer S. Tuttle and Carol Farley Kessler. Columbus: The Ohio State UP, 2011. 180–200.

Einhaus, Ann-Marie. *The Cambridge Companion to the English Short Story*. Cambridge: Cambridge UP, 2016.

Ernst, Jutta. *Amerikanische Modernismen: Schreibweisen, Konzepte und zeitgenössische Periodika als Vermittlungsinstanzen*. Würzburg: Königshausen & Neumann, 2018.

Farquharson, Kathy. "'The Last Walls Dissolve: Space Versus Architecture in *Memoirs of a Survivor*' and the 'The Yellow Wall Paper.'" *Doris Lessing Studies* 28.1 (2009): 4–7.

Felski, Rita. *The Limits of Critique*. Chicago and London: The U of Chicago P, 2015.

Finkelstein, David. "The Role of the Literary and Cultural Periodical." *The Routledge Companion to British Media History*. Eds. Martin Conboy and John Steel. London and New York: Routledge, 2015. 263–272.
Gaskill, Nicholas. *Chromographia: American Literature and the Modernization of Color*. Minneapolis, MN: U of Minnesota P, 2018.
Gilman, Charlotte Perkins. "Excerpt from *The Living of Charlotte Perkins Gilman*." 1935. *Charlotte Perkins Gilman's "The Yellow Wall-paper" and the History of Its Publication and Reception*. Ed. Julie Bates Dock. University Park, PA: The Pennsylvania UP, 1998. 87–88.
Gilman, Charlotte Perkins. *"The Yellow Wall-Paper" and Selected Stories of Charlotte Perkins Gilman*. Ed. Denise D. Knight. Newark: Delaware UP, 1994a. 39–53.
Gilman, Charlotte Perkins. *The Diaries of Charlotte Perkins Gilman*. Vol. 2: 1890–1935. Ed. Denise D. Knight. Charlottesville and London: U of Virginia P, 1994b.
Gilman, Charlotte Perkins [Stetson, Charlotte Perkins]. "The Yellow Wall-Paper." *The New England Magazine*, n.s. 5 (January 1892): 647–656.
Glass, Loren, ed. *After the Program Era: The Past, Present, and Future of Creative Writing in the University*. Iowa City: U of Iowa P, 2016.
Graff, Keir. "A Real Lulu: John Edgar Wideman to Self-Publish," https://www.booklistreader.com/2010/03/05/book-news/a-real-lulu-john-edgar-wideman-to-self-publish/ *The Booklist Reader* 5 March 2010. (20 March 2020).
Hamilton, Caroline. *One Man Zeitgeist: Dave Eggers, Publishing, and Publicity*. New York et al.: Bloomsbury, 2010.
Harris, Donal. *On Company Time: American Modernism in the Big Magazines*. New York: Columbia UP 2016.
Holmes, Timothy, and Miglena Sternadori, eds. *The Handbook of Magazine Studies*. Hoboken, NJ: Wiley, 2020.
Jamieson, Ruth. *Print is Dead. Long Live Print: The World's Best Independent Magazines*. Munich, London, and New York: Prestel, 2015.
Jemisin, N. K. "Introduction." *How Long 'Til Black Future Month?* London: orbit, 2018. vii–xii.
Latour, Bruno. "The Berlin Key or How to Words with Things." *Matter, Materiality, and Modern Culture*. Ed. P.M. Graves-Brown. London and New York: Routledge, 2010. 10–21.
Latour, Bruno. *Reassembling the Social: An Introduction to Actor-Network-Theory*. Oxford: Oxford UP, 2005.
Le Masurier, Megan. "Independent Magazine and the Rejuvenation of Print." *International Journal of Cultural Studies* 15.4 (2012): 383–398.
Le Masurier, Megan. "Slow Magazines: The New Indies in Print." *The Handbook of Magazine Studies*. Eds. Tim Holmes and Miglena Sternadori. Hoboken, NJ: Wiley Blackwell, 2020. 92–104.
Levine, Caroline. *Forms: Whole, Rhythm, Hierarchy, Network*. Princeton: Princeton UP, 2015.
Levy, Andrew. *The Culture and Commerce of the American Short Story*. Cambridge: Cambridge UP, 1993.
Long, Maebh. "The Short Story from Postmodernism to the Digital Age." *The Cambridge Companion to the English Short Story*. Ed. Ann-Marie Einhaus. Cambridge: Cambridge UP, 2016. 128–142.
Lover, Heather. "Close but not Deep: Literary Ethics and the Descriptive Turn." *New Literary History* 41 (2010): 371–391.
Lunden, Jennifer. "'There are things in that paper that nobody knows but me': An Alternative Reading of Neurasthenia." *Charlotte Perkins Gilman: New Texts, New Contexts*. Eds. Jennifer S. Tuttle and Carol Farley Kessler. Columbus: The Ohio State UP, 2011. 161–179.
Lupton, Ellen. *Design is Storytelling*. New York: Cooper Hewitt, 2017.
Lynch, David. *Lynch on Lynch*. Ed. Chris Rodley. London: Faber and Faber, 1997.

March-Russell, Paul. “Writing and Publishing the Short Story.” *The Cambridge Companion to the English Short Story*. Ed. Ann-Marie Einhaus. Cambridge: Cambridge UP, 2016. 15–27.

March-Russell, Paul. *The Short Story: An Introduction*. Edinburg: Edinburgh UP, 2009.

May, Charles. “The American Short Story in the Twenty-First Century.” *Short Story Theories: A Twentieth-First-Century Perspectives*. Ed. Viorica Patea. Amsterdam and New York: Rodopi, 2012. 299–324.

McGann, Jerome. *Textual Conditions*. Princeton: Princeton UP, 1991.

McGettigan, Katie. *Herman Melville: Modernity and the Material Text*. Lebanon, NH: U of New Hampshire P, 2017.

McGurl, Mark. “Everything and Less: Fiction in the Age of Amazon.” *Modern Language Quarterly* 77.3 (2016): 447–472.

McGurl, Mark. *The Program Era: Postwar Fiction and the Rise of Creative Writing*. Cambridge, MA: Harvard UP, 2009.

Melville, Herman (1993). *Correspondence*. Vol. 14. *The Writings of Herman Melville*. Eds. Lynn Horth et al. Evanston: Northwestern UP, 1993.

Miller, Daniel. “Materiality: An Introduction.” *Materiality*. Ed. Daniel Miller. Durham, NC: Duke UP, 2005. 1–50.

Moylan, Michele, and Lane Stiles, eds. *Reading Books: Essays on the Material Text and Literature in America*. Amherst, M.A.: U of Massachusetts P, 1996.

Nagel, James. *The Contemporary American Short-Story Cycle: The Ethnic Resonance of Genre*. Baton Rouge: Louisiana State UP, 2001.

Norrick-Rühl, Corinna. “Short Story Collections and Cycles in the British Literary Marketplace.” *Constructing Coherence in the British Short Story Cycle*. Eds. Patrick Gill and Florian Kläger. New York and Abingdon: Routledge, 2018. 45–68.

O’Connor, Frank. “The Lonely Voice.” *Short Story Theories*. Ed. Charles E. May. Athens: Ohio UP, 1976. 83–93.

Orange, Tommy. “Freyr.” *Zoetrope: All-Story* 22.4 (Winter 2018/2019): 12–19.

Orange, Tommy. “The State,” https://www.newyorker.com/magazine/2018/03/26/the-state *The New Yorker* (March 19, 2018a): n. pag. (April 4 2020).

Orange, Tommy. “What Novels Can Teach Us.” *Time Magazine* 19.2 (2018b): 81.

Orange, Tommy. *There There: A Novel*. New York: Alfred Knopf, 2018c.

Pattee, Fred Lewis. *The Development of the American Short Story: An Historical Survey*. New York and London: Harper, 1923.

Phillips, Philip Edward. “Poe the Magazinist.” *The Oxford Handbook of Edgar Allan Poe*. Eds. J. Gerald Kennedy and Scott Peeples. Oxford: Oxford UP, 2019. 479–498.

Poe, Edgar Allan. “On the Prose Tale: From A Review of Twice-Told Tales by Nathaniel Hawthorne.” 1842. *The Portable Edgar Allan Poe*. Ed. J. Gerald Kennedy. New York: Penguin, 2006. 532–534.

Poe, Edgar Allan. *The Collected Letters*. 3rd ed. Ed. John Ward Ostrom. Vol. 1. New York: Gardian P, 2008.

Reckwitz, Andreas. “Practices and Their Affects.” *The Nexus of Practices: Connections, Constellations, Practitioners*. Eds. Allison Hui and Theodore Schatzki. New York: Routledge, 2016. 114–25.

Roberts, Molly. “‘Cat Person’ is a next step in the #MeToo movement,” https://www.washingtonpost.com/blogs/post-partisan/wp/2017/12/11/why-cat-person-went-viral/. *The Washington Post* (Dec. 11, 2017), n. pag. (27 March 2020).

Roth, Marty. “Gilman’s Arabesque Wallpaper.” *Mosaic: A Journal for the Interdisciplinary Study of Literature* 34.4 (2001): 145–162.

Roupenian, Kristen. “Cat Person,” https://www.newyorker.com/magazine/2017/12/11/cat-person. *The New Yorker* (December 11, 2017), n. pag. (27 March 2020).

Scheiding, Oliver. “Materiality in Twentieth and Twenty-First Century Literary Magazines.” *Routledge Companion to the Literary Magazine*. Ed. Tim Lanzendörfer. New York and London: Routledge, 2022. Forthcoming.
Scott, Heidi. “Crazed Nature: Ecology in ‘The Yellow Wallpaper.’” *Explicator* 67.3 (2009): 198–203.
Seybold, Matt, and Michelle Chihara, eds. *The Routledge Companion to Literature and Economics*. London and New York: Routledge, 2019.
Siewert, Stephanie, and Nils Reiter. “The Explorative Value of Computational Methods: Rereading the American Short Story.” *Amerikastudien/American Studies* 63.2 (2018): 199–230.
Smith, Jennifer J. *The American Short Story Cycle*. Edinburgh: Edinburgh UP, 2018.
Suess, Barbara A. “The Writing on the Wall: Symbolic Orders in ‘The Yellow Wall Paper.’” *Women's Studies* 32 (2003): 79–97.
Tischleder, Babette. *The Literary Life of Things: Case Studies in American Fiction*. Frankfurt: Campus, 2014.
Wheeler, Candace. “The Philosophy of Beauty Applied to House Interiors.” *Houshold Art*. Ed. Candace Wheeler. New York: Harper, 1893. 3–34.
Wideman, John Edgar. *Briefs: Stories for the Palm of the Mind*. Raleigh: Lulu P, 2010a.
Wideman, John. “Guest Author Blog: John Edgar Wideman,” https://blog.lulu.com/john-edgar-wideman/. *Lulu* 14 March 2010b. (20 March 2020).
Wolter, Jürgen. “‘The Yellow Wall-Paper’: The Ambivalence of Changing Discourses.” *Amerikastudien/American Studies* 54.2 (2009): 195–210.

4.2 Further Reading

Evangelisti Allori, Paola, John Bateman, and Vijay K. Bhatia, eds. *Evolution in Genre: Emergence, Variation, Multimodality*. Bern et al.: Lang, 2014.
Hungerford, Amy. *Making Literature Now*. Stanford: Stanford UP, 2016.
McAdams, Dan P. *The Redemptive Self: Stories Americans Live By*. Oxford and New York: Oxford UP, 2006.
Nadiminti, Kalyan. “The Global Program Era: Contemporary International Fiction in the American Economy.” *Novel: A Forum on Fiction* 51.3 (2018): 375–398.
Norris, Sigrid, and Carmen Daniela Maier, eds. *Interactions, Images, Texts: A Reader in Multimodality*. Berlin and Boston: De Gruyter, 2014.
Rajakumar, Mohanalakshmi, and Rumsha Shahzad. “She Needs a Website of Her Own: The ‘Indie’ Woman Writer and Contemporary Publishing.” *Literary Careers in the Modern Era*. Eds. Guy Davidson and Nicola Evans. Basingstoke and New York: Palgrave Macmillan, 2015. 184–195.
Scheiding, Oliver. “The American Short Story.” *Handbook of Transatlantic North American Studies*. Ed. Julia Straub. Vol. 3. *Handbooks of English and American Studies*. Berlin and Boston: De Gruyter, 2016. 234–250.
Senchyne, Jonathan. *The Intimacy of Paper in Early and Nineteenth-Century American Literature*. Amherst and Boston: U of Massachusetts P, 2020.
Thompson, John B. *Merchants of Culture: The Publishing Business in the Twenty-First Century*. Cambridge: Polity P, 2010.
Treharne, Elaine and Claude William. *Text Technologies: A History*. Stanford: Stanford UP, 2020.

Part II: **Close Readings**

Stefanie Schäfer

5 Washington Irving (1783–1859)

Abstract: Washington Irving is generally credited for launching the short story in the United States. He adapted European folklore, sketchy characters, and uncanny stories to an American literary setting, providing the United States with a set of myths that are revived time and again in American popular culture. In keeping with recent scholarship, my contribution reads Irving as a transatlantic author and respondent to Romanticism on both sides of the Atlantic. The chapter's main focus is on Irving's professional self-assertion as author through his narrators, their historic and transatlantic positionalities, and their arrangement of the tales in Irving's collections. As I show in a close reading of "The Adventures of the German Student" and "The Devil and Tom Walker" from *Tales of a Traveler* (1824), Irving's fictional historians and travelers claim the "Americanness" of his writings, while pointing self-critically to the preposterousness of this endeavor of national and cultural distinction.

Keywords: Storytelling, sketch, romanticism, American Revolution, historiography

1 Context

Rip van Winkle, Washington Irving's most widely known character, does something unheard of in American history: He sleeps through the American Revolution. Wandering into the woods, away from his termagant wife, the Dutch colonial Rip falls asleep and returns to his village after the fighting is over and battles have been won. Rip quickly takes up his old habits and is unimpressed by his newly gained status as "free citizen of the United States" (Irving 1983, 783). Ironically, the biggest change for Rip is the passing of his unnerving wife and his escape from the "yoke of matrimony" (783). "Rip van Winkle" uses the German legend of emperor Friedrich Barbarossa sleeping in the Kyffhäuser Mountain to interrogate the cultural meaning of the American Revolution: If Rip managed to sleep all the way through it, then how "revolutionary" was it? Or, put differently, how are revolutions *usually* narrativized – and how does "Rip van Winkle" breach these expectations?

Importantly, Irving's "Rip van Winkle", published in *The Sketchbook of Geoffrey Crayon* (Irving 1983) also marks the beginning of the short story as a genre in the US, showcasing the entanglement of "stories, writers, and themes" in a "transnational network of literary exchanges" (Scheiding and Seidl 2015, 1, 2) and setting a "pattern" (Current-Garcia 1973) that would be developed and theorized by Edgar Allan Poe (↗7 Poe) and Nathaniel Hawthorne (↗6 Hawthorne) until mid-century. In writing, Irving sought out his own trademark style. In 1824, he wrote:

https://doi.org/10.1515/9783110587647-006

> I have preferred adopting a mode of sketches & tales rather than long work, because I have chosen to take a line of writing peculiar to myself; rather than fall into the manner or school or any other writer; and there is a constant activity of thought and a nicety of execution required in writings of the kind, more than the world appears to imagine [. . .].
>
> (qtd. in Scofield 2006, 6)

Irving was the nexus of the Knickerbocker School, a group of New York writers invested in fostering an "American" literature and in establishing New York as a literary center. Named after Irving's fictional historian Diedrich Knickerbocker, the Knickerbockers commented on the developments of the day from their own perceived historical vantage point: As members of the first generation that came of age after the War of Independence, they felt tasked with negotiating the relationship between the United States and the British empire. While Irving tackled this with such unremarkable protagonists as Rip, his contemporary James Fenimore Cooper achieved his literary breakthrough with *The Spy* (1821), a historical romance about the Revolutionary War that applauded American bravery and tapped into the craze around George Washington.

After the revolution, colonial subjecthood had been replaced quickly by citizenship, an exclusive white national manhood tethered to fraternal competition in the public sphere (Nelson 1998, 46). The U.S.'s relation to Britain was told as the story of the prodigal son: the American Brother Jonathan had to leave his father John Bull's farm and fend for himself (Fliegelman 1982). The War of 1812 prompted a revision of the transatlantic kinship narrative: when English troops burned down the White House and the Capitol in 1814, anti-British sentiment ran high. The New England states, however, sought to preserve their economic relations with Britain. The Hartford Convention of 1814/15 even prepared for a separate peace with Britain and secession from the Union. Its efforts failed among public disgrace and in the aftermath, the "Yankees" were described as moneydigging provincials that "calculated the value of the Union," a stereotype that stuck (Greene 1833, 51–56; Schäfer 2021, 200).

Americans in the 1810s thus were faced with various entanglements of the continental, national, and regional identities, between the U.S., Britain, and continental Europe, between the New England elites and their Puritan and Dutch origins. Irving, a member of the New York upper crust, debated these issues in literary clubs and first started commenting on politics and society in the *Morning Chronicle* in 1802, as "Jonathan Oldstyle," a provincial theater critic stuck between "Republican aspirations and federalist vanity" (Burstein 2008, 21). Before turning to the travel narrative format with the *Sketch Book of Geoffrey Crayon, Gentleman*, Irving co-published his own American magazine with his brother William and his brother in-law James Kirke Paulding. *Salmagundi; or The Whim-whams and Opinions of Launcelot Langstaff, Esq. & Others* (1807–1808, Irving 1983) combined society gossip with essayistic satire in the style of the London weeklies *The Tatler* (1709–1711) and *The Spectator* (1711–1712 and 1714). In keeping with the British tradition of ridiculing high society at social events, the fictional commentators of *Salmagundi* themselves prove to be preposterous men of letters, not least by their telling names:

Launcelot Langstaff, Will Wizard, Pindar Cockloft, and Mustapha Rub-a-Dub Keli Khan stylize themselves as authors and critics of New York's or "Gotham's" society. Khan, for instance, is an exoticised traveler to the U.S. modelled on Montesquieu's *Lettres Persanes* (1721). He writes back home, describing the politics and society in the newly minted democracy as a "logocracy,"

> [a] *government of words*. The whole nation does everything [. . .] by word of mouth, and in this manner is one of the most military nations in existence [. . .]. In a logocracy [. . .] there is little or no occasion for fire arms or any such destructive weapons. Every offensive or defensive measure is enforced by *wordy battle*, and *paper war*; he who has the longest tongue or readiest quill, is sure to gain the victory – will carry horrour [sic], abuse, and *ink shed* into the very trenches of the enemy, and without mercy or remorse, put men, women, and children to the point of the –pen! (Irving 1983,144–145, original emphasis)

Hence, the new nation constitutes itself through words only, at the mere inability to employ any other weapon against adversaries that are either spatially removed or (so Khan also suggests) imaginary. A few decades after Khan's fictional travelogue, another traveler to the U.S. found similar words. In *Democracy in America* (1831), Alexis de Tocqueville, commenting on the constitution, found it based on "legal fictions. The Union is an ideal nation that exists so to speak only in the minds, and whose extent and bounds intelligence alone discovers" (Tocqueville 2000, 155). *Salmagundi*'s dictum of the American logocracy thus skeptically interlinks the political and the literary, laying out a key inquiry in American literature about "the act of naming and the politics of representation" (Gustafson 1992, 19).

How this fantastic quality of the "nation" influenced literary production has become a key interest in recent scholarship. For instance, Thomas Koeniges argues that the history of the in the United States needs to be framed as a "history of competing varieties of fictionality" (Koenigs 2021). Similarly, Lara Langer Cohen describes Antebellum literature as propelled by a culture of literary fraudulence (Cohen 2011): It was a "fabrication" in a double sense, both as work in progress and as fabulation, based on "the elevation of printed *literature* to a cultural institution" (Cohen 2011, 6; original emphasis). Mary-Ann Snyder-Körber's reading of Irving's *Sketch Book* similarly argues that the talk about book-making uncovers "pretensions" while also pinpointing Irving's own strategy in "book-making [, which] suggests lack to ultimately claim complexity" (Snyder-Körber 2008, 28). In order to show how Irving's narrator personas appear deficient only on the surface, I want to point to the work of James Kirke Paulding, Irving's lifelong friend. While Irving was only "sort of" (Burstein 2008, 89) invested in the war of 1812, Paulding wrote avidly in its favor and made extensive use of the kinship narrative (see Schäfer 2021, esp. 35–47). Regardless of their political leanings, both Irving and Paulding sought to establish themselves as literati in a highly politicized transatlantic public sphere with similar genres, strategies, and media. Comparied to Paulding's bawdy nationalism, Irving was more nuanced in appropriating European Romantic literary culture and staging celebrity authorship. In England, Irving met Sir Walter Scott and learned about English legends (Rubin-Dorsky 1985, 238); he prepared for his continental travels by studying German "day and night," and his arrival in Dresden in 1822 was a public event (Irving 1991, 1101, and 1103). During his stay, Irving attended the theater almost daily, met German Romantic author Ludwig Tieck, and purchased a copy of the Brothers Grimm's *Deutsche Sagen* (see Reichart 1957, Appendices A and B; cf. Collins 2019).

Irving negotiates transatlantic entanglement on various levels, including the contents, form, and narrative mediation of his writings. On the content level, he taps into themes of his day; on the formal level, he emulates and parodies different authoritative fictional and non-fictional genres; and as author/narrator personas of these different genres, Irving constructs self-conscious *glitterati* who comment on and are a product of the nascent literary industry of the U.S. Through their metanarrative staging of the teller, his tales ask readers to observe the observer, to analyze the analyst. Like other narrative cycles and collections (↗1 Sketches, Tales, and Stories), Irving's writings engage realist depictions (of people and places) while at the same time experimenting with "the production and reception of art" itself (Smith 93). Irving's specialty, I argue, is his parody of professional storytelling by historians, travelers, and authors. After scrutinizing society in *Salmagundi*, Irving turned to historiography, and invented the "Historian" Diedrich Knickerbocker who writes the parodic *A History of New York, from the Beginning of the World to the End of the Dutch Dynasty* (Irving 1983a). Before publishing the *History* in December 1809, Irving built a newspaper hoax around Knickerbocker's alleged disappearance. In keeping with contemporary criticisms of historiography's authority claims (Scheiding 2003, 14), Knickerbocker's American historian-persona thus appears both fraudulent and preposterous.

Like Diedrich Knickerbocker, "Geoffrey Crayon, Gentleman," the narrator of the 1819 *Sketch Book of Geoffrey Crayon* (Irving 1983b) and of *Tales of a Traveller* (1824, Irving 1991), is a funny character in his own right. In the "Author's Account of Himself" at the opening of the *Sketch Book*, he professes that "I knew every spot where a murder or robbery had been committed or a ghost seen" (Irving 1983b, 743). The *Sketch Book* is a mixture between travelogue and fictional sketch (↗1 Sketches, Tales, and Stories) that built on Irving's travels in England between 1815 and 1817. It installs Crayon as a traveler who goes on the "Grand Tour" of Europe to rub elbows with English noblemen, imbibe Europe's cultural heritage, and write a travelogue that would make an intriguing read for Americans back home (see Buzard 2002).

Crayon's penchant for (English) literature, his fascination with books, becomes an enchantment (Sizemore 2013) and offers to American readers in the 1810s a "useable past" that is deliberately "unheroic" (Scofield 2006, 10), humorous, and often fantastic. This fantastic and grotesque past works most successfully in the two tales that brought Irving immediate fame: "Rip van Winkle," as already stated, is a revolutionary tale without the revolution. "The Legend of Sleepy Hollow," the second long-canonized tale from the *Sketch Book*, takes the German legend of a headless rider to a colonial Dutch settlement, where an ambitious newcomer, the Yankee schoolmaster Ichabod Crane, is scared off by the local Dutch boy Brom Bones, who masquerades as the headless rider. Ichabod, the effeminate Yankee schoolmaster, belongs to the New England upstart type; his courtship of the rich heiress Katrina Van Tessel threatens the Dutch community. So the question what happens to Ichabod after his departure from the village is key for poetic justice. Knickerbocker offers competing versions: while the local gossips plead that he was simply "spirited away," according to a farmer, Ichabod 'made it' in New York: he "had kept school and studied law [. . .]; turned politician; electioneered; written for the newspapers; and finally had been made a justice of the Ten Pound Court" (Irving 1983b, 1086).

"The Legend of Sleepy Hollow" and "Rip van Winkle" differ from the other tales in the *Sketch Book* because they are both set in North America and tell about the colonial past; they are attributed not to Crayon but to Knickerbocker's "posthumous writing" (Irving 1983b, 769 and 1058). Knickerbocker's influence therefore lingers and perpetuates Crayon's own collection, and for Irving, this was also a smart marketing trick: After the success of *History of New York*, Irving resurrected the local U.S. historian Knickerbocker (who writes for Americans) to contrast with the travelling book lover Crayon, who writes for Americans and Europeans: In a special addition to the London edition of the *Sketch Book*, Crayon describes his role as "Envoy" to his English readership. He assures them that he will pursue his own path, "half venturing, half shrinking, surprised at his own good fortune, and wondering at his own temerity" (Irving 1983b, 1091). Crayon's gesturing at literary fame abroad, his "surprise" and "wonder" make him seem humble; yet his "venturing" and "shrinking" attempts at writing also prepare readers for the *Sketch Book's* sequel, the 1824 *Tales of a Traveller*, in which Crayon meets many storytellers and sketchy characters.

2 Close Readings

The present section provides a close reading of "Adventure of a German Student" and of "The Devil and Tom Walker," two stories from *Tales of a Traveller* (Irving 1991) that are set in Europe and America respectively, and that address the post-revolutionary moment in the Atlantic world. The dangers and transgressions of this moment are portrayed through the narrators Irving uses, so before delving into the stories themselves, I discern the logic of storytelling and traveling lore in *Tales*. Announced as a "German sequel" to the *Sketch Book*, *Tales* was immediately translated into French and German and had a mixed reception, with English and American reviewers complaining that Irving lacked innovation and was "no longer the exemplary American in Europe [. . .] a symbol of American democracy and a landmark in world literature" (Reichart 1957, 157f.). German readers, however, liked *Tales* because of resonances of the Brothers Grimm's tales, the Faust legend, and contemporary German literature and philosophy. Like Irving's other fictional narrators, Crayon also offers a textual apparatus with preface and epigraphs that tells the reader how the book came about and what (not) to expect from it. In the preface "To the Reader," Crayon insists on being a "mere traveller" who assembles the tales he overheard out of boredom while recovering from a malady (Irving 1991, 380, 383, 384). He is stuck in Mainz, a "frontier town" between France and Germany, in the "Hotel de Darmstadt, *ci-devant* Hotel de Paris, Mentz, *otherwise called* Mayence" (383, 385). The American Crayon blurs the boundaries between the cultures and politics of France and Germany and is generally careless about historical truthfulness and book-making: since "any body is an author today" (383), he needs little professional expertise. He describes his mind as an overstuffed suitcase:

> I am an old traveler. I have read somewhat, heard and seen more, and dreamt more than all. My brain is filled, therefore, with all kinds of odds and ends. In travelling, these heterogeneous matters have become shaken up in my mind, as the articles are apt to be in an ill packed travelling trunk; so that when I attempt to draw forth a fact, I cannot determine whether I have read, heard, or dreamt it; and I am always at a loss to know how much to believe of my own stories. (Irving 1991, 385)

The *Tales* fall into four parts that are less organized by theme but by mood, progressing from "whimsy and fun-poking to [. . .] terror, violence and serious inspection" (Current-Garcia 1973, 335). They also differ in telling scenario and setting: the first three parts are told in Europe by other narrators, but the last part, "The Money-diggers" collects legends Crayon remembers from his childhood in Massachusetts. *Tales* thus mingles European and American folklore and highlights their shared motifs and effects.

The first part, "Tales of the Nervous Gentleman," is a hand-me-down from a person who tells Crayon the stories he overheard one stormy night. The first story discussed here, "The Adventure of the German Student," is therefore wrapped in confusing storytelling layers offered by uncanny narrators (in short: the German students tells a gentleman who tells the nervous gentleman who tells Crayon). The nervous gentleman describes the man he heard the story from, quite fantastically, as "gentleman of the haunted head":

> one side of [the narrator's] face was no match for the other. The eyelid drooped and hung down like an unhinged window shutter. Indeed, the whole side of his head was dilapidated, and seemed like the wing of a house shut up and haunted. I'll warrant that side was well stuffed with ghost stories. (Irving 1991, 393)

The "gentleman of the haunted head" is a grotesque figure whose split mind shows on his face. As a consequence, readers might note that stuffing people with ghosts stories deforms them physically, and makes them crazy. In a few strokes, the narrator here links the haunted house, the skull, and the mind in a hyperbole of the market for stories. That the gentleman claimed he heard "Adventure" from the German student himself, "in a madhouse in Paris" (418, 424), normalizes madness in this day and age.

Through this narrative deformation, "Adventure" articulates a warning against revolutions and intellectualism gone wrong. The protagonist Gottfried Wolfgang goes mad in Paris, during the French Revolution. The opening of the story sets the mood of "tempestuous times" (419) that shake the city to the core. Lightning and thunderclaps resound "La Terreur" (1793–1797), when the elites of the Ancien Regime were guillotined en masse. The German student Gottfried Wolfgang, a "young man of good family," has already been deformed in "mind and body" by "wild and speculative doctrines" (419) when he arrives in Paris to continue his studies. His name points to key thinkers of the German Romantic movement, Johann Wolfgang von Goethe and Gottfried Herder, eminent figures at the Weimar court, theorists of a national culture and body politic (Herder) and of a romantic concept of world literature (Goethe).

Like his namesakes, Gottfried Wolfgang is a "visionary and enthusiastic character" (418) affected by Romantic ardor, a real firehead. Based on his readings of Emanuel Swedenborg, Gottfried Wolfgang has built "an ideal world of his own around him" (419), and now believes that he is doomed. In revolutionary Paris, this belief engenders catastrophe: Gottfried Wolfgang flees from the "popular delirium" and "scenes of blood" (419) to the libraries. There, he trades the revolutionary present for an equally dangerous past: he feasts on "dusty and obsolete" books and becomes a "literary ghoul," whose reading is described as eating dead bodies (419). In the short story, his character declines from intellectualism to melancholy and to madness.

While ingesting "decayed literature" (419), the innocent Gottfried Wolfgang has no relations in the real world. His returning dream of a woman's face becomes an obsession. Passing by the Place de Grève one night, he discovers a hooded figure, a young woman "pale and disconsolate, but ravishingly beautiful" (422), sitting by the guillotine. Gottfried Wolfgang finds his dream come true and takes the young woman home to make love to her.

This initiates a new part of the story: whereas the opening describes Gottfried Wolfgang's Parisian adventure from an external, authorial perspective, his encounter with the dream woman kicks off an internal focalization of feverish deliberations inside his head, up to the climax of the bedroom scene. In free indirect discourse, Gottfried Wolfgang justifies his actions with the help of twisted academic deduction: He tells himself the beautiful woman grows equally "infatuated" with him, since "enthusiasts soon understand each other":

> It was the time for wild theory and wild actions. Old prejudices and superstitions were done away; every thing was under the sway of the 'Goddess of Reason'. Among other rubbish of the old times, the forms and ceremonies of marriage began to be considered superfluous bonds for honorable minds. Social compacts were the vogue. Wolfgang was too much of a theorist not to be tainted by the liberal doctrines of the day. (Irving 1991, 422–423)

This train of thought pits the vocabulary of enlightenment and Romanticism against "sordid form[s]" of custom: for Gottfried Wolfgang, there is no need for courtship and marriage rites. Instead "wildness" (in thinking and doing) and reason, as well as the "social compact" (a nod to Jean-Jacques Rousseau's influential *Contrat Social*, 1762) legitimize his actions. Wolfgang Gottfried pledges himself to the stranger "for ever," and so the path is cleared for sex ("'I am yours,' murmured she, and sank upon his bosom," 423).

After this climax comes a horrifying morning when Gottfried Wolfgang discovers that his conquest "was a corpse," the body of a woman guillotined the day before. When he made love and pledged himself to the woman, he now knows was a demon, Gottfried Wolfgang crossed the line to damnation and madness "for ever." In this sudden turn, Romanticism and Enlightenment ideals become hellish, while the customs of courtship he deemed "superfluous" might have saved him from rash action. Following the logic of correspondences of inside and outside, bodies and histories, the French Revolution has been contorted by *la Terreur*, that "spectre of History" (Bridier 1999, 88). Enlightenment philosophy and the rule of Reason have pulled the German student from the world of the living, made him feast on dead books and have sex with a zombie Parisienne, a headless opposite of his own bodilessness.

Through the voiceover provided by the gentleman of the haunted head, Irving's Geoffrey Crayon warns his readers against Europe's cultural authorities (libraries as "catacombs of departed authors," 419) and its course of history (German Romanticism, Swedenborganism, and the "volcano of human passion" of Revolutionary Paris, 421). German intellectualism and French revolutionary fervor create a deadly union from which no future can be built. Gottfried Wolfgang's Parisian madness links "History" to two reference points: English literature and the American revolution. It clashes with Crayon's own enchantment with English literature, reiterated earlier in the *Sketch Book*'s "Mutability of Literature" (Irving 1983b, 854–864). Whereas English literature creates enchantment as new way for understanding Crayon's own post-revolutionary and transatlantic present (Sizemore 2013, 158), continental literatures, under the sway of the "Goddess of Reason" (422) conjure a state of madness that renders the promising young student useless as a citizen. Read through Irving's own biography, the contrast between English and continental literatures might be informed by Irving's sense of "bewilderment in continental Europe" – and his failed courtship of a young German woman (Reichart 1957, 144).

To readers on both sides of the Atlantic, Irving here offers a commentary on revolutions: the French Revolution incapacitated its elites, while the American Revolution, as we saw with Rip van Winkle, did not change so much on a local level, thus preserving an outlook at prosperity. The palimpsest of haunted storytellers in *Tales* thus points back to Crayon, the traveling American booklover, who finds himself on shifting grounds in the "frontier town" of Mainz. Opposite this healthy old American geezer, all characters of "Adventure," i.e. the deformed storytellers, the beheaded Parisienne and her headless German lover, are sickly European ghosts.

Opposite the mad European revolutionary cityscapes in "Adventure of a German Student," the American rural landscapes portrayed in "The Devil and Tom Walker" call forth different hauntings. The story is also about the past, but it is set in colonial North America and sketches key moments of settlement history that perpetuate the narrative of U.S. exceptionalism. While there is no question that the land needs to be settled, Irving rather asks "how" this has been done, showing his readers what has gone wrong by the 1700s, when "The Moneydiggers" that name this section of *Tales of the Traveller* fall victim to their own greed. "The Moneydiggers" collects local legends from Crayon's childhood days and starts with a view of the haunted landscape around the East River of today's New York city, an estuary that was called Hell Gate. In his description, Crayon shows how the country is literally made by the white men that settle it. In the case of Hell Gate, the makers are the Dutch sailors struggling to navigate the stream's "termagant humor" (Irving 1991, 647):

> At low water, [. . .] it is as pacific a stream as you would wish to see; but as the tide rises, it begins to fret; at half tide it roars with might and main, like a bully bellowing for more drink; but when the tide is full it relapses into quiet, and for a time sleeps as soundly as an alderman after dinner. In fact, it may be compared to a quarrelsome toper, who is a peaceable fellow enough when he has no liquor at all, or when he has a skin full, but who, when half seas over, plays the very devil. [. . .] The Dutch navigators of ancient days [. . .] out of sheer spleen [. . .] denominated it Hellegat and solemnly gave it over to the devil. (Irving 1991, 647)

Crayon's legend personifies the East River as a humorous man of some importance ("alderman") or simply a drunk ("quarrelsome toper"), who roars, bellows, and sleeps soundly at turns. He acts up and hinders the Dutch intrusion and the "founding" of the "city of Manhattoes" (Irving 1991, 647). Irving here invokes the settler colonial narratives of the conquest of the wilderness and the progress of civilization: The old drunk that is the East River can be tricked by adapting to the tidal cycle. Crayon thus acknowledges the hardships and heroism of the first Dutch navigators, but also points to their "spleen": they give the unruly East River "over to the devil" and make it his property – as if it were theirs to give away in the first place.

"Tom Walker" picks up where Crayon's story about Hell Gate's naming left off, in the colonial past of 1727, and it adapts the European Faust-motif: a New England Yankee sells his soul to the devil in exchange for prosperity, and when he feels death coming tries to escape the agreement, in vain. The Devil, "Old Scratch" whisks him away and his fortune is turned to dust. "Tom Walker" was rewritten in American literature and film, among others in Stephen Vincent Benet's 1936 story "The Devil and Daniel Webster." Benet's version inserts the iconic Massachusetts politician and lawyer Daniel Webster into the tale, and changes the ending: In a law trial, Webster outsmarts the devil and saves the poor farmer's soul (see Cooksey 1999).

Irving's comic New England-Faust ridicules the Yankee farmers and the Puritans in various episodes of settlement history: the 1820s present of Crayon's narrative, his childhood days in New York, the Puritan settlement of the early 1700s, and the explorations of the River by Dutch sailors. Tom Walker, the New England farmer is one of the "moneydiggers," impoverished and degenerate colonials who have long forgotten their religious "covenant" with God. Life in New England in the 1700s brings no prosperity, only a daily fight for survival: Tom Walker is the greedy type, "meagre and miserly," his farm has an "air of starvation" and of "sterility" (Irving 1991, 655), the land he wanders is doomed, full of "pits and quagmires," itself half dead, with "trunks of pines and hemlocks [that] lay half drowned, half rotting, looking like alligators, sleeping in the mire" (656).

This rich symbolism conjures a haunted landscape corrupted by the very men who set out to civilize it. The times, "[a]bout the year 1727," are troubling Puritan society, since "earthquakes were prevalent in New England, and shook many tall sinners down upon their knees" (655). By showing rotten people in a rotten landscape, Irving reminds his audience of Puritan mission and thought: the earthquake of October 29, 1727, was perceived as God-sent, for instance by Cotton Mather, who wrote: "[earthquakes] and their causes are still under the government of HIM that is the GOD of Nature" (qtd. in Andrews 1976, 283). By employing the Faust legend and gothic elements, "Tom Walker" shows the height of the crisis which inspired, by the 1730s, the Great Awakening of religious revival and a turn to Calvinism in New England. Echoing the earthquake that brings "tall sinners" before a fall, Tom Walker soon learns that the trees in the woods are stand-ins for the members of his community. The richest among them will pay for their ill-begotten wealth, when Old Scratch cuts them down and takes them to hell (cf. 658).

The story's narrative focus is not so much on the supernatural interactions with the Devil as on Walker's character. Tom Walker is a typical Yankee who recalls the scrawny Ichabod Crane from "The Legend of Sleepy Hollow": Both men are diehard New Englanders who fail to succeed in their communities, to spawn offspring and to secure a prosperous estate. Yet where Ichabod Crane falls victim to his own superstition, Tom, a "hard minded fellow, not easily daunted" (659) is doomed because of his blindness towards "fears" (657). He even "sneers" at the devil (658), so his pride comes before a fall.

The story's descriptive focus is on the swamp where Walker meets the Devil. The North American fauna and flora carries symbolic meaning. The most obvious motif is the hemlock tree (mentioned four times), a North American version of the pine tree named for its smell reminiscent of the hemlock herb, which is associated in classical myth with Socrates's conviction for blasphemy and for influencing Athens' youth. Foster (2014, xxii) observes that in American literature, hemlock is linked to "an exhilarating and foreboding mood to scenes of raw wildness." Socrates's capital punishment was executed by drinking poisonous hemlock, and the overbearing presence of the hemlock in "Tom Walker," "some of them ninety feet high, which made it dark at noonday" (656) tells of the white settlers' sins. The "Black hemlock swamp" (666) forms its own hell gate, a pathway for the Devil.

Irving's gloomy vision of colonial times acknowledges the flaws of settlement history, of Indian removal, and of slavery. The conquered land is sickened by the acts committed on it. Tom's farm is barren, and on the site of an old "Indian fort," he finds a skull with a tomahawk stuck in it, a "dreary memento of the fierce struggles that had taken place in this last foothold of the Indian warriors" (657). Old Scratch claims the fort is on "my grounds." He boasts: "this land belonged to me long before one of your white-faced race put foot upon the soil" (658). He offers Tom the treasure of the legendary seaman and slave trader Captain Kidd. But Tom refuses at first: he will not get a slave ship and join the "black traffick" (662). Instead, the greedy Yankee agrees on working as usurer in Boston, where his business flourishes since economic growth and territorial expansion have spiked a "great speculating fever." In keeping with the Faust legend, the question in "Tom Walker" is not about getting rich, but about getting out of a devilish bargain. Yet Tom's "violent" turn to religion (664) and his carrying a Bible cannot save him. Stubbornness and stinginess are his gravest sins.

Walker's doom, however, engenders little wonder with his contemporaries, the "good people of Boston" who "had been so much accustomed to witches and goblins and tricks of the devil [. . .] from the first settlement of the colony, that they were not so much horror struck as might have been expected" (666). As avid readers, the colonists were used to such stories. Christopher Marlowe's *Doctor Faustus* (1592) was a staple used for didactic purposes and to sustain the colony's ties to England (Mylander 2011). Crayon's story of the sturdy New Englander foregrounds a central motif from the Faust *fabula*: the marriage question. In Marlowe's *Doctor Faustus,* as in Goethe's *Faust* (1808), which Irving read in German, Faust asks the devil for a wife but does not want to abide by the social convention and religious sacrament of marriage; in Marlowe's version, Doctor Faustus says, "marriage is but a ceremonial toy" (Marlowe 1990, 5.148 [53]), and Goethe's Faust seduces young Gretchen and drives her to the grave.

Marriage and its discontents thus pave an inroad for analyzing Irving's storytellers and protagonists. In "Adventure," Gottfried Wolfgang's fate is sealed with his fake marriage vow, when he "pledges himself" to the zombie Parisienne. This might be due, as some scholars have pointed out, to Irving's own experience. After the death of his German fiancée Mathilda Hoffman, Irving never married (cf. e.g. Burstein 2008, Current-Garcia 1973). It may also derive from the conception of the male protagonist/hero in the European folk tale Irving draws from. In "Tom Walker," Tom's (nameless) wife is even worse than her husband. She is "a tall termagant, fierce of temper, loud of tongue, and strong of arm" who repeatedly engages in "wordy warfare" (656) and tries to make her own deal with the devil. Much narrative time is spent on deliberations around the fate of Tom's wife after she takes her most valuable belongings into the swamp and never returns. Various versions of her fate are offered here, but the "most current and probable story" reiterates the misogynist linkage of women's bodies to property (in the 1720s as in the 1820s):

Tom, "anxious about the fate of his wife and his property" (661), discovers her apron and tells himself "[l]et us get hold of the property [. . .] and we will endeavor to do without the woman" (661). Ironically, Tom "consoled himself for the loss of his property with the loss of his wife" (662). The story twice repeats the devaluation of the wife opposite the "property," in a calculation effort that is wrong from the very beginning: The origin of Tom's misery is his barren and loveless marriage, in which the partners "conspire to cheat each other" (655), which leads him astray, to the "deep bosom of the swamp" where he meets the Devil (656).

Looking back at the other henpecked husband that brought Irving fame, Rip van Winkle, it seems that for Irving, marriage is the biggest hindrance to happiness. Paul Downes (2002) reads Dame van Winkle as an embodiment of monarchism vanquished during Rip's sleep in the mountains. Her death by popping an angry vein eliminates female participation in the democratic experiment that is the New Republic. Downes further observes that Dame van Winkle's tongue lashings represent a "noisy aggressive, and excessively embodied female, democratic speech" that is the opposite of "a sympathetic, male economy of honest, disembodied, noble speech" in "Rip van Winkle" (Downes 2002, 158–159).

Yet, as I have shown here, Irving's narrators, albeit all male, are anything but disembodied or noble, for that matter. In recalling the graphic, but "most authentic" storyline of Tom Walker's wife's fate cited above, we have to note that "Rip van Winkle," too, is told by a biased narrator himself, who "[. . .] used to tell this story to every stranger" (Irving 1983b, 783). Pointing back to the "marriage question" in "Adventure," where sexual libertinage leads to taboo-breaking necrophilia, it seems that the way to happy lives and futures in Irving's logic is not refraining from marriage but turning away from sensationalist stories and Old World book culture. Irving's storytellers, provincial theater critics, mysterious historians, overstuffed old travelers, are bachelors all who will ultimately become dead white men. They take to storytelling out of loneliness, literary aspiration, delusion, or boredom. If we read their tales as the first specimens of the American short story, we have to pay more attention to their idiosyncratic fantasies of "America."

3 Theoretical Perspectives

In his day, Irving was a literary celebrity whose biography initially gleaned more academic attention than his oeuvre (Breinig 1972, 11–18). While scholars generally agree that Irving ushers in the short form in the U.S., the poetic reflection on the genre is generally attributed to Poe and Hawthorne and their "development of the romantic impulse" (May 1995, 6); in the nineteenth century, the three appear so closely related that George Parsons Lathrop worked hard to discern their differences in *Scribner*'s in 1876. James Russell Lowell, a humor writer from the Boston literary scene, offers in his 1848 book-length poem "A Fable for Critics" a recipe for cooking up "Irving," a mixture is its own brand, "not wholly deserving a name neither English and Yankee – just Irving" (Lowell 1848, 85).

While he was less popular with his successors (Ralph Waldo Emerson, for instance, bemoaned his lack of "nerve and dagger," qtd. in Giles 2001, 142), Irving the brand has long held its place in U.S. literary history, albeit while sitting on the fence: He is perceived either as representative of eighteenth century literary "sensibility" or an "imperfect precursor to radical romanticists" (Apap and Hoffman 2010, 3). Irving's writings are thus thematically and aesthetically situated at a threshold: the "realistic" parts of his collections may well be attributed to the travelogue and social commentaries on mores and manners, voiced by idiosyncratic observers from Jonathan Oldstyle through Geoffrey Crayon, even if they also count as representatives of eighteenth-century British satire. The "romantic" component in Irving's writing, adapted from his European forerunners, pertains both to the form of the tale and to the gothic, uncanny, and sensationalist contents offered with allusions to oral folklore. Aderman (1986) reads Irving as "Purveyor of Old and New World Romanticism"; and Newman et al. include him in their *Anthology of Transatlantic Romanticism* (2006).

Strikingly, the Romantic debate appears to integrate Irving in a transnational context much earlier than histories of the short story did; as Michael Collins suggests, reading Irving as the founding father of the short story in the U.S. reveals the ideological work of clinging to nationalisms in academia (Collins 2019). Conversely, studying Irving through a broader view of transatlantic literary history shows that the influence goes back and forth: Irving draws from London magazine culture and adapts English and German folklore tales to American settings; his conversations with Walter Scott inspire the latter's *Waverly*-series (1814–1831), and his views of Englishness in the *Sketch Book* inform Charles Dickens's *Pickwick Papers* (1836–1837, see Andrews 2000). In Britain, Irving's stories "set the stage for the publication of short fiction" throughout the nineteenth century (Killick 2016, 2).

Irving's American tales "Rip van Winkle" and "The Legend of Sleepy Hollow" have inspired operas and movie versions that advance their own timely views of U.S. history and culture. "Rip van Winkle" was quickly adapted to the theater by director Dion Boucicault and brought the star actor Joseph Jefferson fame in the U.S. and in London in the 1860s. Jefferson even turned scenes into short films in 1896. Ichabod Crane found his afterlife in Walt Disney's 1949 *Adventures of Ichabod and Mr. Toad.* In Tim Burton's 1999 horror film *Sleepy Hollow* and in the Fox-series *Sleepy Hollow* (2013–2017), Ichabod is a police constable or time-traveler turned private eye.

Recent scholarship pays increased attention to Irving's narrators and to his concepts of authorship (Giles 2001 and 2019, Downes 2002, Snyder-Körber 2008, Sizemore 2013, Schäfer 2021) and addresses "historiography, generational change, early American imperialism and gender and masculinity" (Apap and Hoffman 2010, 3) from a new historicist perspective. A trailblazing example of this was Helmbrecht Breinig's study of authorship and autoreflexivity in Irving's oeuvre from *Salmagundi* through *Tales of the Alhambra* (Breinig 1972). Michelle Sizemore (2013) reads Crayon's enchantment by English literature and landscapes in the *Sketch Book* as a commentary on the passing of time and on historicity; Paul Giles has argued that Irving's stylistics of burlesque "deflates" both authorship and national identity concepts (Giles 2001, 142) and that his "sophisticated self-deprecation" (2019, 84) expresses "skepticism" about the story Americans told themselves about the American revolution (2019, 85). Similarly, Judith Richardson reads *History of New York* as Knickerbocker's "tall tale" of "cosmopolitan-imperial supremacy" (Richardson 2009, 147), a 'History' that self-indulgently writes its capital H into being.

These assessments invite students to take a fresh look at Irving's narrative strategies on the one hand and, on the other hand, at the meaning of his oeuvre for a new, transnational and critical literary and cultural historiography. When we read Irving in a transatlantic network of (literary) relations, the difficulty, playfulness, and nerve-wrecking cultural work of articulating an "American" voice at the beginning of the 19th century becomes obvious. Such a reading also highlights the professionalization of authorship and the habitus attached to this in the U.S. and in Europe. The close linkage of this professionalization to theatrics and (stage) performance culture requires looking beyond the genre and medium of the short story (Collins 2016; Fash 2020). The American short story is part of a multimedial entertainment industry that deliberately indulged anti-intellectualism, play-acting, and make belief (see Schäfer 2021); and Irving's storytellers – haunted, nervous, and weird as they are – play their own intricate parts in it.

4 Bibliography

4.1 Works Cited

Aderman, Ralph M. "The Old and New World Romanticism of Washington Irving." *The Old and New World Romanticism of Washington Irving* Ed. Stanley Brodwin. New York: Greenwood P, 1986. 13–26.

Andrews, Malcolm. "Dickens, Washington Irving, and English National Identity." *Dickens Studies Annual* 29 (2000): 1–16.

Andrews, William D. "The Literature of the 1727 New England Earthquake." *Early American Literature* 7.3 (1973): 281–294.

Apap, Christopher, and Tracy Hoffman. "Prospects for the Study of Washington Irving." *Resources for American Literary Study* 35 (2010): 3–27.

Breinig, Helmbrecht. *Irvings Kurzprosa: Kunst und Kunstproblematik im Erzählerischen und Essayistischen Werk.* Bern: Lang, 1972.

Bridier, Sophie. "*The Adventure of the German Student*: Un Cauchemar Américain à Paris." *Alizés: Revue Angliciste de La Réunion* 17 (1999): 86–99.

Burstein, Andrew. *The Original Knickerbocker: The Life of Washington Irving*. New York: Basic Books, 2008.
Buzard, James. "The Grand Tour and after (1660–1840)." *The Cambridge Companion to Travel Writing*. Eds. Peter Hulme and Tim Youngs. Cambridge: Cambridge UP, 2002. 37–52.
Cohen, Lara Langer. *The Fabrication of American Literature: Fraudulence and Antebellum Print Culture*. Philadelphia: U of Pennsylvania P, 2011.
Collins, Michael J. "Transnationalism and the Transatlantic Short Story." *The Edinburgh Companion to the Short Story in English*. Eds. Paul Delaney and Adrien Hunter. Edinburgh. Edinburgh UP, 2019. 9–23.
Collins, Michael J. *The Drama of the American Short Story, 1800–1865*. Detroit: U of Michigan P, 2016.
Cooksey, Thomas L. "'Talk Not of a Wife': *The Devil and Daniel Webster, Cabin in the Sky*, and *Damn Yankees* – American Contributions to the Faustus Legend." *Journal of Popular Film and Television* 27.3 (1999): 18–27.
Current-García, Eugene. "Irving Sets the Pattern: Notes on Professionalism and the Art of the Short Story." *Studies in Short Fiction* 10.4 (1973): 327–341.
Downes, Paul. "Luxury, Effeminacy, Corruption. Irving and the Gender of Democracy." *Democracy, Revolution and Monarchism in Early American Literature*. Cambridge: Cambridge UP, 2002. 144–164.
Fash, Lydia. *The Sketch, the Tale, and the Beginnings of American Literature*. Charlottesville: U of Virginia P, 2020.
Foster, David R., ed. *Hemlock. A Forest Giant on the Edge*. New Haven: Yale UP, 2014.
Giles, Paul. "Burlesques of Civility: Washington Irving." *Transatlantic Insurrections: British Culture and the Formation of American Literature, 1730–1860*. Philadelphia: U of Pennsylvania P, 2001. 142–163.
Giles, Paul. *American World Literature: An Introduction*. Hoboken, NJ: Wiley and Sons, 2019.
Greene, Asa. *A Yankee Among the Nullifiers*. New York: William Pearson, 1833.
Gustafson, Thomas. *Representative Words. Politics, Literature, and the American Language, 1776–1865*. Cambridge: Cambridge UP, 1992.
Irving, Washington. "Tales of a Traveller." 1824. *Washington Irving: Bracebridge Hall, Tales of a Traveller, The Alhambra*. New York: Library of America, 1991. 379–718.
Irving, Washington. "Salmagundi, or the WhimWhams of Launcelot Langstaff." 1807–1808. *Washington Irving: History, Tales & Sketches*. New York: Library of America, 1983. 45–362.
Irving, Washington. "A History of New York." 1809. *Washington Irving: History, Tales & Sketches*. New York: Library of America, 1983a. 363–732.
Irving, Washington. "The Sketch Book of Geoffrey Crayon, Gent." 1819. *Washington Irving: History, Tales & Sketches* New York: Library of America, 1983b. 731–1092.
Killick, Tim. *British Short Fiction in the Early Nineteenth Century: The Rise of the Tale*. London: Routledge, 2016 [2008].
Koenigs, Thomas. *Founded in Fiction. The Uses of Fiction in the Early United States*. Princeton and Oxford: Princeton UP, 2021.
Lowell, James Russell. *A Fable for Critics*. London: Gay & Bird, 1890 [1848].
Marlowe, Christopher. *Dr. Faustus*. Ed. Roma Gill. New York: Norton, 1990.
May, Charles E. *The Short Story: The Reality of Artifice*. New York: Twayne, 1995.
Mylander, Jennifer. "Fiction and Civility Across the Seventeenth-Century English Atlantic: Teaching the History of Faustus." *Books Between Europe and the Americas: Connections and Communities, 1620–1860*. Eds. Leslie Howsam and James Raven. Basingstoke: Palgrave Macmillan, 2011. 42–63.

Newman, Lance, Joel Pace, and Chris Koenig-Woodyard, eds. *Transatlantic Romanticism: An Anthology of British, American, and Canadian Literature, 1767–1867*. New York: Longman, 2006.
Reichart, Walter Albert. *Washington Irving and Germany*. Ann Arbor: U of Michigan P, 1957.
Richardson, Judith. "Cupola of the World. Diedrich Knickerbocker Constructs the Empire City." *A New Literary History of America*. Eds. Greil Marcus and Werner Sollors. Cambridge and London: Belknap P of Harvard UP, 2009. 145–150.
Rubin-Dorsky, Jeffrey. "The Value of Storytelling: *Rip Van Winkle* and *The Legend of Sleepy Hollow* in the Context of *The Sketch Book*." *Modern Philology* 82.4 (1985): 393–406.
Schäfer, Stefanie. *Yankee Yarns: Storytelling and the Invention of the National Body in 19th-Century US American Literature and Culture*. Edinburgh: Edinburgh UP, 2021.
Scheiding, Oliver. *Geschichte und Fiktion. Zum Funktionswandel des frühen amerikanischen Romans*. Paderborn et al.: Schöningh, 2003.
Scheiding, Oliver, and Martin Seidl, eds. *Worlding America: A Transnational Anthology of Short Narratives before 1800*. Stanford: Stanford UP, 2015.
Scofield, Martin. "The Short Story as Ironic Myth: Washington Irving and William Austin." *The Cambridge Introduction to the American Short Story*. Cambridge: Cambridge UP, 2006. 10–17.
Sizemore, Michelle R. "'Changing by Enchantment': Temporal Convergence, Early National Comparisons, and Washington Irving's *Sketch Book*." *Studies in American Fiction* 40.2 (2013): 157–183.
Smith, Jennifer J. "Collections, Cycles and Sequences." *The Edinburgh Companion to the Short Story in English*. Eds. Paul Delaney and Adrien Hunter. Edinburgh: Edinburgh UP, 2018. 93–107.
Snyder-Koerber, Mary Ann. "Building the Better Book: Conversation, Complication, and the Constitution of the Medium in *The Sketch-Book of Geoffrey Crayon, Gent.*" *American Studies as Media Studies*. Eds. Frank Kelleter and Daniel Stein. Heidelberg: Winter, 2008. 27–37.
Tocqueville, Alexis de. *Democracy in America*. Eds. Harvey C. Mansfield and Delba Winthrop. Chicago and London: U of Chicago P, 2000.

4.2 Further Reading

Aderman, Ralph M., ed. *Critical Essays on Washington Irving*. Boston & Mass: G.K. Hall, 1990.
Antelyes, Peter. *Tales of Adventurous Enterprise: Washington Irving and the Poetics of Western Expansion*. New York: Columbia UP, 1990.
Brooks, Van Wyck. *The World of Washington Irving*. London: Dent, 1945.
LeMenager, Stephanie. "Trading Stories: Washington Irving and the Global West." *American Literary History* 15.4 (2003): 683–708.
McGann, Jerome J. "Washington Irving, A History of New York, and American History." *Early American Literature* 47.2 (2012): 349–376.
Richards, Jason. "Localizing the Early Republic Washington Irving and Blackface Culture." *Imitation Nation: Red, White, and Blackface in Early Antebellum US Literature*. Charlottesville: U of Virginia P, 2017. 159–181.
Smith, Jennifer J. *The American Short Story Cycle*. Edinburgh: Edinburgh UP, 2018.

Philipp Schweighauser

6 Nathaniel Hawthorne (1804–1864)

Abstract: This chapter zooms in on two understudied short stories by Nathaniel Hawthorne that critically probe key processes of modernization that shaped the antebellum era. In "The Celestial Rail-road" (1843) and "Earth's Holocaust" (1844), Hawthorne uses satire to stage a conservative critique of transcendentalist optimism, of many of his contemporaries' belief in technological progress, and of the necessity of radical social reform. In staging this critique, Hawthorne does not merely tap into the Puritan past's power of darkness (as he does in *The Scarlet Letter*); he harnesses that past's theological doctrines and religious narratives for his critique of the present. The theory section suggests two additional, alternative readings through Marxist and ecocritical lenses.

Keywords: Age of Reform, satire, conservatism, Marxism, ecocriticism

1 Context

Along with Edgar Allan Poe (↗7 Poe) and Herman Melville (↗8 Melville), Nathaniel Hawthorne belongs to the bleakest, most skeptical strains of the American Renaissance. If his *Blithedale Romance* constitutes a mildly satirical comment on his own experience with the social wing of transcendentalism, which he experienced firsthand during his short stay at the experimental Brook Farm in 1841, many of his major prose works – from the short stories that he published from 1831 to 1852 to *The Scarlet Letter* (1850) and *The House of the Seven Gables* (1851) – can be read as critical probes of transcendentalism's metaphysical underpinnings informed by his exploration of some of the darkest sites of U.S. history and the human psyche. In Hawthorne (as in Poe and Melville), we can find little of the forward-looking exuberance that characterizes Ralph Waldo Emerson's thought and inspired the poetry and prose of his most ardent literary followers, Walt Whitman and Henry David Thoreau. Melville eloquently captures a major source of Hawthorne's darkness in "Hawthorne and His Mosses," his enthusiastic 1850 review of *Mosses from an Old Manse* (1846): "Certain it is [. . .] that this great power of blackness in him derives its force from its appeals to that Calvinistic sense of Innate Depravity and Original Sin, from whose visitations, in some shape or other, no deeply thinking mind is always and wholly free" (341). Intriguingly, in the two short stories that concern me in the present essay, the lesser-known tales "The Celestial Rail-road" (1843) and "Earth's Holocaust" (1844), Hawthorne goes beyond what Melville discerns. Published first in periodicals and later included in Hawthorne's second collection of tales, *Mosses from an Old Manse*, these two stories forage into the Puritan past not only to tap

https://doi.org/10.1515/9783110587647-007

into its power of darkness but also to harness the theological doctrines and religious narratives of that past for their critique of the present. This is a remarkable fact that should give pause to anyone who reads either his short stories or his major novels – chief among them *The Scarlet Letter* – as unequivocal romantic critiques of the Puritan past.

"The Celestial Rail-road" and "Earth's Holocaust" are also exceptional short stories for a second reason. Together, they constitute a remarkably comprehensive conservative critique of Hawthorne's age that reaches well beyond his repudiation of transcendentalist optimism. These tales' critique is directed against a whole range of technological and social modernizations that shaped the antebellum era. While "The Celestial Rail-road" mainly takes aim at Hawthorne's contemporaries' belief in unfettered technological and industrial progress, "Earth's Holocaust" critically interrogates various social reform movements that emerged during the Age of Reform, from the temperance movement to the movement to abolish capital punishment. In both tales, prevalent beliefs in progress, human perfectibility, spiritual renewal, and social melioration – which were fueled by the Second Great Awakening's post-millennial determination to purify the world in anticipation of the Second Coming – are undermined by persistent older, darker metaphysical forces beyond human control. In staging their critiques of the new, both tales employ satire understood as a literary style that combines humor, distortion, and aggression to subject public figures, institutions, and ideas to a sharp form of cultural critique that is guided by a norm which these figures, institutions, or ideas are perceived to violate or deviate from (see Gaier). In Hawthorne's hands, satire ridicules and diminishes what satire has always ridiculed and diminished – vices and follies – and functions as what it has functioned at least since the Enlightenment: as a social corrective. While these tales' politics of representation are complex (about which, more below), their overall thrust is conservative in that they target various processes of modernization, insist on the reality of human sinfulness, and question the possibility of human perfectibility. In short, "The Celestial Rail-road" and "Earth's Holocaust" testify to Kathleen P. Colgan's perceptive assessment of Hawthorne's political philosophy: "Against reformists' proposals for a socio-political existence which entailed the radical transformation of man and society, against apocalyptic yearnings for some form of world-immanent salvation, Hawthorne consequently registered a vigorous conservative response" (Colgan 2001, 31).

2 Close Readings

First published in the May 1843 issue of the *Democratic Review*, "The Celestial Railroad" updates and modernizes the story of Christian's pilgrimage in the first part of John Bunyan's Puritan classic *The Pilgrim's Progress* (1678/1684) for Hawthorne's mid-nineteenth-century readers. Accompanied by the affable and knowledgeable

Mr. Smooth-it-away (instead of Bunyan's Evangelist), the nameless narrator embarks on an effortless, modern pilgrimage by train "as if the pilgrimage were merely a summer tour" (Hawthorne 1987 [1843], 168). The narrator desires to see the various sites that Christian visited but the train rushes past most of them as Mr. Smooth-it-away scoffs at the outdatedness and inconvenience of traditional pilgrimages. Importantly, the narrator's increasingly ominous companion does not merely embrace the modern technology that gets the pilgrims to the Celestial City quicker and with the least amount of effort; he embraces all things modern, including modern morals, modern fashions, and modern architecture (171). At first, the narrator readily shares these judgments, concluding that he can "very readily dispense with [the] acquaintance" of "Miss Prudence, Miss Piety, Miss Charity" if they are indeed as unfashionable and "prim, starched, dry, and angular" (171) as Mr. Smooth-it-away suggests. Even the fact that Apollyon – the demonic, dragon-like King of the City of Destruction, whom Bunyan's Christian defeats in the Valley of Humiliation – is the train's chief engineer prompts the narrator to exclaim, "Bravo, bravo! [. . .] [T]his shows the liberality of the age; this proves, if anything can, that all musty prejudices are in a fair way to be obliterated" (169). But as the train shoots closer toward the Celestial City, he begins to hear and see the cracks in his companion's narrative as he is startled by the ever more piercing shrieks of the train's engine, perceives the falseness of the light that illuminates the Valley of the Shadow of Death (172), and witnesses several of Vanity Fair's revelers disappearing "like a soap-bubble, and be never more seen of [their] fellows" (178). The narrator begins to understand that there is no shortcut to salvation, and the tale ends with his awakening in two senses of the word: he awakens to the reality that Mr. Smooth-it-away is the devil leading him and his fellow travelers straight to hell, and he wakes up from the dream that is the story he has told.

Hawthorne borrows the allegorical mode from Bunyan and uses it for satirical purposes. Yet what or who are the targets of Hawthorne's satire? Hawthorne's transformation of Christian's arduous pilgrimage to the Celestial City into an effortless train ride to hell aims first and foremost at his liberal contemporaries' belief in the redemptive force of technological and industrial progress. As Cindy Weinstein notes in her reading of the tale, Hawthorne's association of the railroad with hell is fairly conventional:

> Declaring its opposition to the railroad on religious grounds, an Ohio school board called the steam railroad "a device of Satan to lead immortal souls down to Hell," and an 1840 article in *Graham's* magazine entitled "Stage Coaches and Rail Roads" blamed railroad technology for enabling Americans, who "despis[e] every thing but the '*utile*' of life, living only for the sake of making money, and turning from the finest landscape that the Creator ever wakened into beauty," to continue in their "quick, restless, thrifty" ways. (Weinstein 2008, 59)

The quote from *Graham's Magazine* identifies Hawthorne's second major target, the economic system that is fueled by and sustains the nineteenth-century belief in progress: industrial capitalism, which in "The Celestial Rail-road" has its iron produced in the forges of the Valley of the Shadow of Death (Hawthorne 1987 [1843], 172) and whose representatives, "the capitalists of the city," are among Vanity Fair's "largest stockholders" (175). In staging these combined critiques, Hawthorne is far from parodying *The Pilgrim's Progress*, as Anthony Cerulli and Sarah L. Berry have asserted as recently as 2014 (Cerulli and Berry 2014, 119). Instead, he evokes the Calvinist moral codes and values that sustain the original text to judge contemporaneous follies by these old standards. In an essay published in 1991, Michael J. Colacurcio brings out this quality of Hawthorne's text with admirable clarity:

> It is scarcely a question of 'parody,' in the usual sense: nineteenth century religious institutions and attitudes make their own mockery of Bunyan's antique tropes; but in this satire, his old-time standard of religious seriousness is the source of authority rather than the object of ridicule. Taking Bunyan absolutely for granted, Hawthorne revivifies him in order to try conclusions with various forms of American liberalism. (Colacurcio 1991, 4)

Hawthorne's third major liberal target is what an unappreciative reader of Emerson may consider the metaphysical and spiritual corollary of the secular, capitalist belief in progress: the transcendentalist faith in the self's unbounded capacity to renounce the past, throw off the shackles of societal constraints, and reinvent itself on its own terms. Thus, Bunyan's allegorical villains Pope and Pagan, who scare the pilgrims but are, in Bunyan, already unable to hurt them, make way for Hawthorne's "Giant Transcendentalist," who now inhabits their cave at the end of the Valley of the Shadow of Death and "makes it his business to seize upon honest travellers, and fat them for his table with plentiful meals of smoke, mist, moonshine, raw potatoes, and saw-dust" (Hawthorne 1987 [1843], 174). At its core though, Hawthorne's critique of Emersonian transcendentalism here is less metaphysical than noetic and rhetorical: the Great Transcendentalist offers his followers nothing but insubstantial intellectual fare ("smoke, mist, moonshine, raw potatoes, and saw-dust") and incomprehensible diction: "He shouted after us, but in so strange a phraseology, that we knew not what he meant, nor whether to be encouraged or affrighted" (174). Fully in line with its literary model, Hawthorne's critique of transcendentalist intangibility and obfuscation does not, however, constitute a rejection of the quest for transcendence. Quite the contrary is the case: "The Celestial Rail-road" consistently affirms the existence of a world beyond the physical whenever it judges as misguided its narrator's attempts to brush aside signs of such a world, from what the narrator calls the "delusions" and "vain imaginations" (174) of sinful faces appearing in the flames that light the Valley of the Shadow of Death to the supposedly "effervescent pleasures" (175) heaven provides. It is just that this tale's search for transcendence (which it affirms mainly satirically, i.e., *ex negativo*, by presenting fallible figures that either reject this search or eschew the considerable effort it demands) is grounded in a Calvinist instead of an evangelical, Unitarian, or transcendentalist metaphysics.

In line with this, the fourth target of Hawthorne's satire is clergymen who go by names such as "the Rev. Mr. Shallow-deep," "the Rev. Mr. Clog-the-spirit," and "the Rev. Dr. Wind-of-doctrine" (175). These are "Liberal Christians" who "are trying to have the good news of salvation without the bad news of sin and the need of redemption" (Colacurcio 1991, 4). These characters abandon truth-seeking, learning, and theological dogma to appeal to increasingly complacent audiences, who satisfy their spiritual needs with yet another vain exercise (Hawthorne's fifth target): the championing of charitable causes and reform movements, which the denizens of Vanity Fair throw themselves into with gusto but little effort.

Hawthorne merges his critiques of technological progress, spiritual decline, and reformist zealotry when he calls the latter "another species of machine for the wholesale manufacture of individual morality [. . .]. [A] man has merely to connect himself, throwing, as it were, his quota of virtue into the common stock; and the president and directors will take care that the aggregate amount be well applied" (Hawthorne 1987 [1843], 175). In the tale's Vanity Fair episode, these various critiques merge into satirical attacks on commodification and financial speculation that testify to the remarkable modernity of Hawthorne's cultural commentary:

> Some of the purchasers, I thought, made very foolish bargains. For instance, a young man having inherited a splendid fortune, laid out a considerable portion of it in the purchase of diseases, and finally spent all the rest for a heavy lot of repentance and a suit of rags. [. . .] There was a sort of stock or scrip, called Conscience, which seemed to be in great demand, and would purchase almost anything. Indeed, few rich commodities were to be obtained without paying a heavy sum in this particular stock, and a man's business was seldom very lucrative, unless he knew precisely when and how to throw his hoard of Conscience into the market. Yet as this stock was the only thing of permanent value, whoever parted with it was sure to find himself a loser, in the long run. Several of the speculations were of a questionable character.
> (176)

The example of the young man offers an economic lesson about the leveling of qualitative differences under the logic of capitalism: wildly different entities (diseases, repentance, clothes) indeed become exchangeable under the spell of exchange value. Commodification is raised to a new level when Conscience is traded as a "stock" or "scrip." Hawthorne's use of the latter term – which denotes both "[a] small bag or pouch, *esp.* one carried by a pilgrim, shepherd, or beggar" and, since the late eighteenth century, also "a receipt for a share or shares in a loan or a commercial undertaking; a share certificate; (also) a stock, a share" (*OED* 2018) – suggests just how far the logic of capitalism has infiltrated traditional religious pursuits. Hawthorne emphasizes that Conscience is a special kind of scrip though, "the only thing of permanent value," which is forever lost when exchanged in search of financial gain.

Hawthorne's combination of allegory and satire in this mid-nineteenth-century text may strike students of romanticism as antiquated, partly because the type of extended indirect satire he employs had its heyday in eighteenth-century texts such as *The Rape of the Lock* (1712), *Gulliver's Travels* (1726), and *Candide* (1759) but mainly because of Hawthorne's sustained recourse to the allegorical mode. Literary scholars have, after all, opposed the indeterminate, inexhaustible romantic symbol – of which the scarlet letter is a prime example – to the univocal, more heavily conventionalized allegorical style of earlier literary productions. The romantic symbol is "at once infinitely meaningful and incapable of being reduced to any particular meaning" (Halmi 2007, 2), which makes it a major marker of romanticism's modernity. Hawthorne's choice of the older form makes perfect sense for precisely this reason though: he employs it to stress the continuing force of an older, Calvinist religious and moral framework, which provides the story's "moral *a priori*" that "we need to judge the phenomena of present interest" (Colacurcio 1991, 5). It is his use of an old form to express an old set of values and norms that gives his conservative critique such a powerfully archaic feel.

Originally published in *Graham's Magazine* in March 1844 and included in *Mosses from an Old Manse* along with "The Celestial Rail-road," "Earth's Holocaust" shares the earlier story's decidedly conservative attitude toward modernization. In staging its critique, it also taps into the allegorical mode – for instance when a man protesting the destruction of all alcohol in the eponymous bonfire is named "the Last Toper" (Hawthorne 1987 [1844], 185) or when a Bible-reading child is referred to as a "sorely tried Child of Dust" (195) – but it does so in a much less sustained fashion than Hawthorne's adaptation of *The Pilgrim's Progress*. "Earth's Holocaust" does participate with equal force in the satirical mode though, as it moves to the center of attention one manifestation of modernization that "The Celestial Rail-road" critiques only in passing: the antebellum period's social reform movements. The title of the story already announces that there is something dangerously askew with the zeal displayed by the story's fiery reformers, who commit ever more insignia of the past to a rapidly growing pyre that is lit to rid the world of obsolete objects, practices, and ideologies. If permitted full expression, the title suggests, this zeal will result in the earth's wholesale consumption by fire.

The first items of note consigned to the flames are vestiges of a bygone aristocratic world: from "all the rubbish of the Herald's Office" (182) to "the crown-jewels of England" and "the Czar of Russia's sceptre" (184). The narrator has quite a bit of sympathy for this particular expression of revolutionary fervor – the fervor that fueled the American Revolution – as did both Hawthorne and Edmund Burke, the 'father of conservatism,' whom Hawthorne read and admired (Colgan 2001, 157–170). But as we move into Hawthorne's present and more and more objects are consigned to the flames by a succession of reformers that destroy, first, all the world's liquor, coffee, tea, and tobacco (Hawthorne 1987 [1844], 185–186), then the "gowns and petticoats" of women who wish to "assume the garb, together with the manners, duties, offices, and responsibilities, of the opposite sex" (187), then all the planet's weapons, munitions, and instruments of capital punishment (187–189), it becomes clearer and clearer that the tale indicts excessive reformist zeal, denouncing it as a dangerous and destructive force.

Hawthorne employs an intriguing narrative strategy as he takes aim at various social reform movements that sprang up during the Age of Reform (the temperance movement, the women's movement, pacifism, and the movement to abolish the death penalty, among others). His tale provides a stage for a wide range of voices that provide diverging assessments of the proceedings: ardent reformers who, egged on by the common folk, consign ever more items to the flames; conservatives who bewail the passing of old orders yet are impotent to stop it; a "grave observer" (183) constantly by the narrator's side who provides sometimes appreciative but mostly skeptical commentary; and the narrator himself, who applauds some measures – "That was well done!," he exclaims when gallows are thrust into the center of the furnace – and questions the wisdom of others, judging that "they now proceeded to measures in the full length of which I was hardly prepared to keep them company" (191) when marriage certificates, ledgers, legal tender, titles to debts and land, statute books, and written constitutions are thrown into the bonfire. This showcasing of a multiplicity of voices gives Hawthorne's satire a certain Menippean quality (though it is clearly not a pure example of that subgenre).

That the figure of the narrator is a conundrum in itself becomes perfectly obvious once the narrative reaches the burning of books. In this episode, he channels an astonishing variety of voices that cannot be attributed to any one public figure or human type of Hawthorne's time. To make matters even less clear, the narrator is ironized from the episode's start when he attempts to measure the relative merits of various individual writers and their works (from Shakespeare, Milton, and Shelley, to nursery rhymes and popular ballads, to Hawthorne's largely unnamed American contemporaries) by the time it takes to reduce them to ashes and the beauty of the flames their burning produces. Hawthorne next plays a coy game of referring to his own, i.e., Hawthorne's, works as he has his narrator observe that it was "not the writer most frequent in the public mouth, that made the most splendid appearance in the bonfire" (193) only to confess that his texts were "probably changed to vapor by the first action of the heat; at best, I can only hope that, in their quiet way, they contributed a glimmering spark or two to the splendor of the evening" (193). If this leaves any doubts that the narrator's judgments do not always coincide with the implied author's, such doubts are dispersed when the narrator adopts a firmly Emersonian voice as he comforts a "forlorn book-worm" distraught at the destruction of the world's literary production: "Is not nature better than a book?" the narrator asks rhetorically; "Is not the human heart deeper than any system of philosophy? Is not life replete with more instruction than past observers have found it possible to write down in maxims?" (193). Both here and when he meets the burning of clerical garbs and religious symbols with the consoling suggestion that "the wood-paths shall be the aisles of our cathedral, the firmament itself shall be its ceiling" (195), the narrator serves as a vehicle for the author's jabs at, first, Emerson's famous critique of bookish learning, enshrined most notably in his observation that "Books are for the scholar's idle times. When he can read God directly, the hour is too precious to be wasted in other men's transcripts of their reading" (Emerson 1983a, 58–59), and second, the Concord sage's closely related determination to seek truth and revelation in nature, where "no history, or church, or state, is interpolated on the divine sky and the immortal year" (Emerson 1983b, 541). And yet, the narrator is by no means a clear-cut Emersonian, for the conclusion he draws once Bibles are being burned and one of the spectators reveals himself as the devil is neither transcendentalist nor Unitarian, nor evangelical but anti-millennialist, anti-utopian, conservative, and, above all, Puritan.

Before we look at the unexpected turn to Puritan certainties near the story's end, let us pause to consider the effects Hawthorne's choice of narrative strategy has on his satire. In providing a platform for several divergent voices and a narrator whose positionality is exceptionally hard to pin down, Hawthorne's text assumes a powerfully multivocal quality. As such, it bears witness to Dustin Griffin's apt observation that "satire is problematic, open-ended, essayistic, ambiguous in its relationship to history, uncertain in its political effect, resistant to formal closure, more inclined to ask questions than to provide answers, and ambivalent about the pleasures it offers" (Griffin 1994, 5). Yet as we have seen, "Earth's Holocaust" is not an indeterminate text in the post-structuralist sense, for the target of its satirical critique is clear: in this story, it is the unbounded reformist belief in social and political progress that leads to calamity. This is confirmed neither by the many voices that speak in it nor by the narrator's judgments but by the text's narrative trajectory that moves toward what the title announces: the earth's consumption by fire, its holocaust. The satire's main target is the voice that speaks thus:

> "Too cold! too cold!" impatiently exclaimed the young and ardent leader in this triumph. "Let the heart have its voice here as well as the intellect. And as for ripeness, and as for progress, let mankind always do the highest, kindest, noblest thing that, at any given period, it has attained the perception of; and surely that thing cannot be wrong nor wrongly timed."
>
> (Hawthorne 1987 [1844], 190)

As James Hewitson has shown, Hawthorne borrows the tale's apocalypticism from the premillennialist Baptist preacher William Miller, who prophesied that the world would end "sometime between 21 October 1843 and 21 March 1844" (Hewitson 2007, 89), to critique the social reform movements fueled by the Second Great Awakening: "the sheer devastation required by Miller's vision of the imminent Second Coming is made illustrative of the fundamental destructiveness of all organized efforts to ameliorate social ills, implying both that such destruction is inherent to the nature of systematic reform and that it is ultimately futile" (91). In Hawthorne's conservative vision, the reformers themselves beget the destruction foretold by Miller.

What is far less readily identifiable is the norm that founds Hawthorne's satirical critique. Consider the final three paragraphs, where the Calvinist tenet of total depravity is affirmed. Strikingly, the narrator agrees when the devil reveals himself to proclaim that "the human heart itself" is the root of all evil and prophesies that "unless they hit upon some method to purify that foul cavern, forth from it will re-issue all the shapes of wrong and misery – the same old shapes or worse ones – which they have taken such a vast deal of trouble to consume to ashes" (Hawthorne 1987 [1844], 197). "The heart, the heart," the narrator agrees in a moment of recognition, "there was the little yet boundless sphere wherein existed the original wrong of which the crime and misery of this outward world were merely types" (197). Again, we encounter the Calvinist theology that Melville so perceptively locates at the heart of Hawthorne's stories and which fuels, as we have seen, Hawthorne's satire in "The Celestial Rail-road." In "Earth's Holocaust," the validity of that theology is compromised by being first channeled through the devil. Moreover, the narrative does not end here; its concluding sentences further temper the ending's religious severity as the narrator gives expression to the hope that the human heart may, after all, be purified by a type of reform that also aims at the heart. Here are the final sentences of the tale:

> Purify that inner sphere; and the many shapes of evil that haunt the outward, and which now seem almost our only realities, will turn to shadowy phantoms, and vanish of their own accord. But, if we go no deeper than the Intellect, and strive, with merely that feeble instrument, to discern and rectify what is wrong, our whole accomplishment will be a dream; so unsubstantial, that it matters little whether the bonfire, which I have so faithfully described, were what we choose to call a real event and a flame that would scorch the finger – or only a phosphoric radiance, and a parable of my own brain! (197)

A cursory reading of the ending may suggest that the reformers' methods cut too short because they rely solely on the intellect in the sense that they appeal only to human understanding and aim at providing nothing but rational, pragmatic solutions to societal problems. Only if coupled with a reform of human beings' hearts do these movements have a chance to succeed. Yet a closer look shows how strange a conclusion this would provide to a tale that so consistently subjects reformism to satirical critique. Indeed, the narrator's call to combine appeals to the intellect with appeals to the heart echoes the ardent young reformer's appeal quoted above: "Let the heart have its voice here as well as the intellect." More damningly still, it reiterates what the devil says when he tells the narrator that humanity's only hope is to find "some method to purify that foul cavern," the heart. Thus, the narrator's solution, which is actually the kind of solution envisaged by reform movements inspired by the Second Great Awakening, appears as hopelessly compromised. The severe Calvinist theology that raises its specter in the final three paragraphs is compromised too, for one because the devil embraces it. So while I do agree with Agnes McNeill Donohue's assessment that "the weighty and dooming Calvinist view of man's moral nature dominates the theme" (Donohue 1985, 190), Calvinism does not qualify as the unquestioned norm that grounds the tale's satire. What "Earth's Holocaust" does, though, is harness the "power of blackness" of Calvinist theology as the story's most enduring presence to stage Hawthorne's conservative critique of his era's radical reform movements.

3 Theoretical Perspectives

As is the case with any kind of literary text, the critic can make use of a wide variety of theoretical concepts to gain further insights into Hawthorne's short stories. It seems to me, though, that in the case of "The Celestial Rail-road" and "Earth's Holocaust," two theoretical frameworks hold particular explanatory power. For the purpose of the present handbook entry, these frameworks have the additional virtue of opening up the two tales to readings beyond my preceding analyses of Hawthorne's conservative critiques of modernization. The first of these, Marxism, suggests itself based on my discussion of Hawthorne's dissection of capitalist logics in "The Celestial Rail-road." The second, ecocriticism, builds on the results of the first inquiry in that it registers the extent to which Hawthorne's fictional negotiations of industrial capitalism highlight its grounding in the exploitation of natural 'resources.' In "Earth's Holocaust," this recognition is given a decidedly apocalyptical twist.

A Marxist analysis might start by noting just how steeped "The Celestial Rail-road" is in the language of economics. Even the two righteous pilgrims, who choose to travel on foot, use this register when they meet the narrator in Vanity Fair and warn him that the railroad is a road to perdition: "The Lord of the Celestial City [. . .] has refused, and will ever refuse, to grant an act of incorporation for this railroad; and unless that be obtained, no passenger can ever hope to enter his dominions. Wherefore, every man, who buys a ticket, must lay his account with losing the purchase-money – which is the value of his own soul" (Hawthorne 1987 [1843], 178). The upright pilgrims use the language of economics metaphorically here to persuade a subject who understands the world literally in these terms. As we have seen in the close reading section, when Hawthorne stages his critique of commodification and financial speculation, he emphasizes that this economization corrupts not merely subjects' language but life itself as it "resolve[s]," in Karl Marx and Friedrich Engels's words in *The Communist Manifesto*, "personal worth into exchange value" (Marx and Engels 1976, 487). Interestingly though, Hawthorne's description of what happens to the relation between things and ideas in these processes is the opposite of what Marx calls 'commodity fetishism,' where "a definite social relation between men [. . .] assumes, in their eyes, the fantastic form of a relation between things" (Marx 1976, 72). In Hawthorne, capitalism does not charge things with mystical meaning but debases moral ideas, chief among them Conscience, into things. As least in this sense, Hawthorne's critique of capitalism is closer to Emerson's idealism than to Marx's materialism.

A Marxist reading would also notice the division of labor on the train, where the devil's subjects no longer shoot arrows at pilgrims at the Wicket Gate but work at the Station House so that their betters may enjoy a comfortable ride to the Celestial City:

> The Prince's subjects are now pretty numerously employed about the Station House, some in taking care of the baggage, others in collecting fuel, feeding the engines, and such congenial occupations; and I can conscientiously affirm, that persons more attentive to their business,

> more willing to accommodate, or more generally agreeable to the passengers, are not to be found on any railroad. Every good heart must surely exult at so satisfactory an arrangement of an immemorial difficulty. (Hawthorne 1987 [1843], 167–168)

The most immediate reference of the "immemorial difficulty" is to the struggle between Good-will (the keeper of the Wicket Gate) and Beelzebub. This struggle between good and evil is, Hawthorne's narrator suggests at this point, amiably resolved through capitalism's division of labor. Yet when he peers into the depths of the Shadow of the Valley of Death, it becomes clear that those who work there, "the laborers at the forge, and those who brought fuel to the engine" (173) have been consigned to an antechamber of hell: "when they began to draw short breath, [they] positively emitted smoke from their mouth and nostrils" (173). The class politics of such descriptions are ambiguous, especially if we consider that the Valley of the Shadow of Death is also populated by capitalist types such as Mr. Take-it-easy, who light their cigars "at the flame of the crater" (173). These descriptions can be interpreted as much as an indictment of the horrors of industrial labor, which, in Marx's analysis, creates a "mass of misery, oppression, slavery, degradation, exploitation" (Marx 1976, 929), as they can be read as a demeaning othering of those who perform that labor. Considering Hawthorne's conservatism and his insistent critique of a variety of contemporaneous reform movements, one would be hard pressed to read his depiction of laborers' fates in an infernal production setting as a clarion call to improve, let alone revolt against, exploitative, alienating labor conditions. But they certainly do give expression to a conservative variety of critique of the "one single catastrophe which keeps piling wreckage upon wreckage" that Walter Benjamin called "progress" (Benjamin 1968, 257–258) in general and of the 'creative destruction' (see Schumpeter) of industrial capitalism in particular. The figure that leads the narrator to hell, Mr. Smooth-it-away, is, after all, an industrial capitalist himself, "a director of the rail-road corporation" that operates the celestial railroad and "one of its largest stockholders" (Hawthorne 1987 [1843], 166).

Understood in its broad sense as "an umbrella term for a range of critical approaches that explore the representation in literature (and other cultural forms) of the relationship between the human and the non-human, largely from the perspective of anxieties around humanity's destructive impact of the biosphere" (Marland 2013, 846), ecocriticism suggests itself as a second particularly promising theoretical perspective on Hawthorne's two short stories. Already Leo Marx's classic *The Machine in the Garden: Technology and the Pastoral Ideal in America* (1964), which many now consider an early contribution to ecocriticism, briefly discusses "The Celestial Rail-road," calling it "a wonderfully compact satire on the [nineteenth century's] prevailing faith in progress" (Marx 2000, 27). An ecocritical perspective on Hawthorne's tale not only illuminates its critique of progress on the level of the story but also reveals the nature of its intertextual relation to *The Pilgrim's Progress*: by modernizing Bunyan's Puritan allegory, the text itself enacts the intrusion of the machine into Bunyan's Christian garden. An ecocritical mindset further helps us see that several of the phenomena that the narrator of "The Celestial Rail-road" observes allegorize the human exploitation of natural resources as much as they allegorize human sinfulness, from the tunnel that is built to enable passage through Hill Difficulty (Hawthorne 1987 [1843], 171) to the Valley of the Shadow of Death, where "the inflammable gas, which exudes plentifully from the soil" (172) is transformed into an artificial light whose relation to "daylight" is "the same difference as between truth and falsehood" (172). The story's center of industrial production is located in that same valley, at the site that Bunyan still identified as the mouth of hell while Mr. Smooth-it-away assures the narrator that it is nothing but "the crater of a half-extinct volcano, in which the Directors had caused forges to be set up, for the manufacture of railroad iron. Hence, also, is obtained a plentiful supply of fuel for the use of the engines" (172). In Hawthorne's retelling of *The Pilgrim's Progress*, the sources of technological and industrial progress are infernal, i.e., they emerge from the lower regions, in a double sense: their site of production is hell and the resources that fuel them are dug out of the depths of the earth.

More persistently than "The Celestial Rail-road," "Earth's Holocaust" interweaves its exploration of capitalism and social class with the ecological concerns that are at the heart of ecocritical inquiries. The first items consigned to the flames are insignia of aristocracy; in an egalitarian republican spirit, the bonfire is designed to burn away social privilege. It comes as no surprise then that the reformers are joined by an enthusiastic "multitude of plebeian spectators" (Hawthorne 1987 [1844], 183) who desire, among other outcomes, that "the whole soil of the earth revert to the public, from whom it had been wrongfully abstracted" (190–191). From the outset, though, the desire to create a more just social order, which also fuels the burnings that follow, is coupled with a desire to rid the world of its refuse. As we read in the very first sentence, "this wide world had become so overburdened with an *accumulation* of worn-out trumpery, that the inhabitants determined to rid themselves of it by a general bonfire" (181; emphasis added). The problem is, then, one of accumulation, both of (social) capital and of waste. The tale repeatedly stresses that the items given over to the flames are worthless junk, a "heap of condemned rubbish" (181), "articles that were judged fit for nothing but to be burned" (182), "antiquated trash" (182). To twenty-first-century readers at least, it becomes especially clear that there is an ecological dimension to the destruction of the earth's refuse when we learn that, instead of being blown up in the bonfire, the world's gunpowder, "as the safest mode of disposing of it, has already been drowned in the sea" (187). It is at the latest at this point that the reader begins to intimate that the zealous eradication of the earth's waste threatens to lay waste to the earth itself. And indeed, as the bonfire continues to grow and grow, we enter a different ecological scenario: the specter that is now raised is the wholesale destruction of the earth by human activity that recent ecocritical interventions have theorized under the headings of the 'Anthropocene' (see Chakrabarty 2009, 2018; Latour 2014, 2017) and, more radically, 'extinction' (see Colebrook; Morton). In their apocalypticism (or, in Colebrook's case, their 'ethics of extinction'), these theoretical frameworks can help us make sense of a tale that begins with the burning of accumulated waste and ends with the grim possibility, raised by the devil near the ending of the story, that "the earth itself" would be burned "to a cinder" (Hawthorne 1987 [1844], 197). As they allow us to perceive an additional dimension of the apocalypticism that Hawthorne borrowed from William Miller's premillennialist discourse, such frameworks enable students of literature to see what sound theories always enable us to see: hitherto unexplored levels of meaning in literary texts.

I thank Ridvan Askin for his incisive, helpful feedback on an earlier version of this chapter. I also owe thanks to Andrea Wüst and Rahel Ackermann Hui for their diligent proofreading and formatting of the text.

4 Bibliography

4.1 Works Cited

Benjamin, Walter. "Theses on the Philosophy of History [On the Concept of History]." Trans. Harry Zohn. *Illuminations*. Ed. Hannah Arendt. New York: Schocken, 1968. 253–264.

Bunyan, John. *The Pilgrim's Progress*. 1678/1684. A Norton Critical Edition. New York: Norton, 2008.

Cerulli, Anthony, and Sarah L. Berry. "Nathaniel Hawthorne's Warring Doctors and Meddling Ministers." *Mosaic* 47.1 (2014): 111–128.

Chakrabarty, Dipesh. "Anthropocene Time." *History and Theory* 57.1 (2018): 5–32.

Chakrabarty, Dipesh. "The Climate of History: Four Theses." *Critical Inquiry* 35.2 (2009): 197–222.

Colacurcio, Michael J. "Cosmopolitan and Provincial: Hawthorne and the Reference of American Studies." *Studies in the Novel* 23.1 (1991): 3–19.

Colebrook, Claire. *Death of the PostHuman: Essays on Extinction*. Vol. 1. Ann Arbor, MI: Open Humanities P, 2014.

Colgan, Kathleen P. *The Influence of Political Events on Nathaniel Hawthorne's Political Vision and Writings*. Lewiston, NY: Edwin Mellen P, 2001.

Donohue, Agnes McNeill. *Hawthorne: Calvin's Ironic Stepchild*. Kent: Kent State UP, 1985.

Emerson, Ralph Waldo. "The American Scholar." *Emerson: Essays and Lectures: Nature: Addresses and Lectures / Essays: First and Second Series / Representative Men / English Traits / The Conduct of Life*. Ed. Joel Porte. New York: Library of America, 1983a. 53–71.

Emerson, Ralph Waldo. "Nature." *Emerson: Essays and Lectures: Nature: Addresses and Lectures / Essays: First and Second Series / Representative Men / English Traits / The Conduct of Life*. Ed. Joel Porte. New York: Library of America, 1983b. 541–555.

Gaier, Ulrich. *Satire: Studien zu Neidhart, Wittenwiler, Brant und zur satirischen Schreibart*. Berlin: De Gruyter Mouton, 1967.

Griffin, Dustin. *Satire: A Critical Reintroduction*. Lexington: UP of Kentucky, 1994.

Halmi, Nicholas. *The Genealogy of the Romantic Symbol*. Oxford: Oxford UP, 2007.

Hawthorne, Nathaniel. *The Blithedale Romance*. 1852. New York: Norton, 2010.

Hawthorne, Nathaniel. "The Celestial Rail-road." 1843. *Nathaniel Hawthorne's Tales*. Ed. James McIntosh. A Norton Critical Edition. New York: Norton, 1987. 166–181.

Hawthorne, Nathaniel. "Earth's Holocaust." 1844. *Nathaniel Hawthorne's Tales*. Ed. James McIntosh. A Norton Critical Edition. New York: Norton, 1987. 181–197.
Hewitson, James. "'To Despair at the Tedious Delay of the Final Conflagration': Hawthorne's Use of the Figure of William Miller." *ESQ: A Journal of the American Renaissance* 53.1 (2007): 89–111.
Latour, Bruno. *Facing Gaia: Eight Lectures on the New Climatic Regime*. Trans. Catherine Porter. Cambridge, UK: Polity P, 2017.
Latour, Bruno. "Agency at the Time of the Anthropocene." *New Literary History* 45.1 (2014): 1–18.
Marland, Pippa. "Ecocriticism." *Literature Compass* 10.11 (2013): 846–868.
Marx, Karl. *Capital*. Trans. Ben Fowkes. Vol. 1. Harmondsworth: Penguin, 1976.
Marx, Karl, and Friedrich Engels. *Manifesto of the Communist Party*. Collected Works of Marx and Engels. Vol. 6. London: Lawrence & Wishart, 1976.
Marx, Leo. *The Machine in the Garden: Technology and the Pastoral Ideal in America*. New York: Oxford UP, 2000.
Melville, Herman. "Hawthorne and His Mosses." *Nathaniel Hawthorne's Tales*. Ed. James McIntosh. New York: Norton, 1987. 337–350.
Morton, Timothy. *The Ecological Thought*. Cambridge: Harvard UP, 2010.
Oxford English Dictionary Online, https://www.oed.com/. Oxford: Oxford UP, 2018. (20 November 2018).
Schumpeter, Joseph. *Capitalism, Socialism, and Democracy*. New York: Harper & Row, 1942.
Weinstein, Cindy. *The Literature of Labor and the Labors of Literature: Allegory in Nineteenth-Century American Fiction*. Cambridge: Cambridge UP, 2008.

4.2 Further Reading

Bottomore, Tom. *A Dictionary of Marxist Thought*. 2nd ed. Oxford: Blackwell, 1991.
Buell, Lawrence. *The Environmental Imagination: Thoreau, Nature Writing, and the Formation of American Culture*. Cambridge: Belknap P of Harvard UP, 1995.
Colacurcio, Michael J. *The Province of Piety: Moral History in Hawthorne's Early Tales*. Cambridge: Harvard UP, 1984.
Gale, Robert L. *Nathaniel Hawthorne Encyclopedia*. Westport: Greenwood, 1991.
Garrard, Greg. *Ecocriticism*. 2nd ed. New York: Routledge, 2011.
Haslett, Moyra. *Marxist Literary and Cultural Theory*. New York: St. Martin's, 2000.
Kevorkian, Martin. *Writing beyond Prophecy: Emerson, Hawthorne, and Melville after the American Renaissance*. Baton Rouge: Louisiana State UP, 2012.
Michaels, Walter Benn, and Donald E. Pease, eds. *The American Renaissance Reconsidered*. Baltimore, MD: Johns Hopkins UP, 1985.
Reynolds, David D. *Beneath the American Renaissance: The Subversive Imagination in the Age of Emerson and Melville*. New York: Knopf, 1988.
Reynolds, Larry J. *Devils and Rebels: The Making of Hawthorne's Damned Politics*. Ann Arbor: U of Michigan P, 2008.
Zapf, Hubert, ed. *Handbook of Ecocriticism and Cultural Ecology*. Berlin: De Gruyter, 2016.

Gero Guttzeit

7 Edgar Allan Poe (1809–1849)

Abstract: Edgar Allan Poe's literary and critical writings were crucial to establishing the short story as a genre and art form. Two tales by Poe published in 1839, the lesser-known "The Devil in the Belfry" and the famous "William Wilson," exemplify, respectively, the comic and Gothic modes of the emerging antebellum short story. Responding on various levels to Washington Irving, both narratives also adapt European textual models and spaces, coalescing collective and intrapersonal conflicts in semantically rich monsters (devil and doppelganger). Reading them in the context of Poe's seminal reviews of Nathaniel Hawthorne, this chapter demonstrates how Poe's tales abound with metafictional elements, reflecting and shaping a major early phase in the evolution of the American short story.

Keywords: Tale, comic, Gothic, doppelganger, monster, rhetorical poetics, antebellum print culture

1 Context

Many biographical accounts of Edgar Allan Poe begin not with his birth, but with his death. Ever since he died in a Baltimore hospital in the early hours of 7 October 1849, the circumstances have remained mysterious. As such they have become part of the so-called Poe legend (see e.g., Walker 1996), a narrative originated by contemporaries' slanderous biographies (Poe's literary executor Rufus Griswold first among them) and fueled by the ongoing popularity of Poe's morbid tales – the intoxication, madness, and "perverseness" experienced by his narrators oftentimes easily extended to encompass the historical author. Epitomized by Arthur Hobson Quinn's extensive 1941 life of Poe, his later biographers have done much to dispel this mythical narrative and to establish the facts about Poe's life, though the Poe legend continues to be an important reason for Poe's presence in the popular mind.

Edgar Poe was born as the second child of two stage actors in Boston on 19 January 1809. He quickly became an orphan after his father David left the family and his mother Elizabeth died on 8 December 1811. Without ever being formally adopted, Edgar was taken as a foster child into the household of John and Frances Allan in Richmond, Virginia. He grew up in the South before the Civil War, with its political economy and ideology based on the enslavement of African Americans. John Allan later inherited considerable wealth, but he never did much to support Poe's choice to become a poet, and Poe ultimately inherited nothing of Allan's fortune. Poe briefly attended the newly founded University of Virginia in Charlottesville and the United States Military Academy at West Point. At the age of eighteen, he privately published

https://doi.org/10.1515/9783110587647-008

his first collection, *Tamerlane and Other Poems* (1827), of which only twelve copies – among the most valuable American books – are known to be extant. Poe would go on to release many other poems in periodicals and books, such as the perennial favorite "The Raven" (1845), and also to author one novel, *The Narrative of Arthur Gordon Pym* (1838), but he turned to the short story early on to carve out a literary career for himself. Among his literary peers, he became best known for his incisive pieces of criticism and as the editor of such journals as the *Southern Literary Messenger* and *Graham's Magazine*. His life-long dream of owning his own magazine was only partially realized when he assumed ownership of the ailing *Broadway Journal* in 1845, which went bankrupt only a few months later. As one of the first American authors attempting to earn a living solely from their literary works, Poe was often deeply mired in financial difficulties, but by the same token he also became well-acquainted with the highs and lows of the antebellum American literary market. In his afterlife, he became a global icon of American literature.

One of the major reasons why Poe "is in many ways the most impressive figure in the history of the American short story" (Nagel 2015, 62) has to do with the characteristic combination of his creative works and critical statements: not only are Poe's "tales," as he would call them, still regarded as significant for the historical evolution of the form and worthy of being read on their own merits, but his theoretical account of the short story, developed in essays and reviews, also forms a central event in its history. Yet while Poe's definition is said to be "responsible for the birth of the short story as a unique genre" (Patea 2012, 3), its legacy is "double-edged," sometimes proving "to be more of a curse than a blessing for short story theory" (Basseler 2011, 43, 45). Either way, Poe remains indispensable or even inescapable.

Poe's ubiquity does not make it any easier to pinpoint the precise nature and extent of his contribution to the evolution of the genre, but recent criticism by Terence Whalen, Meredith McGill, Eliza Richards, Jerome McGann, J. Gerald Kennedy, Scott Peeples, and others has opened up a variety of vistas onto his oeuvre. What these critics share is the view of Poe as representative of the antagonisms and synergies of antebellum American print culture (↗4 Textual Materiality) with its characteristic medium of the magazine, a culture he described as tending toward "the curt, the condensed, the pointed, the readily diffused – in place of the verbose, the detailed, the voluminous, the inaccessible" (Poe 1984, 1414–1415). Although originally virtually excluded from the paradigm of the American Renaissance as conceived by F.O. Matthiessen in 1941, Poe plays a larger role in revisionist accounts of the canon, such as David Reynolds's *Beneath the American Renaissance* (2011 [1988]). In *Strange Nation: Literary Nationalism and Cultural Conflict in the Age of Poe* (2016), J. Gerald Kennedy has even suggested we analyze the period of antebellum literature as "the Age of Poe," "not because Poe was the dominant cultural presence [. . .] but because his fierce resistance to nationalistic literature signals his awareness of the potential monstrosity of nationalism itself" (Kennedy 2016, xi). Responding to contemporaries such as Washington Irving (↗5 Irving) and Nathaniel Hawthorne

(↗6 Hawthorne), Poe composed his tales using any texts available in antebellum America's varied print culture.

In total, Poe's published short stories, which make up the core of his oeuvre, number about seventy. Beginning with the Gothic tale "Metzengerstein" (1832) and ending with the satire "X-ing a Paragrab" (1849), Poe's short stories formed part of antebellum print culture and were printed (and reprinted) mainly in periodicals. A towering figure in the history of the Gothic tale, he also pioneered the detective story, the weird tale, and the science fiction narrative, all the while developing a revolutionary rhetorical poetics of the short story that featured the effect on the audience and formal unity as its hallmarks.

In the antebellum period, an American literary tradition was still very much in the making – one needs only to think of Sydney Smith's oft-quoted dictum "who reads an American book?" (Smith 1820, 79) – but by the time Poe was writing, Washington Irving had blazed a trail for writers of short narratives. The short story is "arguably the most important literary genre to have emerged in the United States" (Bendixen 2010, 3). But Poe often, particularly in the early stages of his career, downplayed the Americanness of his writings and preferred to be recognized as a cosmopolitan writer, distancing himself from the contemporary "watchword [of] 'a national literature!' – as if any true literature *could be* 'national' – as if the world at large were not the only proper stage for the literary *histrio*" (Poe 1984, 1027).

There were authors commercially far more successful than Poe, such as James Fenimore Cooper, yet Poe's career can be viewed as representative of the vagaries of the antebellum literary market which he traversed virtually through the places of publication of his texts, but also physically, moving from Richmond to Baltimore to New York to Philadelphia, and back to New York, with stints in Boston as well. Poe's description, in a late critical text, of the industry as the "Magazine Prison-House" (1845) told of the difficulties "poor devil authors" had to face in a system that was still a mixture between an earlier, gentlemanly ideal of the man of letters and a commercialized idea of the writer as a producer of the literary text as a commodity (Poe 1984, 1036; see Hayes 2000, 113).

From early on in his career, Poe reflected on the economic dimensions of his own contributions to the marketplace, arguing that successful magazine fiction consisted in "the ludicrous heightened into the grotesque: the fearful coloured [sic] into the horrible: the witty exaggerated into the burlesque: the singular wrought out into the strange and mystical" (Poe 2008, 84). Rather than directly invoking individual genres such as the tale or the novel, he speaks about what John Frow suggests we call "modes": "the extensions of certain genres beyond specific and time-bound formal structures to a broader specification of 'tone'" (Frow 2015, 71). Hence, Frow argues, "[e]xhausted genres such as the Gothic romance may survive in their modal form – quite spectacularly so in the case of the gothic [sic] mode, which passes through early-Victorian stage melodrama into the stories of Edgar Allan Poe and the novels of Charles Dickens, and thence into the vampire novel, the detective novel, and a number of other narrative genres" (72). Notwithstanding that the history of the Gothic romance might be written in different fashion, Poe's description of "the fearful coloured [sic] into the horrible" certainly fits the Gothic mode perfectly (Poe 2008, 84). And "the witty exaggerated into the burlesque" (84) encapsulates the comic mode of Poe's tales. Taken together, these two modes inform a classification that is historically relevant both for Poe and the antebellum short story. Studying *Blackwood's Edinburgh Magazine* and other periodicals, Poe, in the early 1830s, "undertook an extensive-intensive course of familiarizing himself with what constituted best-selling short fiction, which then highlighted either horrifics, derived from antecedent Gothic tradition, or comic themes, or combinations of humor and horror" (Fisher 2008, 4; see Allen 1969). The Gothic and comic modes extend as far back as to the earliest English and American examples of the short story (as identified by Nagel 2015, 6, 12–13): Daniel Defoe's ghost-story "A True Relation of the Apparition of One Mrs. Veal" (1706) and Benjamin Franklin's satire "The Speech of Polly Baker" (1747).

In retrospect, Poe's theory of the tale has come to define the literary historical moment in which he was writing. It exerted its largest influence on later nineteenth- and early-twentieth-century short stories, but it also harked back to earlier rhetoricians' ideas still present in antebellum literary culture (see Guttzeit 2018, 54–123). In sharp contrast with romantic ideas of original genius, Poe's "Philosophy of Composition" (1846), prioritizing the effect on the audience, stays true to the basic principle of Scottish eighteenth-century rhetorician George Campbell's philosophy of rhetoric: "In speaking there is always some end proposed, or some effect which the speaker intends to produce on the hearer" (Campbell, 1988 [1776], 1). As outlined mainly in three reviews of Nathaniel Hawthorne's collections *Twice-Told Tales* and *Mosses from an Old Manse* in 1842 and 1847, Poe's rhetorical poetics of the short story is based on the author's intended effect on the audience, formal unity, and certain assumptions about audience reception. At the heart of Poe's theory is the "unity of effect or impression" rather than romantic inspiration (Poe 1984, 571). This also leads to a strong emphasis on formal unity, which contributed much to an understanding of the short story as a sophisticated form of art:

> A skilful literary artist has constructed a tale. If wise, he has not fashioned his thoughts to accommodate his incidents; but having conceived, with deliberate care, a certain unique or single *effect* to be wrought out, he then invents such incidents – he then combines such events as may best aid him in establishing this preconceived effect. If his very initial sentence tend [sic] not to the outbringing of this effect, then he has failed in his first step. In the whole composition there should be no word written, of which the tendency, direct or indirect, is not to the one pre-established design. (572)

The tale can have certain effects such as truth, terror, passion, and horror, which, in turn, define certain subtypes: e.g., truth is characteristic of the detective story, or as Poe called it, "the tale of ratiocination" (Poe 1984, 573; see also Redling 2011, 96–100). Since "[a]ll high excitements are necessarily transient," the unity of impression "cannot be thoroughly preserved in productions whose perusal cannot be completed at one sitting" (Poe 1984, 571), which leads Poe to elevate the tale over the traditionally more prestigious epic poem as well as the novel. The short story is a "short prose narrative, requiring from a half-hour to one or two hours in its perusal" (572). Based on considerations of the audience's reception of the literary text, Poe's theory of the tale focuses on the author's rhetorical-poetic strategies, presenting an image of the tale-writer as a "skilful literary artist" (572) and ultimately establishing a new sense of the short story.

2 Close Readings

Despite being rarely anthologized, "The Devil in the Belfry," first published in the Philadelphia *Saturday Chronicle* on 18 May 1839, is significant because it presents an archetypal shape of the tale: in formal terms, it is close to the fairy tale, albeit featuring a satirically exaggerated unhappy ending, and it exemplifies Poe's critical insistence on the importance of conclusions in poetic and narrative texts (Poe 1984, 20). Written in the comic mode, it reduces the potential elements of the short story to the bare minimum of narrator, setting, character, and action, to the arrival of a trouble-making outsider in an idyllic community: at noon, the eponymous devil disturbs the peace of the time-obsessed Dutch villagers of Vondervotteimittiss by ringing the village bell thirteen times, throwing people, animals, and objects into disarray.

The first sentence of the tale contains its essence in a nutshell, as it foreshadows the transformation of the idyllic village about to be related: "Everybody knows, in a general way, that the finest place in the world is – or, alas, *was* – the Dutch borough of Vondervotteimittiss" (Poe 2000, 365; original emphasis). This flaunts the story's chief rhetorical device of hyperbole through its assumption of general knowledge and the noncredible utopian quality it attributes to the village ("Everybody knows"; "the finest place in the world"); it mentions what the story will be about (the events that have transformed the village from the state it "*was*" in to the one it "is" in); and it establishes the constitutive dramatic irony that the inhabitants of Vondervotteimittiss, albeit obsessed with time, constantly need to vonder vot teim itt iss: to *wonder* what time it is.

What is also on full display early in the narrative discourse is the highly self-aware nature of the homodiegetic, presumably male narrator. In his version of a *captatio benevolentiae*, he stylizes himself as a neutrally observing and recording historian, promising to "give a history of the calamitous events [. . .] with all that rigid impartiality, all that cautious examination into facts, and diligent collation of authorities, which should ever distinguish him who aspires to the title of historian" (365). His narratorial presence is quite marked in the beginning – this feature might be linked to a common structure in Poe's tales, which often feature a kind of theoretical introduction, as in "The Murders in the Rue Morgue" (1841, see Poe 2000,

527–531) – and he appears to stay true to his purported ideal of impartiality for most of the tale: at the end, however, he is not only positioned as focalizer, saying that the devil raised "such a clatter that my ears ring again even to think of it" (374); but he also makes a rhetorical rather than poetic appeal to his readership to "[l]et us proceed in a body to the borough, and restore the ancient order of things in Vondervotteimittiss" (374). This narrator is thus not as unreliable as many others in the Poe canon (e.g., the murderer of "The Tell-Tale Heart" 1843), but he is certainly not the neutral historian he purports to be.

The narrator's comically inflected unreliability and the setting of the tale add to a strong sense of parody, serving primarily as a device for Poe to respond in literary fashion to contemporaries such as Washington Irving and the Scottish philosopher and writer Thomas Carlyle. Not only was Irving the first major practitioner of the American short story, but he also exchanged letters with Poe; the rural Dutch setting of the tale presents "a parody of the Communipaw landscape, the people, and the sentimental nostalgia of Irving's *History [of New York]*" (Forbes 1982, 93; on Irving and Poe, see Kennedy 2016, 36–72). Based on its link to Dutch speakers in the United States, the story has also been viewed as a satirical send-up of the Dutch-speaking president Martin Van Buren, who served from 1837–1841 (see Whipple 1956, 81). At the time of writing in 1839 in Philadelphia, Poe was in proximity to the Pennsylvania Dutch, 'Deitsch,' or 'Deutsch,' the religious group of German rather than Dutch ancestry. In *The German Face of Edgar Allan Poe*, Thomas S. Hansen and Burton R. Pollin identify the tale as a "persiflage" of certain Pennsylvania Dutch, as it taps into stereotypes about German dipsomania, consonant-clusters, and sauerkraut (Hansen and Pollin 1995, 58–59). As T.O. Mabbott (see Poe 2000, 363) and others have pointed out, Carlyle's parody of German authors such as Jean Paul in *Sartor Resartus* (1833–1834) provides the model for Poe's satirical remarks on the fictional German scholars such as "Kroutaplenttey" (366).

Another lair of transatlantic intertextual relationships beyond Carlyle extends to Gothic writing from Germany translated via Scotland. "The Devil in the Belfry" relates to a short story from *Blackwood's Edinburgh Magazine* that itself capitalized on the demand for German-styled Gothic stories: Robert Macnish's pseudonymously published "The Barber of Gottingen [sic]" (1826; see Mabbott in Poe 2000, 363). The tale features the visitation of a devil upon the eponymous barber, and Poe might also have taken inspiration from phrases such as "Dunder and blixum" (Macnish 1826, 609; see Poe 2000, 366, 373). German Gothic gloom was relevant for Poe's career beyond this tale, particularly in 1839, as comments by his contemporaries attest. One reviewer pointed out that Poe was "deeply imbued with the spirit of German literature" (qtd. in Thomas and Jackson 1987, 275), and critic James E. Heath maintained that Poe was "too much attached to the gloomy German mysticism" (Heath 1839, 708). Poe himself expressed privately in a letter in November 1839 that "little critics [. . .] would endeavor to put me down by raising the hue & cry of *exaggeration* in style, of *Germanism* & such twaddle," trusting that Washington Irving's praise would dispel that idea (Poe 2008, 202). Publicly, in the same year, Poe's preface to his book *Tales of the Grotesque and Arabesque* (copyright date 1840 but published in 1839) was a defense against the charges of Germanism, in which Poe maintained that "terror is not of Germany, but of the soul" (Poe 2000, 473). In the comic mode of "The Devil in the Belfry," German Gothicism has been domesticated and become more of a garden gnome – the text even features a porcelain gnome-like "China man" on the house's mantelpiece (367) – than a haunting ghost or stalking monster.

The monster figure in the tale is the intruding devil, but he is not a very frightening example of his kind. Neither is he an inspiring revolutionary, at a time when many other romantic devils conjured up in the tradition of Milton's Satan in *Paradise Lost* oscillated between villain and revolutionary, between monster and rebel. This devil's rebellion is hardly Miltonic in scope, overcoming only the belfry-man and grotesquely "beating the devil's tattoo" on the latter's "fat" belly with his "hollow" fiddle (372). Though hardly a round character, the devil is described in some detail. What he shares with the villagers is how extremely small they are, adding to the overall impression of hyperbole – he is described as a "little personage," the adult villagers as "little," and the village boys as "two feet in height." They have "red" faces with "double chin[s]" and "big saucer eyes" (368, 369), but he is "finnicky" with "a long hooked nose, pea eyes," and a wide grin, the rest of his "sinister" face obscured by "mustachios and whiskers" (371). In contrast to the garden-gnome German Americans, his description is made grotesque by being inflected with other, African American and Irish American ethnic stereotypes: he is "foreign-looking," his countenance is "of a dark snuff-color" (371) and he plays "Judy O'Flannagan" and "Paddy O'Raferty" (374) on his fiddle to celebrate his triumph over the villagers. This combination is no accident, as it "illuminates the way that the Irish were coded as African American in the antebellum period" (Goddu 2000, 16; see also Kennedy and Weissberg 2001). The biggest contrast to the villagers lies in the fact that the devil saunters, prances, and dances through the village with no idea of "*keeping time* in his steps" (Poe 2000, 371, original emphasis), which could also be seen as tapping into ethnic stereotypes about punctuality and laziness. In contrast, the villagers are obsessed with time to the degree that they all either hold a watch in one of their hands (see 368, 369) or keep one eye on the "great clock" in the belfry (369); they even attach toy watches to the tails of cats and pigs (see 368, 373). Their relationship to time is what defines both the devil and the villagers.

Their common bond to time makes the defining conflict of the tale even stronger. Vondervotteimittiss defines itself through three laws, all of which are broken by the devil: "'That it is wrong to alter the good old course of things,'" "'That there is nothing tolerable out of Vondervotteimittiss,'" and "'That we will stick by our clocks and our cabbages'" (369). The devil then alters the course of time, having come from beyond the hills, and endangers the villagers' "clocks and cabbages." The latter phrase, in its ticking, virtually clock-like alliteration, encapsulates the constricted time and place in which Vondervotteimittiss is imprisoned.

There is even an allegorical transposition of story time into setting. As Jean-Paul Weber (see Weber 1967, 80–81; see also Mabbott in Poe 2000, 364) has shown, the whole village is shaped like a clock, complete with sixty houses standing for minutes, twenty-four cabbages for hours, and seven faces of the great clock for days in a week; the slim devil overpowering the fat belfry-man at noon mimics the overtaking of the hour-hand by the minute-hand at the full hour. Dennis W. Eddings goes so far as to maintain that the "true devil in the belfry is the clock it houses," as it is worshipped by the villagers (Eddings 1994, 3). In any case, in the world of the burghers, time stands so still that it has congealed and solidified into the landscape, buildings, and people.

The story ostensibly features a thoroughly rural setting, which is reinforced by the picture of family life and the limited preindustrial business of carving that produces – yet again – only "two objects – a time-piece and a cabbage" (Poe 2000, 367). However, the theme of time, and in particular the central position of the great clock, evoke an urban context. One of the potential sources for the tale is the idea that Philadelphians put an "excessive emphasis" on punctuality (Mabbott in Poe 2000, 364), and Poe and his family had moved to the city only the year before, in early 1838. More generally, this connects to the mid-nineteenth century urban practice of clockwatching: in Philadelphia, the State House clock, like other clocks across the antebellum United States, "set time standards" for its inhabitants (McCrossen 2013, 43). Hence, the tale can be seen as evidence of "[h]ow closely Poe had observed the habits of city dwellers anxious to regulate their watches" (43). Rather than being merely a story about preindustrial timelessness, then, the story also engages with the exact time of clocks as a defining moment of capitalist modernity. It shows the interconnectedness of these two processes, undercutting the logic of exceptionalism that the narrator had established early on. In opposition to such idyllic imaginings of the countryside as Irving's, Poe's isolated village is already in the thrall of the modern city, also showing – through its typically Poesque logic of inversion – that the city of Philadelphia, perhaps in contrast to Poe's previous place of living, New York City, remained a village.

This contextual relevance does not exhaust the implications of time in the story, which are also thematic and metafictional. As a result of the contrast between the narrator's self-stylization as a historian and its fairy-tale features, the story also engages with the differences between the linear time of history and the circular time of myth. In turn, this engagement shapes the relation between story time and discourse time. For instance, at the climax of the story, discourse time slows down, as the great clock starts to speak:

> 'One!' said the clock. 'Von!' echoed every little old gentleman in every leather-bottomed arm-chair in Vondervotteimittiss. 'Von!' said his watch also; 'von!' said the watch of his vrow, and 'von!' said the watches of the boys, and the little gilt repeaters on the tails of the cat and pig.
> (Poe 2000, 372)

When the clock strikes thirteen, the "big bell" is fully personified, being referred to as "he" – the same pronoun as the devil: "'*Thirteen*!' said he" (373). That this is the devil's doing is recognized immediately and in unison by the villagers in a passage of 'they-narration': "'Der Teufel!' gasped the little old gentlemen, turning pale, dropping their pipes, and putting down all their right legs from over their left knees" (373). The monstrous devil, through his transgressive doings, has taken over time, and the villagers' reaction turns them into virtual doppelgangers of one another: mirroring each other's movements so exactly that their singular behavior can be narrated as the movement of a collective "they," they become one single audience for the devil's mischievous performance.

What exactly defines the resulting devilish temporality? The devil's time is certainly not regarded as correct by the narrator, as his final plea to restore it makes explicit. Yet as Clayton Marsh emphasizes, "instead of stopping to question the clock's unprecedented declaration of a thirteenth hour, [the villagers] simply rush to make up for lost time, credulous that the twelfth hour has somehow expired without their notice" (Marsh 2005, 269). For Marsh, this turns the devil into "a confidence man whose flamboyant self-fashioning proves to be the ultimate manifestation of the villagers' blind obedience to mechanical sources of time and the speed that such obedience engenders" (266–267). A contrasting account, also based on the figure of the devil, has been put forward by Katrina E. Bachinger, who argues that the devil stands for the artist's struggle against the false promises of historical "progress" (Bachinger 1990, 515, 518). It is striking that the devil's time is doubly associated with music – through his dancing (in which he does not keep time) and through his playing the violin (also not in time). Very closely associated with music, Poe's definition of "the Poetry of words" was "the Rhythmical Creation of Beauty" (Poe 1984, 688). Viewed in this light, the dandy devil turns into much more of an Orphic figure, a caricature of the literary writer whom Poe, at various times, described as the "poor devil author" (1036, 1046). Whereas some of the manipulations of time that the devil works appear to not function (out-of-step dancing and violin-playing), he ultimately achieves a "unity of effect" (571) in his audience by stealing an hour of time from the inhabitants of Vondervotteimittiss, roughly the equivalent – in time – of what Poe suggested an audience would spend in perusing a tale. This can be linked to Poe's biographical entry as a writer into time-obsessed Philadelphia, yet the metafictional elements of the tale suggest a more general reflection on the question of what constitutes tale-telling.

The narrator, in his highly self-aware verbal gestures discussed earlier, also appears to pivot, take a step back and then a step forth as he describes his own approach to the story as a historian and etymologist: this, Bachinger argues convincingly, turns the narrator into "the fiddler's *Doppelgänger*, a second devil, if not in ideas then in words" (Bachinger 1990, 531). The purportedly merely passively observed "reverse" (Poe 2000, 370) of fortune in Vondervotteimittiss is, in fact, thoroughly calculated by the narrator, used as a peripety, and placed at a stage in the narrative that guarantees maximum effect – right after a detailed depiction of the previous "happy estate" of the village (370). The narrator's manipulative narrative devilry is just as essential to the tale as that of the devil himself. Without these two there would be no tale to tell, since the inhabitants of Vondervotteimittiss simply observe their iterative routines or look on. What, or rather who, makes a tale are the devil and the narrator – two figures of Poe's "skilful literary artist" – through their singular act of manipulating the time of the audience, spellbinding them and controlling their soul in the thirteenth hour; or, as Poe wrote later in one of the Hawthorne reviews, "[d]uring the hour of perusal, the soul of the reader is at the writer's control" (Poe 1984, 572).

In contrast to "The Devil in the Belfry," the story of "William Wilson" and his haunting doppelganger is well-known and arguably "one of Poe's greatest achievements in prose" (Mabbott in Poe 2000, 422). It was first published sometime between 1 May 1839 and October 1839 in *The Gift* for 1840 (see Mabbott in Poe 2000, 425; Thomas and Jackson 1987, 272) and has garnered an abundance of critical responses (for a recent overview, see Jones 2019). Like "The Devil in the Belfry," it is linked to Washington Irving: in a letter to Irving on 12 October 1839, Poe explained that "William Wilson" was a response to Irving's article "An Unwritten Drama of Lord Byron," in which Irving had detailed the story of Alfonso and his mortal fight with his second self, "an allegorical being, the personification of conscience, or of the passions" (Irving 1835, 143; see Poe 2008, 199). The story is also, like "The Devil in the Belfry" though in a less direct way, connected to German Gothic via the motif of the doppelganger, for which Germany was "a privileged site of formation around 1800" (Murnane 2013, 173), with the term "Doppeltgänger" coined by Jean Paul in his novel *Siebenkäs* (1796–1797). As we move from the comic mode in "The Devil in the Belfry" to the Gothic mode of "William Wilson," the latter tale also features metafictional elements in a much more sustained narrative, yet again asking questions of what the short story is or can be by constructing the figure of a monster.

The trope of the doppelganger is present in a number of Poe's short stories, though nowhere near as obviously as in "William Wilson," one of the classic instances of the genre (see Engel 2012, 80 and Jones 2019, 249n3). The further development of the Gothic double in Poe has been traced by John T. Irwin to the detective story: "if one had to choose a classical model for that innovation in the Gothic double story that Poe created in moving from 'The Fall of the House of Usher' (1839) and 'William Wilson' (1839) to the first Dupin story, one would immediately think of Oedipus, with his double role as culprit and detective" (Irwin 1994, 201). Primarily a story of personal identity, "William Wilson" also raises complex issues of originality and what it means to tell a tale.

The narrative foregrounds issues of identity and doubling right from its opening sentence: "Let me call myself, for the present, William Wilson" (Poe 2000, 426). While the autodiegetic narrator gives himself a pseudonym for moral and perhaps legal reasons, the sentence is also, through its simple yet carefully crafted syntax, indicative of the excessive egocentrism at the heart of the narrator. As many commentators have noted, his choice of a name is quite telling: taking into consideration the shifting potential meanings of its graphemes and phonemes, this 'I' calls himself 'Will I am, Will's son,' emphasizing a patriarchal passing of direction and power from father to son in a virtually biblical phrase that resounds with alliteration. As the title of the story, the name also stands in a doubling relation, as "William Wilson" is, as Nicholas Royle argues, "a text about a double with a title that is thus double, narrated moreover by one who describes his name as a 'fictitious title' [Poe 2000, 431]" (Royle 2003, 190). The opening of the story establishes the oppositional duality of will and conscience, especially when taken together with the motto: "What say of it? What say of conscience grim, / That spectre in my path?" (Poe 2000, 426) (in all likelihood invented by Poe; see Hayes in Poe 2015, 117). The motto also sets the tone for the story and foreshadows its monstrous aspects, obliquely referring to the theme of the doppelganger through its invocation of a ghost. Beyond these thematic and generic aspects, the opening also establishes, in the briefest of ways, a narratee ("Let") and thus connects to the audience in a kind of conversational immediacy. This is reinforced through its emphasis on the "present" act of narration, even while the narrator is writing his story down on the "fair page now lying before me" (426). It thus goes to illustrate the formal demands of the short story – totality, brevity, and economy of words – almost to the point of terseness – exemplifying Poe's definition of the tale and his depiction of the antebellum era as an age of "the curt, the condensed, the pointed" (Poe 1984, 1414–1415).

Despite its brevity, the narrative traces typical steps of a bildungsroman, moving from the Reverend Dr. Bransby's school (modelled on the school Poe attended in Stoke Newington, near London, in his adolescence) to Eton, Oxford, and, in a sort of grand tour across Europe, to Rome. In all these places, the narrator William Wilson is hindered by his doppelganger, a schoolmate also called William Wilson, from committing increasingly immoral acts and crimes: from practical jokes to excessive drinking and swearing to cheating at cards to the seduction of a married woman, only for the second William Wilson to be killed by the first William Wilson.

As "a figure of identity as/in crisis," the motif of the doppelganger generally "renders problematic any fixed sense of individuality or subjectivity" (Murnane 2013, 173), and it is such issues that are dramatized in the relationship between the two main characters in "William Wilson" (to whom I will be referring in the following as Wilson 1, the narrator, and Wilson 2, the doppelganger). The story is most often read as an allegory that makes use of the fantastic: the motif of the doppelganger allegorizes the inner voice of the conscience, Socrates's daimon so to say, as Wilson 2 speaks with "a weakness in the faucial or guttural organs, which precluded him from raising his voice at any time *above a very low whisper*" (Poe 2000, 433; original emphasis). It should be stressed, however, that the story can also be taken literally: as Wittgenstein's concept of family resemblance reminds us, Wilson 1 and 2 might simply be relatives. For one, Kevin J. Hayes points out that the name 'William Wilson' "was quite common in Poe's day" in Philadelphia (Hayes in Poe 2015, 118). Secondly, the case for the two William Wilsons being actual relatives has been made by Lynn Langmade in light of the Minnesota Study of Twins Reared Apart, which showed seemingly unbelievable similarities between twins separated at birth; arguing against "singleton readings" of the story, she points out that "[s]eparated twins [. . .] may not only reunite quite unexpectedly under bizarre circumstances but also lead, inexplicably, 'twin' lives" (Langmade 2012, 5). A literal reading is certainly not favored by the majority of critics; nonetheless, the contrast between the allegorical and the literal, between the fantastic and the realistic, is pertinent with regard to Poe's poetics.

If regarded as an aspect of narrative technique, this ambiguity between the fantastic and the realistic can be explained as a result of Poe's use of what Freud would, eight decades later in 1919, call "the uncanny," *das Unheimliche*, one of the most influential terms in Gothic studies. As Freud originally defined it, "the uncanny is that class of the frightening which leads back to what is known of old and long familiar" (Freud 1999, 219). Jonathan Elmer rightly calls "William Wilson" Poe's "great tale of the uncanny" (Elmer 1995, 56), and the uncanniness of the relationship between the two Wilsons encompasses a long list of features; even leaving out their same birth date and same height, the narrator lists the similarities in an emotionally loaded parallelism: "The same name! the same contour of person! the same day of arrival at the academy!" (Poe 2000, 437). Anticipating Freud's insistence on the importance of early childhood, the narrator mentions "a something" in his double that is "bringing to mind dim visions of my earliest infancy" (436). Interpreted as a personification of conscience, Wilson 2 can easily be identified with the Freudian superego. Two scenes lend themselves particularly well to a reading in terms of Lacanian psychoanalysis and the concept of the mirror stage, the crucial phase in identity formation when one, as a child, identifies with one's mirror image, thus beginning to gain an (illusionary) understanding of oneself as an autonomous subject (for a Lacanian reading of the story, see Jung 2000): when Wilson 1 attempts to play a practical joke on Wilson 2, while the latter is asleep, he is so "[a]we-stricken" at "*what I now saw*" – presumably Wilson 2's "countenance" – that he immediately leaves the school, never to return (Poe 2000, 437; original emphasis). Similarly, at the death of Wilson 2, a "large mirror, – so at first it seemed to me in my confusion – now stood where none had been perceptible before" (447), as Wilson 2 maintains that "*in my death, see by this image, which is thine own, how utterly thou hast murdered thyself*" (448; original emphasis).

The uncanniness of the double extends to many other elements of the story. On the pulpit, the reverend Dr. Bransby has a "countenance so demurely benign," yet as a schoolmaster, he subjects the pupils to corporal punishment "with sour visage," "a paradox [. . .] too utterly monstrous for solution" (428–429). Wilson 1's opponent at the card game in Oxford proposes "to double our already extravagant stakes" (441). What is more, the instances of doubling also extend to the level of style, most clearly in the re-doubled description of the lethal stabbing at the end: "repeatedly through and through" (447), but also in the anaphora of the motto: "What say of it? What say of conscience grim[?]" (426), and in the alliterative description of Wilson 1 being "doubly disgusted" by the arrival of Wilson 2 at the school (434). This extends to full sentences such as "*I fled in vain*," which is repeated on different pages of the story (444, 445). Fleeing in vain from his double, all the narrator can do is to repeat himself in what is yet another act of doubling.

As Paul Christian Jones reminds us, invoking the later examples of Robert Louis Stevenson's Mr. Hyde and *Fight Club*'s Tyler Durden, Poe's "handling of the double's function of depicting human minds as conflicted sites of internal struggles differs significantly from other authors' handling of the device, wherein the double is more commonly manifested as a darker version of the protagonist" (Jones 2019, 237). Rather than projecting physical and moral monstrosity onto the double Wilson 2, the story shows Wilson 1 to be monstrously transgressive as he is also imbued with characteristics of the Gothic villain, such as mania in the family, egocentrism, and the pursuit of innocent women (Poe 2000, 427, 446). In contrast to the devil in the belfry, a comically presented, humanized monster, Wilson 1 is an example of the "monstering of humanity" (Botting 2014, 47), a process that originates in the nineteenth century (see Guttzeit and Bekhta 2019, 6): Wilson 1, seemingly more human than his uncanny doppelganger, is actually the more monstrous of the two. In turn, the monstrous doppelganger himself is humanized when he dies, no longer a "spectre" (Poe 2000, 426) but a bodily human, "no longer" speaking "in a whisper" (448).

In these negotiations of humanity and monstrosity, the story repeatedly foregrounds its own fictional character, especially by allusions to drama. For instance, Wilson 1 remarks that the narrative is nearing "the last eventful scene of the drama" and speaks of it as "the tragedy" (446, 438). He also describes Wilson 2 as an actor who "perfect[s] an imitation of myself [. . .] in words and in actions" (434). The tale thus forms "a self-remarking fiction, a fiction in abyss" (Royle 2003, 190). Royle suggests we read the uncanny story in the comic mode because of its combination of "absurdity" and "implacable earnestness" (190; see also Tally 2014, 120). However, it is precisely the metafictional elements that establish the sense of the Gothic. In particular, the first section of the story, which is set in the schoolhouse, can be seen as a microcosm of the Gothic mode of the short story. The schoolhouse has such a complex, palace-like layout that it becomes difficult "to say with certainty upon which of its two stories one happened to be" (429). This metafictional remark is reinforced in the narrator's depiction of the schoolboys' impression of the mansion as "not very far different from [. . .] infinity" (429). The enclosed space – the grounds of the school are encompassed by a wall "topped with a bed of mortar and broken glass" (428) – is oppressively narrow and constricting but also simultaneously offers "in every creak of its mighty hinges [. . .] a plenitude of mystery" (429). The textual space defined by the story conjures up a fictional "world" of its own (429, 430, 431), complete "with pointed Gothic windows" and "terror-inspiring angle[s]" (430), revisiting the intricate relationship between Gothic architecture and Gothic fiction. While the tale is thoroughly immersed in its uncanny setting, its technique – as in so many later examples of the short story – functions as a "glimpse," to use Katherine Mansfield's formulation (Mansfield 1997, 209; see also McSweeney 2007), into a much larger world shifting between the fantastic and the realistic.

These metafictional elements of the text can also be fruitfully read in conjunction with contemporary print culture or, as Meredith McGill characterized it, "the culture of reprinting" (McGill 2003). Practices of unauthorized reprinting were ubiquitous in the transatlantic antebellum marketplace, in which "[c]irculation outstripped authorial and editorial control" (2). In light of Poe's own (inadvertently self-contradictory) accusation that Nathaniel Hawthorne had plagiarized the ending of "William Wilson" (1839) in Hawthorne's *earlier* short story "Howe's Masquerade" (1838) (of which Poe reviewed the 1842 publication in *Twice-Told Tales*), Jonathan Elmer points out that "William Wilson" enacts Poe's 'perverted' logic of the ascription of plagiarism. This turns "the origin into a repetition of the end, makes the 'original' look merely like a misrecognized 'double'" (Elmer 1995, 75), mirroring the process which "the poet unconsciously performs when he takes as 'his own' a phrase by which he had earlier been 'possessed'" (79). Elmer's argument can be extended and modified if we consider the full implications of the contemporary culture of reprinting: Wilson 1 calls Wilson 2 a "copyist, who, disdaining the letter" gives "the full spirit of the original" (Poe 2000, 435), thus highlighting the anxiety of being plagiarized in a print culture where value is determined by both originality and reprints. Rather than being the conscience of the narrator, then, Wilson 2 is a walking reprint, viewed by Wilson 1 as a plagiarism: an illegitimate, monstrous repetition of himself. The original is haunted by the copy until, at the very end of the tale, he kills him, ending his own moral life, too. Here, the reprint is delineated as essential to the original – that is, stated quite plainly, the author needs to be copied to survive. That Wilson 1 is surprised by his own death of the author shows that he has not mastered the transatlantic culture of reprinting: he attempts to but cannot control the uncontrollable terror of textual circulation.

3 Theoretical Perspectives

Poe's reception has many faces. Even though numerous American and British authors such as William James, T.S. Eliot, and Harold Bloom expressed doubts as to Poe's canonicity, the impact of his critical texts, short stories, and other writings has been felt globally. Much of the canonical ambivalence surrounding Poe in Anglophone contexts – Eliot wrote of him as of a haunting ghost, saying that "one cannot be sure that one's own writing has not been influenced by Poe" (Eliot 1949, 327) – stems from his popularity in a two-fold sense: his use of generic elements of popular literature and his success with mass readerships, highly visible even today, e.g., in the naming of Baltimore's American football team for Poe's poem "The

Raven" (1845). His reception in languages other than English began in 1848, when an imitation of "William Wilson" was published in France (see Vines 1999, 9), and it has extended from Europe to South America and Japan, turning Poe into a figure of world literature.

With regard to the reception of his short stories and short story theory, Poe's momentous impact on American Gothic is hard to overestimate, as is attested by the near synonymy of the term with Poe's best-known tale "The Fall of the House of Usher" (1839, see Perry and Sederholm 2009). His ratiocinative tales about the French private detective Chevalier C. Auguste Dupin established that combination of mysterious crime, homodiegetic narrator, brilliant investigator, and surprising dénouement that Arthur Conan Doyle would later imitate in his Sherlock Holmes stories and that has served as the model for countless other narratives. It is certainly too much to ask of Poe's definition of the tale that it accommodate the variety of literary texts nowadays habitually subsumed under the term "short story." Yet it covers many "well-made" stories written before and after 1900 (see March-Russell 2009, 35), particularly after it had been discussed and formalized in Brander Matthews's *The Philosophy of the Short Story* (1885/1901). For Young and Bailey (2015), Poe is the spearhead of the first of three waves of short story theory that lasted until the 1950s.

Moving from Poe's reception to the theoretical approaches that his short stories have attracted, it is useful to look back on the tales closely read here: "The Devil in the Belfry" has been the subject of more traditional, in particular author-centered methodologies, chief among them influence studies, whereas the reception of "William Wilson" is arguably representative of the development of Poe studies overall. From early on, readings are dominated by psychoanalytic models of the psyche, be they Freudian, Jungian, or Lacanian, a testament to Poe's position as the American short story author most closely intertwined with the history of psychoanalysis. Traced in Scott Peeples's definitive survey of Poe scholarship *The Afterlife of Edgar Allan Poe* (2004), Poe studies moved away from psychoanalysis, first to formalism, then to deconstructionism (see Muller and Richardson's 1988 edition of the famous Lacan-Derrida-Johnson debate on Poe's 1844 Dupin story "The Purloined Letter"), and on to what is still the current phase: socio-historical approaches focusing on questions of gender, race, and class as well as print culture and the nation. Correspondingly, the image of the doppelganger William Wilson changed over the years: from "another ego" on whom the narrator, out of guilt, places responsibility for his actions (Rank's psychoanalytic view 1971 [1914], 76) to "the Romantic individualist for whom the world is nothing but the externalization of the self" (Davidson's formalist view 1957, 201) to the "written self" that displaces "his writing self" (Williams's deconstructionist view 1988, 44). While the doppelganger motif – as suggested in these interpretations – certainly dramatizes an interior conflict, these insights need to be revisited in light of newer paradigms such as antebellum print culture, the literary marketplace, and the nation.

More recent publications on "William Wilson" have reread the story in terms of race and sexuality: as exposing the limits of personhood under the conditions of antebellum law that ideologically construed slaves as things (see Dayan 2001, 114–118) and as the "embodiment of [the] repression" of "[t]he possibility of homoerotic relationship" (Person 2008, 19). Wilson's description of his own name as "the common property of the mob" (Poe 2000, 431) can be connected to aspects of personal identity (see Elmer 1995, 83) or politics and democracy (see e.g., Britt 1995). Wilson's double identity, Paul Giles suggests (see Giles 2001, 189), symbolizes the United States' struggle for cultural self-affirmation after the Declaration of Independence from the United Kingdom, as the story, with an English setting and an American publication venue, makes prominent use of the semantic field of "constitution" (both "physical" and "political"). The "outré" quality of many of Poe's writings, then, shows him to be not just a member but a representative of the "strange nation" (Kennedy 2016) that the United States – politically, yet not culturally independent – represented in the antebellum period.

4 Bibliography

4.1 Works Cited

Allen, Michael L. *Poe and the British Magazine Tradition*. New York: Oxford UP, 1969.

Bachinger, Katrina E. "The Aesthetics of (Not) Keeping in Step: Reading the Consumer Mobocracy of Poe's 'The Devil in the Belfry' Against Peacock." *Modern Language Quarterly: A Journal of Literary History (MLQ)* 51.4 (1990): 513–533.

Basseler, Michael. "Theories and Typologies of the Short Story." *A History of the American Short Story: Genres – Developments – Model Interpretations*. Eds. Michael Basseler and Ansgar Nünning. Trier: WVT, 2011. 41–64.

Bendixen, Alfred. "The Emergence and Development of the American Short Story." *A Companion to the American Short Story*. Eds. Alfred Bendixen and James Nagel. Hoboken, NJ: Wiley-Blackwell, 2010. 3–19.

Botting, Fred. *Limits of Horror: Technology, Bodies, Gothic*. Manchester: Manchester UP, 2014.

Britt, Theron. "The Common Property of the Mob: Democracy and Identity in Poe's 'William Wilson.'" *Mississippi Quarterly* 48.2 (1995): 197–210.

Campbell, George. *The Philosophy of Rhetoric*. 1776. Ed. Lloyd F. Bitzer. Delmar, NY: Scholars' Facsimiles & Reprints, 1988.

Davidson, Edward Hutchins. *Poe: A Critical Study*. 1957. Cambridge: Belknap P of Harvard UP, 1973.

Dayan, Colin (Joan). "Poe, Persons, and Property." *Romancing the Shadow: Poe and Race*. Eds. J. Gerald Kennedy and Liliane Weissberg. Oxford: Oxford UP, 2001. 106–126.

Eddings, Dennis W. *Poe's Tell-Tale Clocks*. Baltimore: The Edgar Allan Poe Society of Baltimore, 1994.

Eliot, T.S. "From Poe to Valéry." *Hudson Review* 2.3 (1949): 327–342.

Elmer, Jonathan. *Reading at the Social Limit: Affect, Mass Culture, and Edgar Allan Poe*. Stanford, CA: Stanford UP, 1995.

Engel, William E. *Early Modern Poetics in Melville and Poe: Memory, Melancholy, and the Emblematic Tradition*. Farnham, Surrey: Ashgate, 2012.
Fisher, Benjamin F. *The Cambridge Introduction to Edgar Allan Poe*. Cambridge: Cambridge UP, 2008.
Forbes, Christopher. "Satire of Irving's A History of New York in Poe's 'The Devil in the Belfry.'" *Studies in American Fiction* 10.1 (1982): 93–100.
Freud, Sigmund. "The Uncanny." *The Standard Edition of the Complete Psychological Works of Sigmund Freud*. 1919. Ed. James Strachey. London: Hogarth P, 1999. 217–256.
Frow, John. *Genre*. 2nd ed. Abingdon, UK: Routledge, 2015.
Giles, Paul. *Transatlantic Insurrections: British Culture and the Formation of American Literature, 1730–1860*. Philadelphia: U of Pennsylvania P, 2001.
Goddu, Teresa A. "Rethinking Race and Slavery in Poe Studies." *Poe Studies/Dark Romanticism* 33.1–2 (2000): 15–18.
Guttzeit, Gero. *The Figures of Edgar Allan Poe: Authorship, Antebellum Literature, and Transatlantic Rhetoric*. 2017. Berlin: De Gruyter, 2018.
Guttzeit, Gero, and Natalya Bekhta. "Introduction." *Monsters and Monstrosity in Nineteenth-Century Anglophone Literature*, special issue of *Anglistik* 30.3 (2019): 5–16.
Hansen, Thomas S., and Burton R. Pollin. *The German Face of Edgar Allan Poe: A Study of Literary References in his Works*. Columbia, SC: Camden House, 1995.
Hayes, Kevin J. *Poe and the Printed Word*. Cambridge: Cambridge UP, 2000.
Heath, James E. "The 'Knickerbocker,' and the 'Gentleman's Magazine.'" *Southern Literary Messenger* 5.10 (October 1839): 708.
Irving, Washington. "An Unwritten Drama of Lord Byron." *The Knickerbocker* 6.2 (August 1835): 142–144.
Irwin, John T. *The Mystery to a Solution*. Baltimore, MD: Johns Hopkins UP, 1994.
Jones, Paul Christian. "Counterparts: Poe's Doubles from 'William Wilson' to 'The Cask of Amontillado.'" *The Oxford Handbook of Edgar Allan Poe*. Eds. J. Gerald Kennedy and Scott Peeples. New York: Oxford UP, 2019. 235–251.
Jung, Yonjae. "The Imaginary Double in Poe's 'William Wilson.'" *Lit: Literature Interpretation Theory* 11.4 (2000): 385–402.
Kennedy, J. Gerald, and Liliane Weissberg, eds. *Romancing the Shadow: Poe and Race*. Oxford: Oxford UP, 2001.
Kennedy, J. Gerald. *Strange Nation: Literary Nationalism and Cultural Conflict in the Age of Poe*. New York: Oxford UP, 2016.
Langmade, Lynn. "The Wilson Duplex: Corporatism and the Problem of Singleton Reading in Poe's 'William Wilson' (or, Why Can't You See Twins?)." *Poe Studies* 45.1 (2012): 5–39.
Macnish, Robert. "The Barber of Gottingen." *Blackwood's Edinburgh Magazine* 20 (October 1826): 604–610.
Mansfield, Katherine. *The Katherine Mansfield Notebooks*. Ed. Margaret Scott. 2 vols. Canterbury, NZ: Lincoln UP, 1997. 2.
March-Russell, Paul. *The Short Story: An Introduction*. Edinburgh: Edinburgh UP, 2009.
Marsh, Clayton. "Stealing Time: Poe's Confidence Men and the 'Rush of the Age.'" *American Literature* 77.2 (2005): 259–289.
McCrossen, Alexis. *Marking Modern Times: A History of Clocks, Watches, and other Timekeepers in American Life*. Chicago: U of Chicago P, 2013.
McGill, Meredith L. *American Literature and the Culture of Reprinting, 1834–1853*. Philadelphia: U of Pennsylvania P, 2003.
McSweeney, Kerry. *The Realist Short Story of the Powerful Glimpse: Chekhov to Carver*. Columbia: U of South Carolina P, 2007.

Muller, John P., and William J. Richardson, eds. *The Purloined Poe: Lacan, Derrida & Psychoanalytic Reading*. Baltimore, MD: Johns Hopkins UP, 1988.
Murnane, Barry. "Doppelganger." *The Ashgate Encyclopedia of Literary and Cinematic Monsters*. Ed. Jeffrey Andrew Weinstock. Farnham, Surrey: Ashgate, 2013. 172–177.
Nagel, James. *The American Short Story Handbook*. Hoboken, NJ: Wiley-Blackwell, 2015.
Patea, Viorica. "The Short Story: An Overview of the History and Evolution of the Genre." *Short Story Theories: A Twenty-First Century Perspective*. Ed. Viorica Patea. Amsterdam: Rodopi, 2012. 1–24.
Peeples, Scott. *The Afterlife of Edgar Allan Poe*. Rochester, NY: Camden House, 2004.
Perry, Dennis R., and Carl Hinckley Sederholm. *Poe, 'The House of Usher,' and the American Gothic*. New York: Palgrave Macmillan, 2009.
Person, Leland S. "Queer Poe: The Tell-Tale Heart of His Fiction." *Poe Studies* 41.1 (2008): 7–30.
Poe, Edgar Allan. *Essays and Reviews*. Ed. G. R. Thompson. New York: Library of America, 1984.
Poe, Edgar Allan. *Tales and Sketches. Volume 1: 1831–1842*. 1978. Ed. Thomas Ollive Mabbott. Urbana: U of Illinois P, 2000.
Poe, Edgar Allan. *The Annotated Poe*. Ed. Kevin J. Hayes. Cambridge, MA: Belknap P of Harvard UP, 2015.
Poe, Edgar Allan. *The Letters of Edgar Allan Poe*. Eds. John Ward Ostrom, Burton Ralph Pollin, and Jeffrey A. Savoye. 3rd ed. Staten Island, NY: Gordian, 2008. 1.
Rank, Otto. *The Double: A Psychoanalytic Study*. 1914. Ed. Harry Tucker. Chapel Hill: U of North Carolina P, 1971.
Redling, Erik. "Tales of Detection and Deterioration: Edgar Allan Poe's 'The Murders in the Rue Morgue' and 'The Fall of the House of Usher.'" *A History of the American Short Story: Genres – Developments – Model Interpretations*. Eds. Michael Basseler and Ansgar Nünning. Trier: WVT, 2011. 95–108.
Royle, Nicholas. *The Uncanny*. Manchester, UK: Manchester UP, 2003.
Smith, Sydney. "Review of *Statistical Annals of the United States* by Adam Seybert." *Edinburgh Review* 33 (1820): 69–80.
Tally, Robert T. *Poe and the Subversion of American Literature: Satire, Fantasy, Critique*. New York: Bloomsbury Academic, 2014.
Thomas, Dwight, and David K. Jackson. *The Poe Log*. Boston: Hall, 1987.
Vines, Lois Davis. "Poe in France." *Poe Abroad: Influence, Reputation, Affinities*. Ed. Lois Davis Vines. Iowa City: U of Iowa P, 1999. 9–18.
Walker, Ian. "The Poe Legend." *A Companion to Poe Studies*. Ed. Eric W. Carlson. Westport: Greenwood P, 1996. 19–42.
Weber, Jean-Paul. "Edgar Poe or the Theme of the Clock." *Poe: A Collection of Critical Essays*. Ed. Robert Regan. Englewood Cliffs, NJ: Prentice-Hall, 1967. 79–97.
Whipple, William. "Poe's Political Satire." *The University of Texas Studies in English* 35 (1956): 81–95.
Williams, Michael J. *A World of Words: Language and Displacement in the Fiction of Edgar Allan Poe*. Durham, NC: Duke UP, 1988.
Young, Emma, and James Bailey. "Introduction." *British Women Short Story Writers: The New Woman to Now*. Eds. Emma Young and James Bailey. Edinburgh: Edinburgh UP, 2015.

4.2 Further Reading

Kennedy, J. Gerald, and Jerome J. McGann, eds. *Poe and the Remapping of Antebellum Print Culture.* Baton Rouge: Louisiana State UP, 2013.

Kennedy, J. Gerald, and Scott Peeples, eds. *The Oxford Handbook of Edgar Allan Poe.* New York: Oxford UP, 2019.

Quinn, Arthur Hobson. *Edgar Allan Poe: A Critical Biography.* 1941. New York: Appleton-Century, 1942.

Reynolds, David S. *Beneath the American Renaissance: The Subversive Imagination in the Age of Emerson and Melville.* 1988. New York: Oxford UP, 2011.

Richards, Eliza. *Gender and the Poetics of Reception in Poe's Circle.* Cambridge: Cambridge UP, 2004.

Savoye, Jeffrey A. *The Collected Works of Edgar Allan Poe: A Comprehensive Collection of E-Texts. The Edgar Allan Poe Society of Baltimore*, https://www.eapoe.org/works/index.html. 1997. (26 February 2021).

Whalen, Terence. *Edgar Allan Poe and the Masses: The Political Economy of Literature in Antebellum America.* Princeton, NJ: Princeton UP, 1999.

Damien B. Schlarb

8 Herman Melville (1819–1891)

Abstract: Known primarily for his novels, Herman Melville also brought to bear his social criticism, wide intellectual and historical compass, and a sardonic sense of humor on the short-fiction genre. The two stories discussed here ponder American techno-modernity and an emerging globalized market economy informed by gender and class divisions. The two-parted story "The Paradise of Bachelors and The Tartarus of Maids" contrasts two scenes of leisure and labor and portrays the compartmentalizing thinking behind transnational capitalism and how it differentiates people by gender, class, and region. "The Lightning-Rod Man" satirizes nineteenth-century American valorizations of science (scientism) and traces the roots of that fervor to the religious revival movements of the day. On the surface, the story allegorizes the emerging rift between science and religion, mass culture, and democratic individualism. Yet Melville shows how, underneath a veneer of cultural competition, religion and technology share discursive and institutional histories.

Keywords: technology, modernity, allegory, transatlantic diptych, religion, scientism, transnational capital, gender

1 Context

On a winter evening in late November 1851, Herman Melville sat down at his desk to reply to a letter he had received the previous day from his friend and mentor Nathaniel Hawthorne (↗7 Hawthorne). Melville's sixth novel, *Moby-Dick*, had just been published, and Hawthorne had praised the book as a major accomplishment, a sentiment that unfortunately only few contemporary readers would share. Sitting at his desk, Melville felt a giddy rush of creative energy at having won the approval of his friend with his "wicked book" (Melville 1993, 212). "I should have a paper-mill established at one end of the house," he exuberantly writes, "and so have an endless riband of foolscap [writing paper] rolling in upon my desk; upon that endless riband I should write a thousand – a million – billion thoughts, all under the form of a letter to you" (213). He fantasized about a writing machine that marries the tireless productivity of an automated paper mill with an equally tireless wellspring of creative inspiration. In this vision, technology and creativity co-create a device of perennial artistic production, in which an endless stream of paper meets a perpetual stream of affectionate, deep thought. Melville and Hawthorne's friendship would not last a decade, and neither would Melville's enthusiasm for the union of mechanical and creative production. Years later, the image of the paper mill reappears in his short fiction. This time, however, the paper mill does not signify a utopian vista

https://doi.org/10.1515/9783110587647-009

of perennial artistic productivity but conjures a grim image of the human cost of an industrialized market economy.

"The Paradise of Bachelors and The Tartarus of Maids" juxtaposes two allegorical scenes: a lavish dinner banquet at an all-male lawyer fraternity at Temple Bar in London and a dour American paper mill populated by an all-female workforce. The two episodes form what Harrison Hayford called in his original edition of Melville's stories a "transatlantic diptych," a two-parted story that indicts various social ills by contrasting two scenes and moralizing that contrast. The story discusses how strands of gender, technology, and socio-economic power relations intertwine two seemingly distinct social milieus. Melville would also publish a scathing indictment of U.S.-American techno optimism with "The Lightning-Rod Man," a story about a travelling salesman who uses fear and quasi-religious fervor to bully his customers into buying his wares. Both stories comment on various crises of modernity: gendered socio-economic inequality, the human cost of industrialization, the contingencies of human communication, and the efficacy of religious language and belief in modern secular society. They juxtapose the hypocrisy of a self-deluding middle class with the suffering of lower-class people, especially women. Time and again, Melville argues that one must confront the anxieties and complexities of modernity head-on. He satirizes ready-made solutions provided by normative belief systems like religious dogma or scientism. In his fiction, Melville lionizes personal independence and agency but acknowledges that both require fierce intellectual commitment to truth-seeking, even at the risk of discovering unpleasantness, e.g. how one's own comfort depends on the suffering of others.

Melville experienced economic hardship first-hand, when his father's failed business ventures forced the family into bankruptcy. Herman had to leave school at age thirteen, taking jobs as a sailor and schoolteacher while writing. The experience made a lasting impression on young Herman. In *Moby-Dick* he would have Ishmael muse "a whale-ship was my Yale College and my Harvard" (Melville 2001, 112). After early successes with adventure novels like *Typee* (1846) and *Omoo* (1847), his career took a downturn. On December 10, 1853, disaster struck when a warehouse of Harper and Brothers that contained unsold copies of his books burned down. The Harpers began billing authors for reprints during that time, which severely reduced his income (Delbanco 2006, 223). Melville had initially scoffed at the idea of writing for American magazines (Sealts 1987, 482), but with debts looming – he owed over $7,000 in loans – he could no longer afford to forego the financial opportunities magazines presented (Delbanco 2006, 141). By the 1850s, the magazine market was booming and skilled authors could earn quick cash compared to the tedious business of procuring advances on novels. In addition, several newer magazines began featuring American authors instead of plagiarizing British writers, a common practice in those days due to the absence of international copyright laws. The catch was that magazines were a disposable medium, requiring a constant supply of new product.

Lasting about three years (1853–1856), Melville's foray into magazine fiction was brief compared to his literary activities as a novelist (1848–1854) and poet (ca. 1860–1891). Yet he left an indelible mark on the genre. Texts like "Bartleby; the Scrivener" (1853), the novella *Benito Cereno* (1855), and the posthumously published, yet unfinished *Billy Budd* (1924) are today considered masterpieces of American literature that have even garnered readerships in disciplines like law and philosophy. Still, Melville's stories remain underexplored in terms of their commentary on the emerging global economy, American technophilia, the magazine market, and material print cultures (↗4 Textual Materiality). "The Lightning-Rod Man" appeared in *Putnam's* in August 1854 and fetched $18.00 (Hayford 1987, 598). In December 1854, "Paradise of Bachelors and the Tartarus of Maids" was published in *Harper's Monthly Magazine*, by then an institution on the magazines market. The story was part of a convolute Melville sold to *Harper's* for $100 (Sealts 1987, 484–485). However, his move to short fiction was not met with universal praise. By early 1855, Melville's long-time editor and friend George W. Curtis condemned his latest short-fiction offerings, commenting that Melville "does everything too hurriedly now" and that his short prose lacked the quality of his earlier novels (in Delbanco 2006, 229). Melville's output during that time (sketches, short stories, diptychs, etc.) shows his penchant for experimentation with literary form. The reputation of his short prose within literary studies has since mended.

The two stories under discussion in this chapter tell us something about Melville's formal, thematic, and moral commitments, specifically his attachment to literary allegory – a preference he shared with other romanticist writers – and his interest in social criticism. That being said, firmly situating Melville in any one literary period or movement is difficult. As this chapter will show, although Melville's short stories display a characteristically romanticist trust in the world-constituting power of language, they also evince elements of literary journalism, extensive intertextual references, and meta-commentary on language, all markers of literary realism and even modernism.

2 Close Readings

The nineteenth century saw an unprecedented euphoria and anxiety about technology. The steam engine (ca. 1712), the cotton gin (1794), the electric telegraph (ca. 1830), the dobby loom (1843), all accelerated production, trade, and travel and sparked the cultural imagination. Industrialization and automation altered notions of time and space. Life itself seemed to speed up at an exhilarating rate. People, goods, and information began to move more rapidly across the country and between continents. Travel by steamer galvanized trade, migration, and recreational travel for those who could afford it. Technological innovations like the lightning rod could be mass-produced and quickly disseminated to customers in remote areas. Edifices like the lightning rod became symbols of human ingenuity and its ability to dominate nature's contingent forces, fostering widespread scientific and technological optimism. Melville eyed critically the quasi-religious fervor with which some valorized these gadgetries.

Industrialized capitalism, it turns out, was indeed a double-edged sword. Eli Whitney's cotton gin, for example, increased productivity but it also fueled the demand for cheap slave labor to pick cotton at an accelerated pace. The "wage slavery" of Northern factories coincided with the horrors of Southern plantation slavery. Those who resided lowest on the socioeconomic totem pole on either side of the Mason-Dixon Line – lower-class children, women, and people of color – bore the brunt of the detrimental effects of capitalism, and literature reflected these ambivalences and discontents. The plight of working women depicted in the "The Paradise of Bachelors and the Tartarus of Maids" (henceforth abbreviated "Paradise" / "Tartarus") foregrounds social inequality as the dark underbelly of both industrialized market capitalism and the fervor of scientism that the lightning-rod man peddles.

The "Factory girls," or "mill girls," in "Tartarus" exemplify the detrimental effects of commodity fetishism as well as the psychological and philosophical implications of industrialization that Karl Marx would diagnose years later. All these effects emerged more clearly as the century rolled on. Urban lower-class women, especially recent immigrants, often found themselves without financial resources, outside sweatshop labor as seamstresses or mill workers. They could not legally own property and had to rely on their husbands or fathers to make a living. Even middle- and upper-class women, although relatively independent and educated, could find themselves entrapped in loveless and even abusive marriages. American cultural history perpetuated this state of affairs by painting an ambivalent picture of women's inner mental and emotional lives: Habitually characterized as feebleminded and childlike, they were simultaneously lionized for their alleged natural moral goodness and morally regenerative influence on men. Women were thus both economically disenfranchised and symbolically over-determined as ethereal, sensual beings in need of constant moral supervision and discipline. This rhetorical double bind served to cordon them off from the supposedly morally corrosive atmosphere of the workplace and the ballot box and relegated them to the home and the schoolroom.

Some middle- and upper-class women circumvented these social and legal restrictions on their professional and intellectual mobility by engaging in social-reform advocacy (e.g. temperance), event planning (author carnivals), and literary authorship. Working women lacked such options and therefore often figure less prominently in our literary and cultural histories of the antebellum period, despite their contributions. Howard Zinn emphasizes how lower-class women of all ethnicities significantly shaped the workers' rights movement in America during the late-nineteenth and early-twentieth centuries, while enduring open hostility and violence

(Zinn 2015, 275, 327). Most canonized literary fiction, Lori Merish argues, portrays female workers as passive, lacking in subjectivity and agency. Melville's rendering of the factory girls follows that typology; however, "Tartarus" also confronts directly the stereotype of passivity as an effect of dehumanizing working conditions. In that respect, it illustrates Merish's claim that the female body was a site of sexualized power relations (Merish 2017, 76).

The short story's two parts speak to these contexts through its narrator – presumably Melville himself, who mediates between upper- and lower-class realities by recounting his experiences of visiting Temple Bar in London and a paper mill in Massachusetts. It was the last of Melville's so-called diptychs, double stories based on materials he had gathered on his trip to England in 1849. The others are "Poor Man's Pudding and Rich Man's Crumbs" (1854) and the unpublished and controversial "The Two Temples" (1924), an acerbic indictment of piety in New York's upper-class Grace Church congregation. Critical readings have mostly focused on the more aesthetically and politically controversial "Tartarus" episode, which focuses on the exploitation of the female body in industrialized New England. Delbanco notes this is to be one of several "social pathology[ies]" that Melville's diptychs address (Delbanco 2006, 224).

Approaches like transnationalism, Marxian world-systems analysis, and new materialism have begun to elucidate how the two episodes portray the consequences of a globalized market capitalism. The story can show us how these lines of inquiry intersect, by allowing us to attend to the way in which profit in a globalized market depends on misogynist, exploitative practices of selective localization, differentiation, and circulation. Bodies and labor must be fixed along the lines of gender, class, and geography in order to become valuable. On the flipside, value becomes profit only when capital can circulate freely. The two seemingly hermetic worlds in the diptych mutually constitute each other, only insofar as they subsist on these geographical and gender differences. The bachelors enjoy their banquet – and their writing paper – *because* the maids suffer in their remote prisonlike paper mill. Melville's short story thus allows us to contemplate the causal relations between these seemingly distinct worlds of creativity and exploitation.

When Melville visited London in 1849, the museums and urban spectacle impressed him greatly, but they also alerted him to the social qualms of an economically stratified and class-based society. On December 19, 1849, shortly before returning to New York, he dined with a group of literati and socialites at Elm Court, Temple, and found the experience highly diverting. The morning after, he writes in his journal while reminiscing about the previous night's merrymaking: "Up on the 5th story we dined. The Paradise of Bachelors" (Melville 1989, 44). The physical elevation of the dining room contributed to the sense of community and fellow feeling he experienced that night. Likewise, in the "Paradise" episode, the narrator sallies forth to a banquet in London's Elm Court Temple Bar, the headquarters of London's elite society of lawyers and legal clerks. Here he partakes in a lavish feast and enjoys the convivial all-male company of the bachelors. After returning home, he is appalled in equal measure when he journeys upstate to pick up writing paper and observes the abysmal working condition for the women who work and board at a paper mill near "Woedolor Mountain" (326), a surreal hellscape of ice and snow populated by tortured mill girls and their grotesque male overseers.

Located high above the streets of London, the bachelors' paradise is a mythical place. Like "the oases in the Sahara" (Melville 1987, 316) it is a sequestered, monastic space that seems to exist on a different plane of existence compared to the rest of the city. Even reaching the place requires skill. One has to "adroitly turn a corner – not a street – glide down a dim monastic way" (316). Making one's way there is enough to "give the whole care-worn world the slip" and become "disentangled" (316) from the profane business conducted on the streets. London's hidden sideways become a space of initiation. Traversing this space mentally and physically prepares the traveler for the realm of the bachelors, just as the arduous journey to the hollow of the Devil's Dungeon presages the site of the maids' misery. Temple Bar stands apart from the rest of the city, even while existing smack in the middle of it. Its halls tend "well up toward heaven" (320). It is a paradoxical space of non-urban urbanity, of sophistication and unobstructed relationality. The bachelor's sheer fantastic conviviality has no place among the grimy streets of London and must itself remain aloof from the everyday struggles of its denizens.

The moral didacticism of all of Melville's diptychs turns on the contrast between the respective episodes. The story's bipartite structure formally signifies this contrast, while its titular metaphor (heaven/hell) connotes social distance and frames it as a matter of class and gender difference. The bachelors' bliss comes at a price, but the toll is levied against the women in the paper mill. On the level of narrative, the distance between the two social milieus, the patriarchal dinner table and the female working-slave factory, achieves the same effect. To this physical remoteness, Melville adds a symbolic one. In terms of the lived experience, the "Paradise of Bachelors" is as far away from the streets below as it is from the "Tartarus of Maids."

Temple Bar is a utopian space of male camaraderie whose "wonderfully unpretending" interior belies its ethereal location (320). Far from being a paradise of pearly gates and golden chalices, it dispenses with "glare and glitter, gimcracks and gewgaws" to foreground the revitalizing quality of human (male) company (320). The narrator imagines the barristers as heirs to the Knights Templar, a religious order who fought to reconquer the Holy Land during the medieval crusades. These modern Templars, however, have been "completely secularized" (317). They have lost their monkish manners and warlike demeanor and excel only in conviviality: "best of comrades, most affable of hosts, capital diner is the modern Templar" (318). Their paradise is a realm of sentimental fellow feeling. They "earnestly express[ed] their sincerest wishes for the entire well-being and lasting hygiene of the gentlemen on the right and on the left" (321). The sheer number of sentimental adverbs and superlatives here conveys simultaneously ironic distance on the part of the narrator and genuineness on the part of his hosts. Their feasting is a function of their sociability, which in turn constitutes a codified symbolic system. Drinking, for instance, expresses fraternity. Having a "lonely, unparticipated glass," therefore, appears "indelicate, selfish, and unfraternal" (321). Melville's coinage "unparticipated" signifies that consumption is never its own end but always communal. One drinks only to share a drink and to express through that act a mutual (male) humanity. Temple Bar, then, is a stage for enacting those sentiments, to instill "[c]omfort – fraternal, household comfort" (322). Even on a tactile level, the bachelors' verbal, warm comfort contrasts the maids' silent, cold isolation.

The Templars are not hedonists but represent the very image of masculine professionalism: "men of liberal sense, ripe scholarship in the world, and capacious philosophical and convivial understanding" (322). Even after several carafes of wine, their "decorum [remains] unassailable by any degree of mirthfulness" (323). Pain and trouble to them seem "monkish fables" (322). They appear genuinely unaware of such profane things as the maids' suffering, and this bliss seems to be a product of their perpetual companionship. The exhilaration of being temporarily liberated from the strains of masculine, performative adversarial competition in the professional realm seemingly reveals what Eve Kosofsky Sedgwick diagnoses as "homosocial desire," which underpins all "compulsory relationships [such] as male friendship" (Sedwick 1990, 186). Still, the barristers voluntarily return to their professional duties, once the evening is done. Appreciating this escapist quality of the bachelors' paradise requires knowledge of "the bugbear styled trouble," which the narrator injects into the story through his journalistic perspective (322). Melville here also celebrates the ability to open oneself up to the paradoxical transience of communal experience for the sake of mutual human comfort. The "forcible light" of the experience prompts him to declare in an outburst of sentimental "admiring candor" that Temple is indeed "the very Paradise of Bachelors" (323). Sentimentality, here, is not a derogatory label. In fact, as I show below, affect is one of the central links between the two stories.

When we rejoin the narrator, he is traveling by horse through the Berkshire Mountains on his way to a paper mill. He wants to procure envelopes and paper for his side venture of producing and shipping seeds to customers around the country, an operation that requires "several hundreds or thousands" of letters to be sent each year (325). He decides to buy paper in bulk, partly "for economy's sake" but also for "the adventure of the trip" (325). Between 1850 and 1863, Melville lived with his family at Arrowhead near Pittsfield and regularly made the arduous commute to the "Old Red Mill" at Dalton (Parker 1996, 810). On 11 January 1851, he made a family excursion of it, and much of what he saw that day must have made its way into the story. The historic Dalton mill is a barometer of the state of industrialization in America and a textbook case for the events Melville relates in his story. "The Old Berkshire Mill," as it was also known, had gone into operation in 1802 (U.S. Census Office 1865, cxxv). Automation of the Massachusetts mills had begun in 1825 with the importation of the Foudrinier machine from England, exponentially increasing productivity almost overnight (cxxvi). By 1850, all but two mills in the United States used automated means of paper production (cxxvii). The paper business was booming in Massachusetts, which at different times housed the greatest number of paper mills in the country. Melville thus reflects on a major economic presence in his immediate surroundings as well as a transformative cultural moment for the United States at large.

While the bachelors' convivial site "lies not far from Temple-Bar" (Melville 1987, 316), the girls' paper mill "lies" deep within the "Devil's Dungeon" (323), an icy hollow within the Berkshires. The double story's parallel opening sentences formally signify that the two sketches compose one conjoined image, but the relationship between the two narratives is one of "inverted similitude" (327), which manifests in several thematic juxtapositions: The warm domesticity of the "Temple" confronts the inhumane cold of "Mount Woedolor." Bachelorhood affords the barristers a state of edenic fraternity, while it condemns the mill girls to perpetual suffering, sterility, and isolation. Even the narrator's emotional responses to the two scenes – his verbose glee at the bachelor's feast and his mute consternation at the maid's suffering – follow that inverted structural logic.

As with the "Temple," reaching "Tartarus" is arduous. The narrator must traverse a supernatural alleyway, turning "from that bustling main road" to the "Mad Maid's Bellow's pipe" (326) a tellingly named, narrow path. Through a "Dantean gateway" (324) he enters the hollow where the paper mill resides. In his long poem *The Divine Comedy*, the medieval poet Dante Alighieri relates how he, led by his teacher Virgil, descends through the nine circles of hell in pursuit of his beloved Beatrice (cf. Book One, "Inferno," Alighieri 2011, 11–20). At the entrance to hell stands a gate that admonishes travelers, "THROUGH ME ONE GOES INTO THE TOWN OF WOE, / THROUGH ME ONE GOES INTO ETERNAL PAIN, / THROUGH ME AMONG THE PEOPLE THAT ARE LOST [. . .] ALL HOPE ABANDON, YE WHO ENTER HERE!" (27; original emphasis). The intertextual reference colors both the landscape and the narrator's quest for paper as otherworldly and foreshadows the story's structure: as Dante must cross the river of the dead, "Acheron," so Melville's traveler passes "Blood River" (Melville 1987, 327), which symbolizes the maids' lifeblood. At the mill, he finds not one but several female phantoms – pseudo-Beatrices – as well as demon-like creatures like the impish "Cupid" (329) and the ominous "Bach" (329). Melville inverts the Dantean journey by having his traveler pass from paradise to hell, thus making the image of suffering the lasting impression of the dyptic story. In *Inferno*, the center of Dante's hell is actually a frozen wasteland. But in place of Dante's three-headed Satan, the center of Melville's hellscape is occupied by the infernal paper mill and its monstrous "machinery" (329).

The story puts in relief the dehumanizing effects of industrialization and the ominous dread of technology that to some was consonant with the transformation of the economy. In contrast to "The Lightning-Rod Man," which I discuss below, there is no celebratory conceit here. The proponents of technology, Cupid and Bach, cater to the sovereign will of the machines they oversee. For instance, the "dark colossal" (Melville 1987, 329) water wheel that powers the factory announces only dread with its "enormous revolutions" (329) that are "grim with its one immutable purpose" (329). To wit, Greek Tartarus is also prison to the Titans, the brute forerunners of the sophisticated Olympians. Like the mythical Tartarus, the factory houses the condemned but also ominous power in the form of a man-devouring machinery. The apparatuses' mechanical movements suggest unreasoning power and singular purpose. The female working bodies in the mill bend to the unfeeling regularity and might of the water wheel. Another machine is fed like an "iron animal" (328) and serviced by the maids like a "Sultan" (328). The machines negate the multiplicity and contradictoriness of human nature and places it under the regime of mechanical regularity, as the narrator muses, "the girls did not so much seem accessory wheels to the general machinery as mere cogs to the wheels" (328). This is antithetical to the free-spirited conviviality of Temple Bar.

When the narrator inspects the machine that presses and combines paper sheets, he finds himself in a room marked by "bloodlike, abdominal heat" (331). Producing a sheet takes nine minutes and the "unvarying punctuality and precision" of this operation impresses on him the notion of "the fulfillment of a mysterious prophecy" (332). The true horror of this scene is not merely that machines control human beings but that they extract and pervert bodily processes. The nine-minute cycle connotes the nine-month duration of pregnancy, just as the redness of Blood River connotes the female menstrual cycle. The point is hammered home by the image of an older maid, a former nurse, who, because the "business" of delivering children "is poor in these parts," now handles "piles of moist, warm sheets, which continually were delivered into the woman's waiting hands" (333). As blood turns to paper and human reproduction into mechanical production, Melville's dream of the writing machine turns into a nightmare.

"Tartarus" criticizes the conjoined ideologies of automation and capitalist value extraction by allegorizing the way female bodies become the raw material that fuels a sexed and gendered political economy. The story can help us attend to the corporeal cost of logics of extraction, to the way workers are silenced while their bodies are used up and turned into exchange value, which is then transferred as profit to remote male "paradises." Marxian-feminist critic Luce Irigaray notes that women make visible the underlying logic of abstraction that informs such an economy. "[B]y submitting women's bodies to a general equivalent, to a transcendent, super-natural value," she notes, "men have drawn the social structure into an ever-greater process of abstraction, to the point they themselves are produced in it as pure concepts: having surmounted all their 'perceptible' qualities and individual differences, they are finally reduced to the average productivity of their labor" (Irigaray 1985, 190).

Melville depicts this logic of abstraction as the gradual transmogrification of female bodies into paper. Upon entering the factory, the narrator beholds, "rows of blank-looking counters [at which] sat rows of blank-looking girls with blank, white folders in their blank hands, all blankly folding blank paper" (Melville 1987, 328). Repeating the adjective "blank" as referring to both bodies and paper obfuscates the difference between the two. Melville here reenacts semantically the chemical process that meshes paper and body. At another station, two women work in tandem, one feeding paper into a ruling machine, the other retrieving the ruled paper. The first woman's face is "young and fair" while the second one is "ruled and wrinkled," as both assume the properties of the paper they are handling (328). In a third instance, he observes blood mixing with pulp as the women get "glued to the pallid incipience of the pulp, the yet more pallid faces of all the pallid girls [. . .] Slowly, mournfully, beseechingly, yet unresistingly, they gleaned along, their agony dimly outlined on the imperfect paper" (334). The paper emerges precisely at the rate that the women's physical frame diminishes, in a process that is both parasitic and vampiric. "Machinery," the narrator notes, "that vaunted slave of humanity – here stood menially served by human beings, who served mutely and cringingly" (328). The motif of "inverse similitude" the narrator spoke of earlier also applies in a wider sense here. The apparatus presents an exact inversion of the automated paper mill Melville had dreamed up in his letter to Hawthorne. Instead of perpetuating human ingenuity by enabling literary production, the machine eats human bodies to produce blank sheets. This ominous blankness connects the story's two episodes. The narrator's transnational and intercontinental movements cause him to muse about "those strange uses to which those thousands of sheets eventually would be put. All sorts of writings would be writ on those now vacant things – sermons, lawyer's briefs, physicians' prescriptions, love-letters, marriage certificates, bills of divorce, registers of births, death-warrants" (333).

In her comparative reading of Melville's machine-related narratives, Branka Arsić argues that the machine gives birth by erasing all marks of life (Arsić 2005, 93). Modern communication and communal life, e.g. the barristers' profession, depends on the maids' silent acquiescence to being gradually erased. Reading "Tartarus" in this manner retroactively turns "Paradise" into a meditation on the hollowness of modern masculinity. The bachelor barristers feast, ignorant of the maids' suffering, nay, feast *on* their suffering. Their paradise becomes a morbid scene of decaying and parasitic masculinity. These new Knights Templar are jovial but flaccid, and their feast is a parody of romantic knighthood. The scene of their great glory is the dinner table, not the battlefield in some distant Holy Land. Just as the maids never seem to eat, the bachelors – like the machines in the paper mill – never stop consuming.

With "The Lightning-Rod Man," Melville satirically contemplates public discourse about modern technological innovation under a market system. In the story, a door-to-door salesman of lightning-rods travels upstate New York during a thunderstorm. He finds shelter at the narrator's house and uses the backdrop of the storm to scare his client into buying his wares. But when he starts ordering his host about the room to avoid being struck by lightning, the narrator mocks the salesman's alarmism and eccentricity. Throughout their conversation, he avers a common-sense theology, culminating in the credo that "the Deity will not, of purpose, make war on man's earth," as he sees professed by the rainbow in the wake of the storm (Melville 1987, 124). Enraged, the salesman charges his host with impiety and threatens to publicly condemn him. The narrator responds by snapping the salesman's rod in half and throwing the unwelcome guest out into the street.

"The Lightning-Rod Man" explores the consequences of such insular thinking as well as the rhetorical continuity between religion and mercantilism that emerge in the nineteenth century. Compared to "The Paradise of Bachelors and the Tartarus Maids," Melville here is less interested in showing the detrimental material effects of automation, but focuses instead on the psychological dimension of scientism, a quasi-religious reverence of science and technology. Protecting oneself against the contingency and harnessing human ingenuity to control natural forces were two ideas that produced immense technological optimism. Benjamin Franklin personified this complex of attitudes like few others in early colonial America. Franklin reports on inventing the lightning rod in *Experimentations and Observations on Electricity* (1747–1753), a collection of letter reports on his experiments. Melville mocks this optimism in the story, for instance through the various appellations the narrator bestows on the salesman. When he stands in the doorway, soaking wet, using his lightning rod as a staff, the narrator greets him sardonically as "that illustrious god, Jupiter Tonans" (Melville 1987, 119). He alludes to the custom of addressing ancient Greek deities in their respective purposive function, here the warlike Jupiter, the "Thunderer." Melville's narrator remains skeptical toward the promise that the natural world could be fully controlled by technology and especially toward the implication that divine Providence could be forestalled by a human-made "pipestem" (124).

The story historicizes religion in America by juxtaposing different versions of belief within the framework of modernist sensibilities: the narrator's nostalgic Christian Protestantism and the salesman's mercantilism. Religion for the salesman amounts to a form of scientist idolatry. "God, for him," as Thomas Werge notes, "is the lightning rod itself and the 'security' and 'omnipotence' it seems to offer the 'modern' man whose fears he [the salesman] seeks to exploit" (Werge 1972, 7). From the start, his own nervousness makes his sales pitch seem alarmist, as is evidenced by their exchange about non-conducting surfaces: "The storm deepens again. Have you a rug in the house? Rugs are non-conductors. Get one, that I may stand on it here, and you too. The skies blacken – it is dusk at noon. Hark! – the rug, the rug!" (Melville 1987, 123). It remains unclear whether the salesman really is a nefarious character. The "sunken pitfalls of eyes ringed by indigo halos" at the very least indicate that his fear-mongering exerts a physical toll on himself (119). The narrator jovially notes that "for one who would arm others with fearlessness," the salesman himself appears "unbeseemingly timorous" (121). As he stands drenched in the narrator's doorframe, his whole demeanor spells desperation rather than malice.

Religious fervor, ignited by fundamentalist groups, spread like wildfire in upstate New York during the Second Great Awakening (ca. 1800–1854). The "Burned-Over

District" saw its share of millennialists, spiritualists, revivalists, and hucksters. For instance, William Miller calculated the Second Coming for 22 October 1844 and assembled followers to wait out what was to become known as the "Great Disappointment." Joseph Smith, after several indictments for fraud, founded the Church of Later Days Saints, based on the alleged discovery of a new Christian gospel. The Fox Sisters attracted sizeable crowds holding "spirit rapping" séances. Melville published a story ("The Apple-Tree Table") about this phenomenon.

Besides drawing on regional history, the story exists in a literary tradition of short prose that negotiates the individual's role in an empiricist universe. Melville drew on a pre-existing American literary tradition of 'thunder-storm' stories, entering into a genre that we would today call disaster or calamity narratives (see Scheiding and Seidl 2011, 67–80). For instance, early New England's foremost theologian Cotton Mather saw thunderstorms as part of God's special providence. The Pittsfield Library Association acquired Mather's *Magnalia Christi Americana* (1702), which includes his essay on thunderstorms, just months before Melville published "The Lightning-Rod Man." In their editorial notes to the Northwestern edition, Harrison Hayford et al. read the salesman's voice as echoing Mather's in his demand to make peace with God in the face of the impending storm (Hayford 1987, 598). The fireside poet William Cullen Bryant followed in this tradition and wrote a story about the providential implications of thunderstorms with "The Whirlwind" (1825). Philip Coleman suggests that in the nineteenth century, the short story genre turned out to be "one of the most effective literary forms for examining the relationship between science and the self" (Coleman 2007, 176).

Melville first published the story anonymously in the August 1854. Together with "I, and My Chimney" (March 1856) and "The Apple-Tree Table" (May 1856), "The Lightning-Rod Man" forms a trilogy of stories that revolve around arguably the same narrator's misadventures in domestic family life. It stands in the satirical tradition and pokes fun at the type of the "Yankee peddler" (see Parker 1964, 156). With "The Lightning-Rod Man," Melville also began experimenting with what he came to see as a quintessentially American type: the confidence man, a type of trickster character, vomited up by a society steeped in the profit-driven ideology of a cosmopolitan market economy that is all about maintaining appearances. Melville would devote a whole novel to this type of character in *The Confidence-Man* (1857). Despite its cultural richness, literary criticism long treated the story as a second-tier piece of writing. Delbanco deems it an "acerbic sketch" of a con man who deals in platitudes and pseudo-science (Delbanco 2006, 228). Josh Matthews argues that the story satirizes "two American types – the rural democrat and the door-to-door salesman," while poking fun at the "rhetoric of consumer marketing in commercial periodicals, while commenting on political and religious divisions in American culture" (Matthews 2010, 57).

"The Lighting-Rod Man" indicts the self-effacing religious rhetoric that works in the service of market capitalism. Allan Emery contests that Melville builds the story around the "dichotomy or science and religion" rather than a conflation of the two

(Emery 1983, 559). However, the salesman builds his ethos not on his scientific expertise but on the opacity of his knowledge. He is more mystic than merchant. When he first enters the house, he immediately begins ordering his host about: "for Heaven's sake, get off the hearth! Know you not, that the heated air and soot are conductors; – to say nothing of those immense iron fire-dogs? Quit the spot – I conjure – I command you," to which the narrator wryly replies, "Mr Jupiter Tonans, I am not accustomed to be commanded in my own house" (Melville 1987, 121). The exchange characterizes the salesman's ethics. Only he knows the safe spots of a house he has never visited before. Because both men are in immediate danger, he has no time to initiate his client through proper ratiocinations; still, he makes time to label the narrator "horridly ignorant" about the layout of his own house in terms of its safe spots during a storm (119). His demeanor is brash, even if we allow for his own anxiety. He demands assent on the basis of faith. For the narrator to purchase a lightning rod means accepting the salesman's arcane authority. However, his authoritarianism countermands the evidentialist intellectual style – based on evidence, persuasion, and assent – that both science and religion had adopted to at this point (Holifield 2003, 4).

The lightning rod, the salesman avers, displaces personal and institutional religion in guaranteeing personal safety. The burnt church steeple in Criggan (an actual historical case) and the girl in Montreal who was struck by lightning at her bed-side while holding her rosary both indict religion as a competing form of institutional protection from calamity (Melville 1987, 120). To drive these competitors out, he claims that "*Mine* is the only true rod" (121). This absolutism bears strange fruit, as the salesman gradually reveals his policy of absolute isolation to his host: "I avoid pine-trees, high houses, lonely barns, upland pastured, running water, flocks of cattle and sheep, a crowd of men" (123). This is the logical end of the salesman's manic wish to insulate himself fully from contingency. Absolute security requires absolute isolation. But the notion that "Man avoid man" in times of need appalls the narrator (123). He refuses to surrender his self-reliance and sociability, huffing at the salesman, "Who has empowered you, Tetzel," he asks, making reference, to Johannes Tetzel, a prominent peddler of indulgences and counterpart to Martin Luther during the German Reformation, "to peddle around your indulgences from divine ordinations?" (124). Rejecting the paralyzing state of distress that security thinking yields, the narrator asserts a fatalism reminiscent of (Dutch) Calvinism. He notes that the days of our lives are numbered, just like "the hairs of our heads" (124). Unfortunately, there is no way of knowing, when those days are up. In the face of such uncertainty, he "stand[s] at ease" and trust that God will not snuff him out unfairly (124). This is both a statement of faith and self-reliance. He appeals to a personal God who is benign and rules the universe in good faith and will not "of purpose" destroy humankind by natural disasters such as lightning (124). Marvin Fisher notes that, the narrator's stance rules out "the kind of God who can be bought or bribed by self-abnegation" in favor of scientific idolatry (Fisher 1977, 123).

3 Theoretical Perspectives

Herman Melville's short fiction shows his romanticist stripes invoking the gothic tradition of Wordsworth, Coleridge, and Hawthorne. At the same time, his narrative perspective carries a journalistic voice, a technique pioneered by Charles Dickens. The two stories I have discussed here illustrate these attributes, but they also show the limitations of each tradition and its literary technologies: e.g., the romanticist allegory cannot go beyond the pathetic lament, "oh! Tartarus of Maids" (Melville 1987, 335). Meanwhile, the realist sketch carries the tint of middle-class voyeurism, fetishizing and dehumanizing the poor twice over with its gaze and lack of commitment to their plight. At best, such generic labels help us contextualize Melville's work in the various overlaying artistic currents of the time and in turn show that literary periodization is never cut-and-dry.

Melville was no revolutionary. His traveler has not come to free the Beatrices from Tartarus, nor does he spoil the barristers' feast in Paradise by pointing out the literal blood on the paper they use. He bears witness. His function is to connect the two episodes. Like Dante, he may "shed tears of sympathy" (Alighieri 2011, 29) for the damned he beholds, but he ultimately heeds Virgil's advice to merely "look, and pass!" (31). He even aggravates the factory girls' plight by patronizing the mill. The factory women, too, seem frozen in place, as they will not or cannot revolt. Fiercer portrayals of feminine agency did exist at the time, for instance in the fiction of E.D.E.N. Southward. Graham Thompson sees the story less as social critique than as a commentary about what literature can do and submits that the story questions "[t]he prospect of correspondence either at the internal level of textual harmony or the external level of social allegory" (Thompson 2012, 526). Later that year, Melville published a short-sketch collection, *The Encantadas* (1854), in which he portrays Hunilla, a native woman of mixed ethnicity who has been stranded on the Galapagos Isles after losing her family in a shipwreck. Hunilla signifies suffering beyond despair that turns into stoic forbearance. The Dalton factory women must have symbolized to him that same indomitable dignity of brokenness.

Critical readings of "The Lightning-Rod Man" have focused on formal elements (esp. allegory) and historicity. Within the historicist camp – the newer and more virulent of the two scholarly conversations – this division breaks down further between those who read the story as commenting on the emergence of commercial culture and those who stress Melville's indebtedness to biblical mysticism in framing the story's central dialogue. By 1854, Protestant exegesis in the Anglophone world had marshaled evidence-based arguments in an attempt to adapt to the new intellectual style in the wake of Enlightenment rationalism, but it could not fully unmoor from mysticism. The lightning rod, as Werge notes, engendered dispute among the New England Protestants in a controversy that "could neither be confined to the churches nor limited in its depth and implications" (Werge 1972, 9). Stephen Frye claims that the story depicts an "alternative modern sensibility that seeks still to find a form of meaning unmitigated by contingency and crass materialism," a meaning rooted in an "enchanted" view of the world (Frye 2006, 116). Meanwhile, Eric Wertheimer sees the lightning rod as a symbol of large-scale secularization efforts, an "emblem of a thoroughgoing skepticism, of all that cannot be satisfactorily secured and underwritten" (Wertheimer 2003, 178). Omitting either history or allegory from our analysis produces an incomplete picture: If we disregard allegory, we miss out on the dialogical discussion of modern personal agency, ethics, and religious form. If we ignore the history of religious fervor that underwrites the salesman's rhetoric, we trivialize the narrator's flight into violence. The difference between the positions may be summed up by a statement by the reverend Josiah Willard Gibbs who noted that, "religion is for the people, [. . .] theology is for the schools" (qtd. in Holifield 2003, 3). The story presents us with a historicizing allegory that exposes the way in which the American traditions of religious language underpin scientific capitalism.

Taken together, the two stories chronicle Melville's changing attitude towards the vision of the writing machine he had developed in his letter to Hawthorne. Both speak to the way technology impacted American communal life and shaped its cultural imagination. Precursors of what we today may call science fiction came to the fore during this time, with Hawthorne and Poe (↗7 Poe) numbering among the genre's forerunners. Even these early representations painted technology and the affects surrounding it in ambivalent colors. Melville's stories follow this trend by articulating visions of awesome and even monstrous technological innovation while musing about the human cost of such progress. As such, they express his disillusionment with techno-modernity.

While early critics pigeonholed Melville as a novelist, recent studies have used the lens of new materialist cultural studies to situate Melville's magazinist writings in the history of the medium and its materiality. Thompson sees Melville's stories as a "literary response" (Thompson 2012, 3) to the way magazines were organized and produced and to the manner in which they shaped and were shaped by reading communities. Katie McGettigan argues that Melville, in all his writings, drew "inspiration from contemporary transformation in the book object" in the context of what she calls "print modernity" (McGettigan 2017, 2). Shining a light on Melville's manifold entanglements in the material cultures of magazines is also part of a broader trend in Literary Studies (↗4 Textual Materiality). Cody Marrs has outlined this program as one of reading Melville "adjacently," i.e., reading with rather than against the grain (Marrs 2019, 7). Such readings harken to post critical models of analysis that focus on the connectivity and enmeshing of literary works in various cultural contexts. Future studies should continue this work of resituating Melville in updated contexts. For example, I argue elsewhere that Melville contributes much to the discourse of postsecularism in literary studies, which questions disciplinary narratives about the wholesale secularization of American society since the nineteenth century (Schlarb 2021, 2). Melville's short fiction complicates this story of American culture by demonstrating that religion and science were never separate discourses. Melville will continue to be a valuable interlocutor for producing this more nuanced account of American cultural history.

4 Bibliography

4.1 Works Cited

Alighieri, Dante. *The Divine Comedy of Dante Alighieri*. 1918. trans. Courtney Langdon. Vol. 1. Cambridge: Harvard UP, 2011.
Arsić, Branka. "Melville's Celibatory Machines-Bartleby, Pierre, and 'The Paradise of Bachelors.'" *Diacritics: A Review of Contemporary Criticism* 35.4 (2005): 81–100.
Coleman, Philip. "'I Am an Experiment': Reflections on Science in American Short Fiction." *On Literature and Science: Essays, Reflections, Provocations*. Portland, OR: Four Courts, 2007. 173–189.
Delbanco, Andrew. *Melville: His World and Work*. New York: Vintage, 2006.
Emery, Allan Moore. "Melville on Science: 'The Lightning-Rod Man.'" *The New England Quarterly* 56.4 (1983): 555–568.
Fisher, Marvin. *Going Under: Melville's Short Fiction and the American 1850s*. Baton Rouge: Louisiana State UP, 1977.
Frye, Steven. "Melvillean Skepticism and Alternative Modernity in 'The Lightning-Rod Man.'" *Poe Studies/Dark Romanticism* 39–40, no. 1–2 (2006): 115–125.
Hayford, Harrison. "Notes on Individual Prose Pieces." *The Piazza Tales and Other Prose Pieces, 1839–1860*. Eds. Harrison Hayford et al. Evanston: Northwestern UP, 1987. 569–793.
Holifield, E. Brooks. *Theology in America: Christian Thought from the Age of the Puritans to the Civil War*. New Haven: Yale UP, 2003.
Irigaray, Luce. *This Sex Which Is Not One*. Ithaca: Cornell UP, 1985.
Marrs, Cody, ed. *The New Melville Studies*. Cambridge: Cambridge UP, 2019.
Matthews, Joshua. "Peddlers of the Rod: Melville's 'The Lightning-Rod Man' and the Antebellum Periodical Market." *Leviathan* 12.3 (2010): 55–70.
McGettigan, Katie. *Herman Melville: Modernity and the Material Text*. Lebanon, NH: U of New Hampshire P, 2017.
Melville, Herman. *Correspondence*. Eds. Lynn Horth et al. Evanston: Northwestern UP, 1993.
Melville, Herman. *Journals*. Eds. Howard C. Horsford et al. Evanston: Northwestern UP, 1989.
Melville, Herman. *Moby-Dick; or the Whale*. 150th Anniversary Eds. Harrison Hayford et al. Evanston: Northwestern UP, 2001.
Melville, Herman. *The Piazza Tales and Other Prose Pieces, 1839–1860*. Eds. Harrison Hayford et al. Evanston: Northwestern UP, 1987.
Kosofsky Sedgwick, Eve. *Epistemology of the Closet*. U of California P, 1990.
Merish, Lori. *Archives of Labor: Working-Class Women and Literary Culture in the Antebellum United States*. Durham: Duke UP, 2017.
Parker, Hershel. *Herman Melville: A Biography, 1819–1851*. Vol. I. Baltimore: Johns Hopkins UP, 1996.
Parker, Hershel. "Melville's Salesman Story." *Studies in Short Fiction* 1.2 (1964): 154–158. Baltimore: Johns Hopkins UP, 1996.
Schlarb, Damien B. *Melville's Wisdom: Religion, Skepticism, and Literature in Nineteenth-Century America*. New York: Oxford UP, 2021.

Scheiding, Oliver, and Martin Seidl. "Early American Short Narratives: The Art of Story-Telling Prior to Washington Irving." *Handbook to the American Short Story: Genres – Classics – Model Interpretations*. Eds. Michael Basseler and Ansgar Nünning. Trier: WVT, 2011. 67–80.
Sealts Jr., Merton M. "Historical Note." *The Piazza Tales and Other Prose Pieces 1839–1860*. Eds. Harrison Hayford et al. Evanston: Northwestern UP, 1987. 457–534.
Kosofsky Sedgwick, Eve. *Epistemology of the Closet*. Berkley, CA: U of California P, 1990.
Thompson, Graham. *Herman Melville: Among the Magazines*. Amherst: U of Massachusetts P, 2018.
Thompson, Graham. "The 'Plain Facts' of Fine Paper in 'The Paradise of Bachelors and the Tartarus of Maids.'" *American Literature* 84.3 (2012): 505–532.
U.S. Census Office. *Manufacturers of the United States in 1860; Compiled from the Original Returns of the Eight Census under the Direction of the Secretary of the Interior*. Washington, D.C.: Government Printing Office, 1865.
Werge, Thomas. "Melville's Satanic Salesman: Scientism and Puritanism in 'The Lightning- Rod Man.'" *Christianity and Literature* 21.4 (1972): 6–12.
Wertheimer, Eric. "Jupiter Underwritten: Melville's Unsafe Home." *Nineteenth-Century Literature* 58.2 (2003): 176–201.
Zinn, Howard. *A People's History of the United States: 1492–2001*. 3rd ed. New York: Routledge, 2015.

4.2 Further Reading

Allen, Tom. "Melville's 'Factory Girls': Feminizing the Future." *Studies in American Fiction* 31.1 (2003): 45–71.
Bush, Harold K., and Brian Yothers, eds. *Above the American Renaissance: David S. Reynolds and the Spiritual Imagination in American Literary Studies*. Amherst: U of Massachusetts P, 2018.
Fisher, Marvin. "Narrative Shock in 'Bartleby, the Scrivener,' 'The Paradise of Bachelors and the Tartarus of Maids,' and 'Benito Cereno.'" *A Companion to Herman Melville*. Ed. Wyn Kelley. Malden: Blackwell, 2006. 435–450.
Scheiding, Oliver. "Subversions of Providential Historiography in Herman Melville's 'Benito Cereno' (1856)." *Re-Visioning the Past: Historical Self-Reflexivity in American Short Fiction*. Eds. Bernd Engler and Oliver Scheiding. Trier: WVT, 1998. 121–141.

Holger Kersten

9 Mark Twain (Samuel L. Clemens, 1835–1910)

Abstract: In the annals of American literature, Samuel L. Clemens ("Mark Twain") is recognized as a major writer of novels and travelogues. Although a handful of his stories enjoy general recognition, he is rarely placed among America's outstanding short-story writers. Seen through the lens of prescriptive short-story theories, Clemens's unconventional work was frequently identified as humorous local-color writing, a category lacking literary prestige. In an endeavor to present a more adequate picture of his achievement, this chapter will show that Clemens produced a wide variety of short texts that elude simple classification. From his early journalistic pieces to the more elaborate writing of his later years, he used short textual units in different ways to articulate his concerns and priorities as a writer. In a close-up look at "A True Story" and "The Man That Corrupted Hadleyburg," two stories that bracket the most productive phase of his career, it will become clear that his linguistic experiments and his intensive engagement with the moral consequences of human behavior elevate his work beyond the confines of traditional genre categories and establish him as an accomplished expert in short-form writing.

Keywords: Storytelling, sketch, short fiction, humor, journalism

1 Context

American literary history accords a special place to Samuel Langhorne Clemens, the writer who became world-famous under his pseudonym, Mark Twain. His reputation rests predominantly on the novels he wrote between 1876 and 1894, most importantly on *Adventures of Huckleberry Finn* (1884), which has been regarded as one of the major achievements of nineteenth-century American literature (Gerber 1985, 1). When it comes to Clemens's short writings, however, it is more difficult to establish a basis for a proper appreciation of his efforts. In his contribution to the handbook *A Companion to Mark Twain*, David E. E. Sloane, author of several authoritative publications on Mark Twain and American humor, places Clemens "among the great short story writers in the world" (Sloane 2010, 78). Five years earlier, Henry Wonham, a scholar with an equally well-established reputation as a Mark Twain specialist, proclaimed that Clemens "was never entirely at home with the short story as a literary form" (Wonham 2005, 357). This chapter attempts to show that writing in short formats was a central element in Clemens's repertoire and can be found in various configurations ranging from brief news items to the famous 17,500-word story "The Man That Corrupted Hadleyburg," and beyond.

https://doi.org/10.1515/9783110587647-010

Clemens did not think of himself as a writer of short stories. For one thing, the term "short story" as a designation for a distinct "genre" was not in widespread use at the time and would not become an object of an intensive debate until Brander Matthews generated broader attention for the issue in his 1884 essay "The Philosophy of the Short-story" (Pattee 1923, 293). Secondly, short forms of writing tended to be judged as an inferior art form, "a convenient, apprentice exercise, a stepping stone to better things" (292). As late as the 1890s, shortly before sitting down to write "The Man That Corrupted Hadleyburg," Clemens regarded a short story as "a novel in the cradle," a project to be further developed (Hirst 1999, 13). Additionally, Clemens had financial concerns: "The short story is the worst paid of all forms of literature," he wrote in response to a request for a contribution from a magazine editor (13). In view of his well-known desire to reap financial rewards from his literary endeavors, a writing activity that did not repay the effort was a disincentive for Clemens to spend time and energy on such projects. Guided by the conviction that the labels "local colorist" and "humorist" constrain an appropriate appreciation of his achievement as a writer of short-form texts, the present chapter will address the complexities involved in the discussion of Clemens as a writer of "short stories." For this purpose, the essay will proceed by sketching the broad spectrum of Clemens's short writings and the circumstances in which they emerged. Anchored in this historical framework, it will then concentrate on two texts, taken from different phases of his career, to illustrate how he produced sophisticated narratives than move beyond the narrow confines of humorous local color writing to involve readers in reflections on the ethical complexities of American life in the nineteenth century and the human condition in general.

Throughout his entire lifetime, Samuel L. Clemens produced an astonishing number of prose texts that, in the words of Edgar Allan Poe, can be "read at one sitting" (Poe 1972, 3; ↗7 Poe). To a certain extent, it can be argued that the majority of his writings, including his episodic travel books and novels, were designed to be consumed in small portions. From this perspective, Clemens undoubtedly deserves attention as an accomplished writer of short-form texts even if the formats he chose do not conform to the strict formal requirements that literary critics have set out in the course of time.

Clemens's path to short prose writing led by way of his exposure to journalism, first as a printer, then as a newspaper reporter in the Far West. During these formative years, he learned to identify interesting subject matter and developed a sense for the skillful and effective use of language. In the early years of his life in the frontier village of Hannibal, Missouri, there was little that predisposed Clemens for a career in literature. During the short period of traditional schooling administered in a small log schoolhouse, Clemens acquired only the rudiments of an education. He learned the basic skills, was praised as a good speller, and acquired a taste for reading. The mandatory visits to Sunday school failed to add much more to this basic repertoire than Bible verses and pious phrases (Wecter 1961 [1952], 82–87). After his father's death, Clemens became an apprentice in a Hannibal print shop in 1848 and pursued this occupation at various printing offices in St. Louis, Philadelphia, and New York. While his years in school probably exposed him to little reading material other than McGuffey's readers, his apprenticeship in print shops afforded him manifold opportunities for reading widely across all kinds of textual genres over a period of ten years (Powers 2005, 26, 47).

Like Benjamin Franklin, Walt Whitman, William Dean Howells, and other American authors who started their literary careers as printers, Clemens experienced the printing office as more than a place of mechanical manual labor. It was here that he discovered the world of writing and it exposed him to writing samples from some of the world's great authors. By requiring intense forms of interaction with texts, the typesetting process put him in a most intimate relationship with a variety of writing styles and subject matter, building up implicit knowledge about textual composition. "One isn't a printer ten years without setting up acres of good and bad literature," Clemens wrote, "and learning – unconsciously at first, consciously later – to discriminate between the two, within his mental limitations; and meantime he is unconsciously acquiring what is called a 'style'" (MT 1992, "Turning," 935).

By the late 1850's, however, Clemens's life took a detour from the written word when a series of coincidences created the opportunity to realize his boyhood dream to become a steamboat pilot on the Mississippi River, a job that promised excitement, prestige, and a generous salary. This interlude ended when the Civil War brought commercial traffic on the Lower Mississippi to a halt in 1861. Clemens then decided to try his luck in the American West where he first shared the hopes and then the disappointment of countless other prospectors who failed to strike it rich in California's and Nevada's gold and silver fields.

A sequence of fortunate events gave his writing career a new impulse when the Virginia City *Territorial Enterprise*, one of the leading newspapers in the Old West, offered him a job as reporter. Since certain misconceptions remain about regions that were remote from the cultural centers of the East, it is worth pointing out that San Francisco and Virginia City were not cultural backwaters, but home to people who enriched the world of the gold- and silver-mining economy with entertainment and intellectual stimulation. The publishers, editors, and writers of the "Sagebrush School" (Berkove 2003) were sophisticated individuals who were well acquainted with the classics of European literature, adhered to enlightened aesthetic standards, and knew how to write in an engaging style on a variety of interesting and timely subject matter.

In this environment, Clemens benefited from the talents and expertise of educated, gifted, and experienced journalists and writers. Under their tutelage, he quickly developed into an attentive and perceptive observer and chronicler of human life in the West. It was in Virginia City that Samuel Clemens "was introduced to ideas and standards that shaped his personality, and to techniques that [. . .] would enable him to become one of the greatest authors of world literature" (Berkove 2005, 168). Writing for newspapers meant keeping the length of a text within circumscribed boundaries. Limited by the four pages available for their daily copy, the *Territorial Enterprise* printed news items and reports about activities in the mines and mills as well as information about the social life of the day in the compact format that is typical for newspaper articles. Consequently, journalists were required to make their writing at the same time concise, effective, and attractive to read – clearly a good training ground for writers of short fiction. The output produced in the years of Clemens's literary apprenticeship at various newspapers in the West was impressive, numbering between 3,000 and 4,000 separate pieces, ranging from short items tossed off in the course of his newspaper work to ambitious imaginative narratives of several thousand words (MT 1979, ET&S1, 2). In this body of texts, he covered a broad thematic range including social and political affairs, topics from the world of literature, theater and the arts, as well as religious and philosophical issues (1).

Although the *Territorial Enterprise* was committed to the principles of accurate journalism, it did not limit itself to factual reporting only. Long before the expression "fake news" was in general use, the paper printed material with a tenuous connection to actual facts. As one historian of the West put it, "[f]ake stories and practical jokes were unquestionably an outstanding feature" (Glasscock 1931, 117). Clemens who, during the course of his employment at the *Enterprise*, adopted the pen-name "Mark Twain" (MacDonnell 2012), found factual reporting less satisfying than the creative use of his lively imagination. Some observers of his journalistic practices spoke of his "indifference to 'news'" which referred to his tendency to shift the focus of a text away from the actual event to the humorous or philosophical perspective which it invited (MT 1979, ET&S1, 21). Clemens actively nurtured the image of "Mark Twain" as a reporter who would embroider his facts and would rarely miss an opportunity to supplant objectivity with fancy (Branch 1969, 5).

Among the text types that fit into the frame of popular periodical publications, the newspaper letter deserves particular attention. It gave a traveling reporter freedom to choose his subject according to his personal interests as long as the material offered a sufficient degree of interest and novelty for the readership at home. Bound by few formal rules, the newspaper letter offered its author much license to combine factual information, personal observations, fanciful invention, and amusing digressions, thus opening up an experimental space for a future writer of tales, sketches, and other forms of short fiction. Whenever the opportunity arose, Clemens eagerly seized upon the creative opportunities this format offered him as a writer.

Clemens's immersion in the world of journalism proved to be an ideal situation for a person with a preference for "controlled incoherence" (MT 1979, ET&S1, 56). It was here that he could experiment with a range of short textual forms – including anecdote, newspaper letter, miniature essay, personal narrative – many of which allowed him to transgress the boundaries of factual news reporting and engage in literary endeavors which blended reality with fictional elements. "Petrified Man" (1862) in which he skillfully combined his ability for writing convincing scientific prose with his penchant for teasing his readers, and his hoax "A Bloody Massacre Near Carson" (1863), whose glaring absurdities first fooled and then infuriated many of his readers, are two early examples that illustrate the irreverent creativity with which Clemens interpreted his job as a journalist at the *Territorial Enterprise.* Building on items like these, Clemens developed a repertoire of techniques in the art of short writing from which he drew generously as his literary career unfolded in the following years.

Samuel Clemens was not a writer who paid much attention to the formal and structural rules of the literature of his time. In fact, he often consciously chose to disregard traditional genres (MT 1979, ET&S1, 56). Although he did care about language and style, he felt free to use whatever form seemed appropriate to his purpose. His allegiances "were not to established literary forms but to imaginative and humorous possibility" (Quirk 1997, 3). There is no evidence to suggest that Clemens thought about the "short story" in any systematic manner. He showed little disposition to classify his work according to accepted genre definitions, and if he used descriptive terms for his short writings, they would be "yarn," "sketch," or "tale." As a newspaperman, a travel writer, and an author of books sold by subscription, his main ambition was to please the general reader, not the arbiters of literary convention. Consequently, his literary repertoire ranged from "the anecdote to the sustained humorous narrative, from the miniature essay to the ambitious set piece" (MT 1979, ET&S1, 52).

The fact that it is hardly possible to detect a clear and consistent pattern in Clemens's shorter texts should not be taken as a sign that he had no sense of form or style. As an heir to the rich tradition of native American humor, Clemens consciously drew on and combined many of the themes and techniques that the writers of the Old Southwest, those of Down East Humor, and the literary comedians had practiced from the 1830s onward (Blair 1960 [1937], 147, 160). These stories followed their own narrative patterns and had distinctive features which contributed to their popularity and influence. The narrative frame, the use of vernacular speech, the deadpan pose, and a taste for incongruous and ridiculous situations were all part of the repertoire of Southwestern humorists, and Clemens often adapted them for his own purposes (Cohen and Dillingham 1975, xx–xxiv).

Although he had very little interest in matters of literary theory, Clemens did have opinions on what constituted good writing. He insisted that "[t]here is only one right form for a story, and if you fail to find that form the story will not tell itself" (Quirk 1997, 5). He held himself to a high standard and would set aside and even destroy large numbers of manuscript pages if he disliked them. Those passages that remained usually underwent an intensive revision process (7). When he criticized the literature of James Fenimore Cooper, Bret Harte, or Jane Austen, he did so on the basis of what he considered their stylistic defects (Kersten 2007, 28, 32). If he had a fixed set of principles for writing good stories, he never expressed it openly. The closest he came to laying down rules for what he called "the humorous story" was in the essay "How to Tell a Story" (1895), originally written for the magazine *Youth's Companion*. With its focus on the strategies of story-telling, on "how a story ought to be told" (MT 1997, "How," 201), the text yields few, if any, insights on the poetics of short-story writing.

With his long-lasting immersion in the world of short texts, it seems obvious that Clemens would have internalized the principles of concise writing. It is easy to imagine how the continuous practice and the routine acquired in the years of his journalism must have revealed to him the benefits of writing in small narrative units, and it is only a small step to adapt the mastery of short forms to the construction of larger projects. Thus, it is no coincidence that Clemens made intensive use of his early travel letters and journalism when he sat down to compose *The Innocents Abroad* (1869) and *Roughing It* (1872), his first full-length books. In fact, Clemens rarely started a new project with a "formal narrative arrangement" but mostly juggled a collection of disjointed notes and separate anecdotes (Hill 1963, 30). If such separate items were reasonably self-contained, they could be published individually, otherwise they could be transformed into chapters of a sustained narrative. In this way they were similar to the "plugs" that he fitted into the "narrative-plank" of his lectures to match "the temper of the audience" (MT 1995, L4, 498).

The episodical and modular nature of much of Clemens's writing is openly visible in the elaborate tables of contents of his books. They spread across several pages and often divide individual chapters into sometimes more than six segments, each one featuring a distinct title. It can in fact be argued that writing in such short segments was Clemens's characteristic way of textual composition. There are many examples where it is hard to draw a solid line separating his long-form publications from his shorter writings. In some instances, Clemens does not even pretend that the segments are organically connected. When he introduces "The Professor's Yarn," which occupies all of chapter 36 in *Life on the Mississippi*, the narrative voice bluntly states: "I insert it in this place merely because it is a good story, not because it belongs here – for it does n't [sic]" (386).

There was also a pragmatic reason for writing in such easily digestible portions: Since Clemens had set his eye on the financially lucrative subscription-book market, his audience was mainly composed of rural readers, "mechanics and farmers," whose busy and exhausting lives gave them few opportunities for sustained reading (Hill 1963, 27). Chapters composed of diverse and self-contained episodes offered instant gratification even to those readers who managed to read only a few pages before they fell asleep after a long day of work and allowed them to appreciate the reading experience without risking to lose track of a stretched-out, intricate plot. In this context it becomes clear that, by delivering entertainment in short reading units in nearly all of his books, Clemens practiced the art of writing in the short form effectively and successfully throughout his entire career.

In a certain sense, most of his books are collections of distinct episodes, scenes, and sketches some of which can be detached from the larger project and published separately (Hook 1985, 104). Clemens himself frequently pulled individual episodes from his book chapters for use in his lectures (Fatout 1960, 242). When he chose twenty items for inclusion in the 1888 anthology *Mark Twain's Library of Humor*, only five of them had been independently published. The remaining fifteen contributions were excerpts taken from five of his books.

This practice has been continued by many editors and critics of Clemens's short writings. Quirk discusses six short stories from *Roughing It* (Quirk 1997, 47) and Wilson gives them separate entries in his *Reader's Guide* (Wilson 1987, 113–130), together with three "interpolated short narratives" from *Life on the Mississippi* (61–66), and two selections from *A Tramp Abroad* (153–162; 217–220). Blair singles out "Jim Baker's Blue-Jay Yarn" as the greatest of Clemens's shorter works and calls it his "other masterpiece" (Blair 1993, 166). From here it is only a small step to argue that a "large number of Twain's best tales did not originally appear as separate short stories" but were embedded "in the long travel narratives or novels" (Gribben 1988, 59).

Blair's appreciation for the "Blue-Jay Yarn" and Gribben's preference for high-quality short writing nested inside of longer texts notwithstanding, there is no consensus among Mark Twain scholars with regard to what might be the author's "best" or "most characteristic" short story. A cursory survey of the relevant literature reveals that discussions of Clemens's short works often single out the "Jumping Frog" story as a significant text whose importance derives both from its formal achievements and its impact on the author's career. It has been used to illustrate the author's narrative ingenuity that "'set all New York in a roar'" (MT 1981, ET&S2, 271). However, the humor inspired by the American West that gave the story its popular appeal has faded in the course of time. As early as 1897, it seemed to be nothing more than "the merest extravagance, as hard to understand as the spirit which prompted the gargoyle on the medieval church" (Thompson 1897, 447). Even Bernard DeVoto, one of the earliest and most vehement defenders of Clemens's literary artistry, conceded that "[n]o one will ever laugh at it again as all America [. . .] laughed at it in the closing weeks of 1865" (DeVoto 1951 [1931], 177). To appreciate the achievement associated with this particular story, modern readers now require the skills of a literary historian. In making a case for a story that is characteristic of Clemens's writing, it is also worth remembering that he was never happy with the label of the "humorist" that remains attached to him to this day. From the early days, when he denounced his first successful story as "a villainous backwoods sketch" (MT 1988, L1, 327) to his old-age reflections on the durability of humor, when he regarded humor as "only a fragrance, a decoration" (MT 1940, "Humorists," 202), Clemens had a complex understanding of the functions and benefits of humor.

Against this background, one might well concede that the "Jumping Frog" has its place in literary history as a marker in Samuel Clemens's career and in the development of the early forms of American literary humor. For an appreciation of Clemens's achievement as a writer of short narratives, however, a fixation on an early expression of his literary talent obstructs the view of a broader perspective. While it is true that his literary career began with the "Jumping Frog," it should not overshadow the fact that, in the decades that followed, his writerly skills and moral vision continued to develop and expand into the mature expression of his literary art. It is for this reason that the present chapter will focus on "A True Story" and "The Man That Corrupted Hadleyburg," two texts that are firmly anchored in the most creative phase of Clemens's career when his experiences in life and literature enabled him to fully realize his potential as a writer.

2 Close Readings

"A True Story" (1874) marks an important landmark in Clemens's literary career. It has certain structural similarities with the "Jumping Frog" but replaces the frivolousness of that story's frontier humor with a considerable measure of moral depth. When he composed the manuscript in 1874, Clemens's literary reputation rested firmly on the material he had written in his early days in Nevada and California which had opened the path for him to become an author of humorous books. Subsequently, the hilarious European travelogue *The Innocents Abroad* became a best-seller in 1869, and the follow-up publication *Roughing It* (1872) made "Mark Twain" a household name in the United States and abroad. The books earned him the title of "the most popular of American humorists" (Taylor 1999, 99). At the same time, however, Clemens was in the process of adapting himself to the new social environment which he had entered by his marriage to Olivia Langdon, the daughter of a wealthy and well-respected coal magnate (Budd 1983, 50–51). When he and his wife settled in Nook Farm, a genteel residential district of Hartford, Connecticut, populated by a community of prominent and influential writers, preachers, and civic leaders (Andrews 1967), Clemens began to explore ways to develop his literary career beyond the confining boundaries of humorous writing. Split between the appeal of the financial rewards flowing out of his writings for a broad popular audience on the one hand, and a budding desire for literary recognition on the other, he began to turn to new subject matter and techniques.

In 1874, Clemens was busy working on a variety of different projects: As he began composing *The Adventures of Tom Sawyer*, he also organized the memories of his time as a steamboat pilot which ultimately developed into the four-part series "Old Times on the Mississippi." These activities were supplemented by "Sociable Jimmy" and "A True Story," two texts whose unconventional format gave them a distinct character of their own. In each case, Clemens chose an African American character as the main narrator and created a linguistic format that allowed them to tell their stories in their own voices. "Sociable Jimmy" appeared in the *New York Times* (29 Nov 1874) and was immediately reprinted by other newspapers. "A True

Story" followed almost simultaneously and became his first contribution to the prestigious *Atlantic Monthly*. Both stories paved the way for the literary innovations that ultimately found their climax in the 1884 publication of *Adventures of Huckleberry Finn* (Fishkin 1993, 15).

As the title indicates, "A True Story" was not pure literary invention. The historical record shows that the material came out of a personal encounter that took place during the summer of 1874. While Clemens and his family spent their summer vacation in Elmira, New York, he listened to the life story of Mary Ann Cord (Jerome and Wisbey 2013, 275–281), a former slave now working as a cook for the family of his sister-in-law. Impressed and moved by what he heard, he first retold the story to some of his friends (MT 2002, L6, 245) before he sat down to give literary shape to "a fragment of [her] history" (222) by adjusting the narrative sequence so that it would begin "at the beginning, instead of the middle" (217). Originating, as it does, from personal experience, "A True Story" offers an interesting angle for exploring biographical aspects of Clemens's attitudes towards African American lives. Its greater value, however, lies in its portrayal of the humanity of African American characters who preserved their dignity throughout the long trials of slavery and the racism of the Reconstruction era.

"A True Story" features momentous scenes from the life story of "Aunt Rachel," an African American woman who was born as a slave in Maryland. Prompted by a casual remark from the narrator, "Misto C—," Rachel responds to the notion that she had "lived sixty years and never had any trouble" (MT 1992, "True," 578). In her uninterrupted and eloquent speech, presented in her distinctive vernacular, she tells the story of how she lost her husband and seven children when they were sold at auction, and how, twenty-two years later, an improbable turn of events ultimately reunited her with her youngest son. Although Rachel concludes her story with an expression of gratitude for this outcome, her final words undercut any notions of a happy ending when she offers an ambiguous and possibly ironic summary: "'Oh, no, Misto C—, I hain't had no trouble. An' no *joy*!'" (582).

The story begins with the words of the narrator who is later called "Misto C—." He sets the scene and introduces "Aunt Rachel," a sixty-year old black servant whom he characterizes as strong, cheerful, and prone to peals of laughter (578). On the following pages, the white narrator reappears only once to highlight the dominant position that Rachel assumes for the remainder of the story. As the language quickly shifts from the white man's standard English to African American vernacular speech, Rachel displays her rhetorical skills when she confronts her listeners with scenes from the harrowing existence of her life as a slave. It is a narrative performance rich with vivid details, moments of suspense, and a colorful language that, at times, is capable of generating brief sparks of amusement in the midst of painful memories. Although Rachel's reunion with her son at the end of her story strikes an emotional note, the text does not surrender to saccharine sentimentality. Its final sentence delivers a searing satirical blow to the ignorant and insensitive white narrator and provides structural closure by reconnecting the text back to the cliché-ridden remark that set the whole narrative in motion. Although some scholars have called the story a "frame story" (Wilson 1987, 268) the term is misleading here. In contrast to many of his other stories where Clemens does indeed use a framework as a literary device, "A True Story" never returns narrative control to the white interlocutor. Especially in dialect stories, the frame often serves as a device to signal the social gap that separates the speaker of the nonstandard variety from the person who represents the accepted linguistic standard. The fact that Aunt Rachel is given the last word in this story is an important clue for an appreciation of the moral position she holds in this narrative.

In contrast to the story of the "Jumping Frog," the contents of Aunt Rachel's story is neither laugh-aloud funny nor even superficially amusing. Although the setup presents a scene of a relaxed end-of-the day conversation among a small group of people enjoying friendly bantering, the atmosphere quickly changes from light-hearted to serious, and ultimately endows the conversation with a moral component. A seemingly innocuous but, in reality, thoughtless and ignorant remark by "Misto C—," the I-narrator of the opening scene, puts an abrupt end to the buoyant laughter and causes an awkward moment of silence before Rachel takes control of the communicative situation. This is conveyed by the fact that 85% of the text appears in her characteristic vernacular. It is reinforced by the description of how her posture changes from "sitting respectfully below our level" (578), as was deemed appropriate for a black servant, to an upright position in which "she towered above us, black against the stars" (579).

The course of the narrative illustrates that this is not a minor detail. As Rachel reports significant moments from her life, her listeners and, by extension, the readers recognize her as a strong and self-confident woman who never allowed herself to be victimized, neither by young rambunctious black men invading her kitchen (581), nor by the sellers and buyers at the slave market. Even under the most humiliating circumstances and in the face of violence, she threatens the white slaveholders that she would kill the man who touches her small son, Henry (579). Despite the fact that she was shackled, she managed to "tear de clo'es mos' off of 'em, an' beat 'em over de head wid my chain" (580). The fearless attitude displayed here is encapsulated in her exclamation "'I's one o' de ole Blue Hen's Chickens, I is!'" – a phrase she had learned from her mother. The expression was originally used by soldiers from Maryland priding themselves on their courage and fighting spirit, and later adopted by the natives of that state. By extension, the phrase has also been used to designate a "'fiery, quick-tempered person'" (Halpert 1951, 196). Appropriated and adapted from the battle cry that originated with white soldiers of the American Revolution, and repeated three times throughout the story, the phrase acquires enough weight to be regarded as a distinctive symbol of Rachel's defiance and resilience. It also carries an undertone of the self-confidence and pride that was familiar to nineteenth-century Americans from the boasts of folk heroes such as Davy Crockett and

Mike Fink who advertised their prowess in animal imagery involving "half-horse, half-alligator" and "red-hot snappin' turtle" (Botkin 1944, 56–57), a tradition which Clemens exploited more fully in chapter three of *Life on the Mississippi*.

With regard to the time of publication, it is remarkable that Clemens decided to portray an African American woman openly displaying her sense of equality. Following in the vein of the Union general who, in her story, tells her "'you's 'mong frens, now'" and listened to her troubles "'jist de same as if I was white folks'" (MT 1992, "True," 580), Rachel informs "Misto C—" that there is no difference between life in black and in white families: Her husband "'was lovin' an' kind to me, jist as kind as you is to yo' own wife,'' and that, as parents, she and her husband "'loved dem chil'en jist de same as you loves yo' chil'en'" (578). As she narrates the reunion scene with her son, Rachel reduces the physical distance to her white interlocutor by pushing his foot, grabbing his hand, and moving her face very close to his. Acting with this sense of self-confidence, she actively resisted and undermined the Black Codes that were designed to keep black people "in their place" and that maintained white supremacy even after the passage of the 14th Amendment. With this historical context in mind, the story can in fact be read as a literary companion piece to the intensive political debate of 1874 that ultimately led to the passing of the Civil Rights Act of 1875 which reasserted "the equality of all men before the law" and expressed that "all persons within the jurisdiction of the United States shall be entitled to the full and equal enjoyment" of the features of public life (Commager 1943, Vol. 2, 86).

Perhaps the most striking formal feature of "A True Story" is its linguistic format. Written almost entirely in nonstandard English, the text attempts to render the protagonist's authentic voice. In his correspondence with William Dean Howells, then the editor of the *Atlantic Monthly*, Clemens expressed how much effort he put into an accurate representation of Rachel's black speech: "I amend dialect stuff by talking & talking & *talking* it till it sounds right [. . .] I want to work at the proofs & get the dialect as nearly right as possible" (MT 2002, L6, 233), an ambition that is visible in the surviving manuscript (Fishkin 1993, 97). With the exception of a few scattered voices (Piacentino 2011, 38–39), the majority of Mark Twain scholars have expressed their appreciation for Clemens's linguistic experiment as an attempt to depict African American speech in a manner that was different from earlier dialect representations that came out of the minstrel tradition and treated black vernacular speech as an index of intellectual and cultural inferiority. Howells's enthusiastic response to the story was partially based on what he regarded as "the best and reallest kind of black talk" (MT 1960, MTHL, 24). Kaplan praised the skillful execution of this linguistic innovation as an "impeccably nuanced but never obscure dialect" (Kaplan 1966, 181), and Fishkin described the vernacular voice as a compelling vehicle that "allows the sheer force of her [Rachel's] character and the concrete truth of her pain to shine through [. . .] with clarity and radiance" (Fishkin 1993, 99).

The fact that "A True Story" differed significantly from the earlier texts Clemens had written, caused a certain amount of confusion among the readers who had pigeonholed him as a humorist. One early critic was "dissatisfied" with him because, in this story, "he is not funny at all" (Moulton 1874). Such defeated expectations notwithstanding, "A True Story" spread widely among America's reading public. After it appeared in the November issue of the *Atlantic*, many daily newspapers immediately reprinted the text, often on their front pages. In the following year it was included in the collection *Sketches New and Old* (1875) and republished in a separate edition together with "The Recent Carnival of Crime" in 1877. Nine years after its original appearance it was included in the January 30 issue of *Swinton's Story-Teller* (MT 1884, "True"), an event which stimulated new reprintings in the local press. In addition to its distribution in printed format, "A True Story" was also featured in public performances by prominent readers at least into the 1880s ("Oak Mound").

From the time of its original appearance to its modern reception, "A True Story" has been regarded as a major literary achievement. Howells publicized his enthusiasm for the story in his review of *Sketches New and Old* in the *Atlantic Monthly* in December 1875. Here, he called it the "most perfect piece of work in the book" and praised it as a "study of character as true as life itself, strong, tender, and most movingly pathetic in its perfect fidelity to the tragic fact" (Howells 1999 [1875], 152). The novelist J.W. De Forest referred to the story as "a really great thing" and "amazingly natural" (qtd. in MT 1960, MTHL, 25). Albert Bigelow Paine, Clemens's official biographer, praised it as a "little masterpiece" (Paine 1912, 515).

Like Clemens's contemporaries, modern critics have appreciated "A True Story" as "one of his finest sketches" (Quirk 1997, 57), "an important stage in Twain's literary development" (Messent 2001, 61), and "an inspiring view of a deep humanity that triumphs over the pain and suffering caused by oppression" (Berkove 2004, xxiii). In the final analysis, the story's impact emerges from the way it evokes the cruelty and inhumanity of slavery, from the determination on the part of the African American characters to assert their dignity even under the most adverse circumstances, and from the way in which it reveals the ignorance and callousness of the white character who initiates the telling of Aunt Rachel's story. Driven by Rachel's "vigorous eloquence" (Fishkin 1993, 99), the story creates an atmosphere of moral seriousness that invites readers to scrutinize and revise stereotypical notions of black experience.

The formal composition of the story offers an innovative use of the traditional frame narrative. With its skillful use of nonstandard language, it marks an important step in the development of dialect writing, moving it away from silly and distractive misspellings and "eye dialect" toward a form of black vernacular speech that would not automatically trigger a humorous response but was suitable for the purposes of serious fiction. In this manner, the story paved the way for the work of Joel Chandler

Harris, Charles W. Chesnutt, Paul Laurence Dunbar and other writers, as well as Clemens's own *Adventures of Huckleberry Finn* (105).

In the quarter-century between 1875 and 1900, Samuel Clemens wrote the books and stories that established his reputation as a major voice in American literature and turned him into a national and international celebrity. This phase is bracketed by "A True Story" as the "career landmark" that "opened up new possibilities for him" (Griffin 2014, 3), and "The Man That Corrupted Hadleyburg" as the high-water mark of his achievement as a writer with an interest in exploring the dilemmas, challenges, and frustrations of leading a moral life. "A True Story" asks readers to take a look at the ordeal of black people during times of slavery and Reconstruction and highlights the role of unthinking whites while simultaneously triggering badly needed reflection and perhaps also stoking a sense of guilt for the inhumane transgressions of white people. "Hadleyburg" opens the focus to ask general questions about moral human behavior in the face of temptations that require a person to balance acting with integrity against the desire to pursue personal material interests. It also highlights the entanglement of people in the moral quandaries that emerge from less than virtuous behavior. In this sense, it has an almost tragic dimension since the original error committed by two people who seemed fairly decent, irrevocably draws them to the fatal ending. In the course of the events, it also implicates other characters who have become trapped in the scheme designed by a malevolent stranger.

"The Man That Corrupted Hadleyburg" was written during Clemens's sojourn in Vienna in late 1898 and published in the December issue of *Harper's Magazine* in 1899. Spread out on 26 printed pages, it was the longest contribution to the issue and featured five illustrations. As was the case with many of his other short works, Clemens reprinted the text in a collection, this time using it prominently in the book's title: *The Man That Corrupted Hadleyburg and Other Stories and Essays* (1900). Benefitting from the story's popularity, the volume was sold out before publication so that a second edition was advertised in late June ("New Publications"). Additional printings manufactured from the first edition plates appeared until at least 1902 (Hirst 1996 [23]). Widely appreciated as one of "the finest of Mark Twain's short stories" (Briden 1993, 486), it also became one of the "most frequently anthologized American works of fiction" (Gribben 1988, 65).

The text is divided into four numbered but untitled sections of uneven length. It is framed by stylized stamp-like renditions of the phrases that are central for the development of the story and, at the end, the original and the revised version of Hadleyburg's motto. The story is told by an omniscient narrator, but offers no clues about a local setting, and refrains from using nonstandard language. Praised for its "shapely plot design" this "well-made story" (Michelson 1995, 178) challenges readers with a complexity that interweaves several narrative twists and turns. It requires them to pay close attention to a number of plot complications which are consecutively introduced into the narrative.

In its bare essentials, the story unfolds the events that happen after a stranger, for the purpose of avenging an offense he once suffered in Hadleyburg, designs a scenario to ruin the town's reputation. The plan revolves around a sack of money intended as a reward for an unnamed individual who once helped the stranger. Providing the citizens with an ambiguous set of clues, he involves them in the search for the deserving recipient while simultaneously creating in them a strong temptation for self-enrichment. In the course of the action, all of the town's leading families lay claim to the treasure only to have their unscrupulous greed exposed publicly at a town-hall meeting. The whole process leads to a series of mutual suspicions, accusations, and defamations, creating an atmosphere which leaves nobody unscathed. In the end, "stripped of the last rag of its ancient glory" (MT 1992, "Hadleyburg," 437), the town gives up its name and eliminates the word "don't" from its long-standing motto "Don't lead us into temptation." In the final sentence, the reader, who has not been given evidence of a wholesale moral reformation of the town's inhabitants, is left to contend with the potential sarcasm of the narrator's assertion that the place "is an honest town once more" (438).

As the story's introduction makes clear, with regard to moral behavior Hadleyburg is no ordinary town, and most certainly not a place filled with hypocrisy and deceit. For three generations, its inhabitants had enjoyed their reputation as an "incorruptible town" which even their neighbors in the region grudgingly conceded (390). With Reverend Burgess, Barclay Goodson, and Jack Halliday it features a number of individuals whose lives are above the low motives of self-interest and self-dealing. The majority of the town's elite, however, the "nineteen principal citizens and their wives" (402), succumb to the lure of money.

Despite the disappointing recognition that none of the townspeople can live up to the reputation that they live in the "most honest and upright town in all the region round about" (390), the story relies on the idea that the main characters, Edward Richards and his wife Mary, do possess a conscience by which they know what would be the right thing to do. They are not depicted as ruthless, corrupt, or evil characters governed only by low instincts. But despite the fact that the town taught "the principles of honest dealing to its babies in the cradle" (390), the couple, like their fellow citizens, are unable to act in accordance with the values that constitute the professed foundation of the town's culture.

The question of why it is possible that a town renowned for its honesty could thoroughly fail what appears to be a test of its moral character lies at the heart of the story. All characters involved in the scheme feel an irresistible attraction to the large sum of money that seems within close reach. The wish for instant and easy wealth is driven by various motives, including a feeling of financial need, the desire for status and, in its worst form, pure greed. For the Richardses, who consider themselves "old and poor," the money promises an end to being "another man's slave" (393). At the other end of the spectrum, Pinkerton, the "mean, smirking, oily" banker (402) and one of the two richest men in Hadleyburg (430), desires the money to satisfy his irrepressible craving for wealth. In the case of Edward and Mary Richards, the reader witnesses the moral struggle they experience as they oscillate between the desire to acquire the treasure and their acute awareness that, by the terms set by the stranger, they are not eligible to receive it. Unable to resist the temptation, they attempt to justify their thoughts and actions by blaming human nature for what they know is dishonest behavior. In a moment of intense inner turmoil, Mary reaches for the line from the Lord's Prayer which serves as the town's motto – "lead us not into temptation" – but is unable to actually speak the phrase. All that remains for her to do is lament her own weakness by pointing to what she perceives as the warped and deplorable design of human nature: "Lord, how we are made – how strangely we are made!" (397–398). In this way, she relinquishes all sense of individual responsibility and agency.

At first sight, it may appear easy to dismiss the behavior displayed in the story as weakness of character. Such a perspective implies that the fall of Hadleyburg would have been avoidable if people had done the "right" thing. However, the impulse to blame people for their morally flawed behavior presupposes that the option of being honest, of making responsible decisions, free of greed and selfishness, was truly available to them. There is evidence to show that Clemens did not believe that this was possible. Already in 1885, fourteen years before the publication of "Hadleyburg," Clemens had determined for himself that "[t]he main-spring of man's nature is just that – selfishness" (MT 1992, "Character," 858). In this narrative, he goes one step further and denies the possibility that education or "training" could remedy this fatal flaw. In an expressive passage early on in the story, Mary has a moment of self-realization and understands that all the training was "weak as water when temptation comes" and that the entire "town's honesty is as rotten" as hers (400).

Struggling with the realization that they are incapable of withstanding the repeated temptations laid in front of them, Mary and Edward Richards conclude that such resistance was beyond the powers that humans possess. For them, the explanation for their dilemma lies in "the designs of Providence" (400). They reassure each other that they are not to blame because they "couldn't help it" since everything "was ordered. All things are" (431). The religious determinism expressed here contains echoes of entrenched Calvinist beliefs in American culture. As Berkove and Csicsila have emphasized, Clemens himself was involved in a lifelong struggle with Calvinism which they consider "a key to his main themes" (Berkove and Csicsila 2010, 13). In a 1904 letter, Clemens admonished himself to be more forgiving to the human race because "it is moved wholly by outside influences, it has no hand in creating the outside influences nor in choosing which of them it will welcome or reject, [. . .] it has no more mastership nor authority over its mind than over its stomach" (MT 1917, L, 764).

With this belief, Clemens was not alone. The end of the nineteenth century saw the rise to dominance of a scientific worldview that – fueled by the path-breaking discoveries in geology, paleontology, and evolutionary biology – undermined many traditional ideas about man's place in nature and eroded illusions about individual choice. As a result, the intellectual climate seemed no longer "appropriate to treat characters as if they were morally accountable" (Mitchell 1988, 525). Consequently, authors such as Stephen Crane, Theodore Dreiser, and Frank Norris began to write about "circumstances that deprived individuals of responsibility" (525). While Clemens shared this perspective, he still found it hard to withhold judgment on morally suspect behavior. Sometimes he denounced humanity as "the damned human race" (Howells 1910, 76) but on other occasions he expressed a degree of pity for people who seemed to be nothing more than puppets helplessly adrift in a world governed by a deity whom he regarded as whimsical, angry, and malevolent (Berkove and Csicsila 2010, 6).

By focusing specifically on the struggles of a well-intentioned and generally honest and moral old couple, "Hadleyburg" dramatizes the conflict between the knowledge of good and right and the actual behavior that disregards these ethical standards. The story puts readers in the uncomfortable position of having to witness how people fail to act according to moral standards although they understand the stakes that are involved in the decisions they face. In his 1885 novel *Adventures of Huckleberry Finn*, Clemens had drawn a hopeful picture of the potential for good moral behavior when the untutored child of a drunkard followed the guidance of "a sound heart" (Fischer et al. 2003, 619) and defied the inhuman laws of a slave-holding society. Fourteen years later, his depiction of Hadleyburg lacks anything resembling a moral core that would enable the town's respectable citizens to overcome their greedy impulses and selfish inclinations.

Although marred by a few philosophical and logical inconsistencies and implausibilities (Quirk 1997, 234), the story has largely received positive feedback from popular and academic audiences. Beginning in late November 1898, local newspapers from Maine to Michigan and Georgia were reprinting a news item which announced the arrival of the Christmas number of *Harper's* magazine containing "The Man That Corrupted Hadleyburg," and predicting that "the majority of readers will probably consider this the best tale Mark Twain has ever written." It was a story not to be missed because it was "one of the things that makes life worth living" ("Fiftieth Year," "Literary News and Notes," 1899). The *New York Daily Tribune* printed a brief review of what it called a "deeply significant" story and praised its artful combination of social satire and the "richest and most spontaneous humor." In it, the reviewer stated, Clemens "analyze[d] some of the great truths of human nature [. . .] with uncommon power" ("The Christmas Magazines," 1899). When it was reprinted in the eponymous collection of stories that came out in June 1900, reviewers renewed their praise. The *Tribune* called the story "a sermon on hypocrisy," and a blend of "tragedy and farce" that turns the narrative into "an unforget[t]able lesson" ("Fiction," 1900). Another reviewer called the story "one of Mr. Clemens's profoundest studies of human nature" and "a revelation of a fundamental fact" ("Stories and Essays," 1999). The *Boston Post* saw the story as evidence that Clemens had now graduated from a mere humorist to "a writer of great power" ("Books of the Day," 1999).

A survey of the scholarly response to "The Man That Corrupted Hadleyburg" largely confirms the early praise accorded to the story. In the one hundred years that lie between Albert Bigelow Paine's verdict that the narrative "takes its place with the half-dozen great English short stories of the world" (Paine 1912, 1068) and Harold H. Kolb's statement that this is "Twain's longest, most coherent, and most carefully plotted" story (Kolb 2015, 302), there is a consensus on the quality of Clemens's literary achievement, making the text "the most critically acclaimed of all Mark Twain's short fiction" (Wilson 1987, 206). While there is agreement about the technical skills displayed in the construction of the story, Wilson's brief survey shows that the interpretations of its meaning vary considerably (206–212). This spectrum includes studies on intertextual connections (see Scharnhorst and Harris), but the majority of critical readings revolve around the philosophical and religious problems raised by the story. One major point of contention is the question of the text's final resolution: Does the final sentence "[Hadleyburg] is an honest town once more" (MT "Hadleyburg," 1992, 438) allow a literal reading and an optimistic interpretation that it constitutes a celebration of the town's moral regeneration (Wilson 1987, 210)? Or is this to be read as an ironic statement which confirms the notion that there is no end to the town's self-delusion? The numerous careful and sophisticated critical readings are vivid testimony to the fact that Clemens managed to write a text that "is rich in interpretive possibility" (Quirk 1997, 103), addressing a host of moral problems involving human virtue, conscience, guilt, and shame – together with fairness and justice. Like all good literature, it exposes readers to a complex situation and challenges them to carefully weigh the options laid out and dramatized in the narrative, inviting them to engage more deeply with the psychological and social consequences of human interaction.

3 Theoretical Perspectives

Anthologies featuring Clemens's shorter writings have been available since he reprinted twenty-seven texts from his days as a journalist in Nevada and California in his 1865 book, *The Celebrated Jumping Frog of Calaveras County and Other Sketches* (Bucci 1996, 1). Practiced throughout the remainder of his life, the routine of reprinting selected material ultimately yielded seven volumes of short prose collections which included historical, philosophical, and scientific essays, political tracts, speeches, sketches, and stories (Messent 2001, 2). After Clemens's death, individual publishers and editors made their own selections from Clemens's shorter writings, referring to them as "short fiction," "short works," or "tales" and "sketches" – an indicator for the general uncertainty regarding genre classification. The first of these anthologies, *The Complete Short Stories and Famous Essays of Mark Twain* (1928), gathered 99 texts on more than 1,000 pages of print. It was based on a loose interpretation of the words "complete" and "short story" that allowed the inclusion of "Tom Sawyer, Detective" (ca. 24,000 words), "Tom Sawyer Abroad" (ca. 34,000 words), "The Mysterious Stranger" (ca. 35,000 words), and "Pudd'nhead Wilson" (ca. 53,000 words), all of them significantly above the word count of many of the best-known stories by Poe and Hawthorne (↗7 Poe; ↗6 Hawthorne) which generally feature fewer than 4,000 words.

Judging from Charles Neider's observation made more than twenty years after its publication, the 1928 anthology did not succeed in drawing much attention to Clemens's shorter writings. In 1957, Neider found himself wondering why the stories of this "very great man, a national monument" (Neider 1990 [1957], xi) had never been collected. He speculated that it may have been due to the fact that Clemens's short pieces were a "hodgepodge" of material that "fluctuate[d] between fiction and fact" (xii), and that his "unconventionality, his insistence on formlessness" (xii) might have prevented potential editors from compiling a consistent body of texts. Dodging the task of providing appropriate criteria for his own selection, Neider chose the opposite path and determined that certain text types should not be considered for inclusion in a volume of "short stories." They included autobiographical material, historical reports, as well as jokes, hoaxes, and moral sermons. At the same time, he felt free to insert into his volume thirteen nonfiction stories, plus the lengthy "The Mysterious Stranger" and material from Clemens's travel books. Regardless of whether these decisions were plausible or not, his project proved to be so popular that it has remained in print until the present day. The book's success also spawned several additional volumes featuring more of Clemens's short writings, making Neider a uniquely successful "popularizer" whose editions have possibly been "more widely read than those of anyone else" (Rasmussen 2001, 3).

From a scholarly perspective, two projects assume special positions among the collections of Mark Twain's shorter writings: The two volumes of tales and sketches edited by Louis J. Budd for the *Library of America* project, and the ambitious, albeit still unfinished, endeavor pursued by the Mark Twain Project at the Bancroft Library. The editorial team at Berkeley has so far published two out of five distinct collections of tales and sketches (Fischer 1993, 496) which cover the period from 1861 to 1865. With their careful headnotes, extensive explanatory notes, and a detailed textual apparatus, these volumes are unrivaled as scholarly editions of the material they present. Budd's two volumes reprint 272 items in chronological sequence, and currently offer the most comprehensive selection from Clemens's short writings. Due to its relative completeness and its solid editorial principles, it is the "single finest collection of Mark Twain's short writings yet published" (Rasmussen 1994).

Although an impressive number of short texts was composed, published, and reprinted during all phases of Clemens's career and in its aftermath, and despite the fact that one or two of them reliably appear in anthologies of American literature and short fiction, his shorter writings have only sporadically received the recognition they deserve. Historical surveys of American short fiction generally do not assign him a prominent place and tend to treat him as a minor contributor to the field. Two years after Clemens's much publicized death, Smith's *The American Short Story* (1912) mentions the author in passing while highlighting Bret Harte, no longer considered a major figure in American literature today, for doing "more than anyone else to give the American short story immediate recognition as a new *genre* both at home and abroad" (Smith 1912, 27). Likewise, Pattee's historical survey, *The Development of the American Short Story* (1923), touches only marginally on the "Jumping Frog," the story that established Clemens's national fame as a humorist, characterizing it as a set of "farcical paragraphs and anecdotes and grotesque sketches" (Pattee 1923, 232) that "rambles in sublime contempt for any short-story canon" (297).

While there has been a growing recognition of Clemens's short writings in recent decades, a comparison with America's most prominent short story writers underlines the limited interest. A quick check of the MLA online bibliography of scholarly publications reveals that there are 437 entries for Melville's "Bartleby" (↗8 Melville), 134 for Hawthorne's "Young Goodman Brown" (↗6 Hawthorne), 117 for Irving's "Rip Van Winkle" (↗5 Irving), and 89 for Poe's "Masque of the Red Death" (↗7 Poe). Among Clemens's short pieces, "The Man That Corrupted Hadleyburg" received the most attention with 34 articles, followed by the "Jumping Frog" story (27), "A Private History of a Campaign That Failed" (18), and "A True Story" (14).

This state of affairs is also reflected in modern handbooks and publications that focus specifically on the American short story. Frequently, the discussion of Clemens's work is conducted in the context of Bret Harte and "local color" fiction, a category which, as Sundquist (Sundquist 1988, 503) and Campbell (Campbell 2003, 93) have argued, tends to diminish the quality of the writings labeled in this way. Similarly, the label "humor" that is reliably attached to Clemens's works, threatens to undermine rather than enhance the value of such texts. Apart from the general prejudice against humor as an indicator of light and minor forms of literature, modern readers often find it difficult to respond to nineteenth-century topics and techniques with the same degree of amusement as Clemens's contemporaries.

A review of the extensive body of research containing a multitude of perceptive insights from scholars past and present illustrates that Clemens's work can profitably be read and analyzed in the context of short-story criticism. At the same time, it is worth noting that his material does not fit easily into the traditional framework of the American short story. The fact that he is most often treated as a "local color writer" and a representative of the "comic short story in America" (Sloane 2010, 78) may be an indication that critics have found it difficult to assign him a proper place in the pantheon of the writers who tend to be seen as the genre's classic representatives. Even Columbia professor and literary critic Brander Matthews, an early enthusiastic champion of Clemens as "a literary artist of a very high order" (Matthews 1999 [1885], 261), did not mentioned him in his influential *The Philosophy of the Short-story* (1901) and left him out of the anthology which he compiled "to exhibit the many varieties possible within the species" (Matthews 1907, 3).

Never interested in following the formal conventions of literary writing and pretending that short pieces were not worth the effort they required, Clemens devoted little effort to a genre that was only beginning to get critical attention at the end of the nineteenth century. At the same time, he did compose a large number of textual units of limited length throughout his life, but there is no evidence that they were motivated by a conscious effort to find a new aesthetic of form. Clemens may have had an intuition that there was a limit to the attention span of the average reader and that the majority among his popular audience could afford to devote only a limited amount of time to reading. Writing texts "in palatable, profitable doses" (MT 1997, ET&S1, 618) may have been a pragmatic move to enhance the reading pleasure of his readership by offering varied and diverse reading matter, and by also providing convenient stopping points which would make it easy for people to pause and resume their reading activity. For him, it was a mode of writing that – from the early news items in the West to the disconnected memories assembled in his autobiographical dictations in his final years – fit his various journalistic and literary projects well and suited his personal disposition.

With its emphasis on the ubiquity of short textual units in the writing of Samuel Clemens, this chapter has suggested that the widespread characterizations of "Mark Twain" as a representative of local color writing and a writer of humorous sketches capture only two aspects of his work. Clemens's formal and thematic range surpasses the limits implied by these labels. As the focused readings of the two stories selected for closer inspection reveal, and as his other writings confirm, he was an individual who was deeply concerned with American life in its multiple facets, and with the moral implications and consequences of human behavior. Consequently, Clemens's work is best understood outside of a normative definition of the "short story" and deserves to be appreciated for its creative manipulation of short textual forms which cover a broad range of themes and ideas in the most variable manner in periodicals, anthologies, and as vital components of his travelogues and novels. Whatever it was that drove him to fit his reading portions to his readers' mental digestive capacity, it proved to be a successful strategy that turned him into "one of the outstanding and most popular figures in the literature of America" (Hart 1961, 150).

4 Bibliography

4.1 Works Cited

4.1.1 Primary Sources

Twain, Mark. "Aunt Rachel's Story: [Mark Twain in Swinton's Story-teller]," https://cdnc.ucr.edu/?a=d&d=SD18840405.2.51.27&srpos=2&e=——-en–20-SD-1–txt-txIN-%22aunt+rachel%22——-1. *Sonoma Democrat [Supplement]* 4 Apr. 1884: 2. (16 June 2021).

Twain, Mark. "A Bloody Massacre near Carson." *Collected Tales, Sketches, Speeches & Essays. 1852–1890*. Ed. Louis J. Budd. New York: The Library of America, 1992. 57–58.

Twain, Mark. "The Character of Man." *Collected Tales, Sketches, Speeches & Essays. 1852–1890*. Ed. Louis J. Budd. New York: The Library of America, 1992. 854–858.

Twain, Mark. *The Complete Essays of Mark Twain: Now Collected for the First Time*. Ed. Charles Neider. Garden City, New York: Doubleday, 1963.

Twain, Mark. *The Complete Humorous Sketches and Tales of Mark Twain. Now Collected for the First Time*. Ed. Charles Neider. Garden City, New York: Hanover House, 1961.

Twain, Mark. *The Complete Short Stories of Mark Twain*. 1957. Ed. Charles Neider. New York: Bantam Books, 1990.

Twain, Mark. *The Complete Short Stories and Famous Essays of Mark Twain*. One Volume Edition. New York: P. F. Collier & Son, 1928 [?].

Twain, Mark. [ET&S1] *Early Tales & Sketches, Volume 1 (1851–1864)*. Eds. Edgar Marquess Branch, Robert H. Hirst, and Harriet Elinor Smith. Berkeley: U of California P, 1979.

Twain, Mark. [ET&S2] *Early Tales & Sketches, Volume 2 (1864–1865)*. Eds. Edgar M. Branch and Robert H. Hirst. Berkeley: U of California P, 1981.

Twain, Mark. "How to Tell a Story." *Collected Tales, Sketches, Speeches & Essays. 1891–1910*. Ed. Louis J. Budd. New York: The Library of America, 1992. 201–206.

Twain, Mark. "Humorists." *Mark Twain in Eruption: Hitherto Unpublished Pages about Men and Events*. Ed. Bernard DeVoto. New York: Harper, 1940. 200–203.

Twain, Mark. "Jim Baker's Blue-Jay Yarn." *The Best Short Stories of Mark Twain*. Ed. Lawrence I. Berkove. New York: Modern Library, 2004. 72–76.

Twain, Mark. "Jim Smiley and His Jumping Frog." *Collected Tales, Sketches, Speeches & Essays. 1852–1890*. Ed. Louis J. Budd. New York: The Library of America, 1992. 171–177.

Twain, Mark. [L] *Mark Twain's Letters: Arranged with Comment by Albert Bigelow Paine*. 2 vols. New York and London: Harper & Brothers, 1917.

Twain, Mark. [L1] *Mark Twain's Letters: Volume 1 (1853–1866)*. Eds. Edgar M. Branch, Michael B. Frank, and Kenneth M. Sanderson. Berkeley: U of California P, 1988.

Twain, Mark. [L4] *Mark Twain's Letters: Volume 4 (1870–1871)*. Eds. Victor Fischer and Michael B. Frank. Berkeley: U of California P, 1995.

Twain, Mark. [L6] *Mark Twain's Letters: Volume 6 (1874–1875)*. Eds. Michael B. Frank and Harriet Elinor Smith. Berkeley: U of California P, 2002.

Twain, Mark. *Mark Twain: Life as I Find It: Essays, Sketches, Tales, and Other Material, the Majority of Which Is Now Published in Book Form for the First Time*. Ed. Charles Neider. Garden City, New York: Hanover House, 1961.

Twain, Mark. *Life on the Mississippi. The Oxford Mark Twain*. Ed. Shelley Fisher Fishkin. New York: Oxford UP, 1996.

Twain, Mark. "The Man That Corrupted Hadleyburg." *Collected Tales, Sketches, Speeches & Essays. 1891–1910*. Ed. Louis J. Budd. New York: The Library of America, 1992. 390–438.

Twain, Mark. *The Man That Corrupted Hadleyburg and Other Stories and Essays. The Oxford Mark Twain*. Ed. Shelley Fisher Fishkin. New York: Oxford UP, 1996.

Twain, Mark, and William D. Howells. [MTHL] *Mark Twain – Howells Letters: The Correspondence of Samuel L. Clemens and William D. Howells, 1872–1910*. Eds. Henry Nash Smith and William M. Gibson. Vol. 1. Cambridge: Belknap P of Harvard UP, 1960.

Twain, Mark. *Mark Twain's Library of Humor*. Eds. Samuel Langhorne Clemens, William Dean Howells, and Charles Hopkins Clark. New York: Charles L. Webster & Company, 1888.

Twain, Mark. "Petrified Man." *Collected Tales, Sketches, Speeches & Essays. 1852–1890*. Ed. Louis J. Budd. New York: The Library of America, 1992. 19.

Twain, Mark. "A Private History of a Campaign That Failed." *Collected Tales, Sketches, Speeches & Essays. 1852–1890*. Ed. Louis J. Budd. New York: The Library of America, 1992. 863–882.

Twain, Mark. *Sketches, New and Old* [1875]. *The Oxford Mark Twain*. Ed. Shelley Fisher Fishkin. New York: Oxford UP, 1996.

Twain, Mark. "Sociable Jimmy." *New York Times*, 29 Nov 1874. Reprinted in Shelley Fisher Fishkin. *Was Huck Black? Mark Twain and African-American Voices*. New York: Oxford UP, 1993. 249–252.

Twain, Mark. *A True Story and the Recent Carnival of Crime*. Boston: James R. Osgood and Company, 1877.

Twain, Mark. "A True Story." *Swinton's Story-teller: A Weekly of Choice Complete Tales* 16 (1884): 22–24.

Twain, Mark. "A True Story, Repeated Word for Word as I Heard It." *Collected Tales, Sketches, Speeches & Essays. 1852–1890*. Ed. Louis J. Budd. New York: The Library of America, 1992. 578–582.

Twain, Mark. "The Turning Point of My Life." *Collected Tales, Sketches, Speeches & Essays. 1891–1910*. Ed. Louis J. Budd. New York: The Library of America, 1992. 929–938.

4.1.2 Secondary Sources

Andrews, Kenneth R. *Nook Farm: Mark Twain's Hartford Circle*. 1950. Seattle: U of Washington P, 1967.

Berkove, Lawrence I. "The Sagebrush School Revived." *A Companion to the Regional Literatures of America*. Ed. Charles L. Crow. Malden, Massachusetts: Blackwell, 2003. 324–343.

Berkove, Lawrence I. "Editor's Preface." *The Best Short Stories of Mark Twain*. Ed. Lawrence I. Berkove. New York: Modern Library, 2004. xix–xxv.

Berkove, Lawrence I. "Nevada Influences on Mark Twain." *A Companion to Mark Twain*. Eds. Peter B. Messent and Louis J. Budd. Malden, Massachusetts: Blackwell, 2005. 157–171.

Berkove, Lawrence I., and Joseph Csicsila. *Heretical Fictions: Religion in the Literature of Mark Twain*. Iowa City: U of Iowa P, 2010.

Blair, Walter, ed. *Native American Humor*. 1937. San Francisco: Chandler Pub. Co, 1960.

Blair, Walter. "Mark Twain's Other Masterpiece: 'Jim Baker's Blue-Jay Yarn.'" *Essays on American Humor: Blair Through the Ages*. Ed. Hamlin Hill. Madison: U of Wisconsin P, 1993. 165–178.

"Books of the Day." *Boston Post*, 1 July 1900. *Mark Twain: The Contemporary Reviews*. Ed. Louis J. Budd. Cambridge: Cambridge UP, 1999. 487.

Botkin, Benjamin A. *A Treasury of American Folklore: Stories, Ballads, and Traditions of the People*. New York: Crown Publishers, 1944.

Branch, Edgar M., Ed. *Clemens of the Call: Mark Twain in San Francisco*. Berkeley: U of California P, 1969.

Briden, Earl. "Man That Corrupted Hadleyburg, The." *The Mark Twain Encyclopedia*. Eds. J. R. LeMaster and James D. Wilson. New York: Garland Publishing, 1993. 484–486.

Bucci, Richard. "Afterword." *Mark Twain. The Celebrated Jumping Frog of Calaveras County and Other Sketches. The Oxford Mark Twain*. Ed. Shelley F. Fishkin. New York: Oxford UP, 1996. 1–18.

Budd, Louis J. *Our Mark Twain: The Making of His Public Personality*. Philadelphia: U of Pennsylvania P, 1983.

Campbell, Donna. "Realism and Regionalism." *A Companion to the Regional Literatures of America*. Ed. Charles L. Crow. Malden: Blackwell, 2003. 92–110.

Cohen, Hennig, and William B. Dillingham, eds. *Humor of the Old Southwest*. 2nd ed. Athens: U of Georgia P, 1975.

"The Christmas magazines," https://chroniclingamerica.loc.gov/lccn/sn83030214/1899-12-19/ed-1/seq-8. *New York Daily Tribune* 19 Dec. 1899: 8. (03 March 2021).
Commager, Henry S. *Documents of American History*. 3rd ed. New York: F.S. Crofts & Co, 1943.
DeVoto, Bernard. *Mark Twain's America*. 1932. Cambridge: Houghton Mifflin Co, 1951.
Fatout, Paul. *Mark Twain on the Lecture Circuit*. Bloomington: Indiana UP, 1960.
"Fiction: A Masterpiece of Humor and Wisdom," https://chroniclingamerica.loc.gov/lccn/-sn83030214/1900-08-04/ed-1/seq-10/. *New York Daily Tribune* 4 Aug. 1900: 10. (03 March 2021).
"The Fiftieth Year of Harper's Magazine," https://chroniclingamerica.loc.gov/-lccn/sn97070614/1899-12-01/ed-1/seq-2/. *The Owosso Times [Mich.]* 1 Dec. 1899: 2. (03 March 2021).
Fischer, Victor. "Mark Twain Project." *The Mark Twain Encyclopedia*. Eds. J. R. LeMaster and James D. Wilson. New York: Garland Publishing, 1993. 495–497.
Fischer, Victor, Lin Salamo, and Walter Blair. "Appendix F: Mark Twain's Revisions for Public Reading, 1895–1896." *Adventures of Huckleberry Finn*. Eds. Victor Fischer, Lin Salamo, and Walter Blair. Berkeley: U of California P, 2003. 617–654.
Fishkin, Shelley F. *Was Huck Black? Mark Twain and African-American Voices*. New York: Oxford UP, 1993.
Gerber, John C. "Introduction: The Continuing Adventures of Huckleberry Finn." *One Hundred Years of Huckleberry Finn: The Boy, His Book, and American Culture*. Eds. Robert Sattelmeyer and J. D. Crowley. Columbia: U of Missouri P, 1985. 1–12.
Glasscock, Carl B. *The Big Bonanza: The Story of the Comstock Lode*. Indianapolis: The Bobbs-Merrill Company, 1931.
Gribben, Alan. "Mark Twain." *American Short-Story Writers Before 1880. Dictionary of Literary Biography* 74. Eds. Bobby E. Kimbel and William E. Grant. Detroit, Michigan: Gale, 1988. 54–83.
Griffin, Benjamin. "Introduction." *A Family Sketch and Other Private Writings*. Ed. Benjamin Griffin. Oakland, California: U of California P, 2014. 1–11.
Halpert, Herbert. "The Blue Hen's Chickens." *American Speech* 26.3 (1951): 196–198.
Harris, Susan K. "'Hadleyburg': Mark Twain's Dual Attack on Banal Theology and Banal Literature." *American Literary Realism* 16.2 (1983): 240–252.
Hart, James D. *The Popular Book: A History of America's Literary Taste*. Berkeley: U of California P, 1961.
Hill, Hamlin. "Mark Twain: Audience and Artistry." *American Quarterly* 15.1 (1963): 25–40.
Hirst, Robert H. "A Note on the Text." *The Man That Corrupted Hadleyburg and Other Stories and Essays. The Oxford Mark Twain*. Ed. Shelley Fisher Fishkin. New York: Oxford UP, 1996. [23].
Hirst, Robert H. "Where Do You 'Find' Mark Twain Letters?," http://digitalassets.lib-.berkeley.edu/bancroftiana/ucb/text/bancroftiana_114.pdf. *Bancroftiana: Newsletter of the Friends of the Bancroft Library* 114 (1999): 13. (03 March 2021).
Hook, Andrew. "Reporting Reality: Reporting Mark Twain's Short Stories." *The Nineteenth-Century American Short Story*. Ed. A. R. Lee. London, Totowa, New Jersey: Barnes & Noble, 1985. 103–119.
Howells, William Dean. "Recent Literature." *Atlantic Monthly* 36 (Dec 1875), 749–751. *Mark Twain: The Contemporary Reviews*. Ed. Louis J. Budd. Cambridge: Cambridge UP, 1999. 151–152.
Howells, William Dean. *My Mark Twain: Reminiscences and Criticisms*. New York: Harper & Brothers, 1910.
Howells, William Dean. "Life and Letters." *Harper's Weekly* 13 (Feb 1897): 155.
Jerome, Robert D., and Herbert A. Wisbey. *Mark Twain in Elmira. With Revisions and Additions by Barbara E. Snedecor*. Elmira, New York: Elmira College Center for Mark Twain Studies, 2013.
Kaplan, Justin. *Mr. Clemens and Mark Twain: A Biography*. New York: Simon and Schuster, 1966.

Kersten, Holger. "'I know grammar by ear only': Mark Twain and the Norms of American Literature." *Amerika und die Norm: Literatursprache als Modell?* Eds. Christopher F. Laferl and Bernhard Pöll. Tübingen: Niemeyer, 2007. 25–36.

Kolb, Harold H. *Mark Twain: The Gift of Humor*. Lanham: UP of America, 2015.

"Literary News and Notes," https://chroniclingamerica.loc.gov/lccn/sn78000873/1899-12-07/ed-1/seq-7/. *The Republican Journal* [Belfast, Maine] 7 Dec. 1899: 7. (03 March 2021).

MacDonnell, Kevin. "How Samuel Clemens Found 'Mark Twain' in Carson City." *Mark Twain Journal* 50.1/2 (2012): 9–47.

Matthews, Brander. "'Huckleberry Finn.'" *Saturday Review* 59 (31 Jan 1855). *Mark Twain: The Contemporary Reviews*. Ed. Louis J. Budd. Cambridge: Cambridge UP, 1999. 260–263.

Matthews, Brander. *The Philosophy of the Short-story*. New York: Longman's, Green and Co., 1901.

Matthews, Brander. *The Short-Story: Specimens Illustrating Its Development*. New York: American Book Company, 1907.

Messent, Peter B. *The Short Works of Mark Twain: A Critical Study*. Philadelphia: U of Pennsylvania P, 2001.

Michelson, Bruce. *Mark Twain on the Loose: A Comic Writer and the American Self*. Amherst: U of Massachusetts P, 1995.

Mitchell, Lee Clark. "Naturalism and the Languages of Determinism." *Columbia Literary History of the United States*. Eds. Emory Elliott, Martha Banta, and Houston A. Baker. New York: Columbia UP, 1988. 525–545.

Moulton, Louise Chandler [L.C.M.]. "Boston. Literary Notices," https://chroniclingamerica-.loc.gov/lccn/sn83030214/1874-10-15/ed-1/seq-6/ . *New York Tribune* 15 Oct. 1874: 6. (03 March 2021).

Neider, Charles. "Introduction." *The Complete Short Stories of Mark Twain*. 1957. Ed. Charles Neider. New York: Bantam Books, 1990. xi–xxii.

"New Publications: Six Books That Sell,", https://chroniclingamerica.loc.gov/lccn/-sn83030272/1900-06-23/ed-1/seq-7/. *The Sun* [New York] 23 June 1900: 7. (03 March 2021).

"Oak Mound Entertainment," https://www.newspapers.com/image/562973002/. *Napa Register [Cal.]* 29 June 1883: 1. (10 March 2021).

Paine, Albert B. *Mark Twain: A Biography: The Personal and Literary Life of Samuel Langhorne Clemens*. 3 vols. New York: Harper & Brothers, 1912.

Pattee, Fred L. *The Development of the American Short Story: An Historical Survey*. New York, London: Harper & Brothers, 1923.

Piacentino, Ed. "Recontextualizing Mark Twain's 'A True Story.'" *Mark Twain Annual* 9 (2011): 31–43.

Poe, Edgar Allan. "Nathaniel Hawthorne: Twice Told Tales. Literary Criticism [1842]." *Die amerikanische Short Story: Theorie und Entwicklung*. Ed. Hans Bungert. Darmstadt: Wissenschaftliche Buchgesellschaft, 1972. 1–8.

Powers, Ron. *Mark Twain: A Life*. New York: Free P, 2005.

Quirk, Tom. *Mark Twain: A Study of the Short Fiction*. New York: Twayne Publishers, 1997.

Rasmussen, R. Kent. "Rev. of *Mark Twain: Collected Tales, Sketches, Speeches, and Essays*," www.twainweb.net/reviews/budd-loa.html. *Twainweb.net*. 10 Oct 1994. (03 March 2021).

Rasmussen, R. Kent. "Charles Neider," https://marktwaincircle.files.wordpress.com-/2019/07/circular-2001-q3.pdf . *Mark Twain Circular: Newsletter of the Mark Twain Circle of America* 15.3 (2001): 2–4. (10 March 2021).

Scharnhorst, Gary. "Paradise Revisited: Twain's 'The Man That Corrupted Hadleyburg.'" *Studies in Short Fiction* Winter 1981.1 (18): 59–64.

Sloane, David E. E. "Mark Twain and the American Comic Short Story." *A Companion to the American Short Story*. Eds. Alfred Bendixen and James Nagel. Chichester U.K. Wiley-Blackwell, 2010. 78–90.

Smith, C. Alphonso. *The American Short Story*. Boston: Ginn, 1912.
"Stories and Essays by Mark Twain." *New York Times Saturday Review of Books and Art*, 23 June 1900. *Mark Twain: The Contemporary Reviews*. Ed. Louis J. Budd. Cambridge: Cambridge UP, 1999. 485–487.
Sundquist, Eric J. "Realism and Regionalism." *Columbia Literary History of the United States*. Ed. Emory Elliott. New York: Columbia UP, 1988. 501–524.
Taylor, Ezekiel D. "Literary Matters / New Books." *Utica [N.Y.] Morning Herald and Gazette*, 23 Feb 1872. *Mark Twain: The Contemporary Reviews*. Ed. Louis J. Budd. Cambridge: Cambridge UP, 1999. 99–100.
Thompson, Charles Miner. "Mark Twain as an Interpreter of American Character." *Atlantic Monthly* 79:474 (April 1897): 443–450.
Wecter, Dixon. *Sam Clemens of Hannibal*. 1952. Boston: Houghton Mifflin Company, 1961.
Wilson, James D. *A Reader's Guide to the Short Stories of Mark Twain*. Boston Mass. G.K. Hall, 1987.
Wonham, Henry B. "Mark Twain's Short Fiction." *A Companion to Mark Twain*. Eds. Peter B. Messent and Louis J. Budd. Malden, Massachusetts: Blackwell, 2005. 357–370.

4.2 Further Reading

Ahrends, Günter. "Mark Twain, Bret Harte, O. Henry und Hamlin Garland." *Die amerikanische Kurzgeschichte*. 5th ed. Trier: WVT, 2008. 88–101.
Branch, Edgar M. *The Literary Apprenticeship of Mark Twain: With Selections from His Apprentice Writing*. Urbana: U of Illinois P, 1950.
Covici, Pascal. *Mark Twain's Humor: The Image of a World*. Dallas: Southern Methodist UP, 1962.
Gibson, William M. "Short Fiction." *The Art of Mark Twain*. New York: Oxford UP, 1976. 72–96.
Kersten, Holger. "The Rise of Local Color Writing in the American West: Bret Harte's 'The Outcasts of Poker Flat' and Mark Twain's 'The Celebrated Jumping Frog of Calaveras County.'" *A History of the American Short Story: Genres – Developments – Model Interpretations*. Eds. Michael Basseler and Ansgar Nünning. Trier: WVT, 2011. 123–136.
Messent, Peter. "Carnival in Mark Twain's 'Stirring Times in Austria' and 'The Man That Corrupted Hadleyburg.'" *Studies in Short Fiction* 35.3 (1998): 217–232.
Messent, Peter. "Character and Agency: Teaching Mark Twain's 'A True Story.'" *Eureka Studies in Teaching Short Fiction* 4.1 (2003): 20–31.
Quirk, Tom. *Mark Twain and Human Nature*. Columbia: U of Missouri P, 2007.
Railton, Stephen. "Mark Twain Tries to Get the Last Laugh: Hadleyburg and Other Performances." *Mark Twain Annual* 3 (2005): 23–36.
Voss, Arthur. "Local Color and Western Humor: Bret Harte and Mark Twain." *The American Short Story: A Critical Survey*. Norman: U of Oklahoma P, 1973. 70–87.
Wuster, Tracy. "'I Wa' N't Bawn in de Mash to Be Fool' by Trash!': Mark Twain's 'A True Story' and the Culmination of Southern Frontier Humor." *Southern Frontier Humor: New Approaches*. Ed. Ed Piacentino. Jackson: UP of Mississippi, 2013. 131–153.

James Nagel

10 Charles W. Chesnutt (1858–1932)

Abstract: By the final decades of the nineteenth century there was an increasing emphasis in American Realism on individual freedom and personal responsibility based on the assumption that people are free agents capable of reason and thus accountable for the key decisions in their lives. One of the African American writers who utilized this idea was Charles W. Chesnutt, a lawyer who brought to fiction a great deal of intellectual and stylistic sophistication. Two of his stories on this theme have come to be regarded as among the most important works of the period: "The Wife of His Youth" and "The Sheriff's Children." The first involves a profound ethical question within the mixed-race community of a fictional city in Ohio, and the second focuses on a white sheriff who once sold a slave and her son down the river and now is forced to confront the consequences of his actions. Both stories deal intensely with a moment of decision that will permanently define the moral standing of the protagonist, and both have important implications for the new society that was forming in the United States after the Civil War.

Keywords: Morality, slavery, short story, realism

1 Context

Charles W. Chesnutt was an extraordinary intellectual presence in the late-nineteenth and early-twentieth centuries. A writer of critically important fiction, a lawyer of impressive ability, a businessman who created a thriving industry in stenography, and an award-winning racial activist who supported civil rights for people of color and became one of the leaders of the National Association for the Advancement of Colored People, Chesnutt stands as a vital figure who helped shape a more just and humane American society before he died in 1932. Although born in Cleveland in 1858, his youth was spent in Fayetteville, North Carolina, where he was educated and began a career teaching before moving into administration.

North Carolina was a particularly remarkable Southern state with a complex history of involvement in the Civil War, having always been divided both topographically and politically. The Coastal Plain in the eastern part was a plantation region with a long tradition of slavery and white supremacy, and that area was quick to support secession and the formation of the Confederacy. The western region deplored the moral compromises of involuntary servitude, resisted the secession, and was reluctant to join the newly formed government of the South. In fact, that area supported the Union army, sending five regiments of troops to join the infantry and two additional units of cavalry as well as an African American regiment

https://doi.org/10.1515/9783110587647-011

of heavy artillery. The eastern part solidly supported the Confederate effort and contributed more men to that cause than any other state, among them many notable generals, including Braxton Bragg and Leonidas Polk. But the fighting within North Carolina itself was somewhat restrained – perhaps in deference to its divided loyalties – and in fact the last major event was the April, 1865, surrender of forces by General Joseph E. Johnston at Bennett Place, Durham County, North Carolina, the largest capitulation on either side during the entire conflict.

It was this region of the country that provided Chesnutt with his knowledge of the legacy of slavery and the gross injustices that persisted even after abolition. Miscegenation, the dislocation of families, violence, and intimidation were part of the psychological matrix of the local culture, even after the war, and the young teacher understood them well. What he portrayed in his fiction written later in his life was not the product of a fanciful imagination but rather the sensitive memories of the people around him as he grew into manhood. When his family moved back to Cleveland, he was exposed to a different climate of racial interactions, one focused primarily on prejudices and attitudes within the mixed-race population. This was the location where he was admitted to the bar and where he acquired considerable wealth with a company devoted to legal records, and it was also where he first began publishing short fiction in the popular literary magazines and books with major publishers in New York and Boston. The appeal of his work cut across racial lines, and his volumes of stories and his novels sold widely, establishing him as the first African American writer to gain a widespread readership in the white community as well as among people of color. His special emphasis was on light-skinned quadroons and octoroons who often found themselves suspended between the two racial groups, neither of which related to them without suspicion and resentment. His emphasis in characterization was the educated quadroon, a person of sophistication, intelligence, and culture who must nevertheless struggle with the vestiges of slavery and miscegenation, often in situations of intense ethical conflict.

2 Close Readings

Chesnutt's first short story celebrated by readers across the ethnic divide was "The Wife of His Youth," which presents the synergy between a racial caste system based on skin color and the use of an ethical crisis at the center of a progressively intense plot, a structure that had become common by the end of the nineteenth century. The emphasis on moral freedom and responsibility central to the movement of American Realism had been at the center of Mark Twain's *Adventures of Huckleberry Finn* in 1885 (↗9 Twain) and *The Rise of Silas Lapham*, by William Dean Howells, published the same year. Multiple short stories, including Sarah Orne Jewett's "A White Heron" in 1886 and Kate Chopin's "Désirée's Baby" in 1893 (↗11 Chopin), expanded the popularity of the plot device in ways that sustained the genre for more than a century. What Chesnutt brought to that tradition was the complexity of racial prejudice within the African American community itself and the deft integrity of his craft, especially with regard to point of view.

The story begins with an emphasis on an unusual form of racial prejudice, a strategy that is unique only because there are no white people in the present action at all. White society is involved only in the background references to an era of slavery and in the indirect legacy of miscegenation, since light skin color becomes a mark of superiority in the black community of Groveland. The narrative stance throughout the explanation of this intra-racial discrimination is one of objectivity tinged with an ironic edge, the narrator possessing a standard of humane conduct that would not allow for superficial judgments.

At the beginning of the first section of the story, the narrator introduces the protagonist, Mr. Ryder, as the "dean of the Blue Veins," a social organization that held that "no one was eligible for membership who was not white enough to show blue veins" (Chesnutt 1899b, "The Wife", 1). The narrator is clearly not comfortable with the slender pretense of such a standard, and he mocks its obviously self-serving justifications, undermining the validity of its discriminatory stance: "By accident, combined perhaps with some natural affinity, the society consisted of individuals who were, generally speaking, more white than black" (1). The purpose of such a practice was quite directly to establish and preserve for posterity the highest standards of personal character and decorum in an exemplary population of free-born African Americans, a group that exemplified the "talented tenth" concept promoted by W.E.B. DuBois (see Feinberg 1990, 219–237). But, the narrator is quick to assert, there are those who "had been known to assail it violently as a glaring example of the very prejudice from which the colored race had suffered most" (Chesnutt 1899b, "The Wife", 2). What becomes clear from the beginning is that the racial conflicts are located within the black community itself, not in its historic struggle with an exploitive white society. The ironic core of this thematic construct is that the Blue Veins, driven by a desire for social elevation and respectability, have repeated the discriminatory exclusion that had oppressed their black ancestors.

Mr. Ryder epitomizes the standards of this exclusive club and is its "recognized adviser and head, the custodian of its standards, and the preserver of its traditions" (3). The crowning elevation of his status is about to take place because, although he was born of the most humble origins in a plantation in the South, he is now about to propose marriage to Mrs. Molly Dixon, a young, wealthy, and light-skinned widow who had taught school in another city. His plan is to reveal this proposal publicly by making it at a ball to which he will invite only the most illustrious members of the Blue Veins. Through this marriage he will thus have solidified his social standing, satisfied his romantic yearnings, and further expanded his considerable financial holdings, which have made him more than comfortable among the well-heeled members of his club.

All of those objectives are made glaringly tenuous in the second section of the story when a tiny woman, elderly and humbly attired, enters the office of Mr. Ryder. She is very dark, toothless, and dressed in old-fashioned clothing: "She looked like a bit of the old plantation life, summoned up from the past" (10). Everything about her suggests the world that Mr. Ryder has struggled so diligently to leave behind, including her dialect, which resonates in sharp contrast to his cultivated discourse. When they first meet, she explains her mission:

> I heerd you wuz a big man an' had libbed heah a long time, an' I 'lowed you would n't min' ef I'd come roun' an' ax you ef you'd ever heerd of a merlatter man by de name er Sam Taylor 'quirin' roun' in de chu'ches ermongs' de people fer his wife 'Liza Jane? (11)

Her question, rendered in conspicuous "eye dialect," establishes her as a lower-class woman of rustic origin, and it becomes progressively more evident that Mr. Ryder himself is the lost Sam Taylor she has been searching for over the last twenty-five years. He immediately recognizes his former wife, and he understands that she embodies the traditional virtues of women who were slaves in the South, especially her "devotion, sacrifice, fortitude, and unaffected dignity" which give her quest and her point of view "a truer standard of judgment than skin color on which to base relationships and obligations to others" (Andrews 1980, 115).

Mr. Ryder's use of language demonstrates how much he has changed in the last twenty-five years. His dialect is that of an educated, polished member of the middle class, a position he has aspired to for many years and which will assuredly be his position once he is married to Mrs. Dixon. His lexicon is in dramatic contrast to that of 'Liza Jane, who seems frozen in the language of her years as a slave. Her comments, however, are entirely without pretension, and the sincerity of her simple requests is poignant. When at the ball Ryder tells the Blue Veins about 'Liza Jane's story, he progressively reverts back to the dialect of the world of his past, a world to which he has, in some dimensions, been forced to return.

Mr. Ryder's predicament is racial only in a specialized sense. The choice he faces at the conclusion involves not the traditional ethnic divisions but a color stratification within the African American community of Groveland, which has established a hierarchy based on gradations of skin color that correlates, to some degree, to social and economic rank. For him to endorse his "slave" marriage to 'Liza Jane involves a retrograde social revision to inferior status. His wedding to 'Liza was not valid after abolition unless both the husband and wife signed a document at the local court house expressing their desire to make it so. Without such a document, Mr. Ryder is no longer married. Thus, were he to choose to marry Mrs. Dixon, his future in the upper echelons of his society would be secure, a status devoutly to be wished.

That is the dilemma he faces in the last section of the story as he prepares to give a speech at what was to have been his engagement party to Mrs. Dixon. At the heart of his decision are wealth and social status, the admiration of his peers, a lifetime with a younger woman of beauty and social graces, and the satisfaction of having "risen" from his humble beginnings. But it would also occasion a rejection of a woman whose devotion to him has seen her through twenty-five years of hardship and privation as she searched throughout the South for her husband and has now expanded her quest to Northern cities that attracted large numbers of freed slaves in the wake of the Civil War. Her selfless commitment to him regardless of hardship, her deeply human confidence that her husband would share her boundless desire to be reunited in their marriage, and her simple humility touch Mr. Ryder deeply, so profoundly that he is willing to forgo "advancement" in favor of honoring his original marriage. As Charles Duncan has pointed out, there are negative dimensions to Ryder's decision: He must now give up "the woman he *currently* loves" and "a reunion with 'Liza Jane represents for Ryder a sort of re-immersion in slavery" (Duncan 1999, 295). This conclusion mirrors that of Silas Lapham in Howells's novel, for he, too, rises morally on the basis of his social and economic fall. What is positive about the outcome is that Mr. Ryder's final judgment reveals a dramatic deepening of his moral fiber, whereas the members of the Blue Veins consider themselves superior simply because of their light color.

Despite the Romantic sentimentality of the ending, "The Wife of His Youth" is important in literary history for a spectrum of reasons, including the fact that the ethical conflict theme that came to dominate Realism is here located entirely within the African American community. No white person appears in the story other than in 'Liza Jane's memories of her life under slavery, which were harshly inhumane in the treatment of her family. The other role of white people is in the miscegenation that brought about the color differential in the Blue Vein Society. In this issue, former slaves and free people of color are revealed to be capable of their own brand of prejudice, one based on skin tone rather than on racial identity. Perhaps the most important aspect of the story is that it is the first important work of fiction to focus on people on the color line, that is, people of mixed race who are socially suspended between a pure-white level of society and a lower level – of entirely African heritage – that they consider beneath them. These are people who have education and sophistication and yet are not accepted into the dominant community, so they have constructed their own elite discriminatory group, which obviously has given them considerable satisfaction. They seem blind to the ethical limitations of their self-congratulation, a lack of moral vision that is dramatically challenged by the appearance of a simple old woman from the South who arrives one day searching for the husband of her youth. By telling the Blue Vein Society about his plight, Mr. Ryder requires them to participate in his moral dilemma, testing the basis for their own caste system based on color; their decision that he should embrace a very dark former slave, of negligible intellect and cultivation, represents a refutation of the core values by which they have conducted their lives.

Chesnutt's first published story about the moral implications of miscegenation was "The Sheriff's Children," which appeared in the November issue of *Independent* in 1889. Here, in a journal supported in large part by a well-educated audience, the author placed the ethical struggle entirely within the mind of a white man, the sheriff of a small town in North Carolina in the early years after the Civil War. As in "The Wife of His Youth," the sexual transgression is many years in the past, when Campbell had a relationship with one of his slaves, Cicely, and produced a son, Tom. When the sheriff fell on hard times financially, he sold Cicely and the boy to a speculator in Alabama, and apparently thought no more of the matter. Now, roughly two decades later, the young man has come back to the town of Troy, searching for his father, and by coincidence ends up being charged with murder. That broad circumstance sets the stage for the sheriff to have to confront the responsibility that has been his all along but that has never entered his consciousness, the duty of a father to his son.

"The Sheriff's Children" and "The Wife of His Youth" are similar in some important respects. In both stories an important character appears at a crucial juncture and occasions the ethical conflict in the protagonist. In both, that character carries vestiges of the era of slavery that challenges the moral integrity of the protagonist, who is a well-respected member of the community. In each, the central character is moved by this intruder toward a new moral realization, a kind of epiphany that demands important changes in his values. However, in "The Wife," that ethical struggle is moved outward from Mr. Ryder to the whole of the Blue Vein Society when he enlists them in making the crucial decision about his former wife. In the second story, that struggle is internalized by the sheriff, who does not betray the intensity of his new realizations with anyone. Other than Tom, only the protagonist knows that he had a son by a former slave, and only he struggles with whether, and to what extent, that fact should be shared with the community. His daughter, Polly, is blissfully unaware that she has a brother, and the vengeful mob that threatens to lynch the prisoner has no idea of the relationship. The ethical plot of "The Wife" moves outward to enlist scores of people in the crucial decision, whereas in "The Sheriff's Children" the movement is inward, toward searing personal realizations of failed responsibility and a final decision to make amends by helping to defend Tom against the charges with which he has been falsely accused.

That fundamental movement lends itself to the actual structure of the story. It opens with a very broad perspective, with an omniscient narrator sketching the history of Branson County, North Carolina, before and just after the Civil War. The general tenor of the area is that it was relatively benign, largely untouched by the violence, although some of the local young men were lost in the wider conflict. The focus then tightens to describe the county seat, Troy, a rather sleepy village with no railroad service and no more than five-hundred citizens. The area is without the great extremes of wealth and poverty often associated with the plantation South, and prestige was more related to the role played in the war than to material possessions. Captain Walker was among the most revered on the basis of his service in the war with Mexico and then later, in the Civil War, having lost an arm at Gettysburg.

The tranquil description of the region ends in roughly 1875 with the dramatic news that Walker, perhaps the most esteemed man in the country, has been murdered, a total disruption of the quiet waters of society.

There are no witnesses to the crime, but there is a report of a "strange mulatto" who was seen near the Captain's house the night before the incident and seen again going out of Troy early the next morning. By evening the suspect has been found and brought into the county jail, which brings a larger crowd than usual into town on a Saturday. Rumors and gossip whip up the assumption of guilt for the arrested man, culminating in the stereotypical call for a lynching. The enthusiasm for this idea overrides any kind of evidentiary rationale, particularly because of the emotional attraction of the "fact" that a white local hero had been murdered by a black man. The fervor of a lynch mentality takes over, and the crowd begins to move toward the jail.

It is only at that point that the protagonist of the story is introduced, as the focus now becomes more limited, offering an objective view of Sheriff Campbell and his domestic situation (Chesnutt 1899a, "The Sheriff's Children," 67). Moving steadily inward to a more restricted concentration, the narration now describes Sheriff Campbell's household, his African American cook and housekeeper, Sis' Nance, and finally Campbell himself. He is presented as tall and muscular with a full beard, and it is clear he is a man of considerable presence. A local black man, Sam, has come to warn the sheriff that a mob has formed and is moving toward the jail. Sam's dialect is important in that it reveals him as lower class, probably uneducated, and from the immediate area: "Shurff, dey gwine ter hang de pris'ner w'at's lock' up in de jail" (68). This lexicon is in dramatic contrast to that of the prisoner, establishing that environment and education, not race alone, is the determinant of personal character and social level. The sheriff is acutely aware of the immediate danger approaching the jail – and he is determined to do his duty and protect the prisoner – but he does not forget his traditional manners, inviting Sam to come in for dinner. As he collects his shotgun and revolver, his daughter, Polly, enters the room, and he tells her that if there is trouble, she should get the horse-pistol from a drawer and use it to defend the house. She later uses it to shoot Tom.

As Campbell is about to depart for the jail and the unruly confrontation with townspeople, the narrator provides more detailed information about his background. He has spent much of his life in Troy and is well known and respected as a man of considerable education and social standing. He graduated from the State University at Chapel Hill before the war and went on to support the Union and oppose secession until it was clear that North Carolina would join the rest of the South. At that point, he joined the Confederate military and distinguished himself, rising to the rank of colonel. In the wake of the war, he returned home, ran for the office of sheriff, and was duly elected. He is widely respected as a man of principle, fully sworn to the faithful execution of his duty, and no one doubts that he will do so. Those character traits are essential for establishing the conflicts that fill the rest of the evening.

The scene in the jail is intensely dramatic. His first action is to confront the lynch mob, all of whom are townspeople he knows. They have already concluded that the arrested mulatto is guilty, and they intend to hang him immediately. Campbell disarms them with humor: "I get seventy-five cents a day for keeping this prisoner, and he's the only one in jail. I can't have my family suffer just to please you fellows" (75). He then threatens to shoot the first man who attempts to break down the door, and the mob knows that he will do so. They have no doubt of his commitment to his duty.

The second line of confrontation is more personal for Sheriff Campbell as he enters the cell of the prisoner to better defend the jail. Tom is described as a "yellow" young man, establishing his mulatto status, and he speaks not the subordinate dialect of Sam but a standard lexicon virtually identical to that of the educated sheriff: "For God's sake, Sheriff, don't let 'em lynch me; I didn't kill the old man" (77). The sheriff unlocks the prisoner's fetters and assures him he will defend him to the best of his ability. At that juncture the lynch mob disbands, and the drama is entirely between the two men in the jail cell. Tom seizes the revolver that Campbell has left on a bench, assuming that no black prisoner would have the courage to take it, and when the sheriff turns away from the window, Tom is pointing the gun at him.

Tom now has the advantage, and this fact has occasioned a stunning transformation in him: "The keen-eyed, desperate man before him was a different being altogether from the groveling wretch who had begged so piteously for life a few minutes before" (81). The sheriff anticipates that Tom will lock him in the cell and make his escape into the night, but he does not do so. Instead, he reveals that he is Cicely's son, and thus Campbell is his father. The sheriff is further astounded when Tom confronts him directly with the historic truth of the situation:

> What father's duty have you ever performed for me? Did you give me your name, or even your protection? Other white men gave their colored sons freedom and money, and sent them to the free States. You sold me to the rice swamps. (85)

Tom was raised in Alabama after Campbell sold him and his mother to a slave trader, and he may well have been aware of the plaçage tradition prevalent in Louisiana, where wealthy white men often established families with women of mixed race. There, however, both law and tradition required that any children issued from the relationship carry the name and, to some extent, the status of the father, not the mother. Such children were often sent to Europe for their education, and they eventually had the right to inherit a third of their father's estate (see Nagel 2014, 9–15). Tom has obviously been educated, given the formulation and content of his comments, and the sheriff realizes that "the mulatto spoke more eloquently and used better language than most Branson County people" (Chesnutt 1899a, "The Sheriff's Children", 86). Campbell is also forced to confront his responsibility in the life of this young man and to see himself as someone who would sell his own child for a profit. The narrator intrudes at this juncture to observe that the conditions of "slavery poisoned the very fountains of life, and created new standards of right. The sheriff was conscientious; his conscience had merely been warped by his environment" (88). At that juncture, without the awareness of either of the two men in the cell, Polly arrives just as Tom is about to shoot Campbell, and she fires the revolver, hitting Tom in the arm. The wound is not fatal. The sheriff and Polly apply a bandage to stop the bleeding, and they leave the jail for the night.

That is the circumstance that leads to Campbell's ethical epiphany as, back at home, alone in his room, he reviews what Tom has said, what he himself has done, and the consequences of his moral blindness: "He saw that he had owed some duty to this son of his, – that neither law nor custom could destroy a responsibility inherent in the nature of mankind" (91). He sees now what he never even considered when it happened that there were better options open to him than selling his own son. Now that the young man was free, he has returned to confront his father, seething with resentment and loathing for the man who had so treated him. Campbell's exploration into this constellation of unaccustomed thoughts leads him to a new realization, that he had failed in his obligations as a father and as a human being. He realizes that he could have spared Tom from slavery by sending him to the North or by keeping him on his plantation. In his attempts to assuage his sense of guilt, he decides to help Tom without publicly acknowledging his paternity. His plan is to hire a lawyer and defend his son in a trial "by which the sheriff might in some degree atone for his crime against this son of his – against society – against God" (93). It is a unique moment in American literature, an instance in which a white character's moral awakening changes his fundamental assumptions about race.

It is this newly transformed Sheriff Campbell who walks into the jail the next morning to begin his new program of mitigation for past sins when he finds that during the night Tom had torn the bandage from his wound and had bled to death (93). The story thus ends in a situation that offers the protagonist no avenue of moral reclamation, and he will have to face his own internal recriminations for the rest of his life. Unlike "The Wife of His Youth," in which the protagonist's ethical struggle is projected outward and fully revealed to the entire community, in this story no one but Campbell has even a hint of the full meaning of what has happened. He must face his own dark past alone.

3 Theoretical Perspectives

The record of critical responses to Chesnutt's stories is a rich and complex commentary on shifting literary tastes, evolving social circumstances, and progressively more insightful readings of his fiction. He was fortunate in that readers across the racial spectrum found his work not only entertaining but humanely enriching, touching on areas previously unexplored in the literature of ethnicity in America, especially the fate of characters of mixed race. He also wrote directly about a forbidden subject, miscegenation, a practice widespread throughout the South but extremely socially sensitive. Chesnutt not only explored the topic but delved into the moral implications of white men who fathered children with their slaves and then abandoned both mother and child to the ways of the world. He created such characters and captured the moments when they examine their own consciences and reproach their own actions, weighing the nature of their culpability.

The biographical approaches to Chesnutt's work focused on his intimate knowledge of small-town North Carolina in the Reconstruction South as well as his personal success in Cleveland, where he not only passed the bar but founded a legal transcription company that became immensely successful. Chesnutt was himself a child of mixed race who could easily have passed for white, but he chose to live among people of his own ethnic background. Doing so seems to have given him a profound insight into the unique struggles of a caste of people excluded from both white and black communities who strove to create a society of their own, as reflected in "The Wife of His Youth." That Chesnutt's personal life led him to play a leading role in the founding of the NAACP and the struggle for civil rights in the early decades of the twentieth century serves as a background for his earlier attempt to portray these matters in his fiction.

The close reading advocated by New Criticism would regard these stories as works of art, searching for aspects of aesthetic unity and design. Chesnutt's handling of plot and character won high praise from such critics, but his uncertain handling of point of view, with its shifts from omniscience to restricted third-person, was thought of as a weakness, as was his use of coincidence. From the perspective of myth criticism, there is much to say about these two stories. For one thing, they both turn on the classic plot device known as "a stranger comes to town," someone from outside the immediate community intrudes so emphatically as to cause a change of direction for the central plot. The sudden appearance of 'Liza Jane and of Tom alters what seems to be a normative situation and brings about the major conflict and final resolution. In literary history, the emphasis on a regional setting with a localized dialect draws from the local color movement, which freed American fiction from the elevated language of the first half of the nineteenth century. In Chesnutt's work the characters speak in the manner of ordinary people in that area, with differentiation by caste and education. The plots of both stories draw on conceivable conflicts, a hallmark of Realism, although both turn on a dramatic coincidence to heighten the conclusion. In terms of regional historiography, these stories capture emerging problems in the era of Reconstruction, when the consequences of miscegenation were becoming manifest, when newly freed slaves attempted to reunite their families.

These issues are also at the center of African American studies approaches to the fiction of the late nineteenth century, especially with regard to the enormous mixed-race population. Their very existence testified to the sexual injustices of slavery and the almost totally ignored culpability of slave owners for the children they had produced, a common theme in all the Confederate states save Louisiana, where the *Code Noir* established rights and protections for such offspring. Slavery is the dominant issue from this critical perspective, although the inception of Jim Crow laws throughout the South gave rise to new dimensions of racial restriction and discrimination, including such basic matters as voting, employment, social interactions, and day-to-day conveniences such as riding a bus or train. "The Wife of His Youth" depicts one approach to such exclusion by establishing an exclusive community within the mixed-race population, although even that was subject to prejudice within the African American community based on skin color. Chesnutt was familiar with that situation during his years in Cleveland, and in his fiction he showed that the slave experience did not inoculate people from their own worst instincts. Such portrayals made Chesnutt's work controversial when it first appeared, but decades of consideration of his fiction, the artistic contribution of his work, and the intellectual reach and personal integrity of his life have made him one of the most admired figures in American literature.

4 Bibliography

4.1 Works Cited

Andrews, William L. *The Literary Career of Charles W. Chesnutt.* Baton Rouge: Louisiana State UP, 1980.

Chesnutt, Charles W. "The Sheriff's Children." *The Wife of His Youth and Other Stories of the Color Line.* Boston: Houghton Mifflin, 1899. 60–93.

Chesnutt, Charles W. "The Wife of His Youth." *The Wife of His Youth and Other Stories of the Color Line.* Boston: Houghton Mifflin, 1899. 1–23.

Duncan, Charles. "Telling Genealogy: Notions of the Family in 'The Wife of His Youth.'" *Critical Essays on Charles W. Chesnutt.* Ed. Joseph R. McElrath, Jr. New York: G.K. Hall, 1999. 281–296.

Fienberg, Lorne. "Charles W. Chesnutt's *The Wife of His Youth*: The Unveiling of the Black Storyteller." *American Transcendental Quarterly* 4.3 (1990): 291–327.

Heermanee, J. Noel. *Charles W. Chesnutt: America's First Great Black Novelist.* Hamden, Conneticut: Archon Books, 1974.

Izzo, David Garrett, and Maris Orban, eds. *Charles Chesnutt Reappraised: Essays on the First Major African American Fiction Writer.* Jefferson, North Carolina: McFarland, 2009.

McElrath, Joseph R., ed. *Critical Essays on Charles W. Chesnutt.* New York: G. K. Hall, 1999.

McWilliams, Dean. *Charles W. Chesnutt and the Fictions of Race.* Athens: U of Georgia P, 2002.

Nagel, James. *Race and Culture in New Orleans Stories.* Tuscaloosa: U of Alabama P, 2014.

Render, Sylvia Lyons. *Charles W. Chesnutt.* Boston: Twayne, 1980.

Simmons, Ryan. *Chesnutt and Realism: A Study of the Novels.* Tuscaloosa: U of Alabama P, 2006.

Wilson, Matthew. *Whiteness in the Novels of Charles W. Chesnutt.* Jackson: UP of Mississippi, 2004.

Wonham, Henry B. *Charles W. Chesnutt: A Study of the Short Fiction.* London: Prentice Hall, 1998.

4.2 Further Reading

Brodhead, Richard H, ed. *The Journals of Charles W. Chesnutt*. Durham: Duke UP, 1993.
Chesnutt, Charles W. *The Colonel's Dream*. New York: Page & Company, 1905.
Chesnutt, Charles W. *The Conjure Woman*. Boston: Houghton Mifflin, 1899.
Chesnutt, Charles W. *The House Behind the Cedars*. Boston: Houghton Mifflin, 1900.
Chesnutt, Charles W. *Mandy Oxendine: A Novel*. Ed. Charles Hackenberry. Urbana: U of Illinois P, 1997.
Chesnutt, Charles W. *The Marrow of Tradition*. Boston: Houghton Mifflin, 1901.
Chesnutt, Charles W. *Paul Marchand, f.m.c.* Jackson: UP of Mississippi, 1998.
Chesnutt, Charles W. *The Quarry*, ed. Dean McWilliams. Princeton: Princeton UP, 1999.
Chesnutt, Charles W. *The Wife of His Youth and Other Stories of the Color Line*. Boston: Houghton Mifflin, 1899.
Ellison, Curtis W., and E.W. Metcalf. *Charles W. Chesnutt: A Reference Guide*. Boston: G.K. Hall, 1978.
Kawash, Samira. *Dislocating the Color Line: Identity, Hybridity, and Singularity in African-American Narrative*. Stanford, CA: Stanford UP, 1997.
Keller, Frances Richardson. *The American Crusade: The Life of Charles W. Chesnutt*. Provo: Brigham Young UP, 1978.
Wright, Susan Prothro, and Ernestine Pickens Glass, eds. *Passing in the Works of Charles W. Chesnutt*. Jackson: UP of Mississippi, 2010.
Ze Winters, Lisa. *The Mulatta Concubine: Terror, Intimacy, Freedom, and Desire in the Black Transatlantic*. Athens: U of Georgia P, 2018.

Ina Bergmann
11 Kate Chopin (1850–1904)

Abstract: Since the rediscovery of her work in the 1970s, Kate Chopin's short stories have been read as feminist texts. "The Maid of Saint Phillippe" (1892) and "The Story of an Hour" (1894) are examples that illustrate the evolution of Chopin's writing from feminine to feminist. They depict female protagonists that strive for freedom from social constraints imposed upon women in the nineteenth century. For these women, freedom can only be found in seclusion from society, be it in the American wilderness or in the confinement of their own home. Solitude warrants liberty. Within society, female freedom remains a dream.

Keywords: Freedom, solitude, feminine phase, feminist phase, new woman

1 Context

"Emancipation. A Life Fable," written in 1869, is the earliest surviving short story by Kate Chopin. The text, only one page long, tells the story of an animal born and raised in a cage. The animal is well cared for and lives comfortably, with food, water, and a bed of straw provided for him. One day, the cage door stands open. The animal is tempted, but first fearingly hesitates to leave. Finally, he cannot subdue his curiousity and takes a leap into the unknown. Henceforth, he lives alone in the outside world without comfort and security. He learns about suffering, seeking, and finding. But his world is richer for the things he sees, smells, tastes, and touches. He does not consider to return to the still open cage because of the joy he feels now, despite the costs of freedom (Chopin 2002b, 659; Nigro 1999, 148).

This story may be read in various ways, for example from a religious perspective, considering its biblical implications; from an ecocritical perspective, regarding the relationship between man and animal; from an ethnic perspective, concerning President Lincoln's "Emancipation Proclamation" of 1863 (Walker 2001, 35–36); or from a feminist perspective, apropos female emancipation (Stepenoff 1987, 452). The story may even be read autobiographically, as it was written at a time when the requirements of the social life of a debutante or 'belle' became a nuisance to the young Chopin (Walker 2001, 36–37). Yet the main subjects of this story are universal: freedom and solitude. The animal revels in his liberty after he has left his cage, although his life was much easier in captivity (35; Saar 1994, 67). He now lives alone and has to rely on himself for his survival. With its primary focus on freedom and solitude, this early story already sets the general themes for Chopin's life and work.

https://doi.org/10.1515/9783110587647-012

Katherine O'Flaherty was born in St. Louis, Missouri in 1850 to "a Celtic father and a Creole mother" (Rankin 1932, 16): her mother came from a French background, her father was an Irish immigrant (Nigro 1999, 148). The most remarkable – and literary relevant – aspect of her childhood is her upbringing in a matriarchy. Her great-grandmother and her grandmother, both widowed, moved in with her mother after the death of Kate's father in 1855. The matrilineal French American influence is central to Chopin scholarship and will be repeatedly referred to below. The Irish American aspects in her work have only recently come into focus (Boren 1992, 16; Shaker 2003, 119; Pierse 2006, 85; Ebest 2013, 25–26; Shaker and Petitt 2013, 385–386; Giemza 2013, 85–109; Shaker, Nigro, and Nestor 2018).

In 1870, soon after her graduation from the Sacred Heart Academy, Kate married Oscar Chopin. They moved to Louisiana, first to New Orleans, and then, when Oscar's business was deteriorating, to Cloutierville. In quick succession, between 1871 and 1879, Kate and Oscar had six children. In her marriage Kate enjoyed more freedom than most women of her time and raised some eyebrows with her behavior and fashion choices. In 1882, Kate's – more or less – conventional life as a wife and mother came to an abrupt end upon Oscar's untimely death. For a while Kate led a relatively carefree existence as a young widow, possibly including a scandalous romance with a married neighbor (Toth 2008, 13–14).

In 1884 Chopin returned to her native St. Louis and started to work seriously on her stories (Toth 2008, 14). The death of her mother in 1885 left Chopin completely on her own (Stepenoff 1987, 454). Grief and loneliness fostered her exploration of change in her writings (Walker 2001, 51). Her career as a published author began in 1889, when she became St. Louis's first professional woman writer (Toth 2008, 14). She published two novels, *At Fault* (1890) and *The Awakening* (1899), and two collections of short stories, *Bayou Folk* (1894) and *A Night in Acadie* (1897). More than fifty uncollected short stories, some published and others unpublished during her lifetime, are available in print (Gilbert 2002, x–xii). She also produced "essays, poems, translations, one play and one polka" (Toth 2008, 14).

In her writing, Chopin pushed the boundaries of freedom of expression (Toth 1992, 28). The portraits of women in her works reflect the free spirit of their author. They are rarely conventional, often controversial, and always compelling. This is perhaps most obvious with Chopin's notorious second novel. In *The Awakening*, she tells the story of Edna Pontellier, a wife and mother, who seeks self-fulfillment and freedom from social conventions. Chopin's working title for the book, "A Solitary Soul," suggests a more negative characterization of the title character's state of being than its final title (Walker 1993, 16). The novel received hostile reviews upon publication and has since had a turbulent reception history, full of ups and downs (Koloski 2008, 161; Stepenoff 1987, 448–449). Because it was ahead of its time, but was taken out of circulation, the novel itself has convincingly been labeled "a solitary book" (Showalter 1993, 169–170).

Freedom and solitude are venerable American ideals. The "Declaration of Independence" (1776) asserts the natural right of liberty most prominently in its well-known second sentence. An individualist understanding of freedom is at the core of U.S. national identity (Smallwood 2014, 111). Personal freedom was the initial motivation for many early settlers to come to the New World. Later, individualism and freedom motivated westward expansion and turned out to be the underlying features of the frontier spirit, as pointed out by Frederick Jackson Turner in "The Significance of the Frontier in American History" (1893). Over the course of American history, freedom was sought in numerous arenas such as religious freedom, economic independence, political autonomy, female emancipation, the abolition of slavery, and the absence of social restrictions, among many others. Reclusiveness can be understood as an extreme manifestation of the ideals of liberty and individualism. The idea(l) of a solitary life in the wilderness is deeply embedded in American culture. Henry David Thoreau's sojourn at Walden Pond, where he famously wrote *Walden* (1854), an account of his retreat to nature, is still the archetypal example (Bergmann 2017, 13).

For women, given the social restrictions in the nineteenth century, freedom and solitude could often only be found by a retreat within their home. Especially for female writers, a private space was crucial. Virginia Woolf's programmatic *A Room of One's Own* (1929) is a seminal work in this context. A secluded room of one's own enables the creative process and has liberating power. The home is reinvented as "herspace," "a stage for a single actor; a house of soliloquy; a studio of rich, internal echoes; a closet of expression; a safe cell for the incubation of ideas" (Malin and Boynton 2003, 144–145). In 1903 Chopin moved into a modest house in St. Louis's West End, "a small house on a city street, where she lived in solitude and freedom" (Stepenoff 1987, 463). In 1904 she died at this place, which contrasted the elegance of her former homes, but perhaps best provided the reclusive setting necessary for her creativity (Stepenoff 1987, 464; see also Bergmann forthcoming).

When considering Chopin's work, it is important to distinguish between the publication and reception history of *The Awakening* and that of her short stories. The short stories were published in two very well-received anthologies and in the most prestigious magazines of the time: *Vogue*, *Century*, *Youth's Companion*, *Atlantic Monthly*, and *Harper's Young People* (Koloski 2008, 161; Stepenoff 1987, 448; Walker 2001, 99, 139). Chopin published stories in family magazines such as *Youth's Companion*, willingly composing her texts with the target audience in mind or meeting the expectations for local color writing (Ewell 1986, 18; Johnsen 2008, 54; Koloski 2008, 161; Saar 1994, 61–62; Shaker 2003). She also published stories with unconventional subjects and daring themes in magazines such as *Vogue*, whose editors and readership accepted controversial ideas (Toth 1999, 172–173; Nigro 1999, 149). With her short stories, Chopin successfully catered to both conservative and progressive audiences (Koloski 2008, 161–162; Johnsen 2008, 54).

The two stories discussed in this chapter fall into these distinctive categories. "The Maid of Saint Phillippe" (1892) has received only sparse critical attention (Saar 1994, 59) and is a story that is presumably unknown to most readers. "The Story of an Hour" (1894) is probably one of Chopin's best-known stories and may, besides *The Awakening*, be her most assessed text. Both stories give expression to Chopin's themes of freedom and solitude.

2 Close Readings

"The Maid of Saint Phillippe" was first published in *Short Stories* in November 1892 (Gilbert 2002, 1056). It is the story of an adolescent girl who faces the challenges of life in the wilderness rather than surrender to social conventions. The story is Chopin's only piece of historical fiction. It is set in the 1760s, the time of her great-great-grandmother. She may have learned about St. Louis's history of that time from her great-grandmother, Victoire Verdon Charleville (Nigro 1999, 148). Her great-great-grandmother may have been a model for the independent Marianne, the protagonist of the story. Chopin's ancestor was the first woman in St. Louis to obtain a legal separation from her husband, operated a highly profitable keelboat line, and gave birth to a child out of wedlock. Although she was widely known as 'La Verdon,' her name was Marianne Richelet Verdon (Toth 1992, 23; Toth 1999, 181–182; Stepenoff 1987, 451–452; Stepenoff 2017, 68–75; Saar 1994, 65).

Marianne Laronce, the focalizer of the short story, is "a French girl of seventeen" (Chopin 2002c, 711). She lives alone with her widowed father Picoté in the small French colonial village of Saint Phillippe, close to the Mississippi River. The first description of the young woman stresses her aptitude for a life in nature and her deviation from the prevailing gender norms: she is "tall, supple, and strong," looks "like a handsome boy," and carries "a gun across her shoulder as easily as a soldier might" (711). Obviously, she is identified as a 'tomboy,' an androgynous girl who enjoys hunting, an activity usually reserved for men. Marianne presents "male strengths in female form," and her appearance is a "sign of difference" (Saar 1994, 66). This is further stressed by the fact that the other women in the village are "already busy preparing the evening meal" (Chopin 2002c, 711) as she returns from a hunting expedition. Interestingly, this is the only reference to other women in the story. The girl lives in a men's world.

Marianne wears "buckskin trappings" and has the "stride" and the "poise" of a "stag" (711). There are only two other characters in Chopin's writing who crossdress: Mamzelle Aurélie in "Regret" (1894) and the eponymous protagonist of "Charlie," written in 1900 (Saar 1994, 67). When the girl walks out of the woods, she momentarily stops on the hilltop to shield her eyes from the descending sun that dazzles her, a gesture that marks her as in command of the land before her. By giving her the identifying accessories and likening her to one of the symbol animals, Chopin is associating her protagonist with the ancient goddess of the hunt, Artemis or Diana.

Upon her return to Saint Phillippe, Marianne is confronted by news that upset her. A young man, Jacques Labrie, brings her up-to-date, offering to escort her – a woman who so obviously does not need male protection – home. During Marianne's absence, the villagers have learned that the English want to take control of their until now inaccessed possessions in Illinois. The French Saint Phillippe is part of this territory. This information ties the story to a specific historical time, 1765, and historical events of the region, which are "indeed true" (Chopin 2002c, 714).

The girl's reaction to Jacques's revelation is twofold. First, she wonders whether the Native Americans living in the region will keep away the English, as they have done in the past. Marianne seems to welcome the protection of the militant tribe living in the wilderness. Learning that this time, the natives will not be able to fight back the English, the girl directs her anger against the French monarch and his remote politics: "[T]hat is a monarch to be proud of! Your Louis who sits in his palace at Versailles and gives away his provinces and his people as if they were baubles!" (712). This scene underlines Marianne's independence as well as her willpower. She has her own opinion and she is not shy to offer it, even if it involves rebellious thoughts. Marianne appears like an American variant of Jeanne d'Arc (1412–1431), the maid of Orleans, who fought against the English in the Hundred Years' War. The title of the short story also hints at this identification.

Jacques relates that most villagers, including himself, will leave Saint Phillippe and settle across the river in Saint Louis. The French soldiers will hand over the nearby fort and their commanding officer, Captain Alexis Vaudry, will return to France. The girl is outraged by the idea that she should leave the place of her birth, and learns that her father has decided to stay by the side of the grave of his wife, Marianne's mother. This also decides the girl's future: "My love belongs to my father. I have but to follow his will; whatever that may be" (714).

As if to affirm this conventional daughterly submission – surprising in a character introduced as strong-willed and independent – Marianne changes into womanly attire. This second introduction of the protagonist, now not as a boyish hunter but as an attractive young woman, serves two functions. On the one hand, the impression of an unwomanly character, which may have appalled the conventional readership of the magazine in which the text was published, is dispelled. The historical setting with its defamiliarization may also add to this intention. On the other hand, the following wooing of Marianne is prepared for by this more conformist description. Yet the fact that "the garments [. . .] had been her mother's once" (714) implies that the fulfillment of the traditional women's role, the submission to her father's will, is only a hand-me-down as well. The role of the "true woman" who practices "piety, purity, submissiveness, and domesticity" (Welter 1966, 152), is not a natural part of the girl's identity. The "tension between the lithe young woman" and the "tradition-bound child of an aging patriarch" (Stepenoff 2017, 76) could not be clearer.

When all villagers have left Saint Phillippe, the dutiful daughter finally dwells there alone with her father. Marianne welcomes the solitude: "Do not ever think that I am sad, father, because we are alone" (Chopin 2002c, 715). Indeed, this new situation bestows freedom and independence upon the young girl, which very much becomes her. "Marianne worked and hunted and grew strong and stronger" (715), while her father becomes more feeble and even child-like. One evening, when Marianne returns from the hunt, she finds her father dead. Although she loved him, she perceives of his death as a release from her obligations and a gift of freedom (Saar 1994, 69). The thought of a future that will be wholly hers takes possession of her: "[A]t once she felt that she was alone, with no will to obey in the world but her own. Then her heart was as strong as oak and her nerves were like iron" (Chopin 2002c, 715–716).

When her father has been buried beside his wife, friends and acquaintances who have returned for the funeral approach the girl one by one. First, the *curé* of Kaskaskia urges her to come with him. He argues that it would be "unseemly" for her to "live alone in Saint Phillippe" (716). But Marianne does not care for conventions. She sends the clergyman on his way, telling him that she will consider his offer and if she decides in favor of his proposition, will come to him in the morning. Jacques, who gently wants to persuade her to begin a new life with him as his wife in Saint Louis, receives the same answer. In the morning, when Marianne leaves her house in her hunting garments, gun across the shoulder, there is a third offer: Captain Vaudry, who has spent the night in the village to watch over her, approaches her with his offer of marriage and an easy life of luxury in France. Marianne is three times tempted to leave Saint Phillippe behind. Not only the number of these approaches, but also the wording of Vaudry's offer reference Christ's temptation in the New Testament, when the devil speaks to Jesus (Luke, 4:1–13; see also Matthew 4: 1–11; Mark 1:12–13). Marianne admits that she sometimes has longed to know what luxury is, instead of living a simple life. Also, she confesses she has occasionally yearned for communion with other people, instead of living in solitude. This scene has been read as a foregrounding of "duality and ambiguity" (Worton 2008, 108–109).

But Marianne remains steadfast in her decision. The young woman's justification of her choice is twofold. She has political reasons, and prefers "wild America" (Chopin 2002c, 717) over European civilization and the political power play of England, Spain, and France, in a stance of "savage resistence" (718): "I have breathed the free air of forest and stream till it is in my blood now" (718). Marianne also has personal reasons to refuse Vaudry. She does not want to be conquered by love, which would leave her "feeble as a little child," like "one who has been stricken" (718). She fears the personal and political unfreedom of marriage and of subserviance that would even effect her offspring: "I was not born to be the mother of slaves" (718). Her final resolution expresses her choices of freedom and solitude: "Freedom is left for me! [. . .] Marianne goes to the Cherokees! [. . .] Hardships may await me, but let it be death rather than bondage" (718). Marianne identifies as a truly independent spirit, an American. And for her, the Native Americans best represent the “possibility of freedom” (Saar 1994, 69). She clings to the nation’s natural rights, as stated in the preamble of the “Declaration of Independence.” Consequently, she chooses “unfettered Darwinian individualism” (Joslin 2008, 79) and liberty. By linking the girl’s crisis with the historical crisis of the nation, Chopin makes “Americanness” (Saar 1994, 63) a central theme of the story.

This is a reading that also ties in with the meaning of the protagonist’s name which, besides its possible autobiographical inspiration, also refers to the French national figure. While the original Marianne metonymically stands for the French Republic and the French Revolution of 1789 that led to its inception, the character in Chopin’s story stands for the American nation and the revolution that let to its founding in 1776, hardly a decade after the time in which the story is set. Like the French Marianne, her American equivalent is a symbol figure of freedom. The topics of freedom and independence are relevant not only on the national level, but also with regard to female emancipation. While the three men who approach the young woman are not at all drawn unsympathetically, they nevertheless are identified as supporters of a patriarchal system that patronizingly supresses women. They believe to know what is best for Marianne and they call her wish to live alone “unseemly” or herself a “child” (Chopin 2002c, 716–717). Marianne opts against the traditional women’s role, against marriage and motherhood, in a patriarchal society. She even chooses the law of the wild, her possible real death, over the ‘civil death’ of marriage, a state in which the wife loses almost all civil rights. The girl decides against social expectations and conventions and chooses nonconformity. It has been argued that the story excells in “would-be feminism,” which is “undercut by the satiric, even flatly funny, ending” (Joslin 2008, 79). However, Chopin cunningly connects the rejection of monarchy with a critique of patriarchy, which in consequence may, as well, be read as un-American. This connection is as smart as Elizabeth Cady Stanton’s linking of the most prominent American feminist text, her “Declaration of Sentiments” (1848), with the “Declaration of Independence.”

“The Maid of Saint Phillippe” is also a coming-of-age text, a typical “story of female initiation” (Bergmann 2011, 307–311). The three male tempter figures aim at making Marianne give up her independence, ‘grow up,’ and accept the conventional role for women. Her psychological journey follows the typical three-part pattern of the traditional American story of initiation, although the “exit” and “transition” are not followed by a “(re-)entrance” (Freese 1986, 52) into society, but by a “denitiation” (Lewis 1955, 115), an initiation away from society: “Marianne [. . .] turned her back upon all of them” (Chopin 2002c, 718). Her isolation in nature is a renunciation of traditional gender roles and the separate spheres for men and women (Bergmann 2003, 160–168).

The story ends with Marianne’s retreat into nature: “With gun across her shoulder she walked up the gentle slope; her brave, strong face turned to the rising sun” (Chopin 2002c, 718). Nature is presented as a “source and symbol of freedom” (Ewell 1986, 80). The positive symbolism of morning, dawn, and the ascendant path affirms a reading of this ending as a new beginning for Marianne. Chopin stresses "the possibility of renewal and rebirth" (Lattin 1982, 223), which ties in with the topics of initiation and emancipation. The ending has been interpreted as a "happy conclusion" (Saar 1994, 67). Marianne's is an "active initiation" (Bergmann 2003, 168). She becomes an agent in her own life and refuses to be reduced to passive submission. Yet it is significant to note that female self-assertion is possible only when away from society.

Of course, the girl's choice of wilderness over civilization is reminiscent of the classic myth of American literature. The protagonist leaves civilization and starts a new life in the wilderness (Saar 1994, 63). This is the romantic promise that in America, a "new land untrammeled by history and social accident" (Baym 1985, 71), a person will be able to achieve complete self-definition. Indeed, Chopin's portrait of Marianne can be read as "more mythical than historical" (Stepenoff 2017, 75). Her retreat is, to refer to only one example here, reminiscent of the protagonist's "light[ing] out for the Territory" (Twain 1985, 369) in Mark Twain's *The Adventures of Huckleberry Finn* (1885) (Saar 1994, 63; Ewell 1992, 163; ↗9 Twain). But to argue that the story (and its protagonist) simply imitate male (plot) patterns (Fluck 1982, 159), that there is "a mismatch between the structure of the myth and the female version of it" (Saar 1994, 64), means to disregard the prominence of the specifically female fear of the dependence of marriage and motherhood.

"The Maid of Saint Phillippe" was a loser in the *Youth's Companion's* folklore contest, for which the author had entered her story (Toth 1999, 125). It was heavily criticized, even by Chopin herself (Saar 1994, 59–61, 72). Twentieth-century critics have labeled it "disastrous" (Seyersted 1969, 82), "weak" (Skaggs 1985, 57), "unconvincing," "clichéd," and even "one of the worst stories Kate Chopin ever wrote" (Fluck 1982, 159). In slightly more benevolent readings, Marianne has been classified as "one of Chopin's most dramatic heroines," her final decision a "melodramatic commitment" to a "romanticised freedom," and the ending "theatrical" (Worton 2008, 108–109). Marianne has been read as "not credible," but "intriguing," the choice of wilderness implying "ambivalent solitude as well as narrative cliché" (Ewell 1986, 80). Chopin's handling of the historical setting has been critiziced as well, but maybe it deliberately aims at obscuring the impact of Marianne's choice of freedom (Ewell 1986, 80), making its radicality more acceptable for her readers through historical defamiliarization.

However, "The Maid of Saint Phillippe" has also been read as Chopin's "own spiritual autobiography," her decision to become a writer as the "equivalent to Marianne's opting to choose freedom" (Saar 1994, 70). It has been argued that the story, Chopin's only historical fiction, was crucial for her career. With it, she tried to position herself within the tradition of the great American authors of historical fiction of the nineteenth century, James Fenimore Cooper and Nathaniel Hawthorne (↗6 Hawthorne) among them (Saar 1994, 62). But she failed. She realized that she needed to write from her own experience and for her own readers, who understood her and were in sympathy with her thoughts. Therefore, the story was an important step in her evolution as a writer (Saar 1994, 72). When set in "the context of American criticism and Chopin's oeuvre," it emerges as a text "that remains relevant as a literary work and gains importance as an antecedent to Chopin's finest works" (Saar 1994, 60). The next story to be discussed here is undeniably among these best writings.

"The Story of an Hour" was first published as "The Dream of an Hour" on 6 December 1894 in *Vogue* (Gilbert 2002, 1056). This magazine allowed Chopin to be "a social critic" (Toth 2008, 21). In the story, a young wife enjoys for one hour her newly-gained freedom as a widow, and instantly dies when she learns that her husband is still alive. Similarly to "The Maid of Saint Phillippe," the story is autobiographically influenced. While the husband in the story is presumably killed in a train accident, Chopin's father actually died in a train wreck in St. Louis in 1855. The protagonist of the story may resemble Chopin's mother Eliza Faris, who had married thirty-nine-year-old Thomas O'Flaherty shortly after her sixteenth birthday, not for romantic reasons, but to solve the Farises' financial problems. At twenty-seven, she was a widow and controlled a large estate. She never remarried (Toth 1992, 25–26; Toth 1999, 10, 182; Giemza 2013, 96). Chopin often wrote about widows, for example in her stories "A Lady of Bayou St. John" (1893) and "A Sentimental Soul" (1895) (Toth 2008, 22).

Mrs. Mallard is a young wife "afflicted with a heart trouble" (Chopin 2002d, 756). Her last name, which is almost homophonous with the French word for ill, *malade*, also hints at this fact (Doloff 2014, 581). The story will reveal that she is not only physically ill, but also emotionally. She is sick at heart. The story starts *in medias res*, with the protagonist being carefully informed by her sister Josephine and her husband's friend Richards, that Brently Mallard has been killed in a "railroad disaster" (Chopin 2002d, 756), news that Richards has learned through two telegrams. Much of the story thus depends on modern transportation and communication technology and the way in which it prompts social change. The railroad and the telegraph, which generally sped up life in the 1890s, become catalysts for the protagonist's emotions. Without them, no incident would have occurred and Mrs. Mallard would not have known of it before the return of her husband. The story can therefore be read as a precursor of later technophobic literary works and as a protomodernist text (Foote 2013, 85–86).

Mrs. Mallard's reaction to the news is not the expected "paralyzed inability to accept" (Chopin 2002d, 756), as a third person narrator, with mainly Mrs. Mallard as focalizer, informs. Rather, she immediately weeps in "sudden, wild abandonment" in her sister's arms, suffers through a "storm of grief" (756), and then equally abruptly, seeks solitude and retreats to her room. Her room, symbolic of female confinement and domesticity, is transformed into a space of transition and anticipation. The meaningfully open window is a threshold to the outside world, the male sphere. The view is of a landscape "all aquiver with the new spring life" (756). A "delicious breath of rain" (756) is in the air, the birds are twittering, and someone in the distance is singing a song, while a peddler in the street is selling his wares. The pastoral scenery is symbolic of the fact that life will go on despite her husband's death and that this moment might be a fresh start, an awakening for her: "There was something coming to her [. . .] creeping out of the sky, reaching toward her through the sounds, the scents, the color that filled the air" (757).

What comes to her is a "moment of illumination," the realization of her new freedom: "free, free, free!" (757). And what makes her "pulses beat fast," warms her "coursing blood," and relaxes "every inch of her body," might very well be "a monstrous joy" (757). But this does not mean that she does not simultaneously mourn her husband. She "had loved him – sometimes" and she is sure that she will cry again when she sees his "kind, tender hands folded in death; the face that had never looked save with love upon her, fixed and gray and dead" (757). But "beyond the bitter moments" (757), she can see a future that will be all hers and she welcomes it with her arms open and spread out. She resolves that she will "live for herself" in those coming years, when there will be "no powerful will bending hers" (757). Mrs. Mallard realizes the "possession of self-assertion" as the "strongest impulse of her being" (757). And again she whispers: "Free! Body and soul free!" (757).

While the hints at the state of the Mallard marriage remain ambivalent, it is nevertheless safe to assert that theirs is a conventional marriage in a patriarchal society. The distribution of power within the marriage and the rule of separate spheres for men and women are not questioned. Mrs. Mallard meets Mr. Mallard's and society's expectations of a "true woman" (Welter 1966, 152). While it is not a romantic union, Brently Mallard is also not depicted as a villain. A Foucauldian reading of the story that positions the Mallard home as a penitentiary and Mrs. Mallard as an infantilized and silenced prisoner (Distel 2018, 66) might exaggerate the general tendency.

The above-described scene is furthermore remarkable because it is loaded with sexual imagery. The "slightly parted lips," "keen bright eyes," fast pulse, "coursing blood" (757), and, finally, the relaxation of her body seem to describe a sexual encounter. Especially the usage of the verb "possess" (757) hints at a sexual union (Deneau 2003, 211–212). But there is no other person in the room, and therefore a variety of more or less convincing readings of this scene have been introduced. One links the passage to the classic tale of Leda and Zeus, who approaches her as a swan, coming from the sky. Another is a Christian interpretation of the Holy Spirit descending upon her and infusing her with knowledge (212). The passage has also been linked to ghost and monster stories (Shen 2009, 121). It is more plausible that her awakening from a numbing existence is described as a sensual experience to elucidate its spiritual as well as physical impact. The usage of the term "possession" (Chopin 2002d, 757) ties in with this. In her upstairs seclusion, she has become a rebellious madwoman, an "angel" turned "monster" (Gilbert and Gubar 2000, 17–20). Her sister Josephine, who wants her to open the door, fittingly fears that she is "making [her]self ill" (Chopin 2002d, 758). It is only then that the reader learns the protagonist's first name. The use of her first name has been read as an indicator of her "new consciousness" (Papke 1990, 63), her individualization from Mrs. Mallard to "Louise" (Chopin 2002d, 758).

When Louise finally leaves the room, she feels as if she has won a fight. She has a "feverish triumph in her eyes" and she carries herself "like a goddess of Victory" (758). This description is over the top and in itself ironic. The narrator stresses the fact that Louise's anticipation of an independent life is naïve. The exaggeration may also function to prepare the steep downfall, indicated also by Louise's descent of the stairway. Suddenly, Brently Mallard, who has not been near the train wreck at all and has not even heard of it, enters the house. Instantly, upon seeing her husband alive, Louise dies: "When the doctors came they said she had died of heart disease – of joy that kills" (758). Chopin employs a surprise ending, a feature that she may have borrowed from Maupassant and that she also uses in "Desirée's Baby" (1892) (Nolan 2008, 121; Walker 2001, 99).

Similar to the criticism occasioned by the ending of "The Maid of Saint Phillippe," this ending has been called "too neat if not too cheap" (Arms 1967, 225). The most common interpretation of this melodramatic ending is that Louise's death results from the shock when she sees her husband, who is still alive (Cunningham 2004, 48). Ironically, though, it is not the joy of finding out that her husband is still alive that kills her. Rather, it is the realization that with Brently Mallard's return, she would have to resume her life as submissive wife. This seems an impossibility to her. She has dreamt of an independent life and has almost tasted freedom. The logical consequence is her instant death, which is made slightly more credible by her previously introduced illness. Given that Louise was just "drinking in a very elixir of life" and wished "that life might be long" (Chopin 2002d, 758), her death is also highly ironic (Shen 2009, 125).

What happens in Mrs. Mallard's room can be described as her "struggle toward selfhood" (Ewell 1992, 160). The dream of female selfhood presented in this story is revolutionary (165), yet for Louise, and for most women in the nineteenth century, it is just that – a dream. Once Louise has found selfhood, she has no place in patriarchal society, no life to lead (Cunningham 2004, 49, 53). Most readings of the story are, of course, feminist. The story has been classified as one of Chopin's "most radical" texts, an "attack on marriage, on one person's dominance over another" (Toth 1990, 252–253). Louise's struggle for selfhood has been read as a rebellion against female "selflessness" (Ewell 1992, 162). There are also analyses that challenge these feminist interpretations. While a reading of Louise as "an immature and shallow egotist" who has to "face the earthly consequence of an impossible dream of her afflicted heart" (Berkove 2000, 158) is hardly convincing, it is true that Chopin abundantly employs irony in the story and does not spare her protagonist (Shen 2009, 123).

"The Story of an Hour" can be read as a "distilled example of the novel of feminist awakening and self assertion" (Wan 2009, 170), a text that can be recognized as leading directly to the achievement of Chopin's *The Awakening*. In "The Maid of

Saint Phillippe," Chopin drew on her own experience as a woman who finds herself alone upon the death of a male relative (Stepenoff 2017, 75). The same holds true for "The Story of an Hour." A "woman's solitude" becomes "a sign of the character's awakening to her own choices" (Toth 1992, 27). Louise Mallard and Marianne Laronce claim freedom for themselves and realize they have to seek solitude in order to be able to enjoy it, even if only briefly. In this, they resemble the unnamed animal in Chopin's earliest story as well as Edna Pontellier in her late masterpiece.

3 Theoretical Perspectives

Chopin's work defies categorization, or, rather, it is very hard to categorize as it fits many labels. There have even been claims that "Chopin's precise position in American literary history remains to be determined" (Stepenoff 1987, 449). Chopin's collections of short stories about rural Louisiana, *Bayou Folk* and *A Night in Acadie*, established her as a writer of the local color school (448). Her contemporaries mainly perceived of her in this way and were shocked by *The Awakening*. As noted earlier, until the publication of this novel, her more radical work had been known only to the readership of certain magazines. Contemporary scholars have dismissed Chopin's categorization as a local color writer on the grounds of the seriousness and skillfullness of her shorter works and their superiority over other Louisiana local colorists (449, 456). The main definitions claim that regionalist writing provides "detailed portraits of [a region's] cultural life and [aims] to represent faithfully the complexities of local character," while local color writing is "presenting quaintness and peculiarity" from an outsider's perspective "for other outsiders" (Fresonke 2018). Chopin's writing fits both labels, but the term 'local color' has been used to disparage the achievements of this style of writing. Labeling Chopin as a regionalist writer instead, positions her writing alongside the parallel traditions of realism and naturalism (Fetterley and Pryse 2002, 4).

Realism and naturalism are, unsurprisingly, labels that various scholars have also attached to Chopin's works. Realism might be defined by William Dean Howells's claim of "fidelity to experience and probability of motive" (Budd 2018). Naturalism has been described as "more alert to science, especially biology, and still more to instinctual drives, especially sex" (Budd 2018). A shorthand might be to identify naturalism as realism plus social determinism. On the one hand, Chopin's description of a sexual encounter in her posthumously published short story "The Storm," written in 1898, has been repeatedly singled out to identify her as a realist (Arner 1970, 3; Gaudé 1975, 1). On the other hand, Edna Pontellier's emergent sensuality in *The Awakening*, and the fact that she has little control over her destiny and her own emotions, has been used to categorize Chopin as a naturalist (Walker 1979, 103). "The Story of an Hour" can similarly serve as an example of Chopin's depiction of social determinism (Emmert 2008, 81).

There are numerous claims that Chopin's writing anticipates modernism or can even be defined as early modernist (Horner 2008, 132–133). This placement is mainly hinged on Chopin's exploration of individualism, and not so much on her mode of representation. Especially Chopin's depiction of the "new woman" (Birkle 1990, 111), the self-reliant and independent female, identifies her writing as pre-modern (Wagner-Martin 2016, 7, 11–13). The labels "new women's writing" (Bhattacharjee and Ray 2018) and "new woman fiction" (Heilmann 2008) have been used to brand Chopin's work.

An innovative categorization of the development of women's writing, created to challenge the periodicity of traditional, male-dominated and -oriented literary history, was introduced in the late 1970s and extended in the 2000s. It identifies four phases of women's writing: the feminine phase (1840–1880), the feminist phase (1880–1920), the female phase (1920–2000), and the free phase (from 2000 onwards) (Showalter 2010, xvii). Chopin's writing chronologically falls into the second category, a phase of protest against the prevailing modes of the dominant tradition and of advocacy of independent rights for women (xvii). Writers of this phase reject the compliant position of femininity (Showalter 2012, 35). The first, feminine phase, is described as an era in which women aimed at equaling the intellectual achievements of the male culture and imitated the prevailing male modes of writing (Showalter 2010, xvii; Showalter 2012, 35). The feminist content of such feminine writing is indirect and can be found between the lines or in the missed opportunities of the texts (Showalter 2012, 35). "The Maid of Saint Phillippe," though actually published in the era of feminist writing, fits the description of feminine writing with its adoption and adaptation of the American male myth. "The Story of an Hour," with its more radical depiction of female rebellion, fits the feminist phase. The two stories discussed here illustrate the evolution from feminine to feminist phase in Chopin's writing.

4 Bibliography

4.1 Works Cited

Arms, George. "Kate Chopin's *The Awakening* in the Perspective of Her Literary Career." *Essays in Honor of Jay B. Hubbell*. Ed. Clarence Gohdes. Durham, North Carolina: Duke UP, 1967. 215–228.

Arner, Robert D. "Kate Chopin's Realism: 'At the 'Cadian Ball' and 'The Storm.'" *The Markham Review* 2.2 (February 1970): 1–4.

Baym, Nina. "Melodramas of Beset Manhood: How Theories of American Fiction Exclude Women Authors." *The New Feminist Criticism*. Ed. Elaine Showalter. New York: Pantheon, 1985. 63–80.

Bergmann, Ina. "Cultures of Solitude: Reflections on Loneliness, Limitation, and Liberation in the US." *Cultures of Solitude: Loneliness – Limitation – Liberation*. Eds. Ina Bergmann and Stefan Hippler. Frankfurt: Peter Lang, 2017. 13–42.

Bergmann, Ina. "Stories of Female Initiation: Flannery O'Connor's 'Good Country People' and Joyce Carol Oates's 'How I Contemplated the World from the Detroit House of Correction and Began My Life Over Again.'" *A History of the American Short Story: Genres – Classics – Model Interpretations*. Eds. Ansgar Nünning and Michael Basseler. Trier: WVT, 2011. 307–322.

Bergmann, Ina. "The Liberty of Solitude: Einsamkeit und Freiheit im Leben und Werk von Emily Dickinson, Louisa May Alcott und Sarah Orne Jewett." *Rückzug. En retrait/e: Produktivität des Solitären in Kunst*, Religion *und Geschlechtergeschichte. La solitude créatice au prisme du genre dans les arts et la religion*. Eds. Jenny Haase, Audrey Lassere, Xenia von Tippelskirch, and Beatrix Trînca. Würzburg: Königshausen & Neumann, forthcoming.

Bergmann, Ina. *And Then the Child Becomes a Woman: Weibliche Initiation in der amerikanischen Kurzgeschichte 1865–1970*. Heidelberg: Winter, 2003.

Berkove, Lawrence I. "Fatal Self-Assertion in Kate Chopin's 'The Story of an Hour.'" *American Literary Realism* 32.2 (Winter 2000): 152–158.

Bhattacharjee, Subashish, and Girindra Narayan Ray, eds. *New Women's Writing: Contextualising Fiction, Poetry and Philosophy*. Newcastle upon Tyne: Cambridge Scholars, 2018.

Birkle, Carmen. "Kate Chopin: 'The Storm': Die Geburt der New Woman aus dem Geiste des Regionalismus." *Die Englische und Amerikanische Kurzgeschichte*. Ed. Klaus Lubbers. Darmstadt: Wissenschaftliche Buchgesellschaft, 1990. 110–119.

Boren, Lynda S. "Introduction." *Kate Chopin Reconsidered: Beyond the Bayou*. Ed. Lynda S. Boren and Sarah deSaussure Davis. Baton Rouge: Louisiana State UP, 1992. 13–20.

Budd, Louis J. "Realism and Naturalism in Literature," https://eas-ref.press.jhu.edu/view?aid=115&from=browse&link=browse%3Fmethod%3Dsubject%26sid%3D7%26type%3D. *Encyclopedia of American Studies Online*. Ed. Simon J. Bronner. Baltimore, Maryland: John Hopkins UP, 2018. (14 June 2019).

Chopin, Kate. "*The Awakening*." *Kate Chopin: Complete Novels and Stories*. Ed. Sandra M. Gilbert. New York: The Library of America, 2002a. 519–655.

Chopin, Kate. "Emancipation. A Life Fable." *Kate Chopin: Complete Novels and Stories*. Ed. Sandra M. Gilbert. New York: The Library of America, 2002b. 659.

Chopin, Kate. "The Maid of Saint Phillippe." *Kate Chopin: Complete Novels and Stories*. Ed. Sandra M. Gilbert. New York: The Library of America, 2002c. 711–718.

Chopin, Kate. "The Story of an Hour." *Kate Chopin: Complete Novels and Stories*. Ed. Sandra M. Gilbert. New York: The Library of America, 2002d. 756–758.

Cunningham, Mark. "The Autonomous Self and the Death of Louise Mallard in Kate Chopin's 'The Story of an Hour.'" *English Language Notes* 42.1 (2004): 48–55.

Deneau, Daniel P. "Chopin's *The Story of an Hour*." *The Explicator* 61.4 (Summer 2003): 210–213.

Distel, Kristin M. "'Free! Body and Soul Free!': The Docile Female Body in Kate Chopin's 'The Story of an Hour.'" *New Women's Writing: Contextualising Fiction, Poetry and Philosophy*. Newcastle upon Tyne: Cambridge Scholars, 2018. 65–78.

Doloff, Steven. "Kate Chopin's Lexical Diagnostic in 'The Story of an Hour.'" *Notes and Queries* 61.4 (2014): 580–581.

Ebest, Sally Barr. *The Banshees: A Literary History of Irish American Women Writers*. Syracuse: Syracuse UP, 2013.

Emmert, Scott D. "Naturalism and the Short Story Form: Kate Chopin's 'The Story of an Hour.'" *Scribbling Women & the Short Story Form: Approaches by American & British Women Writers*. Ed. Ellen Burton Harrington. New York: Peter Lang, 2008. 74–85.

Ewell, Barbara C. "Kate Chopin and the Dream of Female Selfhood." *Kate Chopin Reconsidered: Beyond the Bayou*. Eds. Lynda S. Boren and Sarah deSaussure Davis. Baton Rouge: Louisiana State UP, 1992. 157–165.

Ewell, Barbara C. *Kate Chopin*. New York: Ungar, 1986.

Fetterley, Judith, and Marjorie Pryse. *Writing Out of Place: Regionalism, Women, and American Literary Culture*. Urbana: U of Illinois P, 2002.
Fluck, Winfried. "Tentative Transgressions: Kate Chopin's Fiction as a Mode of Symbolic Action." *Studies in American Fiction* 10 (1982): 151–171.
Foote, Jeremy. "Speed That Kills: The Role of Technology in Kate Chopin's 'The Story of an Hour.'" *The Explicator* 71.2 (2013): 85–89.
Freese, Peter. *The American Short Story I: Initiation. Teacher's Book*. Paderborn: Schöningh, 1986.
Fresonke, Kris. "Regionalism," https://eas-ref.press.jhu.edu/view?aid=562. *Encyclopedia of American Studies Online*. Ed. Simon J. Bronner. Baltimore: John Hopkins UP, 2018. (14 June 2019).
Gaudé, Pamela. "Kate Chopin's 'The Storm': A Study of Maupassant's Influence." *The Kate Chopin Newsletter* 1.2 (1975): 1–6.
Giemza, Bryan. *Irish Catholic Writers and the Invention of the South*. Baton Rouge: Louisiana State UP, 2013.
Gilbert, Sandra M., ed. *Kate Chopin: Complete Novels and Stories*. New York: The Library of America, 2002.
Gilbert, Sandra M., and Susan Gubar. *The Madwoman in the Attic: The Woman Writer and the Nineteenth-Century Literary Imagination*. 1979. New Haven, Conneticut: Yale UP, 2000.
Heilmann, Ann. "*The Awakening* and New Woman Fiction." *The Cambridge Companion to Kate Chopin*. Ed. Janet Beer. Cambridge: Cambridge UP, 2008. 87–104.
Horner, Avril. "Kate Chopin, Choice, and Modernism." *The Cambridge Companion to Kate Chopin*. Ed. Janet Beer. Cambridge: Cambridge UP, 2008. 132–146.
Jefferson, Thomas. "The Declaration of Independence (1776)." *Key Concepts in American Cultural History: From the Colonial Period to the 19th Century*. Eds. Bernd Engler and Oliver Scheiding. Trier: WVT, 2007. 303–304.
Johnsen, Heidi. "Kate Chopin in *Vogue*: Establishing a Textual Context for *A Vocation and a Voice*." *Kate Chopin in the Twenty-First Century: New Critical Essays*. Ed. Heather Ostman. Newcastle upon Tyne: Cambridge Scholars, 2008. 53–70.
Joslin, Katherine. "Kate Chopin on Fashion in a Darwinian World." *The Cambridge Companion to Kate Chopin*. Ed. Janet Beer. Cambridge: Cambridge UP, 2008. 73–86.
Koloski, Bernard. "*The Awakening*: The First 100 Years." *The Cambridge Companion to Kate Chopin*. Ed. Janet Beer. Cambridge: Cambridge UP, 2008. 161–173.
Lattin, Patricia Hopkins. "The Search for Self in Kate Chopin's Fiction: Simple versus Complex Vision." *Southern Studies* 21 (1982): 222–235.
Lewis, R.W.B. *The American Adam: Innocence, Tragedy, and Tradition in the Nineteenth Century*. Chicago: U of Chicago P, 1955.
Malin, Jo, and Victoria Boynton. "Section II: Women's Writing Spaces: Solitude and the Creative Process." *Herspace: Women, Writing, and Solitude*. Eds. Jo Malin and Victoria Boynton. New York: Haworth, 2003. 143–145.
Nigro, Kathleen Butterly. "Kate Chopin." *In Her Place: A Guide to St. Louis Women's History*. Ed. Katharine T. Corbett. St. Louis: Missouri Historical Society, 1999. 148–149.
Nolan, Elizabeth. "*The Awakening* as Literary Innovation: Chopin, Maupassant and the Evolution of Genre." *The Cambridge Companion to Kate Chopin*. Ed. Janet Beer. Cambridge: Cambridge UP, 2008. 118–131.
Papke, Mary E. *Verging on the Abyss: The Social Fiction of Kate Chopin and Edith Wharton*. New York: Greenwood, 1990.
Pierse, Mary S. "Paris as 'Other': George Moore, Kate Chopin and French Literary Escape Routes." *ABEI Journal: The Brazilian Journal of Irish Studies* 8 (2006): 79–87.
Rankin, Daniel S. *Kate Chopin and Her Creole Stories*. Philadelphia: U of Pennsylvania P, 1932.

Saar, Doreen Alvarez. "The Failure and Triumph of 'The Maid of Saint Phillippe': Chopin Rewrites American Literature for American Women." *Louisiana Literature: A Review of Literature and Humanities* 11.1 (1994): 59–73.

Shaker, Bonnie James, and Angela Gianoglio Petitt. "'Her First Party' as Her Last Story: Recovering Kate Chopin's Fiction." *Legacy: A Journal of American Women Writers* 30.2 (2013): 384–396.

Shaker, Bonnie James. *Coloring Locals: Racial Formation in Kate Chopin's* Youth's Companion *Stories*. Iowa City: U of Iowa P, 2003.

Shaker, Bonnie, Kathleen Nigro, and Sarah Nestor. "Catherine O'Flaherty Chopin's Irish Eyes: History, Biography, Fiction, Transatlantic Women 3: Women of the Green Atlantic, The Royal Irish Academy, Dublin, Ireland, June 21–22,2018," https://www.katechopin.org/kate-chopin-panel-in-dublin-21-june-2018/. *KateChopin.org: The Kate Chopin International Society*. (14 June 2021).

Seyersted, Per. *Kate Chopin: A Critical Biography*. Baton Rouge: Louisiana State UP, 1969.

Shen, Dan. "Non-Ironic Turning Ironic Contextuality: Multiple Context-Determined Irony in 'The Story of an Hour.'" *JLS: Journal of Literary Semantics* 38 (2009): 115–130.

Showalter, Elaine. "Towards a Feminist Poetics." *Women Writing and Writing About Women*. Ed. Mary Jacobus. New York: Routledge, 2012. 22–41.

Showalter, Elaine. "Tradition and the Female Talent: *The Awakening* as a Solitary Book." *Kate Chopin: The Awakening*. Ed. Nancy A. Walker. Boston: Bedford, 1993. 169–189.

Showalter, Elaine. *A Jury of Her Peers: Celebrating American Women Writers from Anne Bradstreet to Annie Proulx*. New York: Vintage, 2010.

Skaggs, Peggy. *Kate Chopin*. Boston: Twayne, 1985.

Smallwood, Stephanie. "Freedom." *Keywords for American Cultural Studies*. Eds. Bruce Brugett and Glenn Handler. New York: New York UP, 2014. 111–115.

Stanton, Elizabeth Cady. "Declaration of Sentiments and Resolutions, Seneca Falls Convention (1848)." *Key Concepts in American Cultural History: From the Colonial Period to the 19th Century*. Eds. Bernd Engler and Oliver Scheiding. Trier: WVT, 2007. 331–334.

Stepenoff, Bonnie. "Freedom and Regret: The Dilemma of Kate Chopin." *Missouri Historical Review* 81.4 (July 1987): 447–466.

Stepenoff, Bonnie. "The Real Maid of Saint Phillippe: Kate Chopins French Colonial Heritage." *French Colonial History* 17 (2017): 67–80.

The Bible: Authorized King James Version. Introd. Robert Carroll and Stephen Prickett. Oxford: Oxford UP, 1997.

Thoreau, Henry David. *Walden*. 1854. London: Everyman, 2000.

Toth, Emily. "Kate Chopin Thinks Back Through Her Mothers: Three Stories by Kate Chopin." *Kate Chopin Reconsidered: Beyond the Bayou*. Eds. Lynda S. Boren and Sarah deSaussure Davis. Baton Rouge: Louisiana State UP, 1992. 22–28.

Toth, Emily. "What We Do and Don't Know about Kate Chopin's Life." *The Cambridge Companion to Kate Chopin*. Ed. Janet Beer. Cambridge: Cambridge UP, 2008. 13–26.

Toth, Emily. *Kate Chopin*. New York: William Morrow, 1990.

Toth, Emily. *Unveiling Kate Chopin*. Jackson: UP of Mississippi, 1999.

Turner, Frederick J. "The Significance of the Frontier in American History." *Annual Report of the American Historical Association for the Year 1893*. Washington: Government Printing Office, 1894. 197–227.

Twain, Mark. *The Adventures of Huckleberry Finn*. London: Penguin, 1985.

Wagner-Martin, Linda. *The Routledge Introduction to American Modernism*. London: Routledge, 2016.

Walker, Nancy A. "Introduction: Biographical and Historical Contexts." *Kate Chopin: The Awakening*. Ed. Nancy A. Walker. Boston: Bedford, 1993. 3–18.

Walker, Nancy A. *Kate Chopin: A Literary Life*. Houndmills: Palgrave Macmillan, 2001.
Walker, Nancy A. "Feminist or Naturalist: The Social Context of Kate Chopin's *The Awakening*." *Southern Quarterly* 17.2 (1979): 95–103.
Wan, Xuemei. "Kate Chopin's View on Death and Freedom in *The Story of an Hour*." *English Language Teaching* 2.4 (December 2009): 167–170.
Welter, Barbara. "The Cult of True Womanhood, 1820–1860." *American Quarterly* 18.2 (Summer 1966): 151–174.
Woolf, Virginia. *A Room of One's Own*. San Diego, CA: Harcourt Brace, 1989.
Worton, Michael. "Reading Kate Chopin Through Contemporary French Feminist Theory." *The Cambridge Companion to Kate Chopin*. Ed. Janet Beer. Cambridge: Cambridge UP, 2008. 105–117.

4.2 Further Reading

Beer, Janet. "Introduction." *The Cambridge Companion to Kate Chopin*. Ed. Janet Beer. Cambridge: Cambridge UP, 2008. 1–12.
Berenji, Fahimeh Q. "Time and Gender in Charlotte Perkins Gilman's 'The Yellow Wallpaper' and Kate Chopin's 'The Story of an Hour.'" *Journal of History Culture and Art Research* 2.2 (June 2013): 221–234.
Chopin, Kate. "How I Stumbled on Maupassant." *Biblio* (October 1998): 80.
Diederich, Nicole. "Sharing Chopin: Teaching 'The Story of an Hour' to Specialized Populations." *Arkansas Review* 43.2 (2012): 116–120.
Koloski, Bernard, ed. *Awakenings: The Story of the Kate Chopin Revival*. Baton Rouge: Louisiana State UP, 2009.
Koloski, Bernard. *Kate Chopin: A Study of the Short Fiction*. New York: Twayne, 1996.
Ostman, Heather, and Kate O'Donoghue, eds. *Kate Chopin in Context: New Approaches*. New York: Palgrave Macmillan, 2015.
Petry, Alice Hall, ed. *Critical Essays on Kate Chopin*. New York: G.K. Hall, 1996.
Rippl, Gabriele. "Feminist Voices around 1900: Kate Chopin's 'Desirée's Baby' and Charlotte Perkins Gilman's 'The Yellow Wallpaper.'" *A History of the American Short Story: Genres – Developments – Model Interpretations*. Eds. Michael Basseler and Ansgar Nünning. Trier: WVT, 2011. 157–170.
Sümer, Sema Zafer. "The Birth of a New 'Self' and a New 'Identity' in Kate Chopin's Short Stories." *Selcuk University Social Sciences Institute Journal* 28 (2012): 191–196.

Mirosława Buchholtz

12 Henry James (1843–1916)

Abstract: This chapter begins with an overview of Henry James's career as a short-story writer from 1864 to 1910, and comments on his artistic aims and achievements, his predecessors and followers, and his publishing strategies. The two tales selected for a close reading span twenty-five years of his creative work and illustrate both continuities and changes in the writer's style and theme. The two texts can be classified as American tales in the sense of being set in the United States and exploring American society and culture at the turn of the nineteenth and twentieth centuries. Both tales were published after the expatriate author's increasingly infrequent visits to the country of his birth. "Pandora" is quite clearly an example of James's "international theme," whereas "Crapy Cornelia" shows that a culture shock may be occasioned by differences other than nationality, gender, or class identity, namely, the difference in age and attitude to the past and the future. The juxtaposition of these ostensibly female-centered narratives helps discover the evolution not only in James's perception of the American woman, but also of the man (German or expatriate American) who studies her. The two tales lend themselves to a variety of readings including textual, autobiographical, and political.

Keywords: International theme, American woman, modernity, expatriation, travel, satire, ambiguity

1 Context

Like Nathaniel Hawthorne (↗6 Hawthrone), whom he both admired and criticized, Henry James (1843–1916) preferred to define his short fiction in a more traditional way as "tales" – which in a sense testified to the descent of his literary art from Giovanni Boccaccio in prose and Geoffrey Chaucer in poetry (see Edel 1962, 9). Far more world-savvy than Hawthorne (and "the world" meant to both mostly Western Europe), he availed himself of literary traditions, which he in turn helped to revamp. Some of his longer short narratives are classified as novellas, and are thus heirs to the Italian *novella*, the German *Novelle*, and the French *nouvelle*, all promising newness and trendiness despite their roots deep in the Renaissance. "Short" was a pledge James made to himself while working on novels, but as his literary principles solidified, he faced with increasing dismay the idea and practice of word-count imposed by editors of magazines to which he sent his fiction. To him the difference between a tale and a *nouvelle* was not a matter of the number of words, but of artistic aims and appropriate narrative strategies.

https://doi.org/10.1515/9783110587647-013

According to Leon Edel, only two of James's tales contain around 5,000 words each and come closest to the requirements set by the magazines of his time. On average, his tales run from 10,000 to 20,000 words (9). Although "The Turn of the Screw" reached the figure of some 40,000 words (x), Henry James still called it a tale. Today scholars refer to it as a long short story (see Gale 1989, 680), largely disregarding James's onomastic predilections. The French *nouvelle* was to James a welcome respite from the pressure of word-count; it "has always been a much more elastic form than the English or American 'short story'" (Edel 1962, 8), and its attractiveness to James was precisely in its lack of a prescribed length. At a time of quasi-industrial parameterization of products, including literary works of art, Henry James insisted on the romantic idea of organic form, which exposes one of the unsuspected inner tensions in a self-made professional author of realist fiction.

Instead of counting words, in his Preface to *The Spoils of Poynton* of 1908 (Vol. X of The New York Edition of his fiction), James makes a distinction between "an anecdote" and "a picture": "[a] short story, to my sense and as the term is used in magazines, has to choose between being either an anecdote or a picture and can but play its part strictly according to its kind" (James 1934, 139). He hastens to express his preferences and blurs the line again: "I rejoice in the anecdote, but I revel in the picture; though having doubtless at times to note that a given attempt may place itself near the dividing-line" (139). In his 1908 Preface to *The Reverberator* (Vol. XIII of The New York Edition), James specifies that "[t]he anecdote consists, ever, of something that has oddly happened to some one, and the first of its duties is to point directly to the person whom it so distinguishes" (181). Interestingly, James usually has more to say about the "anecdote" than the "picture," until the 1909 Preface to "The Altar of the Dead" (Vol XVII of The New York Edition), in which he declares in a violent image his authorial power and his real desire: "with the longest and firmest prongs of consciousness, I grasp and hold the throbbing subject; *there* it is above all that I find the steady light of the picture" (258).

Some twenty years before, in a letter of 31 July 1888 to his good friend Robert Louis Stevenson (who was seven years his junior), Henry James, who had just begun work on *The Tragic Muse*, both ambitiously and pragmatically outlined his literary plans and intentions:

> After that, with God's help, I propose, for a longish period, to do nothing but short lengths. I want to leave a multitude of pictures of my time, projecting my small circular frame upon as many different spots as possible and going in for number as well as quality, so that the number may constitute a total having a certain value as observation and testimony. (James 1920, 139)

Leon Edel rightly points to this year and this credo as a kind of caesura. At the age of forty-five, and twelve years into his expatriation in Britain, James had already produced "the equivalent of a lifetime of fiction – some ten novels and half a hundred short stories," and within the following twenty-two years, he "more than doubled the number of his tales as well as of his novels" (Edel 1962, 7). Perseverant productivity was James's response to setbacks and disappointments: two of his social novels published in 1886 – *The Bostonians* and *The Princess Casamassima* – received mixed reviews and did not sell well, while the disastrous end of James's experiment with the theater was soon to come.

Henry James's literary apprenticeship was not so long and probably not so desperate as that of Hawthorne, but he too began by writing short fiction and it took ten years before the short-story writer turned into a novelist. The hierarchy of fiction was taken for granted in the publishing business: for ambitious authors tales were mere stepping stones to more elaborate and substantial forms, just as reviews of others' writing were stepping stones to one's own fiction. James's later use of the *nouvelle* may be seen as an attempt to break out of such hierarchies, and the work of dismantling them was later continued by writers (and readers of James) such as Alice Munro or Mavis Gallant. James's first tale "A Tragedy of Error" appeared without attribution in the *Continental Monthly* in February 1864, some three months before Hawthorne's death. The apparent continuity is deceptive; in fact, James's tale owes much more to French than American authors, especially his favorite of that time, George Sand (see Gale 1989, 664). The first signed tale, "The Story of a Year," appeared in the March 1865 issue of the *Atlantic Monthly*. Before James's first full-fledged novel *Roderick Hudson* came out in 1875, he had published as many as twenty tales, numerous reviews, travel essays, and his first serialized novel *Watch and Ward*, which he soon disowned. His twentieth tale, "The Madonna of the Future" (1873), a story of an expatriate artist who wasted his life and created nothing, may well reflect the fear of the thirty-year-old James himself that the *magnum opus* may never come. Quite typically, it also exemplifies the osmosis between James's fiction and travel writing.

Once he had published his big novel, it seemed easier for his short texts to be noticed, especially if their topics excited controversy. As Edel observes, "it was with a slender tale, about a spirited little American girl in Rome, that he made his enduring reputation" (Edel 1962, 7) in the late 1870s. Edel argues further that James had "at least one masterpiece of the short form for each decade" (7): "Daisy Miller" in the 1870s, "The Aspern Papers" in the 1880s, "The Turn of the Screw" in the 1890s, and "The Beast in the Jungle" in the 1900s. Among the 112 tales forming the canon of James's short fiction as defined by Leon Edel, there are doubtless many more gems, or at least vivid pictures of his times, as promised in the letter to Stevenson.

Henry James "grew up in a culture of literary publication" (Johnson 2009, 426). The mid-nineteenth century United States witnessed the rise of new American periodicals which "played an important role in shaping what eventually became recognized in the 20th century as 'American literature'" (426). Although William James and Henry James benefitted from their father's friendships with the major publishers of the day, "most of the time Henry, Jr., submitted his early stories anonymously and without telling his family, anxiously watching the post to intercept letters of acceptance or rejection" (426). Later he formed friendships with such powerful editors as the "Dean" of American letters, William Dean Howells, who was in charge of the *Atlantic Monthly* from 1871 to 1881. Throughout his creative life, James worked with publishers who often brought out both books and periodicals. The transition from one medium to the other seemed for this reason quite natural (and profitable). His novels were serialized in magazines before they appeared in book form, often simultaneously on both sides of the Atlantic. While periodicals were also the main venue for his short fiction, not all the tales made it into the book-form collections he occasionally published. As Edel puts it, James "never exhausted the back-log he had created in the magazines" (Edel 1962, 7). His first collection of short fiction, *A Passionate Pilgrim and Other Tales* (1875) is a good example; it "contained but six of the twenty-seven tales he had written by that time" (7). More than thirty years later, *The Novels and Tales of Henry James* in twenty-four volumes (with two more added posthumously) contained fifty-two tales, fewer than a half of those he had published before and continued to write in the twentieth century. Remembered as The New York Edition, the volumes appeared from 1907 to 1909 and were meant to become the canon of James's works. He had not only carefully selected, but also thoroughly revised his prose for The New York Edition, which also contains the famous prefaces outlining both anecdotal origins and theoretical underpinnings of his art.

It was Henry James's biographer Leon Edel who first collected the stories in the twelve volumes of *The Complete Tales of Henry James*, which appeared from 1962 to 1964. Placing the tales in chronological order, Edel sought to show James's "development and growth in a form which he found a constant challenge and in which he achieved a singular mastery" (8). While commenting on this output, Edel avails himself of the tripartite division of James's tales into early, middle, and late. He describes the early tales as mostly "apprentice-work" (10), "melancholy and romantic, as young tales can be" (11). Set mostly in the United States, the early tales "have a lugubrious atmosphere, for all their skill" (11). Edel notices in the tales of the middle period, James's "emergence as a brilliant and witty observer of life on both sides of the Atlantic" (11). Apart from "an extraordinary play of intellectual humour as he shows Americans looking at Europe and Europeans at America" (11), Edel also points to James's study of the American woman (12) as the definitive feature in the tales of the middle period. The late tales, Edel declares, "at first glance, might have been written by someone else, so great a change has occurred" (13). James's minute descriptions of manners "give way to a study of states of feeling and dilemmas of existence" (13). Interestingly, David L. Hoover's recent computational studies of James's style have confirmed Edel's claim about the difference between his earlier and late short fiction. The two tales that will be discussed in this chapter are recognizable at first sight as representatives of the middle and the late phases of James's short-story writing.

Apart from the aim of giving testimony to his time and honing his art, James also thought of short fiction as a source of dependable income, though more so in the middle than the late phase. In the 1880s, tales were still his "most profitable publications" on both sides of the Atlantic (Kaplan 1992, 285). For example, James suspended work on *The Bostonians* to write "a couple of tales" – "Georgina's Reasons" and "Pandora" – for lucrative syndication (James 1999, 151–152, editor's note; Kaplan 1992, 285–286). Tales formed his symbolic capital and defined his high status on the literary scene. When cajoling James to contribute to the newly established *Yellow Book: An Illustrated Quarterly*, its young editor Henry Harland, one of many young aficionados, addressed the author as "the supreme prince of short story writers" (qtd. in Kaplan 1992, 410). The first issue of the *Yellow Book* of April 1894 indeed opened with "The Death of the Lion," which bitterly predicted the death of the author, without promising very resolutely the birth of the informed reader. The prophesy proved correct; the *Yellow Book* was a short-lived, though memorable, venture. Unlike Harland, who not only promised a princely remuneration but also gave James freedom to take up as much space as he liked, some of the American editors who solicited contributions from James set limits on both topics and words. James was particularly annoyed when they expected him to write "another 'Daisy Miller'" (qtd. in Kaplan, 1992, 410).

The two tales selected for a close reading in this chapter span twenty-five years of James's creative activity. Both belong to the category of his American tales, and both were written and published after his visits to the country of his birth. "Pandora," following James's travels to the United States between 1881 and 1883, appeared for the first time in New York's *The Sun* (1, 8 June 1884) and was republished in two collections before its canonization in the New York Edition (Vol. XVIII, 1909). "Crapy Cornelia" was published in *Harper's Magazine* (October 1909) and was reprinted in the collection *The Finer Grain* in the following year. It is one of the four American tales James wrote after his return from a tour of the United States in 1904 and 1905, his first visit in twenty-one years. Leon Edel argues that the four (including also "Julia Bride," "The Jolly Corner," and "A Round of Visits") "reflect a sense of shock." "They are," as he puts it, "angry tales" (Edel 1964, 7), voicing the disappointment of an old New Yorker, for whom "in his old age, there seems to have been only the cold bench and the desolation of the metropolitan jungles" (10).

2 Close Readings

"Pandora" is, in a nutshell, the story of a young German count who travels on an ocean steamer from Bremen to New York, via Southampton, to serve as a diplomat in the United States. The microcosm of the ship becomes for him a field of anthropological study. He singles out as a subject of his scholarly pursuits a young American lady, whom he observes, using as his manual a Tauchnitz novel by an American author (see James 1963, 361). As an empiricist, the count not only gathers clues from his book and from a self-appointed American informant, but also relies on his own senses, and eventually engages his subject in an hour-long conversation shortly before their arrival in the United States. Once the journey is complete, the scholar and the studied subject, with her uncouth family in tow, part company, only to meet again eighteen months later at an elegant party in Washington, which is also attended by the recently inaugurated (unnamed) President of the United States. The young American lady, the eponymous heroine of the tale, not only appears at the party but also shines there and carries the day, much to the confusion of the diplomat whose prime concern seems to have been the question of her insecure social standing.

Could a story like this happen today? Intercontinental travel is faster and (usually) better organized in the early twenty-first century than it was in the late nineteenth century. An airplane is a far more frequent choice today than an ocean liner, but travel has become such a common activity of people representing a wide range of social strata and nations that the challenge of understanding humans of other nationalities (let alone social position), in a scientific way, that is, on their own terms, is even more significant today than it was in the 1880s when the story was first published. However, as travel became faster, the opportunity to "study" strangers dwindled, and so did also the curiosity about others along with the need to understand them.

The protagonist of the story, Count Otto Vogelstein, is defined at the outset as "[a]n intelligent young German" (357), straight, elegant, discreet, diplomatic, or, in a word, official. Later the narrator calls him a "candid Teuton" (365), a "theoretic Teuton" (382), and eventually also "an emissary of Prince Bismarck" (399). The count travels to take the post of the secretary of the German legation in Washington and emerges as "a model character for such a purpose – serious, civil, ceremonious, stiff, inquisitive, stuffed with knowledge, and convinced that at present the German empire is the country in the world most highly evolved" (358). He does not have "a high sense of humour," but he is at least aware of this deficiency and its significance, since he knows Americans to be "a highly humorous people" (358). This deficiency may be the reason why Vogelstein, who speaks good English, finds it difficult to catch American jokes (369). He is "still young enough in diplomacy to think it necessary to have opinions" (359) and many of his opinions, especially his political views, are inherited from his noble ancestors. He is, as the narrator soon declares, "a stiff conservative, a Junker of Junkers" who "thought modern democracy a temporary phase, and expected to find many arguments against it in the United States" (359).

When the *Donau*, the steamer on which he travels in the first days of autumn, anchors for several hours in Southampton to receive "many additions" to its "human cargo" (357), he has his chance to observe "curiously, indifferently, vaguely, through the smoke of his cigar" (357) American passengers "absorbed in the huge capacity of the ship" (357). He is far less interested in the many German emigrants on the ship: some of them yellow, some black, and all of them "greasy and matted with the sea-damp" (359). They are physically a contrast to the detached aristocrat, who does not believe that they can improve the quality of the western democracy whose current they are about to swell. Equally mysterious are the numerous Jewish passengers of the ship, "commercial, to a man" (360). Remembering what "a lady in Dresden had once said to him – that America was a country of girls" (360), the count focuses on this species and is soon enough approached by a specimen.

A "slim, brightly dressed, and rather pretty" young lady, who is also repeatedly described as "quick" and "competent" (362) hovers around Vogelstein while he is seated on his sea-chair. The prolonged pantomime (362–363) puzzles Vogelstein until he guesses that she wants to know his name. He stands up, turns the back of the chair to the girl and exposes the superscription in huge elaborate letters: his title and name. The girl expresses gratitude and the short conversation that follows provides a banal explanation: she was looking for one of the sea-chairs that belonged to her family and was lost. The superscription on the back of that lost chair would be much shorter than the Count's: "Our name is Day" (364), she announces. Only later does another American passenger, Mrs. Dangerfield, "a handsome, confidential, insinuating woman" (365–366) reveal that the girl's (probably self-)given name is Pandora. Mrs. Dangerfield is unsparing to Pandora and her kin, dismissing them as provincial people from "the interior" (367).

The informant's emblematic name is, however, ambiguous and detracts from the validity of her claims. She warns, but she may herself pose a danger. By contrast, the name of Pandora Day is melodious to the extent that it would have probably pleased Edgar Allan Poe. Not that James would have been interested in pleasing or imitating Poe. He apparently shared the dismissive opinion of Emerson, who was a friend of James's father, that Poe was a mere "jingle man" (qtd. in McGann 120). James had, evidently, other reasons for picking a melodious name whose rhythm brings to mind "Tandaradei." This is a word from the medieval poem "Unter den Linden" by Walther von der Vogelweide, meant, apparently, to imitate the song of a nightingale. However, Vogel*stein* in James's tale is not the poet von der Vogel*weide*, and the lovely medieval song had since acquired a martial tone when sung by Teutonic Knights at battlefields.

As a student of American "types," Vogelstein keeps comparing Pandora Day and her family with the protagonists of "his volume of Tauchnitz" (James 1963, 365), which "treated of a flighty, forward little American girl, who plants herself in front of a young man in the garden of an hotel" (363). Pandora is "older and not so pretty" (364) as the girl in his book. Like that heroine, she has a mother and a younger brother; unlike the fictional American girl, she has a younger sister and her father is present during the voyage. The count declares that Miss Day has "nothing in common with the heroine of that work, save a certain local quality and the fact that the male sex was not terrible to her" (365). When he finishes reading "his little American story," he can "definitely judge that Pandora Day was not at all like the heroine. She was of quite another type; much more serious and strenuous, and not at all keen, as he had supposed, about making the acquaintance of gentlemen" (368). However, he notices other analogies:

> If his sister did not resemble the dreadful little girl in the tale I have so often mentioned, there was, for Vogelstein, at least an analogy between young Mr[.] Day and a certain small brother – a candy-loving Madison, Hamilton, or Jefferson – who, in the Tauchnitz volume, was attributed to that unfortunate maid. This was what the little Madison would have grown up to at nineteen, and the improvement was greater than might have been expected. (369)

At this point, at the latest, James's reader should remember that the boy's name was Randolph, and not Madison, Hamilton, or Jefferson. Yet James's playfully forgetful allusion not only mimics a careless reading, but also exposes the American habit of naming children, boys in particular, after great men of American politics. James clearly revels in a literary game with his readers, dropping unmistakable hints. Then, to make sure that even the most uninformed reader notices the allusion, he playfully reveals the girl's and the tale's literary antecedent. The hilarious vignettes of Pandora's little sister justify the pronouncement that she "should be a Daisy Miller *enherbe*" (370). "Day," which is also the first syllable of "Daisy," acquires an additional significance as a link between the two tales.

In "Pandora" James kills a few birds with one Vogelstein. He draws portraits of Americans, but to protect himself against accusation of too severe criticism of his native land, he uses as his central reflector a German Count, which, in addition, allows him to share his observations about Germans. "Transcendental" and upper-class, Vogelstein is vaguely fascinated with the "quick, bright, silent girl, who could smile and turn vocal in an instant" (369–370), but he is deeply disturbed by her siblings and her parents, Mr. and Mrs. P.W. Day, whom he describes as "heavy little burghers" (370). Placidly sitting in their sea chairs, carefully tucked in by their energetic daughter, they resemble "a pair of household dogs that expect to be scratched" (371). In the long conversation almost at the end of the voyage, Pandora tells Vogelstein that her family was traveling for two years in Italy, Greece, and Palestine, and has been "everywhere" (373). Vogelstein is supposed to have little sense of humor, so it must be the narrator's American jocularity that shines through the descriptions of Pandora's family as people on whom the European "culture" is not likely to have left any mark (373). Pandora admits further that although they indeed come from Utica, which is "middle-sized" and "horrid," she is now heading for New York, where she works (374).

As it was also in the case of Winterbourne in "Daisy Miller," Vogelstein's interest in Pandora oscillates between the scientific and the amorous, and the reader does not know which may eventually tip the scales. When he praises Pandora as a better type of an American girl than Daisy Miller and admires her intellect (368), is he complementing the American girl on her progress or thinking of Pandora as a prospective wife? When he worries about the social position of her family, who are described dismissively as her "human belongings" (371) or "appendages" (379), is he reflecting on the heterogeneity of American family or considering the presentability of his prospective in-laws in high society? The presence of the American girl on the scene is part of modern life, as the count realizes in his distanced, philosophical manner. He "felt the peril" of "marrying the American girl," but he saw it as "something one had to reckon with, like the rise in prices, the telephone, the discovery of dynamite, the Chassepôt rifle, the socialistic spirit; it was one of the complications of modern life" (370). In a later version of the tale, available online, "the rise in prices" is substituted with "the railway," "the telephone" with "the telegraph," and "Socialist" is capitalized, which all shows James's almost journalistic urge to be up-to date.

By the end of Part I of the tale, which is set on the ship, Vogelstein once again asks Mrs. Dangerfield about Pandora's "social position," to which the woman, despite her previous criticism, responds chidingly: "Poor girl, it isn't fair of you to ask such questions as that" (374). What Vogelstein sees of Pandora when leaving the ship convinces him that she can fend for herself and for her family. At Jersey City, all passengers must go through customs control and then look for further means of transport. Pandora, who seemed afraid of the control, finds a way of convincing the functionary to only superficially inspect their luggage. Her friend Mr. Bellamy was not able to come to greet her, but she uses a letter from him and her own charm to convince the customs officer not to inspect the family's possessions too closely (378–379). Vogelstein has no difficulty with the customs when his English servant (a *deus ex machina* who appears in this scene and never again in the tale) brings the functionary to him. Although the word "victim" surfaces many times with reference to arriving passengers, the count is treated politely to the extent that he even considers tipping the examiner (378). His first lesson in American democracy is a case of high comedy, too. When the servant informs the count that the hack (hackney-coachman) is "beyond everything. He wants thirty shillings to take you to the inn," Vogelstein suggests that the servant should look for a German one. It turns out, however, that the impudent hack is indeed German, and Count Vogelstein has to begin "his career in America by discussing the tariff of hackney-coaches in the language of the fatherland" (380).

It would seem that in Part II of the tale, which is set in the country's capital in April of the second year of Vogelstein's stay in the United States, the young diplomat would feel more secure, but he is still not sure he can judge Americans correctly, which was Pandora's request when they were parting ways (380). He still cannot see the differences in social standing and confuses, for example, Members of the House with "the functionaries engaged for the evening to usher in guests and wait at supper" (384). He repeatedly poses the question of this or that person's social standing, and always receives an evasive answer. It seems unimaginable to him that Pandora Day might be invited to one of the best houses in Washington, that of Mr. and Mrs. Bonnycastle. When he sees her there conversing freely with the President of the United States and "extracting promises from the ruler of fifty millions of people" (389), he is astounded. In an attempt to understand, he dutifully records all the claims to aristocratic privilege that are conceivable in the United States, such as coming from the South, pronounced as "the Sooth" (395), or being a descendant of a participant in the American Revolution (395), but Pandora dismisses all such attempts at myth making (395). It is the host who, once the party is over, takes pity on Vogelstein and explains to him that Pandora is "the latest, freshest fruit of our great American evolution. She is the self-made girl!" (396). Mr. and Mrs. Bonnycastle explain further that the American self-made girl is "not fast, nor emancipated, nor crude, nor loud." She is "simply very successful, and her success [is] entirely personal" (397).

Vogelstein's next meeting with Pandora is planned to take place at a picnic at Mount Vernon, but the diplomat is so curious about the girl who has made herself that he meets her every day. On one occasion he follows Pandora and her companion to the Capitol and takes over the role of a tour guide (399–401). He notices that Pandora seems intensely interested in politics (400–401) and regards the Capitol as "the Acropolis in its prime" (401). The myth she reserves for America (and herself) is clearly older than the American Revolution; it reaches back to antiquity. Pandora's fascination with the American past is evident also during her visit to Mount Vernon. She reminisces about the Civil War (see 402) and avidly explores the home of George Washington. Vogelstein is once again at hand to guide her and misconstrue her historic expedition as an occasion for romance.

In "Pandora" James proves to be quite a diplomat himself; he offers apparently a more flattering image of the American girl than he had done in "Daisy Miller." Pandora is more mature and more desirable to the interested observer than Daisy. Unlike Winterbourne, Vogelstein repeatedly considers marrying Pandora. Instead of flirting with foreigners, however, Pandora remains loyal to her hometown sweetheart, Mr. Bellamy, who, as Pandora's friend reports to Vogelstein, is "one of the leading gentlemen of Utica, and very highly educated. He is a good deal older that Miss Day" (409). He "had some business difficulties," but "has given up business, and [. . .] feels more free" (409). Although Pandora is not a voter herself, she is enough of a celebrity now to influence the newly elected President of the United States and, as it turns out, secure for Mr. Bellamy, whom she soon marries, the post of Minister to Holland (see 412). The inductions (that is, inferences of a generalized conclusion from particular instances) Vogelstein and Mrs. Bonnycastle make about the American self-made girl are proof of her induction into (that is, putting in formal possession of) a position of power in society. Her rise is parallel to the inauguration of the country's president, with a clear hint that the "girl" has power *over* the president, whom she uses to achieve her aims. She also uses the German count, who may have inspired her idea of social promotion through diplomacy. Pandora is not eligible for diplomatic positions herself, but as a proof of "making herself," she also "makes" her fiancé in that she secures for him a position far above that of Count Vogelstein. The better version of Daisy Miller emerges as a (kindly ruthless or ruthlessly kind) manipulator.

A comparative analysis of James's "Pandora" and Goethe's festival play ("Festspiel") *Pandora* is beyond the scope of this reading, and so is also the tracing of Hegel's and Goethe's concept of "the Idea," so elegantly performed in another context by David E. Wellbery (Wellbery 2013, 235). Vogelstein indeed looks at Washington, D.C., with a mind shaped by German philosophy. To him there "was a kind of Hegelian element in it; in the light of these considerations the American capital took on the semblance of a monstrous, mystical *Werden*" (James 1963, 387–388). However, in addition to philosophy, his perception is also influenced by the politics of his social class within the framework of the recently (1871) constructed German Empire (Deutsches Kaiserreich). The tension between philosophy and politics remains just as unresolved as the see-saw between Vogelstein's scientific and amorous pursuits of the American girl.

James may have taken over from Goethe the motif of Pandora's seductiveness, but all other elements of the Greek myth – remolded and put to a variety of uses over the centuries, as Elisabeth Frenzel shows (Frenzel 1992, 617–620) – are ambiguous in the American girl he created. If Pandora Day has all the gifts, is she the one who collects or gives away? Are the gifts innate or acquired? Is it a blessing or a curse to be her husband? What is in her box? And where is the box at all in James's tale? His narrative inventiveness becomes clear also when the reader begins to look for the box and finds a wide range of unexpected "boxes": the steamer named after the river *Donau* and described as "an oblong box" that is afloat for ten days (James 1963, 365), an opera-box of which Pandora cannot even dream in Utica (374), the "battered boxes" of the travelers terrified to open them to the customs officer (376), and finally "box-hedges" ("box" as a small evergreen plant) in the garden of Mount Vernon (405). Vogelstein's mind is at the very beginning conceptualized as a box which contains "several millions of facts, packed too closely together for the light breeze of the imagination to draw through the mass" (358). James plays with the reader's expectations when he refashions the ancient Pandora as someone who wants to open the box of a foreigner's mind while at the same time fearing the inspection of the traveling boxes that belong to her family. Cornelia (in the tale we will consider next) is just as secretive as Pandora, as the modifier "crapy" (draped in crape) suggests in the very title, but nothing is left in her of Pandora's ambition and flamboyance. Some twenty-five years older than Pandora, Cornelia is at best an indicator, a piece of litmus paper, and by no means a (re)agent in the tale named after her, even though she braces herself by the end of the story to demand a role she could play. She is, paradoxically, both a central and a marginal character.

The tale "Crapy Cornelia" is in many ways a contrast to "Pandora." The reader who gets the hang of the convivial light comedy in "Pandora" (a "Festspiel" of sorts, too), will have to recalibrate the analytical instruments to appreciate the more lurid, tortuous, and acerbic style of "Crapy Cornelia." N.H. Reeve notices in James's late tales "a harsher, more aggressive tone of address to the times he was living in" (Reeve 2017a, xxviii). With its total of some 20,000 words, "Pandora" is a diptych:

Part I is set on an ocean liner heading for the New World and Part II takes place in and near the U.S. capital. By contrast, "Crapy Cornelia," which contains over 11,000 words, is divided into five parts, as if it were a proper five-act tragedy (or rather tragicomedy), which it is, in a way. Set in New York, the city of James's birth, in and near Central Park, the tale features a very small cast of three characters. The protagonist, Mr. White-Mason, is an American who, like James, spent his adult life in Europe and returns to the place of his birth to propose to a young and wealthy widow, Mrs. Worthingham. Unlike Vogelstein, who is a young German diplomat, White-Mason is forty-eight and, by his own standards, old. He is an American expatriate of the same social class as his lady friend, facing probably his last chance for marriage. The third unexpected party in the tale is the eponymous Cornelia, who destabilizes what seems to be a clear plan.

What David Wellbery writes with reference to Goethe and Hegel seems unexpectedly relevant to "Crapy Cornelia": "The modern diremption – the split between blind productivity and the melancholy cult of lost beauty – is to be overcome insofar as the Idea, present in itself, finds its imaginative realization in the festival play that knows itself as play" (Wellbery 2013, 235). In Part I of his "late-life" drama, White-Mason (the narrator familiarly dispenses with honorifics) sets out to overcome "the modern diremption" and prepares for the enactment of the nuptial play that "knows itself as play." There is no mention of love or at least lust for the intended at any point in the tale. White-Mason feels inclined to play the game, though the plan causes a discomfort, which he tries to ease by procrastinating: taking a walk and indulging in a quiet meditation. The ludic festival envisioned by Hegel and Goethe – and about to be enacted by White-Mason – discloses, as Wellbery explains, "the world within the horizon of temporality" (235). However, White-Mason has a problem precisely with time. Unlike Vogelstein, who counts days, weeks, months, and years, White-Mason counts only minutes. His is a world in which time is in short supply.

The tale opens with the following image of arithmetical self-discipline: "Three times within a quarter of an hour – shifting the while his posture on his chair of contemplation – had he looked at his watch as for its final sharp hint that he should decide, that he should get up" (James 2017, 318). In the following paragraph, he is reported to have been sitting and meditating "for twenty minutes" (318). It is only at "the single deep note of half-past five borne to him from some high-perched public clock" (319) that he eventually gets up, only to procrastinate some more:

> shaking his garments, settling his hat, his necktie, his shirt-cuffs, fixing the high polish of his fine shoes as if for some reflection in it of his straight and spare and grizzled, his refined and trimmed and dressed, his altogether distinguished person, that of a gentleman abundantly settled, but of a bachelor markedly nervous. (319)

In Central Park White-Mason emerges as the epitome of negativity. The promise of new life is always qualified in one way or another. It is an April day, but already afternoon. The park is filled with "the frolic infants" (318), but "the little girls" represent "the overwhelmingly alien [. . .] populace" (318). When White-Mason formulates his plan, his certainties give way to insecurity, as his word choice, along with the use of a conditional sentence and negations amply show: "the certitude that *if* he did call on Mrs. Worthingham and find her at home he *couldn't* in justice to himself *not put* to her the question that had *lapsed* the other time, the last time, through the *irritating* and persistent, even if accidental, presence of others" (319; emphasis added).

Henry James exposes the physical pain of old New Yorkers in the *new* New York in an ocular-centric manner; they are, to put it mildly, blinded by "excesses of light" (322), "snap-shotted" (321), and left with "a scowl almost of anguish" (321). New York stands for "newness," and to an aging expatriate it is "a world of constant breathless renewals and merciless substitutions" (321). Mrs. Worthingham, a contemporary female Midas of sorts, is undoubtedly part of the changes. Her house, to begin with, has "that gloss of new money, that glare of a piece fresh from the mint and ringing for the first time on any counter, which seems to claim for it, in any transaction, something more than the 'face' value" (320). The idea of marrying her seems tantamount to embracing the new dazzling New York. Can White-Mason bring himself to do that?

In Part II of the tale, White-Mason, like J. Alfred Prufrock a decade later (the similarity was obvious to later readers of the tale, as N.H. Reeve shows, 2017a, xcix–c), pays his visit, again counting minutes ("five" James 2017, 324, "three" 332). He admires "the lustre of Mrs. Worthingham's own polished and prosperous little person," including "the gleam of her lovely teeth, [. . .] her rings and brooches and bangles and other gewgaws, [. . .] her charming complicated yellow tresses" (323), and "her light gemmed hand" with "the prettiest pink finger-nails in the world" (325). He compares all these to the attributes of "the most animated of pink-and-white, of ruffled and ribboned, of frilled and festooned Dresden china shepherdesses with exactly the right system of rococo curves and convolutions and other flourishes, a perfect bower of painted and gilded and moulded conceits" (323). This "scenic extravagance" has its foil in "a third and quite superfluous person, a small black insignificant but none the less oppressive stranger," a "dingy little presence," resembling "some shaggy old domesticated terrier or poodle" (323). This contrast and these "dog" metaphors bring to mind the ones used to describe Pandora and her kin.

The black figure is a mere subhuman intruder at first, but as the head of the woman "crowned with a little sparsely feathered black hat [. . .] came nearer and nearer, while it met his eyes, after the manner of images in the kinematograph," White-Mason recognized in her "the friend of his youth," Miss Cornelia Rasch (326). The recognition seems to rejuvenate White-Mason and enables him to think in terms of years, rather than minutes ("not more than ten [years]" since they last met, 326). The cognitive disjunction on White-Mason's part consists in "addressing the younger and prettier woman," while looking at "his fellow-guest" (327). When he asks if he may visit her, Cornelia "gasps," "wriggles," and "giggles," and gives her permission in an equally disjunctive manner: speaking to him while looking at Mrs. Worthingham. Cornelia leaves immediately, but the remaining "forty minutes" (329) of White-Mason's visit do not end with a marriage proposal. He later recollects his hesitation in a Prufrockian manner: "Shall I now or shan't I? Will I now or won't I? 'Say within the next three minutes, say by a quarter past six, or by twenty minutes past, at the furthest – always if nothing more comes up to prevent'" (332).

When White-Mason retraces his steps to Central Park (in Part III) to consider what he has and has not done, his thoughts about Mrs. Worthingham justify his decision not to propose to her. He condemns her "bright pampered confidence" (329) as well as "the innocent egotism, the gilded and overflowing anarchism, really, of her doubtless quite unwitting but none the less rabid modern note." She may be "in the know" about the ways of the world (*her* world), but she is at the same time "as ignorant as a fish of everything that really and intimately and fundamentally concerned *him*, poor dear old White-Mason" (330). In particular, she has "no instinct for any old quality or quantity or identity, a single historic or social value" (330). Paradoxically, it is the absence of Cornelia that "prevents" this time. The conversation between White-Mason and Mrs. Worthingham circles around the guest who has left. Both are equally surprised that Cornelia is counted among the other's acquaintances. Mrs. Worthingham explains how she met Cornelia in Switzerland two years before: "during twenty days of incessant rain, she was the only person in an hotel of roaring, gorging, smoking Germans with whom I could have a word of talk. She and I were the only speakers of English" (332). When Cornelia was taken ill, Mrs. Worthingham had her servant look after her, for which Cornelia was grateful. To Mrs. Worthingham Cornelia is a "poor little creature" (332) or, as White-Mason assumes, "a little clever queer creature" (333). But what is she to White-Mason? He agrees that she is exactly that, but she nevertheless has some value for him and has something that Mrs. Worthingham obviously lacks.

By the end of his meditation in Central Park, White-Mason makes a conclusion that strikes today's reader as surprisingly up-to-date. He claims that "Mrs. Worthingham had no data" (333). Obviously, he has a different type of "data" in mind from what has become one of the key concepts of the digital world today. It means to him what it had meant from the mid-seventeenth century, "facts given or granted," but the significance of accumulating and storing facts, which White-Mason underlines, appears contemporary. If Mrs. Worthingham in her position of power, guaranteed by wealth, has no data, Cornelia at least *is* "data." She "figures" to White-Mason as "a massive little bundle of data" (333). The paradox of being "massive" and "little" at the same time again brings to mind the wonders of digital age. The social hierarchy is obvious though and related to economic status: Mrs. Worthingham can only *have* and Cornelia can only *be*, at least from White-Mason's perspective. Unlike Mrs. Worthingham, who resists reification at White-Mason's hands, Cornelia becomes a poor old thing, a nostalgic repository. Since "his time and hers had been the same," Cornelia is for White-Mason, "some small extrusion of history" (333).

White-Mason's visit to Cornelia is far more relaxed than his unsuccessful attempt in Mrs. Worthingham's house. His and Cornelia's shared "position as conscious, ironic, pathetic survivors together of a dead and buried society" (334) makes him feel at home; he leans back, stretches his legs, and "after she had let him, above all," he lights "the first of a succession of cigarettes" (336). They meet in Cornelia's "small and quite cynically modern flat" (336), but even though it is her home, she is taken captive by White-Mason, who invents the game of "mak[ing] over and recreat[ing]" their "lost world" (337). She "len[ds] herself, all gallantly, to their game" (336–337) and "after a first inevitable shyness, an uncertainty of wonder, a breathlessness of courage, fall[s] into step and go[es] whatever length he would" (337). Time goes by quickly, and the guest thinks in terms of "half-hours" (337) rather than minutes. Nevertheless, it is obvious that White-Mason treats Cornelia in the same way as Mrs. Worthingham had done two years before in Switzerland: as the only person available in an alien space. White-Mason is fair enough to admit that their youth also brought "modern fever" to "those spacious, sociable, Arcadian days," but it was "so different from any of these arrangements of pretended hourly Time that dash themselves forever to pieces as from the fiftieth floors of sky-scrapers" (338).

The truth is, and White-Mason freely admits it, that they have, in fact, little common past. They had been at no point close friends or lovers (337), but Cornelia had always been "about" and remembers, for example, that White-Mason loved Mary Cardew "in the autumn of '68" (339). Their conversation spills over into Part V of the tale, in which the two discuss Mrs. Worthingham and White-Mason's "situation." He gives his reasons for not proposing to her: "To me her things are awful. They're the newest of the new" (341), and as for herself, "she doesn't know anything" (342). Even though Cornelia is convinced that Mrs. Worthingham adores him, White-Mason dismisses the idea of marrying her. Since he declares so bluntly his love for "old things," Cornelia asks what he thinks of her chance (342), and when he seems to misunderstand, she puts it bluntly: "Do you mean you want to marry *me?*" and then again "Do you want to marry *me?*" (343). He has no such intention, but when Cornelia asks what he can do with her, if he cannot marry her, he offers his companionship: "I can live with you – just this way" (347). He is so pleased with the idea that the scowl disappears from his face. James's male observers, whether German or Europeanized expatriate American – very much like their author – simply will not marry.

3 Theoretical Perspectives

The two narratives discussed above are only a small part of Henry James's achievement as a writer of short fiction, but they nevertheless give an insight into his themes and styles. Both are "American tales" set in (or en route to) the United States and focused on American culture and society. "Pandora" represents more clearly than the other story the "international theme" for which James became famous, but "Crapy Cornelia" exposes the less spectacular though more painful culture shock of an expatriate who returns to the country of his birth. Thus, the clash of cultures emerges as a matter of difference not only in nationality, but also in age, generation, or era. James's characters travel both in space and in time, and sometimes the two kinds of travel converge. In the earlier story, the protagonists travel in space; nostalgic time travel is the shared pursuit of the protagonists in the later tale.

Both tales feature female figures in their titles: the more mature and reasonable Daisy in Pandora and the defeated aging Daisy in Cornelia. Henry James was aware of the popularity of "Daisy Miller" and its consequence in editors' and readers' expectations. The readers demanded "another Daisy Miller" and James delivered, but with an ingenious twist. For example, in "Pandora," James attributes the features of Daisy Miller also to Pandora's younger sister, who is a comic figure. In "Crapy Cornelia," the marginal copies of Daisy seem to belong to the age of mechanical reproduction: one can recognize them in the little beautifully dressed brash girls in Central Park. What is disquieting about them to White-Mason, oppressed by other worries too, is that in spite of being daughters of immigrants, they feel so very much at home in New York, while he – a born New Yorker – feels so very much out of place there (Tambling 2001, 150). James clearly plays in these two tales, and especially in "Julia Bride" (of the same batch as "Crapy Cornelia"), with the figure of Daisy Miller. It seems that the game of looking for "Daisies" in James's fiction never ends. Leon Edel claims that James studied the American woman "with a mixture of affection, awe, and profound mistrust" (Edel 1962, 12). Indeed, none of the female portraits in the two tales is an unqualified glorification.

Despite the feminine titles, these tales (and others, too) backfire on the male protagonist-observers: Winterbourne, Vogelstein, and White-Mason alike. If these three are Henry James's porte-paroles, as they may be seen (James identified himself most openly with White-Mason in a letter to a female friend, qtd. in Herford 2016, 246), then it seems that he was capable of scathing (nearly Puritan) self-scrutiny. In "Crapy Cornelia," "the irritable and complacent White-Mason is as much an object of James's satire as the nouveau-riche Mrs. Worthingham" (Reeve 2017a, xciii) who, and whose house, are apparently modeled on Edith Wharton and her residence (Tintner 2000, 23). In other words, these and other tales open wide the door to the study of James's biography. There are no one-to-one correspondences between his life and the lives of his protagonists, but his fiction gives insights into his experiences and opinions at different stages of his life.

In "Crapy Cornelia," White-Mason, who in some ways resembles Henry James, denounces modernity. Being "snap-shotted" and watching a female head loom ever larger "after the manner of images in the kinematograph," are both unpleasant, threatening experiences that reflect White-Mason's resistance to modern inventions. N.H. Reeve explains that snap-shots were "photographs taken quickly and without preparation by amateur photographers" who used box cameras invented by the Eastman Kodak company in 1900 (Reeve 2017b, 486 note, 355). They were, in that sense, a stark contrast to photographs taken by professional photographers. Around that time, another novelty, the 1895 invention of the Lumière brothers, spread out quickly in Britain. By December 1908, when James wrote "Crapy Cornelia," "there were nearly 200 venues licensed to show films in London alone" (487 note, 362). What was frightening about these inventions was their large scale, infringing on the familiar habits and habitats. Mark Goble notices in James's later tale an "almost archaeological fascination with media technologies at the end of his career" (Goble 2010, 212). From his perspective, James's "Cornelia is at once an almost ghostly remnant of the past and a harbinger of the latest technology" (212). One may add that her "crape" is a fittingly ambiguous attribute that signals both lament for the lost past and unknowability of the future.

It seems that "Pandora" is far more politically inflected than the later tale. It is indeed "satirical about Washington and mildly so about the president (probably Chester A. Arthur)" (Tambling 2001, 160). James remarked in one of his notebooks that the tale "would give me a chance to do Washington, so far as I know it, and work in my few notes, and my very lovely memories of last winter. I might even do Henry Adams and his wife" (James 1987, 23–24). The Adamses appear in James's tale as Mr. and Mrs. Alfred Bonnycastle, and their identity is easily recognizable. There was no need to hide it, as George Monteiro explains, because the portrait is generous, though "lightly satiric" (Monteiro 2016, 62). As a descendant of two American presidents and a historian of the early American Republic, Henry Adams could afford to be even more than lightly satiric about contemporary politics and politicians. His attitude is reflected in James's tale.

In 2002 Lauren Weiner made an attempt to co-opt James's "Pandora" to the study of politics. Following in her footsteps in 2019, Natalie Fuehrer Taylor focuses more specifically on Clover Adams, whom James "once dubbed 'a Voltaire in petticoats'" (Taylor 2019, 188) and who appears in James's tale as Mrs. Bonnycastle. In a manner comparably indifferent to James's ironies and ambiguities, Taylor argues that "Pandora" is "a work of political philosophy, with important contributions to our study of American political thought" (189) and interprets the tale as a fictional equivalent of Tocqueville's eulogy of American women, their freedom, social awareness, and general superiority (193). Taylor unreservedly commends James's praise of Mrs. Adams (Mrs. Bonnycastle) for her "love of American democracy" (197). By contrast, Weiner, who believes James to be "the most Tocquevillian of American authors" (Weiner 2002, 35), rightly observes that neither James nor Tocqueville deplored American women's political inferiority despite their moral and intellectual equality with men (46). Quite surprisingly, however, Weiner reads the scene in which Pandora extracts a promise from the country's president as an example of "the utter openness of the official governing structure in America" (43). In fact, the scene looks much like an example of favoritism and "distribution of spoils" (James 1963, 383), which are hardly commendable in any democracy. It seems that the Bonnycastles think of inviting the unnamed president (in this sense literally a nonentity) to their party as "fun" (383), not because he has such qualities, but possibly because his presence is likely to provoke American girls like Pandora to approach him and ask him for favors. Observing such scenes, the reverse of the "tyranny of the majority," seems to give the Bonnycastles their perverse "fun."

In contrast to "Pandora" with its international theme and the "American girl" motif, "Crapy Cornelia" seems to be all-American, private, and apolitical. White-Mason possesses, however, at least one quality of Alfred Bonnycastle (Henry Adams); he is "not in politics, though politics [are] much in him" (382). White-Mason's withdrawal from the public sphere and his resistance to modernity, symbolized so powerfully by New York and its high society, are political statements, too. His conservatism resembles that of Vogelstein, who is likewise, in a double-edged manner, an object of James's satire. Tambling points to "a repression embodied in" White-Mason's name; his "anti-modernity makes him a builder of that white America that rests upon ignorance of the other" (Tambling 2001, 134). Vogelstein (barely) notices lower-class Germans and Jews during the sea voyage, as well as (tattered) "darkies" in the American open space (James 1963, 403, 410) and a "blackamoor" as a servant indoors (399), but White-Mason who, in his attempt to save his own greatness, creates a secret society (freemasonry of sorts), does not notice them at all. The marginal reference to the self-confident little "aliens" in Central Park serves only to highlight his own uncertainty. The mention of loud Germans in a Swiss hotel who brought Mrs. Worthingham and Cornelia together, however, has its bizarre echo in Cornelia's German family name Rasch, which triggers a myriad of connotations in English.

The study of words – their meanings, polysemies, and sounds – in James's texts is never a waste of time, though rarely conclusive. Equally exciting is the history of revisions and mistakes in the different editions of James's texts. For example, following its publication in *Harper's Magazine* in 1909, "Crapy Cornelia" was included in a collection *The Finer Grain*, which was brought out in London by Methuen and in New York by Scribner's. The Methuen edition is particularly notorious for its many uncorrected proof errors, which surface again in the tale Leon Edel included in *The Complete Tales of Henry James*. One of such errors consists in using "crazy" instead of "crapy" with reference to the eponymous character (James 1964, 347; Reeve 2017b, 488 note 366). One can blame the publisher's lack of care or Henry James's haste while proofreading, but it seems that the choice of the word "crapy" was an invitation to mispronunciations and misprints.

Henry James's tales are a vast and unevenly cultivated area of study. Susan M. Griffin, former editor-in-chief of the *Henry James Review*, admits in an interview that although it is difficult "to find an area of Henry James's writing that has not been studied," some texts appear "relatively neglected and/or open to new approaches." She adds: "Surprisingly (at least to me), many of James's lesser-known short stories (and there are many of them) receive little attention" (Griffin 2017, 177). Interestingly enough, over the past years instead of explorations of the lesser-known texts, there have been attempts to expand the canon of 112 tales established by Leon Edel. These attempts have shown how vast the area really is. For example, in 2004 Floyd R. Horowitz published a selection of James's alleged juvenilia in *The Uncollected Henry James: Newly Discovered Stories*. His stylometric procedures may not be entirely convincing and the twenty-four stories he collected may not have been written by James after all, but the idea of tracing the circumstances of "becoming" the published author Henry James is worthwhile. Family correspondence gives an occasional glimpse of the future author's juvenile creativity. When he was fourteen, his father Henry James Sr. reported in a letter to his mother, Catharine Barber James, that Harry was "a devourer of libraries, and an immense writer of novels and dramas" (qtd. in Moore 1974, 20), which proves that Henry, Jr. found his profession very early and that he was ambitious. There is no mention of shorter forms in this clearly hyperbolic depiction of the young artist's pursuits, but it is by no means certain that the canon of James's tales begins where Edel saw it fit to set the beginning.

Although never a routine activity, writing stories became difficult for James in his later years, and the last tales "cost him quantities of toil and frustration" (Reeve 2017a, xxviii). What is worse, they were not only difficult for James to write but also difficult for his agent to place, so much so that in an unusually short note to his agent J.B. Pinker of 21 July 1909, referring to "The Bench of Desolation," James announced his decision to stop writing stories altogether: "[C]learly I have written the last short story of my life – which you will be glad to know" (James 1999, 483). James kept his word and did not write further stories, but his notebook of 1911 nevertheless "contains several lengthy sketches for possible tales" (Reeve 2017a, xxviii). These inspired ten contemporary authors, including Colm Tóibín and Tessa Hadley, to make stories out of James's ideas. The collection was edited by Philip Horne and came out in 2018 under the title: *Tales from a Master's Notebook: Stories Henry James Never Wrote*. His story-telling impetus did not stop with his last published tale.

4 Bibliography

4.1 Works Cited

Edel, Leon. "General Introduction." *The Complete Tales of Henry James*. Vol. 1. Ed. Leon Edel. London: Rupert Hart-Davis, 1962. 7–16.

Edel, Leon. "Introduction: 1903–1910." *The Complete Tales of Henry James*. Vol. 12. Ed. Leon Edel. London: Rupert Hart-Davis, 1964. 7–11.

Frenzel, Elisabeth. *Motive der Weltliteratur: Ein Lexikon dichtungsgeschichtlicher Längsschnitte*. 8th ed. Stuttgart: Alfred Kröner Verlag, 1992.

Gale, Robert L. *A Henry James Encyclopedia*. New York: Greenwood P, 1989.

Goble, Mark. "Media and Communication Technologies." *Henry James in Context*. Ed. David McWhirter. Cambridge: Cambridge UP, 2010. 203–213.

Griffin, Susan M. "'The Virtues of Inclusiveness': An Interview with Professor Susan M. Griffin", http://dx.doi.org/10.12775/LC.2017.016. *Litteraria Copernicana*, Henry James – muzeum pisarza / The Writer's Museum Vol. 1. (21) 2017: 175–178. (18 February 2021).

Herford, Oliver. *Henry James's Style of Retrospect: Late Personal Writings, 1890–1915*. Oxford: Oxford UP, 2016.

Hoover, David L. "Some Henry James Apocrypha?" Conference of the International Association of University Professors of English, Poznań, Poland, 25 July 2019.

James, Henry. *The Art of the Novel: Critical Prefaces*. New York: Charles Scribner's Sons, 1934.

James, Henry. *Complete Notebooks of Henry James: The Authoritative and Definitive Edition*. Ed. Leon Edel and Lyall H. Powers. New York: Oxford UP, 1987.

James, Henry. "Crapy Cornelia." *The Complete Tales of Henry James*. Vol. 12. Ed. Leon Edel. London: Rupert Hart-Davis, 1964. 335–367.

James, Henry. "Crapy Cornelia," https://ebooks.adelaide.edu.au/j/james/henry/crapy_cornelia/-contents.html. (12 December 2018).

James, Henry. "Crapy Cornelia." *The Jolly Corner and Other Tales: 1903–1910*. Ed. N.H. Reeve. Cambridge: Cambridge UP, 2017. 318–347.

James, Henry. *The Letters of Henry James*. Vol. 1. Ed. Percy Lubbock. London: Macmillan, 1920.

James, Henry. *A Life in Letters*. Ed. Philip Horne. Harmondsworth: Allen Lane, Penguin P, 1999.

James, Henry. "Pandora." *The Complete Tales of Henry James*. Vol. 5. Ed. Leon Edel. London: Rupert Hart-Davis, 1963. 357–412.

James, Henry. "Pandora," http://www.gutenberg.org/files/2299/2299-h/2299-h.htm [eBook#2299]. Transcribed from 1922 MacMillan and Co. *Daisy Miller, Pandora, The Patagonia and Other Tales*, ed. David Price. (12 December 2018).

Johnson, Kendall. "Periodicals." *Critical Companion to Henry James: A Literary Reference to His Life and Work*. Ed. Eric Haralson and Kendall Johnson. New York: Facts on File, Infobase Publishing, 2009. 426–428.

Kaplan, Fred. *Henry James: The Imagination of Genius*. New York: William Morrow and Company, 1992.

McGann, Jerome. *The Poet Edgar Allan Poe: Alien Angel*. Cambridge: Harvard UP, 2014.

Monteiro, George. *Reading Henry James: A Critical Perspective on Selected Works*. Jefferson, North Carolina: McFarland, 2016.

Moore, Harry T. *Henry James*. New York: Thames and Hudson, 1974.

Reeve, N.H. Introduction. *The Jolly Corner and Other Tales: 1903–1910*. Ed. N.H. Reeve. Cambridge: Cambridge UP, 2017a. xxviii–civ.

Reeve, N.H. Notes. *The Jolly Corner and Other Tales: 1903–1910*. Ed. N.H. Reeve. Cambridge: Cambridge UP, 2017b. 425–507.

Tambling, Jeremy. *Lost in the American City: Dickens, James, and Kafka*. New York: Palgrave, 2001.

Taylor, Natalie Fuehrer. "'The Incarnation of My Native Land': Clover Adams in Henry James's 'Pandora.'" *Short Stories and Political Philosophy: Power, Prose, and Persuasion*. Ed. Erin A. Dolgoy, Kimberly Hurd Hale, and Bruce Peabody. Lanham: Lexington Books, 2019. 187–212.

Tintner, Adeline R. *The Twentieth-Century World of Henry James: Changes in his Work after 1900*. Baton Rouge: Louisiana State UP, 2000.

Weiner, Lauren. "Tocquevillian Americans: Henry James, Daisy Miller, Pandora Day." *Seers and Judges: American Literature as Political Philosophy*. Ed. Christine Dunn Henderson. Lanham: Lexington Books, 2002. 33–48.

Wellbery, David E. "The Imagination of Freedom: Goethe and Hegel as Contemporaries." *Goethe's Ghosts: Reading and the Persistence of Literature*. Ed. Simon Richter and Richard Block. Rochester, New York: Camden House, 2013. 217–238.

4.2 Further Reading

Anesko, Michael. *"Friction with the Market": Henry James and the Profession of Authorship*. Oxford: Oxford UP, 1986.
Buchholtz, Mirosława, ed. *Henry James' Travel: Fiction and Non-Fiction*. London: Routledge, 2019.
Gilmore, Michael T. *Surface and Depth: The Quest for Legibility in American Culture*. New York: Oxford UP, 2003.
Hocks, Richard A. *Henry James: A Study of the Short Fiction*. Boston: Twayne Publishers, 1990.
Kraft, James. *The Early Tales of Henry James*. Carbondale: Southern Illinois UP, 1969.
Nyirkos, Tamás. *The Tyranny of the Majority: History, Concepts, and Challenges*. New York: Routledge, 2018.
Vaid, Krishna Baldev. *Technique in the Tales of Henry James*. Cambridge: Harvard UP, [1964] 2013.
Vidal, Gore. *Empire*. New York: Random House, 1987.

Alfred Hornung

13 Jack London (1876–1916)

Abstract: Jack London's success as an important writer of 196 short stories at the turn of the last century is based on his own adventures in exotic places on land and sea and his ambition to emerge from his working-class background in the San Francisco Bay. His experiences of the Klondike Gold Rush in the Yukon and Alaska (1897–1898) and his voyage through the Pacific (1907–1908) provide ample material for thrilling stories in which the author addresses from his socialist perspective political issues of the Anglo-Saxon supremacy and the devastating effects of the colonization of allegedly uncivilized regions with Indigenous populations. London's initial belief in the white man's superiority changes after meeting and marrying Charmian Kittridge. The two stories, "Bâtard" (1902) and "Mauki" (1908), set in the Northland and the Pacific respectively, show the victory of the black underdog over the degenerate white man. They also reflect London's ecological position for an equality of the human and animal world and his eventual vision of a pan-pacific community which prefigures transnational alliances of multiethnic societies in the twenty-first century.

Keywords: Whiteness, racism, underdogs

1 Context

In his short lifetime of forty years, Jack London wrote and published 196 short stories and about sixty books of prose narratives. All of his prolific output is inspired by the journeys of his life full of unusual adventures and by a compulsive writing program of 1.000 words a day to emerge from his working-class background in the San Francisco Bay area and to change the social and political reality of his time (1876–1916), at home and abroad. His first publication, "Story of the Typhoon off the Coast of Japan" (1893), exemplifies this correlation of personal adventures and life writing. As a seventeen-year-old with his experience as an oyster pilot in the Bay he hires on to a sealing schooner bound for the Japanese sea. One of the most dramatic moments of this eight-month trip via Hawai'i is the experience of a terrible typhoon during which he is on deck and which causes the death of a fellow sailor. Inspired by his mother, he writes down in two nights the dramatic weathering of the storm to participate in a writing contest of *The San Francisco Morning Call* and wins the first prize of twenty-five dollars. This literary success becomes the principle of his new business model, and he resolves "to sell no more muscle, and to become a vendor of brains [. . .] a brain merchant" (London 1910, 301, 303).

https://doi.org/10.1515/9783110587647-014

The gradual transition from the exploitation in sweatshops and free enterprise as an oyster pirate to the success of a literary career, which made him the first millionaire American author in 1905, include short phases of formal education at the Oakland High School and a semester at Berkeley, which he quits to begin his own reading of Karl Marx, Charles Darwin, Herbert Spencer, and Friedrich Nietzsche. The acquired knowledge about the Marxist class war, the Darwinian survival of the fittest applied to social evolution by Spencer, and Nietzsche's blond beast and superman become the guidelines for his subsequent journeys in life and in literature. His autodidactic studies are complemented by the help of an amorous attraction for middle-class women who offer a glimpse into the world of the rich without satisfying his political and literary ambitions. The discovery of gold in the Canadian Klondike seems to provide for him the occasion to fulfil all his desires. Rather than striking it rich, the stay in the Klondike Gold Rush and Alaska from 1897 to 1898 becomes a reservoir of stories, heard or experienced in the company of like-minded gold-seekers and adventurers. It repeats a pattern of storytelling, which he had first encountered as a member of the nation-wide hunger march of unemployed on Washington in 1894 and as a hobo and tramp riding illegally on trains across country when the time in traffic was passed by homeless people taking roles in telling stories, which in turn becomes the source of his life narrative *The Road* published only in 1907. But it is the series of adventures connected with his Yukon Gold rush experiences which mark the beginning of Jack London's career as a writer of exotic and thrilling short stories. "It was in the Klondike I found myself," he wrote in a promotional pamphlet. "There you get your true perspective. I got mine" ("Jack London by Himself" qtd. in Labor 1993, xxii).

Apart from a few short pieces and juvenilia published in *The Aegis* of his Oakland High School in 1895 after his road trip across country, most notable among them the "Frisco Kid's Story," serious craftsmanship of a writer showed in the stories derived from his sojourn in the Northland. "To the Man on Trail" in the *Overland Monthly* (1899) and "An Odyssey of the North" in *The Atlantic* (1900) were the breakthrough for his literary reputation. In addition to the thematic originality of the Northland stories, two further items favored London's promotion to a serious writer. On the one hand, the technological innovations of the printing press led to the rise of popular magazines at lower prices for a new readership fascinated with human interest and action-based adventure stories rather than the elitist literature of the genteel tradition. "Cheap paper, new printing technologies, the process of photoengraving, the growth of national advertising, more aggressive promotional methods, better postal rates, and improved distribution made magazines the first truly mass medium" (Labor 1993, xiv; (↗4 Textual Materiality). On the other hand, the literary scene in the first decade of the new century did not really display a range of important American authors (see Knight). Major writers such as Henry James (↗12 James), T. S. Eliot, or Ezra Pound had joined the modernist movement in Europe, leaving the American market to Frank Norris, Theodore Dreiser, and lesser-known writers. In line with his own adventure path, Jack London favored the writing of like-minded authors, such as Robert Louis Stevenson, whose grave on Samoa he visited in 1908 on his sailing trip through the Pacific, Joseph Conrad, or Rudyard Kipling, equally fascinating with their heroes' adventures in unfamiliar exotic settings (London 1986). These situations of extremity in terms of climate and human interactions with unusual forms of the environment and the biosphere also reflect back for Jack London to the social conditions of the United States which he had lived as a subject and explored as an object among the poor and unsheltered and which determined his engagement for socialism and the Muckraking movement with Upton Sinclair (Hornung 1978). Hence his adventure stories often appear to be test cases for the extent and limits of human existence in uncontrollable power relations on Jack London's adventure path. The shift from the Northland stories after his Klondike trip to the Pacific tales is linked to his love life and his move from a conventional family life to his relationship with the bohemian woman Charmian Kittridge, whom he met first in 1900, fell in love with in 1903 and married after the divorce from Beth Maddern in 1905 to live in the agrarian idyll of Glen Ellen in the Californian Sonoma valley rather than in the urban atmosphere. It is Charmian Kittridge with whom he encounters the amenity of Hawai'i and who accompanies him on the sailing trip through the Pacific. Rather than being a mere amanuensis of his manuscripts she shares and influences his political and literary ambitions with her own authorship as evident in *The Logbook of the Snark* (1915) and *Our Hawai'i* (1917; see Hornung 2020).

Jack London's first meeting with Charmian is orchestrated by her aunt Netta Eames, who together with her husband Roscoe provided a home for her and, as the editors of the *Overland Monthly*, promoted her literary interests and acquaintance with writers. In spite of being displeased with London's appearance and dress style she agrees to write a review of *The Son of the Wolf*, the first volume of nine Northland stories published by Houghton, Mifflin and Company in 1900, some of which had appeared in the *Overland Monthly*. This review, which is incorporated into her aunt's biographical essay of "Jack London," notes his "false syntax and flagrant misuse of an occasional word" and recognizes his political message: "Jack London has the avowed belief in an ultimate democracy to be achieved by all peoples whose institutions, ideals, and traditions are ethnologically Anglo-Saxon. 'Not that God has given the earth to the Anglo-Saxon, but that the Anglo-Saxon is going to take the earth for himself,' he declares convincingly" (Eames 1900, 424).

These perceptions form part of the standard evaluation of Jack London in literary histories. He is seen as a macho American writer of successful naturalist animal stories set in extreme climate conditions on land in the Yukon or on sea in the Pacific, where the survival of the fittest is exemplified by the white men of the Anglo-Saxon race. Saving grace of this clichéd perception are his working-class background and his autodidactic reading of Marx, Darwin, and Nietzsche, which, however, resulted in his ambivalent attitude, oscillating between the collective rule of the working class and the individualistic realization of a superman, Nietzsche's "blonde beast" (see Hornung 2016, 2020). It is my assumption that Jack London changed his political and literary perspective after meeting and marrying Charmian, which coincided with his public engagement for socialism and led to their sailing trip to Hawai'i and the Pacific Islands recollected and commemorated in their travel accounts (London 1911; Kittridge London 1915, 1917). The two stories selected for this chapter reflect the gradual transition from the focus on white Anglo-Saxon supremacy to the acceptance of a minoritarian position which Jack London envisioned as a new transnational model of collaboration at the end of his life.

2 Close Readings

The story "Bâtard" is based on Jack London's personal answer to the call of the wild in form of the gold rush in the Klondike. Ending his brief academic education at the University of California at Berkeley after a first disappointing semester, he hoped to strike it rich in the newly discovered gold region in the High North. The new American state of Alaska, bought from Russia in 1867, and the Klondike gold fields in the Canadian Yukon represented a rich field for the author's adventurous mind and literary imagination outside of the familiar civilization. The sparsely populated region spanned from the South of Alaska to the polar region in the North and from the Russian peninsula of Kamtschatka in the West to the Labrador in the East. This vast space represented an unusual colonial contact zone between Indigenous tribes and citizens of the former and new colonial powers of Russia, Great Britain, France, and the United States. Its parameters of existence were male power and animalistic forces in a natural environment of extreme climate conditions. Jack London's excursion into the wild lasted from July 1897 to July 1898. After passing the arduous Chilkoot Trail with the help of Indigenous people he stayed in a hut South of Dawson City, the main area for prospecting. The claim he acquired did not yield any gold. Instead, he derived a great stock of stories from the encounter with fellow adventurers in the saloons during the long polar winters. When he returned to San Francisco he noted bitterly that he had brought back five dollars of gold dust and the scurvy, the arctic lepra disease, but he was also convinced that this stay completed his identity formation (London 1913, 1054). The death of his beloved stepfather makes him the only breadwinner in the family shortly after his return in the fall of 1898 and he is determined to make money by becoming a writer.

The personal hardships and experiences of this year in the Klondike are the material for his first stories and gradual success in having them accepted by the periodicals of the *Overland Monthly* and *Black Cat*. These Northland Stories, which include such classics as "An Odyssey of the North," "The White Silence," "The Law of Life," "To the Man on the Trail," and the 1901 version of "To Build a Fire," depict the physical and psychic stress of isolation in the endless landscape of snow and permafrost, the camaraderie and animosity of fortune hunters and laborers in crowded camps, and the encounter with the Indigenous populations and wild life. The extreme situations of life determined the behavior of human beings, whose survival depended on the adaptation to the laws of life in the unknown environment as practiced by its human and animal inhabitants. These kinds of adventurous stories of exotic places and unusual events were in demand and met with the literary taste of editors and readers. They also raised questions about the progress of humankind, miscegenation, and animalization with respect to the impact of racial ideologies at the turn of the century.

The publication of Jack London's story "Bâtard" in 1902 precedes the author's break-through success of *The Call of the Wild* by only one year. The close interrelation of the abducted Californian family dog Buck with his master John Thornton in the Northland, who liberates him from demeaning mushers, is a portrait of the mutual assimilation and cohabitation of man and animal. Buck, who according to his master lacks only human speech, acts instinctively like a human being who defends his master, avenges Thornton's slaughter by Indigenous people and survives as the leader of a wolf pack, successfully reintegrated into wilderness. The relationship in the previous dog story between Bâtard and his master Black Leclère is equally intimate and close, but totally the opposite in terms of mutual affection. The negativity includes the death of both in the end, which seems to be prefigured from the very first sentences and applies to both animal and man: "Bâtard was a devil. This was recognized throughout the Northland. 'Hell's Spawn' he was called by many men, but his master, Black Leclère was also a devil, and the twain were well matched. There is a saying that when two devils come together, hell is to pay" (729). This devilish pact is made when dog and master first meet, and Black Leclère selects Bâtard as a puppy from a litter and is severely bitten in his hand. He takes this bloody election as a sign of special relation which lasts for five years throughout all of Northland. Covering the whole region outside of the alleged civilization and its biosphere allows Jack London to incorporate in the narrative action familiar patterns commonly associated with the naturalist perspective and writing style, but with decisive differences. First of all, the story deconstructs the common assumption of London's embracement of the Nietzschean superman as represented in the superiority of the Anglo-Saxon race. Instead, he privileges diversity and the equality of animal and man, albeit on the negative side. The very names of the two protagonists appear to be programmatic for the course of the action and prefigure the outcome as the victory of the one over the other. The vicious pursuit of both animal and man to kill the opponent is consciously prolonged in order to relish the desire of mastery by way of mechanical human weapons or animal forces and instinct. It is the initially apparent deficiency of the mongrel nature of the dog expressed in "the shameful name" of Bâtard, which Leclère gives him, but which eventually enables the dog to win.

Thanks to the storekeeper of the Sixty Mile Post, John Hamlin, who sells the puppy to Leclère, Bâtard's progenitors are remembered. The father, unknown to

Bâtard, "was a great gray timber wolf" and the mother "was a snarling, bickering, obscene, husky, full-fronted and heavy chested, with a malign eye, a cat-like grip on life, and a genius for trickery and evil" (730). The difference in Halim's description is significant. Thus, the brief and positive reference to the father's pure wildlife jars with the long and devastatingly evil nature of the mother, a domesticated husky. Hamlin's biased and uneven qualification receives a different evaluation in the narrator's perspective which suggests the puppy's probably successful career of a sled-dog if trained by a benevolent owner, had the history of the dog and his master not been chartered as "a history of war" (730). The mixture of the wild and the tamed sled-dog in Bâtard actually represents for the narrator the particular endowment for the fierce fight: "His were the stealth and cunning of the husky, his mother, and the fierceness and valor of the wolf, his father" (730). While the narrator provides ample information about the dog, the information about the background and heritage of the man is contained in his nationality as a Frenchman. Mono-manically his hatred of the dog is based on "understanding and intelligence," but he ignores that the puppy's initially blind and instinctive hatred develops into a "preternatural intelligence" (730).

The stages of battles and injuries received in beatings and their increasing vehemence permeate the story. They range from a simple ear injury to fights for food and the Darwinian survival of the fittest in the relentless competition with rivals. The predetermined negative filiation does not leave any room for kindness and affection, which when given is turned into its opposite. Aggression is met with more viciousness. Leclère also does not tolerate any kind of intervention, positive or negative by fellow hunters, for it would destroy the script of the devilish warfare between him and Bâtard which he is sure to win. "Their hate bound them together as love could never bind" (731). The battle seems to unroll according to an organic conception of the law of life shared by all parts of the biosphere which Leclère as an outsider understands:

> But Leclère understood. He was a man who lived much in the open, beyond the sound of human tongue, and he had learned the voices of wind and storm, the sigh of night, the whisper of dawn, the clash of day. In a dim way he could hear the green things growing, the running of the sap, the bursting of the bud. And he knew the subtle speech of the things that moved, of the rabbit in the snare, the moody raven beating the air with hollow wing, the baldface shuffling under the moon, the wolf like a gray shadow gliding betwixt the twilight and the dark. And to him Bâtard spoke clear and direct. (732)

When the dog attacks his master in his sleep the fight about life and death is for Leclère the one between Bâtard, whom he recognizes as a "full-bloodied wolf" and himself as "perfect animal" (732), which the narrator connects with "a primordial setting and a primordial scene [. . .] in the savage youth of the world" (733). The devastating effects of this fierce fight leaves both combatants as invalids for quite some time until they recover and allow both to reassert "the supremacy among [their] kind" (734). The hateful relationship includes moments of nurture and mutually interlocked dependence which amounts to forms of human understanding. The new missionary at Sunrise, who at the first meeting is severely bitten by Bâtard in exchange for friendly patting, observes the careful observation between both convalescents and Bâtard's awareness of Leclère's pistol to utter: "'I really believe the brute comprehends'" (735). But the time for the final destruction had not yet come.

To create suspense the narrator delays the final showdown by taking the reader to the actual scenes of action in the vast circumference of the Northland, man and animal out on the trail. The main part of this year-long engagement consists of a series of tortures extended to Bâtard, the worst of which is "the torture of music" (736). Bâtard's impossibility to stand and react to Leclère's playing on an old harmonica, which he consciously uses to direct his hateful fight to the dog's only weakness and to enjoy the dog's suffering expressed in "the long wolf howl" (736). Again, this battle is seen against the background of "the very essence of life":

> At such times he felt impelled to express his own unconquerable essence; and with strong drink, wild music, and Bâtard, he indulged in vast orgies, wherein he pitted his puny strength in the face of things, and challenged all that was, and has been, and was yet to be. (737)

At the highpoint of his utmost pleasure and irrespective of any form of human interdiction, such as Father Gautier's reproof and warning of "instances of concrete perdition," the narrator seems to take sides with forebodings of the end: "But all bad things come to an end as well as good, and so with Black Leclère" (737).

The final scene of action involves a fatal encounter of Leclère and his companion Timothy Brown with two Indigenous men. The laconic and sparse description of the incident leaves much room for interpretation: "It was a simple affair. He had quarrelled with Timothy Brown at McDougall. With Timothy Brown he had left McDougall. Without Timothy Brown he had arrived at Sunrise" (738). In spite of his plausible explanations, which are later confirmed, that the deed was done by two Indians, Leclère is charged with having killed his companion, simply based on his evil record. His dying wish before being hanged is that he sees Bâtard be hanged before him. While preparing Leclère for the hanging with the rope around his neck and his arms manacled on his back standing on a box, the white men neglect to also rope Bâtard. The news of the captured Indians, responsible for the killing, deflects the attention from Leclère's situation, because they want to "teach these damned Siwashes a little manners [. . .] and string up the Beaver for an object lesson" (739). Rather than releasing the doomed Leclère from his awkward position they leave him to his fate "to meditate on [his] sins and the ways of providence" (740). When they return fifteen minutes later, they see that Bâtard has carried out the execution by hurling his body against the box. Even after being shot his teeth hold on to the body of his evil master completing the devilish pact.

From the very beginning of the story, Bâtard and Leclère are outsiders in the Northland. The title of the story introduces the foreignness with the French names of both animal and human protagonist. The topic of a mixed breed in the dog's nature of wolf and husky and the frequent relation of the French to the Catholic religion jar with the characters of the Anglo-Saxon world and their uncontested positions. The first title of the story "Diable – A Dog" in the June 1902 issue of *The Cosmopolitan* uses the French word for "devil" and reinforces the religious connotation along the binary opposition of good and bad without any mitigating mediation. Renaming the story "Bâtard" for the publication in 1904 reveals an interesting change in London's thinking. The publication with the new title comes after his success of *The Call of the Wild* in 1903, in which the benevolence of master Thornton matches the good nature of the dog Buck who survives in the end. The translation of the French word of devil into the French word for "bastard" turns the religious relation into a biological one with a clear discrimination for the diversity of a mixed race. This racial turn receives further prominence by Leclère's name who does not have a first name but goes by the epithet of "Black." The phonetic reproduction of his speech marks him as a Frenchman who deviates from the proper pronunciation of English.

A similar kind of discrimination colors the brief appearance of the Indigenous Siwashes. Obviously, they are the ones who attack the white men on the boat and shoot the Anglo-Saxon Timothy Brown. For the Anglo-Saxon men, Slackwater Charley and Webster Shaw, who represent law and order as storekeeper and musher, the Siwashes are on a lower social rank than the evil Frenchman. Rather than releasing him from his plight and locked-up situation when his innocence in the shooting is affirmed, they feel that the punishment of the "Indians" takes priority and needs immediate attention. Webster Shaw clearly states the unacceptable behavior of the Siwashes as a minor race: "'They're getting fat and sassy, and we'll have to bring them down a peg'" (739). Leclère in view of his own execution regrets that he cannot see the Indian Beaver hanged. What he does not realize yet is that the Indian intermission concludes his own death sentence and allows Bâtard to have his revenge.

Bâtard's final victory is the victory of the discriminated underdog. Rather than attributing the animal's evolution from a mixed puppy of wolf and husky to a fixed biological determinism, I would like to disagree with Earle Labor who sees in "Bâtard" "the naturalistic theme of hereditary and environmental determinism" also used in the other dog stories of *The Call of the Wild* and *White Fang* (Labor 1983, 116). Like Bâtard, White Fang in the 1906 story is a mixture of wolf and husky who reverses Buck's evolution from a domesticated animal to the wild and adapts to the Californian domestic scene after his original wildlife in the Northland. In Bâtard's case, it is precisely the mixture of the wild and the domestic which qualifies him to outdo the evil character of "Black" Leclère. The narrator suggests this positive interrelation for the dog's special capacity which could have been turned into positive ways by a benevolent master. To what extent Jack London's own heritage, his long ignorance of his biological father's identity, who failed to acknowledge his son, as well as his growing up in the household of a mixed African American family as a child when his mother could not take care of him, influenced the conception of

Bâtard and making his discriminated nature a feature of superiority is not proven. But it could be a plausible interpretation.

The equality between animal and man has been a feature of London's stories from early on. The three dog stories published between 1902 and 1906 demonstrate the close relationship and interaction including non-verbal and verbal communication. In this respect, London anticipates aspects of recent research in Animal Studies directed to the languages of animals (Ortiz-Robles 2016). For him, dogs are more than instinct driven. His rebuttal to President Theodore Roosevelt's attack on the fictional treatment of animals in the September 1907 issue of *Everybody's Magazine* caused an animal fakery controversy between nature writers like John Burroughs and popular fiction writers like Jack London (see Fulton 260). In his rebuttal "The Other Animals," London rejected the instinct-driven theory and argued "that certain mental processes in that dog's brain are precisely duplicated in the brain of man" (London 1910, 265). Early on in "Bâtard" the narrator informs the reader that the dog "out of his terrible struggle for life developed a preternatural intelligence" (730). This mental capacity guarantees his survival and his final victory over his partner in crime Leclère to whose hanged body he clings on in conclusion of their devilish pact. Hanging on to the hanged man after being shot documents in the final picture their intricate interrelation and the equality of man and animal. This became Jack London's dominant topic toward the end of his life in Hawai'i when he remembers a childhood conversation with his African American nanny Jennie Prentiss in which she misunderstood his question about gorillas in Africa as guerillas which caused a confusion and moment of distrust. He relates this memory of Jennie's political experience on a plantation in the South in a speech, which he gave to the Hands-Around-the-Pacific Organization in Honolulu in 1915, and in which he envisioned a pan-pacific union of all pacific islanders with a universal language (London 1915, 119). In this speech, London even suggests that animal languages share such a universal language, as becomes insistently evident in his preoccupation with animals in the final years of his life in Hawai'i and on his Beauty Ranch in the Sonoma Valley of California. Two dog stories appeared posthumously in 1917: *Jerry of the Islands* and *Michael, Brother of Jerry*. When his favored horse Neuadd died in 1914, he immediately wanted to compensate the loss by writing a novel about the horse. Restating his conviction that "horses were like men" also justified the high price of ten thousand dollars he had paid for Neuadd: "I wasn't buying a horse. I was buying a man, and I think I've bought a whole lot of man" (qtd. in Labor 2013, 374).

Most Jack London critics agree that his project of sailing the world changed his outlook on life and on writing. The origin of this project is closely connected to London's relationship to Charmian and their life on his second wife's summer place in Glen Ellen from 1905 on where the plans for the trip were conceived. The San Francisco earthquake in 1906 delayed the construction of the sailing ship and the start of the voyage until April 1907. After the short trip to Hawai'i repairs of the ship were needed, forcing the party to stay on the island for five months. This delay turned out to be basically very happy for the Londons' marital life, and the embracement of the pleasant Hawaiian lifestyle prepared them for the eventual trip through the South Seas, recorded in London's *The Cruise of the Snark* (1911; 1986) and Charmian's *The Log of the Snark* (1915). While they appreciated the amenities of the Polynesian landscape and the exotic life of Indigenous populations, they soon found their expectations destroyed when they wanted to recuperate Herman Melville's exotic experiences on the Marquesas islands, described in his autobiographical novel *Typee* (1846; ↗8 Melville). Actually, London registers here the same kind of intrusion of Western civilization into the pacific paradise which he had encountered during his stay in Hawai'i and had made the topic of a new type of short story critical of the European and American colonizers. In "The House of Pride" (1907) he defends the Indigenous life style against the corrupt influences of American missionaries and sugar cane colonizers who exploit the islands; in "Koolau the Leper" (1908) he incorporates his very positive perceptions of leper patients during his visit to the quarantined patients on the Island of Molokai and represents the historic fight of the infected protagonist and his fellow patients against the intrusions of the American army as part of the dethronement of the last Hawaiian Queen Lili'uokalani and the U.S. annexation of Hawai'i in 1893. This new anti-colonial attitude informs his writings on the voyage and shatters his erstwhile beliefs in the prerogative of the Anglo-Saxon race and the dominance of the white man, once his ship arrives in the Melanesian islands. Short stories like "The Terrible Solomons" (1908) or the ironic "The Inevitable White Man" (1908) give a rather bleak account of the mostly black population on these islands and the devastating effects of the blackbirding system, a form of slavery practiced for recruiting workers on the plantations operated by white masters. Part of the bleak picture conveyed in these stories as well as in his essays and his autobiographical travel account matches the miserable situation on board the Snark, which had turned into a hospital ship full of patients ridden with tropical diseases. Jack London seems to have suffered most from a number of persistent illnesses which seriously impeded his energy and creativity. The assumption that his diminished physical strength may have contributed to his move away from the belief in the superiority of the white man does not really conform to the new attitude visible in his Hawaiian stories and their obvious criticism of the ravages of colonialism. Based on my interpretation of "Bâtard," I would actually like to argue that London's alliance with the (under)dog and its innate superiority underlies his whole life and becomes dominant after his marriage to Charmian and his discovery of the native environment in the Pacific. It is the topic of "Mauki" (1908), sharing the defining criteria of blackness and the devil.

The short story begins with the dominant role of the title hero and his most significant features: "He weighed one hundred and ten pounds. His hair was kinky and negroid, and he was black. He was peculiarly black. He was neither blue-black nor purple-black, but plum-black. His name was Mauki, and he was the son of a chief" (London 1908, 1532). His superior status in his Indigenous community is defined by his blackness, intensified by his mother's coloring his teeth black with a powdered mineral which turned them "deep black, or, perhaps better, *lamp*-black" (1532). The landscape of his place of origin, Port Adams, provides the material for this color and matches the bleak picture: "Port Adams is a salt-water village on Malaita, and Malaita is the most savage island in the Solomons – so savage that no traders nor planters have yet gained a foothold on it" (1532). The problematic interaction between the white men and the frightening aspect and ruthless behavior of the Indigenous population underlies the course of the action which follows the story of Mauki's evolution from a happy childhood in his family, his abduction and enslavement by various masters, and his eventual victory over his white molesters and the courageous return to his native community. The narrator's long depiction of the hero's physical features and unusual peculiarities focus on the special items needed for the story of his liberation from slavery to unroll. These include three taboos, which forbid him to "shake hands with a woman," to eat clams and touch any part of a crocodile (1532), the piercing of his ears "in a couple of dozen places" (1532–33), which serve as pockets for personal items, as do his hair for a knife and his nose for a ring of a turtle-shell with the handle of a china cup. In an almost racist statement, the narrator gives Mauki "a nice face [. . .] and for a Melanesian it was a remarkably good-looking face" (1533). Last but not least, he is an excellent swimmer and in the line of his father's birth as a salt-water man, half amphibian. This constitutes all the qualities and competences needed to endure and survive the series of mistreatment and torture in the forced service of cruel masters.

Stolen by bushmen at the age of seven he becomes "the slave of old Fanfoa, head chief over a score of scattered bush-villages on the range-lips of Malaita" (1533). For ten years Mauki is in the service of Fanfoa and experiences several militant interventions of white men, including the system of blackbirding as well as the recruited men's revenge who "chopped off the two white men's heads, killed the boat's crew, and burned the ketch" (1534). To satisfy Fanfoa's desire for tobacco Mauki is traded as a slave and has to sign on "to toil for three years on the plantations of the Moongleam Soap Company" (1535). Commenting on the devastations wrought by the military equipment, the narrator states that "white men must be terrible [. . .] and were possessed of such devil-devils – rifles that shot very rapidly many times" (1534). Mauki's fate he compares in a Biblical image to "a lamb led to the slaughter" (1534). Backed by "the power and all the warships of Great Britain" (1535) the employees of the Company take him away from his familiar environment and displace him to New Georgia where he lives the miserable life of a slave in all kinds of laborious jobs. In his alienated existence he has to learn the makeshift language of the bêche-de-mer English to communicate with the white men and all the recruits who spoke different kinds of dialects (1535). To escape from this hard labor, Mauki makes several attempts, alone and in company, to return to his home in Malaita, but is caught, severely beaten and returned to the plantation increasing the tenure of his recruitment. It is only when he is made a house-boy that his situation changes considerably because he gets access to weapons and ammunition for yet another attempt to escape together with ten Malaita boys and one boy from San Cristoval. The latter serves to introduce the topic of cannibalism, for the Malaita boys kill him, save his head, cook and eat the rest of him. Further escapes and recaptures take the reader through the vast domain of the Melanesian Islands showing the extent of the area and the reach of white colonization. In this process, the time of Mauki's slavery adds up to eight and a half years, when the company decides to take him to Lord Howe Atoll and deliver him into the mercy of Max Bunster for the final showdown, after all other measures of punishment for "Solomon Islands' incorrigibles" had failed (1538).

Before the narrator can introduce the character of Max Bunster, a German trader in the pay of the Moongleam Soap Company, Jack London provides the geographical and cultural background for the setting of Lord Howe, which intensifies the dramatic and violent encounter between the vicious white man and the freedom-loving enslaved man from Malaita. In contrast to the elevated islands of the Melanesian Solomons, Lord Howe is an atoll, a ring of sand with a mighty lagoon whose people and language are Polynesian. The large extent and the remoteness of this location, untouched even by white missionaries, allows the five thousand natives to live a peaceable and primitive life until one fatal incident in which they attack an intruding bark and kill all on board except for the second mate. The revenge instigated by the surviving mate with three trading schooners leads to havoc and destructive violence "to preach the white man's gospel that only white men shall kill white men and the lesser breeds must keep hands off" (1539). Max Bunster had also been displaced to this "most out-of-the-way place to be found" in an effort of the Moongleam Soap Company to get rid of him (1539). It seems to fit his character: "He was a strapping big German, with something wrong in his brain. Semi-madness would be a charitable statement of his condition. He was a bully and a coward, and a thrice-bigger savage than any savage on the island" (1539). Depictions of Bunster's cruel mistreatment of the natives, who flee before him and hide like the king and his prime ministers, set the stage for the life-and-death battle between him and Mauki, sentenced to toil for eight and half years under him. The physical weight of the white master's two hundred pounds meets with Mauki's one hundred and ten, a dominance which correlates with the narrator's moral assessment: "Bunster was a degenerate brute. But Mauki was a primitive savage" (1540). Jack London's life-long interest in boxing, his own practice with sparring partners including Charmian as well as his reports of professional matches for American newspapers, is relevant for the analysis of the confrontation of a heavy-weight white man with the seemingly inferior light-weight black opponent. As such, this constellation prefigures the author's acceptance and appreciation of the superior qualities of a black boxer over a white one, which he actually experienced and reported on for the *New York Herald* at the end of his aborted voyage on boxing day in Sydney, where he recuperated from treatment in the hospital, and after his return to the United States in Las Vegas in 1910 (see Hendricks/Shepard 1970, 258–301; Wöll 2018). The reported acknowledgement of the victory of a black man over the white man also applies to the fight between Mauki and Bunster. It reflects London's reassessment of the Darwinian belief in the survival of the fittest according to the natural selection of the favored races by opting for Thomas Huxley's alternate biological assessment which favors the life of Indigenous people over that of the so-called civilized ones (London 1982, 180). This attitude also justifies the means needed to overcome the physical imbalance and to defeat the superior white man.

The first rounds of the battle go to Bunster, who exercises his role as master of the house over his cook and general house-boy and squashes all attempts of opposition by using the knowledge about Mauki's taboos and inferior status as an enslaved native. Mauki, in turn, is aware of the uneven fight and suffers all forms of punishment for some time.

> One of the trader's favorite tricks was to catch Mauki's kinky locks and bat his head against the wall. Another trick was to catch Mauki unawares and thrust the live end of a cigar against his flesh. This Bunster called vaccination, and Mauki was vaccinated a number of times a week. Once, in a rage, Bunster ripped the cup handle from Mauki's nose, tearing the hole clear out of the cartilage. (1542)

Enumerating all forms of facetious mistreatments and torture, the narrator passes the judgment: "Bunster was a devil" (1541). The worst and most painful treatment is the use of the skin of a ray fish which functions like a rasp on Mauki's body removing the skin from its sensitive parts. Mauki endures this repetitive treatment, which for Bunster is a form of amusement, knowing that his time of fighting back would come. When the master falls ill one day and develops blackwater fever, Mauki acts out his revenge ending in a deadly knock-out. After having prepared for his escape and return home, he reassures his final verdict with the king and the authorities of the atoll as well as with the boat-boys and then takes care of Bunster. Roles reversed, he applies the ray fish mitten on Bunster, eventually removing all of the skin of the white man's body. As Mauki is about to leave well equipped with weapons and supplies, "a hideous, skinless thing came out of the house and ran screaming down the beach till it fell in the sand and mowed and gibbered under the scorching sun" (1544). The final sign of defeat and victory is Mauki's removing Bunster's head and taking it as a trophy. It empowers him to take revenge on his former masters like Fanfoa and to make his way to the security of the bush-villages making himself the chief over them. More than the fear of the British government for having taken a white man's head he fears the all-powerful Moongleam Soap Company and therefore reimburses it for the eight and one-half years of labor on demand. His protection in the bush against all kinds of dangers and enemies is Bunster's head, "the most powerful devil-devil on Malaita, and to the possession of it is ascribed all of Mauki's greatness" (1545). As such, it matches the weapons and ammunition of the colonial powers.

The critical reactions to Jack London's "Mauki" as well as to his other stories written during his pacific voyage address personal issues, literary demands of the periodical market in the United States and the author's attitude toward the exploitation of the Indigenous population and their resources by the white man and Western civilization. As in many other stories, London's personal experiences figure importantly in this short story. Thus, he devotes the last three chapters of *The Cruise of the Snark* to their stay in the Solomons which lasts a total of five months. Here he records their travels to various islands including Malaita on board of ships in the service of blackbirding, recruiting natives as plantation laborers. Rather than using his own sailboat, they rely on the navigation skills of experienced captains like Captain Jansen and the German Captain Keller, who support each other when in peril: "[. . .] the white man, the inevitable white man, coming to a white man's rescue" (London 1986, 267). Compounded by his medical problems, the difficult maneuvers in unknown seas and the frightening interactions with the natives and wayward recruits on the ships determine his perception of "The Terrible Solomons" in the eponymous short story, summed up in *The Cruise of the Snark*: "If I were a king, the worst punishment I could inflict on my enemies would be to banish them to the Solomons." But the following sentence, "on second thought, king or no king, I don't think I'd have the heart to do it," reverses the verdict and actually reveals the author's ambivalent attitude which eventually takes the side of the natives in the islands (261). It actually allows him to move from the depiction of the strangely exotic features of "true head-hunting cannibals" on board Captain Jansen's *Minota* (246) to the representation of Mauki as the hero of his story whose features are couched positively. Hence, Jeanne Campbell Reesman calls it "a story of true heroism and the achievement of freedom" (Reesman 2006, 45). According to Russ Kingman's research, Jack and Charmian actually met with the young cook Mauki on September 28, 1908, whom they had heard about at the home of the Australian trader Harold Markham in Lord Howe Atoll (Kingman 1992, 93–94). In a comparison of Mauki's crimes and those of the plantation manager, the Londons side with the native cook as described in Charmian's *The Log of the Snark*:

> The third and last member of Markham's household is a mild-faced Solomon Island cook, who, despite his deceptive weak prettiness, is deservedly serving an aggregation of sentences that cover eight years, for murders, escapes in handcuffs, thefts of whaleboats – a history of blood-curdling crimes and reprisals too long to go in here, but which so tickles Jack's fancy that he intends making a short story of it, to be called "Mauki," and including it in his collection *South Sea Tales.* (Kittridge London 1915, 445)

Charmian contrasts this predilection for Mauki with Markham's behavior whose mistreatment of his native wives she also records. This position is shared by most postcolonial critics who weigh London's depiction of the violence of the islanders against "the crimes of the white invaders" (Riedl and Titze 2006, 24) and "the weird brutality that has everywhere accompanied European efforts to 'civilize' the remote corners of the world" (61). London's biographer Earle Labor maintains that the author "was scarcely blind to the immoral stupidities of the white race in its treatment of darker-skinned people" (Labor 1983, 214).

For the readers of short stories in Jack London's time at the beginning of the twentieth century and the editors of literary periodicals different standards applied. On the one hand, the editors of critical Muckraking journals favored the author's acclaimed socialist agenda and used London's stories to display the imperialist attitudes of the United States and of European powers leading to the destruction of Indigenous cultures in the South Seas. At the same time, they also catered to the exotic gaze of their readers with the sensational topics of cannibalism and headhunting. While Jack London and Charmian relate their accounts of cannibalism among the Melanesian population, which they heard and were afraid of, cultural historians deny the existence of cannibalism and qualify it as a myth prevalent at the turn of the century. Thus, Keith Newlin questions why Jack London would expose himself and his wife to the alleged dangers of "the savage Other," if it were not to fulfill his desire for adventures raised by his readings of travel reports. Hence, he argues that the Londons' "failure to find what they had traveled specifically to see – evidence of cannibalism, eye-witness observations of headhunting in action – is that neither was present in Melanesia at the time of their voyage" (Newlin 2018, 7). In support he cites William Arens' *The Man-Eating Myth* (1980) who sees "little empirical evidence of actual cannibalism; rather, cannibalism is largely a projection of Europeans who sought to justify their exploitation of Indigenous people by depicting them as the savage Other, as animals beyond the pale of civilization" (in Newlin 2018, 7), and Gananath Obeyesekere's *Cannibal Talk* (2005) who argues "that cannibalism is mostly 'talk' by both colonizers and the colonized, intended both to intimidate and to defend against exploitation" (in Newlin 2018, 7). Finally, Newlin juxtaposes the Indigenous practice of using an enemy's head as a trophy in internecine battles, such as Mauki's taking Bunster's head, to Jack London's search for literary topics and touristic treasures. "In his travels through Melanesia," he writes, "Jack London operated as a metaphorical headhunter, collecting artworks and taking photographs, trophies to signify his presence in an exotic land and his skill at negotiation, just as Melanesian headhunters collected heads. [. . .] London's attitude toward the black islanders did not undergo a shift to perceiving their shared humanity; rather, it remained little different from what he had been reading" (21). The focus of Newlin's reading on the topics of cannibalism and headhunting during Jack London's five-month sojourn in Melanesia recognizes London's eyewitness experience of blackbirding on Captain Jenkin's *Minota* and the recruited "islanders' violent response to colonial labor practices" and he quotes Judith A. Bennett's figure of one hundred thousand Melanesians who were recruited as indentured laborers between 1863 and 1914 (12–13; Bennett 1987, 169–170). That London conveyed this piece of historical knowledge to his readers when American and European imperialist ventures were in vogue, is no small achievement. What Newlin entirely ignores is London's rewriting of the colonial story making the historical Mauki into a literary hero who wins the fight over the cruel white man. Taking into account the context of Jack London's own biography and the final years of his life, the change in his recognition of native difference is tantamount.

3 Theoretical Perspectives

Jack London was a very prolific writer at the turn of 1900 who became an important voice for the representation of precarious and endangered lives in urban slums, in remote places outside of Western civilization at home and abroad. His own life, in which he successfully overcame his working-class background through unusual adventures and exotic experiences on land in North America, Europe, and Asia and on sea mostly in the Pacific, often underlies schematically his novels, essays, lectures, and short stories. Sharing the literary ambitions of international writers like Joseph Conrad, Rudyard Kipling, and Robert Louis Stevenson, he shaped his career as a truly transnational writer, critic, and socialist activist addressing a great variety of topics which found an avid readership not only in the United States, but also in Europe, the Soviet Union, and Asia. Earle Labor sums up alphabetically the impressive range of subjects in his introduction to *The Complete Short Stories of Jack London*:

> aging, alcoholism, boxing, bullfighting, child labor, ecology, extraterrestrial fantasy, gambling, gold mining, hoboing, love (primitive and atavistic as well as romantic and ideal), mental retardation, myth, naturalism, Platonic idealism, political corruption, psychology (human and animal), racial and sexual exploitation, revolution, scientific experimentation, seafaring, slum dwelling, socialism, suicide, war, wildlife, and the writing game [. . .].
>
> (Labor 1993, vol. 1, xiii)

In merely twenty-five years of literary production and authorial presence he not only turned out to be a commercial success, but also influenced and shaped the course of American literature in the first decade of the twentieth century until America's entry into the First World War. For Jeanne Campbell Reesman, he is "one of the most radically experimental writers of his day" (Reesman 2010, 1). It is especially the late phase of his career connected with his discovery of the peoples and cultures in the pacific islands which many critics credit with his innovative insights and techniques. Thus, Lawrence Berkove believes that "the short fiction of London's later years is intellectually sophisticated and surprisingly subtle, and his work soared to new artistic heights" (Berkove 2010, 1). While critics agree upon the wide range of subjects and praise "his ability to enter into the minds of very different people" by presenting the perspectives of "the 'other' in gender, race, class, mental ability, morality, and philosophy" (Reesman 2010, 10), they attribute his writing to different genres and styles. The most common classifications are realism and naturalism.

Jack London's love of life and his encounter with a great variety of different people in his pursuit of all kinds of adventures on land and sea make for an autobiographical background of his writing. Starting with his first short story "Story of the Typhoon off the Coast of Japan," continued with his experiences on the road, in the Northland and the Pacific, his stories often oscillate between realistic descriptions of the landscape and people and fictional elements. This realistic vein also is apparent in his journalistic pieces as war correspondent in the Russo-Japanese war or his reportage of the misery of precarious lives in East London of *The People of the Abyss*. Depending on the brutality of the topic or the emotional commitment he calls this form of writing "disgusting Realism" or "impassioned realism" (1). The classification of his stories as naturalistic fiction is connected both to the French influence via Emil Zola, whose novels he read and quoted, and the philosophical sources of naturalism in general and London's own in specific. Jack London's autodidactic reading of evolutionary biology in Charles Darwin, Herbert Spencer, Thomas Huxley, the message of Karl Marx, and Friedrich Nietzsche's idea of the blond beast give his writing at times a fatalistic design and pessimistic spirit or liken it to racist theories of the Anglo-Saxon superiority and white supremacy. Next to fatalistic pessimism Jack London also presents a form of "naturalistic romance" which Jacqueline Tavernier-Courbin attributes to the canine hero Buck in *The Call of the Wild* (1994). The setting of stories and life in a harmonious natural environment or in a savage wilderness of hopeless situation will make the difference.

The two short stories under discussion here certainly display a fated setting from the very beginning with an unexpected ending. Both "Bâtard" and "Mauki" are set in a place which are peopled with devilish figures situated in a form of hell. The intense blackness of the protagonists intensifies this feeling of what Labor with reference to the Melanesian Islands calls an "inferno" as opposed to the Polynesian paradise (Labor 2013, chs. 21–22). Although the actions in both stories follow a deterministic script, the outcome reverses the readers' expectation by presenting the victory of the black underdog over their white masters. To what extent this constitutes in "Mauki" a form of "racial satire" – as Reesman has it – or a form of "a grim poetic justice" in Labor's reading depends on the readers' judgment (Reesman 2006, 45; Labor 1983, 214). These evaluations transcend the superficial classification of Jack London's stories as examples of racist ideologies rampant at the turn of the century. Instead, this chapter emphasizes the autobiographical subtext of London's own life, his engagement for the status of animals as part of all organic life, and his alliance with the concerns of Indigenous people as part of his pacific experiences and the embracement of a Hawaiian Kanaka identity, accorded him by the Hawaiians during his final stays on the Islands when he proposed a pan-pacific community of all Pacific Islanders. It also corresponds to London's overall ecological concerns for organic life, promoting alternative agrarian principles on his Beauty Farm in California.

In terms of style and technique, critics classify Jack London with the rising California school of writers prefiguring the works of John Steinbeck. The short laconic sentences as well as ironic elements precede Ernest Hemingway's short story technique (↗16 Hemingway), as does the thematic arrangement in Northland stories, pacific stories, or South Seas tales. Giving the underdog and underprivileged characters a heroic position in his stories is reminiscent of his own background and life-long relationship with his African American nurse, and it prefigures William Faulkner's similar preference for black characters as the voices for a future supplanting the degenerate white race (↗18 Faulkner). In line with recent criticism these stories show that a revision of the simplistic perception of Jack London as a writer of thrilling adventures stories is urgently needed. London's emergence from the hard labor of his working-class milieu and his resolve "to sell no more muscle, and to become a vendor of brains" (London 1905, 301) has been likened to the author's philosophy of life "as a counterindustrial force in his work that transforms a mechanistic discourse of efficiency into one of self-actualization" (Dorson 2019, 166). This counternarrative to the discourse of his time presenting the self-actualization of his underprivileged characters also includes an ethical dimension which Reesman quotes from Donald Pizer's definition of naturalism as "reflect[ing] an affirmative ethical conception of life [. . .] endowing the lowest character with emotion and defeat and with moral ambiguity, no matter how poor or ignoble he may seem" (Pizer 1984, 11–12) in her analysis of Mauki. To Reesman's evaluation of Jack London's literary achievement as "one of the writers of his day who led the revolt against sentimentalism and literary propriety" (Reesman 2010, 3), I would like to add his vision of demographically changing societies in which the underprivileged manifest their innate capabilities in affirming their position in a white world. In his stories Jack London anticipates some of the major concerns of life in the twenty-first century worldwide, making him a truly transnational writer whose "fight for a better America" (Tichi) extends into our own time.

4 Bibliography

4.1 Works Cited

Bennett, Judith A. *Wealth of the Solomons: A History of a Pacific Archipelago, 1800–1978.* Honolulu: U of Hawaii P, 1987.

Berkove, Lawrence I. "Jack London's 'Second Thoughts.'" 2002. Rpt. *Short Story Criticism.* Ed. Jelena O. Krstovic. Vol. 133. Detroit: Gale, 2010. 1–22.

Dorson, James. "Flow Management: Jack London and Efficient Living in the Progressive Era." *Fictions of Management: Efficiency and Control in American Literature and Culture.* Eds. James Dorson and Jasper J. Verlinden. Heidelberg: Winter, 2019. 163–187.

Eames, Ninetta. "Jack London." *Overland Monthly* 35.209 (May 1900): 417–425.

Fulton, Maurice G., ed. *Roosevelt's Writings: Selections from the Writings of Theodore Roosevelt.* New York: The Macmillian Company, 1920.

Hendricks, King, and Irving Shepard, eds. *Jack London Reports.* Garden City, Ney York: Doubleday, 1970.

Hornung, Alfred. *Narrative Struktur und Textsortendifferenzierung: Die Texte des Muckraking Movement (1902–1912).* Stuttgart: Metzler, 1978.

Hornung, Alfred. *Jack London: Abenteuer des Lebens*. Darmstadt: Lambert Schneider, 2016.
Hornung, Alfred. “Jack London’s Journeys of Life.” *The Journey of Life in American Life and Literature*. Ed. Peter Freese. Heidelberg: Winter, 2015. 91–100.
Hornung, Alfred. “Charmian Kittredge London’s *Our Hawaii*.” *Women and U.S. Politics: Historical and Contemporary Perspectives. Essays in Honor of Hans-Jürgen Grabbe*. Eds. Julia Nitz and Axel R. Schäfer. Heidelberg: Winter, 2020. 199–212.
Kingman, Russ. *Jack London: A Definitive Chronology*. Middletown: David Rejl for Jack London Research Center, 1992.
Knight, Grant C. *The Strenuous Age in American Literature*. Chapel Hill: U of North Carolina P, 1954.
Labor, Earle. “Jack London’s Mondo Cane: ‘Bâtard,’ *The Call of the Wild*, and *White Fang*.” *Critical Essays on Jack London*. Ed. Jacqueline Tavernier-Courbin. Boston: G.K. Hall & Co., 1983. 114–130.
Labor, Earle. “Jack London’s Pacific World.” *Critical Essays on Jack London*. Ed. Jacqueline Tavernier-Courbin. Boston: G.K. Hall & Co., 1983. 205–222.
Labor, Earle, Robert C. Leitz, III, and I. Milo Shepard, eds. *The Complete Short Stories of Jack London*. 3 vols. Stanford: Stanford UP, 1993.
Labor, Earle. “Introduction.” *The Complete Short Stories*. Stanford: Stanford UP, 1993. xiii–xxxvi.
Labor, Earle. *Jack London: An American Life*. New York: Farrar, Straus and Giroux, 2013.
London, Charmian Kittredge. *The Log of the Snark*. New York: Macmillan, 1915.
London, Charmian Kittridge. *Our Hawaii*. New York: Macmillan, 1917.
London, Jack. “Story of a Typhoon off the Coast of Japan” (1893). *The Complete Short Stories*. Vol. I. Stanford: Stanford UP, 1993. 1–5.
London, Jack. “A Diable – A Dog.” *The Cosmopolitan* 33 (June 1902): 218–226.
London, Jack. “Bâtard” (1902). *The Complete Short Stories*. Vol. I. Stanford: Stanford UP, 1993. 729–741.
London, Jack. “Mauki” (1908). *The Complete Short Stories*. Vol. II. Stanford: Stanford UP, 1993. 1532–1545.
London, Jack. *Revolution and Other Essays*. New York: The Macmillan Company, 1910.
London, Jack. “What Life Means to Me” (1905); rpt. *Revolution and Other Essays*. New York: The Macmillan Company, 1910. 293–309.
London, Jack. “The Other Animals” (1908); rpt. *Revolution and Other Essays*. New York: The Macmillan Company, 1910. 237–266.
London, Jack. *The People of the Abyss*. 1903. *Novels and Social Writings: The People of the Abyss, The Road, The Iron Heel, Martin Eden, John Barleycorn, Essays*. Ed. Donald Pizer. The Library of America. New York: Viking, 1982. 1–184.
London, Jack. *The Cruise of the Snark: A Pacific Voyage*. 1911. New York: KPI, 1986.
London, Jack. *John Barleycorn*. 1913. *Novels and Social Writings: The People of the Abyss, The Road, The Iron Heel, Martin Eden, John Barleycorn, Essays*. Ed. Donald Pizer. The Library of America. New York: Viking, 1982. 933–1112.
London, Jack. “The Language of the Tribe.” *The Mid-Pacific Magazine* 10.2 (August 1915): 117–120.
London, Jack. *Novels and Social Writings: The People of the Abyss, The Road, The Iron Heel, Martin Eden, John Barleycorn, Essays*. Ed. Donald Pizer. The Library of America. New York: Viking, 1982.
Newlin, Keith. “Among Cannibals and Headhunters: Jack London in Melanesia.” *Journeys* 19.1 (2018): 1–26.
Ortiz-Robles, Mario. *Literature and Animal Studies: Literature and Contemporary Thought*. London: Routledge, 2016.
Pizer, Donald. *Realism and Naturalism in Nineteenth-Century American Literature*. Rev. ed. Carbondale: Southern Illinois UP, 1984.

Reesman, Jeanne Campbell. “Fictional Experimentation.” 1999; rpt. *Short Story Criticism*. Ed. Jelena O. Krstovic. Vol. 133. Detroit: Gale, 2010. 1–22.
Reesman, Jeanne Campbell. “Rough Justice in Jack London’s ‘Mauki’.” *Studies in American Naturalism* 1.1/2 (Summer/Winter 2006): 42–69.
Riedl, Gary, and Thomas R. Tietze, eds. *Jack London’s Tales of Cannibals and Headhunters. Nine South Seas Stories by America’s Master of Adventure*. Albuquerque: U of New Mexico P, 2006.
Tavernier-Courbin, Jacqueline. *The Call of the Wild: A Naturalistic Romance*. New York: Twayne Publishers, 1994.
Tichi, Cecelia. *Jack London: A Writer’s Fight for a Better America*. Chapel Hill: U of North Carolina P, 2015.
Tietze, Thomas R., and Gary Riedl. “‘Saints in Slime’: The Ironic Use of Racism in Jack London’s South Sea Tales.” *Thalia: Studies in Literary Humor* 12.1–2 (1992): 59–66.
Wöll, Steffen. “Bleeding Borders: Space, Blackness, and Hybridity in Jack London’s Representations of the American Southwest.” *Amst* 63.1 (2018): 5–28.

4.2 Further Reading

Berkove, Lawrence I., ed. *Critical Insights: Jack London*. Pasadena: Salem P, 2011.
Bloom, Harald, ed. *Bloom’s Modern Critical Views: Jack London*. New York: Infobase Learning, 2011.
Brandt, Kenneth K. *Jack London: Writers and Their Work*. Devon: Liverpool UP, 2018.
Hodson, Sara S., and Jeanne Campbell Reesman, eds. *Jack London: One Hundred Years a Writer*. San Marino: Huntington Library, 2002.
Lasartemay, Eugene, and Mary Rudge. *For Love of Jack London: His Life with Jennie Prentiss – A True Love Story*. New York: Vantage, 1991.
Newlin, Keith, ed. *The Oxford Handbook of American Literary Naturalism*. Oxford: Oxford UP, 2011.
Reesman, Jeanne Campbell. *Jack London’s Racial Lives: A Critical Biography*. Athens: U of Georgia P, 2009.
Williams, Jay, ed. *The Oxford Handbook of Jack London*. Oxford: Oxford UP, 2017.

Birgit Däwes

14 Zitkala-Ša (1876–1938)

Abstract: The first female Native American writer in the English language to be published without the collaboration of an editor, Zitkala-Ša has often been read as torn between her Yankton Dakota heritage and the mainstream American culture into which she was assimilated. More recently, however, critics have increasingly pointed out the aesthetic strength and strategic resistance of her work. While "The Soft-Hearted Sioux" and "A Warrior's Daughter" accommodate non-Native audiences by including stereotypical motifs and melodramatic or sentimentalist passages, this chapter argues that they also celebrate polysemy and pluralism, highlight the performativity of subject positions, and elude the fixity and passivity of "Indian" victimhood through dynamic tactics of textual representation. Through their focus on Native American agency, image control, and traditional patterns of knowledge, they promote Indigenous cultural sovereignty. Simultaneously, because of their cleverly ambivalent narrative strategies, they might well be recontextualized as constitutive examples of a tradition of Indigenous modernism.

Keywords: American Indian Stories, Native American modernism, sovereignty, transmotion

1 Context

In June 1876 – the same year Zitkala-Ša was born on the Yankton reservation in South Dakota – several thousand Lakota, Yankton, Santee, northern Arapaho, and northern Cheyenne gathered in southeastern Montana to organize a protest against the violations of the Fort Laramie treaty of 1868 and the ensuing U.S. policy of forcing the Plains tribes onto reservations. Resisting the U.S. Presidential "order that all Sioux bands must come into their assigned agencies by January 31, 1876, or be considered hostile and subject to capture or killing" (Nies 1996, 283), they successfully defeated the Seventh Cavalry in the Battle of the Greasy Grass, as the Lakota called it. More broadly known as "the Battle of the Little Bighorn," and mythologized into a heroic "last stand" of a Civil War veteran and commanding Major General, George Armstrong Custer, this encounter marked one of the last successful Indigenous attempts at resisting dispossession and annihilation, as Cherokee writer Thomas King reminds us: "Five companies under Custer's command – 258 soldiers – were wiped out, along with 7 civilians and 3 Arikara scouts" (King 2012, 11). The success of defending Native sovereignty did not last, however, and by 1877, most Sioux bands had surrendered. The narrative of Little Bighorn was appropriated into a tale of heroism and bravery for the Seventh Cavalry, and although a National Cemetery,

https://doi.org/10.1515/9783110587647-015

which was later expanded into a National Monument, was dedicated to the U.S. soldiers in 1879, "[t]here were never any figures on Northern Cheyenne and Lakota casualties" (11; Greene 2008). After the massacre of Wounded Knee in 1890, during which the Seventh Cavalry killed between 250 and 300 Lakota civilians (see King 160), mostly women and children, the dispossession of Indigenous Americans became widely effective, and the policy of assimilation that many critics have come to term "cultural genocide" (e.g., Alvarez 2016, 135–136) built full momentum. Within this strategy, in 1879, General Richard Henry Pratt founded the Carlisle Indian Industrial School, the most widely known federal boarding school for Native American children (see Lomawaima and Ostler 2018). Pratt's motto of "kill[ing] the Indian to save the man" (qtd. in Krupat 2018, xvii) was followed, on a federally legislative level, by the General Allotment Act of 1887 and the Curtis Act of 1898, which fractionalized Native American lands and replaced communal land ownership by individual property, ultimately reducing Indigenous territory possession by almost two thirds, "from 147 million acres in 1887 to 55 million acres by 1934" (Isakson and Sproles 2008, 70).

Zitkala-Ša's life and literary work are, in many ways, exemplary of these times and tensions, not only because she served as a teacher at Pratt's Carlisle Indian School herself. She was a classically trained violinist and a political activist for Native American rights and citizenship, the founder and president of the National Council of American Indians, the editor-in-chief of the *American Indian Magazine* and a secretary on the Uintah and Ouray reservation in Utah, an ethnographer and the co-composer of an opera, a wife, mother, poet, and writer of fiction. Educated by her Dakota mother on the Yankton reservation, at White's Manual Labor Institute, a Quaker missionary school, and later at Earlham College (Indiana) as well as the New England Conservatory (Rappaport 1997; Lewandowski 2016), she lived in a number of places between South Dakota, Indiana, Utah, and Washington, D.C., and was clearly well-versed in shifting social and cultural registers. Her name serves as an illustrative cipher of her multiple affiliations and loyalties: born Gertrude Simmons in 1876, and first engaged to a Yavapai medical doctor named Carlos Montezuma, she married Raymond Bonnin, a Yankton Nakota administrative clerk in 1902 (Lewandowski 2016, 56–61; Rappaport 1997, 101), took his last name to become Gertrude Simmons Bonnin, and changed her name to Zitkala-Ša (Lakota for "Red Bird") around the turn of the century. "[A]s I grew I was called by my brother's name – Simmons," she wrote to Montezuma. "I bore it a long time till my brother's wife – angry with me because I insisted upon getting an education said I had deserted home and I might give up my brother's name 'Simmons' too. [. . .] Then, I choose to make a name for myself" ("Letter [ca. August 1901]" 64–65; see also Hafen 2001, xvii–xviii).

It is under this chosen name that most of her writing is published, signifying its distinctly Dakota cultural context to a largely non-Native audience. In 1901, Ginn & Company published *Old Indian Legends,* a volume of fourteen traditional stories that she had collected and translated from Dakota elders, and which were illustrated by Winnebago artist Angel deCora, a colleague of Bonnin's at Carlisle. As she writes in the preface, "[t]he old legends of America belong quite as much to the blue-eyed little patriot as to the black-haired aborigine" (Bonnin 1901, "Preface," 5). For both of these, her aim is "to transplant the native spirit of these tales – root and all – into the English language, since America in the last few centuries has acquired a second tongue" (5). In addition to preserving and mediating the stories to a non-Native target audience, Zitkala-Ša clearly uses the book to make a statement about colonialism: English is, after all, only America's "second tongue" – an acquired cultural identity that has dominated but not replaced the continent's original languages and people. This stance pervades all of her work more or less evidently, as "The Soft-Hearted Sioux" and "A Warrior's Daughter" illustrate. In addition, Zitkala-Ša wrote essays, speeches, and poetry for the *American Indian Magazine*, a journal she edited for the Society of American Indians.

American Indian Stories, which contains most of her short fiction, is arguably her best-known work, published by Hayworth in Washington, D.C., in 1921, where Bonnin had served as secretary of the Society of American Indians and become increasingly involved in campaigns for Native American rights and citizenship. Seven of the ten prose texts in the volume had previously been published elsewhere: "Impressions of an Indian Childhood," "The School Days of an Indian Girl," and "An Indian Teacher Among Indians" appeared in the highly influential *The Atlantic Monthly* in 1900; "The Great Spirit" was originally published as "Why I Am a Pagan" (1902), also in *The Atlantic Monthly*. "The Soft-Hearted Sioux" and "The Trial Path" were published in *Harper's Magazine* in 1901, and "A Warrior's Daughter" appeared in *Everybody's Magazine* in 1902. This context is important, not least because of the journals' impact on public opinion and their influence on the shaping of a literary canon. As Barbara Chiarello and Ryan Burt point out, *The Atlantic Monthly* "reverberated with slights against the Indian" (Chiarello 2005, 7), and thus "pushed readers to acknowledge, embrace, and even celebrate, a 'new' imperial reality" (Burt 2010, 61). Zitkala-Ša's stories thus substantially challenged, and offered convincing counterpoints to, the common turn-of-the-century stereotypes of the (conveniently disappearing) "noble" or "ignoble savage." As "The Soft-Hearted Sioux" and "A Warrior's Daughter" specifically show, her texts criticize American assimilation policies, especially the mission school system, and emphasize the need for Indigenous political and cultural sovereignty. The fact that they do so in the English language, and through mainstream media of public opinion, while cleverly teasing – and ultimately subverting – readers' nostalgic or exoticist expectations about tipis (or "tepees," in Zitkala-Ša's spelling), beaded deerskins, and medicine men, places her resistant textual moves at the very heart of imperialist and settler colonialist discourse.

As I argue here, "The Soft-Hearted Sioux" and "A Warrior's Daughter" not only serve as literary vessels to mediate Indigenous cultural sovereignty to non-Native readers, but their emphasis on performative and cultural rather than fixed biological notions of identity, their play with semantic polyvalence, their engagements of temporality, their structures, and their narrative strategies powerfully merge into an aesthetics of what Gerald Vizenor has termed "literary transmotion" (Vizenor 2015, 18): the elusive concept of "a movement, though not necessarily linear, across imagined boundaries" (Gamber 2008, 222). Drawing from the Anishinaabe word *manidooke,* which "means in translation 'to have a spiritual power' and 'to conduct a ceremony'" and which "describes a visionary sense of motion and presence" (Vizenor 2015, 18), transmotion – much like his influential concept of "survivance" – celebrates an Indigenous-centered worldview in opposition to any discourse of settler colonialism that freezes Native American agency into absence, stagnation, or fixed denominators of the past. Operating as literary acts of "resistance to manifest manners, the commercial modalities of monotheism and tragedy, and the cruel legacy of victimry" (18), Zitkala-Ša's stories overturn stereotypes through multiple subjectivities, textual ambiguity, and subtle shifts of meaning, moving across semantic, narratological, and cultural boundaries. Even though markers of the canonical literary contexts of realism, regionalism, or sentimentalism can be found in these texts, my reading of Zitkala-Ša's stories accordingly places them instead in a tradition of a distinctly Native American form of modernism.

2 Close Readings

Narrated from a homodiegetic, first-person point-of-view, and subdivided into five sections, "The Soft-Hearted Sioux" revolves around the return of a boarding-school educated Native American man to his home reservation. Even though arranged in linear order, the story's plot unfolds on two time levels: in the first section, the nameless narrator is sixteen years old and embedded in the safe domestic space of his family home; the subsequent four sections follow his development after his return from the mission school – where he "hunted for the soft heart of Christ" (Zitkala-Ša 2003, 119) – nine years later. When he comes home with "the white man's Bible in my hand, and the white man's tender heart in my breast" (120) in order to teach Christianity to the Sioux, he finds that his education has alienated him from his community and his traditions: "I did not grow up the warrior, huntsman, and husband I was to have been" (119). He first encounters an "old warrior," who leads him to his father's wigwam, then a "medicine-man," who looks like "eternal death" (120) to him, and finally his father, "a brave man's body broken by sickness" (121). Sitting down amidst his parents, the young man no longer "feel[s] at home," because "far apart in spirit our ideas and faiths separated us" (121). He is openly antagonized by the community's medicine man, who calls him a "traitor to his people" (122) and eventually expels him and his family from the band, moving on to "untainted ground" (122) with the rest of the villagers. More importantly, both his mother and father blame him for his failure to provide for his "aged parents" (123) and urge him to “bring [them] meat” (123) in order to keep them from starvation. After several unsuccessful days, the narrator eventually finds a domesticated herd of cattle near the village and kills one of the twenty animals. When the non-Native owner runs after him and confronts him, the Sioux man loses control of the situation and ends up killing “the paleface” (125). His delivery of the “life-giving food,” however (124), comes too late: he returns to his parents’ wigwam only to find that his father has already died. The story ends four days after the narrator “gave [him]self up to those who were searching for the murderer” (125) and awaits his death sentence in a prison cell. Haunted by the “shrieking winter winds” (125), who blame him in the voice of his father, he wonders about the spiritual consequences of his actions in the afterlife: “I wonder who shall come to welcome me in the realm of strange sight. Will the loving Jesus grant me pardon and give my soul a soothing sleep? or [sic] will my warrior father greet me and receive me as his son?” (125). When the guards come during sunrise, he concludes by describing his “strong” heart and his “dry and eager” hands: “Serene and brave, my soul awaits the men to perch me on the gallows for another flight. I go” (126).

Most critics have read the story in the context of Zitkala-Ša’s “disillusionment with missionary schools” (Ruoff 2005, 165), as an indicator that the non-Native education system leads only to death and despair (Furer 2010, 204). According to Jeanne Smith, the story “presents in unequivocal terms the threat that the cultural displacement of a mission education poses to the individual and to the community” (Smith 1994, 51) and “shows the crippling psychological effect of mouthing the ‘bigoted creed’ of an alien language and culture” (52). Or, as Lucy Maddox puts it, “[t]he story thus contains an implicit critique of the Carlisle philosophy that separating Indian children from the old ways of their families is necessary for the survival of the children and, ultimately, their families” (Maddox 2005, 147). The criticism of the “educational arm of the allotment reformation process” (Burt 2010, 62) is undisputed here; it is slightly counterbalanced, but not undermined, by the fact that “The Soft-Hearted Sioux” was published in written form, and in English. However, the reduction of the story’s conflict to a binary cultural divide would be too simple. Similarly, the implicit tragedy of the protagonist’s situation – rejected and expelled by his community, unsuccessfully searching for the identity of who he “was to have been” (119), inadvertently becoming a murderer by trying in vain to save his father, and facing premature death through non-Native executive institutions – is not reducible to a dichotomous problem. A closer look at the representation of the characters, the communal implications, the story’s title, and the symbolism of the heart, reveals that “The Soft-Hearted Sioux” contains a much more complex significational agenda.

To begin with, the protagonist is the first-person narrator, focalizer, and thus empathic center of the story, but other characters are provided with a voice as well. Tellingly, much in line with the matriarchal society of the Dakota and their traditions of orality, the first spoken words are uttered in the opening section by the narrator’s grandmother, who diagnoses his liminal situation on the verge of adulthood: "My grandchild, you are tall and are no longer a little boy" (118). As Ruth Spack notices, the grandmother's representation here "links her with the Lakota creation story of White Buffalo Calf Woman" (Spack 1997, 35), who offers the "truth" (35) in the form of a ceremonial pipe to the tribe. Her question about his plans to get married is followed by practical advice by the adolescent's mother: "My son, be always active. Do not dislike a long hunt. Learn to provide much buffalo meat and many buckskins before you bring home a wife" (Zitkala-Ša 2003, 119). The narrator reacts with "confusion" (119), silence (119), and "refusing to smoke my grandmother's pipe because my heart was too much stirred by their words" (119). He leaves the tepee, and the story resumes in section II, when he has spent nine years away from his family. His activity is now limited to "preach[ing] Christianity" to his tribe (120). Similarly, large sections of direct speech come from the medicine-man, who tries to heal the narrator's father. This privileging of an Indigenous voice – who charges the protagonist with treason for having adopted a religion that helps neither to "defend his people" nor to "keep off starvation" (122) – is further expanded by the narrator's own linguistic framing of the medicine-man. At first, the latter is seen as a rival and competitor in the project of saving "my father's soul" (121), then as a force with "the poisonous fangs of a serpent" who alienates him from his community (122), and eventually as a victorious leader who "disbanded the people" (122) and abandons the narrator's family. Despite this internalized demonization, however, the narrator also refers to the "cunning magician" (122) as a "sacred man" (121), and since the latter, in turn, terms the Christianized son's heart a "nest of young snakes" (122), and is quoted directly, their voices and mutual charges are balanced.

Furthermore, their conflict of interest is undermined by the relativization of the son's actual shortcomings, since it does not necessarily seem to be his "fea[r] to kill" (118) that keeps him from bringing home food, as Julianne Newmark and others suggest (Newmark 2002, 252); it is also his failure to find it: "I started forth again to hunt food for my aged parents. All day I tracked the white level lands in vain. Nowhere, nowhere were there any other footprints but my own!" (Zitkala-Ša 2003, 120). Colonization has clearly taken its toll; the resources of the "*white* level lands" (123, emphasis added) have been exploited, the buffalo brought to near extinction, so that the only nourishment left "upon the white country" (124) and on "the white prairie" (124) is "the white man's cattle" (124). This larger, clearly color-coded discourse of settler colonial dispossession overshadows his attempts, relativizing his family's and community's blame of individual failure. The narrator's father's statement that "I can not live without the medicine man" (121) thus also proves true, since he dies just after the "sorcerer of the plains" (118) has left. In a larger reading, therefore, the son's failure to provide food is only secondary to the spiritual deprivation of traditional knowledges, and the "warrior father" (119) – as a representative of Indigenous Americans at large – has turned into "an old gray-haired skeleton lying rigid and stiff" (125) because of these larger dispossessions.

On a narratological level, this effect is reinforced by the apparent loss of agency during the murder scene. After the young man has killed the white farmer's animal, he describes the ensuing chase and confrontation in remarkably passive terms:

> Suddenly like thunder an angry voice shouted curses and threats into my ear! A rough hand wrenched my shoulder and took the meat from me! I stopped struggling to run. A deafening whir filled my head. The moon and stars began to move. Now the white prairie was sky, and the stars lay under my feet. Now again they were turning. At last the starry blue rose up into place. The noise in my ears was still. A great quiet filled the air. In my hand I found my long knife dripping with blood. At my feet a man's figure lay prone in blood-red snow. The horrible scene about me seemed a trick of my senses, for I could not understand it was real. (124–125)

The short sentences and abundance of exclamation marks indicate the increasing speed of the plot development as well as the narrator's confusion and emotional involvement. The impressionist tumbling of the world around him corresponds to the actual movement of rolling on the floor while fighting his pursuer, but it also indexes the upending of the Sioux homeland through non-Native interference. According to Carmen Birkle, the stream-of-consciousness technique draws readers "into the murder, into the psyche of the Sioux, and thus into the inevitability and almost justification of his reaction" (Birkle 2004, 180). This effect is additionally underlined by the lack of intentional movements: a passive observer of a situation spun out of control, the narrator does not describe any deliberate movements of his knife. As if waking from a dream, he finds the dead body on the ground, shocked by the "horrible scene" (125) and unable to understand its causes. Once more, a closer reading of the scene from an Indigenous perspective is useful here: if we take Ruth Spack's clue about the grandmother's similarity with White Buffalo Calf Woman and look into the Lakota creation story more closely, we find that in that story, two young men encounter White Buffalo Calf Woman approaching their camp. One of the men desires her, and in spite of his companion's warnings, he tries to take hold of her. According to a Lakota elder, "[h]is companion saw him attempt to embrace her and there was a cloud closed about them so that he could not see what happened. In a short time, the cloud disappeared and the woman was alone" (qtd. in Walker 1991, 109). The only remains of the young man are his "bare bones" on the floor (qtd. in Walker 1991, 110). The parallels are remarkable: as if caught in a cloud, the narrator cannot capture what happens, but when he regains his cognitive skills, he inexplicably finds a dead body at his feet. From this angle, the murder scene in "The Soft-Hearted Sioux" not only embeds the protagonist in a larger oral tradition, but it substantially frames the short story within an Indigenous-centered cosmological context. The fact that he turns himself over to the authorities and accepts his fate demonstrates a keen sense of individual justice and heightens readers' sense of empathy, but the story also adds a question mark to the larger issue of cause and effect and ultimately inverts the hierarchy of cultural traditions.

This critique of colonialist dispossession in combination with an Indigenous world view is conspicuously embedded in the story's title as well. Elizabeth Wilkinson explains that "Bonnin is a member of the Yankton branch of the Dakota peoples. The Dakota consist of three branches, distinguished by language: the Dakota, Lakota, and Nakota. Within each of the three larger, language-based groups are smaller bands" (Wilkinson 2013, 53n.). Instead of "Yankton" or "Dakota," however, the title references "Sioux," which is the French (mis)adaptation of an Anishinaabe term later assimilated into English (cf. Hassrick 1964, 6; Johnson 2012, 5). The lack of a more differentiated rendering is probably due to the target readership and market, as Jane P. Hafen notes: "Zitkala-Ša's choice of 'Sioux' in the title rather than 'Lakota' or 'Yankton' or 'Nakota' immediately acknowledges the dominant culture of her audience" (Hafen 1997, 32). If the primary addressees of Zitkala-Ša's text are non-Native, her critique of dispossession and assimilationist policies is even more precisely placed. Furthermore, "Sioux" could be singular or plural, so that the title also leaves open who "the" soft-hearted Sioux really is – or are. While it seems evident at first to associate the protagonist, it is equally possible to consider other candidates, especially since the title's "heart" is not exclusively used with reference to the narrator, but also in connection with his father (Zitkala-Ša 2003, 119), or the Sioux community and camp at large (120, 122). If we thus allow once more for a communal perspective, the text exposes not only the fluidity of the signifier "Sioux," but also our expectations of reducing this signifier to the singular. By using "Sioux" in her title, Zitkala-Ša thus cleverly re-appropriates the French moniker for her own purposes, catering to audience expectations but also doubling the central message of the plot: that the European politics of unilateral assimilation and dominance are marked primarily by their devastating consequences.

Just as the title highlights the generic claim of the story's political message, so does the signifier of the "heart," which undergoes multiple symbolic shifts in the course of the narrative. The word "heart" is used 20 times in the story, including the title, and it is distributed across the five sections in remarkable symmetry, with a culmination of six occurrences in the central section III, in which the narrator is confronted by the medicine-man. The word is furthermore used three times each in sections I (including the title) and V, and four times each in sections II and IV. The first reference is to the father's heart ("I have been counting in my heart the bravest warriors of our people," 119), the second reference to the narrator's, whose "heart was too much stirred by their words, and sorely troubled with a fear lest I should disappoint them" (119), illustrating – in this juxtaposition – the story's central conflict between the son and his father. The use of the heart as a seat of emotions and character traits also reappears most dominantly in sections III, IV, and V, exclusively with reference to the narrator himself: "My heart beat hard and irregularly" (121), "A tremor played about my heart" (122), "my honest heart" (122), "a strange warmth in my heart" (124), and the final diagnosis that "[m]y heart is strong" (125). Other uses of "heart" are spatial ("the heart of the camping ground," 120) and temporal ("With every heart-throb I grew more impatient," 120). Throughout the story, seven instances refer to the "soft heart of Christ" (119, 121), the "soft-hearted Son of God" (123), the "white man's tender heart" (120), or the narrator's "soft heart"

(122–125), respectively, so that next to the title, a total of nine references to the "soft heart" establish the context of a Christian affiliation. Yet, of the twenty occurrences throughout, only twelve references are to the narrator's own heart: four are linked to Christ's or the white man's heart, and the other four refer to Indigenous persons or locations: the father (119), the campground (120), and – quite remarkably, at the end of section III – the "hearts of the people," won by the medicine-man (122). The interpretation of the title's "Soft-Hearted Sioux" as an exclusive reference to the protagonist and narrator is thus not entirely sound. If we follow the reading of the title as a plural, referring to the community at large, the narrator's fate no longer seems tragic, since, in the final scene, he concludes that "my heart is strong" (126). This could be read in direct opposition to the leitmotif of the "soft heart" and thus indicates a recovery of agency. Whereas the population of the village, led by the medicine-man's ostracizing of the protagonist, has passively accepted the choice to simply leave, the narrator – having taken on the full responsibility for his actions – has been restored to a position of action, empathy, and respect.

Aesthetically, the references to the symbol of the heart work like expressive brush strokes that apply similar semantic colors in apparently identical movements, but with variations or differences – very much like the famous technique Gertrude Stein used for her literary portraits, which – like visual portraits on canvas – she modeled on modernist painting and, taking into account the seriality of the movement, on cinema (see Stein 1998, 294–295). "The Soft-Hearted Sioux," however, is more than merely an Indigenous adaptation of European modernist strategies. In its resistance to static signifiers, its embrace of multiple meanings and ambivalence, its privileging of a communal perspective, and its unique merging of an individual's story with Lakota oral traditions, the story establishes its own version of modernism as a manifestation of what Vizenor terms "transmotion" (Vizenor 2015, 18). "Native survivance," Vizenor writes, "is natural motion, and the sentiment is contrary to dominance, that continuous murmur of state surveillance and the ethnographic containment and closure of cultures in federal archives. Survivance creates a sense of presence, and a singular resistance to manifest manners, the commercial modalities of monotheism and tragedy, and the cruel legacy of victimry" (18). This sense of presence is underlined by the use of present tense in the last section of the story, and by the narrator's last words: "I go" implies choice, willfulness, and bodily motion (Zitkala-Ša 2003, 126). In this sense, "The Soft-Hearted Sioux" is (and are) not a victim, but an embodiment of survivance as a principle, someone whose "eyes are dry and eager for new scenes" (126), but who retains his sense of dignity and self.

Like "The Soft-Hearted Sioux," "A Warrior's Daughter" (1902) opens with a family scene in a tepee during the protagonist's childhood and largely takes place on two time levels. Anthologized in Paula Gunn Allen's influential collection *Spiderwoman's Granddaughters* (1983), the text is often considered "[o]ne of her most interesting stories" (Brown Ruoff 2005, 165), and especially celebrated by feminist critics. Told from a limited, third-person narrative perspective, the story centers on

Tusee, a young Dakota woman, who – using courage, cleverness, and costumes – rescues her lover from captivity at an enemy's camp. With the exception of the opening, the story is written in present tense, which not only adds to the effect of immediacy and suspense, but which also underlines a policy of representation that focuses on Native American survivance, presence, and agency.

In the first quarter of the story, Tusee is introduced as an eight-year-old girl, growing up in the family of "the chieftain's bravest warrior" (Zitkala-Ša 2003, 132) as the pride and "great joy" of her father (132). The description of Tusee focuses both on her size, as a "wee" (132) and "little" (133, 134) daughter with a "small body" (132), "a wee hand" (132), a "small forefinger" (134), and "small, brown hands" (134), and on her resemblance to and close relationship with her "proud" father (133, 134), who "tenderly placed a strong arm about her" (133). She is performing a traditional dance to her father's chanting, when a former captive and slave, who has been adopted into Tusee's family, joins the scene in a foreshadowing of events to come. His "stranger's accent" announces his embodiment of difference: "His features were of the Southern type. From an enemy's camp he was taken captive long years ago by Tusee's father. But the unusual qualities of the slave had won the Sioux warrior's heart, and for the last three winters the man had had his freedom" (133). Tusee addresses him as "uncle" and even speaks his language, "for he had taught her many words and phrases" (134). Since the girl has been promised a pony for her first dance, she wants her "uncle" to go and catch the animal the next day. Even though her mother questions her request, the man expresses his loyalty and obedience, to which Tusee reacts with gratitude. The narrator, who stays largely in the background throughout the rest of the story, comments on this scene heterodiegetically: "Her childish faith in her elders was not conditioned by a knowledge of human limitations, but thought all things possible to grown-ups" (134). This exposition not only serves the characterization of Tusee as a "buoyant spirit" (134) full of "joy" and "glee" (134), and as a foreshadowing of her possibilities when she grows up, but it also explains how Tusee will later be able to use the antagonist's language strategically for her purposes, and – even more importantly – it precludes any fixed dichotomy between the Dakota and their enemies.

The next section of the story resumes at an unspecified later time, when Tusee is old enough to get married, and her "lover" tells her that he has asked her father "for his only daughter" (135). The father, however, sets a task for him: "Naught but an enemy's scalp-lock, plucked fresh with your own hand, will buy Tusee for your wife" (135). The next day, a war party including her lover rides south "to avenge a former wrong" (135), and even though Tusee promised him to "wait and watch for [his] return" (135), she joins the "faithful women" who follow them. At the enemy's camp, the time between dawn and dusk – and the decisive battle – is narrated in only five sentences, with a much stronger focus on the consequences: before they return home, the warriors realize that two members of their group are dead, while another "is a captive to the foe" (136). Whereas the Dakota leave for their own camp,

Tusee single-handedly "plans to cheat the hated enemy of their captive" (136), the execution of which covers the entire second half of the story. She begins by approaching the camp "with a panther's tread and pace" (136), asking the Great Spirit for assistance: "All-powerful Spirit, grant me my warrior-father's heart, strong to slay a foe and mighty to save a friend!" (137). As a next step, she seductively lures her lover's captor away from the camp, announcing in his language that "I wish to tell you who I am" (138). When he eagerly follows her, she keeps her promise: "like a panther for its prey," she attacks, and tells him that "I am a Dakota woman" (139) just as the stabbed enemy "falls heavily at her feet" (139).

Intensifying the suspense of the story, the following section is focalized through Tusee's nameless lover, who witnesses what looks like an old woman walking around the camp. When she approaches, he realizes it is Tusee, who cuts him loose and tells him to follow. However, since he is "numb and helpless" (140) and unable to walk, she transforms one more time: "The sight of his weakness makes her strong. A mighty power thrills her body. Stooping beneath his outstretched arms gasping at the air for support, Tusee lifts him upon her broad shoulders. With half-running, triumphant steps she carries him away into the open night" (140).

Ending on these words, the story emphasizes Tusee's identity by reiterating her name, her Dakota affiliation, her gender, and her physical strength, which becomes all the more striking through the contrast with the initial emphasis on her "small" and "wee" body. Whereas, in the opening's encounter with the former-enemy-turned-uncle, she stood "in front of the strong man" and "clapped her small, brown hands with joy" (134), the roles have now been reversed, but not necessarily along binary lines. In the climactic confrontation with the enemy who captured her lover, she benefits from a "warrior's strong heart" (139), given to her by "the Great Spirit" (139) in response to her prayers, so she can defeat him. The final scene, however, juxtaposes her with a member of her own tribe, whose "weakness makes her strong" (140). Critics have widely evaluated this scene as a feminist expression of triumph over male dominance: Zitkala-Ša's biographer Tadeusz Lewandowski reads it in the context of the author's struggle with her fiancé Carlos Montezuma "over her overt refusal to adopt a submissive position within their proposed marriage" (Lewandowski 2016, 60), and Andrew J. Furer writes: "On the one hand, Tusee exemplifies the superiority of Native American women to overcivilized, hyperfeminized Euro-American women, especially in their alternative sex role as warrior women; on the other, Tusee is a model for what all American women can achieve if they break free of Victorian notions of femininity" (Furer 2010, 205). Carmen Birkle emphasizes the "phallic knife" (Birkle 2004, 223) as well as the "three generations of women" through which Tusee has gone in her masquerade (223), and intriguingly compares the motif of the weak husband and the triumphant wife to Charlotte Perkins Gilman's "The Yellow Wall-Paper" (224). For a feminist approach, it is also particularly relevant that the last sentence describes Tusee carrying her future husband "away into the open night" (Zitkala-Ša 2003, 140) instead of back to their camp and thus into established social structures and domestic surroundings. In contrast to the fiancé's original expectation – that she patiently wait for his return – she has to actively liberate and literally sustain him, clearly inverting their traditional gender roles. Moreover, since Tusee underlines her identity ("I am a Dakota woman!" 139) at the climactic moment of killing the enemy, the mixture of "racial pride and women's rights" (Furer 2010, 204) seems prominent.

It is, however, not evident at first glance which language she uses at this moment. Like the rest of the story, the sentence is rendered in English for the mainstream audience. Since Tusee lured her victim away from the group by "speak[ing] to him in his own tongue" (Zitkala-Ša 2003, 138), she is likely to use his language in this situation as well, since she provides him with an explanation for her action that she probably wants him to understand. The Dakota had a number of enemies before European contact, including the Crow to the west, and the Anishinaabe to the east and north. Since the foes described in this story are located "a day's journey" (136) to the South, they are probably Pawnee, the "hereditary enemies" (Rodriguez 2002, 274) of the Sioux. Thus, Tusee likely uses the Pawnee language to express her Dakota origin. It is remarkable that, while occasional Dakota words or phrases are inserted into the text, such as Tusee's father's affirmative "Hechetu!" ("It is so," 134; see Lakota Language Consortium 2018) or her fiancé's "Hoye" ("All right," 135; see Lakota Language Consortium 2018), these are not used by the protagonist herself. She is thus not firmly located within a specific language but seems to be flexibly multilingual. She may not use a lot of words in the first place, "only nod[ding] assent, for mere words are vain" (135), yet her ability to easily shift between languages and registers ultimately ensures her triumph.

Most importantly, "A Warrior's Daughter" cleverly relocates the question of cultural difference – the boundaries of which Zitkala-Ša's contemporaneous readership would place between Native American and European-American cultures – into a different arena. As Ruth Spack notes, there are no references whatsoever to Europeans in this story (Spack 2001, 190). Instead, the plot introduces us to an America untainted by colonialist influence, except for the language in which the story itself is rendered. In this context, the argument that Spack makes elsewhere – that the story's plot cleverly echoes and inverts "Pocahontas's seventeenth-century rescue of John Smith" (Spack 1997, 35) – becomes all the more prominent, since the rescued male is "Indian, not white" (36) and thus effectively "Indianiz[es]" (36) the stereotypical legend, highlighting Native American agency and power. Furthermore, Indigenous sovereignty, territory, and ownership are underlined by the warrior father's initial perception: "His head was so poised that his eye swept easily the vast level land to the eastern horizon line" (Zitkala-Ša 2003, 132). While the "white country" of "The Soft-Hearted Sioux" (124) has become limited in extent and resources by the intrusion of white farmers and missionaries, Tusee's territory is exclusively Native American, allowing her to move freely and strongly "into the open night" (140) in the end.

In the absence of Europeans, the cultural boundaries around which the story revolves run between the Dakota and their enemies, foregrounding for early twentieth-century readers the original social and cultural diversity of the North American continent. Even though the enemies remain nameless, and their language unspecified, there is no essentialist condemnation on the basis of ethnicity or cultural affiliation: the "Southern type" of elderly rider introduced early in the story has been able to change his social status from former captive to family member because of his "unusual qualities" of character (133). Cultural conflicts are addressed in the traditional Dakota way – through warfare – but they do not operate by ideologies of racism, nor do they aim at the annihilation of the Other. The title underlines this further: even though the protagonist's Dakota background becomes evident, the generic reference to "a" (not "the;" not "a Dakota") warrior's daughter – much like the definite article in "The Soft-Hearted Sioux" – allows for a broader application of Indigenous sovereignty.

In this exclusively Indigenous universe, Tusee is the only character given an individual name and an explicit, self-declared identity position ("I am a Dakota woman!" 139). At the same time, she is also the character most capable of shifting between different roles, languages, and costumes: as a child whose "eyes brimmed over with light" (134), a "loyal daughter" (134), a "handsome woman smiling" seductively (138), a "fleeing woman" (139), or a "bent old woman [. . .] with a bundle like a grandchild slung on her back" (139), she performs different roles in a strategy to claim her place in society and to defeat her opponents through flexible adaptation. Jeanne Smith has read this versatility as a "trickster strateg[y]" (Smith 1994, 53): "Tusee is a powerful trickster/savior in this story. Her disguise as a bent woman with a bundle on her back echoes a Lakota trickster tale in which Iktomi throws grass into his blanket and carries it as if it is a great burden in order to trick and catch some ducks" (53–54). As in "The Soft-Hearted Sioux," therefore, the story employs traditional Dakota knowledge and oral traditions to provide intertextual backgrounds in subtle gestures toward Indigenous empowerment and agency.

In addition to the diegetic advantages of Tusee's shapeshifting and bilingualism in outwitting her Native opponents, her skillful performance of different roles involves an important statement on cultural identity as a concept. Instead of being fixed and limited to her respective social status as daughter, future wife, or even future grandmother, she acts out these roles – in "easy movements" (132), like her initial dance – in strategic performances, effortlessly crossing any boundaries between them. Identity itself is thus not a static given, but fluid, multiple, and contingent – a situational possibility among others. Thus coding subjectivity as performative, "A Warrior's Daughter" not only effectively undermines notions of Native American identity as static, homogeneous, or frozen in the past, but also promotes a concept of subjectivity that was later considered a signature element of European-American modernism – and, through Judith Butler's work on performative subjectivity, of post-modernism.

In combination with the narrative techniques of shifting focalizers and stream-of-consciousness, Zitkala-Ša's use of Dakota cultural heritage, of multiple languages, semantic pluralism and representational flexibility, and her notions of fluid and multiple subjectivities, are clearly exemplary, if not constitutive, of a distinctly Indigenous version of modernism. Her style is as far from the realist, exoticist, or naturalist expectations of her time (see Okker 1994, 92) as it is from what Louis Owens has termed (with reference to N. Scott Momaday's *House Made of Dawn*) the "unmistakable signs of privileged and presold modernist discourse to be recognized by a jury from that same arena" (Owens 1998, 64). Just as Tusee disrupts the Pawnee community from within through a clever adaptation of their ways, Zitkala-Ša uses the English language to infuse a mainstream audience with Dakota oral traditions, strong and sympathetic Native American characters, claims to Indigenous sovereignty, as well as "the marvelous collage of ancestral remembrance and cultural survivance" that Vizenor sees at the heart of transmotion (Vizenor 2015, 25).

3 Theoretical Perspectives

The earliest criticism of Zitkala-Ša's short stories came from her contemporaries, and it was not particularly kind. When "The Soft-Hearted Sioux" was first published in *Harper's Magazine*, its critique of assimilationist policies in general and boarding schools in particular enraged Richard Henry Pratt, who – in the Carlisle newspaper – reviewed the story as "morally bad" (qtd. in Katanski 205, 128). "All that Zitkalasa [sic] has in the way of literary ability and culture," he writes, "she owes to the good people, who, from time to time have taken her into their homes and hearts and given her aid. Yet not a word of gratitude [. . .] has ever escaped her in any line of anything she has written for the public" (qtd. in Katanski 2005, 128). Especially with regard to "The Soft-Hearted Sioux," Pratt claims that the author "harms the educational work in progress for the race from which she sprang" (qtd. in Katanski 2005, 128). This is an understandable reaction given that Pratt, "whose considerable energies were given to urging the breaking up of reservations" (Maddox 2005, 145), felt directly criticized by Zitkala-Ša's writing, but it also shows the broad reception of the story and the impact Pratt feared it must have had on public opinion. Zitkala-Ša did not seem to be particularly impressed by these attacks: "Certainly people are welcomed to their opinions," she wrote to her fiancé in 1901, but "I must cling to my own" ("Letter [March 5, 1901]" 24). Ironically, as Ruth Spack points out, Pratt's criticism would "have the unintended effect of increasing Zitkala-Ša's literary output" (Spack 2001, 188), since she felt all the more motivated to proceed with the publication of her *Old Indian Legends*.

After the immediate reception of her work during her lifetime (she passed away in 1938), it took more than four decades for critics such as Dexter Fisher, Paula Gunn Allen, Mary Stout, or William Willard to engage with Zitkala-Ša's texts again, and in 1994, Patricia Okker noted that "few scholars have studied Zitkala-Ša's work in much detail" and "literary analysis of her work remains scanty and partial" (89). It was primarily Dexter Fisher who, in his 1979 article on "Zitkala-Ša: The Evolution of a Writer," acknowledged her contribution to "progressive reform for Indians" (Fisher 1979, 237) and her merit in "struggl[ing] toward a vision of wholeness in which the conflicting parts of her existence could be reconciled" (237), but he ultimately considers her literary output flawed by "a model of ambivalences, of oscillations between two diametrically opposed worlds" (237). This strand of criticism, based on a dichotomous understanding of a cultural divide between Native and non-Native Americans, has been remarkably dominant in the reception of Zitkala-Ša's fiction to the present day. Whereas Fisher speaks of her "curious blend of civilized romanticism and aggressive individualism" (236), later critics phrased this positioning less harshly, but to the same outcome: in an article brimming with factual errors (e.g., Zitkala-Ša's date of birth), Mary Stout considers her as suffering from "culture shock" (Stout 1984, 70) and concludes that "[s]he did not have the skill of a Leslie Silko or an N. Scott Momaday" (74). Because of the frustration expressed in the author's letters, her political activism, and her rhetoric in essays such as "Why I am a Pagan" (1902) or "America's Indian Problem" (1921), many critics have evidently focused on Gertrude Simmons Bonnin's position "between" her Yankton and white heritage (through her absent white father, but also through her education), and her "liminal position between the Native and white worlds" (Diana 1997, 156), reading her situation as "irreducible to either culture and alienated from each" (Carpenter 2004, 1). Attila Silkü organizes his analysis around the binary juxtaposition of "natural phenomena – ground squirrels or ice – with unnatural – for example, the glass marbles of the missionaries – as if *representing the two sides of her hybrid character*" (Silkü 2002, 118; emphasis added), and even Jeanne Smith seems to read Zitkala-Ša's multicultural development as a deficit, writing that she "moved in many cultural worlds, belonging fully to none" (46).

Even in positive assessments of her achievements, dualist structural orders prevail: Franci Washburn, for instance, divides Bonnin's life into two periods (her "early career" as a writer, and her "later career" of political activism) and concludes that she "took the torn fragments of her life and stitched them, at times painfully, into a functioning, productive bicultural identity, dedicated to protecting and advancing the well-being of Native peoples" (Washburn 2008, 273–274). This emphasis on the interdependency between her life and her work is shared by many critics (Beardslee 2001, Newmark 2002, Irvin 2006) and was reinforced by the proliferation of approaches to *American Indian Stories* as "part autobiography, part storytelling" (Newmark 2002, 336), even though the texts "appeared with no generic designation" (Spack 1997, 28). Furthermore, whereas many critics, including Diana (2008),

Furer (2010), Silkü (2002), or Irvin (2006), praise her work from a feminist or postcolonial point of view, others, such as Robert Warrior, argue for a critical and "less celebratory" assessment (Warrior 1995, 10), underlining that because of "the presence of Bonnin and other SAI figures in the antipeyote camp" (9), their status is complicated and at least partially counterproductive with regard to traditional "forms of Native existence" (10).

Going beyond these binary or biographically tinted interpretations focusing on Bonnin's political life, I would instead follow Jane P. Hafen (1997), Ruth Spack (1997), Gary Totten (2005), Ryan Burt (2010), Elizabeth Wilkinson (2013), Kiara Vigil (2017), and others in emphasizing the "subversive tactics" (Spack 1997, 26) of Zitkala-Ša's texts and her "strategic [. . .] ability to negotiate shifting expectations and representational politics based on both Yankton communal values and American modernity" (Vigil 2017, 132). As I have argued here, Zitkala-Ša's short stories are prime examples of transmotion, the fluid, transversal movement across boundaries of space and time, of languages, cultures, and meanings. Firmly rooted in Dakota storytelling traditions and a Dakota worldview, her complex engagements of multiple subjectivities and semantic ambivalence contribute to the formation of a Native American literary modernism that is both an expression of Indigenous literary sovereignty and a political and aesthetic claim to a place in the American canon.

According to Gerald Vizenor, rock paintings around the Great Lakes demonstrate that the "elements of form, perspectives of natural motion, and visionary dimensions in ancient Native American art anticipated the movements of impressionism, cubism, surrealism, and abstract impressionism" (Vizenor 2015, 21). These "movements of impressionism and cubism were honored elsewhere, but Native Americans were restricted by exhibition curators to the dubious romance of simulated traditional scenes" (23). Along similar lines, Bill Anthes deplores "postwar America's systematic denial of subjectivity to nonwhites, women, and homosexuals": "Indeed," he writes, "it was assumed that modernist subjectivity was unattainable for artists of color in the twentieth century due to a critical double-standard" (Anthes 2006, 146). Whereas, more recently, Native American modernism has increasingly been recognized as an independent movement in literature and the visual arts, including John Joseph Mathews, D'Arcy McNickle, George Morrison, Allan Houser, or Jaune Quick-to-See Smith, Zitkala-Ša's work reminds us that these delimitations of an established canon are in continuous need of revision. Through their narrative strategies and conflicted subjectivities, her short stories deserve a place not only in anthologies of Native American literature or of what Christopher Schedler calls "Border Modernism" (Schedler 2003), but of American literature as a whole.

4 Bibliography

4.1 Works Cited

Allen, Paula Gunn, ed. *Spider Woman's Granddaughters: Traditional Tales and Contemporary Writing by Native American Women*. New York: Ballantine, 1983.

Alvarez, Alex. *Native America and the Question of Genocide*. Lanham: Rowman & Littlefield, 2016.

Anthes, Bill. *Native Moderns: American Indian Painting, 1940–1960*. Durham: Duke UP, 2006.

Beardslee, Karen E. *Literary Legacy, Folklore Foundations: Selfhood and Cultural Tradition in Nineteenth- and Twentieth-Century American Literature*. Knoxville: U of Tennessee P, 2001.

Birkle, Carmen. *Migration – Miscegenation – Transculturation: Writing Multicultural America into the Twentieth Century*. Heidelberg: Winter, 2004.

Burt, Ryan. "'Death Beneath this Semblance of Civilization': Reading Zitkala-Ša and the Imperial Imagination of the Romantic Revival." *Arizona Quarterly* 66.2 (Summer 2010): 59–88.

Carpenter, Ron. "Zitkala-Ša and Bicultural Subjectivity." *Studies in American Indian Literatures* 16.3 (Fall 2004): 1–28.

Chiarello, Barbara. "Deflected Missives: Zitkala-Ša's Resistance and Its (Un)Containment." *Studies in American Indian Literatures* 17.3 (Fall 2005): 1–26.

Diana, Vanessa Holford. "'Hanging in the Heart of Chaos': Bi-Cultural Limbo, Self-(Re)Presentation, and the White Audience in Zitkala-Ša's *American Indian Stories*." *Cimarron Review* 121 (Oct. 1997): 154–172.

Diana, Vanessa Holford. "Zitkala-Ša's and Sui Sin Far's Sketch Collections: Communal Characterization as Resistance Writing Tool." *Scribbling Women and the Short Story Form: Approaches by American and British Women Writers*. Ed. Ellen Burton Harrington. New York: Peter Lang, 2008. 98–111.

Fisher, Dexter. "Zitkala-Ša: The Evolution of a Writer." *American Indian Quarterly* 5.3 (August 1979): 229–238.

Furer, Andrew J. "From 'Water Drops' to General Strikes: Nineteenth- and Early Twentieth-Century Short Fiction and Social Change." *A Companion to the American Short Story*. Eds. Alfred Bendixen and James Nagel. New York: Wiley Blackwell, 2010. 187–214.

Gamber, John. "Tactical Mobility as Survivance: *Bone Game* and *Dark River* by Louis Owens." *Survivance: Narratives of Native Presence*. Ed. Gerald Vizenor. Lincoln: U of Nebraska P, 2008. 221–246.

Greene, Jerome A. *Stricken Field: The Little Bighorn Since 1876*. Norman: U of Oklahoma P, 2008.

Hafen, Jane P. "Introduction." *Dreams and Thunder: Stories, Poems, and* The Sun Dance Opera. By Zitkala-Ša. Ed. Jane P. Hafen. Lincoln: U of Nebraska P, 2001. xiii–xxiv.

Hafen, Jane P. "Zitkala-Ša: Sentimentality and Sovereignty." *Wicazo Sa Review* 12.2 (Autumn 1997): 31–41.

Hassrick, Royal B. *The Sioux: Life and Customs of a Warrior Society*. Norman: U of Oklahoma P, 1964.

Irvin, Amanda. "Out of Body Experiences in the Texts of Zitkala-Ša and Gertrude Bonnin." *Pembroke Magazine* 38 (2006): 80–86.

Isakson, Hans R., and Shauntreis Sproles. "A Brief History of Native American Land Ownership." *Indigenous Peoples and Real Estate Valuation*. Eds. Robert A. Simons, Rachel Malmgren, and Garrick Small. Berlin: Springer, 2008. 63–76.

Johnson, Michael G. *Tribes of the Sioux Nation*. New York: Bloomsbury, 2012.

Katanski, Amelia V. *Learning to Write 'Indian': The Boarding School Experience and American Indian Literature*. Norman: U of Oklahoma P, 2005.

King, Thomas. *The Inconvenient Indian: A Curious Account of Native People in North America.* Toronto: Anchor Canada, 2012.

Krupat, Arnold. *Changed Forever: American Indian Boarding-School Literature.* Albany: SUNY P, 2018.

Lakota Language Consortium, https://www.lakotadictionary.org/phpBB3/nldo4.php. *New Lakota Dictionary Online.* Ed. Jan F. Ullrich. (9 February 2021).

Lewandowski, Tadeusz. *Red Bird, Red Power: The Life and Legacy of Zitkala-Ša.* Norman: U of Oklahoma P, 2016.

Lomawaima, K. Tsianina, and Jeffrey Ostler. "Reconsidering Richard Henry Pratt: Cultural Genocide and Native Liberation in an Era of Racial Oppression." *Journal of American Indian Education* 57.1 (Spring 2018): 79–100.

Maddox, Lucy. *Citizen Indians: Native American Intellectuals, Race, and Reform.* Ithaca, New York: Cornell UP, 2005.

Newmark, Julianne. "Writing (and Speaking) in Tongues: Zitkala-Ša's *American Indian Stories.*" *Western American Literature* 37.3 (2002): 335–358.

Nies, Judith. *Native American History.* New York: Ballantine, 1996.

Okker, Patricia. "Native American Literatures and the Canon: The Case of Zitkala-Ša." *American Realism and the Canon.* Eds. Tom Quirk and Gary Scharnhorst. Newark: U of Delaware P, 1994. 87–101.

Owens, Louis. *Mixedblood Messages: Literature, Film, Family, Place.* Norman: U of Oklahoma P, 1998.

Raheja, Michelle H. "Visual Sovereignty." *Native Studies Keywords.* Eds. Stephanie Nohelani Teves et al. Tucson: U of Arizona P, 2015. 25–34.

Rappaport. Doreen. *The Flight of the Red Bird: The Life of Zitkala-Ša.* New York: Puffin Books, 1997.

Rodriguez, Junius P. "Pawnee." *The Louisiana Purchase: A Historical and Geographical Encyclopedia.* Ed. Junius P. Rodriguez. Santa Barbara, California: ABC Clio, 2002. 273–274.

Ruoff, A. LaVonne Brown. "Pre-1968 Fiction." *The Cambridge Companion to Native American Literature.* Eds. Joy Porter and Kenneth M. Roemer. New York: Cambridge UP, 2005. 161–171.

Schedler, Christopher. *Border Modernism.* New York: Routledge, 2003.

Silkü, Attila. "Fiddler on the Threshold: Cultural Hybridity in Gertrude Bonnin's *American Indian Stories.*" *Journal of Postcolonial Writing* 40.1 (2002): 115–124.

Smith, Jeanne. "'A Second Tongue': The Trickster's Voice in the Works of Zitkala-Ša." *Tricksterism in Turn-of-the-Century American Literature: A Multicultural Perspective.* Eds. Elizabeth Ammons and Annette White-Parks. Hanover: UP of New England, 1994. 46–60.

Spack, Ruth. "Dis/engagement: Zitkala-Ša's Letters to Carlos Montezuma, 1901–1902." *MELUS* 26.1 (Spring 2001): 173–204.

Spack, Ruth. "Re-Visioning Sioux Women: Zitkala-Ša's Revolutionary *American Indian Stories.*" *Legacy* 14.1 (1997): 25–42.

Stein, Gertrude. "Portraits and Repetitions." *Stein: Writings, 1932–1946.* Vol. 2. Eds. Catharine R. Stimpson and Harreit Chessman. New York: Library of America, 1998. 287–312.

Stout, Mary. "Zitkala-Ša: The Literature of Politics." *Coyote Was Here: Essays on Contemporary Native American Literary and Political Mobilization.* Ed. Bo Schöler. Special Issue of *The Dolphin* 9 (1984): 70–78.

Totten, Gary. "Zitkala-Ša and the Problem of Regionalism." *American Indian Quarterly* 29.1–2 (2005): 84–123.

Vigil, Kiara M. "Expanding Interpretations of Native American Women's History." *Great Plains Quarterly* 37.2 (Spring 2017): 131–144.

Vizenor, Gerald. "Literary Transmotion: Survivance and Totemic Motion in Native American Indian Art and Literature." *Twenty-First Century Perspectives on Indigenous Studies.* Eds. Birgit Däwes, Karsten Fitz, and Sabine N. Meyer. New York: Routledge, 2015. 17–30.

Walker, James R. *Lakota Belief and Ritual*. Eds. Raymond J. DeMallie and Elaine A. Jahner. Lincoln: U of Nebraska P, 1991.

Warrior, Robert Allen. *Tribal Secrets: Recovering American Indian Intellectual Traditions*. Minneapolis: U of Minnesota P, 1995.

Washburn, Franci. "Zitkala-Ša: A Bridge Between Two Worlds." *Their Own Frontier: Women Intellectuals Re-Visioning the American West*. Eds. Shirley A. Leckie and Nancy J. Parezo. Lincoln: U of Nebraska P, 2008. 269–302.

Wilkinson, Elizabeth. "Gertrude Bonnin's Rhetorical Strategies of Silence." *Studies in American Indian Literatures* 25.3 (Fall 2013): 33–56.

Zitkala-Ša (Gertrude Simmons Bonnin). "The Soft-Hearted Sioux." 1901. *American Indian Stories, Legends, and Other Writings*. By Zitkala-Ša. Eds. Cathy N. Davidson and Ada Norris. New York: Penguin, 2003. 118–126.

Zitkala-Ša (Gertrude Simmons Bonnin). "A Warrior's Daughter." *American Indian Stories, Legends, and Other Writings*. By Zitkala-Ša. Eds. Cathy N. Davidson and Ada Norris. New York: Penguin, 2003. 132–140.

Zitkala-Ša (Gertrude Simmons Bonnin). "Letter to Carlos Montezuma, March 5, 1901." *Letters, Speeches, and Unpublished Writings, 1898–1929*. By Zitkala-Ša. Ed. Tadeusz Lewandowski. Amsterdam: Brill, 2018. 23–25.

Zitkala-Ša (Gertrude Simmons Bonnin). "Letter to Carlos Montezuma, ca. August 1901." *Letters, Speeches, and Unpublished Writings, 1898–1929*. By Zitkala-Ša. Ed. Tadeusz Lewandowski. Amsterdam: Brill, 2018. 64–65.

4.2 Further Reading

Cox, James H. *The Political Arrays of American Literary History*. Minneapolis: U of Minnesota P, 2019.

Justice, Daniel Heath. *Why Indigenous Literatures Matter*. Waterloo: Wilfrid Laurier UP, 2018.

Taylor, Melanie Benson. *The Cambridge History of Native American Literature*. Cambridge: Cambridge UP, 2020.

6.2 Further Reading

Nassim Winnie Balestrini

15 Sherwood Anderson (1876–1941)

Abstract: In his short story cycle *Winesburg, Ohio* (1919), Sherwood Anderson depicts – on the most obvious level – Midwestern small-town life in the 1890s. According to him, his stories provide insight into the inner lives of late-nineteenth-century American youths who are struggling to understand who they are. Underneath a realist surface, Anderson pursues an expressionist trajectory by employing an omniscient narrator and by depicting individuals in relation to others. For his characters, understanding one's own self remains as difficult or even impossible as forming functioning long-term relationships. In tune with an evolving modernist outlook, the stories leave much to be interpreted and determined by readers willing to explore the intrinsic multi-dimensionality of Anderson's diction, symbolism, and allusions. Recent Anderson scholarship reinterprets the stories from extrinsic, contextual perspectives such as Anderson's response to neoliberal capitalism and the First World War.

Keywords: Modernism, realism, expressionism, relationality, symbols

1 Context

Sherwood Anderson (1876–1941) is most prominently remembered for *Winesburg, Ohio*, his book of short stories published a century after Washington Irving's 1819 *Sketch-Book* (↗5 Irving). By 1919, American fiction had definitely come into its own, and twentieth-century American authors could look back at decades during which their predecessors had been developing and diversifying the short story. The genre had swiftly moved from Irving's use of the picturesque and satirical modes to Edgar Allan Poe's gothic horror and his tales of ratiocination (↗7 Poe), to Nathaniel Hawthorne's narratives that scrutinize the ponderous legacy of Puritan New England in light of mid-nineteenth-century concerns regarding national identity, science, belief, and art (↗6 Hawthorne), to Herman Melville's critique of capitalism and imperialism (↗8 Melville), to the upsurge of magazine publications focused on postbellum trends that came to be called 'local color' and 'regionalist' fiction, which prominently featured women writers like Mary E. Wilkins Freeman, Harriet Beecher Stowe, Kate Chopin (↗11 Chopin), Alice Dunbar-Nelson, and Rose Terry Cooke, and to the harrowing naturalist stories by authors like Ambrose Bierce and Stephen Crane. Additionally, so-called short story cycles – such as Sarah Orne Jewett's *The Country of the Pointed Firs* (1896) – demonstrated the novel-like potential of convincingly weaving multiple stories into larger patterns of plot, character, and thematic development.

Beginning to write during the transition from Naturalism to Modernism, Anderson was among those writers who strove to combine well-established forms with

https://doi.org/10.1515/9783110587647-016

bold experimentation and innovation. Accordingly, Anderson anchors *Winesburg, Ohio* in his re-definition of the term 'grotesque.' Rather than focusing on the outward features of the centuries-old grotesque tradition of ornately and unnaturally intertwining plant, animal, and human shapes, the opening text of *Winesburg, Ohio*, which is entitled "The Book of the Grotesque" (Anderson 1996b, 5–7), explains that human beings who firmly believe in what they regard as a singular truth become so obsessed with this persuasion that they shrivel into absurdly grotesque beings (7). The volume's subtitle, *A Group of Tales of Ohio Small Town Life*, may lead potential readers to expect vignettes that nostalgically review possibly eccentric and yet lovable representatives of an America that was quickly vanishing during the second age of industrialization and the sweeping changes brought about by mass immigration and urbanization. The definition of grotesqueness, however, points Anderson's readers in the direction of an expressionist project – an attempt to narrate inner, mental worlds (see Stouck 1990, 31). Thus, "The Book of the Grotesque" presents the figure of an elderly male writer who is metaphorically "pregnant" with "a woman, young, and wearing a coat of mail like a knight" (Anderson 1996b, 5). Her symbolically impenetrable outfit, then, is only important in the sense that the narrator defines the writer's province as exploring the inner lives of fictional characters: "The thing to get at is what the writer, or the young thing within the writer, was thinking about" (5). At the end of the short text, the narrator claims that the 'pregnant' writer never published his book about the grotesques who, in turn, never got to realize how "the truth" they "embraced became a falsehood" (7). The writer's reason was that he feared "becoming a grotesque" himself (7).

The omniscient narrator's superior knowledge about the old writer as well as the seeming disconnectedness of this opening text and the stories that follow throws the reader into a quandary regarding authorship, control over one's subject matter, and the process of trying to understand and assess ideas either as "thoughts" or "truths" (6). With this prefatory text, Anderson reverses Geoffrey Crayon's ostensible function as something like a narrating tour guide to, admittedly, not necessarily typical tourist spots and important places in Irving's *Sketch-Book* (↗5 Irving) and leaves the reader in limbo regarding the significance of the unnamed failed writer. Or, rather, to do Irving's satirical bent justice, Anderson's narrator does not even pretend to think that claiming irrefutable truths will not result in distortions that will do the fictional human characters an injustice. Also, "The Book of the Grotesque" raises the question as to whether the subsequent stories are supposed to be read as the unnamed failed writer's tales which the anonymous narrator decides to present to the world without openly admitting it.

Anderson began working on the *Winesburg, Ohio* stories in 1915. He had already published two novels (*Windy McPherson's Son* [1916]; *Marching Men* [1917]) and several of said stories before the short story cycle appeared as a book in 1919. Five novels followed until 1936 (*Poor White* [1920]; *Many Marriages* [1923]; *Dark Laughter* [1925]; *Beyond Desire* [1932]; *Kit Brandon* [1936]) as well as three further short story collections (*The Triumph of the Egg* [1921]; *Horses and Men* [1923]; *Death in the Woods* [1933]). Anderson's oeuvre also includes two volumes of poetry (*Mid-American Chants* [1918]; *A New Testament* [1927]), the volume *Plays: Winesburg and Others* (1937), and various volumes of memoirs, essays, and journalistic work. His final book publication before his death in 1941 was a collection of prose sketches with photographs, *Home Town* (1940). His extensive publication history notwithstanding, Anderson is mostly remembered for his short stories and especially for *Winesburg, Ohio* as an early modernist masterpiece that offered an assessment of the 1890s and their legacy for the era of the First World War.

The 1890s as a decade of brewing social tensions in the United States and imperialist ambitions in U.S. foreign policy as well as the 1910s as a decade of global war and mass destruction may have induced Anderson's retrospectively rose-tinted characterization of the time, place, and concomitant sense of self in the mid-1870s Midwest, as described in his memoirs: "Fortunate to have been born an American in what may well turn out to be America's happiest period, to have been born poor and in a small town, when community life was intimate and close, to have had to work as a laborer both in factories and on farms, to have known thus from whence comes the food that has nourished my body and with what toil it is produced [. . .]" (R.L. White 1969a, 11). In contrast to this nostalgia for the non-metropolitan lower classes, Anderson depicts the impact of the First World War as being encapsulated within the devastating notion of "lives [. . .] thrown away like worn out shoes" (14) – a notion that he counters with affirming the human "hunger for life, the hunger for the affection that gives life, [which] is the oldest and strongest hunger in the world" (15).

Living in a Chicago boarding house and earning a meagre living in advertising, Anderson characterizes innovative developments in American literature as agents of change, as "a kind of renaissance, in the arts, in literature [. . .]" (317; also see 447). Anderson's focus on what he sees as a general human urge, which he voices in the preface to the memoirs he was working on towards the end of his life, coheres with the ways in which the author depicted his work as a writer of literature. While pro-business journalism and advertisements tend to glorify surfaces and sell ideas to potential customers (see 18), his brand of modernist realism ostensibly grows out of the above-mentioned young woman in knightly armor that resides in the writer's mind. Accordingly, Anderson claims that when he wrote his "first authentic tale" for *Winesburg, Ohio* in a rush of inspiration ("at one sitting" [237]), this story embodied "authentic[ity], [. . .] structure, beauty, strength" and emerged "out of that strange, more real life into which I have so long been trying to penetrate and that is the only real reality" (238). I will come back to the question of what constitutes 'reality' in the next section of this chapter.

Literary historians have suggested quite an array of genre designations for what is most commonly called the short story cycle, and they have suggested various ways in which such a set of short stories published as a book can create coherence (see Birkenstein 2010, 485; Kennedy 1995, vii, ix, xiv; Whalan 2007, 15). While Anderson did not set out to write the stories with such a cycle in mind, he subsequently did regard *Winesburg, Ohio* as initiating a new genre (glibly ignoring Jewett's work and other possible predecessors). During his work on the play adaptation of *Winesburg, Ohio*, Anderson wrote to Arthur Barton in a letter dated 26 November 1932: "Now to me it seems that the theme of the *Winesburg* book, the thing that really makes it a book – curiously holding together from story to story as it does – is just that there is a central theme. The theme is the making of a man out of the actual stuff of life" (Modlin 1984, 153). Whether this idea of an individual's development and, possibly, maturation through encountering 'life' implies a storyline reminiscent of stories of initiation, of the *Bildungsroman*, and of the *Künstlerroman* has remained a mainstay of critical debate about Sherwood Anderson's stories in *Winesburg, Ohio*.

As indicated, Anderson saw himself as an innovator in artistic matters and as a pioneer who smoothed stylistic and thematic paths for younger contemporaries like Hemingway (↗16 Hemingway). In the spirit of critiquing the prevailing American preoccupation with financial success, Anderson applauded the pursuit of new artistic directions. He interpreted the transition from being severely sanctioned for stories that were deemed vile and uncouth to being considered a leading short story writer as the triumph of a modernist willingness to face unpleasant realities over the desire for formulaic stories of social affirmation: "I think that it is now generally recognized that the little book [i.e., *Winesburg, Ohio*] did something of importance. It broke the O. Henry grip, de Maupassant grip. It brought the short story in America into a new relation with life" (R.L. White 1969a, 349–350). Anderson's assessment reflects the work of literary historians who have numbered him among the writers who introduced "the 'new realism'" (Weston 2010, 278).

Beyond contributing to the development of the short story's thematic scope and versatility, Anderson's method of representing the character of George Willard not by focusing on his individuality but rather by depicting his relationships with others (see Anderson qtd. in R.L. White 1969b, xxiv) prefigures a relational approach to narration that Anderson also uses in his memoirs – decades before 'relationality' entered the vocabulary of life-writing studies (Smith and Watson 2010, 216). As Anderson argues, "[m]an cannot think clearly of self, cannot see himself, except through others. The self you seek, the true self you want at last to face, is hidden away" (R.L. White 1969a, 20). The fact that Anderson considers relational structures worth exploring indicates that, in contrast to modernists who disavowed the possibility of understanding one's self and others, the quest for meaning is worth the effort, even though guarantees for a successful completion of the journey cannot be provided.

In the two stories that will be discussed in the second section, Anderson puts his particular spin on long-standing themes and stock characters in the American short story tradition: a young single woman waiting in vain for her prospective husband's return from having made it financially so that marriage becomes possible in "Adventure" and a young man's ostensible maturation into adulthood in "Sophistication." Alice Hindman, the protagonist of "Adventure," finally realizes that she may both spend her life and ultimately experience death while consistently feeling lonely and disconnected from others. George Willard and Helen White in "Sophistication" share a brief moment of essential joy amidst social predicaments focused on frustration and fear, but this moment of release is less understandable to them than to the narrator of the story. Discussed in conjunction with one another, the narratives confirm Anderson's interest in exploring the human psyche relationally. The interlaced features of the two texts also make a case for the writer's attempt to avoid strictly dichotomous depictions of gender differences in favor of addressing a more general state of inhibited communicative abilities and of lacking self-awareness.

2 Close Readings

"Adventure" pits the Olympian perspective of an omniscient narrator against the protagonist's limited range of emotions and rational capabilities. The heterodiegetic and covert voice allows for the narrative strategy of retrospection and, at the closing of the story, for a prospective evaluation of Alice Hindman's inner and outer experience. Most strikingly, the narrator's awareness of social processes in the United States of the 1890s clashes with the protagonist's inability to pursue newly emerging options for women. Stuck in the past, Alice feels destined to be a 27-year-old single woman whose purview centers around the shocking notion "that many people must live and die alone, even in Winesburg" (Anderson 1996a, 64).

Numerous stories depict young women whose male love interest leaves a constricting social environment in order to strike it rich and subsequently come home to get married. These men then either never return or return too late, be it because they fail financially or because they meet other women away from home. In possibly the best-known late-nineteenth-century American short story in this vein – Mary E. Wilkins Freeman's "A New England Nun" (1891) – the patiently waiting betrothed develops eccentric traits and eventually transfers her emotional attachment from her fiancé to unchanging delicate objects and meticulous habits. At the end of the story, she somewhat gladly relinquishes her fiancé to another woman. While she can continue undisturbed in her set ways, the narrator points out the narrowness and the isolation of becoming an oddity within her unchanged social environment. This "New England Nun," who will eke out her existence without close human associates, can do so only because she is a woman of means. Similarly, William Faulkner's "A Rose for Emily" (1930; ↗18 Faulkner), which postdates Anderson's "Adventure," somewhat compensates for the titular character's loneliness by upholding her social status to an extent that preserves at least a modicum of dignity. Even more strongly, class as a factor of allowing a woman's singleness figures prominently in Henry James's "The Beast in the Jungle" (1903; ↗12 James). May Bartram consciously indulges John Marcher's pervading sense of doom – which makes him an Andersonian grotesque before the fact – so fully that she voluntarily lives a life of unrequited love. She does so, however, from a position of financial security and independence from a husband/provider.

In contrast to both Freeman's and Faulkner's spinsters, Anderson's Alice Hindman is caught between her perception of the static social mores of Winesburg and the narrator's hints at possibly changing options for women in U.S. society. Except for a brief moment, Alice cannot imagine exploring such options, especially not without male permission. She lacks the education to consider modified social attitudes towards women as well as the social status to break with traditional gender roles without fearing sanctions. Having fallen in love with Ned Currie, the sixteen-year-old becomes his lover the night before he leaves for Cleveland in the hopes of a journalistic career. Alice regards herself as Ned's designated wife and does not question his contradictory behavior: he first forbids her, for reasons of decency, to follow him to Cleveland and to work for a living before getting married; but then, in the emotional heat of having sex with her in a field before his departure, Ned simply "forgot" (60) his previously stated desire "to protect and care for her" (59–60). The narrator drily depicts both the night of passion and Ned's waning interest in Alice as being directed by forces not subject to the fictional characters' control. For instance, the narrator describes a rather hackneyed atmospheric prelude to making love: "[t]he moon came up and they found themselves unable to talk" (60); when Ned is subsequently "caught up by the life of the city," he meets other women, as a result of which "he forgot Alice" except for brief moments when he sees "the moon shining on the grass as it had shone that night on the meadow by Wine Creek [. . .]" (60). A naturalist sense of determinist subjection to physical urges goes along with Ned's inability to feel deeply and to transform a pleasant memory into a personal commitment.

Complementing the moon as an ultimately meaningless emblem of romance, the story includes further symbols of restricted and clichéd social roles and relations. These symbols refer to im-/mobility, existential needs, and the Bible. Thus, they closely comment on the topic of hopes connected to relocating to a better place in order to achieve prosperity and happiness, and on the claims that material pursuits, social relations, and metaphysical beliefs make on the protagonist.

Anderson's use of symbols exemplifies how he manages a realist surface – that is, representing a small town in the United States of the 1890s in convincing material and social particularity – while also pursuing his expressionist trajectory of externalizing his characters' inner worlds. We learn that Alice's father, who died when she was twenty-two years old, "owned a harness repair shop" (60), whereas her stepfather was "a carriage painter" (59). On the one hand, these details fit into a rural environment in which harnesses were needed both for farm animals and for horses that pulled carriages, which were still the prevailing mode of transportation. On the other hand, the harness as a means of restricting and guiding movement alludes to the social strictures that dominate Alice's thought and behavior. The carriage takes on further significance because such a vehicle transports Alice and Ned to the countryside when they become lovers.

Right after the reference to Alice's father's passing, the narrator mentions that the widowed mother bought a loom and began to earn a living as a weaver. A loom also contains a harness, which directs the course of warp threads during the process of weaving. This cements a symbolic reading of real-life objects, and it possibly implies that the women in Winesburg are complicit in upholding the restrictions imposed on them. This perspective certainly applies to Alice herself who, metonymically linked to her mother, earns a living in a dry goods store, that is, a store that sells textiles and other non-food wares. Just as the horse-drawn carriage was being replaced as a means of transportation, the dry goods store was beginning to lose its importance in the age of the rising department store.

Reminiscent of Freeman's "New England Nun," Alice transfers her affection from human beings to material objects in the course of transitioning from the role of the patiently waiting wife-to-be to that of the emotionally drained and lonely near-hermit. Her inability to "bear to have anyone touch the furniture of her room" (61) precedes her insight that "for her the beauty and freshness of youth had passed" (62) and that her body, by implication, will remain as untouched by others as her furniture. The ability to purchase furniture does not instill pride in having earned money for her own sake. It simply fosters her parsimonious urge to offer Ned her riches and to spend money not for herself but strictly for and with her husband. Significantly, her dreams do not center around a house and other material objects that will stay put. She rather imagines that they "can travel together all over the world" (61) in order to fulfil Ned's (and ostensibly not her) penchant for traveling. Ironically, she does not link his desire for mobility to the fact that his departure from Winesburg went along with abandoning her physically and emotionally.

Just as the symbolic implications of material objects allow the narrator to communicate a reading of Alice's psyche and of her limited perspective on her own life, biblical allusions provide an expressionist subtext to a story about a lonely woman. Early in the story, the narrator describes Alice Hindman as follows: "Her head was large and overshadowed her body. Her shoulders were a little stooped and her hair and eyes brown. She was very quiet but beneath a placid exterior a continual ferment went on" (59). Curiously, her immense head overwhelms her slight frame; the size of her head along with her brown hair and eyes may even evoke similarity to a horse. More importantly, the discrepancy between her serene (and, by implication, socially conformist) appearance and her bubbling internal world makes her resemble a volcano. This resemblance is strengthened in the narrator's depiction of how, when the uncontrolled need to converse takes hold of Alice and Ned, her emotions pierce the calm outer shell: "He became excited and said things he did not intend to say and Alice, betrayed by her desire to have something beautiful come into her rather narrow life, also grew excited. She also talked. The outer crust of her life, all of her natural diffidence and reserve, was torn away and she gave herself over to the emotions of love" (59).

The metaphor of "[t]he outer crust of her life" gains further significance in the section of the story which represents what the narrator terms Alice's "adventure" (63). Having come home from work, she takes off her clothes, listens to the rain outside, and – inexplicable to herself – is overcome by "a strange desire" so that,

> [w]ithout stopping to think of what she intended to do, she ran downstairs through the dark house and out into the rain. As she stood on the little grass plot before the house and felt the cold rain on her body a mad desire to run naked through the streets took possession of her.
>
> She thought that the rain would have some creative and wonderful effect on her body. (63)

During her unpremeditated actions, she is controlled by the seemingly insane impulse "to leap and run, to cry out, to find some other lonely human and embrace him" (63). She approaches a man walking on the sidewalk "without stopping to consider the possible result of her madness" (64) and shouts that "[w]hoever you are, you must wait" (64). When the old man, who is hard of hearing, responds after all, "Alice dropped to the ground and lay trembling. She was so frightened at the thought of what she had done that when the man had gone on his way she did not dare get to her feet, but crawled on hands and knees through the grass to the house" (64). Alice's irrational behavior obviously transgresses social norms of decency to a degree that drives home how her sense of extreme isolation spurs a desperate and hopeless act of establishing human communication. Her retreat back into her house turns her into something like a shy animal that silently escapes from a danger zone.

Anderson's use of verbs describing Alice's actions ("to leap and run, to cry out") and the fact that the 'adventure' occurs in the rain from which Alice expects some magical restoration of her wholeness create the impression that this passage is to evoke Psalm 42. This allusion provides the reader with a specific perspective on the text rather than representing Alice's own understanding of the events. A biblical subtext also makes sense within the realist level of narration, for Alice "joined the Winesburg Methodist Church [. . .] because she had become frightened by the loneliness of her position in life" (62). Thus, becoming a church member fulfills a social and psychological function rather than one of seeking spirituality through religious affiliation.

The opening verse of Psalm 42 presents an analogy between a male deer's thirst and a believer's desire for God: "As the hart panteth after the water brooks, so panteth my soul after thee, O God." Water as life-giving in the physical sense for the animal and in the spiritual sense for the believer recurs in Anderson's story through setting Alice's experience during a rainy night. The biblical text dispels the believer's sense of dejection at possibly having been forgotten (verses 6 and 9) through the confirmation of belief in divine assistance and in God as the source of the soul's health (verse 11). Alice, however, comes to the devastating conclusion that she will remain isolated until her death (see Anderson 1996a, 64).

Naming her Alice Hindman contributes to reading her 'adventure' in light of the biblical text and its apparent lack of significance for the bearer of the telling name. As a hind is a female red deer, the second half of her last name sits uneasily with the first half, if we read '-man' as a male gender marker rather than as a generic reference to a human being. The first name, Alice, is a short form of Adelheidis or Adelaide (literally, "the noble type") which evokes (as its best-known fictional representative in an English-language literary context) the character of Alice in Wonderland who does anything but live in reality.

The closing representation of Alice Hindman's worst fear, to "live and die alone, even in Winesburg" (64), disillusions anyone who might hope that small-town society guarantees social connectedness. The phrase may even be a reference to Anne Brontë's *Agnes Grey* (1847) in which the narrator depicts unfulfilled love and the inability to actively communicate with a potential loved one in the sardonic statements: "the fly must seek another mate, the worm must live and die alone" (Brontë 1998, 139). In Alice's case, however, the prospect of finding a different partner than the one originally desired remains an impossibility, and she finds no comfort in Protestant piety and humility either.

Long before Alice's story ends with her crushing sense of loneliness, she is depicted unmistakably as a character in whose case relational self-definition has gone awry. Even though Alice eventually lets go of her obsession with marrying Ned, she does not develop independent agency but always waits for things to happen to her: "Although she sometimes thought of Ned Currie, she no longer depended on him. Her desire had grown vague. She did not want Ned Currie or any other man. She wanted to be loved, to have something answer the call that was growing louder and louder within her" (Anderson 1996a, 63). She does not realize that only she can perceive her inner voice and act upon it, just as much as she never assumes that she could use her financial independence as the material basis of social agency or individual satisfaction.

The second story to be discussed here, "Sophistication," places George Willard, the young man who works for the Winesburg newspaper and who serves as listener-confidant for various characters in the course of the short story cycle, centerstage. At the same time, the narrative's contemplation of having attained intellectual maturity also extends to Helen White, George's love interest and the local banker's daughter. The story is structured like a frame plus an appended 'pathway' leading out of the frame, as Anderson embeds George and Helen's reunion within the community-focused account of the final afternoon and evening of the Winesburg County Fair during which "[. . .] an American town worked terribly at the task of amusing itself" (Anderson 1996c, 130). Instead of offering a folkloristic and nostalgic outlook on the fair, the description rather bespeaks dirt, grime, and a certain degree of forced jolliness induced by the occasion. Thus, the opening description of the setting includes a sunset whose flaming redness stems from excessive amounts of dust in the air (130). Similar to this utterly de-romanticized sunset, George and Helen enjoy the post-fair silence that envelopes the anything but impressive "half decayed old grandstand" (134). Rather than offering a stage show, this location provides an elevated vantage point for surveying the corn fields and the lights of the town. The noisy late-night entertainment in Winesburg is contrasted with the spell-like stillness (135) shortly before the story ends. In the final two paragraphs, George and Helen first see each other as young adults who deserve mutual "respect" (136); but then their "embarrassment" makes them behave like "excited little animals" that nudge and pull at each other, rolling down the hill and acting silly. Instead of resolving the contradictory references to juvenile and mature features in either of the characters, the narrator closes by claiming that "[m]an or boy, woman or girl, they had for a moment taken hold of the thing that makes the mature life of men and women in the modern world possible" (136). As Helen and George oscillate between dignified notions of respect that link mature adults and childish behavior expressive of their love of life, the enigmatic elixir of survival may be the balance between serious sophistication and simple joys.

The story title raises the question as to whether the narrative contains a definition of sophistication. The narrative does not include strong examples of education-based higher knowledge, but rather reflects on instances of sophistry and disillusioned urbanity. Although the narrator attributes an affirmative, encouraging experience to George and Helen, the enjoyment of this special recipe for survival as adults "in the modern world" is characterized as fleeting ("for a moment"). Related to this short-lived and enigmatic quality, the story as a whole represents sophistication as a central element of maturity and adulthood. The state of having achieved sophistication is, however, unmasked as a primarily subjective (and thus unreliable and volatile) feeling about one's own or another person's development. In contrast to the often socially affirmative trajectory of realist tales, Anderson's expressionist story depicts the outside events realistically but leaves characters' inner worlds opaque and inconclusive.

The narrator introduces George as a nervous and impatient observer who tries to find Helen in the crowd. George's impression of being world-weary ("He felt old and a little tired" [130]) and of having developed the ability critically to consider his own memories exceeds an objective realization of the passing of time: "To his mind his new sense of maturity set him apart, made of him a half-tragic figure" (130). Willard as a journalist/writer knows how to fictionalize his own self. His self-perception as "half-tragic" implies what historian Hayden White would later call "emplotment" (H. White 1978, 84) because he borrows the pattern and concomitant interpretation of his life story from a well-known literary genre, that is, tragedy. As a form associated with classical antiquity, with Shakespeare, and with 'high art' in general, tragedy's emplotment is closely linked to George's sense of self-worth and his potential (or maybe even deserved) status in Winesburg society. At the same time, he only adopts this outlook in part ("half-tragic"), for he is emotionally insecure, lacking in self-awareness, and unable to grapple with his grief: "He wanted someone to understand the feeling that had taken possession of him after his mother's death" (130). Contrary to expectation, George's idea of being sophisticated does not provide a serene self-assuredness but rather unsettles him and makes him yearn for a partner who will provide "understanding" (131), a slippery term in its own right.

In contrast to Alice Hindman in "Adventure," a female character whose degree of insight into her own self does not exceed a shocking awareness of how loneliness impacts her personality and experience, Helen White is depicted as going through the same changes as George (see 131) and as realizing the upsides and downsides of strategic public displays of ostensible sophistication. Having brought one of her college instructors from Cleveland to the county fair, she grows tired of his verbose foregrounding of his educatedness (131). But she does not recognize the faultiness of her own notion of maturation with which she intends to impress George, as indicated by the fact that she ascribes this process to her exposure to city life ("the months she had spent in the city, the going to theatres and the seeing of great crowds wandering in crowded thoroughfares, had changed her profoundly" [132]) and not to internal growth.

George's "half-tragic" self-image and Helen's faulty reasoning then raise suspicions as to whether these two characters have matured at all. Assessing their maturity hinges upon the juxtaposition between their shared memory of a summer evening several years before and the fall evening that they spend in the present depicted in "Sophistication." The narrator dismisses said summer evening as, "when looked at quite sensibly, rather stupidly spent" (132): adolescent George had bragged to Helen about his maturity already then. In retrospect, George is highly embarrassed by this memory of his confidently trumpeted, yet unfounded claims. Bothered by this unpleasant memory, he gets so upset at another man's bragging that he runs, falls, tears his trousers, and decides to confront Helen at her parents' house (133). This sequence of rather childish emotional impulses and the playground-like incident of torn clothing calls the narrator's references to George's newly found maturity into question and reveals what they are: George's wishful thinking about his own development.

During their excursion to the fairground grandstand, Helen serves as a mirror for George's emotional processes (134). Contrary to earlier on, George now interprets the silence surrounding them as indicating "his own insignificance in the scheme of existence" (135). Thus, the bragging or – to use a synonym – grandstanding of years ago is replaced by a shared moment of emotional union and humility: "In that high place in the darkness the two oddly sensitive human atoms held each other tightly and waited. In the mind of each was the same thought. 'I have come to this lonely place and here is this other,' was the substance of the thing felt" (135). Both characters experience mutual understanding as overcoming their existential loneliness – but their companionship, as discussed above, only exists during this brief phase of a "spell" (135) which gives them a sense of mutual "respect" (136) without offering unequivocal or lasting explanations.

The diction in this pivotal scene invites discussion of Anderson's claims regarding his stylistic choices. In his memoirs, he describes his own limited vocabulary and lack of education in the literary arts as an asset in his attempt to capture the ostensible reality of his subject matter. He seemingly recalls the following conversation with himself:

> But should I use in my writing words that were not part of my own everyday speech or my everyday thought?
>
> I did not think so.
>
> "No," I had long been telling myself. "You will have to stay where you have put yourself." There was the language of the streets of American towns and cities, the language of the factories and warehouses where I had worked, of laborers' rooming houses, the saloons, the farms. (White 1969a, 339)

Whether the language of farmers and laborers would have included a metaphorical use of "atoms" for George and Helen or an encapsulation of their being "oddly sensitive" as representing the "substance" of their experience remains a matter of debate. Similarly, Anderson's use of symbols and allusions enriches the superficially homespun characteristics of his vocabulary with modernist subtexts and interpretational leeway. Again, the subtleties of Anderson's style undermine the seemingly plain realism and facilitate insight into the inner emotional workings and, thus, into the Andersonian 'real reality' of his characters.

The shifting implications of "Sophistication" and the contrasting emotional states contribute to not perceiving George and Helen as Andersonian grotesques that cling to a two-dimensional understanding of a single truth which will guide them through life. The gamut between their sense of individual insignificance and isolation, on the one hand, and their enjoyment of mutual understanding, wordless communication, and uninhibited and childlike joy, on the other, indicates the difficulty and complexity of what the narrator sketches as the requirements of the so-called modern world. Whereas "Adventure" closes with a grotesqueness-inducing and overarching fear, "Sophistication" allows at least a fleeting experience of relief.

In a manner that prefigures modernist writers like F. Scott Fitzgerald, Anderson's new realist style serves the purpose of depicting everyday events alongside the emotional upheavals that the characters experience but that they – more often than not – cannot articulate in a socially acceptable or intersubjectively comprehensible manner. Cultural symbols that may still have carried shared depths of meaning have, as in the case of biblical examples, lost their power to comfort individuals like Alice Hindman or to make her feel connected to other characters and to a higher being. The omniscient narrator represents another stylistic throwback to nineteenth-century fiction, but Anderson's narrative voice does not provide clarity and disambiguation. On the contrary, as will become increasingly prevalent in modernist stories, the reader is left to figure out a meaningful path through the narrative.

3 Theoretical Perspectives

Critics have been debating in which ways the stories in *Winesburg, Ohio* participate in well-established genres or whether and how they modify extant forms. Most frequently, such studies refer to narratives about transitioning from adolescence to adulthood and, more broadly, of maturing (or not) as an individual – and, thus, to narratives that incorporate elements of the *Bildungsroman* and *Künstlerroman* traditions. Mark Whalan describes the texts as "elegiac, coming-of-age stories" (Whalan 2007, 2). Linking Anderson and Hemingway (↗16 Hemingway), George Monteiro finds that both follow "a pattern of initiation that leaves the boy more experienced in the raw ways of the world but barely more perceptive" (Monteiro 2010, 226). Focusing on the issue of entering adulthood follows what Anderson himself claimed as the pivot of his stories while he was in the process of writing them. In a November 14, 1916, letter to Waldo Frank, he describes the projected book of narratives as "suggest[ing] the real environment out of which present-day American youth is coming" (Anderson 1996d, 141), thus claiming national significance despite the focus on a fictional small town in the Midwest.

According to Marcia Jacobson, George Willard's inability to mature into adulthood, to find a spouse, and to launch a fruitful writing career precludes reading *Winesburg, Ohio* as coherent with the *Bildungsroman* and *Künstlerroman* traditions (Jacobson 1990, 66, 68). Jennifer J. Smith (2018), on the other hand, argues that Anderson appropriates and modifies the *Bildungsroman* tradition and uses metatextual strategies in a modernist manner. Thus, recent work on Anderson takes into consideration that, again as in Hemingway's case, Anderson does not simply eschew earlier literary styles but rather adapts them to his purposes. For instance, Aaron Ritzenberg (2010) cogently shows that Anderson's "Sophistication" reinterprets the eighteenth-century sentimental trope of physical touch within a modernist context. The story thus shows that the lack of shared notions about the human body, about gender, and about the implications of such physical contact disallows clearly delineated interpretations.

A larger trend in twenty-first-century Anderson scholarship focuses on contextualizing his fiction more succinctly within sociopolitical movements and developments. In his study of Anderson's *Winesburg, Ohio* and Jean Toomer's *Cane*, Whalan proposes that Anderson's stories need to be understood with the First World War, race, gender, demographic change, economic developments, and the Fordist organization of industrial processes in mind (Whalan 2007, 27; see Colton 2018). Equally transcending the claim that Anderson influenced, in particular, Hemingway and Faulkner, scholars have also regarded *Winesburg, Ohio* as having impacted works by non-white American writers throughout the twentieth century (see, for instance, Xu 2010, 439). The upsurge in Anderson criticism since the 1990s has opened up the discussion of his work to perspectives influenced by the reconceptualization of U.S. literary history of the last three decades. Rather than reading his works through his biography and within the context solely of other white male modernist authors, scholars have been exploring the much more complex and variegated landscape of short story writing and reading practices.

4 Bibliography

4.1 Works Cited

Anderson, Sherwood. "Adventure." Winesburg, Ohio: *Authoritative Text, Backgrounds and Contexts, Criticism*. Eds. Charles E. Modlin and Ray Lewis White. New York: Norton, 1996a. 59–64.

Anderson, Sherwood. "The Book of the Grotesque." Winesburg, Ohio: *Authoritative Text, Backgrounds and Contexts, Criticism*. Eds. Charles E. Modlin and Ray Lewis White. New York: Norton, 1996b. 5–7.

Anderson, Sherwood. "Sophistication." Winesburg, Ohio: *Authoritative Text, Backgrounds and Contexts, Criticism*. Eds. Charles E. Modlin and Ray Lewis White. New York: Norton, 1996c. 130–136.

Anderson, Sherwood. "Letters." Winesburg, Ohio: *Authoritative Text, Backgrounds and Contexts, Criticism*. Eds. Charles E. Modlin and Ray Lewis White. New York: Norton, 1996d, 141–151.

Bendixen, Alfred, and James Nagel, eds. *A Companion to the American Short Story*. Malden, MA: Wiley-Blackwell, 2010.

Birkenstein, Jeff. "'Should I Stay or Should I Go?' American Restlessness and the Short-Story Cycle." *A Companion to the American Short Story*. Eds. Alfred Bendixen and James Nagel. Malden: Wiley-Blackwell, 2010. 482–501.

Brontë, Anne. *Agnes Grey*. Oxford: Oxford UP, 1998.

Colton, Aaron. "Metafiction, Literary History, and the Limits of Industrial Identity in *Winesburg, Ohio*." *Studies in American Fiction* 45.1 (Spring 2018): 61–89.

Jacobson, Marcia. "*Winesburg, Ohio* and the Autobiographical Moment." *New Essays on* Winesburg, Ohio. Ed. John W. Crowley. Cambridge: Cambridge UP, 1990. 53–72.

Kennedy, J. Gerald. "Introduction: The American Short Story Sequence: Definitions and Implications." *Modern American Short Story Sequences: Composite Fictions and Fictive Communities*. Ed. J. Gerald Kennedy. Cambridge: Cambridge UP, 1995. vii–xv.

Modlin, Charles E., ed. *Sherwood Anderson: Selected Letters*. Knoxville: U of Tennessee P, 1984.

Monteiro, George. "The Hemingway Story." *A Companion to the American Short Story*. Eds. Alfred Bendixen and James Nagel. Malden: Wiley-Blackwell, 2010. 224–243.

Ritzenberg, Aaron. "Holding on to the Sentimental in *Winesburg, Ohio*." *Modern Fiction Studies* 56.3 (Fall 2010): 496–517.

Smith, Jennifer J. "Locating the Short Story Cycle." *The American Short Story Cycle*. Edinburgh: Edinburgh UP, 2018. 12–36.

Smith, Sidonie, and Julia Watson. *Reading Autobiography: A Guide for Interpreting Life Narratives*. 2nd ed. Minneapolis: University of Minnesota P, 2010.

Stouck, David. "Anderson's Expressionist Art." *New Essays on* Winesburg, Ohio. Ed. John W. Crowley. Cambridge: Cambridge UP, 1990. 27–51.

Weston, Ruth D. "Eudora Welty and the Short Story: Theory and Practice." *A Companion to the American Short Story*. Eds. Alfred Bendixen and James Nagel. Malden: Wiley-Blackwell, 2010. 277–294.

Whalan, Mark. *Race, Manhood, and Modernism in America: The Short Story Cycles of Sherwood Anderson and Jean Toomer*. Knoxville: U of Tennessee P, 2007.

White, Hayden. “The Historical Text as Literary Artifact.” *Tropics of Discourse: Essays in Cultural Criticism*. Baltimore: Johns Hopkins UP, 1978. 81–100.

White, Ray Lewis, ed. *Sherwood Anderson’s Memoirs: A Critical Edition: Newly Edited from the Original Manuscripts*. Chapel Hill: U of North Carolina P, 1969a.

White, Ray Lewis. Introduction. *Sherwood Anderson’s Memoirs: A Critical Edition: Newly Edited from the Original Manuscripts*. Ed. Ray Lewis White. Chapel Hill: U of North Carolina P, 1969b. xv–xxxix.

Xu, Wenying. “The Asian American Short Story.” *A Companion to the American Short Story*. Eds. Alfred Bendixen and James Nagel. Malden: Wiley-Blackwell, 2010. 436–449.

4.2 Further Reading

McKenzie, Precious, ed. *Sherwood Anderson’s* Winesburg, Ohio. Leiden / Boston: Rodopi, 2016.

Nagel, James. “The Twentieth Century: A Period of Innovation and Continuity.” *A Companion to the American Short Story*. Eds. Alfred Bendixen and James Nagel. Malden: Wiley-Blackwell, 2010. 217–223.

Nagy, Peter. “The Woman in the Man: Male Modernism and Cross-Gender Identification in Sherwood Anderson’s *Winesburg, Ohio*.” *College Literature* 45.4 (Fall 2018): 773–800.

Ross K. Tangedal

16 Ernest Hemingway (1899–1961)

Abstract: Two of Ernest Hemingway's Nick Adams stories represent a crucial key to understanding the writer's canon. "Indian Camp" and "The Battler" (both 1925) feature protagonist Nick at two critical points in his youth: first, as a young boy traveling with his physician father to deliver a baby at the Indian Camp up in Michigan; and second, as a young adult on the run heading north, where he meets a disgraced boxer and his pal Bugs in the dark after being thrown from a train. These are stories about the senses, how Nick (at two different points in his youth) experiences life beyond his control. He finds out about life and death in the visceral "Indian Camp," and he escapes a strange, indefinable danger in "The Battler." Hemingway puts us in the middle of the action, where we participate in the story rather than observe it. We experience through Nick Adams the horrors, dangers, and troubles of growing up.

Keywords: Trauma, innocence, race, violence, multisensory

1 Context

Ernest Hemingway's first short story collection, *In Our Time* (1925), gave us "Indian Camp," "The Doctor and the Doctor's Wife," "The End of Something," "The Three-Day Blow," "The Battler," "Cross-Country Snow," and his masterpiece, "Big Two-Hearted River." In *Men Without Women* (1927), he gave us "In Another Country," "The Killers," "Ten Indians," "An Alpine Idyll," and "Now I Lay Me." Finally, in *Winner Take Nothing* (1933) he gave us "The Light of the World," "A Way You'll Never Be," and "Fathers and Sons." These stories feature Nick Adams, Hemingway's first great character who mirrors much of the author's early life and experiences. Other Nick Adams stories, as they are called, have been released in various forms, including "The Last Good Country" and "Summer People," and Don Daiker has argued that "A Very Short Story" from *In Our Time* is in fact a Nick Adams story. Though the stories do provide a mural of Nick's life, specifically his youth and young adulthood, they do not appear in a linear fashion.

The Nick Adams stories distill much of Hemingway's best work into the form he may very well have mastered as a professional writer. Though F. Scott Fitzgerald sold several stories to *The Saturday Evening Post* and *Cosmopolitan* for top dollar by the end of the 1920s (he made as much as $4,000 a story by 1929), Hemingway made a modest sum from his short story output. Many of them were published in *Scribner's Magazine* and *Cosmopolitan*, yet none netted the kind of sums that his friend and rival Fitzgerald was receiving. Hemingway made his money writing novels, and he had two bestsellers (*The Sun Also Rises* and *A Farewell to Arms*) by the time he was

https://doi.org/10.1515/9783110587647-017

thirty. But his stories were his great art form, the form that he recounts most fondly in his Paris memoir, *A Moveable Feast* (1964):

> It was a pleasant café, warm and clean and friendly, and I hung up my old waterproof on the coat rack to dry and put my worn and weathered felt hat on the rack above the bench and ordered a café au lait. The waiter brought it and I took out a notebook from the pocket of the coat and a pencil and started to write. I was writing about up in Michigan and since it was a wild, cold, blowing day it was that sort of day in the story, I had already seen the end of fall come through boyhood, youth and young manhood, and in one place you could write about it better than in another. That was called transplanting yourself, I thought, and it could be as necessary with people as with other sorts of growing things. But in the story the boys were drinking and this made me thirsty and I ordered a rum St. James. This tasted wonderful on the cold day and I kept writing, feeling very well and feeling the good Martinique rum warm me all through my body and my spirit. (Hemingway 1964, 17)

Hemingway is writing "The Three-Day Blow," a Nick Adams story where Nick and his friend Bill, after Nick has broken up with Marjorie in "The End of Something," proceed to get drunk and talk about baseball, literature, and marriage. Though stories like "Indian Camp," "The End of Something," and "Big Two-Hearted River" garner the most attention from critics and readers in *In Our Time*, Hemingway's story about two drunken boys talking shop forms the basis for the writer's essay on remembrance and experiential writing. I find this not only intriguing, but somewhat agitating, given that a story with a role as a bookend to (some would argue) a much better and more transcendent story ("The End of Something") is given such primacy in Hemingway's treatment of writing. Yet he chooses this story to highlight his tangible connection to the place of writing, the feeling of writing, and the act of writing. It is the sensation he is after, and he expects us to accept his tale as if it were our own.

Two Nick Adams stories represent a crucial key to understanding the work of Ernest Hemingway. "Indian Camp" and "The Battler" feature Nick at two critical points in his youth: first, as a young boy traveling with his physician father to deliver a baby at the Indian Camp up in Michigan; and second, as a young adult on the run heading north, where he meets a disgraced boxer and his pal Bugs in the dark after being thrown from a train. I choose to focus on these stories for their attention to the senses, how Nick experiences life and death in "Indian Camp," and how Nick escapes an unknowable danger in "The Battler." Hemingway was a master of control, attempting to get at what he called "the fourth and fifth dimension of writing" (Hemingway 1935, 27) in which the reader became a part of the story, as if we are there in that place in his time, with him as a participant rather than an observer. "Indian Camp" and "The Battler" put readers in the broken-down shanty up north with the Indians and next to the putty-faced fighter by the fire with Nick Adams, who serves as an unwitting guide to the horrors and dangers of growing up in a violent, modern present.

2 Close Readings

When I read the Nick Adams stories, I focus on how old Nick is at any given time. His age means everything, as we justify our interpretations to suit a ten-year-old, a fifteen-year-old, or a thirty-year-old. My reasoning comes from a quote by early Hemingway scholar Carlos Baker, who argued that "Hemingway's aim in the Nick Adams stories is always the aim of an artist. He is deeply interested in the communication of an effect, or several effects together, in such a way as to evoke the deep response of shared human experience" (Baker 1952, 131). Life and death play a major role in "Indian Camp," the second story in his 1925 debut collection, *In Our Time*. According to Robert Paul Lamb, with "Indian Camp" "Hemingway became Hemingway, the story in which he originally brought into play many of the technical innovations of his craft" (Lamb 2013, xvi). Though he revised his original opening to the story (the excised portion was published posthumously as "Three Shots"), Hemingway was able to evolve out of his original indebtedness to Gertrude Stein and Sherwood Anderson, from whose influence his earliest stories sprang. "Indian Camp" features minimal dialogue, with Nick saying very little. His father, Dr. Adams, performs a crude Caesarean section on a pregnant Indian woman, whose husband lays in the bunk above with an axe wound on his foot. The story is about the terror of newness, especially when we are young and inexperienced, but it is also about the inevitability of growing up. Nick Adams is forever changed when he leaves the Indian camp, but what in him is different than when the story began? And how does Hemingway come about communicating an effect that, undoubtedly, affects us?

Judging by dialogue, Nick is roughly ten years old, which immediately shifts our focus to reading the story through the eyes of a child. Nick sits huddled in his father's arms, cold in the morning on the water, as the Indians row them to the camp (Hemingway 1925, 15). It is dark, wet, cold, and early. Once at the shanty, a broken down structure filled with would-be horrors, Nick immediately perceives two things. As they make their way further into the shanty, Hemingway writes, "The room smelled very bad" (16). Nick's sense of smell is bombarded first, followed closely by his sense of hearing. But the focus first is on smell. Just what does this room smell like? And why wouldn't Hemingway use more detailed words for "very bad" like vile, awful, grotesque, or disgusting? It is here that the reader must recognize Nick as the primary filter for the story. A ten-year-old has a limited vocabulary, some may even argue an immature vocabulary, due to their age and lack of experience. What does Nick actually smell? Feces, blood, sweat, tobacco, disease (from the father's gangrenous foot), vaginal fluids, urine, dog hair/feces/urine (several dogs rush out to them as they walk up to the shanty), vomit, etc. A woman has been laboring for forty-eight hours and a man's foot has succumbed to infection from an axe-wound, not to mention the multiple witnesses standing about smoking, eating, drinking, and sweating. How could Nick possibly define the smells of things he has never experienced fully? For him, the room smells "very bad," and for us, the room is Nick's personal hell well before the main trauma visits him. How do you explain labor to a ten-year-old in this circumstance? How would you describe the surroundings? Hemingway's restrictive word choice forces readers to create the environment that Nick inhabits, and we feel for Nick because we have done so.

Nick also hears. The Indian woman is screaming, causing Nick to nervously retreat from his father, who is clearly trying to involve his son in the process, going so far as to call him an "interne" (17). There is concern as to why a father would bring his child to a setting like this, leading us to wonder about Nick's circumstances: Where is his mother? Is he an only child? Where were Nick and Dr. Adams when they were called to come to the camp? The doctor's semi-genuine attempts to include Nick are both necessary (he needs more hands to help with the operation) and circumstantial (Nick is there, so his father may as well try and teach him something). Nick's mind cannot disregard the screams, which removes any enthusiasm from him once his father explains himself: "'Listen to me. What she is going through is called being in labor. The baby wants to be born and she wants it to be born. All her muscles are trying to get the baby born. That is what is happening when she screams'" (16). Nick appears to understand, replying "'I see'" (16). It is a lot for a ten-year-old child to process, let alone understand. But Nick wants to understand. His father tells him what is happening, and he responds with "I know" and "I see," when he neither knows nor sees. Ten-year-olds do not understand, nor should they understand, what being in labor is, yet Nick wants his father to know that he cares, that he is trying. That he is a man.

Just then, the woman screams out again, causing Nick to return to a state of fear: "'Oh, Daddy, can't you give her something to make her stop screaming?'" (16). The doctor's response is both cold and assured, paternal and clinical: "'No, I haven't any anesthetic,' his father said. 'But her screams are not important. I don't hear them because they are not important'" (16). Much has been made of the doctor's response. Is he a cold-hearted misogynist denying the pain of a woman in labor, denying her the space to feel real pain in security rather than dismissal? Or is he a good doctor, blocking out the screams to focus on what will be a major operation, something that must remain emotionless and clinical? The key here is Nick, who, regardless of his father's intentions, cannot understand anything other than the terror of the sound of screaming. "Indian Camp," as I have explained, is about the senses. In this moment, Nick's ears become the primary sense that he associates with the operation.

His father has him hold a basin filled with water, which is used as a place for bloody implements to be placed, like his father's jackknife post-surgery. Nick's reactions during the surgery amplify his sensory experience. The following sequence shows Hemingway's craft of omission, his ability to provide great detail and channel complex emotions with essential word choice rather than extraneous adjectives or adverbs:

> Nick held the basin for his father. It all took a long time. His father picked the baby up and slapped it to make it breathe and handed it to the old woman.
> 'See, it's a boy, Nick,' he said. 'How do you like being an interne?'
> Nick said, 'All right.' He was looking away so as not to see what his father was doing.
> 'There. That gets it,' said his father and put something into the basin.
> Nick didn't look at it.
> 'Now,' his father said, 'there's some stitches to put in. You can watch this or not, Nick, just as you like. I'm going to sew up the incision I made.'
> Nick did not watch. His curiosity had been gone for a long time. (17)

How long is a long time? To Doctor Adams? To the woman? To Nick? A standard c-section performed in ideal circumstances today, with no complications and using modern instruments, can take anywhere from forty-five minutes to one hour (from first incision to final suturing and bandaging). The initial incision to delivery can be anywhere from two minutes to ten minutes, depending on the health of the mother and the health and position of the baby.

Given that the story takes place in the early 1900s, the doctor is using a jack-knife with no instruments or anesthetic, and the woman is being restrained by witnesses rather than straps, we could assume that from initial incision to delivery, the surgery took longer than fifteen or twenty minutes. But does Nick watch what his father is doing? Some scholars note that Nick *watched* the baby being delivered, *watched* his father perform a c-section, and *saw* a woman give birth. The story suggests that Nick did not see anything, his eyes presumably clenched shut while his father cuts open a young woman screaming with child. "He was looking away so as not to see what his father was doing" pertains not only to the post-delivery "it's a boy" moment but also from the moment he takes hold of the basin for his father. He holds it with his eyes shut firmly, though one sense cannot be blocked out. Nick would hear screaming, crying, the sound of the knife cutting flesh, the sound of her head hitting the floor as she passes in and out of consciousness, the first cries of the baby boy, the sound of hands splashing in the basin water, organs moving around in the womb, the placenta being removed and placed in the basin, the silence of an unconscious mother who has yet to meet her son. And what is the "it" of "it all took a long time"? Nick would not use the terms *birth* or *c-section* or *surgery* to describe this, since he has not experienced any of them. For him, the "it" is the multisensory trauma of hearing the screams along with a series of new macabre sounds. The birth did not take a long time; *it all took a long time*, the eternity of listening and imagining rather than seeing the horror (the miracle?) of how life begins. Which is worse?

It would make sense for Hemingway to end the story there. Nick has been traumatized in multiple ways, having yet to visually witness horror. He has smelled it, heard it, and imagined it; he does not have to see it. But he does, and what he sees (what he actually sees) elevates this story from one of youthful inexperience and initiation into one of infinite pathos for a little boy bearing witness to life, death, and mortality in the shadow of his father's approval. Upon finishing the surgery, Dr. Adams jubilantly takes to the father in the upper bunk, hoping to congratulate him. He leans in with his lantern, turning the father's head toward him. His hand comes away wet with blood, the father having slit his throat from ear to ear. Nick, having finished with the basin, stands in the doorway of the kitchen area, seeing clearly the horror of the upper bunk. Having not seen the delivery of the baby, Nick instead bears witness to the result of the father's suicide, a bloody scene akin to the blood of the surgery. Dr. Adams does his best to remove Nick, as the next scene finds them back on the water rowing back to the mainland. He apologizes to his son, agreeing that "It was an awful mess to put you through" (18). The "mess" could be read as any number of things, from the surgery and its instruments to the overall condition of the shanty and its inhabitants, to the upper bunk sagging with fresh blood. We are meant to feel Nick's terror, for the mess of his experience is now our mess. But we have the luxury of experience to mitigate the trauma; Nick is a child with no such luxury.

In the boat, the night now dawn, Nick begins to ask his father a series of questions, though they are not in the order we would expect. Whether he is in shock or not, his questions speak to a psyche complicated (for the first time) by the violent new experience of life and death.

> 'Do ladies always have such a hard time having babies?' Nick asked.
> 'No, that is very, very exceptional.'
> 'Why did he kill himself, Daddy?'
> 'I don't know, Nick. he couldn't stand things, I guess.'
> 'Do many men kill themselves, Daddy?'
> 'Not very many, Nick.'
> 'Do many women?'
> 'Hardly ever.'
> 'Do they ever?'
> 'Oh, yes. They do sometimes.'
> 'Daddy?'
> 'Yes.'
> 'Where did Uncle George go?'
> 'He'll turn up all right.'
> 'Is dying hard, Daddy?'
> 'No, I think it's pretty easy, Nick. It all depends.' (19)

Aside from the questions regarding Uncle George, whose role in the story is the point of some controversy (is he the father of the mother's child?), Nick's questions carry an existential weight that most ten-year-olds would never carry. I find it striking that his first question has nothing to do with the suicide (the trauma he actually *saw*), but rather his initial query has to do with "ladies" having a "hard time having babies." Even in his trauma, he feels the need to impress his father by asking the medical question first. He wants his father to know that he was listening when the doctor told him about the screams, and when the doctor called him an "interne."

There is a deep sadness to Nick's choice of question, in that he suppressed the question about suicide in favor of the question he thought his father would want him to ask. Life and death are difficult concepts for anyone, and Nick displays the curiosity of a child in the face of his trauma. Why does he ask that question first? The suicide is clearly on his mind, as he asks an additional five questions focused on death and dying. Yet, in his youthful desire to say the right thing for his father's approval, Nick shows the early signs of trauma: suppression of the primary experience.

Nick's hand trails in the water, the water warm to the touch in the cool morning. Unlike the beginning of the story, Nick is now seated apart from his father, removed from his strong supportive arms. Nick feels "quite sure that he would never die" (19). When he asks his father if dying is hard his father responds, "'No, I think it's pretty easy, Nick. It all depends.'" Nick's sole experience with death is witnessing the result of a suicide, where he knows that the man took his own life. A ten-year-old would have no real experience with death, and here Nick shows his youthful ignorance, as he can assure himself (perhaps) that he will never die, not *that way*. His father knows that with death, it all depends on the circumstances; the many shades of death are known to an adult: accidents, deliberate killings, suicide, illness, etc. But at this point in his life, Nick thinks that death is something to be controlled; death does not happen to you, rather you control death. What we want for Nick is to process his trauma in a safe space, but in the final image of the story Nick is alone (apart from his father's embrace), his hand feeling warm in the cold water as he convinces himself that death shall have no dominion over him. The multisensory experience of Nick Adams up north with his father becomes more than a coming-of-age tale; in Hemingway's hands, a child is stripped of his innocence by the sights, sounds, smells, and touch of a dark morning at the Indian camp, with a father unable to protect his son from neither the horrors of life nor the mechanics of death.

Paul Smith (1989) traces the composition of "The Battler" first to December 1924, with the manuscript comprised of four handwritten pages and fifteen typed pages with holograph revisions. Once hearing of publisher Horace Liveright's issues with the story "Up in Michigan" and "Mr. and Mrs. Elliot," Hemingway revised the story sometime between early March and March 31st, at which point he returned his signed contract to Liveright (see Smith 1989, 115–116). Hemingway insisted in his letter to Liveright that his new story "makes the book a good deal better. It's about the best I've ever written and gives additional unity to the book as a whole" (Spanier, DeFazio, Trodgon 2013, 295). Of course, the young writer may simply be trumpeting his own work in order to assure his new publisher of the story's strength (given his avowed love for "Up in Michigan" throughout his early career). However, he may also be speaking truthfully, for the story being replaced would certainly attract undo attention for the time period, and Hemingway knew that. He wrote John Dos Passos on 22 April, "This Battler story is a hell of a swell story and better than Up In Mich. Altho I always liked Up In Mich. Altho Some did not. I suppose if it was called Way Out In Iowa Mencken would have published it if the fucking would have been changed to a community corn roast" (323). Five years later, when preparing the collection to be reprinted by Charles Scribner's Sons, Hemingway wrote editor Max Perkins that *In Our Time* was "a damn fine book – Especially The Battler – Cat In the Rain – Out of Season and the first couple of stories – Those three I couldn't write any better now" (Bruccoli 1996, 149). The three main characters in "The Battler" represent different sensations – Nick's isolation, Ad's craziness, and Bugs's hospitality – which create more than a coming-of-age moment for Hemingway's protagonist. Though he certainly comes of age, we must consider exactly what Nick learns, what Nick speculates, and finally, what Nick experiences as a result of his run-in with Ad Francis and his pal Bugs in the woods at night. Further, what does Hemingway intend for us to learn, to speculate, and to experience when reading "The Battler"? Given the compositional relationship to "Up in Michigan" and Hemingway's assertion that the story provides greater unity to the collection, we may read "The Battler" as a controlled, subtly realized, and ultimately effective story of sexual danger and rape as a the "ultimate in evil" Philip Young attempts to locate in many of Hemingway's stories (Young 1966, 39). If so, then perhaps Horace Liveright missed the point of "The Battler" altogether, and frankly, maybe we have been missing the point of the story as well.

The relationship between Bugs and Ad Francis is crucial in recognizing the rape subtext throughout the story. Though some scholars deny that the two are homosexual, I find it compelling and necessary for the story. Bugs's name could be short for "buggery," which points to the physical relationship between the two. Young acknowledges that Bugs's nature makes for some kind of ultimate evil (39), and Kenneth Lynn sees a sadistic quality in Bugs, describing him as one of Hemingway's "dark mother figures" (Lynn 1987, 272). But few discuss that same evil in Ad Francis. Due to his mental instability, vulnerability (after being blackjacked by Bugs), and almost childlike final words ("I got an awful headache, Bugs," Hemingway 1930, 62), Ad is frequently cast as a victim, the victim of too many hits to the head, too many punches, and too much unpleasantness. Bugs tells Nick that Ad is basically a kept man, since his ex-wife "'sends him money'" (61), and once knocked out Ad ceases to control his own story. His past is filtered through Bugs's telling, which we and Nick apparently trust, and the sexual ambiguity surrounding the fighter's wife/manager/sister never gets resolved, even by Bugs. Therefore, it is Bugs, not Ad Francis, who controls the narrative, which leads Nick and readers to form our own judgments about the fighter without his influence. But the way Bugs tells that story, the inflections he uses, and the way Hemingway names Bugs at various moments, points to a textual pattern crucial in recognizing the story's central relationship. Scholars have pointed out this pattern, where Bugs is named in three ways depending on his role at that moment. When named "Bugs," the character is humanized and given a sort of agency; when named "negro," he is a mother-figure to Ad and a hospitable host to Nick; when named "nigger," he performs the role of a wife – he cooks, cleans up, soothes Ad's wounds, and speaks almost lovingly about his companion. George Monteiro (1986) suggests that Bugs sees the beauty in Ad's features when he describes his friend's sister as a "mighty fine woman" (Hemingway 1930, 61), which contrasts with Nick's earlier description of Ad's face as "dead looking in the firelight" (55; Monteiro 1986, 180). Bugs's low, soft, and polite voice tells Nick: "I'd have warned you about him but he seemed to have taken such a liking to you and I thought things were going to be all right" (Hemingway 1930, 62), and he assures Nick that "I hate to have to thump him and it's the only thing to do when he gets started. I have to sort of keep him away from people" (62). The uncovering of Bugs and Ad's relationship as potentially more than platonic leads to a more complex reading of the story, for as Michael Reynolds argues, "dark things happened on that peninsula in Hemingway's imagination" (Reynolds 1999, 280).

One of those "dark things" may appear at the beginning of the story. The exchange between Ad Francis and Nick early on in "The Battler" extends the almost dreamlike quality of the story. Nick comes upon Ad with his own wounds, a black eye and ripped pants. After some small talk, Ad asks, "'You're a tough one aren't you?' 'No,' Nick answered. 'All you kids are tough.' 'You got to be tough.' Nick said. 'That's what I said.' The man looked at Nick and smiled. In the firelight Nick saw that his face was misshapen. His nose was sunken, his eyes were slits, he had queer-shaped lips. Nick did not perceive all this at once, he only saw the man's face was queerly formed and mutilated. It was like putty in color. Dead looking in the firelight" (Hemingway 1930, 55). Nick finds grotesquery in Ad's smiling face, a smiling face "taking a liking" to Nick. All these kids are tough, kids like Nick, and the kids he used to fight, and the kids he met in prison. He asks an embarrassed Nick, "Don't you like my pan?" before taking off his cap. "He had only one ear. It was thickened and tight against the side of his head. Where the other ear should have been there was a stump" (55). This makes Nick "a little sick" (55). Ad appears to be bragging to Nick, showing him his toughness, his wounds, and his ability to "take it.": "'I could take it. [. . .] Don't you think I could take it?" Nick's "You bet!" is more survivalist than laudatory. He is embarrassed, afraid, and sick to his stomach. Ad let's Nick know that he's crazy. "'You're all right,' [Nick] said. 'No I'm not. I'm crazy. Listen, you even been crazy?' 'No,' Nick said, 'how does it get you?' 'I don't know,' Ad said. 'When you got it you don't know about it'" (56). This craziness is open to interpretation. Philip Young traces Ad Francis to the real-life Ad Wolgast, the "Michigan Bearcat" or "Michigan Wildcat" and lightweight champion of the world who was judged legally insane by 1917 due to too many hits to the head (Young 1966, 37). Ad's craziness certainly takes a cue from the real Ad's nervous breakdown reported in the *Leavenworth Weekly Times* on 22 February 1917 (1). However, that same craziness may be a cue to his relationship with Bugs. Judging by Bugs's account of Ad's relationship with his wife, his craziness may be his subdued homosexuality. He brags "'they all bust their hands on me [. . .] they couldn't hurt me'" (Hemingway 1930, 56), then looks at

Nick. This look, along with his defensive state, leads me to believe that Ad was badly injured in prison, and those injuries certainly go deeper than beatings. He and Bugs met in prison, he for "busting people all the time" and Bugs for "cuttin' a man" (61). Is his craziness a result of his own sexual abuse, his role as a homosexual in a culture (sport) unable to fathom such an abomination? If readers are willing to interpret the relationship between Bugs and Ad as more than platonic, than they should be willing to see non-boxing abuse as part of Ad's own history.

But to this point nothing has truly cued the possibility of an attack. Nick is embarrassed, sick, and scared, but Ad has yet to make any moves. However, one act in particular stands out, and it occurs immediately prior to Bugs's appearance. He tells Nick that he is Ad Francis, which Nick begrudgingly believes. "'Come on,' the man took hold of his hand. 'Take hold of my wrist. Put your fingers there.' The little man's wrist was thick and the muscles bulged above the bone. Nick felt the slow pumping under his fingers" (56). Nick literally feels Ad's power, his blood and his skin. "Feeling the slow hard throb under his fingers Nick started to count. He heard the little man counting slowly, one, two, three, four, five, and on – aloud. 'Sixty,' Ad finished. 'That's a minute. What did you make it?' 'Forty,' Nick said. 'That's right,' Ad said happily. 'She never speeds up'" (57). Just then Bugs appears. What happens if he does not show up? Nick goes along with Ad's odd requests in order to appease him. However, what does he stand to lose if he does not follow Ad's orders? After finishing the story, we know several things: First, Ad has money, therefore he wouldn't plan on robbing Nick, who has nothing anyway. Second, though we determine Ad is mentally unstable, he does not feel insulted by Nick until after Bugs arrives, which leads us to read Ad as relatively controlled and calm during the early exchange, not wild-eyed and crazy. Third, and most importantly, Ad has done this all before. Bugs tells Nick after blackjacking Ad, "'He won't remember nothing of it. I have to do it to change him when he gets that way,'" and he's seen Ad like this "'plenty of times before'" (60). So many questions revolve around these responses. How many times has Ad been like this, and where were they? Who was with them? What does the result of Ad's craziness look like when Bugs is not there to tap him on the head with his blackjack? Does Ad take Nick's arm after the second count and attack him? Has Bugs stopped Ad from attacking and assaulting others before? He does tell Nick to leave before Ad wakes up, "'I don't like to not be hospitable, but it might disturb him back again to see you. I hate to have to thump him and it's the only thing to do when he gets started. I have to sort of keep him away from people'" (62). Young and Lynn see this as completely self-serving on Bugs's part, since he has a financial stake in Ad not being jailed, but is Bugs also protecting Ad from his own urges, the darkness, the ultimate evil, the unpleasantness?

Early on Hemingway writes that Nick "must get to somewhere" after being thrown from the train (54); and near the end "he found that he had a ham sandwich in his hand and put it in his pocket" (60). He doesn't even recognize the sandwich in his own hand until he is walking away. We, along with Nick, do not know what we have seen, what we have experienced, and what we have escaped. Liveright approved of the story, apparently condoning the sister/wife incest subtext in "The Battler" over the rape subtext in "Up in Michigan." But what if he missed the subtext? It is possible that Ad Francis, in his damaged state due to past events, had planned on raping Nick once he got the young man to take hold of his wrist. This is hard to prove, since Bugs comes in at the opportune moment, effectively saving Nick from something. From the pieces Hemingway gives us, it is clear that "The Battler" represents a dreamlike state-of-being in Nick's continuing development. Though the fire illuminates Ad's pugilist features, it also shines a light on the danger from which Nick escaped. We can never really know what Nick Adams escaped from that night in upper Michigan near Mancelona, but maybe it is best that we keep asking questions about the dark places of Hemingway's imagination.

3 Theoretical Perspectives

Early interpretations of the Nick Adams stories focused partially on their autobiographical nature. Carlos Baker compared Hemingway's experimentation with Nick Adams to Henry Adams's *The Education of Henry Adams*, suggesting that the stories may even be arranged under the title "The Education of Nicholas Adams" (Baker 1952, 128). Baker stressed that Hemingway "was always prepared to invent people and circumstances, to choose backgrounds which would throw his people in three-dimensional relief, and to employ as symbols those elements of the physical setting, which could be psychologically justified by the time and place he was writing about" (129–130). Philip Young wrote extensively on the similarities between Nick Adams and Mark Twain's *Huckleberry Finn* (↗9 Twain), noting that "the parallel is pretty complete. For twenty-five years Hemingway went on telling Twain's story, as it was lived by Twain's protagonist, as both heroes came up against the vices and false values of their societies, were struck down and limped off [. . .] where Twain found his boy too complex and let him go, Hemingway exploited the condition and raised him to complicated manhood" (Young 1966, 234). Verna Kale believes "that neither Hemingway's greatest fans nor his harshest critics can absolve him of having simply set his own life into prose is a testament to his success in creating emotional truths too real to be disbelieved" (Kale 2016, 51).

While Young noted, "the adventures of the generic Nick Adams are the adventures of Huckleberry Finn in our time" (Young 1966, 234), Carlos Baker argued that "the story of Adams is a presented vision of our time" (Baker 152, 131). Baker figures that "to record for posterity another chapter in his own fictional autobiography does not interest [Hemingway] at all" (131). With regard to "Indian Camp," Robert Paul Lamb lists the many elements at play in the story:

> childbirth, father-son relationships, hunting/fishing, northern Michigan, physical mutilation, competence, and powerlessness. They came together in the mind of a writer whose great and pervasive theme was man's efforts to control his fate, efforts marked by small successes and ultimate failure. The emotions evoked by such thematic concerns – exhilaration, hope, fear, anxiety, and despair – colored the terrain of his imagination. These emotions can be found at the start in this, his first great story. (Lamb 2013, 10)

Here we see the autobiographical and the artistic merge into a unique evocation of the senses, a technique Hemingway would use in all his best work. Aside from autobiography, Hemingway's attention to the Native American population provides fertile ground for analysis. Amy Strong's excellent analysis of race "serves as a metaphor for the ways Hemingway's story has been read; readers have also trailed their hands in the river of forgetfulness, ignoring the Indians' role not only in this story, but in the making of American identity" (Strong 2008, 17). Controversy also exists over whether Dr. Adams should be culpable in the trauma visited upon his son. Lisa Tyler argues that the story, "is a work about men's responses to violence and their capacity for empathy," where young Nick chooses not to empathize with the Indian woman, but rather his father's removed approach, which "damns him to a cold isolation and instrumental rationality in which other human beings are regarded as objects rather than subjects in their own right" (Tyler 2006, 39). Donald Daiker disagrees, calling for readers to instead sympathize with Dr. Adams, who patterns for his son, "the power of compassion and love and, above all, the example of grace under pressure" (Daiker 2016, 55). Within this disagreement is another opportunity for readers to engage with the text. Is the doctor a good doctor and, if they are relatable, is he a good father? And how would a ten-year old see his doctor father – as a doctor, or as a father? Our reading of Nick's responses to his father provides us with ample opportunities to relate to him, Hemingway's chief goal.

A similar balance is struck between the senses and reality in "The Battler." Frederic Svoboda contends that Hemingway created a "mood of wilderness isolation" in the story (Svoboda 1995, 20), and Joseph Flora notes Nick's "metaphysical isolation [. . .] alone, with much of the day spent" (Flora 1982, 85). This isolation can be compared to Jim Gilmore and Liz Coates in "Up in Michigan," especially when Liz closes the story looking over the water. In fact, Philip Young classifies "The Battler" as an isolated Hemingway story in its own right: "People who complain about the sordid nature of many of Hemingway's stories seldom if ever cite this one, perhaps because the unpleasantness is more in the undertones and in things not said than in the outer events which, though not happy, are not entirely extraordinary in our time" (Young 1966, 36). He then concludes that "if the subtleties are drawn out and examined, 'The Battler' is as unpleasant as anything its author ever wrote" (36). The unpleasantness refers to Bugs's characterization of Ad Francis's relationship with his wife/sister/manager, but Young correctly classifies the story as much darker than meets the eye. This darkness takes on a variety of manifestations. Young argues that Bugs's alleged homosexuality is a clearly malevolent force, though Nick fails to fully recognize it (39). Flora disagrees, seeing no clear homosexual context, but rather a “nonsexual camaraderie” which furthers Nick’s notion that “even if the world is hostile and fierce there are caring hands and caring voices” (Flora 1982, 92–93). Smith concludes that “the episodic narrative of Nick Adams needed a predictive moment when the callow kid who claims he is tough can catch a portentous glimpse of someone who has been so beaten his memory goes awry” (Smith 1989, 120). These three manifestations – sexual, social, and mental – represent much of the interpretive criticism of the story, and each can be used to recognize the inherent violence within. Regardless, one overriding constant in the interpretations of “The Battler” settles on the unpleasantness, the mystery, and the danger at the heart of one of Hemingway’s darkest stories.

In *The Sun Also Rises* (1926), Jake Barnes recalls how “at noon of Sunday, the 6th of July, the fiesta exploded. There is no other way to describe it” (Hemingway 1926, 156), imbuing one word, “exploded,” with all the sights, sounds, tastes, and feelings of a colorful fiesta. In *A Farewell to Arms* (1929), Frederic Henry mentions that “abstract words such as glory, honor, courage, or hallow were obscene beside the concrete names of villages, the numbers of roads, the names of rivers, the numbers of regiments and the dates” (Hemingway 1929, 185). Those places were imbued with the sense of dignity, a sense readers must understand in order to understand the novel. In *Green Hills of Africa* (1935), Hemingway argues that “nothing else matters” other than the honest attempt to place in readers a feeling of lived experience, a sense of something that will not “go bad afterwards” (Hemingway 1935, 27). For Ernest Hemingway, the way to a reader was multisensory, and he honed his skills first in stories like “Indian Camp” and “The Battler.” With his short stories, he was able to evoke particular emotions in ways that few writers before or since could, thanks in large part to his dedication to craft and the ability to see clearly.

4 Bibliography

4.1 Works Cited

“Ad Wolgast Suffering from Nervous Breakdown,” https://www.newspapers.com/newspage-/94659110/. *Leavenworth Weekly Times*. 22 February 1917. 1. (9 July 2016).

Baker, Carlos. *Hemingway: The Writer as Artist*. Princeton, NJ: Princeton UP, 1952.

Bruccoli, Matthew J., ed., with the assistance of Robert W. Trogdon. *The Only Thing That Counts: The Ernest Hemingway-Maxwell Perkins Correspondence*. Columbia, South Carolina: U of South Carolina P, 1996.

Daiker, Donald A. “In Defense of Hemingway’s Doctor Adams: The Case for ‘Indian Camp.’” *Hemingway Review* 35.2 (2016): 55–69.

Flora, Joseph M. *Hemingway’s Nick Adams*. Baton Rouge, LA: Louisiana State UP, 1982.

Hemingway, Ernest. *A Farewell to Arms*. New York: Scribner’s, 1929.

Hemingway, Ernest. *Green Hills of Africa*. New York: Scribner’s, 1935.

Hemingway, Ernest. *In Our Time*. 1925. New York: Scribner's, 1930.
Hemingway, Ernest. *A Moveable Feast*. 1964. Restored Edition. New York: Scribner's, 2011.
Hemingway, Ernest. *The Sun Also Rises*. New York: Scribner's, 1926.
Kale, Verna. *Ernest Hemingway*. London: Reaktion Books, 2016.
Lamb, Robert Paul. *The Hemingway Short Story: A Study in Craft for Writers and Readers*. Baton Rouge, LA: Louisiana State UP, 2013.
Lynn, Kenneth S. *Hemingway*. New York: Fawcett Columbine, 1987.
Monteiro, George. "This Is My Pal Bugs: Ernest Hemingway's 'The Battler.'" *Studies in Short Fiction*, 23.2 (1986): 179–183.
Reynolds, Michael S. *Hemingway: The Paris Years*. 1989. New York: W.W. Norton & Company, 1999.
Smith, Paul. *A Reader's Guide to the Short Stories of Ernest Hemingway*. Boston: G.K. Hall & Company, 1989.
Spanier, Sandra, Albert J. Defazio, and Robert W. Trogdon, eds. *The Letters of Ernest Hemingway Volume 2: 1923–1925*. New York: Cambridge UP, 2013.
Strong, Amy L. *Race and Identity in Hemingway's Fiction*. New York: Palgrave Macmillan, 2008.
Svoboda, Frederic J. "False Wilderness: Northern Michigan as Created in the Nick Adams Stories." *Hemingway: Up in Michigan Perspectives*. Eds. Frederic J. Svoboda and Joseph J. Waldmeir. East Lansing: Michigan State UP, 1995.
Tyler, Lisa. "'Dangerous Families' and 'Intimate Harm' in Hemingway's 'Indian Camp.'" *Texas Studies in Literature and Language* 48.1 (2006): 37–53.
Young, Philip. *Ernest Hemingway: A Reconsideration*. University College: Pennsylvania State UP, 1966.

4.2 Further Reading

Dudley, Marc Kevin. *Hemingway, Race, and Art: Bloodlines and the Color Line*. Kent: Kent State UP, 2012.
Moddelmog, Debra A., and Suzanne Del Gizzo, eds. *Ernest Hemingway in Context*. New York: Cambridge UP, 2013.
Paul, Steve, Gail Sinclair, and Steven Trout, eds. *War + Ink: New Perspectives on Ernest Hemingway's Early Life and Writings*. Kent: Kent State UP, 2014.
Pottle, Russ. "Hemingway and *The Journal of the American Medical Association*: Gangrene, Shock, and Suicide in 'Indian Camp.'" *Hemingway Review* 35.1 (2015): 35–52.
Reynolds, Michael S. *Critical Essays on Ernest Hemingway's In Our Time*. Boston: G.K. Hall, 1983.
Trogdon, Robert W. *The Lousy Racket: Hemingway, Scribners, and the Business of Literature*. Kent: Kent State UP, 2007.
Wagner-Martin, Linda. *Hemingway's Wars: Private and Public Battles*. Columbia: U of Missouri P, 2017.

Erik Redling

17 Zora Neale Hurston (1891–1960)

Abstract: Zora Neale Hurston is generally known for her novel *Their Eyes Were Watching God* (1937) and the frequently anthologized short story “Sweat” (1926) which both support the dominant interpretive practice of reading Hurston’s work through the dual lens of race and gender. The two rarely examined short prose texts selected for the “Closed Reading” section, however, depart from the prevailing conventional explanatory pattern of Hurston’s oeuvre. While her first published short story, “John Redding Goes to Sea,” focuses on a young artist’s discovery of the resources of imagination, the second episodic text, “The Eatonville Anthology” (1926), consisting of fourteen very short humorous stories, participates in the Classical literary tradition of character sketches established by Theophrastus’s book *Characters*. Both short prose texts support readings that transcend the dominant interpretive pattern that limits her work to interpretations along the lines of identity categories. Both can suggest a broadened view of Hurston as a citizen of a global republic of letters.

Keywords: Imagination, metaphor, allegory, romanticism, sketch, republic of letters

1 Context

Zora Neale Hurston, a prominent figure of the Harlem Renaissance, is best known for her novel *Their Eyes Were Watching God* (1937), which is generally regarded as her masterpiece. Besides *Their Eyes*, Hurston’s extensive oeuvre includes three other novels, two books of folklore, an autobiography, various plays and poems, numerous essays and articles in magazines and newspapers, and, published between 1921 and 1951, more than twenty short stories (see Cronin 1998, 1). Yet, despite her substantial body of work and the central role she played in the Harlem Renaissance, she was largely forgotten in the 1950s. As M. Genevieve West writes, “by the time of her death in 1960,” she had become “a literary has-been, a marginalized, out-of-print outsider” (West 2005, 1).

Hurston’s rediscovery in the 1970s is closely associated with the Pulitzer-prizewinning writer Alice Walker (↗27 Walker), who, in her endeavor to establish a tradition of African American women writers, adopted Hurston as a “model” for her own life and work, and memorably said of *Their Eyes* that “*There is no book more important to me than this one*” (Walker 1984, 7, 12–13, 86). In her essay “Looking for Zora” (originally published as “In Search of Zora Neale Hurston” in the March 1975 issue of *Ms.* magazine), Walker recounts her trip to Florida in 1973 to locate Hurston’s unmarked grave in the segregated Garden of Heavenly Rest cemetery at Fort Pierce and, on her supposed burial site, arranged for a headstone to be placed with the following inscription:

https://doi.org/10.1515/9783110587647-018

ZORA NEALE HURSTON
"A GENIUS OF THE SOUTH"
1901 — — — 1960
NOVELIST, FOLKLORIST
ANTHROPOLOGIST

(Hemenway 1980, 3)

This essay and several others on Hurston in Walker's collection *In Search of Our Mothers' Gardens*, together with essays by Helen Mary Washington, June Jordan, and other Black feminist critics, along with Robert Hemenway's seminal biography *Zora Neale Hurston: A Literary History* (1980), sparked a fresh interest in Hurston's life and writings and gained her canonical status in the 1980s (see West 2005, 2; ↗2 Canon Formation). In his influential study *The Signifying Monkey: A Theory of African-American Literary Criticism* (1988), Henry Louis Gates, Jr. states that "Zora Neale Hurston is the first writer that our generation of black and feminist critics has brought into the canon, or perhaps I should say the canons. For Hurston is now a cardinal figure in the Afro-American canon, the feminist canon, and the canon of American fiction" (Gates 1988, 180).

Hurston's life and works inspired a younger generation of African American women writers including Alice Walker, Gloria Naylor, and Toni Morrison and attracted much critical attention over the past five decades. Her books are all in print. A volume of her short stories and a two-volume Library-of-America edition of her writings appeared in 1995, and the intermittent rediscovery of previously published but mostly neglected texts and the publication of archival material confirm a continuing keen interest in Hurston. A 2010 special issue of *Amerikastudien/American Studies*, edited by Glenda R. Carpio and Werner Sollors, reprinted five early short stories by Hurston that originally appeared in the *Pittsburgh Courier* (four in 1927, one in 1933), which, as the guest editors point out, "provide fuller insight into Hurston's engagement with urban black life in Harlem" (Carpio and Sollors 2010, 557); and Hurston's manuscript on the African ex-slave Oluale Kossola (a.k.a. Cudjo Lewis), rejected by contemporary publishers, was finally published as *Barracoon: The Story of the Last Slave* in 2018. Both of these recent publications, together with several others (e.g., *Zora Neale Hurston: Collected Plays*, 2008), shed new light on Hurston's multifaceted work.

Hurston was born on January 7, 1891 (rather than in 1901 as the headstone indicates; Hurston provided different birth dates in various phases of her life) in Notasulga, Alabama, and her family moved three years later to Eatonville, Florida, an "all-black" incorporated community in the United States, where she was a student at the local Hungerford School between the ages of nine to thirteen (Wall 1995, 1013, 1014; Boyd 2003, 17, 19, 38). After her mother Lucy Potts Hurston died on September 18, 1904, Hurston attended schools in Jacksonville (1904–1905) and Baltimore (1917–1918) before she moved to Washington, D.C. in the summer of 1918 and found work as a waitress (Wall 1995, 1014–1015; Boyd 2003, 49, 50, 51, 76–78). In December Hurston enrolled at

Howard Academy and graduating from that prep school in May 1919 entered Howard University in the fall, taking courses in Greek, English, and Spanish while working as a manicurist to support her college education (Wall 1995, 1015; Boyd 2003, 80–81). Her literary ambitions received a vital impulse when she became a member of the university's literary club called the Stylus, which was co-founded by Alain Locke and his colleague Montgomery Gregory. Her first two literary publications, the poem "O Night" and the short story "John Redding Goes to Sea," appeared in the May 1921 issue of the club's magazine *The Stylus* (Hemenway 1980, 19; Wall 1995, 1015; Boyd 2003, 84–85). Hurston's second published short story, "Drenched in Light," debuted to a national audience in the December 1924 issue of the journal *Opportunity* and, following the counsel of its editor Charles S. Johnson, she moved to New York in January 1925, "where the New Negro movement was in full flower" (Boyd 2003, 92, 93; Wall 1995, 1016).

The "Negro Renaissance," as Alain Locke in the signal collection *The New Negro: An Interpretation* (1925; e.g., Locke 1925a, xi) called what we know today as the Harlem Renaissance (1918–1937), "was the first significant literary and cultural movement in African American history" (Bernard 2007, 28). Featuring "an unusual outburst of creative expression" (Locke 1925a, xi) in the African American arts (e.g., literature, music, and visual arts), the New Negro movement derived its name from Locke's manifesto "The New Negro" (published in the eponymously titled anthology) in which he distinguishes between generational types: "The Old Negro [. . .] was a creature of moral debate and historical controversy. His has been a stock figure perpetuated as an historical fiction partly in innocent sentimentalism, partly in deliberate reactionism" (Locke 1925b, 3). Yet, as Locke declares, "[t]he day of 'aunties,' 'uncles' and 'mammies' is [. . .] gone. Uncle Tom and Sambo have passed on" (5). Diametrically opposed to such "negative stereotypes" (Bernard 2007, 29), he posits the "New Negro" who evinced characteristics such as "renewed self-respect and self-dependence" and creative "self-expression" (Locke 1925b, 4, 5). "The world of African American arts of the 1920s," Emily Bernard writes, "was intoxicated with the idea that it had invented itself, not only in terms of its creative ambitions, but also as a locus of a new black identity, namely the New Negro. Black intellectuals of the Harlem Renaissance invested in the ideology of the New Negro all of their ambitions to liberate black people – psychologically, socially, and even politically – from the denigration of the slave past" (Bernard 2007, 29–30). Hurston's move to New York brought her into close contact with Black artists and intellectuals of the burgeoning Harlem Renaissance.

Her nascent literary career received a significant boost when she won second-place awards for her short story "Spunk" and her play *Color Struck* in a 1925 *Opportunity* literary contest, and at the magazine's awards dinner in New York, met such luminaries as the co-founder and trustee of Barnard College, Annie Nathan Meyer, and the prizewinners in poetry, Langston Hughes and Countee Cullen (Hemenway 1980, 20–21; Wall 1995, 1016; Boyd 2003, 96–97). Meyer helped Hurston obtain a scholarship to Barnard College where she enrolled in the fall as "Barnard's only black student at the time" (Hemenway 1980, 21). After her first term, she studied anthropology under the guidance of the distinguished cultural anthropologist Franz Boas along with his colleagues Melville J. Herskovits, Ruth Benedict, and Gladys Reichardt; she received a BA degree in May 1928 (Hemenway 1980, 21; Wall 1995, 1016, 1018; Boyd 2003, 100–102).

Hurston's Harlem years were characterized by a flurry of literary activities. In 1926, for instance, she collaborated with Hughes, Wallace Thurman, Bruce Nugent, Gwendolyn Bennett, and Aaron Douglas to create a new magazine entitled *Fire!!* (subtitled "A Quarterly Devoted to the Younger Negro Artists") to which she contributed the play *Color Struck* and the short story "Sweat" (Hemenway 1980, 43; Wall 1995, 1016–1017; Boyd 2003, 122–123). Although *Fire!!* was planned as a quarterly, its first and only issue appeared in November 1926. In the same year Hurston published "The Eatonville Anthology" in three installments in *The Messenger*. Hurston also worked with Hughes on a play, *Mule Bone: A Comedy of Negro Life*, which was neither staged nor published in its entirety during their lifetimes (written mostly in 1930, it was posthumously published in 1991). Friction from the collaboration on *Mule Bone* effectively ended their friendship (Hemenway 1980, 136–137; Wall 1995, 1019; Boyd 2003, 198–217).

Between February and March 1927, and with the support of Boas, Hurston set out on the first of several extended field trips to the South (with stops in Eatonville and Jacksonville in Florida) to collect African American folklore. Sponsored by benefactor Charlotte Osgood Mason, a wealthy white New Yorker who subsidized several Harlem Renaissance figures, and whom Hurston addressed as "Godmother" in her letters, Hurston embarked on further folklore-gathering expeditions between December 1927 and March 1931, which took her not only to Eatonville but also to a lumber camp in Polk County, Florida, to New Orleans, and to Nassau in the Bahamas, and where she collected a wealth of folktales, work and blues songs, hoodoo remedies, and other ethnographic materials (Wall 1995, 1017–1019; Boyd 2003, 156–197).

Hurston attained the peak of her literary career in the 1930s and early 1940s. She published her first novel, *Jonah's Gourd Vine*, in 1934 and turned her rich folklore discoveries into the widely acclaimed *Mules and Men* (1935). A Guggenheim fellowship granted to Hurston in March 1936 enabled her to conduct research in Jamaica and Haiti, where, in late 1936, she wrote her second novel *Their Eyes Were Watching God* (published in 1937; Boyd 2003, 286–287, 293–295). She subsequently converted the ethnographic material she collected in the Caribbean into the book *Tell My Horse: Voodoo and Life in Haiti and Jamaica* in 1938. She then published her third novel *Moses, Man of the Mountain* in 1939, her autobiography *Dust Tracks on a Road* in 1942, and her last novel, *Seraph on the Suwanee*, written largely during her stay in Honduras, in 1948 (Boyd 2003, 385–386).

Hurston's literary career faded by the late 1940s. Shortly before the publication of *Seraph on the Suwanee*, she was charged with child molestation on September 13, 1948 (Boyd 2003, 387). "The charges," as Valerie Boyd stresses, "were groundless and Hurston categorically denied them" (387). Though the charges were dismissed on March 1949 (400), the experience slowed her progress and threatened her reputation. Living in Miami at the time, Hurston continued to work on a book-length manuscript entitled "The Lives of Barney Turk," but, like drafts of other prospective novels that she worked on in the 1950s, "The Golden Bench of God" and "Herod the Great," it was not published (Hemenway 1980, 328; Boyd 2003, 407–408, 412, 422–423). In her final decade, she wrote several political articles for newspapers and magazines (e.g., "What White Publishers Won't Print," 1950, and "Mourner's Bench, Communist Line: Why the Negro Won't Buy Communism," 1951), and took several jobs, working as a maid in Miami, a librarian in Cocoa, and a substitute teacher in Fort Pierce. Impoverished and virtually unknown, she died in a segregated nursing home in Fort Pierce on January 28, 1960 (Hemenway 1980, 325, 328; Wall 1995, 1028–1032; Boyd 2003, 403, 431).

The prevailing context of Hurston's rediscovery was the convergence of the feminist and the Black Civil Rights Movement in the 1970s, which manifested itself in gender- and race-dominated readings of Hurston's work, the bulk of criticism devoted to *Their Eyes Were Watching God* (1937). A few examples of the criticism and theory devoted to Hurston's work over the last five decades suffice to illustrate the tendency to examine her writings (most notably *Their Eyes*) through the lenses of race and gender. Three early representative samples are Mary Helen Washington's signal essay "The Black Woman's Search for Identity" (1972), S. Jay Walker's article "Zora Neale Hurston's *Their Eyes Were Watching God*: Black Novel of Sexism" (1974), and Missy Dehn Kubitschek's "'Tuh De Horizon and Back': The Female Quest in *Their Eyes Were Watching God*" (1983). Washington argues that "[t]he Black frame of reference is achieved in Ms. Hurston's novel [*Their Eyes Were Watching God*] in three ways: 1) the language is the authentic dialect of Black rural life; 2) the characters are firmly rooted in Black folk culture; and 3) Janie's search for identity is an integral part of her search for Blackness" (Washington 1972, 68–69). Kubitschek attests that Washington's essay "has led the way in reasserting the centrality of Janie's search for identity and her connections to her community" (Kubitschek 1983, 109). Drawing on the model of a hero's journey that Joseph Campbell delineated in his study *The Hero with a Thousand Faces* (1949), Kubitschek proposes that the novel's protagonist, Janie Crawford, is a "questing heroine" (111) who experiences "communal acceptance" (114). "Through Janie," Kubitschek writes, "Hurston merges the quest pattern with the Afro-American call-and-response to form a new experience, a group quest or ascent" (114). Walker claims that the novel "deals far more extensively with sexism, the struggle of a woman to be regarded as a person in a male-dominated society, than [with] racism, the struggle of blacks to be regarded as persons in a white-dominated society" (Walker 1974, 520).

Developing the notion of the "quest," Henry Louis Gates, Jr. regards Janie Crawford's search for identity also as a "quest to become a speaking black subject" (Gates 1988, 181) in his highly influential chapter, "Zora Neale Hurston and the Speakerly Text," from his path-breaking critical study *The Signifying Monkey: A*

Theory of African-American Criticism (1988). He perceives Hurston's literary rendering of Janie Crawford's black vernacular speech as "the voice of the black oral tradition" and claims that "Hurston's text is the first example in our tradition of 'the speakerly text,' by which I mean a text whose rhetorical strategy is designed to represent an oral literary tradition" (181). A little more than a decade later, Susan Edwards Meisenhelder reads most of Hurston's book-length works, including *Their Eyes*, through the dual lens of race and gender in her often-quoted book *Hitting a Straight Lick with a Crooked Stick: Race and Gender in the Work of Zora Neale Hurston* (1999). Acknowledging her debt "to the work of many black feminist critics who have answered Barbara Smith's 1977 call in 'Toward a Black Feminist Criticism' for an examination of the interlocking politics of race and gender in the works of black women writers" (Meisenhelder 1999, 10), she contends that "[b]uried in each of the books [. . .] lie stories of power and dominance, acquiesence [sic] and resistance to the racial and gender hierarchies Hurston saw in the dominant world" (13). More recent publications on Hurston also follow this interpretive pattern of examining her writings in light of race and gender (e.g., Haddox 2001, Miller 2004, Brooks 2010, and Bailey 2016) and together have provided valuable insights into Hurston's oeuvre.

Hurston studies have been partially blinded by a presumption that her work is primarily definable by its treatment of the identity categories race and gender. Scholarship on subjects other than race and gender in Hurston's oeuvre, such as folklore and ethnography, humor, local color, and, environmentalism, has been considerably less forthcoming. Several studies on Hurston and folklore/ethnography have appeared (e.g., Hemenway 1976, Adams 1991, Lawrence 2000, Mehring 2011). Hemenway, for instance, argues that "Hurston was serious about collecting folklore" in his 1976 article "Folklore Field Notes from Zora Neale Hurston" (Hemenway 1976, 39). He regards her as a serious if unconventional folklorist whose "unorthodox collecting techniques, characterized by a close identification with her informants, enabled her to discover lore inaccessible to other collectors" (39). John Lowe examines the role of humor in Hurston's short fiction and particularly in her novels in his significant study *Jump at the Sun: Zora Neale Hurston's Cosmic Comedy* (1994). "Hurston's co(s)mic prose," he states, "would play a useful function, creating a kind of systolic movement to her books, like the beating of a great heart" (35). Geneva Cobb-Moore inscribes Hurston into the local-color genre (together with Bret Harte, Mark Twain, William Faulkner, and Eudora Welty; Cobb-Moore 1994, 26; ↗9 Twain, ↗18 Faulkner) in her article "Zora Neale Hurston as a Local Colorist" (1994) and claims that "Florida's rich topography, the Eatonville community, and Joe Clarke's store porch are permanent features in Hurston's local colorist works" (27). More recently, a few critics have interpreted her works in terms of environmentalism (e.g., Davis 2006, Hicks 2010). Scott Hicks, for instance, argues that "Zora Neale Hurston grants readers a revolutionary, liberating vision of environmentalism in her autobiography, fiction and non-fiction" (Hicks 2010, 114). Yet such critical readings are exceptions to the rule.

Hurston criticism overall is dominated by discussions of race and gender in her works. She is routinely portrayed as a regional Southern black woman writer whose prose is firmly grounded in the black oral tradition and features folkloric material (e.g., folktales, songs, and hoodoo remedies), oral storytelling performances, and literary renderings of black vernacular speech. In this dominant account, her work is mostly set in the fictional village of Eatonville, Florida and revolves around strong African American women who challenge conventional gender roles and overcome obstacles, such as patriarchal oppression, domestic violence, and white exploitation, on their way toward independence and self-realization.

This well-entrenched interpretive pattern has also affected the reception of Hurston's more than twenty short stories and in particular her widely lauded short story "Sweat," perhaps her most frequently anthologized text. Like *Their Eyes*, "Sweat" has been the subject of much critical discourse because it reflects several aspects of the prevailing interpretative pattern (e.g., Lupton 1982, Seidel 1991, Lindenberger 2002, Carter 2014): Delia Jones is physically abused by her husband Sykes, an oppression that outweighs the racial tension in the background of the story. The two rarely discussed short prose texts that I have selected for the "Close Readings" section below, however, depart from the conventional wisdom on Hurston's oeuvre. "John Redding Goes to Sea" (1921), her first published short story, features a young male protagonist who dreams of traveling to the horizon but tragically dies before he can fulfill his dream. The story centers on a subject that connects Hurston to English Romanticism as much as to Twentieth-Century racial and sexual paradigms: the artist's discovery of the resources of imagination. Likewise, "The Eatonville Anthology" (1926), consisting of fourteen very short humorous chapters, may examine gender relations and racial types, but also participates in the genre of character sketches that is rooted in a classical tradition established by the Greek philosopher and rhetorician Theophrastus's little book *Characters*. Both short prose texts support close readings that transcend the dominant focus on the representation of, or quest for, identity and point to an ampler consideration of Hurston as a citizen of a global republic of letters.

2 Close Readings

Although "John Redding Goes to Sea" exhibits several of the features often emphasized in Hurston's literary art, such as the "rural, Eatonville-like village" setting (Hemenway 1980, 19), African American folklore (e.g., John's mother believes in conjure), literary renderings of black speech, and plot elements similar to the ones expressed in the novel *Their Eyes Were Watching God* (e.g., a young person overcomes obstacles and desires to travel to distant horizons), the short story has been mostly ignored in critical discussions of Hurston's writings. Only a few critics have examined it at length. Hemenway, for instance, considers it the work of an "apprentice," and describes it as "a groping, stumbling attempt to capture the folk ethos, overladen with sentimentality" (19). "What makes the story worth reading," he declares, "is the folkloric texture" (65). Valerie Boyd asserts a biographical interpretation: "The story contains a hodgepodge of autobiographical elements: As a child, Zora always dreamed of the horizon and felt the urge to wander, which her mother blamed on travel dust sprinkled on the doorstep by an enemy. And the story's names are directly from Zora's family tree: John was her father, Alfred was his father" (Boyd 2003, 85). She also points out that Hurston "flipped the genders" in the story (85): "John finds understanding in his father, while Zora found encouragement in her mother. The other parent – the one of the opposite sex as the child – is the naysayer: In Zora's life, it was her father; in 'John Redding,' it's the mother, who curiously, has the same name as Zora's reviled stepmother" (85). For Boyd, the short story demonstrates "Zora's earliest attempt to make sense of her father's life, and death, through fiction" (85).

While both critics afford valuable insights into folkloric and biographical facets of "John Redding Goes to Sea," they neglect to address the main issue of the story: a young artist's discovery of the resources of imagination. Hurston identifies with the young male protagonist John and, imagining her future through the eyes of a young boy (and later adult), explores various resources of the imagination, including metaphor, personification, simile, and allegory, and the process of creating art. The reversal of gender destabilizes the concept and challenges the critical view that Hurston's literary works focus primarily on African American women, who, like Janie Crawford in *Their Eyes Were Watching God* (1936), escape the conventional gender roles enforced by patriarchy and thereby gain their independence and their voice. It indicates that gender represents only one factor in a story that instead emphasizes the imagination. "John Redding Goes to Sea" is certainly the work of an apprentice writer, but it showcases the tremendous creative potential of this young author exploring the resources of the imagination available to her as an artist in her very first published short story.

Before the story turns to its main topic, a young boy's imagination and a young artist's invention of original metaphors, "John Redding Goes to Sea" introduces the topic of the individual and the community. The first sentence of the story, which is told from a third-person perspective, places the name "John Redding" at its syntactical center and emphasizes the individuality of the male protagonist, highlighting his separation from the village community: "The Villagers said that John Redding was a queer child" (Hurston 1995a, 925). The "Villagers" (with a capital "V") compare themselves with the young boy and, stigmatizing him as "queer," reinforce their collective identity and "normality" vis-a-vis the "queer child."

A further distinction, belief or disbelief in black folk magic, divides John's family. Sharing the opinion of the "Villagers," John's mother, Matty, remarks sorrowfully to his father, Alfred, that John has been conjured: "Alf, it's too bad our boy's got a spell on 'im" (925). She believes (as the readers learn later) that "travel dust" is the cause for John's unceasing wish to leave his hometown and his mother:

"The very night John wuz bawn, Granny seed ole Witch Judy Davis creepin outer dis yahd. You know she had swore tuh fix me fuh maryin' you, 'way from her darter Edna. She put travel dust down fuh mah chile, dat's whut she done, tuh make him walk 'way fum me" (927). John's father, by contrast, repeatedly denies the existence of conjure magic ("Ah keep on telling yuh, woman, tain's so," 925) and provides a different explanation for John's desire to travel: "Matty, a man doan need no travel dust tuh make 'im wanter hit de road. It jes' comes natcheral fuh er man tuh travel" (927). Their disagreement about the reliability of supernatural evidence is not based on gender or race. From the very beginning, then, the story centers on a young male – not a female – African American protagonist, who is discriminated against by the village and whose father and mother hold to diametrically opposed views on folk magic and superstitious belief.

Following the initial dialogue between Matty and Alfred, the third-person narrator acknowledges that "[p]erhaps ten-year old John *was* puzzling to the simple folk there in the Florida woods," because "he was an imaginative child and fond of day-dreams" (925). What follows is Hurston's literary rendering of a sudden creative impulse of imagination that transforms a literal description of the St. John River and southern Florida's flora into a metaphorical exemplification of John's 'lush' imagination:

> The St. John River flowed a scarce three hundred feet from his back door. On its banks at this point grow numerous palms, luxuriant magnolias and bay trees with a dense undergrowth of ferns, cat-tails and rope-grass. On the bosom of the stream float millions of delicately colored hyacinths. (925)

Whereas the first sentence has a straightforward literal meaning, the notion of lushness expressed in the phrase "numerous palms, luxuriant magnolias and bay trees" of the second sentence signals figurative meanings and metaphorically insinuates the rich and bountiful imagination of John, whose creativity has a verdant foundation ("with a dense undergrowth of ferns, cat-tails and rope-grass"). Additionally, the same notion of vast abundance in the phrase "millions of delicately colored hyacinths" "floating" in the third sentence also signifies the fluid ("float") and highly creative ("millions of" ideas) quality of his budding ("delicate") fantasy. The two sentences display a parallel syntactical construction – "On its banks [. . .]" and "On the bosom [. . .]" – that not only links them to each other but metaphorically points to a surface ("On its banks [. . .]") and a deeper level ("On the bosom [. . .]") of John's imagination that rhetorically suggests the flow of the St. John River. The reference to the spring-blooming "hyacinths," a flower with multiple symbolic meanings, signifies the nature metaphor's participation in a symbolic tradition (see Ferber 2017, 108). Intratextually, it also points to a later passage (a passage that also prefigures the pear-tree metaphor in *Their Eyes Were Watching God*) in which Hurston modifies the metaphor in her extended figurative translation of fresh love and erotic intimacies between John and his newly wed wife, Stella. This time, the hyacinths "lie" "on the surface of the river" instead of floating at the bottom of the river, indicating a reversal: "The miles of hyacinths lie like an undulating carpet on the surface of the river" (Hurston 1995a, 930). To further emphasize John's ingenuity, the narrator also presents the young boy's recurring "day-dreams" (925) of leaving his rural hometown in southern Florida in favor of discovering "distant lands" (925) as variations on the trope "riding away to the horizon" (925): for instance, "Sometimes in his dreams he was a prince, riding away in a gorgeous carriage [. . .]" (925).

The story then zeroes in on the inventive resources of the ten-year-old John's imagination. Initially, John utilizes basic devices of verbal creativity – metaphors, personifications, and a simile – to understand and describe his experience of the world around him. In one incident, John personifies weeds and screams at them for trapping a few of twigs that he had thrown into the river to float as ships along the current:

> One day his father came upon him scolding the weeds for stopping his sea-going vessels.
>
> "Let go mah ships! You ole mean weeds you!" John screamed and stamped impotently. "They wants tuh go 'way. You let 'em go on!" (926)

His father reveals to him that he also wanted to leave and travel the world when he was young but was unable to fulfill his dream: "Well, son, when Ah wuz a boy Ah said Ah wuz goin' too, but heah Ah am. Ah hopes you have bettah luck than me" (926). Observing a change in his father's mood, John tries steer him away from thinking about this missed chance in life. He plays a game with his father through which he can showcase his singular imagination:

> "Pa, Ah betcha Ah seen somethin' in th' wood-lot you ain't seen!"
>
> "Whut?"
>
> "See dat tallest pine tree ovah dere how it looks like a skull wid a crown on?"
>
> "Yes, indeed! said the father looking toward the tree designated. "It do look lak a skull since you call mah 'tention to it. You 'magine lotser things nobody else evah did, son!"
>
> "Sometimes, Pa dat ole tree waves at me just aftah th' sun goes down, an' makes me sad an' skeered, too." (926)

Alfred Redding genuinely admires John's imaginative comparison of a tall pine tree to the image of "a skull wid a crown," which may suggest the common vanitas image and memento mori motif of a skull to which (pre-)Romantic artists returned in the eighteenth century (see, for instance, Richard Bentley's woodcut illustrations of the "graveyard poet" Thomas Gray's poems collected in the 1753 book entitled *Designs by Mr. R. Bentley, for Six Poems by Mr. T. Gray*). Simultaneously, it recalls the crown of thorns worn by Jesus Christ at his crucifixion. Further, John's personification of the "ole tree" "wav[ing]" at him "just aftah th' sun goes down" demonstrates once more his imagination's creative resources and, together with Hurston's artistic and religious allusions, foreshadows John's death at the end of the story.

The middle section of the story displays another significant reversal of the conventional interpretive pattern that dominates the critical response to Hurston's work. Typically, men represent obstacles for Hurston's female protagonists (e.g., Janie Crawford in *Their Eyes* and Delia Jones in "Sweat"). In *Their Eyes Were Watching God*, for instance, Logan Killicks and Jody Starks hold Janie back from gaining her independence and her voice. Both men, as critic Wilfred D. Samuels points out, "claim the prerogative of owning 'their' women like they own their land or dry-goods store" (Samuels 1998, 243). In "John Redding Goes to Sea," by contrast, the roles are reversed (243). While Alfred Redding is an understanding father who supports his son's dreams and aspirations because he had never realized his own ("Ah wants mah John tuh go an' see cause Ah wanted to go mah self," Hurston 1995a, 927), John's mother, Matty, and, later, his wife, Stella, prove to be obstacles to John's ambitions and desires. Both women fail to comprehend John's urge to travel and want him to settle down and stay in his hometown. Three times Alfred and/or John asks Matty and/or Stella to allow John to travel abroad, and three times Matty and/or Stella refuse to bless his plans. As Samuels succinctly writes, "In 'John Redding Goes to Sea' father and son are imprisoned by women" (Samuels 1998, 244).

The question-and-answer pattern (Alfred and/or John ask for Matty's and/or Stella's consent and Matty/Stella refusing to give their blessings) frames the two main topics of this middle section: the impact of literacy on John's creative mind and his development into a young man with a complex poetic intellect. After Alfred Redding first informs his wife about their ten-year old son's future plans to go to sea ("He 'lows he's goin' to sea when he gits grown an' Ah reckon Ah'll let 'im," Hurston 1995a, 927), Matty Redding suffers "a fit of weeping" (927), which ends the conversation about that issue. John stays at home and, like Hurston, attends not only the local village school but also a high school in Jacksonville, "where none of the villagers went" (928). The formal education powerfully changes the way John speaks and thinks and, at the same time, provides him with new resources for his imagination. When Alfred mentions his now-grown-up son's plans to travel for the second time to his wife, he is joined by John, who implores his mother to let him follow his dream:

> "Mama," John began slowly, "It hurts me to see you so troubled over my going away, but I feel that I must go. I'm stagnating here. This indolent atmosphere will stifle every bit of ambition that's in me. Let me go, Mama, please. What is there here for me? Give me two or three years to look around and I will be back here with you and papa, and I'll never leave you again. Mama, please let me go. Why sometimes I get to feeling just like a lump of dirt turned over by the plow – just where it falls there's where it lies – no thought or movement or nothing. I wanter make myself something – not just stay where I was born." (929)

Beginning with this passage, John speaks Standard English rather than African American vernacular throughout the rest of the story. His deliberate expression ("slowly"), his elevated style ("It hurts me to see you so troubled [. . .]"), his adherence to normative grammatical rules, his adoption of a Latinized vocabulary (e.g., "stagnating," "indolent," and "stifle"), and his use of "fine" rhetorical devices (e.g., emphatic repetition and the rhetorical question "What is there here for me?") demonstrate the transformative effect of literacy on his previously oral mind. He retains his imaginative bent and a trace of his former vernacular language ("wanter") across the cognitive-linguistic boundary, but his creative processes have changed: the simile is now the product of a literate mind that compares an inner feeling to an agricultural – and thus rural – image ("Why sometimes I get to feeling just like a lump of dirt turned over by the plow"), and he expresses it in measured Standard English.

John's imagination gains in poetic complexity over time, and he metamorphoses into a highly articulate poet. He joyously informs his mother about an opportunity to join the Navy and see the world and, after she harshly and cruelly refuses to give her consent ("Naw, John, Ah'll nevah give mah consent. I know yous hardheaded jes'l-ak yo' paw; but if you leave dis place ovah mah head, Ah nevah wants you tuh come back heah no mo," 932), he and his father walk down to the river where John reiterates the same simile ("I get to feeling just like a lump of dirt") and translates it into a simpler but fuller metaphor ("I feel that I am just earth"):

> "You know, papa, sometimes – I reckon my longing to get away makes me feel this way . . . . I feel that I am just earth, *soil* lying helpless to move myself, but *thinking*. I seem to hear herds of big beasts like horses and cows thundering over me, and rains beating down; and winds sweeping furiously over – all acting upon me, but me, well, just soil, *feeling* but not able to take part in it all." (933)

Then John displays the full development of his imaginative powers:

> "Then a soft wind like love passes over and warms me, and a summer rain comes down like understanding and softens me, and I push a blade of grass or a flower, or maybe a pine tree – that's the ground thinking. Plants are ground thoughts, because the soil can't move itself. Whenever I see little whirls of dust sailing down the road I always step aside – I don't want to stop 'em 'cause they're on their shining way – moving! Oh, yes, I'm a dreamer . . . . I have such wonderfully complete dreams, papa. They never come true. But even as my dreams fade I have others." (933–934)

This lyrically Romantic passage, intensely invested in the pathetic fallacy, illustrates how far John has travelled on his journey towards becoming a gifted poet and artist. John has gradually changed from an imaginative ten-year-old boy, who randomly perceives twigs as ships and compares a pine tree to a skull with a crown, into a young man whose schooling has transformed his language, thoughts, and creative resources to become the inspired poet whose extended metaphor blurs the line between poetry and prose. To argue that John is no longer a "Black" character because he speaks Standard English and not what critics consider a hallmark of Hurston's "oral" writings – "authentic" black dialect – would miss the point of the story. Looking through the eyes of a male character, Hurston imagines her own development from a gifted child into a school- and college-educated literary writer whose creativity does not depend on racial or sexual identity.

The final part of "John Redding Goes to Sea" evinces a further reversal of the conventional wisdom about Hurston's writings. Whereas African American women such as Janie Crawford manage to fulfill their dream of reaching the horizon ("Ah done been tuh de horizon and back," says Janie; Hurston 1995c, 332), John joins a team of workers to help fortify an unfinished bridge across the St. John River before an impending storm and tragically dies without ever having a chance to follow his dream of "riding away to the horizon." Only his dead body floats towards the horizon at the end of the story.

Near the conclusion, in a conversation between Alfred Redding and John and the white bridge-builder Mr. Hill, who is looking for additional men to secure the bridge, Hurston briefly addresses the issue of race relations. Upon Mr. Hill's straightforward greeting ("Howdy John, Howdy Alf. I'm mighty glad I found you. I am in trouble."; Hurston 1995a, 934), Alfred Redding gives him a guarded but friendly answer: "'Well now, Mist' Hill,' answered Alfred slowly but pleasantly. 'We'se glad you foun' us too. What trouble could *you* be having now?'" (934). The italicization of the pronoun "you" indicates that Alfred Redding is aware of the white man's presence and deliberately, if subtly ironically, expresses his surprise that a "white man" like Mr. Hill could have a problem. John's reply, by contrast, does not indicate any change in demeanor and tone: "'I'll go, Mister Hill,' said John with a great deal of energy. 'I don't want papa out on the bridge – too dangerous'" (935). Unlike his father, John apparently takes his equality for granted.

Hurston subsequently devotes the final pages of the story to the exploration of her own resources of imagination and establishes an allegory with the help of Biblical images. In search of his son after the storm has subsided, Alfred Redding asks a surviving worker, Fred Mimms, about the whereabouts of his son and in what way the bridge had been wrecked by the storm (938). Mimms's account reveals two correspondences with the Bible. His first statement – "Den dat rain commenced – an' Ah nevah seed such a downpour since de flood. We set dere and someone begins tuh pray. Lawd how we did pray tuh be spared!" (983) – may remind readers of the Biblical flood as related in the Book of Genesis (and the flood in *Their Eyes Were Watching God*, which itself evokes the Biblical story). His second statement – "Comin' down stream, as fas' as 'press train wuz three big pine trees. De first one wuzn't fohty feet from us and there wasn't no chance to do nothin' but pray" (938) – encourages recognition of a similarity between "the three big pine trees" and the three crosses standing on "Golgotha" or "Cavalry" in the Gospels. Finally, the narrator's description of John's dead body floating on a piece of wood invites readers to remember Christ's crucifixion:

> He lay prone upon his back. His arms were outstretched, and the water washed over his brogans but his feet were lifted out of the water whenever the timber was buoyed up by the

> stream. His blue overalls were nearly torn from his body. A heavy piece of steel or timber had struck him in falling for his left side was laid open by the thrust. A great jagged hole wherein the double fists of a man might be thrust, could plainly be seen from the shore. The man was John Redding. (939)

Hurston, in this example, has embedded three markers in her text that foster an allegorical reading of Mimms's narrative and the narrator's portrayal of John's body (see Hemenway 1980, 19). She has generated a Biblical allegory.

In her first short story, Hurston imagines her own future through the eyes of John and wonders what will happen to her, the gifted and imaginative young artist. Will her exceptional creative talent – which she showcases in the story – be recognized? Will she be acknowledged as an innovative member of the literary Stylus club and the "New Negro Movement"? Will she realize her dreams and ride away to the horizon? Or will she die, her dreams unfulfilled, her creative potential unrecognized. Her autobiographical projection onto a male character, along with her neutralization of his race by assigning him a "white" language and education, challenges the view of Hurston as an author who is mainly interested in issues of gender and race and, instead, draws attention to the topic of a young artist's discovery of the resources of the imagination. She explores this topic by experimenting with ever more complex metaphors, by presenting in the final part of the story correspondences with the Bible that justify an allegorical reading, and by charting the development of John's creative potential from his "natural" use of basic rhetorical devices, such as metaphor, personification, and simile, through schooling and literacy into a powerful poetic imagination with which he can speak for others. The story points to a complexity in Hurston's writings that needs further investigation.

"The Eatonville Anthology" demonstrates another facet of Hurston's complexity: her formal experimentation, which is influenced by the sketch tradition. Published as a mini-series in the September, October, and November 1926 issues of the African-American monthly magazine *The Messenger*, "The Eatonville Anthology" is quite unusual in that it challenges the convention of regarding a short story as a single narrative text. While the first two installments offer thirteen short sketches about a group of characters living in the small town of Eatonville, Florida, the third installment presents a short folktale about Brer Rabbit, Brer Dog, and Miss Nancy Coon. (Note: Since recent editions of "The Eatonville Anthology" reproduce the punctuation and italicizations of the original three installments inaccurately and fail to indicate the serial character of the story collection, I recommend – if possible – the use of the original publication.)

Hurston's "The Eatonville Anthology" has been subject to scant critical attention. Critics tend to read the compilation of altogether fourteen short texts as a literate transcription of oral compositions (e.g., Hemenway 1980, 70; Boyd 2003, 140, 141; Frever 2006) and through the dual lens of gender and race (e.g., Boyd 2003, 139; West 2014; Jones 2016). Hemenway, for instance, claims that "[t]he fourteen parts have no thematic, structural, or imagistic relationship beyond their general identification with Eatonville in a bygone age" (Hemenway 1980, 70) and concludes that "[t]he anthology is just that, a series of self-contained stories as they might be told in a night's lying session at Joe Clarke's store or at a meeting of the Niggerati at Hurston's apartment" (70). "If 'Sweat' is the decade's finest example of her written art," he declares, "'The Eatonville Anthology' is the best written representation of her oral art" (70). Likewise, Boyd sees Hurston as "steeped in the oral tradition where stories were meant to be enjoyed in a communal setting" (Boyd 2003, 141) and detects common thematic concerns – e.g., gender and domestic violence – that link a few short narratives together: "In several of the tales, for example, women are routinely beaten by their husbands with no comment from the community (or from Hurston)" (139). Jill C. Jones also considers Hurston a "storyteller" in her essay titled "Taking the Axe to Babylon: Zora Neale Hurston's 'Lost' Caroline Stories, Gender, Place, and Power" (Jones 2016, 486) and examines the already known and the recently rediscovered 'Caroline story'-versions of the sketch "Pants and Cal'line" from "The Eatonville Anthology," which, as Jones states, "revolves around a woman who will not be beaten, literally or figuratively. Whether she's in rural, small-town Florida or the city of New York, Caroline, like other Hurston heroines and Zora Neale Hurston herself, prevails" (484).

While such critical readings afford significant insights into Hurston's "anthology," critics have neglected to read her collection of sketches within the frame of the American sketch-tradition (↗1 Of Sketches; see also Fash 2020), not to mention the even wider frame of the Classical tradition of the character sketch. "The Eatonville Anthology" participates in a character-sketch tradition that goes back to Theophrastus's notable book *Characters* (late 4th century BC) in which the Greek philosopher and rhetorician presents, as J. W. Smeed writes, "thirty short prose sketches of imaginary but typical individuals (the Coward, the Boor, the Chatterer, and so on). These 'characters' are all of bad or foolish types" (Smeed 1985, 3). In "The Eatonville Anthology," Hurston's humorous sketches about individual Eatonville inhabitants (including a dog called Tippy) exhibit a structure similar to Theophrastus's *Characters*. The classically educated Hurston introduces a Theophrastian structure in the first sketch of "The Eatonville Anthology" and explores the creative instability of the Theophrastian structure in her subsequent character sketches, which she contrasts, in turn, with the African American folktale about Brer Rabbit, Brer Dog, and Miss Nancy Coon of the third installment. Echoing the topic of the individual and the community introduced in "John Redding Goes to Sea," Hurston juxtaposes a "community" of sketches that participate in a "Western" literary tradition of the character sketch with an "individual" African American folktale that participates in a specifically African American tradition of oral sketches.

Besides a rhetorical style that is "unique in Greek literature" (Rusten 2002, 9), each of the thirty sketches in Theophrastus's work *Characters*, as Jeffrey Rusten observes, exhibits a tripartite – and in a few cases a fourfold – structure:

1) *Title*: a single-word personality trait, always ending in -ία;
2) *Definition* in abstract terms of this quality;
3) *Description*, the longest part of each chapter, introduced with the formula "the X man is the sort who . . . ," and continuing in a series of infinitives giving characteristic actions;
4) *Epilogue* (in some cases) in a more rhetorical style, with moralizing generalizations. (8)

Rusten points out that "the setting [of Theophrastus's *Characters*; ER] is [. . .] unmistakably the Athens of the last few decades of the fourth century B.C., whose customs, institutions, and prejudices form the backdrop of every character's actions" (9) and that "the descriptions are equally distinctive as literary portraiture. They are never generalizations, but catalogues of vivid detail" (9). Rusten adds that Theophrastus's sketches are firmly rooted in ethics ("the Greek word for 'character' is usually ἦθος," 12–13; Engl. "ethos"): "Basic to his [Theophrastus's; ER] whole enterprise is the notion that individual good or bad traits of character may be isolated and studied separately" (13). The character sketches, Rusten notes, also are comical: "For all their ethical basis, Theophrastus' sketches – especially in extended scenes like 'Idle Chatter' (3), 'Rumor-Mongering' (8), or 'Cowardice' (25) – quite obviously have comic affinities as well" (16).

The tenth sketch of Theophrastus's *Characters*, "Pennypinching," illustrates many features of the model:

10. Pennypinching

(1) Pennypinching is an immoderate sparing of expense. The pennypincher is the sort (2) who stipulates the repayment of a half-cent "within the month, to his house." (3) When he is sharing a dinner he reckons up how many glasses each has drunk; his initial offering to Artemis is smaller than any other at the table. (4) When someone has bought goods for him at a bargain price and presents his bill, he says they are too expensive, and rejects them. [. . .]. (12) When he goes shopping, he returns home without buying anything. (13) He forbids his wife to lend out salt, or a lampwick, or cumin, or oregano, or barley groats, or garlands, or sacrificial cakes, maintaining that these small items add up to a lot over the course of a year.

(14) [In general, pennypinchers like to see their money boxes moldy and the keys to them rusty . . . .] (Theophrastus 2002, 83–85)

The single-word title "Pennypinching," which specifies the single fault that will be the dominant topic of the character sketch, is followed by a general definition of the trait ("Pennypinching is an immoderate sparing of expense."), the formula ("The pennypincher is the sort who . . . "), and a series of when-clauses (with main clauses) that illustrate the typical activities of the "pennypincher" (e.g., "When he goes shopping, he returns home without buying anything."). Rich in details of the daily life of ancient Athens, the description includes a phrase spoken by a pennypincher in a particular situation, a small list of kitchen items, and a reference to the custom of giving an initial offering to a mythological god or goddess at the beginning of a meal.

The structure of Hurston's sketches in "The Eatonville Anthology" displays close similarities to Theophrastus's model of character delineation. Both numbered their sketches. A brief excerpt from the first short story will illustrate the structure of the character sketch:

> I
>
> THE PLEADING WOMAN
>
> Mrs. Tony Roberts is the pleading woman. She just loves to ask for things. Her husband gives her all he can rake and scrape, which is considerably more than most wives get for their housekeeping, but she goes from door to door begging for things.
>
> She starts at the store. "Mist' Clarke," she sing-songs in a high keening voice, "gimme lil' piece uh meat tuh boil a pot uh greens wid. Lawd knows me an' mah chillen is SO hungry! Hits uh SHAME! Tony don't fee-ee-eee-ed me! (Hurston 1926, 261)

Similar to Theophrastus's sketches, Hurston's first sketch is prefaced by a title that announces a character trait ("The Pleading Woman") before it continues with a statement that a specific character exhibits the particular trait ("Mrs. Tony Roberts is the pleading woman.") and two sentences that explain the quality in abstract terms ("She just loves to ask for things. [. . .]"). What follows is an example of Mrs. Tony Roberts's typical activity of pleading. She begins her begging tour at the local store and visits several other people, including a woman with the name Mrs. Pierson, until she has gathered all the ingredients she needs for a meal she has planned: "[. . .] and the pleading woman moves on to the next place, and on and on. The next day, it commences all over" (261). As in Theophrastus's sketches, the description also provides detailed information on the setting, its inhabitants, and customs: we learn, for instance, that Eatonville has a store and that the owner of the store, Mr. Clarke, knows Mr. Tony Roberts, the husband of the pleading woman, and we learn her exact words ("Tony don't fee-ee-eee-ed me!"). Like Theophrastus, Hurston also evokes humor. The stark discrepancy between Mrs. Roberts's actual prosperity and her relentless begging for food, her persistent invocation of the Christian God, and the stoic reaction of Mr. Clarke, who is unperturbed by her unceasing nagging, is comical. Yet Hurston's model differs from Theophrastus's, for the title "The Pleading Woman" consists of three words and denotes a character trait specific to a female character, whereas Theophrastus always describes activities performed by men.

The second sketch exhibits Hurston's exploration of the creative instability of the initial structure established in the first sketch. Here is the complete second sketch:

II
TURPENTINE LOVE

> Jim Merchant is always in good humor – even with his wife. He says he fell in love with her at first sight. That was some years ago. She has had all her teeth pulled out, but they still get along splendidly.
>
> He says the first time he called on her he found out that she was subject to fits. This didn't cool his love, however. She had several in his presence.
>
> One Sunday, while he was there, she had one, and her mother tried to give her a dose of turpentine to stop it. Accidentally, she spilled it in her eye and it cured her. She never had another fit, so they got married and have kept each other in good humor ever since. (261)

The title "Turpentine Love" consists of two words and refers to a love relationship made possible through the use of turpentine rather than to a character trait of a person. The first part of the opening sentence follows the formula "X [Name of an Eatonville resident] is . . . " and presents the notion of "good humor" as a character trait which is attributed to Jim Merchant ("Jim Merchant is always in good humor – even with his wife."). But instead of ending the initial seven-word main clause of the second sketch with a period, as she did in the seven-word opening sentence of the previous sketch, she adds a phrase (separated by a dash) that turns a likely character sketch about Jim Merchant's good humor into a supposedly comical husband-and-wife joke. What follows is a description of a humorous incident that ends in a happy marriage (with the feeling of a fairy tale-like ending). In contrast to the first sketch, Hurston does not provide the exact words of the main character Jim Merchant.

A brief glance at the other nine sketches in the first installment of "The Eatonville Anthology" reveals Hurston's experimentation with the creative instability of the sketch's structure. While all the sketches feature Roman cardinal numbers (though without periods), some have titles and others not. The number of words in a title varies. While the title of the third sketch has none, the fourth title has one ("Tippy"), and the fifth title eight ("The Way of a Man with a Train"), and so on. With the exception of the eleventh sketch, all initial sentences begin with a name, but not all follow the formula "X [Name of an Eatonville resident] is . . .," and the attributed "character trait" varies greatly: e.g., "Sewell is a man who lives all to himself." (VIII, no title) and "Mrs. Clarke is Joe Clarke's wife." (IX, no title).

The second of the three installments of "The Eatonville Anthology" juxtaposes the sketch "The Head of the Nail," which is a complete story (featuring a beginning, middle, and end), with the deliberately unfinished sketch "Pants and Cal'line" that invites readers to participate in the storytelling, like the porch sitters, and imagine a humorous end. This time the Roman numbers that preface the two short segments have periods ("XII." and "XIII."). These observations suffice to showcase Hurston's creative play with the instability of the character sketch's structure.

Hurston contrasts the thirteen character sketches of the first two installments, which participate in the Theophrastian tradition of character sketches, with the final installment's African American folktale about the animal characters Brer Rabbit and

Brer Dog and Miss Nancy Coon that breaks from the literary structure established at the outset of "The Eatonville Anthology." Surprising to the readers who may expect another Theophrastian character sketch due to the serial 'logic' of the story collection, the folktale features no Roman numeral, no dot or title, and the first sentence does not begin with a name but nevertheless introduces a characteristic trait of the animals in earlier days: "Once way back yonder before the stars fell all the animals used to talk just like people" (Hurston 1926, 332). The narrator then relates a rivalry between Brer Rabbit and Brer Dog, who try to win the favor of Miss Nancy Coon: "In them days dogs and rabbits was the best of friends – even tho both of them was stuck on the same gal – which was Miss Nancy Coon" (332). Both propose to her and, seemingly favoring Brer Dog because he can sing, she asks them to come back after a week, when she will announce her decision. Brer Rabbit promises Brer Dog to improve his singing and, having invited him to "meet [him] t'morrer in de huckleberry patch" (332), asks him the next day to stick out his tongue and splits it with a knife: "The dog has been mad at the rabbit ever since" (332). The story explains the etiology of a character trait of dogs. With this folktale Hurston distinguishes between the Classical literary tradition and the African American oral tradition. She establishes a difference between the "community" of the first thirteen Theophrastian character sketches and the "individual" African American folktale.

"The Eatonville Anthology" demonstrates that Hurston was an experimental short fiction writer who was interested in issues of literary structure. Not only does it contest the convention of perceiving a short story as a single narrative, but it also challenges readings that describe the fourteen short chapters as written transcriptions of oral compositions. In contrast to Hemenway's claim that the fourteen episodes are "self-contained stories" without any "thematic, structural, or imagistic relationship" (Hemenway 1980, 70), for instance, the close reading of "The Eatonville Anthology" has shown that the first time- and geographically bound thirteen sketches about early-twentieth-century inhabitants of Eatonville participate in the Classical literary character-sketch tradition that is rooted in Theophrastus's *Characters*. Featuring a Theophrastian structure, the first sketch is followed by twelve sketches that display Hurston's creative exploration of the structure's instability, and, thus, all sketches are structurally linked to each other. The fourteenth sketch, the African American folktale about Brer Rabbit and Brer Dog and Miss Nancy Coon, by contrast, is not tied to a specific time and place and participates in the specifically African American oral tradition, and yet the etiological explanation of a dog's character trait links the story to the other character sketches presented in the first two installments of "The Eatonville Anthology."

3 Theoretical Perspectives

The theorization of Hurston so far has limited her to a corner that is too restrictive. Critics have tended to examine her oeuvre (especially *Their Eyes* and "Sweat") primarily through the lenses of race and gender, and for good reason: the Civil Rights Movement and the Feminist Movement are self-justifying rationales. However, it is time to treat Hurston like the canonical world writer she should be. No critic reduces writers of Faulkner's or Welty's stature almost exclusively to gender, race or folklore categories. Flannery O'Connor (↗19 O'Connor) is in the process of being liberated from decades-long interpretations of short stories and novels almost exclusively within a Christian or Southern-regional framework. Recent critical approaches to her work include formalism, poststructuralism (e.g., Lacanian psychoanalysis), ecocriticism, gender theory, Whiteness studies, and media studies (film and photography). Likewise, Hurston studies need to open up to critical approaches other than ones that focus on race and gender. The "Close Readings" section of this chapter presented examples of alternatives: a young artist's imagination in "John Redding Goes to Sea," a subject that can be linked to English Romanticism (e.g., Samuel Taylor Coleridge's discourse on imagination articulated in his works, especially in his two-volume autobiography *Biographia Literaria*, 1817), and the sketch genre, particularly the character sketch genre that goes back to Theophrastus's *Characters*, in "The Eatonville Anthology."

Theoretical approaches to Hurston's writings could include intertextuality, reception theory, transnational studies, transmedial studies, poststructuralism (e.g., Foucault, Lacan, etc.), postcolonialism, just to name a few. For instance, critics tend to mention Milton's *Paradise Lost* when they discuss the rattlesnake in Hurston's short story "Sweat," a literary text replete with Biblical allusions, but her work points to many other intertextual references to both canonical and lesser known literary texts. In Hurston's debut novel *Jonah" Gourd Vine*, for instance, the character Lucy Potts recites three verses from an unnamed poem: "'A chieftain to the highland bound, cried 'Boatman do not tarry' – to the final. 'My daughter, oh my daughter'" (Hurston 1995b, 35). These three verses derive from the poem "Lord Ullin's Daughter" by the late eighteenth- and early nineteenth-century Scottish poet Thomas Campbell (see Campbell, 1907, 165, 167). A further intertextual allusion to Shakespeare may be detected in the thirteenth sketch of "The Eatonville Anthology." In "Pants and Cal'-line," Cal'line catches a glimpse of her philandering husband carrying a box and thinks to herself, "Very much like a shoe-box. So!" (Hurston 1926, 297), which may be a barely veiled adaptation of the famous line – "Very like a whale." (*Hamlet* 3.2.373) – spoken by Polonius in Shakespeare's renowned play *Hamlet*. The same sketch stops in medias res and asks readers to imagine, like the porch sitters in the unfinished story, a humorous end to the conflict between the axe-swinging Cal'line and her husband. Relevant to the reading activity Hurston's work inspires is Wolfgang Iser's reception theory outlined in *The Act of Reading: A Theory of Aesthetic Response* (1978) in which readers fill in textual gaps during the act of reading. Hurston's writings offer many possibilities for the application of a full range of theoretical approaches to literature.

Critics need to liberate Hurston studies from identity-based frameworks (Southern-Black-oral-feminist) that misleadingly create the impression that Hurston's imaginative horizons were mostly limited to the all-black village of Eatonville, Florida. We must recognize her as the canonical author she truly is and explore other facets of her work than race and gender to further nuanced understandings of her writings. Indeed, her work invites research in many areas that will demonstrate her membership in a global republic of letters.

4 Bibliography

4.1 Works Cited

Adams, Amelia Marie. "Zora Neale Hurston: Native Anthropologist." *All about Zora: Views and Reviews by Colleagues and Scholars, at the Academic Conference of the First Annual Zora Neale Hurston Festival of the Arts*. Ed. Alice Morgan Grant. Winter Park, FL: Four-G Publishers, 1991. 35–42.

Bailey, Amanda. "Necessary Narration in Their Eyes Were Watching God." *The Comparatist* 40 (2016): 319–337.

Bernard, Emily. "The Renaissance and the Vogue." *The Cambridge Companion to the Harlem Renaissance*. Ed. George Hutchinson. Cambridge: Cambridge UP, 2007. 28–40.

Boyd, Valerie. *Wrapped in Rainbows: The Life of Zora Neale Hurston*. London: Virago, 2003.

Brooks, Daphne. "Sister, Can You Line It Out?: Zora Neale Hurston and the Sound of Angular Black Womanhood." *Amerikastudien/American Studies* 55.4 (2010): 617–627.

Campbell, Thomas. *The Complete Poetical Works of Thomas Campbell*. Ed. J. Logie Robertson. London: Henry Frowde, Oxford UP, 1907.

Carpio, Glenda R., and Werner Sollors, eds. "Five Harlem Short Stories by Zora Neale Hurston." *Amerikastudien / American Studies* 55.4 (2010): 557–560.

Carter, Catherine. "The God in the Snake, the Devil in the Phallus: Biblical Revision and Radical Conservatism in Hurston's 'Sweat.'" *Mississippi Quarterly: The Journal of Southern Cultures* 67.4 (Fall 2014): 605–620.

Cobb-Moore, Geneva. "Zora Neale Hurston as Local Colorist." *Southern Literary Journal* 26.2 (Spring 1994): 23–34.

Coleridge, Samuel Taylor. *Biographia Literaria or Biographical Sketches of My Literary Life and Opinions. The Collected Works of Samuel Taylor Coleridge*. Eds. James Engell and W. Jackson Bate. Vol. 7. Princeton: Princeton UP, 1983.

Cronin, Gloria L. "Introduction: Going to the Far Horizon." *Critical Essays on Zora Neale Hurston*. Ed. Gloria L. Cronin. New York: Hall, 1998. 1–29.

Davis, Cynthia. "The Landscape of the Text: Locating Zora Neale Hurston in the Ecocritical Canon." *Florida Studies: Proceedings of the 2005 Annual Meeting of the Florida College English Association*. Eds. Steve Glassman, Karen Tolchin, and Steve Brahlek. Newcastle upon Tyne: Cambridge Scholars P, 2006. 149–156.

Fash, Lydia G. *The Sketch, the Tale, and the Beginnings of American Literature*. Charlottesville, VA: U of Virginia P, 2020.
Ferber, Michael. "Hyacinth." *A Dictionary of Literary Symbols*. 3rd edition. Cambridge: Cambridge UP, 2017. 108.
Frever, Trinna S. "'Mah Story Ends,' or Does It?: Orality in Zora Neale Hurston's 'The Eatonville Anthology.'" *Journal of the Short Story in English* 47 (Autumn 2006): 16 paragraphs. Web. 20 March 2021.
Gates, Henry Louis, Jr. "Afterword." *Zora Neale Hurston: The Complete Stories*. Eds. Henry Louis Gates, Jr. and Sieglinde Lemke. New York: Harper Perennial, 1996. 285–294.
Gates, Henry Louis, Jr. *The Signifying Monkey: A Theory of African-American Literary Criticism*. New York: Oxford UP, 1988.
Gray, Thomas. *Designs by Mr. R. Bentley, for Six Poems by Mr. T. Gray*. London: Printed for R. Dodsley, in Pall-Mall, 1753.
Haddox, Thomas F. "The Logic of Expenditure in 'Their Eyes Were Watching God.'" *Mosaic: An Interdisciplinary Critical Journal* 34.1 (2001): 19–34.
Hemenway, Robert. "Folklore Field Notes from Zora Neale Hurston." *The Black Scholar* 7.7 (April 1976): 39–46.
Hemenway, Robert. *Zora Neale Hurston: A Literary Biography*. 1977. Urbana, IL: U of Illinois P, 1980.
Hicks, Scott. "Zora Neale Hurston: Environmentalist in Southern Literature." *The Inside Light: New Critical Essays on Zora Neale Hurston*. Ed. Deborah G. Plant. Santa Barbara, CA: Praeger, 2010. 113–125.
Hurston, Zora Neale. *Barracoon: The Story of the Last Slave*. Ed. Deborah G. Plant. London: HarperCollinsPublishers, 2018.
Hurston, Zora Neale. "The Eatonville Anthology." Three installments. *The Messenger* 8.9–11 (September, October, and November 1926): 261–262, 297 and 319, 332.
Hurston, Zora Neale. "John Redding Goes to Sea." *Zora Neale Hurston: Novels and Stories*. Ed. Cheryl A. Wall. New York: Library of America, 1995a. 925–939.
Hurston, Zora Neale. *Jonah's Gourd Vine. Zora Neale Hurston: Novels and Stories*. Ed. Cheryl A. Wall. New York: Library of America, 1995b. 1–171.
Hurston, Zora Neale. *Their Eyes Were Watching God. Zora Neale Hurston: Novels and Stories*. Ed. Cheryl A. Wall. New York: Library of America, 1995c. 173–333.
Iser, Wolfgang. *The Act of Reading: A Theory of Aesthetic Response*. Baltimore: Johns Hopkins UP, 1978.
Jones, Jill C. "Taking the Axe to Babylon: Zora Neale Hurston's 'Lost' Caroline Stories, Gender, Place, and Power." *Mississippi Quarterly* 69.4 (Fall 2016): 481–499.
Kubitschek, Missy Dehn. "'Tuh De Horizon and Back': The Female Quest in *Their Eyes Were Watching God*." *Black American Literature Forum* 17.3 (Autumn 1983): 109–115.
Lawrence, David Todd. "Folkloric Representation and Extended Context in the Experimental Ethnography of Zora Neale Hurston." *Southern Folklore* 57.2 (2000): 119–134.
Lindenberger, Laura. "'Sweat': Through the Lens of Womanness." *The Corinthian* 4.1 (2002): n.p. Web. 20 March 2021.
Locke, Alain. Foreword. *The New Negro: An Interpretation*. New York: Albert and Charles Boni, 1925a. ix–xi.
Locke, Alain. "The New Negro." *The New Negro: An Interpretation*. New York: Albert and Charles Boni, 1925b. 3–16.
Lowe, John. *Jump at the Sun: Zora Neale Hurston's Cosmic Comedy*. Urbana, IL: U of Illinois P, 1994.
Lupton, Mary Jane. "Zora Neale Hurston and the Survival of the Female." *Southern Literary Journal* 15.1 (Fall 1982): 45–54.
Mehring, Frank. "Performing Anthropological Authenticity: Ethnic Identity Politics from Minstrel Realism to Zora Neale Hurston's Transatlantic Blues." *Cornbread and Cuchifritos: Ethnic*

Identity Politics, Transnationalization, and Transculturation in American Urban Popular Music. Eds. Wilfried Raussert and Michelle Habell-Pallán. Trier: WVT, 2011. 229–252.
Meisenhelder, Susan Edwards. *Hitting a Straight Lick with a Crooked Stick: Race and Gender in the Work of Zora Neale Hurston*. Tuscaloosa, AL: U of Alabama P, 1999.
Miller, Shawn E. "'Some Other Way to Try': From Defiance to Creative Submission in 'Their Eyes Were Watching God.'" *The Southern Literary Journal* 37.1 (2004): 74–95.
Rusten, Jeffrey. Introduction. *Theophrastus* Characters, *Herodas* Mimes, *Sophron and Other* Mime Fragments. Ed. and transl. by Jeffrey Rusten and I. C. Cunningham. Loeb Classical Library. Cambridge, MA: Harvard UP, 2002. 5–44.
Samuels, Wilfred D. "The Light at Daybreak: Heterosexual Relationships in Hurston's Short Stories." *Critical Essays on Zora Neale Hurston*. Ed. Gloria L. Cronin. New York: Hall, 1998. 239–253.
Seidel, Kathryn Lee. "The Artist in the Kitchen: The Economics of Creativity." *Zora in Florida*. Eds. Steve Glassman and Kathryn Lee Seidel. Orlando: U of Central Florida, 1991. 110–120.
Shakespeare, William. *Hamlet*. Ed. Harold Jenkins. London: Routledge, 1982.
Smeed, J. W. *The Theophrastian 'Character': The History of a Literary Genre*. Oxford: Clarendon P, 1985.
Theophrastus. *Characters. Theophrastus* Characters, *Herodas* Mimes, *Sophron and Other* Mime Fragments. Ed. and transl. by Jeffrey Rusten and I. C. Cunningham. Loeb Classical Library. Cambridge, MA: Harvard UP, 2002. 45–175.
Walker, Alice. *In Search of Our Mothers' Gardens: Womanist Prose*. London: Women's P, 1984.
Walker, S. Jay. "Zora Neale Hurston's *Their Eyes Were Watching God*: Black Novel of Sexism." *Modern Fiction Studies* 20.4 (1974): 519–527.
Wall, Cheryl A. "Chronology." *Zora Neale Hurston: Novels and Stories*. Ed. Cheryl A. Wall. New York: Library of America, 1995. 1013–1032.
Washington, Mary Helen. "The Black Woman's Search for Identity." *Black World* 21.10 (1972): 68–75.
West, M. Genevieve. *Zora Neale Hurston and American Literary Culture*. Gainesville, FL: UP of Florida, 2005.
West, M. Genevieve. "'Youse in New Yawk': The Gender Politics of Zora Neale Hurston's 'Lost' Caroline Stories." *African American Review* 47.4 (Winter 2014): 477–493.

4.2 Further Reading

Carby, Hazel V. "The Politics of Fiction, Anthropology, and the Folk: Zora Neale Hurston." *New Essays on Their Eyes Were Watching God*. Ed. Michael Awkward. Cambridge: Cambridge UP, 1991. 71–94.
Casanova, Pascale. *The World Republic of Letters*. Transl. Malcolm DeBevoise. Cambridge: Harvard UP, 2005.
Cole, Jean Lee, and Charles Mitchell, eds. *Zora Neale Hurston: Collected Plays*. New Brunswick, NJ: Rutgers UP, 2008.
D'haen, Theo, David Damrosch, and Djelal Kadir, eds. *The Routledge Companion to World Literature*. Abingdon: Routledge, 2014.
Giles, Paul. *American World Literature: An Introduction*. Hoboken, NJ: Wiley and Sons, 2019.
Giles, Paul. *The Global Remapping of American Literature*. Princeton: Princeton UP, 2011.
Kaplan, Carla, ed. *Zora Neale Hurston: A Life in Letters*. New York: Anchor Books, 2003.
Scheiding, Oliver, and Martin Seidl, eds. *Worlding America: A Transnational Anthology of Short Narratives before 1800*. Stanford: Stanford UP, 2015.

Timo Müller

18 William Faulkner (1897–1962)

Abstract: William Faulkner's short stories, like the rest of his oeuvre, are intertwined with the history of the South that they chronicle from the first settlers to the mid-twentieth century. At the same time these stories are among the most renowned works of the international modernist avant-garde. This chapter traces the interplay between local subject matter and avant-garde aesthetics in two of Faulkner's most accomplished stories, "Dry September" (1931) and "Barn Burning" (1939). Following a brief introduction to Faulkner's life and works, the essay addresses central topics and concerns in both stories, in particular their depiction of Southern racial and class tensions respectively. It examines the aesthetic strategies Faulkner employs in probing these tensions and concludes with a survey of critical readings of the stories from the New Criticism to the present.

Keywords: South, lynching, racism, class, narrative perspective

1 Context

William Faulkner was born into a prominent northern Mississippi family in 1897. His great-grandfather had been an officer in the Civil War; the family legacy and that of the Old South would shape Faulkner's life and works. Mississippi in the early twentieth century was torn between idealizing memories of the antebellum past and the changes brought about by modern technology and business. The stark social hierarchies of the agrarian South persisted but were increasingly challenged by the risks and opportunities of the capitalist marketplace. In many respects, racism remained the only common denominator among the white population. This meant increasing segregation and oppression for African Americans, who made up the majority of Mississippians until the 1930s. Faulkner's oeuvre revolves around the all-pervasive history of the South and the conflicts it creates for later generations. His literary writings offer so powerful a testimony of these issues that they have themselves become part of Southern identity and are featured prominently in many history books about the region (Pascoe, Leathem, and Ambrose 2005). At the same time, his stories and novels made Faulkner one of the most esteemed members of the international literary avant-garde. How did this local writer become an outstanding figure in the history of the American short story?

As a young man, Faulkner did not think of himself as a local figure. He admired the European writers of the *fin de siècle*, made plans to travel to Italy and France (which he carried out in 1925), and enlisted in the Canadian Royal Air Force to become a pilot in World War I. While he never made it beyond ground training, he

https://doi.org/10.1515/9783110587647-019

remained fascinated by airplanes and played up his military experience after the armistice. It was this fascination that inspired his early efforts as a writer. His very first published story, "Landing in Luck" (1919), is about a young air force pilot and his instructor. Faulkner also wrote several war stories such as the memorable "Turnabout" (1932), in which an American air force pilot and a British naval gunner accompany each other on combat assignments.

By the mid-1920s, however, Faulkner realized that his deepest imaginative powers were inspired by his hometown of Oxford, Mississippi, and its surroundings – the "little postage stamp of native soil" that he would immortalize as Yoknapatawpha County (Cowley 1958, 127). He began to write a novel based on his family history that was published as *Sartoris* in 1929. The project branched out into a monumental oeuvre of fourteen novels and dozens of stories that tell of distinguished families, memorable events, and notorious figures in Yoknapatawpha from the early nineteenth century through the Civil War to the arrival of modern technology and business. Rather than one systematic history of the county, these novels and stories present a multitude of interrelated histories, whose subjectivity is highlighted by the prominent role of the narrator. Many of the characters appear in several works, whether as protagonists or in passing references; events briefly mentioned in one work may become the central concern in another. All of these stories and novels were modeled on Faulkner's hometown, of which he remained a resident almost all his life, farming his own land and riding in the countryside.

Faulkner's second novel of Yoknapatawpha County turned out to be his first masterpiece. *The Sound and the Fury* (1929) tells the story of the Compson children, who also figure prominently in the short stories "That Evening Sun" (1931) and "A Justice" (1931). The Compsons are one of the oldest families in the county, but their fortunes are in decline, and the children wrestle with this decline in their different ways. The artistic achievement of *The Sound and the Fury* lies in its language and narrative structure. Rather than presenting a straightforward family saga, the novel successively delves into the minds of four central characters: the three Compson brothers, whose emotions revolve around their sister Caddy in entangled ways, and the old house servant Dilsey, a black woman whose benign wisdom sets a counterpoint to the brothers' obsessions. The very different styles of these four sections, three of which are internal monologues, convey a striking sense of the characters' unique personalities, perspectives, and limitations.

This conjunction of local history and modernist experimental narrative became Faulkner's groundbreaking contribution to world literature. In various manifestations it shapes the major novels he published over the following decade and a half: *As I Lay Dying* (1930), *Light in August* (1932), *Absalom, Absalom!* (1936), *The Hamlet* (1940), and *Go Down, Moses* (1942). Since none of these novels sold very well – his only bestseller was *Sanctuary* (1931), a gruesome tale of rape and murder in Yoknapatawpha – Faulkner had to keep working away at more popular kinds of writing. His main source of income during this period was scriptwriting for the Hollywood studios, but he also made money from short stories he wrote for national magazines such as *Harper's Weekly*, *Scribner's*, and the *Saturday Evening Post*. These magazines paid well and reached a large audience, even if their ideas of good writing did not always square with Faulkner's. The strategies he devised to break the history of Yoknapatawpha County down into individual anecdotes for magazine publication resulted in a large, intriguing body of work that established Faulkner as one of the masters of the American short story.

His main achievements in the genre, in regard to both quantity and quality, came in the 1930s. Between the appearance of his most famous story, "A Rose for Emily," in 1930, and his elegiac masterpieces "Delta Autumn" and "The Bear" from 1942, he published over sixty stories and wrote many more. Among the most impressive are "Red Leaves" (1930), about a slave hunt in the Native American community of Yoknapatawpha County; "That Evening Sun" (1931), about a black servant stalked by her former lover; "Mountain Victory" (1932), a tragic episode in the wake of the Civil War; and "Mule in the Yard" (1934), a hilarious tale of peasant wit and double dealing. Faulkner's stories are seldom narrated in a straightforward manner, nor are they organized around a central theme or motif. Titles such as "Red Leaves" or "That Evening Sun" give no indication of what is going to happen in the story, and even when the title announces a central motif – as in "The Leg" (1934) or "The Brooch" (1936) – that motif tends to remain in the background of the narrative. Most of Faulkner's stories are long-winded anecdotes told by characters who either witnessed the events firsthand or heard about them from others. The stories often switch from one perspective to another and go back and forth in time. As in Faulkner's novels, the focus is not so much on the events themselves as on the ways in which they are passed on, judged, and imagined by others.

This chapter will examine two of Faulkner's masterpieces in the genre, "Dry September" (1931) and "Barn Burning" (1939). Not only do these stories frame the most productive decade of Faulkner's writing career; they also show the scope of his aesthetic strategies and the breadth of his thematic vision: one investigates the New South, the other the survival of the Old. In fact, one could argue that from the mid-thirties, the short story gradually superseded the novel in Faulkner's output. Most of the books he published after *Absalom, Absalom!* were either collections of short stories – *Knight's Gambit* (1949) and *Collected Stories* (1950) – or novels composed of existing stories and related anecdotes. This is true of *The Unvanquished* (1936), a series of episodes about the Sartorises' Civil War experience; *The Wild Palms* (1939), which consists of two separate novellas; *The Hamlet* (1940), which incorporates several previously written stories wholesale or in part; *Go Down, Moses* (1942), a history of the McCaslin family in seven episodes, five of which had been published as stories before; and to an extent *The Town* (1957) and *The Mansion* (1959), which continue the history of the Snopes family begun in *The Hamlet* (Creighton 1977; Ferguson 148–182; Carothers 1985, 25–56).

Faulkner's powers as a writer declined in the 1940s as marriage problems, struggles with alcoholism and depression, and increasing public interest in his personal life combined to draw energy from his writing. Nevertheless, he continued to publish novels and stories until his death in 1962. Among the more successful of his later stories are "Two Soldiers" (1942), told by a farmer's boy whose older brother enlists for World War II; "Shingles for the Lord" (1943), a humorous anecdote that ends with the protagonists unintentionally setting the local church on fire; and "A Courtship" (1948), one of the best of Faulkner's Indian tales. The 1940s were also the time when Faulkner became internationally recognized as one of the greatest writers of the twentieth century, winning the Nobel Prize in 1949 and the National Book Award for his *Collected Stories* (from which the quotations in this chapter are taken) in 1951. Those stories not incorporated in the *Collected Stories*, incorporated many unpublished during his lifetime, are available in paperback as *The Uncollected Stories of William Faulkner* (1979).

Faulkner continues to be ranked among the major figures in American and twentieth-century literature. He is primarily regarded as a novelist, but several of his stories have become classics of American literature and are widely read in high schools and universities. Faulkner himself spoke highly of the short story on several occasions and even said that it required greater artistic skills than the novel because its compact form demanded greater awareness of details (Faulkner 1980, 217; Ferguson 8–12). Today he is regarded as one of the masters of the American short story, both because of his innovative handling of language and narrative technique and because his stories capture the complexities of social conditions in the modern South in a nuanced, farsighted manner. Faulkner's stories cover a broad range of topics and social milieus. The largest group of stories is set in Yoknapatawpha County. Its scope includes the town and the country, the upper and lower classes, and the various ethnic groups that live in the area. Another group of stories revolves around airplane pilots, especially the British air force in World War I. A few additional stories are set in Europe, Mexico, and coastal cities in the United States. Recurrent topics across these groups include initiation into adult life, disappointed or deferred love, and the tension between individuals and claims in the community. Some of Faulkner's stories are comical anecdotes, others tell tragic stories of disillusionment, betrayal, and death. While few of them directly reference historical or political events, the turmoil of Southern life since the Civil War reverberates throughout the stories and at times disrupts the routines of daily life, most dramatically perhaps in "Dry September."

2 Close Readings

Set in Jefferson during a heat spell of "sixty-two rainless days" (Faulkner 1950, 169), "Dry September" traces the psychological dimensions of a racist lynching. A lynch mob forms in the local barbershop after a middle-aged townswoman, Minnie

Cooper, has leveled an unspecified accusation against Will Mayes, a respectable black worker. Even though no one seems to believe Minnie, the men in the barbershop start rumors of sexual assault, organize into a mob, and drag Mayes to an abandoned brick kiln outside town. The story follows these events through the eyes of Hawkshaw, one of the barbers, who is convinced of Mayes' innocence and leaves the group on the way to the lynching. A second plot line centers on Minnie and the lonely, repressed life she leads after rejecting potential husbands out of youthful arrogance. Long past her prime, Minnie enjoys the undivided attention of the town once more when she goes to the movies after the lynching, but she breaks down halfway through the show and has to be taken home by her friends. The story as a whole suggests – but never confirms – that Minnie's accusation was baseless and that the leader of the lynch mob, John McLendon, murdered an innocent man.

The central concern of the story is not the lynching itself but the social conditions that enable and sanction such irrational violence: racial prejudice, sexual repression, coercive group dynamics, and a twisted notion of Southern life and values. With the exception of Hawkshaw, none of the white townspeople cares to find out whether Mayes is actually guilty. Several of them hint that Minnie cannot be trusted but they proceed with the lynching anyway. Their main motivation is to prevent a threat of black men molesting white women that only exists in their own imagination. Faced with Hawkshaw's reasonable defense of Mayes, the mob leaders resort to racial prejudice, claiming that a black man's word should never be taken above a white woman's and that blacks need to be taught a lesson no matter what happened. The story suggests that such fantasies of black sexual threat are nurtured by sexual repression among the whites themselves. Shut up in a "small frame house with her invalid mother and a thin, sallow, unflagging aunt," Minnie is prevented from following or even admitting her sexual urges. McLendon is married but sleeps alone, brutally rejects his wife's solicitude, and shows more interest in his pistol than in his wife, all of which suggests that violence might be his outlet for sexual repression or even impotence (Putzel 1985, 226–228). Another of the mob leaders, an overexcited boy called Butch, is the only single teenager in a story that otherwise stresses the romance and sensuality of teenage love.

The relationship between individuals and the community is another important topic of the story, which has been called a "study in group violence" (Gerstenberger and Gaber 1969, 406–407). The opening section traces how the diverse individual opinions about Minnie's claims form into a collective resolve to assault Mayes. The lynch mob is depicted as an artificial community forced in line by the aggressive McLendon, who relies on racial prejudice and thinly veiled threats of physical violence to divide the men into those "that're with me" and those that allegedly side with black people against whites (Faulkner 1950, 172). When Mayes is confronted by the mob, he recognizes this dynamic and tries to save himself by individualizing the group members. The mob consists of seven or eight local men, and Mayes addresses up to five of them by their names to find out what is happening. He does not receive a single answer: the men now act as an anonymous collective rather than as a community of responsible, accountable individuals. The coercive force of the group becomes obvious when Hawkshaw wants to leave and McLendon refuses to slow down the car.

The story situates the lynch mob within the larger community of the town, and implicitly within the imagined community of the white South. Some of the lynchers try to keep their plan secret, especially from the sheriff, which suggests that the larger community disapproves of vigilante violence. The only explicit approval of the lynching comes from "coatless drummers" in front of the hotel, that is, from unrespectable outsiders (180). This continues a theme from the opening section, where another drummer who "looked like a desert rat in the moving pictures" joins McLendon in working up the lynch mob (170). While these references imply that racism infiltrates the community from outside, the townspeople implicitly support the lynching when they become "deferent, protective" toward Minnie afterward (181). Her friends note in thrilled voices that not a single black person is to be seen in the town square.

Given their doubts about Minnie's character, the townspeople seem to rally not so much around her as around an abstract ideal of pure Southern womanhood. This aligns them with the lynchers, who justify their plans as a defense of "our mothers and wives and sisters" (172). The fact that much of this rhetoric comes from the rat-like drummer, who presumably hails from another part of the South, embeds the lynching into the history of the region as a whole. The South had long imagined itself in terms of honor, purity, and chivalry when the actual common denominator was often racial prejudice and violence (Wyatt-Brown 1982). The rhetoric of chivalry is exposed as an empty fraud when McLendon returns home from punishing imagined violence against "a white woman" only to perpetrate actual violence against his own wife (Faulkner 1950, 171). Similarly, one of the more cautious men at the barbershop is accused of Northern sympathies by the drummer even though he was born and raised in the town while the drummer is a temporary visitor.

All of these dimensions of the lynching unfold in the dry heat that pervades the story from beginning to end. The climate has an obvious effect on the characters, who have grown restless and irritable under the continual heat. It also has a range of symbolic overtones, one of which emphasizes the townspeople's racism. The main culprits in the lynching are described as heated throughout: McLendon and Butch are sweating heavily, the drummer looks like a desert rat, and Minnie is "feverish" when she goes to the movies (180). The recurrent image suggests that these characters are mentally heated too: that they lack self-control, rational thought, and responsible behavior – the very values on which Southern whites based their sense of racial superiority. As in many of his works, Faulkner ironically exposes the absurdity of such hierarchies by ascribing these values to the African Americans instead of the whites. In "Dry September" it is Will Mayes who remains controlled and rational as he confronts the mob. He addresses the whites respectfully, inquires what they accuse him of, and keeps his wits together as he is handcuffed and forced into the car. The contrast is enhanced by the complementary motif of ice. The coolness of ice is associated with Mayes (he works at the ice plant) and contrasted with Minnie, who needs to be cooled with ice packs after her mental breakdown at the movie theater.

In between the publication of "Dry September" and "Barn Burning," the Great Depression shifted attention from race to class in public debates around the South. The survival of old customs and hierarchies had kept the South an underdeveloped agricultural region where not only African Americans but also the majority of whites lived in relative poverty. Many of them worked as sharecroppers, small farmers dependent on the landowners who loaned them farmland and equipment. The plight of these sharecroppers received wide media coverage in the 1930s and gradually took center stage in Faulkner's work as well (Atkinson 2006). "Barn Burning" is perhaps the most famous outcome of this development. Its title indicates that this story too is structured by motifs of heat and coldness. It continues to explore the instability of Southern identity narratives and the tensions between individual and community. Rather than race, however, "Barn Burning" puts the focus squarely on class hierarchies, which are represented by the wealthy landowner Major de Spain on the one side and the sharecropper Abner "Ab" Snopes on the other.

The story opens with Ab standing trial for burning an owner's barn. Barn burning was among the most serious of crimes in rural communities, where a family's livelihood depended on the grain, livestock, and equipment stored in their barn. Ab escapes conviction but is told to leave the county. He moves his family to de Spain's land, only to provoke the new owner right away by ruining an expensive rug in his mansion. After another trial – this time Ab takes the landowner to court for fining him for the rug – Ab resolves to burn de Spain's barn as well. As in "Dry September," much of the drama is created by the narrative perspective. "Barn Burning" is focalized through Ab's younger son, the ten-year-old Sarty, whose moral development emerges as the main interest of the story.

Like Hawkshaw in "Dry September," Sarty acts as both a center of consciousness and a center of conscience. The story takes his side and sympathetically follows the evolution of his moral awareness. The conflict he struggles with is between two communities: family and society. He cannot remain loyal to both because his father defines family in opposition to society. Abner sees society as oppressive because its socio-economic system keeps him and his family poor despite their hard work. While the societal authorities do not question this class hierarchy, however, they do show acts of kindness toward the Snopeses. De Spain fines them much less than the rug was worth, and the justices in both trials seem to be upright and considerate. Such acts of kindness are lacking within the Snopes family, where Abner reigns over his wife and children with a cold, detached brutality. This contrast may be the key factor in Sarty's eventual decision to side with society over family and warn de Spain that Abner is about to burn the barn. The story ends with Sarty running away from the fire and hearing shots fired at his father and older brother. He has lost his family and is now on his own, in a society that holds little in store for people like him.

The tension between individual and community thus plays out on two levels. In following his own principles rather than those of the family, Sarty reenacts his father's behavior toward society at large. Abner is the supreme individualist in the story. He pursues his course of action independently of either institutional authorities or social customs. He remains unconcerned by the authority of the court, the power of wealthy landowners, but also by communal values such as honesty, loyalty, or human sympathy. Burning barns and lying about it afterward runs counter to all of these values and makes him an outcast in consequence. The story points out that he has behaved in this ruthlessly independent manner all his life. During the Civil War, when most Southerners made great sacrifices for their country, he stole army supplies and sold them to the enemy. His decision to name Sarty after one of the local war heroes, Colonel Sartoris, shows that he has always used his family to cover up his crimes. At the same time, the name indicates that Sarty has been caught between family and society, between his father and the moral authorities of the local community, from the very beginning.

Rather than celebrating individualism, family, or society, "Barn Burning" remains ambivalent about all of these options. While the story stresses Abner's cold, at times brutal behavior toward his family, it also emphasizes the instinctive love and loyalty Sarty offers his father. It also gives Abner an opportunity to defend his behavior toward the landowners. In speaking to his son, Abner points out that the landowners have amassed wealth by exploiting sharecroppers like himself, so that they can afford imported rugs from Paris while Sarty and his siblings have to sleep on the floor. The story implicitly supports this argument at several points, for example by noting that even de Spain's servants wear better clothes and ride better horses than the Snopeses could ever afford. Besides criticizing class hierarchies, the story also questions the communal values and ideals of the Old South. When Sarty is asked to testify at the first trial, the justice says, "I reckon anybody named for Colonel Sartoris in this country can't help but tell the truth" (Faulkner 1950, 4). This might indicate that the supposedly neutral courts are often prejudiced in favor of influential personalities – a group that would include de Spain, whose mansion Sarty compares to a courthouse. The comment also shows that the efficiency of the legal system is hampered by the widespread glorification of the Old South. The judge is ready to take Sarty's word simply because of his name, while Sarty is about to lie in order to cover up a crime.

It is this self-satisfied indulgence in social myths that Abner was playing on when he chose Sarty's name. The justice's comment reveals that the trick is working. This sets the tone for another main theme underlying the story: the decline of the Old South. Like "Dry September," the story exposes retrogressive notions of Southern identity as false and self-defeating. In both stories these notions tear apart the very community they are supposed to hold together. In "Dry September" the result is immediate physical violence; in "Barn Burning" they enable Abner to burn another barn and presumably get killed in the process. On a structural level, "Barn Burning" marks the decline of the Old South and the rise of a new social order, one dominated not by the heroes of the past but by individualists out for their own gain. This shift would become the dominant theme in Faulkner's oeuvre after "Barn Burning," and the Snopes family remained its unerring representative. Abner may be killed at the end, but he has forced his landlord out of the aloof unconcern of his privileged position. At court over the rug, de Spain "wore on his face an expression not of rage but of amazed unbelief which the boy could not have known was at the incredible circumstance of being sued by one of his own tenants" (18). The religious vocabulary underscores that the landowner is no longer an untouchable, vengeful god but a common citizen. Abner has challenged not only his social superiority but the entire system of shared, unspoken certainties on which the old hierarchy was founded.

One of the most striking differences between "Barn Burning" and "Dry September" is the narrative setup. While "Barn Burning" is told from the perspective of a young boy, with the narrative voice staying close and sympathizing with him, "Dry September" adopts a distant tone and mostly an outside perspective on the events. Where one story dives into the mind of one character, the other studies the psychology of a group of characters in a detached, almost scientific manner. This range is characteristic of Faulkner's stories, which employ an extraordinary variety of narrative voices and perspectives. Short stories usually operate with a single narrator and a limited number of perspectives, but Faulkner often multiplies narrative voices and perspectives within the same story. This serves different purposes: it may indicate that the story is local folklore that has been told and retold across generations (as in "A Justice"); it may create dramatic tension by letting the reader know things the characters do not know (as in "That Evening Sun"); or it may enable a more nuanced, comprehensive view of the incidents in the story. "Dry September" and "Barn Burning" fall into the last category.

In "Dry September" the shifts of focalization between Hawkshaw and Minnie create two complementary plot lines that develop the key concerns of the story. The story draws numerous connections between these plot lines through shared motifs (heat, ice, movies) and through the striking parallels between the main culprits. McLendon and Minnie are both described as isolated, repressed, "furious," and overheated in a physical and mental sense (172). Both were widely respected in their youth – McLendon was decorated for valor in World War I – and have now lapsed into obscurity; both have rejected offers of love and feel lonely as a result; both live in small, imprisoning houses. These parallels illuminate the complex psychological motivations behind the lynching and show that responsibility is shared by many members of the community, no matter if they were present at the lynching or not.

Another shared feature of the plot lines is selective omniscient narration. The narrator follows Hawkshaw and Minnie around but seldom enters into their minds. This limits the reader's insights into the characters' motivations, and it highlights that the characters are confronted with the same problem. They do not fully understand what is going on even when they are in the middle of it. Hawkshaw is left alone in the dark at key points in the plot: when the lynchers arrive at the ice plant and after he has jumped out of the car. In both scenes the focalization stays close to him, emphasizing his lack of orientation. Similarly, when Minnie goes to the movies the narrator adopts her perspective to emphasize the other visitors' unsettling glances and the contrast between the "silver glare" of the screen and the happy couples entering the theater (181). The repeated references to Minnie as being "feverish" stress the unreliability of her senses and thus of her perception (180).

The story contrasts these instances of limited perception with long passages of detached observation by the heterodiegetic narrator. In these passages the story resembles a scientific study, and the narrator emphasizes the analytic setup by repeatedly describing the characters as if they were in a laboratory: "the sky was as clear as the inside of a brass bell" (175); "they seemed to breathe, live, in a bowl of molten lead" (177). In this laboratory the characters become animals trapped in their cages, where the reader can see them following their drives and instincts rather than thinking rationally. The main perpetrators of the lynching are all described in animalistic terms. McLendon's house is "trim and fresh as a birdcage and almost as small" (182); the drummer in the barbershop "looked like a desert rat" with "his frothy beard" (170); Butch resembles an ape given his hulking figure, his sweaty armpits, and his position hanging from the side of the car. Even Hawkshaw bears an animal in his name, which one of his colleagues shortens to "Hawk" as if anticipating the barber's half-participation in the lynching. All of these devices provide clues to the characters' motivations rather than to the events themselves. This complements the narrator's detached report of the events and involves readers the more deeply in the psychological dimension of the story.

The distant narrative voice and the scene cuts between Hawkshaw and Minnie also contribute to another important aesthetic strategy of the story: its use of film as motif and technique. Not only do the characters visit a movie theater, but the narrator draws on cinematic terms to describe characters (such as the visiting salesman in the barbershop) and to portray the protagonists' behavior as artificial and unreliable. Minnie regards the movie theater as "a miniature fairyland with its lighted lobby and colored lithographs of life caught in its terrible and beautiful mutations." This confusion of the movies with real life is repeated when the film starts and Minnie thinks "life began to unfold" on the screen (181). It also seems to inspire her gaudy dress and dramatic behavior, which make her an outsider in the small, rural town.

Here too Minnie and McLendon are paralleled. The final scene of the story shows McLendon looking at the "dusty screen" of his porch, which separates him from the world (183). At an earlier point his entry into the barbershop is described, with overtones of parody, as the arrival of a western hero in the saloon:

> The screen door crashed open. A man stood in the floor, his feet apart and his heavy-set body poised easily. His white shirt was open at the throat; he wore a felt hat. His hot, bold glance swept the group. His name was McLendon. [. . .]
>
> "Well," he said, "are you going to sit there and let a black son rape a white woman on the streets of Jefferson?" (171)

McLendon too lives in a dream world, one where he is a military leader and his aggressiveness translates into heroic resolve. No one has mentioned rape before; by all appearances this version of the events has formed in his own imagination. The lynching thus becomes a movie script directed by and starring McLendon, who restages an absurdly distorted version of the war in which he was decorated for valor. Another script he draws on, the scholar William Bache points out, is that of the Hollywood romance where the dashing hero avenges the honor of a beautiful young girl (Bache 1954, 53–57). The parodic overtones of McLendon's entry in the barber shop indicate how inappropriate these scripts are. Even by his own standards, ganging up on a defenseless individual hardly qualifies as valor or even combat. By the ethical standards of the story, he callously murders an innocent man.

Whereas "Dry September" achieves complexity and dramatic effect through analytic distance to the characters, "Barn Burning" does so by presenting the story through the eyes and mind of its main character. It is this internal focalization that establishes Sarty's conflictive relationship with family and society as the main theme of the story. Yet "Barn Burning" also introduces additional layers of perception. The narrative voice alternates between rendering Sarty's thoughts and impressions in the first person, summarizing them in the third person, and occasionally adding to them from an outside perspective. Most of the story falls into the second category, but even here the story maintains a distinction between Sarty and the narrator by contrasting their stylistic repertoire. Where Sarty uses simple words, faulty grammar, and rural dialect – the language of a sharecropper's child – the narrator commands an extensive vocabulary and often launches into grandiose rhetoric that takes the story far beyond its rural setting. These contrasts are established in the very opening lines:

> The store in which the Justice of the Peace's court was sitting smelled of cheese. The boy, crouched on his nail keg at the back of the crowded room, knew he smelled cheese, and more: from where he sat he could see the ranked shelves close-packed with the solid, squat, dynamic shapes of tin cans whose labels his stomach read, not from the lettering which meant nothing to his mind but from the scarlet devils and the silver curve of fish – this, the cheese which he knew he smelled and the hermetic meat which his intestines believed he smelled coming in intermittent gusts momentary and brief between the other constant one, the smell and sense just a little of fear because mostly of despair and grief, the old fierce pull of blood. He could not see the table where the Justice sat and before which his father and his father's enemy (*our*

> *enemy* he thought in that despair; *ourn! Mine and hisn both! He's my father!*) stood, but he could hear them, the two of them that is, because his father had said no word yet[.]
> (Faulkner 1950, 3)

Except for the brief italicized passage, the opening scene is in the voice of the third-person narrator. The narrator renders Sarty's perceptions throughout, but his comments on the meaningless letters and the "old fierce pull of blood" reveal that the narrator knows more than Sarty or any other single character. Moreover, the narrator's language clearly distinguishes him from the illiterate child. Not only does he use words such as "hermetic" and "intermittent," but he speaks in the complex, at times monumental diction that Faulkner developed to mastery in his longer novels but often struggled to fit into the limited space of the short story. "Barn Burning" is among his best stories not least because it puts his trademark style to convincing use. The narrative voice introduces a layer of reflection that turns a simple crime story or children's tale into a complex meditation on the individual's sense of place in the social world.

This broader significance is indicated by the opposition the opening scene sets up between Sarty's father and the justice of the peace. By referring to the latter as "the Justice" the story lends him an allegorical quality. Abner seems to be struggling not merely against the social authorities (before whom he is made to stand like a schoolboy) but against the principle of Justice itself. In the course of the story this allegorical layer is elaborated by frequent depictions of Abner in Satanic terms. This motif too is introduced in the very first sentence, where the narrator takes care to point out that among the food are "scarlet devils" (a type of canned meat). The motif reappears in Abner's "stiff foot" and "curled claw," in the "clocklike finality" of his actions, in his continual association with fire, in his wife's religious invocations – "Abner! No! No! Oh, God. Oh, God. Abner!" (20) – and above all in his futile revolt against the god-like power of the landlord, or Lord of the land, whose mansion is described as a gated, flowery paradise. The religious allegory reinforces the impression that "Barn Burning" is not simply the story of a farmer's boy but a reflection on the human condition itself.

In the opening scene, Sarty's bottom-up perspective and his father's schoolboy attitude in front of the justice's desk also establish the stark class hierarchies characteristic of the rural South. The theme is introduced by several related motifs, which reappear throughout the story. Sarty's craving for the canned food reveals the extreme poverty in which his family lives even though food is available in the community. This poverty drives them to instinctive, irrational action, an impression the story sustains with pervasive animal imagery. At the trial Sarty sits "crouched" at the back and later flies at the boy who calls him a barn burner in a "red haze" (5). Abner beats his sons with the same detached brutality with which he strikes his mules. Sarty marvels at his father's "wolflike independence"; elsewhere he compares him to a wasp and a fly (7). Sarty's sisters are twice described as "bovine," and one of them says that they live like hogs in their shack (9). The motif of class barriers is literalized in the many physical barriers the Snopeses confront. The de Spain mansion is guarded by "fence massed with honeysuckle and Cherokee roses," a barrier that serves both to exclude the lower classes and to assert economic and symbolic power through ornamentation (10). Abner in turn resists barriers whenever he can. The initial trial results from his unwillingness to fence in his hog, and afterward, when the family camps in a grove, he uses a "nearby fence" to build his fire instead of getting wood from the trees. Yet the class barriers prevail in the end. When Sarty runs to warn his father that de Spain is coming after him, the fence around the mansion is too high for him to jump over. The barrier secures not only the landlord's possessions but his ability to dole out punishment at the same time – a conflation of powers foreshadowed by Sarty's initial confusion of the mansion with a courthouse.

In adopting Sarty's perspective on these events, the story foregrounds the gradual development of the boy's consciousness and conscience. Like "Dry September," "Barn Burning" is less interested in the crime itself than in the psychological circumstances that lead up to it. The very title of the story gestures toward this shift of interest. It may refer to an actual burning but also to a social and psychological phenomenon: the habit of burning barns that Abner acquires and enacts in the course of the story. Just as "Dry September" describes neither the alleged molestation nor the lynching, "Barn Burning" does not describe any actual burnings. Instead it explores the long, complex process that results in the deed, so that in the limited space of a short story it nevertheless unfolds a comprehensive panorama of the rural South.

3 Theoretical Perspectives

Faulkner's short fiction was read more widely than the novels in his lifetime, yet it has received less consideration from scholars. The first book-length studies did not appear until the 1980s (Skei 1981; Carothers 1985), and surveys of the American short story continue to treat Faulkner as primarily a novelist even while they acknowledge the excellence of his stories (Stevick 1984, 126). Nonetheless, the productivity of Faulkner scholarship in general has resulted in many useful readings of individual stories within broader studies of the work. Many of these readings have been collected for easy reference in two guides to critical readings of his stories (Jones 1994; Towner and Carothers 2006). Readers new to the stories will find helpful introductions in Edmond Volpe's *A Reader's Guide to William Faulkner: The Short Stories* (2004). The main trends in Faulkner criticism are reviewed in Tyler Hagood's *Following Faulkner: The Critical Response to Yoknapatawpha's Architect* (2017).

Readers of "Dry September" and "Barn Burning" generally agree that both stories are among Faulkner's best, and that both encourage ethical and psychological interpretations. "Dry September" has inspired long discussions of who bears what degree of guilt for the lynching. Critics have scrutinized not only the obvious candidates, Minnie and McLendon, but also the men at the barbershop, including Hawkshaw, who could be held accountable for not preventing the lynching more decisively. Another much-debated question has been whether the social constraints and repressions to which the characters are exposed alleviate their guilt (Carey 1964, 27–30; Crane 1985). Scholarship on "Barn Burning" centers on ethical questions as well, especially on Sarty's conflict of loyalty and on Abner's justification for his crimes. Judgments of Abner have ranged all the way from rejection and outright contempt (Howe 1975, 85; Cackett 1989) to appreciation of his social criticism and even his fatherly role (Zender 1989; Atkinson 2006, 196–197). Broadly speaking, there has been a shift from the former to the latter position over time. Sarty has consistently been regarded as the moral hero of the story because he tries to maintain his loyalty and integrity against all odds. Some critics have pointed out, however, that the story can also be read as a dramatization of Sarty's erroneous judgments, from his perception of the first landlord as an enemy to his belief in Abner's stories of the war (Carothers 1985, 60–64).

The prominence of Sarty's inner life in the story motivated a range of psychological interpretations, which often addressed ethical questions as well. In one of the earliest sustained psychological readings of "Barn Burning," Irving Malin regards Sarty as an example of the process of Freudian ego formation. Malin argues that Sarty forms his value system in conflict with various superegos, including his father, whom he eventually rejects, and the "truth and justice" represented by the court (Malin 1957, 81). In a similar vein, Warren Beck describes the story as a "miniature *bildungsroman*" (Beck 1976, 279). Given the different narrative setup of "Dry September," which largely depicts the protagonists from an outside perspective, psychological interpretations of this story have foregrounded the impact of the social environment on character formation and decision making. In a typical assessment, Beck points to the "latent racial hysteria in certain whites which can erupt into mass psychosis" (173; cf. Jenkins 1981). The lack of evidence for Minnie's claims has also encouraged psychological speculation. Critics have offered explanations ranging from "wish-fulfillment" (Malin 1957, 8) and "psychoneurotic" frustration (Carey 1964, 27–30) to issues of power and control (Douglas 1981, 162). Given the parallels between Minnie and McLendon, it is not surprising that the latter's actions have been interpreted in very similar terms (Carothers 1987, 67; Dowling 1989, 146).

The focus on ethics and psychology is representative of early Faulkner scholarship, most of which followed the New Critics in seeking to reveal the deeper meanings of the texts through close readings of their language, structure, and characterization. The interest in structure shows in the many discussions of such aspects as family relations in "Barn Burning" (Trilling 1967; Bradford 1981; Hiles 1985), the parallels between Minnie and McLendon in "Dry September" (Faulkner 1973, 47–50; Beck 1976, 173), and the organization of narrative voice and perspective. In James Ferguson's

(1991) study of Faulkner's short fiction, "Dry September" is one of the most frequently cited examples for structure and point of view. "Barn Burning" has been the subject of narratological analyses such as Wayne C. Booth's classic study of unreliable narration (Booth 1961, 308–309).

Another structural feature that caught the attention of early critics were the intertextual relationships among the novels and stories Faulkner set in Yoknapatawpha County. Such comparative readings often looked for complementary information across different works and noted contradictions between specific texts. "Dry September," for example, has strong links to the story "Hair" (1931), which appeared in the same year and recounts Hawkshaw's long courtship of a local girl. While some critics drew on "Hair" to illuminate the barber's motivations in "Dry September," others pointed out that his personality differs strongly between the two stories (Jones 1994, 143–144). Similarly, "Barn Burning" was often compared to *The Hamlet*, which summarizes the events related in the story and gives us a detached, amused outside perspective very different from Sarty's (Carothers 1985, 109–124; Jones 1994, 8–9). Many of these intertextual readings downplayed the numerous inconsistencies in the oeuvre and sustained the assumption that Faulkner's works formed a coherent 'saga' of Yoknapatawpha County (Cowley 1946, 6–9).

The notion of a coherent saga reflects the New Critics' assumption that great works of literature were unified by the author's creative genius. This emphasis on unity and authorial control came under suspicion from the 1960s onward, when French poststructuralism shifted scholarly attention to the disconnections and misunderstandings within texts, and also to the sociopolitical debates in which texts participate. This shift arrived in Faulkner criticism about two decades later and foregrounded previously neglected elements of his fictional world: the oppressed, marginalized, and disabled people of Yoknapatawpha County, but also those parts of the oeuvre that had previously been disregarded because they did not meet the New Critics' criteria for modernist masterpieces. "Dry September" benefited more from this shift than "Barn Burning." The latter had already been recognized as a masterpiece, and it pays little attention to the favorite topics of the new political criticism, race and gender. One of the few readings in this vein is Jennie Joiner's (2010) comparative essay on the construction of black masculinity in "Barn Burning" and Flannery O'Connor's story "The Artificial Nigger" (1955). Joiner argues that Abner Snopes, like other Southern men, casts Sarty in the role of a racial other against which Abner can define his white masculinity.

The social construction of race and gender plays a more prominent role in "Dry September," as we have seen, and in the 1990s much of the debate on the story revolved around Faulkner's depiction of these dynamics. Anne G. Jones argued that the story depicts an "ideology of penetration," where male sexuality constitutes itself in "the act of entering an 'other'" (Jones 1995, 129). James Heffernan read the story as a manifestation of the "forgotten trauma of lynching" (Heffernan 1999,

801). Faulkner witnessed a lynching himself as a child, in 1908, when a black convict called Nelse Patton murdered a white woman and was lynched by a crowd of several thousand people in Oxford, Mississippi. The shift in literary scholarship toward historical contextualization brought new awareness of this event, which featured prominently in Joel Williamson's *William Faulkner and Southern History* (Williamson 1996, 157–162) and has since been mentioned regularly in critical readings of "Dry September" (Atkinson 2006, 157; Bleikasten 2017, 12). The recent reception of "Barn Burning" has also stressed historical context, particularly that of the Great Depression period in which Faulkner was writing the story. This has resulted in a revalorization of Ab Snopes, who is now often read as a socialist critic of class hierarchies in the New South (Godden 1997, 123–125; Lessig 1999; Atkinson 2006, 194–220). The wide variety of critical readings highlights the range of topics, approaches, and perspectives that Faulkner managed to incorporate in both "Dry September" and "Barn Burning" – owing not least to his innovative experimentation with narrative technique. It is this complexity of form and content that has made both stories enduring classics of American short fiction.

4 Bibliography

4.1 Works Cited

Atkinson, Ted. *Faulkner and the Great Depression: Aesthetics, Ideology, and Cultural Politics.* Athens: U of Georgia P, 2006.
Bache, William B. "Moral Awareness in 'Dry September.'" *Faulkner Studies* 3 (1954): 53–57.
Beck, Warren. *Faulkner: Essays by Warren Beck.* Madison: Wisconsin UP, 1976.
Bleikasten, Andre. *William Faulkner: A Life Through Novels.* Bloomington: Indiana UP, 2017.
Booth, Wayne C. *The Rhetoric of Fiction.* Chicago: Chicago UP, 1961.
Bradford, M. E. "Family and Community in Faulkner's 'Barn Burning.'" *Southern Review* 17.2 (1981): 332–339.
Cackett, Kathy. "'Barn Burning': Debating the American Adam." *Notes on Mississippi Writers* 21 (1989): 1–17.
Carey, Glenn O. "Social Criticism in Faulkner's 'Dry September.'" *English Record* 15.2 (1964): 27–30.
Carothers, James B. *William Faulkner's Short Stories.* Ann Arbor: UMI Research, 1985.
Carothers, James B. "'I Ain't a Soldier Now': Faulkner's World War II Veterans." *Faulkner Journal* 2 (1987): 67–74.
Cowley, Malcolm. "Introduction." *The Portable Faulkner.* Ed. Cowley. New York: Viking, 1946. 1–26.
Cowley, Malcoln, ed. *Writers at Work: The* Paris Review *Interviews.* London: Secker & Warburg, 1958.
Crane, John K. "'But the Days Grow Short': A Reinterpretation of Faulkner's 'Dry September.'" *Twentieth Century Literature* 31 (1985): 410–420.
Creighton, Joanne V. *William Faulkner's Craft of Revision: The Snopes Trilogy,* The Unvanquished, *and* Go Down, Moses. Detroit: Wayne State UP, 1977.
Douglas, Ellen. "Faulkner's Women." *A Cosmos of My Own: Faulkner and Yoknapatawpha, 1980.* Eds. Doreen Fowler and Ann Abadie. Jackson: UP of Mississippi, 1981. 149–167.

Dowling, David. *William Faulkner*. New York: St. Martin's P, 1989.
Faulkner, William. *Collected Stories of William Faulkner*. New York: Random House, 1950.
Faulkner, William. *Lion in the Garden: Interviews with William* Faulkner, *1926–1962*. Eds. James B. Meriwether and Michael Millgate. Lincoln: U of Nebraska P, 1980.
Faulkner, William. *The Uncollected Stories*. Ed. Joseph Blotner. New York: Random House, 1979.
Ferguson, James. *Faulkner's Short Fiction*. Knoxville: U of Tennessee P, 1991.
Godden, Richard. *Fictions of Labor: William Faulkner and the South's Long Revolution*. New York: Cambridge UP, 1997.
Hagood, Taylor. *Following Faulkner: The Critical Response to Yoknapatawpha's Architect*. Rochester: Camden House, 2017.
Heffernan, James. "The Simpson Trial and the Forgotten Trauma of Lynching: A Response to Shoshana Felman." *Critical Inquiry* 25.4 (1999): 801–806.
Hiles, Jane. "Kinship and Heredity in Faulkner's 'Barn Burning.'" *Mississippi Quarterly* 38.3 (1985): 329–337.
Howe, Irving. *William Faulkner: A Critical Study*. Chicago: Chicago UP, 1975.
Jenkins, Lee. *Faulkner and Black-White Relations: A Psychoanalytic Approach*. New York: Columbia UP, 1981.
Joiner, Jennie. "Constructing Black Sons: William Faulkner's 'Barn Burning' and Flannery O'Connor's 'The Artificial Nigger.'" *Flannery O'Connor Review* 8 (2010): 31–47.
Jones, Anne G. "Desire and Dismemberment: Faulkner and the Ideology of Penetration." *Faulkner and Ideology: Faulkner and Yoknapatawpha 1992*. Eds. Donald M. Kartiganer and Ann Abadie. Jackson: UP of Mississippi, 1995. 129–171.
Jones, Diane B. *A Reader's Guide to the Short Stories of William Faulkner*. New York: Hall, 1994.
Lessig, Matthew. "Class, Character, and 'Croppers: Faulkner's Snopeses and the Plight of the Sharecropper." *Arizona Quarterly* 55.4 (1999): 79–113.
Malin, Irving. *William Faulkner: An Interpretation*. New York: Gordian, 1957.
Pascoe, Craig S., Karen T. Lethem, and Andy Ambrose, eds. *The American South in the Twentieth Century*. Athens: U of Georgia P, 2005.
Stein, William Bysshe. "Faulkner's Devil." *Modern Language Notes* 76.8 (1961): 731–732.
Stevick, Philip. *The American Short Story 1900–1945: A Critical History*. Boston: Twayne, 1984.
Taylor, Walter. *Faulkner's Search for a South*. Urbana: U of Illinois P, 1983.
Towner, Theresa, and James B. Carothers. *Reading Faulkner: Collected Stories*. Jackson: UP of Mississippi, 2006.
Trilling, Lionel. *The Experience of Literature: A Reader with Commentaries*. New York: Holt, Rinehart, 1967.
Volpe, Edmond L. "'Barn Burning': A Definition of Evil." *Faulkner: The Unappeased Imagination*. Ed. Glenn O. Carey. Troy: Whitston, 1980. 75–82.
Volpe, Edmond L. *A Reader's Guide to William Faulkner: The Short Stories*. New York: Syracuse UP, 2004.
Williamson, Joel. *William Faulkner and Southern History*. Oxford: Oxford UP, 1996.
Wyatt-Brown, Bertram. *Southern Honor: Ethics and Behavior in the Old South*. New York: Oxford UP, 1982.
Zender, Karl F. "Character and Symbol in 'Barn Burning.'" *College Literature* 16.1 (1989): 48–59.

4.2 Further Reading

Blotner, Joseph. *Faulkner: A Biography*. New York: Random House, 1974.
Brooks, Cleanth. *William Faulkner: First Encounters*. New Haven: Yale UP, 1983.

Howe, Irving. *William Faulkner: A Critical Study*. Chicago: U of Chicago P, 1981.
Kerr, Elizabeth M. *Yoknapatawpha: Faulkner's 'Little Postage Stamp of Native Soil.'* 1969. New York: Fordham UP, 1976.
Oates, Stephen. *William Faulkner: The Man and the Artist*. New York: Harper & Row, 1987.
Parini, Jay. *One Matchless Time: A Life of William Faulkner*. New York: HarperCollins, 2004.
Skei, Hans. *William Faulkner: The Novelist as Short Story Writer: A Study of William Faulkner's Short Fiction*. Oslo: University Press, 1985.

Mark S. Graybill

19 Flannery O'Connor (1925–1964)

Abstract: Flannery O'Connor's short fiction is generally regarded as "southern grotesque," a label she encouraged. This essay explains the term as used by O'Connor and other theorists and extends its meaning via O'Connor's notion (borrowed from Thomas Mann) that the "grotesque is the true anti-bourgeois style." This idea makes her stories, particularly "Good Country People" and "A View of the Woods," interpretable as critiques of philistinism, and further suggests intertextual connections with the work of Vladimir Nabokov, whom she named as a possible influence. The tales also suggest the promise and peril of applying disabilities studies and ecocritical perspectives to O'Connor's work.

Keywords: grotesque, philistinism, *künstlerroman*, disabilities studies, ecocriticism

1 Context

Grotesque: this is the default descriptor of Flannery O'Connor's work. Indeed, Robert C. Evans, with only a little hyperbole, claims that "if there is one term used by practically every commentator on O'Connor from the mid-1950s to the early twenty-first century, 'grotesque' is that term" (Evans 2016, 55). In its nascent, sixteenth-century European form, it implied an "ornamental style suggested by antiquity, understood not only [as] something playfully gay and carelessly fantastic, but also [as] something more ominous and sinister in the face of a world totally different from the familiar one" (Kayser 1963, 21). While artists in subsequent periods and movements update the grotesque in certain ways, it always involves, says Wolfgang Kayser, "an attempt to invoke and subdue the demonic aspects of the world" (188). Geoffrey Harpham offers his own definition, even while calling it "the slipperiest of aesthetic categories," having evolved far from its origins in the Renaissance, when the term described ancient murals discovered in "caves [*grotto*] near Rome" (Harpham 1976, 461). Its basic features are strangeness/uncanniness, incongruity, and the subversion of norms and expectations; it "depends not only on physical conditions the deformity of which most people would recognize, but also on our conventions, our prejudices, our commonplaces, our banalities, our mediocrities" (463). A genuinely grotesque work "must arouse [. . . l]aughter and astonishment [. . . and] either disgust or horror" (463). Among a short list of literary artists who employ it, Harpham mentions O'Connor, though without specific analysis of her fiction. His sketch, however, accords surprisingly well with her work, including the nonfiction, which frequently invokes the grotesque directly or allusively.

https://doi.org/10.1515/9783110587647-020

In "Some Aspects of the Grotesque in Southern Fiction," O'Connor defines this mode with more circumspection and nuance than elsewhere (e.g., "The Fiction Writer and His Country"). The grotesque connects "concrete" reality with something ineffable and mysterious which is "just as real to [the writer . . .] as the [reality] that everybody sees"; authors employing it "use the concrete in a more drastic way" than traditional realists – which is to say, theirs is "the way of distortion" (O'Connor 1969, 42). Such fiction is "wild [. . .] violent and comic, because of the discrepancies that it seeks to combine" (43). O'Connor also approvingly cites Thomas Mann's claim that "the grotesque is the true anti-bourgeois style," but suggests that its power is largely lost on American readers, who have "managed to connect [it] with the sentimental." When "used in a legitimate way," the grotesque will give "the intellectual and moral judgments implicit in it [. . .] ascendency over feeling" (43). Thus, her basic aesthetic and heuristic assumption becomes clear: readers should judge, not identify with, her characters, and the grotesque – if both she and her audience have the correct sense of it – will lead to this result. O'Connor aims to compel readers to set sentimentality aside and see her characters – and humanity itself – as spiritually incomplete, maimed, deformed, and both shockingly and hilariously ugly.

Her concern with the audience's response echoes Kayser's observation that "the grotesque is experienced only in the act of reception" (Kayser 1963, 181). The resonances between Harpham's and O'Connor's conceptions of the grotesque, however, are even clearer. In order to be moved to "intellectual and moral judgements," the reader must experience seemingly contradictory experiences of "[l]aughter and astonishment [. . . and] disgust or horror" that Harpham posits as requisite – the only reasonable reactions, O'Connor would say, to such incongruous or discrepant admixtures. In fact, Harpham's schema defines the grotesque as either primarily "comic or terrible" (Harpham 1976, 464). O'Connor's comments on her fiction often suggest she saw herself as a practitioner of the former; for her, the "wild" characters, objects, and scenarios her imagination produces are funny first and foremost: "I read [my stories] over and over and laugh and laugh" (O'Connor 1979, 80–81). The stimulus for shock and laughter surely involves the recognition of physical or spiritual "deformity," to use Harpham's word, or, in O'Connor's synonymous term, "distortion." O'Connor's invocation of Thomas Mann is interesting. The phrase "anti-bourgeois style" echoes Harpham's observation that the grotesque represents not just the extraordinary and fantastic – though it certainly does that – but also exposes and interrogates "our conventions, our prejudices, our commonplaces, our banalities, our mediocrities." By "our," Harpham presumably means American, middle class, (likely white) readers. Relatedly, while O'Connor is usually labeled a southern artist, an identity she embraced, her citing Mann's implicit critique of bourgeois values illustrates her surprisingly cosmopolitan tastes. In fact, the quotation actually comes from Mann's introduction to Joseph Conrad's "The Secret Agent" (Conrad 1933, 241). Aside from Henry James (another influence of international renown; ↗12 James), few fiction writers are cited in O'Connor's nonfiction more frequently than Conrad. Though she calls herself "peculiarly possessed of the modern consciousness [. . . which psychologist Carl] Jung describes as unhistorical, solitary, and guilty" (O'Connor 1979, 90), like her modern literary heroes, her imagination – including her fascination with the grotesque – was sparked by the larger historical context. She deliberately observed, reflected, and evaluated the regional, national, and even international anxieties of mid-century America: she knows, for example, that "the bomb that was dropped on Hiroshima affects life on the Oconee River, and there's nothing [the artist] can do about it" (O'Connor 1969, 77). O'Connor would likely agree with Harpham's statement that "no longer is the grotesque a method of portraying only the distorted inner landscapes of the diseased or neurotic imagination [. . .]: in a bomb-dominated, anxious time, objective reality, revealed to man by his most reliably 'realistic' methods of observation, provides the stimulus for the grotesque" (Harpham 1976, 467).

O'Connor's imagination is complex, reflecting her engagement with the South, Realism, Modernism, New Criticism, and even the debilitating Systemic Lupus Erythematosus (SLE), from which she suffered most of her life, and which led to her death at thirty-nine. But it was her religion that the "cradle Catholic" O'Connor regarded as definitive. "I see from the standpoint of Christian orthodoxy," she proclaims bluntly in "The Fiction Writer and His Country" (O'Connor 1969, 32). Measured against the perfection of God, in a modern, largely secular society that has, in her view, willfully suppressed the memory of that perfection, human beings nearly always appear to her as incomplete, fractured, deformed – i.e., ugly – in their thoughts and behavior. What they lack inside often shows externally, in facial expressions, gestures, and other physical features. As one "who see[s] by the light of [her] Christian faith," O'Connor implicitly situates herself among those artists who possess "the sharpest eyes for the grotesque, the perverse, and for the unacceptable" (33); she felt compelled to create vivid, arresting images and scenarios for an audience unable to apprehend that which they should find disturbing in modern life. "When you can assume that your audience holds the same beliefs you do," she explains, "you can relax a little and use more normal means of talking to it; when you have to assume that it does not" – which, in O'Connor's case, is nearly always – "then you have to make your vision apparent by shock – to the hard of hearing you shout, and for the almost-blind you draw large and startling figures" (34). To do this, she had to go "the way of distortion," and her characters' bodies frequently bear the brunt of it.

Literally large figures, mostly women, abound in the short stories: the mountainous Mrs. Shortley in "The Displaced Person," the inflated (outside and in) Ruby Turpin in "Revelation," the hulking Joy/Hulga Hopewell in "Good Country People" (who I will call Hulga hereafter, for ease of reference). In contrast, Sarah Ruth of "Parker's Back" is outsized in her starkness, her taut body and face indicating the austerity of her fundamentalist religious piety, "the skin on her face [. . .] thin and drawn as tight as the skin of an onion and her eyes [. . .] gray and sharp like the points of two icepicks" (O'Connor 1988, 65). The "icepick" eyes, which suggest

O'Connor's own hyper-perceptive vision (a somewhat misleading detail about this ultimately rather myopic character), also recall Mrs. Freeman, an important secondary character in "Good Country People": her eyes "never swerve [. . .] to left or right [. . .] as if they follow [. . .] a yellow line down the center" (263). Other characters stand out for their impaired senses and/or damaged, sometimes absent, body parts: Hulga, who relies on a prosthetic leg and suffers from both a heart condition and near-sightedness; Tom Shiftlet ("The Life You Save May Be Your Own"), who is missing half of his left arm; Shiftlet's deaf and intellectually challenged bride, Lucynell Crater; and in Rufus Johnson ("The Lame Shall Enter First"), a delinquent who wields his club foot "like a weapon" (601). But some of the figures whom readers are likely to label "grotesque" because of physical disabilities or other "abnormalities" seem created to highlight the flaws of those around them. With varying degrees of violence (and success), Johnson, Mary Grace ("Revelation"), and, most uniquely, the hermaphrodite in "A Temple of the Holy Ghost," all shock the protagonist into an awareness of his or her own moral disabilities.

What O'Connor finds most disturbing in her characters, and in the real-life human beings they reflect, is pride. All the protagonists suffer from some form of it. As Michael Raiger observes, "[t]he grotesque in [. . .] O'Connor's work is sin reduced to its essential form in pride; the grotesque is the idol of pride, the distortion of the soul in its transgression against its own natural form; it is the creature in its limitations attempting to become a god" (Raiger 1998, 266). This most pervasive of sins makes them vulnerable to the psychological, and hence spiritual, deformities of delusion and obsession, ushering them inevitably into extreme, violent, and potentially perspective-altering situations. Harpham writes that in the modern grotesque, "the neurotic himself, or at least the outsider, has come to feel himself custodian of the height of reason [. . .] the standard of sanity [. . . and] the world has itself become more and more hallucinatory" (Harpham 1976, 467). In her novels, O'Connor regards figures such as Hazel Motes (*Wise Blood*) and Old Tarwater (*The Violent Bear It Away*) as the "standard of sanity" in a world gone mad. For secular readers, she says that the former's "integrity lies in his trying with such vigor to get rid of the ragged figure [of Christ] who moves from tree to tree in the back of his mind," but for her, "Hazel's integrity lies in his not being able to do so" (O'Connor 1988, 1265). Similarly, about the raging fundamentalist Mason Tarwater, O'Connor says, "the old man [. . .] speaks for me" (O'Connor 1979, 350). In her short fiction, by contrast, the main characters usually regard themselves as gatekeepers of reason, but pay handsomely – or rather, unhandsomely – for their delusional arrogance. Such is the case, for example, with Mrs. May ("Greenleaf"), Mr. Fortune ("A View of the Woods"), and one of O'Connor's most well-known characters, the Grandmother ("A Good Man is Hard to Find"), whose generic moniker implies the near universality of her (self-)destructive pride. Each enacts some form of self-worship, usually motivated by a powerfully ingrained sense of racial and/or class superiority. The violence they experience is often less grotesque than the ridiculous behavior that gives rise to it.

Still, her tales are known for their endings, where the result of such behavior becomes manifest. A white woman dies of an apparent heart attack after a black woman whom she has insulted hits her in the head with a purse ("Everything That Rises Must Converge"). A farm owner is gored by her neighbor's bull ("Greenleaf"). Members of a vacationing family are assassinated, one-by-one, by a group of escaped convicts, with the grandmother being shot three times through the chest ("A Good Man Is Hard to Find"). A young boy seeking the Kingdom of God drowns in a river ("The River"). Another neglected boy hangs himself in order to join his dead mother in heaven ("The Lame Shall Enter First"). Such bleak endings raise a question with which every O'Connor critic must wrestle at some level: of what value, ultimately, is the grotesque in her hands? Staking out what might be called the "negative" position early on was the novelist, John Hawkes, a friend of O'Connor's with whom she argued extensively about the meaning of her work. Hawkes sees in her a great satirist who, like her predecessor, Nathanael West, possesses an "inverted attraction for the reality of our absurd condition," and unsparingly "demolish[es] 'man's image of himself as a rational creature' [. . .] exploit[ing . . .] the syntax of the devil" to do so (Hawkes 1962, 396, 397, 398). In this most "vigorously moral of writers," he argues, "the actual creation of fiction seems often to depend on immoral impulse" (398). Her vision, then, seeks not so much "to invoke and subdue the demonic aspects of the world" (Kayser 1963, 188), but to let them loose. Josephine Hendin (1970, 29), less charitably than Hawkes, notes the emotional, psychological, and spiritual flatness of O'Connor's characters and her revulsion by the human body (which she compares to that of William S. Burroughs). There "is at least a possibility," writes Hendin, "that her religion provided a legitimate sanction for violent and destructive impulses [. . .] which became acceptable when [. . .] directed at the 'godless'" (16). Similarly, Claire Kahane (née Katz) underscores the "sadistic quality of [O'Connor's] narrator, who acts as an archaic superego" that allows the author "to express precisely those demonic impulses which she must punish in them" (Kahane 1974, 57), suggesting "a psychological demand which overshadows her religious intent" (56). Taking more account of O'Connor's humor, André Bleikasten argues that it "seldom provides comic release" (Bleikasten 1985, 138). And grotesqueness goes beyond characters to infect the whole landscape, the entire geographical universe, they inhabit; with their "fierce, fiery suns [. . .] blank or blood-drenched skies [. . . and] ominous woods," they "are landscapes of nightmare." While "O'Connor defended her use of the grotesque as a necessary strategy [. . .] one is left with the impression that in her work it eventually became the means of a savage revilement of the whole creation" (142). This last claim, echoed by other critics, O'Connor herself would have rejected as a misreading of her incarnational imagination – that is, her Catholic habit of tracing God's omnipotence and omnipresence in every aspect of creation, no matter how seemingly lowly, vile, disturbing, disgusting. Regarding the violence in her work, O'Connor felt it "strangely capable of returning [her] characters to reality and preparing them to accept their moment of grace" (O'Connor 1969, 112).

More critics agree with O'Connor than with Hawkes and company. In various ways, Richard Brinkmeyer (1989), John Desmond (2010), Richard Giannone (1999), Christina Bieber Lake (2005), and Ralph Wood (2004), to name a few, have argued for the broadly redemptive function of O'Connor's stories. Gilbert H. Muller, who authored the first book-length study of O'Connor and the grotesque, belongs to this group. In his view, her fiction exaggerates an already distorted world in which characters travel "dark and impenetrable mazes which seemingly lead nowhere, but which unexpectedly reveal an exit into Christianity's back yard." (Muller 1972, 18) The stories unflinchingly address the jarring "contradictory aspects" O'Connor witnessed, turning the whole scene into a quasi-realistic representation that ultimately "embodies a transcending principle of order" (18). Delma Eugene Presley, another early defender of O'Connor's "moral" use of the grotesque, regards the author's Catholicism and southernness as paramount. Elaborating on O'Connor's oft-quoted description of the South as "Christ-haunted" (O'Connor 1969, 44), Presley contends that the region, with its strong tradition of evangelical Protestantism, allows a native such as O'Connor to "bear witness about human possibilities in the light of God's grace as well as the inhuman possibilities of lives lived without that light" (Presley 1972, 39–40). Her faith in O'Connor's ability to harmonize her aesthetic intentions with her art is unshakable: she likens the author to a tailor working with a three-way mirror wherein "each section is broken and often pieces are completely missing" – a nifty metaphor for the fractured, distorted, modern world. Yet the author-tailor can "use the mirror with no problems because [she] remembers what it was before it was broken and has confidence about what the whole man is supposed to resemble" (39). Thus, it is not O'Connor who distorts, but the world she describes.

Two later, more sophisticated, formulations of O'Connor's "positive grotesque" are offered by Marshall Bruce Gentry and Anthony DiRenzo. Starting with O'Connor's comment that the grotesque might sometimes be thought of as pointing to "the good [. . .] under construction," Gentry argues that her characters struggle to uncover that good in the unconscious: "[t]he typical O'Connor protagonist is oppressed by degradations of [. . .] society's ideals: by the economy of the [s]outh, by ignorance, by physical deformity and disease, by systems of class and race, and by the strictures of race" (Gentry 1986, 14–15). On the conscious level, characters succumb to these forces, but their unconscious – which, viewed through a Christian lens, leads to God – often asserts itself in order to "transform that grotesquerie into a force of redemption" (15). Thus, Hulga unconsciously "desires her betrayal by Manley Pointer" (115), the crooked Bible salesman, so that she may begin the process of jettisoning her consciously created, ugly identity (Hulga) and returning to her true self (Joy). Yet Gentry acknowledges, directly or tacitly, that some characters do not undergo this process of redemption: In "The Lame Shall Enter First," Sheppard –

an intellectual like Hulga – neglects his suffering son in favor of Rufus Johnson, but unlike her, "his unconscious maneuvers are consistently self-serving rather than self-redeeming" (155). Tellingly, Gentry includes no discussion of an apparently irredeemable figure such as Mark Fortune in "A View of the Woods." DiRenzo mounts a more complex and spirited defense. Drawing heavily upon Mikhail Bakhtin's theory of the carnivalesque, he traces O'Connor's use of the grotesque to medieval folk art, wherein representation of the human body is "gay, triumphant, and at the same time mocking, deriding" (DiRenzo 1993, 65). This paradigm attempts to account properly for an unappreciated tonal complexity in O'Connor's fictional universe, which "is a weird, threatening [one], by turns bewildering and delightful [. . .] we find ourselves in an enclosed passage between the sacred and profane, the chapel and the square, heaven and hell, our eyes endlessly tracing the surface of startling, twisted figures" (15). One such figure is Hulga, whom DiRenzo calls (debatably, I think) "[u]ndoubtedly [. . .] O'Connor's most grotesque creation" (74). He likens the tale in which she lives to a medieval fabliau that mixes disparate elements in a garishly comic way consistent with O'Connor's belief in Catholic sacramentalism. Here and elsewhere, O'Connor "stresses the carnality of the human body to express divinity, without resorting to sublimation" (81). Like the author herself, DiRenzo "laughs and laughs" at a character such as Hulga. It is presumably in this "positive" response that redemption might lie. Yet, here is another theory of O'Connor's grotesque that seems ill-equipped to account for darker stories such as "The Lame Shall Enter First" and "A View of the Woods," both of which merit hardly a mention. Thus, O'Connor's use of the grotesque continues to trouble.

2 Close Readings

O'Connor saw herself as a "thirteenth century" kind of Catholic (Gooch 2007, 23), so DiRenzo's medieval-centric approach would win her approval. I propose, however, a reading of two stories informed by historical and literary contexts a bit closer to her home/time: namely, mid-twentieth century America, a milieu characterized (from O'Connor's perspective) by shallowness and arrogance – not only among the elitist, secular audience she so overtly targeted in her essays, but also among the seemingly religious middle class whose banality and moral complacency she finds just as unacceptable.

Discussing "A Good Man Is Hard to Find," DiRenzo passingly mentions one of O'Connor's most famous literary contemporaries. The title story of her first collection, he suggests, "is a religious parable that refuses to transcend *poshlost*, what Nabokov called the essential banality or gaucheness of ordinary human existence" (DiRenzo 1993, 146). *Poshlost* – which, roughly translated into English, means "philistinism," punningly transformed by Nabokov into "posh-lust" – deserves further attention, especially as it relates to "Good Country People," the tale with which

O'Connor chose to end that collection. The protagonist, Joy Hopewell (who has rechristened herself Hulga), has a host of physical ailments, including poor eyesight, a heart condition, and, most prominently, a prosthetic leg. The 32-year-old Ph.D. lives on a farm, where she endures the constant chirpiness of her mother and the prying eyes and general coarseness of Mrs. Hopewell's tenant, Mrs. Freeman. When a seemingly innocent Bible salesman named Manley Pointer shows up, Hulga decides to seduce him with her nihilistic worldview – only to learn, quite painfully, that he is a more committed nihilist than she. After an abortive romantic tryst in the barn, he takes her spectacles and her wooden leg – and, ultimately, her dignity, hissing as he leaves her helpless in the hayloft, "you ain't so smart. I been believing in nothing ever since I was born" (O'Connor 1988, 283).

A few years after writing "Good Country People," O'Connor advised her frequent correspondent, Betty Hester, "I tell you if you don't know Nabokov, you ought to" (O'Connor 1979, 339). About the controversial author of *Lolita* (1958), she also wrote to Hawkes, "I have always liked Nabokov, beginning with the one I read called *Bend Sinister*. That was a long time ago and I have forgotten everything about it except that I was impressed, even possibly influenced" (343–344). Many aspects of the Russian émigré's work may have appealed to O'Connor, but I suspect his disdain for philistinism is the most meaningful. Ole Nyegaard sheds light on Nabokov's use of this concept, explaining that philistinism is "the dogma of preconceived ideas [. . . in which] communion is not with the Holy Spirit, or any such concept, but with the spirit of the group" (Nyegaard 2004, 342–343). By embracing without question the conventional ideas of a group to whom they belong (or fervently wish to), philistines speak in clichés, ritualized language emptied of its meaning. The hollow expressions of the philistine bespeak an unearned arrogance that distinguishes, sometimes subtly, between those in the club and those outside. Nabokov himself associates philistinism with the terms *genteel* and *bourgeois*. He writes that

> [g]enteel implies the lace-curtain refined vulgarity which is worse than simple coarseness. To burp in company may be rude, but to say "excuse me" after a burp is genteel and thus worse than vulgar. The term *bourgeois* I use following Flaubert, not Marx. Bourgeois in Flaubert's sense is a state of mind, not a state of pocket. A bourgeois is a smug philistine, a dignified vulgarian. (Nabokov 1981, 309)

Such smugness is bad enough kept to oneself, but the true philistine imposes his or her "dignified vulgarity" on others. Placing the term in the context of church history, Nyegaard notes that "[p]hilistines not only refuse genuine religious insight, but also deny others access to any understanding" (Nyegaard 2004, 344).

Such observations call to mind O'Connor's interest in the grotesque as the "true anti-bourgeois style" (O'Connor 1969, 43). She cites Mann's phrase several times in her nonfiction, and her fiction has no shortage of characters seeking membership in the bourgeois club, even if they lack the vocabulary to describe their desires. Often, they are the class-conscious, middle-aged women who populate her best-known stories. Mrs. Hopewell certainly fits this mold, but so does Hulga, who rejects her mother's bourgeois values. In her own, more grotesque way, Hulga demonstrates a more consequential refusal of "genuine religious insight." Blindness – to one's own flaws, to the dangerous predilections of others, to God's presence in the world – is a persistent theme in O'Connor. Hulga stands as the most famous blind character this side of *Wise Blood*'s Hazel Motes. Among the many things that she cannot see is that, as much as she would like to separate herself from her mother, she suffers from a similar myopia. With her penchant for clichés and her blithe acceptance of appearances, Mrs. Hopewell cuts a most Nabokovian figure. Her pretension to gentility O'Connor subverts by making her blind to the absence of it in the Freemans, whom she categorizes undiscerningly as "good country people." She comically calls Mrs. Freeman "a lady," and while to the reader Mrs. Freeman may appear the baser person, because, for example, she brings up in polite conversation "how many times [her pregnant daughter] had vomited since the last report" (O'Connor 1988, 264), Mrs. Hopewell is the philistine. Recall Nabokov's pithy statement quoted earlier: "[t]o burp in company may be rude, but to say 'excuse me' after a burp is genteel and thus worse than vulgar" (Nabokov 1981, 309). Mrs. Freeman may burp, but Mrs. Hopewell excuses her.

Mrs. Hopewell also preaches common sense – a concept Nabokov detests. She wishes that her intellectual daughter possessed the down-to-earth sense of Mrs. Freeman's girls, Glynese and Carramae (the sourness of whom she sweetens, as is her habit, by calling them Glycerin and Caramel): "there were not many girls with Glynese's common sense. [Mrs. Hopewell] said what she admired in those girls was their common sense" (O'Connor 1988, 274). Nabokov contends that "the biography of common sense makes nasty reading," complains that it "has back-kicked dirt at the loveliest of queer paintings because a blue tree seemed madness to its well-meaning hoof," calls it "fundamentally immoral, for the natural morals of mankind are as irrational as the magic rites that they evolved since the immemorial dimness of time" (Nabokov 1980, 372). Though she would not state it quite the same way, O'Connor also disdains the kind of conventional, unimaginative thinking that murders mystery. In Nabokov's canon, there are many Mrs. Hopewells (though never with the rural, southern inflection of O'Connor's character). In *Pnin*, one of O'Connor's favorites, two stand out: Betty Bliss, a former student and potential love interest of professor Timofey Pnin's, and Judith Clyde, who introduces his lecture to her Cremona Women's Club. While these women's patina of education seems to put them in a different class than Mrs. Hopewell, beneath it lies the same empty politeness. The former bears a name that surely delighted O'Connor, who usually devised similarly transparent or ironic appellations. Like Mrs. Hopewell, Betty Bliss is an unfailingly cheerful, conventional, ignorant woman with a "servant maid's mind" (Nabokov 1985, 152). Described as having "myopic gray eyes [that] peered at you with [. . .] ingenuous sympathy" (151–152), she reacts to a decorative bowl of Pnin's with the same depthless gaiety we might expect from Mrs. Hopewell: "My, what a lovely thing!" (153). Mrs.

Clyde's introductory speech echoes hollowly with the same type of meaningless expressions that riddle Mrs. Hopewell's conversations. "He hardly needs any introduction, of course, and we are all happy to have him," she says witlessly about the guest speaker. "I am sure you will all be delighted to know that there is a grand surprise in store for all of us" she declares, referring to the lecturer who will follow Pnin (26–27). As Nyegaard explains, "[t]his presentation is devoid of content, being purely ritual. What may sound like utter nonsense serves as the opening prayer before a social communion" (Nyegaard 2004, 342).

O'Connor deplores such ritualized language, precisely because it has been separated from religious meaning and purpose. So, when Mrs. Hopewell tells Manley Pointer that "good country people are the salt of the earth!" (O'Connor 1998, 271), she speaks as a philistine: "salt of the earth" is divorced from its original context in Christ's Sermon on the Mount, which is itself more complex than Mrs. Hopewell can dream. Salt was an invaluable commodity in the ancient world, so much so that Roman soldiers received it as payment for their work, and in his sermon, Christ uses it both to motivate and admonish his disciples: "Ye are the salt of the earth: but if the salt have lost his savour, wherewith shall it be salted? It is thenceforth good for nothing, but to be cast out, and to be trodden under foot of men" (Matthew 5:13). Like so many of Christ's metaphors, this is somewhat open to interpretation: Does salt's white color symbolize purity? Is Christ referring to its purifying properties, or perhaps its efficacy as a (moral) preservative? By "savour," does he suggest that those who speak the Word of God must make it appealing (i.e., tasty) to potential converts? Salt's association with true religious insight, however, is clear. If it loses that quality, it becomes worthless, and must be thrown into the road as refuse. Surely, Mrs. Hopewell does not understand the significance of her words, especially given that her Bible is "in the attic somewhere" – a fact about which she lies to Pointer (O'Connor 1988, 271). The phrase is doubly ironic, considering that Pointer, the fake Bible salesman and true unbeliever, consciously disobeys Christ's directive, and that Mrs. Hopewell herself professes a belief that has "lost [its] savour" due to neglect, and is therefore "good for nothing."

O'Connor intends for the reader to judge Hulga even more harshly, because her blindness is even more severe, exacerbated by the education that she attempts to weaponize. If Mrs. Hopewell demonstrates a desire to belong to the club of genteel ladies (she has to convince herself that Mrs. Freeman is not "trash," because she would not socialize with such people), her daughter makes plain that she is a member of the intellectuals (or as O'Connor sometimes snidely pronounced it, "interleckchuls"). O'Connor tends to be much harder on her learned characters than Nabokov is on his. Still, whatever affection Nabokov seems to harbor for Pnin, *Pale Fire*'s Charles Kinbote (1962), *Ada*'s Van Veen (1970), and even (or especially) *Lolita*'s Humbert, ultimately their flaws are clearly damnable. As Nyegaard points out, Lolita's mother, Charlotte Haze, "fits the description of the philistine" because she demonstrates (and here he quotes Nabokov) a "passionate urge to conform, to belong, to join" (Nyegaard 2004, 357). But Humbert, the artist and intellectual with the supposedly unassailable aesthetic sensibility, remains the target of moral judgment, no matter how vigorously he strives to deflect it. He "fails to overcome his limitations because he perceives people according to the images he creates" (361). So it is with Hulga, who also views herself as a kind of artist.

Critics do not generally see Hulga as an artist figure. Tara Powell, for example, categorizes O'Connor's intellectual characters as "students, Educationists, [or] pseudointellectual artists," and deems Hulga "[t]he definitive example" of the student (Powell 2012, 30, 32). Yet Hulga uses the language of the artist to describe her intellectual aims. She sees changing her name from Joy, for example, as "her highest creative act" (267) – until, of course, she meets Manley Pointer, whom she regards as a character to be authored by her. Pointer will become her greatest artistic triumph, the product of her "true genius" (O'Connor 1988, 276). Perhaps like Humbert, Hulga is an artist *manqué* who creates the Manley Pointer she wants to see, whose solipsism and overweening intellectualism ultimately renders her even more conventional than her mother, making her the worst philistine – or as Nabokov would put it, the worst *poshlyak* – of all. Her most consequential disability is not her lost leg, but her blindness to details, to which O'Connor alludes at story's end, as Hulga, now spectacle-less, sees the blurry image of Pointer walking away across the field, which appears a "green speckled lake" (283). Though she would hardly place herself in their class – or they in hers – Hulga, like Mrs. McKintyre ("The Displaced Person"), Turpin ("Revelation"), May ("Greenleaf"), and Cope ("A Circle in the Fire"), and a host of others, pays a heavy price for her philistinism. Left without a "leg to stand on," both literally and figuratively, Hulga sits in the loft crying, "her churning face" metonymically representing the intellectual, emotional, and spiritual distortion she has embodied all along. Of course, the word "churn" here is also a metaphor whose ingenuity Nabokov might appreciate. Its earliest meaning in English (around the fifteenth century) referred either to the "vessel in which cream or milk is agitated to separate it and make butter" or the act of such agitation. In the seventeenth century, it came to mean "shaking or agitating" more broadly. Here, both meanings apply. Hulga is clearly "agitated" by her humiliation at Manley Pointer's hands. Even more meaningfully, the final image of her face connects it to the farm life upon which she has habitually looked down. The reader is given one last opportunity to laugh, and to sit in judgement of her pride.

O'Connor's most frequently anthologized story, then, seems a good example of the negative grotesque. The author herself bluntly says in a letter that "[n]othing 'comes to flower' [a phrase her correspondent, Betty Hester, apparently used] [. . .] except [Joy's/Hulga's] realization in the end that she ain't so smart" (O'Connor 1979, 170). Is it possible, though, to read it as an instance of the positive grotesque? Another trait that O'Connor's best work shares with Nabokov's is circumspection, especially when it comes to the theme of redemption. While O'Connor was a devout Catholic, and thus believed in an afterlife, and while Nabokov, though not religious in any orthodox sense, claimed faith in an "otherworld" and contended that his status as a creative writer made him take seriously the notion "[t]hat human life is but a first installment of the serial soul and that one's individual secret is not lost in the process of earthly dissolution" (Nabokov 1980, 377), neither attempted to represent *too* directly in their fiction the transcendent realm they felt to be real. They simply provided the details and hoped – usually against hope – that their readers would shrug off their own *poshlost* and know what to make of them. Thus, following O'Connor's own lead (as almost every critic is compelled to do by the force of her essays and letters), perhaps we should see in Hulga not unadulterated ugliness, but "good under construction." One way to do this is to approach the story as a kind of embryonic *künstlerroman*.

In such a reading, perhaps Hulga is not an artist manqué in the mold of Humbert Humbert, but a genuine artist-in-the-making. The similarities between character and creator are well-documented: O'Connor actually called the intellectual, sharp-tongued, physically disabled woman living on a farm with her mother her "heroine" and a "projection" of herself (O'Connor 1979, 106). Though a philosopher by training, an apparent devotee of scientism (which places "too high a value on natural science in comparison with other branches of learning or culture," Sorrell 1991, 2), and "a dilettante in nihilism" (Edmondson, 2003, 66), Hulga nevertheless displays the artist's active imagination and the impulse to create, albeit in a nascent, perhaps even unconscious, form. The first thing she brings into existence is a new self, which necessitates, of course, a name other than the blandly optimistic "Joy" with which her mother saddled her:

> She had a vision of the name [Hulga] working like the ugly sweating Vulcan who stayed in the furnace and to whom, presumably, the goddess had to come when called. She saw it as the name of her highest creative act. One of her major triumphs was that her mother had not been able to turn her dust into Joy, but the greater one was that she had been able to turn it herself into Hulga. (O'Connor 1988, 266–267)

While O'Connor likely expects us to laugh at Hulga's sourness and cynicism here, suggesting that her creative energies are misplaced, the woman who went by the unusual middle name, "Flannery," and went to great lengths in her fiction and nonfiction to forge a strong authorial identity, must at some level have appreciated her character's attempt at self-fashioning. It is also telling that Hulga draws on classical mythology, something that O'Connor, despite her denials, does on occasion (see Scouten, 1973; Shields, 1981; see also Dumas and Wilson, 2013).

As discussed earlier, things go awry when Hulga attempts to author someone else's identify for him. Elizabeth Hubbard eloquently describes the essential character flaw in how Hulga deals with Manley Pointer and everyone else. This artist-in-the-making engages in the "objectification and reduction of others [. . .] from her relationship with her mother to her relationship with Manley Pointer, whom she believes she has successfully pigeonholed as 'good country people' – the kind of person who can be easily seduced and then tormented by guilt and whom she will be able to take in hand for her own ends." Perhaps, then, "[h]er inability to recognize the alterity in another" is a major part of her grotesqueness (Hubbard 2011, 56). If Hulga's inability to recognize otherness – which is, of course, a kind of figurative blindness that many O'Connor characters exhibit – has its ground in modern philosophy, her fantasy of remaking Manley Pointer in her own image draws on the vocabulary of the artist:

> During the night she had *imagined* that she seduced him. She *imagined* that the two of them walked on the place until they came to the storage barn beyond the two back fields and there, she *imagined*, that things came to such a pass that she very easily seduced him and that then, of course, she had to reckon with the remorse. *True genius* can get an idea across even to an inferior mind. She *imagined* that she took his remorse in hand and changed it into a deeper understanding of life. (O'Connor 1988, 276; emphasis added)

The repetition of the word "imagined," and the reference to "true genius," situate Hulga in the time-honored tradition of the Romantic artist. At some level, Hulga fancies herself a poet with a deep understanding of human thoughts and emotions, but at the story's end, we find out how undeveloped her artistic sensibility is. Like O'Connor's most fully drawn artist figure, Asbury Fox in "The Enduring Chill," Hulga gets a lesson in humility that requires the loss of her sight. At the story's start, O'Connor establishes Hulga's figurative lack of vision: she has "the look of someone who has achieved blindness by an act of will and means to keep it" (265). At the end, though, she is reminded of her literal near-sightedness, as Manley Pointer confiscates her spectacles, leaving her to squint, through tears of humiliation, at the blurry image of his "blue figure struggling successfully over the green speckled lake" (283). Earlier, with glasses on, Hulga touts her ability to "see through to nothing"; now, without them, she can finally, however weakly, begin to see through to *something*. Frederick Asals deems this moment of (in)sight a "mock miracle" (Asals 1982, 71), but what she sees is strangely beautiful, nearly like an impressionist painting – suggesting, perhaps, a modicum of growth. Though often celebrated or vilified for leading her characters to violent deaths, O'Connor lets Hulga live. Those O'Connor characters with creative impulses, even if flawed – Asbury, O.E. Parker ("Parker's Back"), and the unnamed (and clearly autobiographical) narrator in "A Temple of the Holy Ghost" – fare better than those with no hint of artistic vision. Hulga, acutely cognizant of the physical, mental, emotional, and spiritual limitations of her disease and disabilities – but also now possibly aware of how her condition uniquely empowers her to take advantage of her creative gift – perhaps, like O'Connor herself, will become a powerful artist.

About 18 months after writing "Good Country People," O'Connor completed "A View of the Woods," which is not among her most celebrated works. Suzanne Morrow Paulson speaks for many critics when she calls it "an inferior work [. . .] weak in philosophical and psychological richness" (Paulson 1988, 25). The story is worth considering, however, because it emblematizes, differently than its more heralded predecessor, the tonal and thematic challenges in O'Connor's handling of the grotesque. Like most O'Connor stories, it features an intergenerational conflict, this time not between a mother and daughter, but a grandfather, Mark Fortune, and his granddaughter, Mary Fortune Pitts. The two start out with little "spiritual distance between them" (O'Connor 1988, 526), as they both enjoy surveying land that Fortune has sold for development being transformed by machinery. When Fortune decides to sell the lot across the road from Mary's house, however, she turns on him, calling him "the Whore of Babylon" and ultimately attacking him physically (532, 533). There are few of the snicker-inducing moments of "Good Country People" here. Given its pervasive grimness, one wonders if this is one of the works that made O'Connor "laugh and laugh."

Two haunting images stand out: the opening and closing scene of the monstrous bulldozer "gorging itself" on clay (546), and the nine-year-old Mary clinging to a tree trunk as her father "beat[s] her around the ankles with his belt," doing so "as methodically as if her were whacking a bush with a sling blade" (530). In both cases, violence occurs with disturbing efficiency, to either disorienting or stomach-turning effect. In the latter, Mary's face "contort[s] into a puzzle of small red lumps [. . . with] her nose and eyes [. . .] running" (530). In the former, the narrator personifies the excavator – illustrating Kayser's claim that in the grotesque, inanimate objects assume "an ominous vitality" (Kayser 1963, 183) – whose actions suggest repulsion against its own actions: gulping chunks of clay into its "disembodied gullet," it emits "the sound of a deep sustained nausea and a slow mechanical revulsion," then rotates around to vomit up the earth (O'Connor 1988, 525). These scenarios create a feeling of deep dread absent from "Good Country People," and the response is well-founded: perhaps no O'Connor tale features a more "wild, violent" ending than "A View of the Woods." Fortune, having sold the plot of land for the construction of a gas-station, is beset by Mary, who comes on him like "a pack of small demons" (545). With her atop him, pummeling his chest – inside of which resides a bad heart, both in the literal and figurative sense – he grabs her throat and bashes her head against a rock, killing her. After telling her corpse, "[t]his ought to teach you a good lesson" (546), the exhausted old man apparently dies of a heart-attack – thus closing the deadliest of all O'Connor tales. Only "A Good Man Is Hard to Find" has a higher body count but contains nowhere near this level of graphic violence.

Even though Asals deems "A View of the Woods" a "lesser" work than "Good Country People" (Asals 1982, 99), it is no mere catalog of physical brutality: like its earlier counterpart, it mounts a critique of philistinism, which here takes the form of absolute devotion to economic progress. The aptly named Mr. Fortune engages in conspicuous consumption by driving a Cadillac, though the car's "battered" body and gauche "mulberry" color diminishes its impressiveness as a status symbol (O'Connor 1988, 525). More significantly, he regards himself as a god of economic progress, and his self-aggrandizement has disconnected him from the true God, of course, but more visibly, from nature and his own family. Fortune's devotion to an abstraction makes him most ridiculous. He subscribes unthinkingly to a particularly American brand of "groupthink," a crass capitalism bent on an ever-expanding commercial infrastructure: "He wanted to see a paved highway in front of his house with plenty of new-model cars on it, he wanted to see a supermarket store across the road from him, he wanted to see a gas station, a motel, a drive-in picture-show within easy distance" (527). Fortune soothes whatever doubts he harbors about his master plan by repeating to himself the readily received rhetoric of American economic nationalism: "All men were created free and equal. When this phrase sounded in his head, his patriotic sense triumphed and he realized that it was his duty to sell the lot, that he must insure the future" (539). Fortune exhibits an absurdly simplistic understanding of that founding statement, engaging with the potent idea as nothing more than a cliché. In this, he reminds us not only of Mrs. Hopewell, who may love empty phrases more than any other O'Connor character, but also her daughter, whose superficial devotion to science, nihilism, and even her mother's own "salt of the earth" values – "aren't you just good country people?" she incredulously asks Pointer upon seeing his hollowed-out Bible filled with pornographic images and condoms – illustrate her susceptibility to abstraction.

Though she winds up fighting her mentor, Mary also internalizes class-coded abstraction. At the outset, she watches the excavator – an obvious metaphor for a destructive, isolating social and economic system – with "complete absorption" (525). Like Hulga, Mary wears glasses, suggesting both literal and figurative vision problems: "[h]er pale eyes behind her spectacles followed the repeated motion of [the bulldozer] again and again" (525). She has inherited his bourgeois concept of "land owner" (although Fortune exercises the power that comes with it by following the mantra, "sell, sell, sell!"). When one of the construction workers gets too close to the "stob" demarcating Fortune's land from sold property, she warns, "[i]f you don't watch him . . . he'll cut off some of *your dirt*," then "shake[s] her fist at the worker" (528; emphasis added). But Mary has also absorbed her mother's attitudes about land ownership as an indicator of class, as becomes clear when the young girl begins to question her grandfather:

> The Fortune house was set back about two hundred feet from the road and it was this two hundred feet that [Mr. Fortune] intended to sell. It was the part that his daughter airily called "the lawn" though it was nothing but a field of weeds.
>
> "You mean," Mary Fortune said after a minute, "the lawn?"
>
> "Yes mam!" he said. "I mean the lawn," and he slapped his knee. (531–532)

O'Connor's repetition of "the lawn" is significant. Mary's mother (and Mark's daughter) embraces what Virginia Jenkins calls "the front-lawn aesthetic" embraced by middle-class Americans in the twentieth century. As Jenkins explains, by the 1930s,

> the American elite transmitted their preference for front lawns to the larger society through the example of their own estates, public parks, horticulture shows, community garden contests, fix-up and clean-up campaigns, golf courses, and articles and advertising in popular magazines. (Jenkins 1994, 91).

While the economic upheavals of the Great Depression and the Second World War interrupted its spread, this aesthetic flourished again by the 1950s, when the story is set. That Mary's mother, and under her influence, Mary herself, refers to the lawn "airily" indicates, once again, the abstract thinking that O'Connor abhors. Interestingly, however, the source of judgement in this passage is ambiguous. Does it emanate from Mr. Fortune, who "slap[s] his knee" as he laughs at Mary, calling her "mam" perhaps not as a mere colloquialism, but to mock her middle-class pretensions? Or does it come from the narrator, more likely than Mr. Fortune to use a word such as "airily," but less likely than he to dismiss the space as "nothing but a field of weeds"?

Whatever the case, while young Mary exhibits a nascent philistinism, Mark Fortune's own tendencies are clearly worse. He quotes the Bible (in this case, Matthew 7:1) with a similar lack of understanding to, but a considerably meaner spirit than, Mrs. Hopewell, yelling at his granddaughter, "Jedge not [. . .] lest ye be jedged" (O'Connor 1988, 532). In his self-delusion and arrogance, however, he more closely resembles Hulga. Like her, he has a heart problem that bespeaks a physical and moral condition. Relatedly, he is blind to the inherent (O'Connor might say divine) value of the natural world. Similarly to the way Hulga goes "for walks but [. . .] [doesn't] like dogs or cats or birds or flowers or nature," Mr. Fortune peers "across the 'lawn'" and sees nothing of value – "not a mountain, not a waterfall, not any kind of planted bush or flower, just woods" (268, 538). The word "planted" indicates his belief that anything not put there by man is worthless. Looking into those woods for the (religiously symbolic) third time on the evening before he sells the land to the suitably named Mr. Tilman, however, he sees something different: after he "stare[s] for some time," he senses "an uncomfortable mystery" via a "hallucination" in which the forest appears "bathed in blood" (538). Elsewhere, O'Connor writes that "[t]he beginning of human knowledge is through the senses," and sight is undoubtedly the most important of them, such that "[t]he writer should never be ashamed of staring" (O'Connor 1969, 67, 84). Her oeuvre teems with characters, including Tom Shiftlet, Mrs. May, O. E. Parker, Asbury Fox, and, of course, Hulga Hopewell, who stand on the threshold of a vision that, were it to deepen and sharpen, might develop into something akin to the artist's eye, and thus the prophet's vision. Mr. Fortune's values are so hopelessly warped, however, that once "this unpleasant vision" passes, he wills himself to ignore it, even though in his mind the "hellish red trunks r[i]se up in a black wood" when he closes his eyes at night (O'Connor 1988, 538). Whereas Hulga's vision at the end of her story suggests the possibility of an earthly future, Mr. Fortune's obsessive devotion to materialism – embodied in the last thing he sees, the "yellow monster" of a bulldozer – blinds him permanently, insuring his death.

Perhaps Fortune's greatest sin is his attempt to re-create his granddaughter in his own image, making him a parodic God. Hulga attempts something similar with Pointer, of course, though with different results. Asals asserts that of all of O'Connor's stories, "A View of the Woods" "most overtly makes use of the double motif" (Asals 1982, 99). In trying to make his granddaughter into a version of himself, though, he also recalls Nabokov's Humbert. Just as he "safely solipsize[s]" Lolita (Nabokov 1991, 60), so Fortune transforms his little beloved into a version of himself. Her face appears "a small replica of the old man's" and he is sure that "she [is] like him on the inside too" (O'Connor 1988, 525, 526). The comparison of Mark and Mary Fortune to Humbert and Lolita is discomforting, given the implication that O'Connor might be adding to the already sordid phenomenon of pederasty the equally disgusting possibility of incest. But while the story features erotic undertones that would not be out of place in *Lolita* – in, for example, the images of Fortune waking in the morning to find Mary "sitting astride his chest" the way she had "straddled the hood" of his Cadillac earlier in the story (535, 525), and her sitting on the bumper of the car "with her bare feet on his shoulders" (528) – no actual sexual relationship exists between the old man and the girl. Rather, the story's subliminal erotic energy serves as a metaphor for Fortune's own perversity, which is not sexual but spiritual. It makes sense, then, that in the story's climactic battle scene, Mary kicks her grandfather "in the crotch," causing him to "thr[o]w himself on the ground and [. . .] roll like a man on fire" (545). Like her counterpart in Nabokov's novel, Mary is unable to escape the destruction wrought by her solipsizer, though her death is immediate, not many years in the offing, as is Lolita's. Humbert's blindness to what makes her a unique human being ultimately destroys them both. The same might be said about Mark Fortune and Mary Fortune Pitts. In O'Connor's essays, letters, and stories, intellectual characters often seem the most grotesque. One of the earliest commentators on O'Connor's work contends that these figures "are incomplete as human beings and cannot act successfully in the real world outside of their minds [. . . because] they have abused God's most generous gift to man – rational intelligence – and have committed the most heinous of sins" (McCarthy 1966, 1144). Yet most of O'Connor's intellectuals, including Hulga, Asbury, Julian, and Sheppard – while they certainly pay a painful price for their arrogance – stand some chance of redemption. Not so Mark Fortune, whose end suggests there are sins O'Connor found even more troubling than intellectualism.

3 Theoretical Perspectives

While O'Connor criticism has remained focused on religious themes since the earliest scholarly publications in the 1960s, it also engages with developments in the larger evolution of critical/literary theory. (*The Flannery O'Connor Bulletin*, initiated in 1972 and renamed *The Flannery O'Connor Review* in 2000, has been a significant scholarly outlet.) Most of it builds on the fascination with violence and the grotesque demonstrated by her earliest reviewers but apprehends these elements through a variety of theoretical lenses. Numerous monographs and articles use Bakhtin's theories to conduct narratological analyses, paying particular attention to the role of the third-person narrator (see Gentry 1986; Brinkmeyer 1989; DiRenzo 1993; Hayward and Magistrale 1996; Askin 2007). A growing body of gender-conscious criticism explores and/or interrogates O'Connor's attitude, as manifested in her short fiction, toward conventions of southern patriarchy and womanhood (see Westling 1985, especially 133–174; Gordon 2000; the essays in Caruso 2004, which focus almost exclusively on the short stories; Miller 2017, 123–146). Another recurring topic of interest is O'Connor's treatment of race, especially whiteness, doubling, and the civil rights movement, issues that arise more frequently in the stories than the novels (see Wood 2004, 93–119; Caron 2008; and Fowler 2013, especially 72–87, 2015, and 2016). O'Connor studies has also seen a burgeoning interest in political history and theory, beginning in earnest with Bacon (1993) and intensifying recently with the essays in Edmonson (2017). Perhaps the most promising, yet problematic, interpretive paradigms are disabilities studies and ecocriticism, which can be (and have been) applied to numerous O'Connor stories, but are particularly relevant to "Good Country People" and "A View of the Woods," respectively.

Unsurprisingly, O'Connor's (and her critics') investment in the grotesque poses a significant challenge for reading O'Connor through critical disabilities theory. The most basic step has been to move from a focus on "deformity" to "disability" (Libow 2017, 386). Another related issue pertains to O'Connor's method of employing physical disability as a metaphor for spiritual flaws. Critics such as Sara Hosey and Jess Libow address this by focusing on the materiality of disability. The latter's analysis of O'Connor's use of "assistive technology" – e.g., Hulga's glasses and prosthetic leg – and specifically the way "her characters repurpose their devices, reconstructing and reimagining embodiment," is particularly cogent and original (Libow 2017, 388). But many critics, including Kathleen Patterson, Carl Horner and Elena V. Barnes, Laura Behling, and Timothy Basselin, remain essentially untroubled by disability-as-metaphor. For Patterson, stories such as "Good Country People" provide "no definitive answers" about the "relationship between disability and identity," but it is enough that they aim "to astonish [. . .] readers with the full range of motion and emotion [. . . disabled characters] are capable of" (Patterson 2003, 99, 102). In their reading of "A Good Man Is Hard to Find," Horner and Barnes (an M.D.) explore the "connection between the self-attacking, self-dismantling, killing violence of SLE and the Misfit who emerges on one level of O'Connor's ambiguous literary universe as a metaphor of the baffling disease" (Horner and Barnes 2005, n. pag.), but seek to compensate for the danger that disease and disability become abstractions by highlighting specific clinical details of the condition. Behling argues that characters such as Hulga and Rufus reveal to the non-disabled their true selves and, indeed, the true nature of reality: "it is not the physically disabled character who needs redemption or repair," but the non-disabled, who fail to "understand that being human is not reliant on being whole" (Behling 2006, 98). Similarly, Basselin, author of the only monograph to consider disability in O'Connor's work, argues that "[d]isability, especially when understood theologically, helps imagine alternative visions of being that are dependent on limitation and mutual vulnerability" (Basselin 2013, 8). Following (again) the idea that the grotesque can be "good under construction," O'Connor saw in the limitations engendered by disease and disability "gateways to reality" (8), and Basselin explores the ways her own experiences with SLE shaped, along with the catalyzing element of Catholic belief, her understanding of the human condition. O'Connor largely rejected Enlightenment thought, especially the tendency to regard human suffering as a problem to be solved. Ultimately, Basselin admits the difficulty of employing disability studies/theory to O'Connor's work, because it "tends to critique any use of disability-as-metaphor, which O'Connor uses regularly" (109). His addressing this issue at the end of his study seems an attempt to downplay it. As Ben Rhodes notes in his theology-centric review, "Basselin [also] fails to come to grips with the legitimate worry about equating disability with the grotesque" (Rhodes 2014, n. pag.). Disabilities studies theorist Rosemarie Garland-Thomas complained more than 20 years ago that "critics seem unable to go beyond" the grotesque (Garland-Thomas 1997, 168), and in some ways, not much has changed.

Though O'Connor's work seems similarly inhospitable to ecocriticism, a fairly robust body of such scholarship is emerging. Commentators have long acknowledged the importance of the landscape in her fiction (see Pyron 1983; Westling 1985, 156–174; Martin 1987). Nature, especially the woods, often vacillates between exhibiting a "menacing, surrealistic" vitality and an "indifferen[ce]" to the "grotesque violence" that unfolds among the characters (Pyron 1983, 558–559), but the narrative focus remains squarely on the human environment. Indeed, nature itself is personified, as "[t]he land, the trees, the sky, and the relentless eye of the sun [. . .] become some of [the] most powerful characters" (Westling 1985, 6). This approach seems contrary to the ecocritical perspective. Lawrence Buell observes that anthropomorphism poses risks for the environmentally sensitive writer, because "[t]he ethics of care may be deflected back into narcissism" (Buell 1995, 218). Yet perhaps O'Connor's overarching goal is to expose and ridicule narcissism (i.e., pride), and to do so, she frequently underscores her characters' retrograde attitude toward nature, particularly their unwillingness to acknowledge their connection to the natural environment and the rest of humanity – both of which, at least from O'Connor's perspective as a Catholic, embody God's ultimate goodness. Elsewhere, I have explored the resonances between O'Connor's fiction, including "A View of the Woods," and deep ecology, a philosophy first articulated by the Norwegian Arne Naess that argues for the notion of "'biocentric equality' between humanity and nature" (Graybill 2011, 2); this paradigm also points to similarities (either biographical or ethical) between O'Connor and *Silent Spring* author Rachel Carson and, more significantly, the eco-theologist Pierre Teilhard de Chardin, whom O'Connor acknowledged as an influence (the collection containing "A View of the Woods," *Everything That Rises Must Converge*, takes its title from a Teilhard quotation). Interestingly, several ecocritical readings (like some of the disabilities-sensitive interpretations discussed above) integrate feminist theory, demonstrating, for example, how in "A View of the Woods," O'Connor provides her own take on the motif in American literature wherein women embody the "virgin land" ideal and "serve metaphorically as an analogue to the power of the land to control and consume men's lives [. . .] mak[ing] freedom possible only to those men who overpower, desert, or destroy the very ground under their feet" (Hewitt 2004, 130). Thus, Mary stands caught between two men, her grandfather and her father, who wage a battle over her body – which is also to say, over the land. While O'Connor's understandable inclination to metaphor remains a point of debate in the "eco-critical conversation" on her work, the tendency toward healthy nuance (see Bewes 2005; Achilles, 2015), coupled with an increasingly historicist sensibility (see Flanagan 2017), suggest that this will be a particularly fruitful approach in the future.

4 Bibliography

4.1 Works Cited

Achilles, Jochen. "Real Machines in Imaginary Gardens: Leo Marx's *The Machine in the Garden* and Environmental Liminality in Flannery O'Connor's 'A View of the Woods.'" *Rural America*. Eds. Antje Kley and Paul Heike. Heidelberg, Germany: Universitätsverlag, 2015. 207–225.

Asals, Frederick. *Flannery O'Connor: The Imagination of Extremity*. Athens: U of Georgia P, 1982.

Askin, Denise T. "Carnival in the 'Temple': Flannery O'Connor's Dialogic Parable of Artistic Vocation." *Christianity and Literature* 56.4 (2007): 555–572.

Bacon, Jon Lance. *Flannery O'Connor and Cold War Culture*. New York: Cambridge UP, 2003.

Basselin, Timothy J. *Flannery O'Connor: Writing a Theology of Disabled Humanity*. Waco, TX: Baylor UP, 2013.

Behling, Laura L. "The Necessity of Disability in 'Good Country People' and 'The Lame Shall Enter First.'" *Flannery O'Connor Review* 4 (2006): 88–98.

Bewes, Timothy. "What Is a Literary Landscape?: Immanence and the Ethics of Form." *differences: A Journal of Feminist Cultural Studies* 16 (2005): 63–102.

Bleikasten, André. "The Heresy of Flannery O'Connor." *Les Américanistes*. Eds. Ira D. and Christiane Johnson. N.p.: Kennikat P, 1978. Reprinted in *Critical Essays on Flannery O'Connor*. Eds. Melvin J. Friedman and Beverly Lyon Clark. Boston, MA: G. K. Hall, 1985. 138–158.

Brinkmeyer, Robert. *The Art and Vision of Flannery O'Connor*. Baton Rouge: Louisiana State UP, 1989.

Buell, Lawrence. *The Environmental Imagination: Thoreau, Nature Writing, and the Formation of American Culture*. Cambridge, MA: Harvard UP, 1995.

Caron, Timothy. "'The Bottom Rail Is on the Top': Race and 'Theological Whiteness' in Flannery O'Connor's Short Fiction." *Inside the Church of Flannery O'Connor: Sacrament, Sacramental, and the Sacred in Her Fiction*. Ed. Joanne Halleran McMullen. Macon, GA: Mercer UP, 2008. 138–166.

Caruso, Teresa, ed. *"On the Subject of the Feminist Business": Re-reading Flannery O'Connor*. New York: Peter Lang, 2004.

DiRenzo, Anthony. *American Gargoyles: Flannery O'Connor and the Medieval Grotesque*. Carbondale: Southern Illinois UP, 1993.

Dumas, Jacky and Jessica Hooten Wilson. "The Unrevealed in Flannery O'Connor's 'Revelation.'" *Southern Literary Journal* 45 (2013): 72–89.

Edmondson III, Henry T., ed. *A Political Companion to Flannery O'Connor*. N.p.: U of Kentucky P, 2017.

Edmondson III, Henry T. "'Wingless Chickens': 'Good Country People' and the Seduction of Nihilism." *Flannery O'Connor Review* 2 (2003): 63–73.

Evans, Robert C. "Flannery O'Connor's Short Fiction: Major Trends in Critical Commentary." *Critical Insights: Short Fiction of Flannery O'Connor*. Ed. Robert C. Evans. Ipswich, MA: Salem P, 2016. 53–69.

Flanagan, Christine. "Social Distortion: Displaced Landscapes and the Machines of Progress in Flannery O'Connor's 'The Displaced Person' and 'A View of the Woods.'" *Flannery O'Connor Review* 15 (2017): 18–34.

Fowler, Doreen. "Aligning the Psychological with the Theological: Doubling and Race in Flannery O'Connor's Fiction." *Flannery O'Connor Review* 13 (2015): 78–89.

Fowler, Doreen. "Death, Denial, and the Black Double: Reading Race in Flannery O'Connor's Fiction." *Mississippi Quarterly* 69.3 (2016): 303–326.

Fowler, Doreen. *Drawing the Line: The Father Reimagined in Faulkner, Wright, O'Connor, and Morrison*. Charlottesville: U of Virginia P, 2013.

Gentry, Marshall Bruce. *Flannery O'Connor's Religion of the Grotesque*. Jackson: UP of Mississippi, 1986.

Gilbert, Susan. "'Blood Don't Lie': The Diseased Family in Flannery O'Connor's *Everything That Rises Must Converge*." *Literature and Medicine* 18.1 (1999): 114–131.

Gooch, Brad. "Thirteenth-Century Lady." *Flannery O'Connor Review* 5 (2007): 23–34.

Gordon, Sarah. *The Obedient Imagination*. Athens: U of Georgia P, 2000.

Graybill, Mark S. "O'Connor's Deep Ecological Vision." *The Flannery O'Connor Review* 9 (2011): 1–18.

Harpham, Geoffrey. "The Grotesque: First Principles." *Journal of Aesthetics and Art Criticism* 34.4 (1976): 461–468.

Hawkes, John. "Flannery O'Connor's Devil." *Sewanee Review* 70 (1962): 395–407.

Hayward, Elizabeth Mahoney, and Anthony Magistrale. "Inverting Bakhtin: The Carnivalized World of Flannery O'Connor." *Literature and Belief* 16.1 (1996): 83–98.

Hendin, Josephine. *The World of Flannery O'Connor*. Bloomington: Indiana UP, 1970.

Hewitt, Avis. "'Ignoring Unmistakable Likeness': Mark Fortune's Miss-Fortune in 'A View of the Woods.'" *"On the Subject of the Feminist Business": Re-reading Flannery O'Connor*. Ed. Teresa Caruso. New York: Peter Lang, 2004. 129–154.

The Holy Bible. Authorized King James Version. N.p.: Thomas Nelson Bibles, 1987.

Horner, Carl S., and Elena V. Barnes. "Misfit as Metaphor: The Question and the Contradiction of Lupus in Flannery O'Connor's 'A Good Man Is Hard to Find'", https://dsq-sds.org/-article/view/607/784. *Disabilities Studies Quarterly* 25.4 (2005). (16 July 2019).

Hosey, Sara. "Resisting the S(crip)t: Disability Studies Perspectives in the Undergraduate Classroom." *Teaching American Literature* 6.1 (2013): 23–44.

Hubbard, Elizabeth. "Blindness and the Beginning of Vision in 'Good Country People.'" *Flannery O'Connor Review* 9 (2011): 53–68.

Jenkins, Virginia. *The Lawn: A History of an American Obsession*. Washington, D.C.: Smithsonian Books, 1994.
Katz, Claire. "Flannery O'Connor's Rage of Vision." *American Literature* 46.1 (1974): 54–67.
Kayser, Wolfgang. *The Grotesque in Art and Literature*. Trans. Ulrich Weisstein. New York: McGraw Hill, 1963.
Lake, Christina Bieber. *The Incarnational Art of Flannery O'Connor*. Macon, GA: Mercer UP, 2005.
Libow, Jess. "Prosthesis Repurposed: Gender and Rehabilitation in Flannery O'Connor's Fiction." *Journal of Literary and Cultural Disability Studies* 11.4 (2017): 385–401.
Mann, Thomas. "Conrad's 'The Secret Agent.'" *Past Masters and Other Papers*. Trans. H. T. Lowe-Porter. London: Martin, 1933.
McCarthy, John F. "Human Intelligence versus Divine Truth: The Intellectual in Flannery O'Connor's Works." *English Journal* 55.9 (1966): 1143–1148.
Miller, Monica. *Being Ugly: Southern Women Writers and Social Rebellion*. Baton Rouge: Louisiana State UP, 2017.
Muller, Gilbert H. *Nightmares and Visions: Flannery O'Connor and the Catholic Grotesque*. Athens: U of Georgia P, 1972.
Nabokov, Vladimir. *The Annotated Lolita*. Revised and updated. Ed. by Alfred Appel. New York: Vintage Books, 1991.
Nabokov, Vladimir. *Lectures on Literature*. Ed. Fredson Bowers. Intro. by John Updike. New York: Harvest/HBJ, 1980.
Nabokov, Vladimir. *Lectures on Russian Literature*. Ed. Fredson Bowers. New York: Harvest/HBJ, 1981.
Nabokov, Vladimir. *Pnin*. New York: Vintage International, 1985.
Nabokov, Vladimir. *Strong Opinions*. New York: Vintage International, 1973.
Nyegaard, Ole. "*Poshlust* and High Art: A Reading of Nabokov's Aesthetic." *Obis Litterarum* 59 (2004): 341–365.
O'Connor, Flannery. *Collected Works*. New York: Library of America, 1988.
O'Connor, Flannery. *The Habit of Being*. Ed. Sally Fitzgerald. New York: Farrar, Straus and Giroux, 1979.
O'Connor, Flannery. *Mystery and Manners: Occasional Prose*. Eds. Sally Fitzgerald and Robert Fitzgerald. New York: Farrar, Straus and Giroux, 1969.
Patterson, Kathleen A. "Negotiating Elevators and Haylofts: Disability and Identity in Flannery O'Connor's Short Fiction." *Literature and Sickness*. Ed. David Bevan. Amsterdam: Rodopi, 1993. 95–103.
Paulson, Suzanne Morrow. *Flannery O'Connor: A Study of the Short Fiction*. Boston: Twayne, 1988.
Powell, Tara. *The Intellectual in Twentieth-Century Southern Literature*. Baton Rouge: Louisiana State UP, 2012.
Presley, Delma Eugene. "The Moral Function of Distortion in Southern Grotesque." *South Atlantic Bulletin* 37.2 (1972): 37–46.
Pyron, Virginia. "'Strange Country': The Landscape of Flannery O'Connor's Short Stories." *Mississippi Quarterly* 36.4 (1983): 557–568.
Raiger, Michael. "'Large and Startling Figures': The Grotesque and the Sublime in the Short Stories of Flannery O'Connor." *Seeing into the Life of Things: Essays on Literature and Religious Experience*. Ed. John L. Mahoney. New York: Fordham UP, 1998. 242–270.
Rhodes, Ben. "Disability Theology and Flannery O'Connor." *reformation* 21 September 2014. Web. 17 July 2019.
Scouten, Kenneth. "The Mythological Dimensions of Five Flannery O'Connor Works." *Flannery O'Connor Bulletin* 2 (1973): 59–72.

Shields, John C. "Flannery O'Connor's 'Greenleaf' and the Myth of Europa and the Bull." *Studies in Short Fiction* 18.4 (1981): 421–431.
Sorrell, Tom. *Scientism: Philosophy and the Infatuation with Science*. New York, NY: Routledge, 1991.
Westling, Louise. *Sacred Groves and Ravaged Gardens: The Fiction of Eudora Welty, Carson McCullers, and Flannery O'Connor*. Athens: U of Georgia P, 1985.
Wood, Ralph C. *Flannery O'Connor and the Christ-Haunted South*. Grand Rapids, MI: William B. Eerdmans, 2004.

4.2 Further Reading

Cofer, Jordan. *The Gospel According to Flannery O'Connor: Examining the Role of the Bible in O'Connor's Fiction*. New York: Bloomsbury, 2014.
Gooch, Brad. *Flannery: A Life of Flannery O'Connor*. New York: Little, Brown: 2009.
Prown, Katherine Hemple. *Revising Flannery: Southern Literary Culture and the Problem of Female Authorship*. Charlottesville: U of Virginia P, 2001.
Shloss, Carol. *Flannery O'Connor's Dark Comedies: The Limits of Inference*. Baton Rouge: Louisiana State UP, 2012.
The Letters of Flannery O'Connor and Caroline Gordon. Ed. Christine Flanagan. Athens: U of Georgia P, 2018.

Steven Tracy

20 James Baldwin (1924–1987)

Abstract: Best known for prose works such as *Go Tell It on the Mountain* (1953), *Notes of a Native Son* (1955), and *The Fire Next Time* (1963), James Baldwin is also a brilliant short story writer. The paired reading of the short stories "Sonny's Blues" and "Going to Meet the Man" from his 1965 collection *Going to Meet the Man* brings together two very distinct short prose texts and emphasizes their common themes and issues. Both stories feature a main character who embarks on a quest to meet "the man" (an ambiguous phrase the stories explore in different ways) and undergoes a rite of passage. Both stories also topicalize the relationship between the individual and the community as well as the interplay of light and darkness. Central to African American culture, music (blues, jazz, and gospel music) plays an important role in Baldwin's short stories and expresses his musical aesthetic which, following a West African philosophy, views such music as a sacred-secular continuum, rather than as a sacred-secular dichotomy. In both short texts, Baldwin examines racism and its horrific consequences that inevitably result in black suffering and pain.

Keywords: Quest narrative, rite of passage, music, racism

1 Context

James Baldwin is known as a passionate polemicist, prose writer *par excellence*, and an occasional poet who reveled in his role as one spokesman for the African American community in the second half of the twentieth century. A brilliant orator, political essayist, dramatist, and nuanced novelist, he has been recognized for such prose classics as *Go Tell It on the Mountain* (1953), *The Amen Corner* and *Notes of a Native Son* (1955), *The Fire Next Time* (1963), *If Beale Street Could Talk* (1974), and *The Evidence of Things Not Seen* (1985). But his contributions as a short story writer are just as stellar. Lean, intellectual, visceral, polished, insightful, and inciting, Baldwin shows in his short stories "Sonny's Blues" and "Going to Meet the Man" from the collection of the latter title (1965) the multi-faceted lives of African Americans from the inside: the problems and dilemmas from the perspectives of race, class, and gender; cultural strategies for coping with those problems and dilemmas; and the spectrum of outcomes from being swamped, to limping through, to transcendence.

The eponymous story essentially begins, ends, and casts its shadow over the entire collection, being the titular first words one encounters, the last story in the volume, and the journeys undertaken in the works in-between, including in "Sonny's Blues," which occupies the center and heart of the volume. Both stories describe quests to meet "the man," though there is some ambiguity in that phrase. In one

https://doi.org/10.1515/9783110587647-021

sense, "the man" may be an authority figure from outside the black community who is frequently perceived as an enemy. However, the perceived enemy may be from within the black community as well. Alternatively, the man may be a positive role model in the community, a teacher to some protagonist's tyro. Or, indeed, the elder of the two may be meeting a boy who is being initiated into manhood, either seriously or ironically. In any case, the person must embark actively upon a journey to encounter this "man," apparently not someone previously encountered in his full (in)humanity, either literally or figuratively, to gain some perspective on the world. In each of the stories, Baldwin describes multiple encounters that may fulfill the action of the title, suggesting a deliberate ambiguity and mutual reciprocal directionality that assist in nuancing the title.

Music is, Baldwin recognizes, an essential component of African American culture, therefore it is central to his portrayal of his people struggling with the various forms of discrimination and the resulting (earthly) pessimism and (heavenly) hope that arise in both blues and gospel music. That is, the "sun's gonna shine in my back door someday" (House) of the blues lyric and "I'm gonna cross the river of Jordan some of these days" (Coleman) of the gospel song both hold out hope in the face of devastating racial and human tribulations. Baldwin's world is not merely a black and white world, but a world between those North and South poles, which the entire globe inhabits, and which explores the grey hues hovering tentatively between ebony and alabaster, stygian and transcendent. Thus, music plays a central role in Baldwin's universe. It is important to understand that Baldwin's musical aesthetic for his fiction follows a West African philosophy that establishes a sacred-secular continuum as opposed to a sacred-secular dichotomy or opposition. This roots it squarely in a historical and racial ground that honors the ancestors and creates for Baldwin a sense of continuity, pride, and persistence. Even though he realizes that the black community sometimes recognizes and acts upon those perceived "oppositions," seeing the blues as the devil's music and gospel as God's music, the viewpoint of the story from the author's point of view is frequently that there is a higher reasoning or philosophy that binds the two together, envisions them as positively overlapping to provide a holistic and unifying home or community. In a sense, there is no secular: it all is sacred, though some of it can be considered more profane than other elements.

2 Close Readings

In "Sonny's Blues" we clearly have, in its "Bildungsroman" theme, in Sonny's case a "Künstlerroman": two brothers who mature and develop an understanding of each other over a period of time that is contextualized by the father's tragic loss of his own brother; the mother's sense of the environment as potentially threatening even to those who are "good"; the death of the older brother's daughter and his difficult sojourn in a job that depresses him in body and spirit; and Sonny's drug addiction and need to be heard, listened to, and accepted for who he is. Once Sonny embraces the jazz community, he comes into conflict with a brother whose understanding of Sonny's avocation is limited by a restrictive social and moral sense. To Sonny's brother, jazz folks are bad folks, associated with bars, alcohol and other drugs, promiscuity, and a distorted sense of reality. To him, Sonny was admirable: "When he was about as old as the boys in my classes his face had been bright and open, there was a lot of copper in it; and he'd had wonderfully direct brown eyes, and great gentleness and privacy" (Baldwin 1988b, 86). But he has come from a family that has produced a middle-class brother who is an educator. Sonny's brother knows, after all, that his students' "heads bumped abruptly against the low ceiling of their possibilities" (87), but he had managed to channel his own frustrations into a middle-class paradigm for existence. He is unable to "find any room for [Sonny's addiction] anywhere inside [himself]" (87). Jazz and Sonny, he feels, don't mix, shouldn't mix. He is aware of his role as the older brother and therefore caretaker for and advisor to Sonny, as expressly emphasized by their mother, but he has not been good at carrying out his obligations. He is not very "fatherly" until he loses his daughter, but even then, it is more about his needs than Sonny's. Baldwin's strategy not to name the brother and force us to think of him and call him "Sonny's brother" is effective in that even though the brother is in the important role of narrator, his identity and importance in the story are really determined by his relationship to Sonny. It is as Sonny's brother that he has his importance, because this is primarily Sonny's story that illuminates the brother and the community.

Baldwin's story is extremely aural, not only in its use of music, but also in its expression of the necessity to attend to each other's words and actions for the sake of survival: "You can't talk it and you can't make love with it, and when you finally try to get with it and play it, you realize *nobody's* listening. So *you've* got to listen. You got to find a way to listen" (115). Unfortunately, he hears but does not listen to Sonny's problems and pleas because he has his own pre-fabricated middle-class understanding of Sonny's life, as well as solutions to Sonny's problems: Sonny should get his education, get married and settle down, have kids, and avoid any music that is more dangerous than his understanding of Louis Armstrong's. By this time in his career, Armstrong had in some quarters a reputation as a smiling, minstrelsy-soaked Uncle Tom because of his ubiquitous smile and grandfatherly demeanor. Actually, beneath that exterior was still a revolutionary performer and a sometimes-outspoken critic of American racism. Because Sonny's brother does not keep up with jazz, he is in fact nearly two generations behind in his knowledge of the performers in the idiom, demonstrating his incomplete cultural contemporaneity. And his first thought when Sonny says he wants to be a musician is "do you want to be a concert pianist, you want to play classical music and all that [. . .]?" (102). Classical music, to him, is the only music he feels is good enough for Sonny to play.

The emphasis on hearing comes in a Christian context in the story. Baldwin, while treating jazz, blues, and pop music on the "secular side" of the coin, also employs gospel music in the text to set up his aesthetic continuum. From his own religious upbringing, Baldwin understands the importance of hearing in the Bible: "In Isaiah 22:14 and 50:4, the prophet gives precedence to hearing over seeing, and Deuteronomy 5:1 and 6:30 and Jeremiah 2:4 invoke the Jewish faith with the phrase 'shema Ysrael' –Hear, O Israel" (Tracy 2015, 165). Baldwin juxtaposes all these musical genres to demonstrate the continuity that they have in the African American community, and the necessity of hearing them in that way for the survival of family, community, and culture. David Leeming credits Beauford Delaney as being crucial for Baldwin's understanding of the relationship of these various genres: "Delaney was to reconcile for his *protegé* the music of the Harlem streets with the music of the Harlem churches [. . .]" (Leeming 1994, 34). We should note that this collection is dedicated to Beauford Delaney. Baldwin intertwines various tunes from these genres in a skillful way to elucidate how they each deal with similar subject matter, counselling faith and perseverance in dealing with earthly problems.

Such well-known gospel tunes as "The Old Ship of Zion," "God Be With You 'Til We Meet Again," "If I Could Only Hear My Mother Pray Again," and "This Little Light of Mine" – dealing with the story's main themes of life's journey, separation, family, and the employment of spirit to illuminate the world and become spiritually transcendent – sit side by side with such transfiguring jazz figures as Louis Armstrong and Charlie "Bird" Parker, nuggets of wisdom in blues lyrics from traditional songs such as "Boogie Woogie," "Some Cold Rainy Day," and "Lonesome Day Blues," and the pop song/torch song "Am I Blue?" The latter is actually an amalgamation of characteristics from various genres: it was notably performed in Baldwin's lifetime by Billie Holiday, known as a jazz singer, with references to the blues, but in a pop format. Thus, the genres can exist not only separately, but may be profitably combined, as they are by Sonny, into a historical and socio-political rendering on the story of African Americans in America. Previous literary critics Saadi Simawe (2000) and Clarence E. Hardy III (2003) make clear that Baldwin associates blues and gospel music in his play *The Amen Corner* for example; Hardy further asserts that the "blues achieves a kind of sacred status" (Hardy 2003, 55) in Baldwin's last novel *Just Above My Head*. But earlier, in "Sonny's Blues," Baldwin makes a nuanced and artful argument for such a close possible connection.

The brother's sense of propriety limits his responses to Sonny's needs. He mistakenly views the community as fragmented, the jazz part on the bad side of town, the gospel on the good. This is because he hears jazz, but does not listen to it, the same as his wife Isabel's family. They are tortured by what they hear as Sonny endlessly practices his playing on their piano, and it drives them crazy in the way that in old Tarzan movies the unrelenting drums would drive the people on safari crazy. "But if [they were] to listen" (Hughes 1994a, 412), as Langston Hughes advises in his poem "Lady's Boogie," about a "dicty" African American woman who classes off from boogie woogie, they would hear the connection, in polyrhythms, percussiveness, syncopation, antiphony, and other musical elements, between gospel music and jazz. You not only have to *hear*, but to know *how* to listen. Not just through the ear, but through the mind and the gut; not just vibrating membranes, but visceral moments of vivid enlightenment. What Sonny recognizes is the common source of the musical genres: he hears the musical elements of both, but he also hears the suffering and transcendence expressed in both genres, and he *listens* to them, *bridges* them, combines them in a holistic world view that animates his appreciation for his African American culture as well as his own musical and humanistic spirit. His appreciation of the gospel musicians in the street outside his brother's window, his contribution with the tambourine, demonstrate his recognition of the value of the music, even if the people who play it may conceive of themselves as in conflict with the jazz world in which Sonny lives. But he feels the kinship, recognizes the brotherhood and sisterhood of the performers. In this sense they are closer spiritually to Sonny than his *actual* brother, suggesting the fragility of their family ties when juxtaposed to traditional community ties. Because the brothers are not in sync in terms of cultural community, they have a fractured relationship.

Thus, Sonny, who has gone to meet the white man in his court and prison experiences, also has to go to meet his brother, the man whose underdeveloped sense of the community aesthetic limits him from understanding Sonny's life. And so Sonny goes to meet the men in the jazz community, and benefits from the brotherhood that the brother's ideas prevented between the two of them, Creole especially. But at the same time, for the two brothers to reconcile, the older brother must go to meet Sonny's bandmates, but more importantly, must go to meet the man, Sonny. It is only when the brother is able to show Sonny the respect that a man deserves, only when he not only hears but listens to jazz – and therefore listens to Sonny – that the two can understand and reconcile.

This understanding and reconciliation is staged as a ritualistic rite of passage at the climax of the story. Sonny's brother has finally agreed to go to hear Sonny play in a club, and their entry into the jazz palace – they treat Sonny like royalty, there in "his kingdom [. . .] it was not even a question that his veins bore royal blood" (Baldwin 1988b, 118); their approach to the carefully-lighted and respectfully-treated "altar" that is the stage; the brotherly treatment of Sonny, like a special little boy come for his baptism and communion; and the pride in Sonny's maturation as a musician all mark this out as a spiritual ceremony, even though church folks would not consider it to be a very sacred place. Baldwin makes it clear that they must not be able to control the mode of hearing and listening to the community. While Baldwin does not criticize those connected to the church traditions in the story, he advocates for a broader acceptance of various elements of the community and a holistic view of the spirituality that imbues those various elements. Brought along carefully, nurtured in a holistic approach to music and community, Sonny is about to shine. He has gone to meet the men in the band; he has gone to meet the mature Sonny; he has gone to meet the man who is his brother. The man does not refer to white folks here: they are peripheral to the community healing itself and setting its things in order.

Significantly, the brother is placed by Creole "at a table in a dark corner" (118). The spotlight is on Sonny, the musician/communicator who offers a lesson in history and aesthetics and culture. In this Sonny is preparing to serve as a griot or jali, West African terms for a bardic figure who serves a variety of different purposes – social, political, religious, historical – as he or she reports news, gossip, culture, and other things in an artistic display of community traditions. The brother's role is not a passive one: he must actively hear and listen to the message and respond to its informing and healing power. The call and response elements that are in the music are activated and mirrored in the leader-audience call and response. Once again, Baldwin affirms aesthetic and social connections to the West African and earlier African American traditions, demonstrating how contemporary music retains and nurtures its continuity with tradition.

Creole is the essential figure for Sonny in this ritual of emergence into adult male identity and griot status. The word 'creole' itself refers to a mix of languages or cultures, a combination or blending that implies much about his character and function. He encourages a blend of the variety of cultural, especially musical, traditions through his reverence for the traditions, the community, and Sonny, through his ceremonious yet necessary preparing of the way for Sonny's emergence. He is father and colleague, spiritual advisor with a heart of jazz. He handles things and encourages improvisation, controls them, and sets them free: "He had them on a short rein" and "[He] wanted Sonny to leave the shoreline and strike out for deep water" (120). Creole determines when Sonny's time has come, accompanying Sonny to the liminal area from which Sonny will set out on his inaugural journey into musical, cultural, and spiritual manhood.

Becoming a griot involves a long period of study, comparable to the achievement of a PhD, with a mentor who imparts the nuances of becoming and being a griot to his student. Fulfillment is a cause for celebration, a transformation that Sonny's brother sees in Sonny as he struggles closer to attainment:

> And the face I saw on Sonny I'd never seen before. Everything had been burned out of it, and, at the same time, things usually hidden were being burned in, by the fire and fury of the battle which was occurring in him up there. (120)

This, Sonny's trial by fire, was not a mechanical exercise, "so much wood and wires and little hammers and big ones, and ivory" (120). Rather, like the piano player in Langston Hughes's poem "The Weary Blues," he transforms the instrument, the music, and himself by placing ebony hand on ivory key and making the piano "moan with melody" (Hughes 1994b, 50), a cultural transformation most people did not dream it could do, except for the musicians and audiences that heard and listened. Now he tells a story, a triumphal expression of tragedy racial and personal, historical and political, imposing and exposing:

> Creole began to tell us what the blues were all about. They were not about anything very new. He and his boys up there were keeping it new, at the risk of ruin, destruction, madness, and death, in order to find new ways to make us listen. For, while the tale of how we suffer, and how we are delighted, and how we may triumph is never new, it always must be heard. There isn't any other tale to tell, it's the only light we've got in all this darkness.
>
> (Baldwin 1988b, 121)

This little light of Sonny's, now not just a son but a brother, illuminates the man that Sonny has become and now met among the family at the bar. And not only has Sonny met the man, but Sonny's brother has met Sonny the man, and has met himself with his new consciousness of Sonny's maturity, and his own. The brother realizes the importance of developing on and developing past Louis Armstrong, telling a contemporary and ancient story with contemporary language and style that is based on what has gone before. Finally, Sonny's brother understands that Sonny "could help us be free if we would listen, that he would never be free until we did" (122).

Because Sonny's brother sees the child and adult that is Sonny before him, the child that is father to the man, he sends Sonny a highly symbolic, if gustatorially unusual, drink to congratulate him: Scotch and milk. The alcohol is for the man he has become, the milk for the child from which he has grown. And as Sonny's brother recognizes Sonny's triumph, he too exhibits a maturity towards which he has been developing in the story. The "Bildungsroman" and "Künstlerroman" plots complement each other; the various types of music complement each other; and the various parts of the community complement each other. And we hear and listen to the wisdom of the ages as the little child, now man, shall lead us.

The final story in the collection *Going to Meet the Man* shares its title with the entire volume, clearly demonstrating the dominance of the "Bildungsroman" theme and the various characters whose "maturation," or, comparatively, failure to mature, unify the community and the story. As in "Sonny's Blues," Baldwin employs flashback time switches to emphasize the "child is father to the man" theme, helping his readers to see how elements of the Southern gothic past, heir to the sins of slavery, impinge upon subsequent times at public and private levels, reverberating historically from Jesse's innocent childhood early in the twentieth century through a trial by fire to a graceless, impotent, proxy adulthood in the 1950s. Once again, there is a question about who "the man" is, and who is going to meet him. The ambiguities call into question the values of the characters, the nature of the communities, and the role that music plays in forcing them to face each other.

The limited omniscient narrator does not narrate the story in chronological order. Through time shifts, we are first exposed to the white man, grown into pathetic lawman Jesse, being queried by his wife Grace as to "What's the matter?" (Baldwin 1988a, 198) when his impotence in bed becomes obvious. This comes before we meet the child Jesse years earlier on his way to his first lynching. This is in part to emphasize that Jesse is going to meet himself as a child, which exposes who he is as a man. Clearly there is a great deal that is "the matter" with respect to this scene with his wife, from dysfunctional genitals to dysfunctional family to dysfunctional town to dysfunctional world. Impotent patriarchy in the southern white community in the Jim Crow era is highlighted in a number of ways in the story, not only sexually, but most emphatically so.

Baldwin leads off the story this way to foreground his analysis of what had produced the current state of race relations in the South in America. Clearly Baldwin appeals to the southern gothic past for explanation: "[. . .] he had the feeling that he had been caught up in a nightmare, a nightmare dreamed by a child; perhaps one of the nightmares he himself had dreamed as a child" (203). Feeling as if the elements of his existence "had been subtly and hideously displaced" (203), he seems on an unconscious level to understand that he had been the creator of his own world, even if he prefers to see himself as a victim in a world he did not understand: "What had the good Lord Almighty had in mind when he made the niggers?" (199). It emerged from the fractured psyche of the white South, mixed with his own learned psychic fears and weaknesses to produce a world for which he does not wish to take responsibility. The sins of the fathers are not only visited upon the sons; here, the sins of the sons are visited upon the fathers, through the agency of de-chronologized time, to make Baldwin's point that it is an endless cycle that must be understood politically, socially, and psychologically before it can be changed. This is why Baldwin makes the choice of limited omniscient point of view: he can both observe and describe Jesse, but he can also explore his sometimes unspoken prejudices, weaknesses, fears, insecurities – what he hides from the people he does not want to judge him.

His sexual power queried by his wife, Jesse offers explanations from being tired to working too hard out loud, but internally he reveals a deeper need for the hypersexualized figure of an African American woman to prod him to potency in a way that he could not imagine a "lady" like his wife doing. His need to dominate and control his sexual victim is what activates his sexual responses, indicating a pathetic streak of cowardice masked by the appearance of competence and confidence. After his failure to attain an erection, he himself "felt something he had never felt before" (199):

> He felt that he would like to hold her, hold her, hold her, and be buried in her like a child and never have to get up in the morning and go downtown to face those faces, good Christ they were ugly! And never have to enter that jailhouse again and smell that smell and hear that singing; never again feel that filthy, kinky, greasy hair under his hand, never again watch those black breasts leaping against the cattle prod, never hear those moans again or watch that blood run down or the fat lips split or the sealed eyes struggle open. (199–200)

It is, of course, his wish to return to the womb, to escape all responsibility, as well as all contact with the black world he has savaged, all the tawdriness of his mixed feelings of repulsion and attraction swirling in his subconscious mind in a chaos of evangelical, social, and sexual confusion. Ironically given the title of the story, the man is going to meet himself in the form of a child, one whose past will greatly illuminate who Jesse the adult is. So, he goes to meet the child, and encounters him, but also encounters himself, as well as white southern racist patriarchy and all that produced it. The "nightmare dreamed by a child" he wishes to displace with a womb-like serenity that is nonetheless invoked in all the language of repulsion he can muster as a way of portraying how inextricably they are bound in his mind and life. What scared him as a child was his inability to maintain control over himself and his life, and a fear that he was incompetent to maintain the semblance of superiority over those he had been taught are inferior to him, and whom he must subject to his power in order to justify the long racist past and continuing racist present. It is a nightmare generated out of his own subconscious mind, a self-fulfilling prophecy that he grapples with, with fear, guilt, a sense of inadequacy, and hubris, the sum of which is beyond his ability to understand. It leads to nemesis.

Jesse very deliberately begins the self-empowering narrative of his punishment of an African American male attempting to register to vote, even as that narrative is intertwined with dissatisfaction and despair. Best to flee from the moment of his failure and his present life to a site of power and control. Here his weapon of choice is phallic: "I put the prod to him [. . .]" (201). The extra prodding that he needed from his wife and desired from an African American woman, an indicator of his impotence, he turns into a weapon. It is an electric shock to the man's system, akin to the activating shock Jesse needs to be a "man." Besides infantilizing the prisoner – "He'd pissed his pants already" (201) –, he keeps prodding him into his own urine and excrement until finally reaching the locus of power that he wishes to challenge: "The boy rolled around in his own dirt and water and blood and tried to scram again as the prod hit his testicles [. . .]" (202). But while Jesse is busy mastering the young black man in the cell, he is being over-mastered by his own emotions, which shames him: "And now he was shaking worse than the boy was shaking. He was glad no one could see him" (202). Deep inside, Jesse realizes that the power is not his, but a surrogate power that he wields in a manner that violates basic human values of decency. It shakes him to his core, as if with the prod he is shocking himself, and thus explains his impotence in bed with his wife. What he thinks demonstrates his own power produces self-inflicted shocks that serve to lessen his own virility.

Jesse refuses to take responsibility. As with many other elements in his life, he scapegoats African Americans as the cause of his, the community's, and the country's deterioration from the Biblical way things are supposed to be to the wretched present, that made the entry into an African American house "like the entrance to a cave [. . .]" (203). It was his sense that there was a "secret place where the darkies lived" (206) that unnerves him, makes him respond in fearful, violent ways. Jesse recognizes things about the black community – and therefore the white community and therefore himself – that he does not want to admit. These elements are exposed in the remembrance of his first lynching. It is described in interesting terms, in terms of a common male rite of passage, terms sometimes used to describe a first sexual encounter: "[T]hey had taught him what it meant to be a man" (205). Because what it takes to be a man, technically, is to have a penis and testicles (unless, of course, some accident or different choice or development has rendered them absent). At this point, the young child Jesse is going to meet the man, the white community's version of the black man and how he must be treated. In meeting the black man, Jesse must also reckon with himself and the rest of his life. It is a visceral experience. And because the black captive is himself going to meet the man, the white man, he is another of the characters implicated by the title of the story. It is his encounter with, deep down, what the white man is about, and it is shocking (like the prod), captivating, enthralling, horrifying – a climactic event in the development of the boy into the man.

That encounter is represented at both stages of Jesse's life by another common element of Baldwin's corpus, including "Sonny's Blues": music, especially religious music. "Going to Meet the Man" is replete with music, a veritable soundtrack that acts as a lifeline for African Americans, as well as a tortuous conscience for Jesse, who virtually writhes under the sound of the songs sung by African American protesters. It was actually his responsibility to make the music cease: "they were still singing and I was supposed to make them stop" (201), and he is desperate to do so as he tortures his prisoner: "[. . .] and he screamed again, 'You had enough? You going to make them stop singing now?'" (202). What was it that bothered Jesse so much about the singing? For one thing, it reminded him of their problematic presence, more than a nuisance to him and his community. Then there was the primal crime of the Bible he proclaimed to love so much. How does one escape the valley of the shadow of that story in the modern South, the guilt of fratricide and the shame of lying, or pretending to lie, to one's God? And how does one escape the biblical story of Moses and the move to the Promised Land when applied to African Americans? Jesse was refusing to follow his own scripture.

Beyond that, it was not only the words to the songs, which sought "mercy" and the hope "to go to heaven," that weighed so heavily on him. He sometimes thought "they were singing for mercy for his soul, too" (204), which implicitly means that he is aware of not only his guilt but his sham Christianity as well. But the meaning goes beyond religious implications, what happens inside the church. The singing is taking it to the streets, uprooting the biblical meaning and applying it to contemporary life, as the typology does. He thought of himself as a protector of the white folks' Christianity from the "niggers," while he was protecting the "niggers from themselves" (205), blaspheming without self-consciousness by using the shameful epithet while he states his own Christian case. But he recognized that their songs were a subjective and stylistic rebuttal to all the unholiness he held holy: "He knew that some of the young people had changed some of the words to the songs [. . .]. They had not been singing black folks into heaven, they had been singing white folks into hell" (205). The style produces a mixture of feelings in Jesse, both sadness/fear and a strange pleasure: "'Even when they're sad, they sound like they just about to go and tear off a piece'" (208) – which is, interestingly, a sexual reference to having sexual intercourse, another way Jesse sexualizes the African Americans in the story even as their sexuality makes a visceral appeal to him. It produces in him the same type of "icy fear" (204) that Sonny's brother felt in "Sonny's Blues" as the latter fears that Sonny's life is lost to drugs and the brother helplessly, guiltily, and cluelessly clings to his middle-class insularity (Baldwin 1988b, 86, 91). But the lyric that follows applies to both the African Americans and to Jesse, so in a sense it unites him with the people from whom he consciously tries to be most alienated: "*Oh Lord! Come on and ease my troubling mind!*" (208). The lyrics from the gospel song *Ease My Trouble in Mind* recorded by such groups as the Spirit of Memphis Quartet refer not only to the wounds inflicted by the whites on the black community, but also to the guilt, confusion, and anger felt by Jesse and, presumably, at least some other whites pulled along by fear and racism into the maelstrom of monstrous wickedness.

The renowned song from which the lyrics "I stepped in the river at Jordan" likely derive, "Wade in the Water," were well known in the black community through frequent repetition and recordings by groups such as the Golden Gate Quartet. The words deal with the notion of achieving salvation through crossing over to the promised land (207–208). Frequently the slaves who made and sang the song would connote by the wading into water that the hounds were on the trail of escapees, and thus one must wade into the water to eradicate the trail (207). In the scenario in the story, the hounds are actually the white folks, viciously and animalistically pursuing the *de facto* still-enslaved blacks whose *de jure* freedom is undercut by almost a century of Jim Crow law and custom. Jesse knows at some level that these songs are leveled at him and the white community, and he knows the songs well enough through repetition – "All that singing," he complains (204) – that he realizes it overtakes his own consciousness: "When I begin to feel it, I gets kind of musical, too" (208). And he realizes that the songs reflect not only biblical ties, but contemporary strife as well. Another hint for Jesse that they do is his realization that "the young people had changed some of the words to the songs" (205). A common tactic in the civil rights era, traditional folk songs were also given appropriate variants that adapted them to the current situation – which itself was consistent with folklore practice. This kind of improvisation kept the songs current while still rooted in centuries old group tradition, indicating that an abiding faith and expression drove the musical expression in new times of trouble. The confrontation between the hateful tradition and the reverently perseverant tradition produces the confused responses in Jesse, who states his prejudices in such an argumentative way, while reluctantly conceding an identification with the African Americans underneath the layers of hatred.

This situation is reflected by the Manichaean interplay of light and darkness. Baldwin steps up his interplay of light and darkness that he had introduced earlier in the story (198, 199, 200, 203) as we move inexorably to the climactic lynching scene. Just as in "Sonny's Blues," the notion of introducing a bit of the light of understanding into the darkness of unknowing forms a central part of the meaning of "Going to Meet the Man." The "river of Jordan" lyric arises abruptly to confront Jesse "[o]ut of the darkness of the room. Out of nowhere" (207). As Jesse looks out, he hears the singing come "across the dark fields. There were no lights anywhere" (208). As they drive to the lynching in the darkness, in the dark Jesse's father's voice sounds, "in the darkness of the car [. . .] small and accusing" (209). In his adulthood, Jesse feels this smallness as he accuses the blacks of not knowing their places, his smallness being an analogue to his sexual feelings about himself. Jesse's father "switched off the lights" (209) as they awaited the time to join the other white folks on the way to the lynching.

Out of all this darkness, Jesse and the other whites arrive at the rite of passage, played and re-played so many times because everyone in the white community must be indoctrinated, and the scene must be re-enacted to convince and re-convince the whites of what they claim to know and be. As the caravan departs to the lynching, "bright dust filled the air" (211), Baldwin suggesting that though there is some illumination, it is obscured and dirtied by the purpose of the hateful hegira. His father, who seems to grow "immense" (213) in Jesse's eyes, has a look that is strange to Jesse. In particular, the father's eyes look as they do "in the sunlight," though Jesse is unsure whether it is the sunlight or some internal electrical prod that makes them look as they do, something Jesse "had never seen before" (213). When they arrive at the scene, "the sunlight hit [Jesse's] face, the trees moved away from them, and they were there" (213–214). In this brilliant sense Baldwin announces the arrival at the unholy/holy place, providing revelation through the emergence of sunlight and the removal of obscurities, bringing them to the "clearing" a word used six times within the space of twenty-four lines (214). It is here that Jesse's life will be transformed, will be made tragically unclear, for him and for the black community.

Jesse goes to meet the man, his father, who is here, immense, seemingly proud and strong, hoisting Jesse up on his back so Jesse can see and learn from the spectacle. Here he meets, from a distance, though very personally through intimate violence, the lynched man, Jesse searching desperately for an explanation: "*What did the man do? What did he do?*" (215). Baldwin even has Jesse call him "*the man*" in acknowledgment of the man's humanity, in Jesse's eyes at that time. Jesse's confusion in his encounter with himself logically conflicts with the scene, which Baldwin describes brilliantly with "a terrible odor across the clearing" (215). Jesse *sees* the light of the "gleaming" body (216), and the flash (216) of the phallic knife and *hears* his father tell him he "wasn't never going to forget this night" (217) but will never quite understand it. He has learned the behavior, been indoctrinated into the "club," been given "a great secret which would be the key to his life forever" (217), but it will thwart him the rest of his life. Jesse's response to his father's reminder that he had told him he would never "forget *this* picnic" (217) is key: "I reckon" (217). In the vernacular, "reckon" means 'suppose' or 'guess,' hardly a strong affirmation. It is 'maybe' with a vaguely positive leaning. Furthermore, it does call to mind Jesse's moment of *reckoning*, the point at which he will struggle mightily, as did Jacob with the angel, to understand his intuitive horror but communal acceptance of the acts committed there. Baldwin makes a peripheral and subtle reference to the biblical story when, as time changes back to impotent present, Grace does as the angel did to Jacob: "Grace stirred and touched him on the thigh" (217). When the angel touches Jacob on the thigh, he becomes impotent, and so does Jesse's father, and Jesse, and all the other lynching men, whether conflicted or not about what they do with their Bible. What a strange "grace" this is for the southern white men in the story. It is no salvation for sinners or the unmerited blessings of God, but rather suffering of the innocent blacks and the merited struggling damnation of the whites. The story ends "in a moonlight that has now grown cold as ice" (218). Cold in a prevalent darkness, but with a little illumination that allows us as readers to see a little bit in all this darkness, just as Jesse does.

In these two stories, Baldwin uses certain common motifs and themes to construct two very different scenes. By emphasizing the timeless themes of negotiating the rite of passage, reconciling the interplay of darkness and light, and emphasizing the relationship of individual and community, and focusing in part on music as an important arbiter of these themes, Baldwin gives us two nuanced stories with roots and branches. That is we can see the evidence of things past, and can wrestle with contemporary manifestations of the difficult struggles with which we are faced in this world. Baldwin does not provide easy answers; there are no such solutions to problems so unfortunately timeless. But he does provide some moonlight in all this darkness. As the narrator says in "Sonny's Blues," "it's the only light we've got in all this darkness" (Baldwin 1988b, 121). That is, the songs and stories that confront the darkness.

3 Theoretical Perspectives

Scholarship on Baldwin's work has, in recent decades, become so complex and extensive that it has itself become a field of research (James 2019; Francis 2014; Francis 2015; Magras 2016). James Baldwin was a prolific writer of fiction in various genres (poetry, drama, short stories, novels, children's literature, and screenplays) and an outspoken public intellectual who not only participated in public debates and was a constant media presence but who also published numerous essay collections analyzing the political realities of the day on both an abstract and a personal level. This is reflected in the critical evaluation of his works. Critical and theoretical perspectives on his oeuvre tend to read his works in light of his political activism and the question of what it means to be a black and gay man in the U.S. of the mid-20th century on the one hand (e.g., Field 2011; Sinitiere 2016), or, on the other hand, evaluating his texts – both fictional and non-fictional – as epitomizing an autonomous aesthetic and philosophy, reflecting on the human condition in a rather universalist manner (e.g., Burrell 2012; Golden 2012; Mambrol 2018; Martinez 2007; Jones 2015). In Baldwin's works private and public, fictional as well as non-fictional dimensions are however intricately connected. Reading Baldwin's works is therefore most productive when regarding these different perspectives not as mutually exclusive but instead acknowledging their interconnectedness and the complex character of Baldwin's work which defies conventional categorizations (cf. e.g., Schwarz and Kaplan 2011; Miller 2019a; Elam 2015; Schulenberg 2018).

Short prose texts were a part of Baldwin's writing in his childhood and already then he experienced infringement from without. In the opening essay to *A Native Son* he writes: "[T]he first effort of mine to be seen in print, occurred at the age of twelve or thereabouts, when a short story I had written about the Spanish revolution won some sort of prize in an extremely short-lived church newspaper. I remember the short story was censored by the lady editor, though I don't remember why, and I was outraged" (Baldwin 1998a, 5). He furthermore wrote short stories as a student working for *The Magpie*, a student magazine at Frederick Douglass Junior High School (1998b, 845–846). Later in life, Baldwin was painfully aware of the demands and concomitant restrictions placed on him as a black writer, being, "in effect, prohibited from examining my own experience too closely by the tremendous demands and the very real dangers of my social situation" (Baldwin 1998a, 8), restrictions which also translated into his private life: "I don't like people who like me because I'm a Negro; neither do I like people who find in the same accident grounds for contempt" (9).

The collection *Going to Meet the Man* was published in 1965 "at the height of the civil rights movement," (Dudley, 82) encompassing short stories written since 1948 ("Previous Condition") and also his most-anthologized short story "Sonny's Blues" (published in the *Partisan Review* in 1957). Immediately after the publication of *Going to Meet the Man*, reviewers complained about the "[t]he triteness of the writing" (Kauffman 1965) or the stories being "crude and unconvincing" (*Kirkus Review* 1965). In these reviews, Baldwin's short stories are continually measured against his other writings, and in particular his essays, when, for instance, Kauffman claims that "[v]ery little of this writing shows the incisiveness of prose, the rhythmic control, the attractively serpentine and dramatic structure, the simple care, that mark his other writing" or when Donoghue says "[w]ritten at intervals since 1948, the short stories can be read as footnotes to the novels, but this is unfair. Some of them are slight things, hardly more than sketches, unofficial essays prophesying fire next time. But there are two stories, 'Sonny's Blues' and 'This Morning, This Evening, So Soon,' which are far better than anything else Mr. Baldwin has done in fiction." The reception of Baldwin's short stories has undergone a transformation towards an emphatically positive evaluation since the collection's first publication in 1965. Dudley, for instance, interprets Baldwin's short stories as texts where "what he was able to do with the novel he distilled nearly perfectly in the shorter form," adding that "'Sonny's Blues' and ['Going to Meet the Man'] [. . .] are [. . .] stories of alienation and a special brand of racial pain. In the former, that pain is quite personal, and literally familial. In the latter, it is of an empathetic kind. In both instances, though, Baldwin examines the destructive nature and human collateral of institutionalized racism" (Dudley 2020, 11).

For approximately three decades now, there has been a "reinvigorated interest in Baldwin" (Magras 2016, 105) inside and outside of academia, which has intensified in the wake of the Black Lives Matter movement (Joyce et al. 2015). This renewed interest is manifested, for instance, in workshops and conferences (e.g., at the 2015 ALA or 2019 MLA conventions; cf. Jackson et al. 2019), in honorary celebrations (e.g., when New York declared 2014 "the year of James Baldwin"; Dudley 2020, 3), in memes or publications published in light of the recent civil rights debates (e.g., Eddie S. Glaude's *Begin Again* (2020), Raoul Peck's documentary *I Am Not Your Negro* (2017), the Netflix adaptation of the novel *If Beale Street Could Talk* (2018), the founding of the journal *James Baldwin Review* in 2015, or the special issue "James Baldwin and Philosophy" of the journal *Critical Philosophy of Race*; cf. James 2019). *Going to Meet the Man* furthermore inspired a Jazz compilation by the Metta Quintet by the same title (2002). It is telling that more than three decades after Baldwin's death, the critical reception of his work spans the spectrum between private and public, fictional and non-fictional, and appeals to transnational academic and non-academic audiences. In particular, the recent engagement with his work underlines simultaneously the timelessness and timeliness of James Baldwin's writing which, one can hope, will continue to reverberate with future audiences.

4 Bibliography

4.1 Works Cited

Baldwin, James. "Autobiographical Notes." *James Baldwin: Collected Essays*. Ed. Toni Morrison. New York: Library of America, 1998a. 5–9.

Baldwin, James. "Chronnology." *James Baldwin: Collected Essays*. Ed. Toni Morrison. New York: Library of America, 1998b. 845–855.

Baldwin, James. "Going to Meet the Man." *Going to Meet the Man*. 1965. New York: Laurel-Dell, 1988a. 198–218.

Baldwin, James. "Sonny's Blues." *Going to Meet the Man*. 1965. New York: Laurel-Dell, 1988b. 86–122.

Baldwin, James. *Going to Meet the Man*, www.kirkusreviews.com/book-reviews/james-baldwin/going-meet-man/. *Kirkus Review*. 15 June 1965, (February 24, 2021).

Burrell, Vernita. "Keep Your Hands Off My Ethics: The Ethics of Reading in *The Bluest Eye* and 'Sonny's Blues.'" *ASEBL Journal* 7.1 (2012): 1–4.

Donoghue, Denis. *Blues for Mr. Baldwin*, www.nybooks.com/articles/1965/12/09/blues-for-mr-baldwin/. The New York Review 9 December 1965, (February 24, 2021).

Dudley, Marc K. *Understanding James Baldwin*. Columbia, SC: U of South Carolina P, 2020.

Elam, Michele, ed. *The Cambridge Companion to James Baldwin* Cambridge: Cambridge UP, 2015.

Field, Douglas. *James Baldwin*. Horndon, UK: Northcote, 2011.

Francis, Conseula. "Reading and Theorizing James Baldwin: A Bibliographic Essay." *James Baldwin Review* 1.1 (2015): 179–198.

Francis, Conseula. *The Critical Reception of James Baldwin, 1963–2010: "An Honest Man and a Good Writer."* Rochester, NY: Camden House, 2014.

Glaude, Eddie S. *Begin Again: James Baldwin's America and its Urgent Lessons for Our Own*. New York: Crown, 2020.

Golden, Timothy Joseph. "Epistemic Addiction: Reading 'Sonny's Blues' with Levinas, Kierkegaard, and Nietzsche." *The Journal of Speculative Philosophy* 2.3 (2012): 554–571.

Hardy III, Clarence E.. *James Baldwin's God: Sex, Hope, and Crisis in Black Holiness Culture*. Knoxville: U of Tennessee P, 2003.

Hughes, Langston. "Lady's Boogie." *Collected Poems of Langston Hughes*. Ed. Arnold Rampersad. New York: Knopf, 1994a. 412.

Hughes, Langston. "The Weary Blues." *Collected Poems of Langston Hughes*. Ed. Arnold Rampersad. New York: Knopf, 1994b. 50.

I Am Not Your Negro. Directed by Raoul Peck, 2017. Netflix.

If Beale Street Could Talk. Directed by Barry Jenkins, 2018. Netflix.

Jackson, Robert, et al. "James Baldwin: Interventions." *James Baldwin Review* 5.1 (2019): 197–218.

James, Jenny M. "Trends in James Baldwin Criticism, 2015–16." *James Baldwin Review* 5.1 (2019): 143–159.

Jones, Meta DuEwa. "Baldwin's Poetics." *The Cambridge Companion to James Baldwin*. Ed. Michele Elam. Cambridge: Cambridge UP, 2015. 41–55.

Joyce, Justin A., et al. "Baltimore is Still Burning: The Rising Relevance of James Baldwin." *James Baldwin Review* 1.1 (2015): 1–9.

Kauffman, Stanley. *Going To Meet the Man,* archive.nytimes.com/ www.nytimes.com/books/98/03/29/specials/baldwin-going.html?r=1&oref=slogin. *The New York Times* on the Web 12 December 1965 (February 24, 2021).

Leeming, David. *James Baldwin: A Biography*. New York: Knopf, 1994.

Magras, Lydia Brown. "Review: Conseula Francis. *The Critical Reception of James Baldwin 1963–2010: 'An Honest Man and a Good Writer.'" Reception: Texts, Readers, Audiences, History* 8.1 (2016): 105–108.

Mambrol, Nasrullah. "Literary Criticism of James Baldwin." *Literary Theory and Criticism*. 25 Apr. 2018, literariness.org/2018/04/25/literary-criticism-of-james-baldwin/ (February 2, 2021).

Martinez, Inez. "'Sonny's Blues' and Cultural Shadow." *JJSS: Journal of Jungian Scholarly Studies* 3 (2007): 1–8.

Miller, D. Quentin. "Introduction." *James Baldwin in Context*. Ed. D. Quentin Miller. Cambridge: Cambridge UP, 2019. 1–4.

Miller, D. Quentin. *James Baldwin in Context*. Cambridge: Cambridge UP, 2019.

Schulenberg, Ulf. "'Speaking Out of the Most Passionate Love' – James Baldwin and Pragmatism." *EJAS: European Journal of American Studies* 2.2 (2007): 1–19.

Schwarz, Bill, and Cora Kaplan, eds. *James Baldwin: America and Beyond*. Ann Arbor: U of Michigan P, 2011.

Simawe, Saadi. "What Is in a Sound? The Metaphysics and Politics of Music in *The Amen Corner*." *Reviewing James Baldwin's Things Not Seen*. Ed. D. Quentin Miller. Philadelphia: Temple UP, 2000. 12–32.

Sinitiere, Phillip Luke. "James Baldwin: Biographical Dispatches on a Freedom Writer." *James Baldwin Review* 2.1 (2016): 140–152.

Tracy, Steven. “Sonny in the Dark.” *The James Baldwin Review* 1.1 (2015): 164–178.
Turner, Mark, Kurt Rosenwinkel, Mark Gross, George Colligan, Joshua Ginsburg, and H. Benjamin Schuman. “Going to Meet the Man / Metta Quintet.” New York, NY: Koch Jazz, 2002.

4.2 Further Reading (Discography)

Boyd, Jim. “This Little Light of Mine.” The Library of Congress.
Coleman, Jaybird. “I’m Gonna Cross the River of Jordan.” *Jaybird Coleman and the Birmingham Jug Band*. Document DOCD 3140.
Count Basie. “Boogie Woogie.” *America’s #1 Band*. Columbia Legacy C4X 87110.
Golden Gate Quartet. “Wade in the Water.” *Swing Down, Chariot*. Sony 47131.
Hallsway High School Quartet. “The Old Ship of Zion.” *Negro Religious Field Recordings*. Document DOCD 5312.
House, Son. “Delta Blues.” The Complete Library of Congress Sessions, 1941–1942. Travelin’ Man 02.
Hill, Chippie. “Some Cold Rainy Day.” *Bertha “Chippie” Hill*, 1998a. 5–9. Document DOCD 5330.
Holiday, Billie. “Am I Blue?” *The Lady: Complete Collection*. Sony OODP 570–577.
Jackson, Mahalia. “God Be With You ’Til We Meet Again.” *The Forgotten Recordings*. Acrobat 3006.
Jackson, Mahalia. “If I Could Only Hear My Mother Pray Again.” *Gospels, Spirituals, and Hymns*. Columbia 65594.
Johnson, Rev. Anderson. “If I Could Only Hear My Mother Pray Again.” *1950s Gospel Classics*. Document DOCD 5464.
McTell, Blind Willie. “Lonesome Day Blues.” *Blind Willie McTell, Vol. 2*. Document DOCD 5007.
Plantation Echoes. “God Be With You ‘Til We Meet Again.” The Library of Congress.
Spirit of Memphis Quartet. “Ease My Trouble in Mind.” *Happy in the Service of the Lord*. Acrobat ADDCD 3007.
Weaver, Curley. “Some Cold Rainy Day.” *Curley Weaver*. Document DOCD 5111.

Michael Grimwood

21 Shirley Jackson (1916–1965)

Abstract: Shirley Jackson's signature story "The Lottery" (1948) anticipates the agoraphobia that later immobilized her. "Biography of a Story" (1960), in which she recounts the origins of that story, replicates its basic plot. She may have written "The Lottery" partly in response to the New England tradition of town meeting, perhaps one meeting in particular, and under the influence of the Congressional anti-Communist investigations of the late 1940s. Its narrative structure relies upon an ironic management of information. The reader's position both inside and outside the narrative point of view accentuates the story's focus upon the benefits and liabilities of membership. Game theory illuminates the lottery's capacity to generate communal solidarity. Public-sphere theory affords further insight into the lottery's social functionality. Girard's attribution of the origins of culture to ritual sacrifice can also advance understanding of the story's meaning. The rollcall that occupies the central third of the story may be as significant as the story's surprise ending. "Biography" revises "The Lottery" as a "scene of circulation" that imagines the entrance of writers and readers into the communal space of publication as trauma.

Keywords: Agoraphobia, town meeting, Red Scare, game theory, public sphere, scapegoat

1 Context

Few writers' reputations have been dominated by one short story as much as Shirley Jackson's by "The Lottery." Although she published many other stories, six novels, and several other books, she remains identified with the work that brought her to public notice early in her career. Her two final novels and a few of her other stories have remained in circulation, but "The Lottery" is still the linchpin of her oeuvre. It so shadowed her subsequent efforts that she felt compelled in 1960, a dozen years after she wrote it and five years before her death, to account for it. In "Biography of a Story," a talk she prepared for the few audiences she consented to address, she described the story's composition, publication, and reception in departures from the truth sufficiently misleading to constitute a parallel work of short fiction (Franklin 2016, 222–223). She claimed that she had invented "The Lottery" during a routine outing from her home in North Bennington, Vermont, on a "warm morning" a few weeks before *The New Yorker* published it on June 26, 1948. She had walked into town to fetch groceries and mail, pushing her daughter in a stroller, composed the story in her mind while returning, and immediately typed it almost verbatim. When

https://doi.org/10.1515/9783110587647-022

the story was published, she claimed, she had followed the same route into town while performing the same daily ritual and had been snubbed by the postmaster because of the loads of hostile mail the story had generated. She then provided a profusion of excerpts from "the three-hundred odd letters" she said she had received in response to "The Lottery," which showed, she argued, that "people who read stories are gullible, rude, frequently illiterate, and horribly afraid of being laughed at." Some of the writers had been "positive that *The New Yorker* was going to ridicule them in print" and had forbidden publication, Jackson said – as she defied their will by publishing, often out of context, laughable portions of their letters (Jackson 2010, 787–790).

Jackson's retroactive identification of "The Lottery," in which Tessie Hutchinson leaves home to suffer a fatal encounter with her neighbors, with her own angry reception by the postmaster and her readers suggests that the story sprang from autobiographical origins, though not as she described them. "Biography" is inaccurate: the story's typescript bears an inscription indicating she had written it by mid-March; it reveals numerous revisions; and the response from readers was more confused than hostile (Franklin 2016, 223–244). But the story about the story may tell an inner truth: "The Lottery" expressed Jackson's fear of public space – not only the literal space represented by the town "square" in the story and by Main Street in North Bennington, but also the virtual space of publication.

Long before a psychiatrist diagnosed as agoraphobia the panic attacks and domestic self-imprisonment Jackson suffered in the early 1960s, she had displayed symptoms of the disorder both in her life and her fiction. She would confide to a diary in 1963 or 1964, "i [*sic*] think all my books laid end to end would be one long documentation of anxiety" (Oppenheimer 1988, 249–252; Franklin 2016, 426–427, 456–458, 469–478). Having grown up in a suburb of San Francisco in the 1920s and early 1930s, she had moved to progressively denser urban environments – suburban Rochester during high school, central Syracuse for college, and New York City, to which she and her classmate and then husband, Stanley Edgar Hyman, moved when they graduated in 1940 (Oates 2010, 805–807). New York was an intellectual magnet that provided the resources they needed to start writing careers – hers in fiction, his in journalism – and literary criticism, but it also activated in Jackson a phobic temperament that had probably always been latent. Although she appreciated the pleasures of urban life, she increasingly minded its perils. This descendant of architects "was convinced she was going to be killed by something dropping from a building," said an acquaintance. "She pretty much convinced me it was dangerous to walk down the street" (Oppenheimer 1988, 109). Her stories from the early 1940s frequently concern urban hazards. In "The Intoxicated," for example, a seventeen-year-old girl sobers a guest at her parents' party by describing an apocalyptic vision: "I think of the churches as going first, before even the Empire State building. And then all the big apartment houses by the river, slipping down slowly into the water with the people inside" (Jackson 2010, 7). Partly because of her difficulty in managing urban menace (and partly because of a teaching opportunity for Hyman at Bennington College), the couple moved to Vermont with the first of their four children in 1945 (Oppenheimer 1988, 109–112).

Jackson continued to write stories about frightening cityscapes while she was turning her attention to superficially bucolic New England village settings. In "The Daemon Lover," which she probably wrote in 1947 (Franklin 2016, 239, 542), a young woman goes to an apartment building to meet James Harris, her fiancé, but she cannot find his name on the mailboxes. She enters buildings, knocks on doors, and asks people where to find him. In the end, she believes she has traced him to a probably empty apartment, to which she returns and returns, without avail, in the coming days (Jackson 2010, 10–25). In "Pillar of Salt," also written in 1947 (Franklin 2016, 542), a married couple who live in New Hampshire revisit New York after having, like Jackson and Hyman, moved away. Disaster dominates their sight-seeing: an airplane crash site, rides in defective taxis, a fire alarm, a dismembered leg, crumbling buildings (Jackson, 2010, 184–193). Finally, after shopping at a store, the wife becomes too paralyzed by traffic and pedestrian commotion to cross the street: "A crowd of people formed around her suddenly; [. . .] and she had a sudden feeling of being jammed in the center and forced out into the street when all of them moved as one with the light changing." She makes her way back to the store, back to the curb, and again back to the store, where she calls her husband to beg him to rescue her (Jackson 2010, 197). A case study in agoraphobia, "Pillar of Salt" is a pure version of a plot that pervades Jackson's fiction: a young woman leaves the relative security of home, goes out into a dangerous world, and suffers trauma. Though not set in a city, "The Lottery" follows the same plot. No less than the urban stories, it simulates an anxiety attack, albeit surrealistically and symbolically. Jackson may not yet have failed to control her own panic in public, but she was already reporting the phenomenology of the condition. "The Lottery" not only describes Tessie's panic and the vertigo that publication may have caused Jackson; it seeks to induce the same experience in its unprepared reader.

"The Lottery" may have been inspired not only by Jackson's general fear of public encounters but also by a particular occasion of institutionalized social pressure. The typescript of the story that she sent to her agent, Eleanor Kennedy, and that Kennedy then submitted to *The New Yorker*, is dated March 16, 1948 (Franklin 2016, 223). No proof exists that this was the date of the story's composition, but if any element of "Biography" rings true, it is her claim that she wrote the story quickly, with minimal revision, and dispatched it immediately. March 16 was the date of town meetings in Vermont in 1948. A foundation of political culture in New England still widely practiced there, town meeting is a local governance conducted by all of a community's citizens rather than their representatives. After a decline at the end of the nineteenth century, it enjoyed a revival in the early twentieth, and its prestige as a national institution of "pure democracy" grew through mid-century. Its reputability was enhanced by the production, beginning in 1935, of a popular radio program called *Town Meeting of the Air*, on which experts discussed national issues

(Bryan 2004, 25–40). During the 1930s and 1940s, the radio-reinforced mystique of town meetings was frequently deployed against dictatorship (Gould 1940b). Following World War II, the United Nations was sold to the American public partly as the "Town Meeting of the World" and "the safeguard of our civilization" (Bonsal 1947, 44). "The Lottery" emerged near this crest of town meeting's esteem.

The story exhibits many conventions of what might be called the "town-meeting narrative." Once the tradition became a point of pride for Americans of all regions, newspapers and magazines across the country regularly ran accounts of particular meetings to advertise the quaint virtues of this living relic of American democracy. These articles exhibited generic features, many of which "The Lottery" shares. They are usually narrated by an outsider, often guided by an insider. They attend to the weather, though most town meetings occur in wetter, colder seasons than "The Lottery." They set the assembly inside a building rather than outside, but like them "The Lottery" describes a venue (25). Like "The Lottery," they often lavish description on the citizens' arrivals (25). They spotlight women's participation, even before the Nineteenth Amendment, and children's activities though the children never vote. The officers of the meeting, the "moderator" and the town clerk, take oaths and conduct the proceedings according to arcane rules of voting, just as people in "The Lottery" take their chances according to arcane rules enforced by Mr. Summers, "the official of the lottery," and Mr. Graves, who administers the "proper swearing-in" to Summers (26). The officers must produce a correct roster of eligible voters; Summers and Graves must produce accurate lists "of heads of families, heads of households in each family, members of each household in each family" (26). Just as "The Lottery" details the setting-up of the "black box" from which villagers extract their lots (25), some meeting narratives detail the setting-up of the meeting's furniture, including the ballot box. Just as "The Lottery" focuses attention on Old Man Warner (27–28), many meeting narratives dwell on such stereotypical "Yankee" personalities as the curmudgeon who bemoans the current generation's deviations from the tried and true. Like "The Lottery" (25–26), town-meeting narratives often rehearse the history and durability of the institution; like "The Lottery" (27), they often mention threats to its continuity or experiments with other procedures. The subtext, increasingly close to the surface, of town-meeting discourse during the 1930s and 1940s was the desirability of "participation," of discharging one's "civic responsibilities." "The Lottery" may have startled readers in 1948 partly because it embodied this consensus with ironic fullness (Brackett 1892; Barron 1903; Googins 1906; Chase 1910; Holden 1921; Waugh 1925; Shute 1928; Robbins 1934; Dickinson 1938; Gould 1940a; Trott 1941; Wilburr 1942; Faught 1942; Webster 1945; Lagemann 1947; Bachman 1947).

Although the town meeting is usually lauded as a repository of communal amity, hostility frequently disrupts the harmony, as it comes to do in "The Lottery." Town-meeting narratives frequently acknowledge quarreling and even violence (Googins 1906, 564; Waugh 1925, 389; Shute 1928, 35, Robbins 1934, SM6). Occasionally the friction verges on victimization. A town meeting can be a lions' den for anyone fitting the role of Daniel. Sometimes, individuals must go into battle against strong majorities (Barron 1903, 406–408; Wilburr 1942, 25–26). Sometimes towns that have trouble filling offices impose them on hesitant candidates (Brackett 1892, 578; Robbins 1934, SM16). In "The Lottery," the town has decided to assign its most undesirable "office" through random selection. Like some other reluctant office-holders, Tessie is an unhappy "winner," in whose dilemma one can recognize an exaggerated form of the agoraphobic pain that political scientist Jane Mansbridge reported from her studies of town meetings (Mansbridge 1980, 60–65).

Jackson did not have to read articles to know that town meetings could be stressful. The March 1948 meetings in Bennington and North Bennington were momentous and contentious. North Bennington's meeting decided not to merge with the much larger town of Bennington. Bennington's meeting replaced the village president with a younger, more conservative man. Both meetings were controversial enough to attract record turnouts ("Plumb Wins" 1948; "North Village" 1948). Only three years after she had arrived in North Bennington, and long before she became a virtual shut-in, Jackson was still trying fitfully to integrate herself and her family into the community (Oppenheimer 1988, 182–185, 195; Franklin 2016, 200–202, 266–268), so it is possible that she attended one of the meetings. Whether she did or not, she would have heard about them during the period when she wrote "The Lottery" – through conversations with friends, through gossip overheard at the post office and the grocery, by reading the local newspaper. If she attended the North Bennington meeting, where she was eligible to vote, she probably witnessed a painful confrontation between one person and everyone else. Someone had proposed the town's effectual dissolution. The next day's *Bennington Banner* reported a unanimous "standing vote" of "[n]early one hundred persons" against the idea. But the vote cannot have been *quite* unanimous. No one rose to defend consolidation, but its proponent was surely there. That person must have suffered a humiliating chastisement by a crowd of friends and neighbors ("North Village" 1948).

Jackson's attentiveness to the discomfort a meeting could generate coincided with a general American anxiety about public space and group loyalty in the 1940s. The United States' participation in World War II was followed by a rivalry with the Soviet Union that upended the nation's politics. In November 1946, the end of the hot war and the beginning of the Cold War precipitated the victory, in Congress and many states, of conservative Republicans, who immediately began investigating the U.S. government's alleged failure to resist Soviet aggression. President Harry S. Truman, to forestall more extreme measures by the new Congress, established a Loyalty Program to gauge every federal employee's allegiance, partly through a compulsory oath and partly through an invitation of "derogatory information" from anonymous sources (Caute 1978, 268–270; Navasky 2003 [1980], 21, 29). Particular accused government workers became *causes célèbres*: William Remington in April 1947, "Mr. Blank" in June 1947, Dorothy Bailey and Edward Condon in March 1948 (Caute 1978, 276–277, 287, 291–292, 470–471). Meanwhile, the House

Committee on Un-American Activities ("HUAC") launched investigations of whole sectors of American life, including entertainment, labor, and education (Goodman 1968, 186–189). In the fall of 1947, HUAC's interrogation of the "Hollywood Ten" set a pattern for public trials of citizens' loyalty (Navasky 2003 [1980], 79–83; Goodman 1968, 207–225). These cases dominated national news during the year before Jackson wrote "The Lottery."

"The Lottery" deploys two devices upon which the emerging U.S. security state depended: the loyalty test and the list. The HUAC hearings, particularly, created a genre of public ordeal that could overwhelm any citizen with a leftist history. Government bureaucrats, Hollywood professionals, union officials, academics, journalists, and others the new Congress distrusted could find themselves in danger of losing their careers, livelihoods, reputations, freedom. Subpoenaed and sworn, they would be asked about their membership in various left-wing organizations and required to "name" the "names" of other members, not because the committee needed the information, which it had usually already acquired, but because only the betrayal of friends, through the ritual "naming of names," could prove that suspects had broken with the past and resumed their patriotic reliability (Navasky 2003 [1980], ix, 28, 314–323). The citizens of Jackson's village attend their own "naming of names": they pass their loyalty test not only by offering up their own lives but by demonstrating a readiness to betray every single other villager.

The reading of the town's names evokes a primary rhetorical form of the Red Scare: the list. The loyalty boards compiled lists of suspect employees ("Text of Truman's Order" 1947). Congressional committees called lists of witnesses. Witnesses were pressured to produce lists of names. Names were compiled into "blacklists" of people an industry would avoid hiring or into intermediate "graylists." The Attorney General's list of subversive organizations, to which lists of suspect people belonged, was first published in March 1948. The FBI maintained a Security Index, a list of subversives to be rounded up in the event of an emergency. When Senator Joseph McCarthy assumed leadership of the repression in 1950, he claimed to brandish lists of names of "known Communists" working in particular agencies (Navasky 2003 [1980], 22–23, 85–88). Lists of names were the medium through which Americans managed their suspicion of other Americans. In "The Lottery" as in the nation, the list is a comprehensive heuristic for monitoring the loyalty of the citizenry. In the informant culture of Lotterytown, the enlisted have come to include everyone.

Jackson and Hyman had reason to fear that their names were on lists and that their loyalty might be tested. Hyman had been an enthusiastic member of the Young Communist League in college, and he had persuaded Jackson to attend a few meetings. They consorted with other shady characters, including Walter Bernstein, Hyman's high school classmate who had become a blacklisted screenwriter. Both as a Bennington professor and as a *New Yorker* staff writer, Hyman was susceptible to scrutiny. The couple's vulnerability was confirmed later when, probably unbeknownst to them, the FBI opened an investigation after a house mover reported some of the titles in their huge library when he helped move them to Connecticut for two years in 1949 (Franklin 2016, 81–82, 85, 100, 310–312). When Jackson wrote "The Lottery," she must already have known that she and her family were among the multitudes subject to the inquisition.

2 Close Readings

Harold Bloom has opined that, although "The Lottery" is "crisply written and cunningly plotted," it "scarcely bears rereading, which is [. . .] the test for canonical literature" (Bloom 2001, 9). He recognizes that this is a story one can experience fully only if one has no previous knowledge of its outcome. But it is also a story one cannot understand, sentence by sentence, unless one has already read it. To appreciate "The Lottery" more than once, one must always read it as if for the first time. To appreciate "The Lottery" even the first time, one cannot understand any sentence in it without anticipating the ending.

The New Yorker would not carry tables of contents for another two decades, so its readers in 1948 would learn what was in an issue only by reading or browsing through it. They might well have assumed at first that "The Lottery" was journalism, as it withholds evidence of its fictionality for as long as it can. "The Lottery" is not called "The Stoning." The very same story, except with that title, would be a very different story indeed, whose readers would brace for its final mayhem from the beginning. The "great pile of stones" the boys collect (Jackson 1948, 25) would already make ominous sense and proclaim their affinity to bloodshed. The actual title not only diverts attention from the violent conclusion; it constrains the meaning of the story from the start by setting expectations that are quasi-generic, albeit mistaken. A story entitled "The Lottery" will probably follow some trajectories rather than others. Insofar as the title stipulates a plot in which one person will win and others will lose, it may augur winning rather than losing, because losing a lottery is probably too ordinary an event to merit narration, while a spectacular win is newsworthy. Yet a story about a lottery is probably not only about winning. What story could be more unsatisfying than one in which a character has a problem, buys a lottery ticket, and – presto! – wins the solution? So a story entitled "The Lottery" will probably be about a "win" that turns out to be a "loss," just as a story entitled "The Diamond as Big as the Ritz" or *The Treasure of the Sierra Madre* is likely to end with emptied pockets or dreams. Besides, a lottery, even before someone writes a story about it and before anyone draws a lot, is already a narrative event. People who play the lottery are buying counterfactual plots, alternative existences, fantasies about fortune, fame, travel, big houses, fast cars.

The narrator of "The Lottery" appears to be "omniscient," but she dispenses her knowledge selectively. Despite her privileged awareness, she opens the story by misleadingly describing an idyllic setting: "The morning of June 27th was clear and sunny, with the fresh warmth of a full-summer day; the flowers were blossoming profusely and the grass was richly green." The citizens of an unnamed town – let's call it Lotterytown – are assembling: they "began to gather in the square, between the post office and the bank, around ten o'clock; in some towns there were so many people that the lottery took two days and had to be started on June 26th, but in this village, where there were only about three hundred people, the whole lottery took only about two hours, so it could begin at ten o'clock in the morning and still be through in time to allow the villagers to get home for noon dinner" (25). Nothing in this introduction prepares readers for the horror to come, though the narrator must already know about it. At least the *story* knows how it will end, because the remainder of the text already lies waiting beyond this page. In a way, Tessie will be correct when she complains that the lottery is corrupt, for the fix *is* in: no matter how many times one reads the story, she will suffer a catastrophe. The lottery may not be rigged against her, but "The Lottery" is. On a second reading, we know the narrator has systematically withheld essential information.

"The Lottery" depends more than most stories on its narrator's exploitation of the gulf between what she knows and what she can expect readers to know. Sometimes she pretends that readers are already inside the community so she does not need to mention the ritual's true purpose. The narrator does *not* begin, "Every year, in late June, the citizens of a certain village conduct a lottery through which they select one of their number for the others to stone to death." Ostensibly, she foregoes such a sentence because we already know about this lottery. Actually, of course, Jackson withholds it precisely because we lack that information and she wants to keep us in the dark until as close as possible to the end. When we are told that "The children assembled first, of course" (25), the phrase "of course" implies that we must already know why the children gather first. It appears to endow us with a portion of the narrator's omniscience, though we cannot know until after we see the story's whole trajectory what she is hinting about it. On a first reading, when "Bobby Martin had *already* stuffed his pockets full of stones" (25, emphasis added), he simply seems eager to gather stones, as boys do. On subsequent readings, Bobby is preparing for the murder of a neighbor or family member, as he has done every year since he was a toddler. The narrator's reticence actuates irony, since the words seem to signify one meaning but actually signify another.

Yet at other times the narrator acknowledges that readers stand outside the community and must be provided inside information. When Dickie Delacroix appears in the second paragraph, the narrator reveals that "the villagers pronounced his name 'Dellacroy'" (25), because otherwise we would have no way of knowing, as his neighbors do, that they have anglicized the Francophone name. The tension between the assumption that we know and the assumption that we are ignorant, between being citizens and being aliens, is crucial to the story's meaning. By drawing *New Yorker* readers into the village circle, by making urban sophisticates feel at home within the pastoral town, the story invites them – us – into provisional complicity with local practice. At the same time, by excluding readers from the town's lore, the story casts us as ethnographers witnessing a remote culture's exotic custom or incriminates us as foreigners illicitly spying. The tension between the insider status and the outsider status of the reader communicates the anxiety the townspeople turn out to suffer from their own fellow membership.

"The Lottery" makes its readers simultaneously participants in and strangers to the community of its narration. Membership, and therefore exclusion, is its deepest, darkest subject. From the outside, until near the end, the story appears to celebrate the mundane pleasures of small-town life. Once one has absorbed the ending, and fallen inside, the story exposes the horror beneath the village's superficial geniality. Outside, the "boisterous play" (25) of the children is roughhouse; inside, it is brutality that promises savagery. Outside, the demeanor of adult males when they arrive – "surveying their own children, speaking of planting and rain, tractors and taxes," telling jokes and smiling rather than laughing (25) – bespeaks New England reserve. Inside, it attests to dread, and their observation of their children amounts to panoptic control, not just benign care. Not only have they been potential victims of the lottery, vulnerable to sudden execution year after year; they have been, year after year, executioners.

For residents of Lotterytown, participation in the lottery – citizenship – is a birthright. Just as they belong to their families and thereby to the village, they belong to the lottery. They do not enroll in its lists; the lists enroll them. But Tessie's late arrival at the proceedings suggests that any individual in the town might choose to drop out. On a first reading, her tardiness may seem casual and unintentional: "Clean forgot what day it was," she says to Mrs. Delacroix, joining the crowd, "and then I looked out the window and the kids was gone, and then I remembered it was the twenty-seventh and came a-running." And Mrs. Delacroix's response – "You're in time, though" (26) – may imply a minor negative consequence for lateness, perhaps an eligibility deadline for a prize. On a second reading, she may be citing a harsher penalty, for the ending surely suggests that residents who fail to attend the lottery might pay a high price. And we know on a second reading that June twenty-seventh is a date no one in Lotterytown can have "clean forgot," even temporarily. Once inside the community of the story, we recognize that Tessie's delinquency can result only from reluctance or even rebelliousness, not just from a dutiful preoccupation with housework. The real subject of this passage, the real subject of the townspeople's small talk, is the possibility that Tessie, or any of her neighbors, including even the belated "official," Mr. Summers, might not show up.

The middle third of "The Lottery" is dominated by a rollcall, the usual purpose of which is to establish who is attending and who is not. But for this lottery, strangely, that determination happens *before* the rollcall. Summers begins the proceedings with an informal headcount to establish that everyone is present. Apparently preoccupied with the box and the papers, he has already missed Tessie: "Thought we were going to have to get on without you," he tells her (26), either casually on a first reading or more darkly on a second, as she arrives. "Anybody ain't here?" he then asks the crowd (26), knowing that everyone has been registering the arrivals of everyone else. When people inform him that Clyde Dunbar is absent because he broke his leg, the rollcall cannot begin until Mr. Summers certifies Dunbar's excuse and confirms that a substitute, his wife, will draw for him, because "it was the business of the official of the lottery to ask such questions" (26). A precondition of the lottery is that everyone be present or accounted for. On a first reading the reason for this may seem to be to ensure that everyone has a chance to win a prize. But even on a first reading that reason frays: if a prize were imminent, would not those present be happy for their neighbors to stay home, reducing competition? The effort to ensure full participation is instead an early sign that the villagers seek a maximal risk pool. On a second reading, the pre-rollcall headcount, like the children's gathering of stones and the men's subdued mood, clearly foreshadows the revelation that the lottery's winner is actually its loser. Nevertheless, Summers is not only determining Clyde's whereabouts in case the Dunbar family "wins" and Clyde must be summoned. Paradoxically, he is also serving a social-welfare function, ascertaining that the Dunbar family is provided for during Clyde's convalescence. He extends similar solicitude to the Watson family, who have lost a breadwinner, and to Warner, the oldest resident, whose attendance must be a matter of uncertainty (26). The lottery looks out for the needy, including survivors of lottery victims (but all the townspeople are lottery survivors), by determining whether their needs are being met. The lottery is a mechanism through which the community both kills one of its members and monitors all of their circumstances. Its sinister function is the counterfoil to a benefit; the sacrifice is the instrument of a public good.

When "The Lottery" reaches the rollcall, its forward motion as a narrative is highjacked by an inherently non-narrative ordering principle. Temporarily, an organization of information according to causes and effects, unfolding relationships among individuals and between individuals and groups, is displaced by the bureaucratic logic of the alphabet. A story gives way to a list. The story "names" the "names" Adams, Allen, Anderson, Bentham, Clark, Delacroix, Dunbar, Graves, Harburt, Hutchinson, Jones, Martin, Overdyke, Percy, Summers, Warner, Watson, and Zanini (26–27). But the rollcall is only the most mechanical component of a foreordained agenda that has driven the story from its beginning. Like town-meeting narratives, the story of the lottery this year follows a routine much the same as last year's or next year's. "The Lottery" fuses "news" about a unique event with cyclical repetition of recurrent information. Thus, the behavior of the villagers, as they listen for their names, turns out to be just as generic and scripted as the rollcall itself. The lottery is an annual theater in which all the residents must play the part of neighbor. Not only must Mr. Adams force a grin when he draws the first "ticket" (26); he must simulate patience and indifference through the duration of the tally, as must everyone else. Everyone must pretend this event is humdrum: "Seems like there's no time at all between lotteries any more," Mrs. Delacroix says to Mrs. Graves. "Time sure goes fast," Mrs. Graves replies (27), and the reader can supply the unspoken ironic supplement: when you're having fun. Whatever one's apprehension – and the nervous tension that permeates the assembly signifies that many, perhaps all but Warner, suffer stage fright – one summons the resources with which to soldier through. If one can pretend to be a citizen on lottery day, one can probably play the role every other day. The lottery teaches participants how to pretend to be citizens of Lotterytown year-round. Awaiting the climax schools them in self-control, fortitude, sportsmanship, and other virtues that will sustain them through the year. At first, despite her late arrival, Tessie appears to perform her composure as effectively as her neighbors do. When her family's name is called, she prods her husband to "Get up there, Bill," and those near her laugh. But eagerness may be as indecorous a sign of anxiety as reluctance. When Mrs. Dunbar says, "I wish they'd hurry," twice, her older son curbs her impatience. "They're almost through," he says, though they are still in the P's (27).

Old Man Warner, who has survived the lottery seventy-seven times, is its strongest advocate: "Listening to the young folks, nothing's good enough for *them.* Next thing you know, they'll be wanting to go back to living in caves, nobody work any more, live *that* way for a while. Used to be a saying about 'Lottery in June, corn be heavy soon.' First thing you know, we'd all be eating stewed chickweed and acorns. There's always been a lottery" (27). For Warner, the lottery is eminently just and efficient. It guarantees an agricultural order more advanced than hunting and gathering. It marks the community as civilized rather than primitive. From his perspective, indeed, the "winner" may be a true winner rather than a loser. As the master of a potlatch is whoever wastes the most possessions by giving them away or destroying them, so the honor of the lottery may accrue to the competitor who makes the fullest sacrifice. Warner may in fact have been disappointed every year for seven decades. Naturally, he would lament the ingratitude of a winner who acts as if she is a loser. His neighbors agree with him sufficiently that they maintain the ritual, because they appear to value its benefits more than they fear its risks.

The lottery consists of at least two drawings, not one. First, a family is chosen from the pool of all families; then a member of that family is chosen from its members. An additional layer may be involved in some cases: the lists that Summers and Graves must "make up" include a list of heads of families and a list of heads of households (26). When the Hutchinson family are selected, Summers "consult[s] his next list," the second of three, and asks Bill Hutchinson whether "You got any other households in the Hutchinsons?" Much to Tessie's chagrin, the answer is no, because her married daughter Eva will draw "with her husband's family" (27). Had Eva been a son, an intermediate drawing between the two Hutchinson households would have improved Tessie's odds. Instead, only one additional lottery is necessary – among Bill; Tessie; Bill, Jr., an awkward teenager; twelve-year-old Nancy; and "little Dave" (28). The division of the lottery into stages motivates population growth by encouraging large families (Oehlschlaeger 1988, 259). The logic of increasing population by killing people is counterintuitive, but the process works: The population of the town, "about three hundred" near the beginning of the story, is "more than three hundred and likely to keep on growing" a few paragraphs later (Jackson 1948, 25). If one's family is chosen, one's chance of survival increases the larger it is. Thus Warner, a family of one, is more vulnerable than Tessie, from a family of five. If Warner is a bachelor, his survival of seventy-seven lotteries in a town that probably has fewer than seventy-seven households is remarkable.

The multiple structure of the lottery also encourages betrayal. However ably Tessie may enact her civic duty during the rollcall, she breaks from her performance once the first round concludes with her husband holding the fatal "ticket": "You didn't give him time enough to take any paper he wanted. [. . .] It wasn't fair!" she shouts at Summers. Her husband tries to manage her: "Shut up, Tessie." Mrs. Delacroix admonishes her: "Be a good sport, Tessie." Mrs. Graves reminds her that "All of us took the same chance." But the floodgates of Tessie's dissent have opened, and out pours her disloyalty not only to the community but to her family. "There's Don and Eva," she yells. "Make *them* take their chance!" By trying to include them in the risk pool for the second drawing (27), she justifies her own selection. She is the appropriate victim because, more clearly than anyone else, she has withdrawn from the communal consensus. By the reluctance she expresses through her late arrival, through her declaration of the lottery's unfairness, and through her "unsporting" attempt to jeopardize her daughter, she earns her punishment. Tessie will finally consent to the process by at least drawing her own lot: "She hesitated for a minute, looking around defiantly, and then set her lips and went up to the box. She snatched a paper out and held it behind her" (28). But her consent is "defiant." Her demise is an object lesson for everyone else: the way to avoid her fate next year is to report promptly, play one's part, accept one's fate. The way to survive is to subordinate oneself to the group, suppressing one's loyalty to mere self or to family and friends. Still, far from discouraging betrayal, the lottery is perfectly designed to induce it. Tessie's children are as ready to forsake her as she them: when Nancy and Bill, Jr., open their unmarked "tickets," they "beam[ ] and laugh[ ], turning around to the crowd and holding their slips of paper above their heads" (28). After Tessie is "chosen," her husband and children participate in her execution along with her friends and neighbors. Perhaps the story's most horrifying detail is that "someone" gives "a few pebbles" (28) to little Davy so that he too can slay his mother.

The execution to which the lottery leads is by stoning. Most methods of capital punishment delegate the killing to one specialist or a representative group. Decapitation is best conducted by a headsman. Drawing and quartering, electrocution, hanging, guillotining, and lethal injection accommodate relatively few practitioners; others must assume the passive role of spectators. Most firing squads consist of a delegated sub-group, one member of which fires a blank so that everyone may imagine that he or she is not culpable. Stoning, perhaps uniquely, involves more-or-less equal participation by everyone. The killing in "The Lottery" appears to require the unanimous participation of all 300 townspeople, all of whom are visible to each other and none of whom can throw blanks. Stoning is the one method of execution that commits everyone in a community to direct responsibility for the act.

The crowd that is "upon" Tessie in the story's last sentence is unconventional. People who join mobs are usually anonymous and fungible, but the mob at the end of "The Lottery" consists of people whom the story has carefully labeled by name and by position within the kinship structure that constitutes the community's prevailing self-definition. At no point do the individuals turn into interchangeable parts of a brutal mass. When Mrs. Delacroix arms herself with "a stone so large she had to pick it up with both hands," whether fiendishly to maximize pain or compassionately to end Tessie's misery, the text makes clear that she is the same Mrs. Delacroix who welcomed Tessie earlier. The woman carrying smaller stones who replies to her, "You'll have to go ahead and I'll catch up with you," is specifically Mrs. Dunbar (28). What is most shocking about "The Lottery" is not that the pleasant community it depicts "turns into" a vicious mob but that the townspeople remain as individuated and as civil as they have been all along – or, rather, that the polite community has been a mob all along. Nonetheless, the crowd has undergone a dramatic double transformation for the reader. The townspeople are first benign participants in a quaint ritual; then, each of them turns out to be a prospective victim; then, each of those victims, save one, becomes an executioner. People who have stood under threat of being stoned avenge themselves upon that vulnerability by stoning a representative of the people who would have thrown stones at them. The simplest algebra of justice is served. A primary reality of the story is that it reveals the grim truth beneath a facile masquerade.

"Biography of a Story" bears enough similarities to "The Lottery" to constitute a revision as much as an explanation. A mother leaves the shelter of her home to enter a specific public space inhabited by other members of her community, on a specific date during a specific season. She confronts an official, one of whose jobs is to maintain a public record of the community's membership. From a "box," she receives inscription-bearing paper. The delivery of this marked paper prompts a violent confrontation between one individual and a hostile group of other people. The approximately three hundred citizens of Lotterytown prepare to throw stones; the approximately three hundred people who purportedly wrote letters to *The New Yorker* (did the number derive from Jackson's memory of the story rather than from an actual count of the letters?) launch hurtful judgments. In each case, the mother exposes a daughter to danger. In "The Lottery," a woman's entrance into literal public space results in a physical attack. In "Biography," sending "The Lottery" into the virtual public space of publication results in a verbal attack. Both "stories" demonstrate the risks of engagement with one's community.

3 Theoretical Perspectives

Although the critical response to "The Lottery" has included feminist (Oehlschlaeger 1988; Carpenter 1988; Hall 1993, 48–53) and Marxist (Kosenko 1985) readings, folkloristic and anthropological studies (Nebeker 1974; Allen 1980; Gibson 1984) have predominated. Critics have ignored or relied minimally on some resources that offer strong leverage for discerning the lottery's function within its community. Game theory, public-sphere theory, and René Girard's theory of sacrificial violence afford overlapping insights that can illuminate the story. All of these discourses suffer from defects. Game theorists depend unrealistically upon the rationality of human "players." The philosophers of public space disagree with each other profoundly. Girard practices an anthropology without direct observations or informants. There is no evidence that Jackson was aware of game theory, and she cannot have known in 1948 about political and religious ideas that were yet unformulated. But, taken together, these perspectives provide a powerful lens through which to read – or, *pace* Bloom, to reread – "The Lottery." Within their combined framework, the stoning matters less than the lottery, and the lottery matters less than the assembly.

Although Jackson was probably unfamiliar with the work of John von Neumann, Oskar Morgenstern, John Nash, and other pioneers of what was to become a substantial discipline within mathematics and the social sciences (Poundstone 1992, 1–99), she was herself a game-player and sports fan (Oppenheimer 1988, 108, 121–22, 144, 195; Franklin 2016, 8, 260–261), attuned to the increasing indispensability of game-playing to American life. While the game theorists were transforming military and geopolitical strategy, market analysis, and business regulation, Jackson was devising in "The Lottery" a game that clarifies the relationship we all "enjoy" with the communities we inhabit. The fundamental game-theory paradigm is "Prisoner's Dilemma." Police separately interrogate two criminal accomplices, offering them reduced sentences if each will betray the other. If the prisoners "cooperate" with each other by refusing to confess, they can go free. But most people do not trust each other, so the "dominant strategy" in Prisoner's Dilemma is plea bargaining, or "defection": both suspects reduce their liability by betraying their partners (Poundstone 1992, 116–125). Jackson's lottery also forces its participants to choose between their own interests and those of their fellows, but it relies upon an opposite logic and a different game paradigm – the "Stag Hunt," in which a party of hunters at posts in a forest must each decide whether to remain vigilant for the group, to help ambush a large prey that will feed everyone, or give in to the temptation to bag a passing squirrel or rabbit that might fill one family's larder. Mathematically, the dominant strategy of the Stag Hunt is cooperation rather than defection (Poundstone 1992, 218–221). Likewise, the dominant strategy of the lottery is to subordinate oneself to the common good rather than pursue one's own. All residents of Lotterytown annually commit themselves to sacrificing not only themselves but their families, friends, and neighbors to the whole community that consists of those same people. The headcount before the rollcall secures such "cooperation" by ensuring – *contra* Prisoner's Dilemma – that nobody need proceed until everyone knows that no one else is "defecting." Everyone runs the same risk of turning out to be the "stag"; everyone participates equally in slaying it.

Game theory lends to "The Lottery" a logicality that contradicts the horror most readers of the story experience. If Old Man Warner were right and the lottery were necessary to Lotterytown's prosperity, no fairer procedure seems possible. If the community were to select its weakest, most unpopular, or least productive member, its life would be twisted by perverse incentives. Even Warner might not want to have grown old in a Lotterytown that always selected its oldest member. And a rotation system would be unworkable: surely a family who knew that this year is their "turn" would strongly consider moving elsewhere before June. By comparison the lottery embodies democratic equity. Even if Warner is wrong, and the lottery does not guarantee the corn yield, it produces the real blessings of solidarity. Once a year, the citizens gather to renew their mutual commitment, not to each other but to the community. Against the benefits of belonging to a social polity, they balance the mathematically negligible risk (on average, one in about three hundred) of their own demise. The function of the lottery is to induce people to prove their fidelity to the community in the most dramatic possible way, not by swearing fealty to it but by volunteering to die and murder for it. Or rather: the function of the lottery is precisely to tempt its participants not to participate, to lure them into revealing any disloyalty they might harbor. It requires no one to say a word, except "Here," but as a trial of faith it is surer than any oath. It generates dissent for the purpose of eradicating it, a test of each player's loyalty not just to the game that is the lottery but to the game that is village life.

Following World War II, many Western intellectuals, appalled by Europe's turn towards totalitarianism, devoted urgent attention to the idea of free discourse in "public space." In *The Open Society and Its Enemies* (1945), for example, Karl Popper blamed the fascist disaster of the Thirties and Forties on Plato, Hegel, and Marx, and he recommended a course correction to fortify liberal democracy. In *The Human Condition* (1958), Hannah Arendt extolled the virtues of a "life devoted to public-political matters" as they had been conducted in the Greek *polis*. Citizenship consists of courageously "leaving one's private hiding place" and entering the *agora*, the common, where one gains agency and identity in the "constant presence of others." To live one's life "without speech and without action," by contrast, is to be "literally dead to the world." Without intending to, "The Lottery" anticipates Arendt's logic perversely: The ritual annually enables citizens to confirm their moral identities by entering their common space together. But Tessie's entrance into public space is precisely what opens her to the "mob rule" Arendt fears (Arendt 1998 [1958], 12, 51, 176, 186, 188, 203).

Jürgen Habermas, in *The Structural Transformation of the Public Sphere* (1962), provides a more historically rigorous theory of the origins of the "public sphere," which he traces less to ancient Greece than to the developing market economy of seventeenth-century Britain and eighteenth-century France, where courtly authority was displaced by bourgeois institutions such as coffee houses, salons, newspapers, postal service, and public libraries. While for Arendt the public sphere consists of visibility to others' eyes, Habermas asserts that "a public consisting of private persons" requires a strong "private sphere" to engender them. But, beginning in the late nineteenth century, the "public sphere" has disintegrated due to a government-fostered blurring of the boundary between the public and the private realms. For Habermas, the intrusion of HUAC into the friendships, associations, and reading habits of private citizens would exemplify "the interferences of the state in the private sphere." The coercion of Tessie and her neighbors to leave their homes and assemble in the common would exemplify not the triumph of the public sphere but the "deprivatization" of the private sphere. The lottery may reinforce family structure, but only to subordinate people's loyalty to parents, siblings, and children to their loyalty to the town. It illustrates that "a public made up of rationally debating private citizens has collapsed" (Habermas 1989 [1962], xvii–xviii, 27–51, 55, 79, 142–143, 146, 152, 247). The most specific subject of "The Lottery" is the particular degradation of the American public sphere caused by the Red Scare in the late 1940s.

Postmodernist contributors to public-sphere theory (Fraser 1990 and Warner 2002, for example) have rejected both Arendt's ancient-Athenian and Habermas's bourgeois-European public spheres as too exclusive. Whereas Arendt defines politics "restrictively as an active stance, verbal and physical, that takes place within a clearly demarcated public sphere," Judith Butler, in *Notes Toward a Performative Theory of Assembly* (2015) makes room for alternative modes of assembly and speech. Whereas Habermas identifies public space with increasingly obsolete classes and institutions, Butler hopes to make it available to marginalized people who have been denied access to the *polis*: women and gays, slaves, migrants, the disabled, the "precarious." She does not mention agoraphobes, but they are perhaps the definitional case of the excluded. If the right to speak implies a right to remain silent, perhaps the right of assembly implies a right *not* "to appear" (Butler 2015, 55, 58, 205). The ultimate task for an "open society" might be to open the *agora* to the agoraphobe.

For Butler, assembling entails performance. Regardless of any messages a crowd delivers by words or banners, the fact that bodies have gathered with other bodies is itself a statement. Even if they stand silently, the assembled assert their right to assemble, which may be the primary message they communicate. They may, in short, stand together simply to affirm their *standing*. A civil-rights demonstration can "speak" first about the "right to have rights" before it addresses a particular policy or outrage (Butler 2015, 18, 26, 40, 48–49). The people of Lotterytown, with strangely similar bravery, *perform* their citizenship. Villagers appear at the assembly to perform their willingness to subject themselves to bodily risk and moral hazard. The ritual has no discursive content beyond the reading of names, because its primary purpose is to occasion the participants' theatrical nonchalance, their enactment of the role of participant, the annual renewal of their silent pledge of allegiance. And, of course, unbeknownst to first-time readers, the performance includes their willingness not only (perhaps) to die for the town but also (certainly) to stone to death a friend, a family member, a neighbor.

Although that stoning has prompted scholars to provide a steady supply of ethnographic analogues for "The Lottery," from the Bible, ancient Greece and Rome, Frazer's *Golden Bough*, and other sources, little attention (Lamerson 2010, Bailey 2014) has been paid to René Girard's theory of "sacrificial violence" to explain the lottery's utility to Lotterytown. In *Violence and the Sacred* (1972), *Things Hidden Since the Foundation of the World* (1978), and *The Scapegoat* (1982), Girard hypothesized that the function of violent ritual sacrifice is to prevent worse violence. According to these books, all human communities must in their "primitive" pasts have suffered cycles of violence and "mimetic" reprisal that jeopardized their existence. Such "sacrificial crises," by reducing everyone to the same aggressive stance, threatened the "cultural order," which is only "a regulated system of distinctions." The selection of a scapegoat restored distinction by defining a difference between one citizen and all others. Subsequently, the commemorative sacrifice of a victim, whether a human who stands for all members of the community or an animal who stands for the human substitute, "secur[es] the safety of the group by checking the impulse for revenge" (Girard 1977 [1972]), 9–10, 15, 21, 23, 31, 49, 85). Through "unanimous violence," "the opposition of everyone against everyone else is replaced by the opposition of all against one" (Girard 1977 [1972], 93; 1987 [1978] 24). The sacrifice occurs at the climax of a "festival," a holiday from normality during which work is suspended to make room for a contest, game, or sporting event. Typically, the selection of the scapegoat is "arbitrary," often by lot (Girard 1977 [1972], 119, 257, 311–314). Stoning is a frequent method of execution (Girard 1977 [1972], 94, 248; 1986 [1982], 170–171, 176, 178, 182). In "The Lottery," the annual ritual simultaneously erases distinctions by reducing everyone to the status of killer and killed and restores them by organizing the populace as a set of families.

In Girard's telling, the "victimage mechanism," at the "threshold of hominization" (Girard 1987 [1978], 95), not only underlies "all ritual and all religion" (Girard 1977 [1972], 302) and "dominates world mythology" (Girard 1986 [1982], 71). It is the root of linguistic signification and, thereby, of human thought. The mission of sacrificial violence, to restore social distinctions, must have overlapped with the development of linguistic structure, because the symbolic displacement that constitutes ritual sacrifice is the source of signification itself: "The origin of symbolic thought lies in the mechanism of the surrogate victim." The scapegoat "gives birth to language" (Girard 1977 [1972], 235, 306) because it is the first object that stands for something it is not. Girard attributes the emergence of this first sign, before the existence of a system of other signs, to the "uniquely dynamic" "model of drawing lots" as "the simplest" and "most rudimentary form of symbolism" and the basis for "[a]rticulated language." To ensure the maintenance of the "miraculous peace" produced by the original "act of collective murder," humans developed a "penchant for reproducing the sign, [. . .] for reproducing the language of the sacred by substituting, in ritual, new victims for the original victim" (Girard 1987 [1978], 73, 100–101, 103).

By convening to select and stone a victim, the citizens of Lotterytown not only perform their solidarity with the town; they also perform their adherence to the symbolic order of written language. First the lottery causes them to gather according to patterns of pleasure, attraction, and affinity; then it pulls each head of family out of this jumble of lived relationships and inserts him into the one-dimensional, unidirectional, irrational master order of the Roman alphabet, from which there can be no appeal. At its heart, "The Lottery" consists of the community's ritual obeisance to the rule of literacy. Each individual demonstrates, through commitment to the sacrifice for which the roster-reading is a prelude, a willingness to subordinate his or her very life to the alphabetical regime. Indeed, the alphabetical list may be less an instrument of the sacrifice than the sacrifice is an instrument of the list, lending it a gravity appropriate to its import.

Thus, the text of "The Lottery" interrupts itself midway to rehearse the elements of its own communicative technology. What plot device could be more self-reflexive, more meta-literary, for a published text than a recitation of the alphabet? The story that has introduced a miscellaneous array of characters pauses for them to match themselves against the story's *dramatis personae*. This matching of people with their names bears a "family resemblance" to the "language games" that Ludwig Wittgenstein proposed in *Philosophical Investigations* (1951) (2009 [1951], 35–36 [¶64–¶67]) and to a cluster of works of fiction from the Forties and Fifties. In Saul Bellow's "Looking for Mr. Green" (1951), for example, a government worker searches Chicago tenements for the titular "Tulliver Green," carrying a welfare check with that name, like the Prince seeking a foot for Cinderella's slipper. Near the climax of Ralph Ellison's *Invisible Man* (1952), the anonymous narrator wanders the streets of Harlem after donning dark glasses and finds himself mistaken for "Rinehart," whose identity shifts as he is incorrectly recognized by various members of the crowd (Ellison 1995 [1952], 479–499). In Jackson's own story "Nightmare," unpublished during her lifetime, a Manhattan office worker named Toni Morgan, sent by her boss to deliver a package elsewhere in the city, finds herself pursued by the producers of a radio game show, the object of which is to "Find Miss X." As she traverses the streets, a sound truck follows her describing Miss X by listing Toni's features but changing the description to fit Toni's efforts to disguise herself. Finally, she acquiesces in her new identity, or non-identity, pleased to have *become* Miss X (Jackson 1997, 37–51). In "The Daemon Lover," when the protagonist cannot find James Harris's name on any of the mailboxes at his apartment building, she spends the rest of her day searching for him throughout the confusing and hostile city. She never secures him, but his name proliferates throughout the remainder of *The Lottery; or, The Adventures of James Harris* (1949), Jackson's first short story collection. To unify the volume,

Jackson modified many of the stories (though not "The Lottery") by changing a male character's name to James, or Jim, or Mr. Harris (Jackson 2010, 1–239). A dozen years later, at the end of Stanley Kubrick's film *Spartacus* (1960), written by the blacklisted Dalton Trumbo, a member of the Hollywood Ten using his own name for the first time in years, the name-matching game results in one of the most iconic rollcalls of the Cold War era. To foil the Roman effort to find the leader of a slave revolt so that he can be crucified, all of his fellow slaves volunteer for victimization: "I am Spartacus," they respond to the blacklist of one name.

Whereas the signifiers in these festivals of "language on holiday" float free, the language game in "The Lottery" is remarkably pat: the names are called and their human referents step forward one by one to reconfirm their places in a nomenclature, to accept their assigned discursive identities. Perhaps it is inevitable that one of the men presiding over this ceremony of literacy is the fittingly named postmaster, Mr. Graves. His regular job is likewise to connect people with letters, to reduce humans to addresses and their relationships to correspondence.

It is appropriate too that, in "Biography of a Story," Jackson framed her composition of "The Lottery" between encounters with a real postmaster. "Biography" constitutes what public-sphere theorist Michael Warner calls a "scene of circulation" (Warner 2002, 69, 74). In the print culture of the mid-twentieth century, the postmaster is literally a conduit of publication. After Jackson composed "The Lottery," she indeed mailed it to her agent, who mailed it to *The New Yorker*, which communicated with Jackson primarily through the mail. When the story was published by *The New Yorker*, it was delivered to its readers primarily through a postal-subscription *list* of names and addresses. When those readers responded to their experience, they mailed letters to the editor, who collected them in bundles and mailed them to Jackson, who received them through the post office and later assembled them into a *list* of illustrative excerpts. To at least a few of these correspondents, she sent responses through the mail; and *The New Yorker* responded to all the mail it received with letters (Shirley Jackson Papers, Box 32 Scrapbook; Box 48, Folder 20). In "Biography," Jackson "explains" "The Lottery" by placing it within the "feedback loop" (Warner 2002, 71) of its own consumption. But "The Lottery" already depicts that system in its primal depiction of the consequences of entering public space. The selection and stoning of Tessie in Lotterytown mirror "Biography"'s singling out of Shirley Jackson for verbal persecution by the community of readers of "The Lottery." (Let's call that community "Lottery"-town.) As she reports the letters of her readers, they testify to their own, *our* own, sensation of trauma. "The Lottery"'s conjunction of the alphabet with ritual violence, of written words with retribution, may express Jackson's complex attitude toward her own literary vocation as much as it expresses her attitude toward the contemporary politics of scapegoating and interrogation. By constructing an alphabetical lure to entice her readers out of our own private spaces into the initially attractive public sphere of "The Lottery," and then by assaulting us with the shock of its conclusion, she replicates her own agoraphobic panic in us.

4 Bibliography

4.1 Works Cited

Allen, Barbara. "A Folkloristic Look at Shirley Jackson's 'The Lottery.'" *Tennessee Folklore Society Bulletin* 46.4 (December 1980): 119–124.

Arendt, Hannah. *The Human Condition*. 1958. 2nd ed. Chicago: U of Chicago P, 1998.

Bachman, John W. "The Significance of the Junior Town Meeting Movement to the Evolving Secondary School Curriculum." *Education* 67 (May 1947): 571–577.

Bailey, Ted. "Sacred Rule in Shirley Jackson's *The Lottery*." *British and American Studies* 20 (2014): 37–42.

Barron, Alexander J. "A New England Town Meeting of To-Day." *Outlook* 75 (17 October 1903): 405–409.

Bellow, Saul. "Looking for Mr. Green." *Commentary* 11.3 (March 1951): 251–261.

Bloom, Harold. "Introduction." *Shirley Jackson (Bloom's Major Short Story Writers)*. Ed. Harold Bloom. Broomall, PA: Chelsea, 2001. 9–10.

Bonsal, Stephen. "Town Meeting of the World." *Collier's Weekly* 4 January 1947: 44.

Brackett, Anna C. "The Aryan Mark: A New England Town Meeting." *Harper's Monthly* 85 (September 1892): 577–585.

Bryan, Frank M. *Real Democracy: The New England Town Meeting and How It Works*. Chicago: U of Chicago P, 2004.

Butler, Judith. *Notes Toward a Performative Theory of Assembly*. Cambridge, MA: Harvard UP, 2015.

Carpenter, Lynette. "Domestic Comedy, Black Comedy, and Real Life: Shirley Jackson, A Woman Writer." *Faith of a (Woman) Writer*. Eds. Alice Kessler-Harris and William McBrien. Westport, CT: Greenwood P, 1988. 143–148.

Caute, David. *The Great Fear: The Anti-Communist Purge Under Truman and Eisenhower*. New York: Simon and Schuster, 1978.

Chase, Sidney M. "Town Meeting Day." *Scribner's* 48 (November 1910): 554–563.

Dickinson, S. L. "First Monday in March." *Scribner's* 103 (March 1938): 57–59.

Ellison, Ralph. *Invisible Man*. 1952. New York: Vintage, 1995.

Faught, Millard. "Town Meeting." *Rotarian* 61 (October 1942): 35–37.

Franklin, Ruth. *Shirley Jackson: A Rather Haunted Life*. New York: Norton-Liveright, 2016.

Fraser, Nancy. "Rethinking the Public Sphere: A Contribution to the Critique of Actually Existing Democracy." *Social Text* 25/26 (1990): 56–80.

Gibson, James M. "An Old Testament Analogue for 'The Lottery.'" *Journal of Modern Literature* 11.1 (March 1984): 193–195.

Girard, René. *Violence and the Sacred* [*La Violence et le sacré* 1972]. Trans. Patrick Gregory. Baltimore: Johns Hopkins UP, 1977.

Girard, René, in collaboration with Jean-Michel Oughourlian and Guy Lefort. *Things Hidden Since the Foundation of the World* [*Des choses cachées depuis la foundation du monde* 1978]. Trans. Stephen Bann and Michael Metteer. Stanford, CA: Stanford UP, 1987.

Girard, René. *The Scapegoat* [*Le Bouc émissaire* 1982]. Trans. Yvonne Freccero. Baltimore: Johns Hopkins UP, 1986.

Goodman, Walter. *The Committee: The Extraordinary Career of the House Committee on Un-American Activities*. New York: Farrar, Straus and Giroux, 1968.
Googins, George E. "The Town-Meeting." *Outlook* 82 (10 March 1906): 561–565.
Gould, John. *New England Town Meeting: Safeguard of Democracy*. Brattleboro, VT: Stephen Daye, 1940a.
Gould, John. "No Place for Dictators." *New York Times* 7 Apr. 1940b: 106.
Habermas, Jürgen. *The Structural Transformation of the Public Sphere: An Inquiry into a Category of Bourgeois Society*. [Strukturwandel der Öffentlichkeit 1962.] Trans. Thomas Burger and Frederick Lawrence. Cambridge: MIT P, 1989.
Hall, Joan Wylie. *Shirley Jackson: A Study of the Short Fiction*. New York: Twayne, 1993.
Holden, Raymond. "Town Meeting – 1921." *Survey* 45 (26 March 1921): 917.
Jackson, Shirley. "The Lottery." *New Yorker* 26 June 1948: 25–28.
Jackson, Shirley. *Just an Ordinary Day*. Eds. Laurence Jackson Hyman and Sarah Hyman Stewart. New York: Bantam, 1997.
Jackson, Shirley. *Novels and Stories: The Lottery, The Haunting of Hill House, We Have Always Lived in the Castle, Other Stories and Sketches*. Ed. Joyce Carol Oates. New York: Library of America, 2010.
Kosenko, Peter. "A Marxist/Feminist Reading of Shirley Jackson's 'The Lottery.'" *New Orleans Review* 12.1 (Spring 1985): 27–32.
Kubrick, Stanley, director. *Spartacus*. Screenplay by Dalton Trumbo. Performances by Kirk Douglas, Laurence Olivier, Jean Simmons. Bryna/ Universal, 1960.
Lagemann, John Kord. "Democracy in Our Town." *Collier's Weekly* 119 (8 March 1947); 13–15, 43.
Lamerson, John M. "Everyone Wins This 'Lottery.'" *Lamar Journal of the Humanities* 35.1 (Spring 2010): 25–31.
Mansbridge, Jane J. *Beyond Adversary Democracy*. New York: Basic, 1980.
Navasky, Victor S. *Naming Names*. 1980. 3rd ed. New York: Hill and Wang, 2003.
Nebeker, Helen E. "'The Lottery': Symbolic Tour de Force." *American Literature* 46.1 (March 1974): 100–107.
"North Village Disapproves Consolidation," *Bennington Evening Banner* 17 March 1948: 1.
Oates, Joyce Carol, ed. "Chronology." *Novels and Stories: The Lottery, The Haunting of Hill House, We Have Always Lived in the Castle, Other Stories and Sketches*, by Shirley Jackson. New York: Library of America, 2010. 805–813.
Oehlschlaeger, Fritz. "The Stoning of Mistress Hutchinson: Meaning and Context in 'The Lottery.'" *Essays in Literature* 15.2 (Fall 1988): 259–265.
Oppenheimer, Judy. *Private Demons: The Life of Shirley Jackson*. New York: Putnam's, 1988.
"Plumb Wins By 16-Vote Edge Over Gibney; Defeat Parking Meters; Base Tax Same; Favor Highway Pay Increase, Police No," *Bennington Evening Banner* 17 March 1948: 1, 5.
Popper, Karl. *The Open Society and Its Enemies*. 2 vols. 1945. Princeton: Princeton UP, 1962.
Poundstone, William. *Prisoner's Dilemma*. New York: Random House-Anchor, 1992.
Robbins, L. H. "Town-Meeting Time Is Here Once More." *New York Times Magazine* 28 October 1934: SM6, 16.
Shirley Jackson Papers. Washington, D.C.: Library of Congress.
Shute, Henry A. "A Special Town Meeting." *Saturday Evening Post* 200 (31 March 1928): 35, 93–94.
"Text of Truman's Order to Test Loyalty of U.S. Employees." *New York Times* 23 March 1947: 49.
Trott, Harlan. "It's Town-Meeting Time." *Christian Science Monitor* 1 March 1941: 4.
Warner, Michael. "Publics and Counterpublics." *Public Culture* 14.1 (Winter 2002): 49–90.
Waugh, Frank A. "Town Meeting." *Survey* 54 (1 July 1925): 389.

Webster, Clarence M. *Town Meeting Country*. New York: Duell, Sloan and Pearce, 1945.
Wilburr, Harriette. "New England Town Meeting." *Common Ground* 2 (Spring 1942): 19–26.
Wittgenstein, Ludwig. *Philosophical Investigations* [*Philosophische Untersuchungen* 1951.] Trans. G. E. M. Anscombe, P. M. S. Hacker, and Joachim Schulte. 4th ed. Chichester, West Sussex: Wiley-Blackwell, 2009.

4.2 Further Reading

Camus, Albert. *The Plague*. [*La Peste* 1947.] Trans. Stuart Gilbert. 1948. New York: Random-Vintage, 1991.
Friedman, Lenemaja. *Shirley Jackson*. Boston: G. K. Hall-Twayne, 1975.
Hattenhauer, Darryl. *Shirley Jackson's American Gothic*. Albany: State U of New York P, 2003.
Hutchinson, George. *Facing the Abyss: American Literature and Culture in the 1940s*. New York: Columbia UP, 2018.
Jackson, Shirley. *The Magic of Shirley Jackson*. Ed. Stanley Edgar Hyman. New York: Farrar, Straus and Giroux, 1966.
Jackson, Shirley. *Come Along with Me*: *Part of a Novel, Sixteen Stories, and Three Lectures*. Ed. Stanley Edgar Hyman. 1968. New York: Penguin, 1995.
Jackson, Shirley. *Let Me Tell You: New Stories, Essays, and Other Writings*. Ed. Laurence Jackson Hyman and Sarah Hyman DeWitt. New York: Random, 2015.
Miller, Arthur. *The Crucible: A Play in Four Acts*. 1953. New York: Penguin, 2003.
Orwell, George. *1984*. 1949. New York: Houghton Mifflin-Harcourt, 2017.

Sabrina Völz

22 Bernard Malamud (1914–1986)

Abstract: Bernard Malamud is considered a fine novelist, a prominent figure of the Jewish Renaissance, and a master of short fiction. Much of the literature on Malamud's short stories has focused on his first two collections, *The Magic Barrel* (1958) – for which he won the National Book Award – and *Idiots First* (1963). This chapter heeds Michael Basseler and Ansgar Nünning's call to "strike a balance between well-known, oft-anthologized stories and some lesser known ones," to avoid the "incestuous" repetition of scholarly inquiry into only a handful of short stories (Basseler and Nünning 2011, 4). While any attempt to classify Malamud's fifty-five short stories written between 1940 and 1984 will have its obvious limitations, this study pairing "The Jewbird" (1963) – a well-researched story – with "My Son the Murderer" (1968) – one of Malamud's most overlooked stories (Marx 2014) – seeks to shed new light on the artistic and thematic span of Malamud's short story writing produced during a decade raging with turmoil. Themes, such as father-son relationships and suffering, will also be explored. "The Jewbird" treats the interplay between elements of the fantastic and ordinary as well as characters who elicit both humor and pity, topics familiar to both the humanist's readership and scholars. In contrast, "My Son the Murderer" does not fit into scholarly categorizations of Malamud's short stories, such as those dealing with New York Jews, poor immigrants, artists and authors, Italians in Rome, or the relationship between African Americans and Jews. "My Son the Murderer" explores the breakdown in communication between father and son, their diverging views on life's purpose, and the despair of a young man called to serve his country in a war that he does not endorse.

Keywords: Suffering, magical realism, father-son conflict

1 Context

> My work, all of it, is an idea of dedication to the human. Malamud (Frankel 1991 [1966], 21)

With regard to the various streams of Judaism, Malamud might best be characterized as a secular Jew inasmuch as his father was not religiously observant, and the boy's home life was devoid of lessons in Jewish religious rituals and the Torah. He did, however, grow up in a neighborhood surrounded by other Jews, read Yiddish literature, and regularly attended Yiddish theater as a child. Only as an adult did Malamud become formally acquainted with Jewish religious thought. When asked about Jewish tradition in his work, he remarked:

https://doi.org/10.1515/9783110587647-023

> I would say that my subject matter mixes the universal and the particularly Jewish. Some borderline figures in my work act under the influence of their Jewish background. I write about Jews because I know them. I'm comfortable with them. The interest in ethicality by the Jews has always exited me. The struggle against self is a basic struggle. Almost anything one reads touches on this. Also, the history of the Jews is a great drama. Obviously, it's not one of pure darkness. I also like Jewish history as a metaphor for the fate of all men. (Leviant 1974, 49–50)

"The Jewbird" is indicative of the interplay between the universal and the particular and the struggle against the self as a metaphor for humankind, as will later be shown. The "pure darkness" to which Malamud alludes likely references the Eastern European pogroms, the diaspora, and the Holocaust. Only a few of Malamud's stories deal directly with the Holocaust, e.g. "The Loan," "The German Refugee," and "The First Seven Years." His work instead revolves around the general concept of suffering deeply rooted in that experience and in Jewish culture in general.

For Malamud, suffering is neither religiously sanctioned nor senseless as many viewed it during the Holocaust. Malamud instead considers it to be an integral part of the human condition that requires the conscious endurance of personal anguish and/or recognition of, empathy toward, and alleviation of the suffering of others. As Sheldon N. Grebstein already observed in 1973, "[Malamud's] heroes all suffer deeply, but they are also secular men whose suffering is not always voluntary, undertaken wholly for exalted reasons, or blessed by great rewards. Malamud's real concern is for the social and moral aspects of suffering as they impinge upon personality" (Grebstein 1975 [1973], 21). His short stories generally show that each individual has the opportunity through suffering, the accepting of responsibility as well as the performing of ethical acts of charity to evolve as a person (Halio 1994, 336). Malamud seems to prefer success to failure, commenting that "[a] Malamud character is someone who fears his fate, is caught up in it, yet manages to outrun it" (Shenker 1991 [1971], 34). For Malamud to only write stories about characters able to resolve their conflicts or escape their predicaments, however, would have been naive. It is, therefore, interesting to examine two stories that explore the conditions under which and the reasons why some Malamudian characters fail.

Malamud belongs to a diverse group of experimental American writers of Jewish heritage who moved Jewish literature away from the margins to the heart of American letters between 1945 and 1965. Previously, Jewish literature produced in the United States from 1930 to 1945 was largely influenced by its European roots and Jewish literary traditions. Much of it was concerned with matters of European emigration, e.g. "flight from persecution, arrival in New York, a life of struggle, and a spiritually impoverishing success" (Fried 1988, 34). Malamud, son of Russian Jewish immigrants, forged a link between his contemporaries and the previous generation of Jewish writers. In addition to including elements of Old World culture and the shift from emigrant to immigrant experience into his narratives, Malamud's short stories contain elements of Jewish folklore, humor, and Yiddish. In contrast to his literary predecessors of Jewish heritage, most of Malamud's stories – though many of his characters are Jewish – treat universal themes. At the same time, Malamud – like his contemporaries, Saul Bellow and Philip Roth – did not shy away from using his writing to directly criticize American society, including Jewish behavior and traditions, when he saw fit. These three prominent writers – who all rejected the label 'Jewish American writer' as restrictive – explore similar themes in their writing but hold, in part, diverse views about Jewishness as well as styles of and approaches to writing. They ranged from "emotional-synthetic and rational-analytic to comical-satiric" (Hornung 1997, 318; translation is mine). An article of this scope cannot afford a detailed comparative analysis of the three great writers of the Jewish Renaissance. Yet it is necessary to call attention to "Eli, the Fanatic," the final story in Roth's short story collection, *Goodbye, Columbus* (1959), since it is relevant to both of Malamud's stories investigated here. Eileen H. Watts has, for example, noticed the "political, social, and psychological fallout of [the] assimilated Jew as good tenant, unassimilated Jew as bad" in Roth's "Eli, the Fanatic" as well as Malamud's "The Jewbird" (Watts 1996, 159). The 'good' tenant is one who does not make waves and adheres to WASP rules. I would like to take her argument a step further, paying particular attention to both authors' shared view of suffering as an integral part of being Jewish, of being human.

Appearing just four years before Malamud's "The Jewbird," "Eli, the Fanatic" deals with the fear assimilated Jewish American suburbanites had of Orthodox Jewish immigrants from Eastern Europe. Hired as a lawyer to represent the interests of an affluent, assimilated Jewish community, Eli Peck writes a letter to Mr. Tzuref, the head master of a yeshiva for displaced persons set to open in their residential area. In the letter, the protagonist underlines the importance of upholding zoning laws and complying with accommodation strategies, such as the wearing of American clothing. In other words, yeshiva and its Orthodox residents, students, and staff would be tolerated as long as they adhere to rules and remain invisible or blend in. As an ominous warning of sorts, Eli blames European Jews for contributing to their own persecution:

> Woodenton, as you may not know, has long been the home of well-to-do Protestants. It is only since the war that Jews have been able to buy property here, and for Jews and Gentiles to live beside each other in amity. For this adjustment to be made, both Jews and Gentiles alike have had to give up some of their more extreme practices in order not to threaten or offend the other. Certainly such amity is to be desired. Perhaps if such conditions had existed in prewar Europe, the persecution of the Jewish people, of which you and those 18 children have been victims, could not have been carried out with such success – in fact, might not have been carried out at all. (Roth 2006 [1959],195)

Even in a work of fiction, Eli's ludicrous and cruel accusation was bound to provoke a backlash outside the realm of fiction, particularly because the blame placed on Holocaust victims and survivors is voiced by a Jewish character in a story written by a Jewish author less than two decades after the end of the Holocaust. Within the story, Mr. Tzuref ignores the provocative remarks and instead suggests that that resolution to their conflict might best be brought about if the community or Eli, at the very least, tries suffering: "'Stop with the law! You have the word suffer. Then try it. It's a little thing'" (198).

The long-bearded yeshiva caretaker who up until that point had only worn traditional black clothing and a "wide-brimmed Talmudic hat" (190) complies with Eli's demands after the lawyer – realizing that "the greenhorn" has no means to buy new clothing – donates his best suit to the less fortunate refugee. The turning point in the story takes place when Eli curiously puts on the caretaker's discarded Orthodox clothing after it appears on his doorstep. Breaking his own rules and donning the tattered black suit, Eli then walks outside for all residents to see and repeatedly begs the Orthodox man to let him make amends (216). Eli literally and metaphorically walks in the other man's shoes, learning the meaning of "sacrifice through love, which for Philip Roth appears to be the essence of Judaism" (Halio 1992, 34). According to the community's logic, anyone empathizing with the Orthodox Jews to a point in which that person would be seen in their old, worn-out black Orthodox clothes must be a fanatic or at least mentally unstable. The story concludes with a doctor inserting a needle into Eli's skin: "The drug calmed his soul, but did not touch it down where the blackness had reached" (Roth 2006 [1959], 221). The blackness, now inseparable from Eli's soul, signifies a real and lasting epiphany. Instead of being rewarded, Eli pays a heavy price for his sacrifice and insight, a widespread theme in Jewish literature that can also be found in Malamud's work.

Certainly, the long history of anti-Semitism as well as discrimination against and hostility toward Jews in the United States propagated by the Ku Klux Klan, Henry Ford – who idolized Adolf Hitler – and Father Charles Coughlin in the 1930s had lingering effects on American Jews even as their prosperity increased and anti-Semitism as well as social, structural, and economic discrimination began to decline after World War II (Dinnerstein et. al. 2003, 184–188, 206). In that delicate environment, some Jewish community members expressed their stark disapproval of *Goodbye, Columbus* (1959). They believed it damaged the progress that Jews had made by 'airing their dirty laundry' in public. Roth was even accused of "anti-Semitism, or Jewish self-hate" (Halio 1992, 24), a further topic which Malamud explores in "The Jewbird."

In contrast to Roth's 'confrontational' variety of social satire, Malamud was more interested in exploring general questions of morality and the survival of humanity: "'The purpose of the writer,' says Malamud, 'is to keep civilization from destroying itself. But without preachment. Artists cannot be ministers. As soon as they attempt it, they destroy their artistry'" (Wershba 1991 [1958], 7). As such, the sociopolitical criticism which can nevertheless be found in Malamud's stories is less biting and more a byproduct of that journey rather than its primary aim. The continuous state of turmoil in the U.S. during the 1960s must have brought a renewed urgency to Malamud's stated mission as a writer.

The generational conflict that began in the 1950s with the rebellion of America's youth against unchallenged traditions and the values of the older generations continued well into the 1960s in the form of student activism, the counterculture, and the anti-war movement that sought to put an end to the armed conflict in Vietnam, a war spanning over two decades that was never formally declared. The Vietnam conflict was also the first televised war. TV coverage brought battlefield round-ups and policy stories from Washington into the living rooms of Americans on a daily basis, but the main emphasis was placed on "the visual and above all the personal: 'American boys in action' was the story, and reports emphasized their bravery and their skill in handling the technology of the war" (Hallin 2004, 2447). With the Tet Offensive in January 1968, U.S. citizens became painfully aware that their government had misrepresented 'the facts' about the Vietcong's resolve and the war's so-called progress. The media responded with increased investigative journalism, photographs, and an onslaught of reports detailing the human costs of the war. In the same year, the U.S. Supreme Court ruled 7–1 in the *United States v. Paul David O'Brien* that the constitutional right to free speech does not apply to the burning of draft cards. And with the assassination of Martin Luther King, Jr. in 1968 and the seemingly death of his dream, the bitterly divided nation became consumed by destructive race riots. And if that is not enough, 1968 also witnessed the shooting of Andy Warhol, Robert Kennedy's assassination, and the brutal squashing of the anti-war protests at the Democratic National Convention. The end of the 1960s saw the belief in progress, which had already started to falter with the slayings of John F. Kennedy in 1963 and Malcom X in 1965, go down in dust.

Since a number of these significant events are not directly mentioned in Malamud's stories and interviews, one can only speculate how deeply Malamud was affected by them. We do know, however, that he was horrified by the Vietnam War, placing it on the same level as Stalin's Moscow Purge, the Holocaust, and the atomic bombing of Hiroshima and Nagaski:

> I was born just before the profuse blood letting of World War I. I have not forgotten the bleak hungry depression. [. . .] Nor how Stalin's Moscow Trials were for years passionately defended by men of intellect and good will. How Hitler, a fanatic with a stench for a soul, came to power in Germany. A line of Jews more than a thousand miles long was gassed to death, their bodies plundered of shoes, teeth fillings, human fat for soap. [. . .] World War II, begun in a graveyard of sinking ships, ended with atom bombs that should have never been dropped, destroying Hiroshima and Nagasaki. [. . .] Overnight we were fighting a new war in Korea. A young American President had the back of his head shot off into the hands of his wife. And now, in Vietnam, a small, gentle faced people, caught between two armies, have become living torches. After what one lives through who runs from content. (Malamud 1969, 12)

Obviously concerned about the fate of humanity and the atrocities committed against humans by humans, Malamud was moved to write "My Son the Murderer." Instead of focusing on society and its struggles as a whole, Malamud's stories concern themselves with the human condition on a micro level, reflected through external and internal conflicts, interpersonal communication and relationships. Individuals can change; individuals can make a difference. But sometimes it is complicated.

2 Close Readings

Reminiscent of the poem "The Raven" by Edgar Allan Poe (↗7 Poe), "The Jewbird" opens in medias res with the entrance of a frazzled-looking, undernourished, and slightly cross-eyed black bird named Schwartz into Harry Cohen's kitchen window: "That's how it goes. It's open, you're in. Closed, you're out" (Malamud 1997a, 322). Obviously in need of food and shelter, the fate of the bird and other refugees like him hinges on the luck of the open window and the charity of others (Hanson 1993, 360). Immediately Schwartz informs Harry Cohen, his wife, and their son, Maurie, that he is "'flying but [. . .] also running'" from "'Gevalt, a pogrom,'" from "'anti-Semeets [. . .] of any kind, [. . .] including eagles, vulture, and hawks. And once in a while some crows will take your eyes out'" (Malamud 1997a, 323). In other words, anti-Semites come in all shapes and sizes, including fellow Jews. Although the Cohen family is somewhat surprised by the Yiddish-speaking bird who looks "like a dissipated crow" (323), they immediately accept the situation as should implied readers. While trying to make sense of the odd circumstances, the Jewish family asks Schwartz if he is a "ghost or dybbuk" (323) or if he might "'be an old Jew changed into a bird by somebody.'" It is a claim that the bird does not exactly reject: "'Who knows?' answered Schwartz 'Does God tell us everything?'" (324). Instead of vacillating between the real and the fantastic, the two occur simultaneously and are "virtually indistinguishable" (Abramson 1993, 133). Once the surprise about the bird's ability to speak has subsided, Schwartz's supernatural abilities are accepted as a normal part of daily life (Sío-Castiñeira 1998a, 57). A short time later, Schwartz conveys to the others that he is a Jewbird, a revelation causing Cohen to breakout in hearty laughter. The lighthearted atmosphere begins to sour, however, when the cold-hearted frozen-foods salesman realizes the bird's intentions to stay (Malamud 1997a, 321). Politely but insistently, the Jewbird asks for compassion: "Where there's charity I'll go" (324). In doing so, Schwartz refers to the Jewish tradition, requiring assistance to those in need.

The Jewish practice of *tzedakah*, derived from the Hebrew root *tzedek*, is often translated as charity – as the Schwartz does – but it literally means justice: "In other words giving to the poor is not merely charitable; it is the fulfillment of basic justice. [. . .] *Tzedahak* is not only about personal discretion, it is also about the obligation and responsibility to effectively deal with poverty" (Goldstein 2006, 385). Even so, Cohen must be begged by his family to even give Schwartz the minimum he requires. Instead of bringing home the herring the refugee repeatedly requests, the miserly Cohen buys him a bird feeder full of dried corn which Schwarz rejects: "Not for my digestion. [. . .] Cramps. Herring is better even if it makes you thirsty. At least rainwater don't cost anything" (Malamud 1997, 325). Schwartz has his rough edges and is sarcastic at times. Yet Edie, who functions as a contrast character, serves Schwarz herring "with an occasional piece of potato pancake, and even a bit of soupmeat when Cohen wasn't looking" (325). Trying to avoid conflict with her patriarchal husband, Edie acts in secret but at least fulfills her responsibility to a fellow creature.

In return, Schwartz behaves as a "traditional Jewish uncle" (Shaw 2000, 139) or as any caring parent might do. He silently suffers on Maurie's behalf by patiently helping the "not overly bright," 10-year-old boy with his homework (Malamud 1997a, 322), listening "to him practice his screechy violin" (325), and reading him comic books "though he personally disliked them" (326). Instead of following Schwartz's altruistic example, Cohen ignores his son, only taking an interest in him when – as a result of Schwartz's care – Maurie's grades improve to a C-. It is at that point that Cohen starts to selfishly dream about the boy's acceptance into an Ivy League university. The father's sole interest in his son is implicitly situated in Cohen's desire to climb the social ladder, to obtain a piece of the American Dream through his son. Comically, Schwartz provokes Cohen by shattering his host's foolish dream: "'He [Maurie] won't be a shicker [drunk] or a wife beater, God forbid, but a scholar he'll never be, if you know what I mean, although maybe a good mechanic. It's no disgrace in these times'" (326).

This comment marks the beginning of Cohen's all-out assault against the "goddam pest and freeloader" (327). Cohen's first name, Harry, exposes his true nature. The verb "to harry," means to "persistently carry out attacks on (an enemy or an enemy's territory)" or to "persistently harass" (*Lexico*, 2019). Cohen forces Schwartz to sleep on the balcony during the winter as well as endure herring mixed with cat food and the popping of paper bags when Schwartz is sleeping. And finally, in hope of scaring the 'unwanted' house guest away, Cohen buys an adult cat. As days turn into weeks, the conflict continues to escalate. Acting as a mediator, Edie encourages Schwartz to accommodate her husband, telling the bird that he smells and needs to take a bath, a request to which he replies: "'Everybody smells. Some people smell because of their thoughts or because who they are. My bad smell comes from the food I eat. What does his [Cohen's] come from?'" (Malamud 1997a, 327). Using the corrective function of humor, Schwartz exposes Cohen's unacceptable social behavior. Schwartz openly speaks out what others, including Edie, know but remain silent about. In this role, Schwartz is also "the compound image of opportunist and saint who tests to the extreme the humanity and compassion of others" (Richmann 1986, 95). In Jewish terms, the black bird is clearly a schnorrer, defined as a "mendicant who often goads his benefactors to charitable actions" (Kremer 2001, 125), i.e. Schwartz provides the Cohen family patriarch with the opportunity to make amends.

The only person who seems disturbed by Schwartz is the head of the family. Cohen tolerates the crow as long as it suits his purpose. The news of his mother's death and Maurie's zero on a math test, however, sends Cohen into a frenzy. Seeking both an outlet for his anger and revenge against the bird, a comical fight scene ensues that at first glance appears similar to those found in the cartoon *Tom and Jerry*. Yet it tragically ends with Cohen first whirling the bird rapidly around his head, then flinging the dizzy Jewbird, who latches on to Cohen's large nose for dear life, out of the apartment "into the night" and presumably to his death (Malamud 1997a, 329).

Similar to the conflict depicted in Roth's "Eli, the Fanatic" between assimilated Jews and their Eastern European counterparts, "The Jewbird" portrays Cohen's prejudice against the Orthodox crow, who – at the beginning of the story – refers to himself as an "old radical" and starts dovening or praying when asking for mercy (323). After all, Jewbird is "a clever variation of 'blackbird' which highlights the potentially racial connotation of the term" (Shaw 2000, 138). Jewbird also brings the racial slur "Jewboy" to mind (Wirth-Nesher 2003, 117). In contrast to Roth, however, Malamud places the humorous, but sharp-tongued Schwartz, Yiddish for black, in the role of the Eastern European Orthodox protagonist of the story. Malamud accords Schwartz with greater status than American writers – even after Malamud – usually give to animals in fiction. Maria Moss, for instance, calls attention to the function of animals in John Updike's "The Cat" and Richard Ford's "The Puppy." In those stories, the animals seem to be "mere 'byproducts,' reflecting mainly the protagonist's emotional instabilities" (Moss 2015, 122). While Malamud does use Schwartz to show Cohen's emotional instabilities and while Schwartz does ultimately become a victim and die a terrible death, the bird's roles are not limited to those functions alone. He is speaker of truths as well as Maurie's companion and mentor.

Cohen, whose family name ironically signifies the priestly status of the *cohanim*, is relegated to the role of antagonist who tries to turn the attention away from his own anti-Semitism by accusing Schwartz of being a "number-one troublemaker" (Malamud 1997a, 328). Cohen has assimilated to the point that he has lost touch with his priestly heritage, community, and humanity. Instead of trying to empathize with, show compassion, and take responsibility for the refugee as Eli eventually does in Roth's story, Cohen remains self-centered, loathing his "subconscious double" (Hershinow 1980, 125). Schwartz is a lasting reminder of Cohen's own uneasy ethnic and religious heritage as well as his status as an outsider which he longs to overcome. Like Poe's raven, Schwartz haunts Cohen's conscience (Hansen 1993, 363; Charles 2012, 1119). The assimilated head of the house is forced to look within himself, a process laden with great suffering: "Schwartz was driving him [Cohen] bats by being there always, even in his dreams" (Malamud 1997a, 328). It is a reading that fits well to the black bird's general symbolic tradition. As a literary symbol, the raven or crow represents "the dark side of the psyche but can function positively if the individual in question is able to come to terms with the issues involved there" (Biedermann 1994, 281). Upon Edie and Maurie's return, Cohen attempts to deceive his family about Schwartz's whereabouts but ultimately he only deceives himself: "'I threw him [the Jewbird] out and he flew away. Good riddance'" (Malamud 1997a, 329). Just as the severe injuries to Cohen's nose obtained in the final battle with

Schwartz will surely leave a scar, the killing of the Jewbird will not likely lift Cohen's self-imposed imprisonment. As Sheldon J. Hershinow elaborates, "Harry is a trapped and tortured character; his failure to accept the Jewbird represents a rejection of himself" (Hershinow 1980, 125).

Having already achieved closure, "The Jewbird" could well have ended with Cohen's murder of Schwarz – especially since the implied reader knows the truth – but Malamud, the humanist and moralist, drives the message home in the story's final paragraphs. In the spring, "moved by a memory" (Malamud 1997a, 330), Maurie sets out to look for his lost companion and mentor. The boy finds a "dead black bird in a small lot by the river, his two wings broken, neck twisted, and both bird-eyes plucked clean" (330). Cyclical in nature, the story returns to Schwartz's prophetic words uttered at the beginning of the story: "And once in a while some crows will take your eyes out'" (323). Responding to the boy's question about who killed him, Edie replies "Anti-Semeets" (330). By imitating the Jewbird's Yiddish diction, she underlines her empathy with the creature. Cohen had tried to hide his hatred because "Maurie hated violence and Cohen didn't want to leave a bad impression" (328), but Edie sees through the façade and overcomes her silent complicity.

Stubbornly refusing to accept *tzedakah* and Schwartz's markers of difference, Cohen's failure is one of his own making. The cold-hearted frozen foods salesperson neither shows remorse for his actions, nor does he respect the deceased. Instead of giving "the brokenhearted bird" a decent burial, Schwartz's cadaver was left to rot and fall prey to scavengers. The anti-Semitic Cohen is rotten to the core. And even if readers are unfamiliar with Jewish cultural references in the text, "The Jewbird" can still be read as a general cautionary tale or a timeless call for action to support the less fortunate, refugees, and other marginalized groups, topics that are just as relevant for readers today as they were when "The Jewbird" first appeared.

Malamud distances himself from the fantastic, at least for a time, only five years after the appearance of "The Jewbird." This is not to say that "My Son the Murderer" (hereafter "My Son") represents a general rejection of Malamudian hope, but it does illustrate the magnitude of the impact of the Vietnam War on an author who was not known for commenting on current geo-political realities in his writing. Malamud would not "run from content," the mass killing in Vietnam. The origins of that conflict were so firmly embedded in American culture that they also infiltrated everyday life and family relations. Malamud used his pen to create a powerful literary rendition of that dilemma, to hold up a mirror to society and expose one son's existential crisis while exploring a father's sheer helplessness.

"My Son" opens in media res with narrative focalizations that shift between son and father, focalizations that unfold before the eyes of the reader without the direct interference of a narrator:

> He wakes feeling his father in the hallway, listening. He listens to him sleep and dream. Listening to him get up and fumble for his pants. He won't put on his shoes. To him not going to the

> kitchen to eat. Staring with shut eyes in the mirror. Sitting an hour on the toilet. Flipping the pages of a book he can't read. To his anguish, loneliness. The father stands in the hall. The son hears him listen.
>
> My son the stranger, he won't tell me anything. (448)

The narrative style is markedly different from the cohesive, orderly narration found in "The Jewbird." In that story, the perspectives of several characters are presented to readers through tagged dialogues, attributed to the appropriate speaker. At the beginning of "My Son," readers will, however, need a moment to catch their bearings since framing and attribution are missing. This method of 'showing' allows characters to speak through the written word directly to the readers in the present tense instead of reporting their thoughts and speech acts through a narrative voice: "My son the stranger, he won't tell me anything" (448). The present tense draws readers into the story on a more emotional level. About half way through the short story, the narrator begins to consistently frame the narrative and uses the more traditional past tense (451).

The son has become a stranger who no longer adheres to social conventions, e.g. participating in conversations, putting on one's shoes as a part of getting dressed, or eating in the kitchen with the family. As hard as the father tries to listen and discern what ails his son and what his son is doing behind closed doors, the son reciprocates with silence. Likewise, the son's conversational and thematic silence may have partially been caused by the father's suffocating listening techniques as well as the metaphorical invasion of the son's private spaces: the son's dreams or the bathroom where the son sits on the toilet. Just as Schwartz haunts Cohen, the father haunts his son. Both want the other to change. The main difference between the two stories, thematically speaking, is that the beneficial potential of suffering in "The Jewbird" is entirely absent from "My Son." Yet suffering, at its most basic level, is something that all humans share: humans live; humans love; humans suffer.

Regarding style, the two stories also reveal Malamud's vast writerly toolkit. The short story master makes use of the mode of narration which Grebstein refers to as "selective omniscience." It is the ability of the writer to retain

> the objectivity, the freedom to move through time and space, and the power to know all [. . .]; yet by refraining from editorial intrusions and maintaining the focus on a single character or a few characters, the writer can shift into interior monologue or take a stance [. . .] through the character's eyes without any obvious break in the narrative seam or detection by the reader. (Grebstein 1975 [1973], 33–34)

This narrative style increases the story's tension and actively engages readers. The narrator places the perspectives of both father and son side by side on an equal footing, requiring readers to become active. In short stories, it is less common for the narrator to withhold descriptions of the characters' outer appearances. Malamud has acknowledged his indebtedness to Ernest Hemingway's "ability to compress"

(Tyler 1979; ↗16 Hemingway). By omitting character descriptions and severely limiting the number of adjectives used in "My Son," Malamud strips the story in Hemingwayesque fashion of all potential distractions in order to focus on the father-son conflict and compel the reader to co-construct the story's meaning. Important in this context is the minimalist tendency to give credence to not only what is said but also what is not said, e.g. the reasons why the son refuses to engage in a deeper conversation with his father. As Malamud described in an interview with Joseph Wershba in 1958, the writer "creates the architecture, the form and content. [. . .] The artist cannot legislate the reader's interpretation. [. . .] Imagination is required" (Wershba 1991, 5).

Another interesting feature of the story and one that engages the readers' imagination is Leo's renaming of Harry: "My Son the Murderer" (Malamud 1997b, 448), the short story's title; "my son the stranger" (448); "my temporary son" (449); "my son the prisoner" (449); "my temporary son" (449); "my son with the dead hand" (449); and "my son who made himself into a lonely man" (453). Each of these represents one aspect of the son's troubles that the father repeats, either aloud or in his thoughts, as if he is trying to make sense of them. Taken together, these monikers emphasize the complexity of "Harry's obsessions and Leo's equally compulsive concern" (Solotaroff 1989, 127). This list of obsessions sheds light on the power dynamics in the story. Through the repetition of the possessive adjective 'my' as well as the family role 'son' and not the first name 'Harry' – the pattern that Philip Roth chose for the title of his short story "Eli the Fanatic" – Harry is repeatedly defined in terms of his relationship to his father. By repeatedly naming his son's obsessions in that manner, the father reveals his own obsession with Harry, an obsession from which the son rebels through his silence.

When the story opens, the twenty-two-year-old Harry has been at home for several months after graduating from college the previous summer. He seems to be suffering "from a mental breakdown that prevents him from communicating with his mother and father" (Sío-Castiñeira 1998b, 92). Indeed, Harry shows signs of psychological paralysis and depression, but mental health was still a taboo topic in society during the 1960s and perhaps, therefore, not obvious to the father especially since he is close to his son. Nevertheless, Harry has directly or indirectly communicated some of his concerns with his parents. First, Harry cannot concentrate; he flips the pages of the book "he can't read" (Malamud 1997b, 450). Second, the son feels hopeless and confused by a world in which nothing lasts or has meaning: "The goddamn world is temporary" (449). Moreover, he does not see any value in his college education, telling his father "you know where you can stick that" (449). A college education, once viewed as a key to a successful career and a good life, will not save Harry from the draft that can only be deferred while actively pursuing a college degree. A third factor that points to Harry's depression is agitation. The son is easily moved to strong emotions, threatening his father, for instance, with murder after Harry catches Leo invading his privacy by reading his mail (452), another statement that readers along with Leo must interpret. Did he mean it literally or not? The narrator abstains from taking a position. Fourth, Harry has withdrawn from his family and society. He seems to have no friends, and no reason to get up in the morning. He is neither interested in hearing happy childhood stories (450), nor does he take part in any meaningful activities, including applying for jobs or beginning his career, acts which social conventions would dictate after a person graduates from college. Likewise, Harry does not want to distract himself with meaningless routines: "I don't want temporary work. I want the opposite of temporary, but where is it? Where do you find it?" (449). These questions are addressed to both father and readers. The son's only task is to wait for the letter from the draft board that he believes will inevitably arrive. At one time, Harry proactively wrote letters to the draft board (452), but when the story opens, he has already hit rock bottom, silenced himself, and chosen to suffer alone within the confines of his "cell-like jail that his parents' house has come to represent" (Tayyar 2008).

While Harry does not know how to discover life's meaning, he does know, however, that it cannot be found in actively contributing to the war effort, by killing others and placing his life on the line for a meaningless war:

> At night I watch the news programs. I watch the war from day to day. It's a big burning war on a small screen. It rains bombs and the flames roar higher. Sometimes I lean over and touch the war with the flat of my hand. I wait for my hand to die.
>
> My son with the dead hand.
>
> I expect to be drafted any day but it doesn't bother me the way it used to. I won't go. I'll go to Canada or somewhere I can go. (Malamud 1997b, 449)

Halfway around the world, the war has invaded the homes of Americans by directly sending graphic images of the destruction of Napalm bombs and the war into the homes of private citizens. The war is so toxic that Harry waits for his hand, a potential agent of death, to die when he touches the screen so that he does not have to literally use it to kill others and/or potentially harm himself: "Harry must confront the awful truth of his existence: not only was he born to die, but to kill, and there will be no impromptu moment of magic to change his fate" (Tayyar 2008). Perhaps Harry's son with the dead hand also touches the television screen to show his solidarity with all the young men whose personal stories as soldiers are reported daily on the news. Until the Vietnam War, it was taken for granted that patriotic young men would willingly rise to the occasion and perform their duties as soldiers when called upon in the name of liberty and justice for all. No questions were asked. The conflict between older and younger generations who began to question most American values during the 1960s is implicit in Harry's reference to his potential exodus to Canada. And although Harry remarks that he is not as bothered by the thought of being drafted as he used to be, readers will not likely take his remarks at face value. Fleeing to Canada might help him survive and protect him from the horrors of war, but it would not likely end his problems at home or help him understand life's deeper meaning.

From Leo's perspective, he is doing everything he can to help Harry. He believes that worrying about someone else is "the worst kind [. . .] because if he won't tell you, you can't get inside of the other person and find out why. You don't know where's the switch to turn off. All you do is worry more" (Malamud 1997b, 450). Leo has taken vacation from work to be with his son even though it is "hard to deal with a person who won't reciprocate to you" (450). The father can manage his own worries, but failing to know what ails his son drives him in frustration to make insensitive remarks – "Harry, don't worry so much about the war" (450) – and take desperate actions, such as following him around when Leo leaves the house (451) or reading his son's mail (452), actions which ironically push Harry even further away. Leo might not recognize Harry's depression, but he does encourage him to get outside help and seek a doctor: "Maybe you ought to go to a doctor? [. . .] It's not my health. Whatever it is I don't want to talk about it" (449). The father attempts to show his affection for his son by sharing fond memories from the boy's childhood with him or by doing small favors for him, such as cooking him an egg (450–451).

The father lacks both an effective support system and the ability and distance to piece together relevant clues about Harry's behavior and concerns. Harry's mother worries about her son but is busy taking care of her daughter during her difficult fourth pregnancy and watching her daughter's three children (448–449). The father does have a friend, Moe, that he can confide in, but he is most concerned about distracting Leo from his troubles. Leo feels so helpless that he fantasizes about communicating with Harry through letters: Leo "opened the mailbox. [. . .] He looked to see if one of them [letters] was, by any chance from his son to him. My dear father, let me explain myself. The reason I act as I do. [. . .] There was no such letter" (451). Leo feels he has no alternative but to follow his son around like a lost puppy, hoping for some kind of attention. After the letter from the draft board arrives and Leo is caught opening Harry's mail, Harry needs some space and tries to escape to Coney Island, a once proud amusement paradise for New York's middle and upper classes that symbolically stands forlorn in the midst of a winter landscape. Leo rushes after his flesh and blood, hoping to apologize and to keep Harry from doing anything rash:

> Harry, it was a mistake, excuse me, I'm sorry I opened your letter.
>
> Harry did not move. He stood in the water, his eyes on the swelling deaden waves.
>
> I'm frightened of the world, Harry thought. It fills me with fright.
>
> He said nothing.
>
> A blast of wind lifted his father's hat and carried it away over the beach. [. . .] Leo chased after his hat. He chased it one way, then another, then toward the water. The wind blew the hat against his legs and he caught it. By now he was crying. (453)

Leo's anguish and helplessness has brought the grown man to tears. For Harry, who conceals his inexpressible existential fear (Shaw 1996, 330), the time for communication has passed. The father chases aimlessly after his hat in an environment which he cannot control. The ocean with "deaden" or foreboding waves is certainly not a place of renewal. Likewise, the son has had his foundation ripped out from underneath him and is drifting aimlessly through life. The last lines of the story shift back into the present tense and free indirect discourse, leaving readers to make sense of the basic plot of the story without narrative comment:

> My father listens in the hallway. He follows me in the street. We meet at the edge of the water.
> He runs after his hat.
> My son stands with his feet in the ocean. (Malamud 1997b, 454)

While the story does technically come to a conclusion, it still leaves many questions open: Why can't father and son help each other? Has the father finally given up? Will Harry return to shore and leave for Canada? Will Harry commit suicide? The end has been criticized as too abrupt (Sío-Castiñeira 1998b, 97), but its ambiguity is actually powerfully unsettling. It is likely to accompany implied readers for some time after they finish reading the story. "As a writer," Malamud remarked, "I want uncertainty. It's part of life. I want something the reader is uncertain about. It is this uncertainty that produces drama" (Frankel 1991 [1966], 20). And certainly "My Son the Murderer" is one of the most dramatic of all of Malamud's short stories. Parents often worry about their children or even suffer as a result of circumstances in the lives of their children beyond their control. Just as some children, even as adults, need to free themselves from their parents' smothering love. In such cases, each of the two knows how the other will act, but both parent and adult child may not be able to control their responses or behavior. It is a vicious circle and an age-old, universal story.

3 Theoretical Perspectives

The majority of scholars has come to view anti-Semitism or Jewish self-hate as a dominant theme in Malamud's "The Jewbird." Watts is no different. In her study, she points to the symbolism in the story, e.g. the empty bird cage and the escaped tenant when the story opens as a hint of anti-Semitism to come, anti-Semitism from which escape might be possible: "Cohen is still a Jew, whose frozen heart need only be thawed" (Watts 1996, 159). New to the discourse, more importantly, is Watts's application of Sander Gilman's concept of Jewish self-hatred to Malamud's story. For Gilman, Jewish self-hate is comprised of the dialectic relationship between "the liberal promise" – which is achieved when the group of outsiders assimilate and become like the reference group and thus accepted – and "the conservative curse" which is a reminder of the outsiders' loss of power as they give up their own identity in an effort to be accepted by the mainstream (Gilman 1986, 2). This double bind can only be endured when repressed, as Cohen does: "He has absorbed the Gentile's airs of superiority and directs them toward the embarrassing 'greenhorns.' He is probably ashamed of and looks down on the behavior and speech of low-class, eastern European Jews – in other words, we presume, the behavior of Cohen's own parents" (Watts 1996, 161). Watts further likens Cohen's increased harassment of Schwartz to Jewish pogroms (Watts 1996, 162) and Edie's charge of anti-Semitism at the end of the story, anti-Semitism which still exists in altered forms, e.g. JAP jokes, today.

Another topic that has occupied a central position in scholarship on "the Jewbird" is Malamud's fusion of fantasy and reality. It is, however, not until 1998 that a Malamud scholar finally addresses the looming question whether the term "magical realism," first used with the new Latin American novel, can be applied to "The Jewbird." Begoña Sío-Castiñeira convincingly applies the three components of magical realism, as outlined by Amaryl B. Chanady, to Malamud's story: "the juxtaposition of the natural and the supernatural on the same textual level, the resolution of the antinomy between both words, and the authorial reticence" (Sío-Castiñeira 1998a, 57). Malamud positions fairy tale tropes, e.g. an unpleasant visitor whom another character must learn to love in a realistic setting, to uncover the "hidden, ambiguous, and mysterious" side of reality that extends beyond the realm of logic (59). For Sío-Castiñeira, the narrator lends credibility to the fantastic by treating Schwartz exactly in the same manner as the human characters are treated and by employing several focalizers: "a hating frozen-food dealer, an unbiased wife, an innocent child, and an implied author who does not contradict the observations of the characters" (60).

In a further key essay entitled, "Animals Speaking in the Fiction of Jin and Malamud," Matt Prater applies critical animal theory to "The Jewbird." Doubly Othered, Schwartz as a Jew and a bird embodies both Othered human animal and non-human animal perspectives. In addition to viewing Schwartz's mistreatment as a reflection of the way in which humans treat each other, as conventionally scholars have done, Prater also considers the value animals have in their own right, apart from their relationship to human animals, arguing against limiting Schwartz to the role of a symbol (Prater 2013, 2). Sensitive to the "particularities of Schwartz's bodily life as bird addressing the fears, needs, and desires which come of out the bird experience," Malamud succeeds in taking up the topic of anthropomorphism through the lens of magical realism – the bird can speak and pray as the pun "dovening" suggests – without making the story anthropocentric (3). Reminding readers of the anti-Semitic employment of animal metaphors to dehumanize Jews as well as to justify discrimination of and even violence toward them in the past, Prater further notes that Cohen's maltreatment of Schwarz as a bird turns into the anti-Semitic abuse of a Jewbird, picking up on Watts's reading of Cohen's self-hatred (5). In the end, Schwartz must die as an expendable victim. It seems the death or murder of non-human animals inevitably evokes strong human emotions that ironically the death of human animals may not.

Most of the scholarship on "My Son the Murderer" ranges in length from a sentence to a few paragraphs (Solotaroff 1989, 257; Abramson 1993: 135; Shaw 1996,

330–331; and Aarons 2016, 9–11). In contrast, Sío-Castiñeira is one of the few scholars to devote an entire scholarly essay to "My Son." She situates it in the context of Expressionist paintings with its focus on emotional experience and "interior human degradation" (Sío-Castiñeira 1998b, 92). For her, one of the expressionist characteristics in "My Son" is the constant "change of focalization that has no precedent in Malamud's stories" (93). Malamud abandons the calm, clear progression of thought into an "aggressive subjectivity" that develops into a "portrait of emotional turmoil" (93). For this reason, Sío-Castiñeira sees the narrative more as a poem than a "narrative instance" (96).

Sío-Castiñeira then elaborates on the father-son conflict as the key expressionist characteristic found in the literature during WWI, a time in which hostility between generations and murder were omnipresent (93). In the son's "existential confusion," he cannot overcome social structures and norms, two aspects over which he has no control (94). The son's locked door "symbolically prevents his father from accessing his son's inner world" (95). The genuinely concerned father envisions Harry's "seven independent and imaginary metamorphoses," using Kafkaesque terminology to express the repressed factors hidden deeply within the son (96). These labels the father attaches to his son in an attempt to understand him are also accompanied by "different instances of indignation in different degrees, depending on the quality of attack he feels threatened with" (101). The more Harry becomes detached from society, the more his father becomes lonely, unprotected, and insecure (102). Harry becomes "the metaphysical being par excellence, immobilized and paralyzed, the man who can never find what he is looking for" (104). In the end, he is left to stand at the water and contemplate suicide.

All in all, the scholarship on Bernard Malamud has been slowing down for the past few years, a development that may have to do with the pigeonholing of Malamud as a Jewish American writer of mainly historical interest. This essay has hopefully shown the timeless, aesthetic quality of Malamudian stories. "My Son the Murderer" with its complex narrative style does not overtly have much to do with Judaism or Jewish identity. It can be read in terms of the Vietnam War, but it can also be understood as a story with a universal theme, e.g. father and son or generational conflict. The references to war are both specific and ambiguous at the same time. Malamud's artistic range makes his writing interesting for diverse audiences and continues to inspire budding writers. Hanif Kureishi, a short story writer and novelist of Pakistani and English descent, seems to make that very point with his story, "My Son the Fanatic" (1994). Its opening as well as the father-son conflict serve as a tribute to Malamud's story: "Surreptitiously, the father began going into his son's bedroom. He would sit there for hours, rousing himself only to seek clues" (Kureishi 2010, 9). In "My Son the Fanatic," the father is bewildered by his son's actions and becomes obsessed with trying to understand Ali's changes in behavior and views, his "sudden transformation from a seemingly happy black British adolescent to a secluded and increasingly aggressive Muslim fundamentalist" (Jansen

2018, 118). Similar to Roth's "Eli the Fanatic," "My Son the Fanatic" questions who the fanatic actually is: the helpless father or the son, a devout Muslim. While Kureishi's homage in that story to Philip Roth has not gone unnoticed (see Jansen 118), it would seem that Kureishi's debt to "My Son the Murderer" has not been recognized in scholarship until now, begging the question what other gaps in Malamud's short stories are still left for scholars to discover.

4 Bibliography

4.1 Works Cited

Aarons, Victoria. "'Moved by a Memory': Bernard Malamud's Literary Landscape." *Bernard Malamud: A Centennial Tribute*. Eds. Victoria Aarons and Gustavo Sánchez Canales. Detroit: Wayne State UP, 2016. 3–16.

Abramson, Edward A. *Bernard Malamud Revisited*. New York: Twayne, 1993.

Basseler, Michael, and Ansgar Nünning. "Introduction: Features, Objective, and Premises of *A History of the American Short Story*." *A History of the American Short Story: Genres – Developments – Model Interpretations*. Trier: Wissenschaftlicher Verlag Trier, 2011. 3–20.

Biedermann, Hans. *Dictionary of Symbolism: Cultural Icons and the Meanings Behind Them*. Trans. James Hulbert. New York: Meridian, 1994.

Charles, Hal. "Bernard Malamud." *Critical Survey of Short Fiction*. Ed. Charles E. May. Vol. 3. 4th ed. Ipswich: Salem P, 2012. 1114–1122.

Dinnerstein, Leonard, Roger L. Nichols, and David M. Reimers. *Natives and Strangers: A Multicultural History of Americans*. New York: Oxford UP, 2003.

Frankel, Haskel. "Bernard Malamud." 1966. *Conversations with Bernard Malamud*. Ed. Lawrence M. Lasher. Jackson: UP of Mississippi, 1991. 19–21.

Fried, Lewis. "American Jewish Fiction, 1930–1945." *Handbook of American-Jewish Literature: An Analytical Guide to Topics, Themes, and Sources*. Ed. Lewis Fried. New York: Greenwood P, 1988. 35–60.

Gilman, Sander. *Jewish Self-Hatred: Anti-Semitism and the Secret language of the Jews*. Baltimore: John Hopkins UP, 1986.

Goldstein, Warren. *Defending the Human Spirit: Jewish Law's Vision for a Moral Society*. Jerusalem: Feldheim, 2006.

Grebstein, Sheldon Norman. "Bernard Malamud and the Jewish Movement." 1973. *Bernard Malamud: A Collection of Critical Essays*. Eds. Eslie A. Field and Joyce W. Field. Englewood Cliffs, N.J.: Prentice-Hall, 1975. 18–44.

Halio, Jay L. "Malamud, Bernard." *Reference Guide to Short Fiction*. Ed. Noelle Watson. Detroit: St. James, 1994. 335–337.

Halio, Jay L. *Philip Roth Revisited*. New York: Twayne, 1992.

Hallin, Daniel C. "Vietnam on Television." *Museum of Broadcast Communications: Encyclopedia of Television*. Ed. Horace Newcomb. 2nd ed. Vol. 4. New York: Fitzroy Dearborn, 2004. 2446–2448.

Hanson, Philip. "Horror and Ethnic Identity in 'The Jewbird.'" *Studies in Short Fiction* 30 (1993): 359–366.

"Harry", https://www.lexico.com/definition/harry. *Lexico*. Oxford University Press, 2019. (03 March 2021).
Hershinow, Sheldon J. *Bernard Malamud*. New York: Frederick Ungar, 1980.
Hornung, Alfred. "Postmoderne bis zur Gegenwart." *Amerikanische Literaturgeschichte*. Ed. Hubert Zapf. Stuttgart: Metzler, 1997. 304–375.
Jansen, Bettina. *Narratives of Community in the Black British Short Story*. Cham: Springer International Publishing, 2018.
Kremer, S. Lillian. "Reflections on Transmogrified Yiddish Archetypes in Fiction by Bernard Malamud." *The Magic Worlds of Bernard Malamud*. Ed. Evelyn Avery. Albany: State U of New York P, 2001. 123–138.
Kureishi, Hanif. "My Son the Fanatic." *Collected Stories*. London: Faber and Faber, 2010. 116–127.
Leviant, Curt. "My Characters Are God-Haunted." 1974. *Conversations with Bernard Malamud*. Ed. Lawrence M. Lasher. Jackson: UP of Mississippi, 1991. 47–53.
Malamud, Bernard. "The Jewbird." *The Complete Stories*. Ed. Robert Giroux. New York: Farrar, Straus, and Giroux, 1997a. 322–329.
Malamud, Bernard. "My Son the Murderer." *The Complete Stories*. Ed. Robert Giroux. New York: Farrar, Straus, and Giroux, 1997b. 448–454.
Malamud, Bernard. "Theme, Content, and the 'New Novel." *Page 2: The Best of 'Speaking of Books' from The New York Times Book Review*. Ed. Francis Brown. New York: Holt, Rinehart, and Winston, 1969. 9–13.
Marx, Bill. "Fuse Interview [with Philip Davis]: Jewish-American Writer Bernard Malamud at 100 – Appreciating the Beauty of the Ethical", https://artsfuse.org/102166/-fuse-interview-jewish-american-writer-bernard-malamud-at-100-appreciating-the-beauty-of-the-ethical/. *The Arts Fuse*. 23 March 2014. (9 July 2018).
Moss, Maria. "'Their deaths are not elegant' – Portrayals of Animals in Margaret Atwood's Writing." *Zeitschrift für Kanada-Studien* 35 (2015): 120–135.
Prater, Matt. "Animals Speaking in the Fiction of Jin and Malamud." CLCWeb 15.4 (2013), http://docs.lib.purdue.edu/clcweb/vol15/iss4/1. (27 October 2021).
Richman, Sidney. "The Stories." 1966. *Bernard Malamud: Modern Critical Views*. Ed. Harold Bloom. New York: Chelsea House Publishers, 1986. 71–100.
Roth, Philip. "Eli, the Fanatic." *Goodbye, Columbus and Five Short Stories*. 1959. London: Vintage, 2006. 187–221.
Schenker, Israel. "For Malamud It's Story." 1971. *Conversations with Bernard Malamud*. Ed. Lawrence M. Lasher. Jackson: UP of Mississippi, 1991. 32–34.
Shaw, Martin Urdiales. *Ethnic Identities in Bernard Malamud's Fiction*. Oviedo, Spain: Universidad de Oviedo, Servico de Publicationes, 2000.
Shaw, Martin Urdiales. "Conflicts of Communication and Perception in Malamud's Stories about Madness." *Many Sundry Wits Gathered Together*. Santiago de Compostela: Universidad de Santiago de Compostela, 1996. 329–335.
Sío-Castiñeira, Begoña. "'The Jewbird': Bernard Malamud's Experiment with Magical Realism." *Short Story* 6.1 (1998a): 55–64.
Sío-Castiñeira, Begoña. *The Short Stories of Bernard Malamud: In Search of Jewish Post-Immigrant Identity*. New York: Peter Lang, 1998b.
Solotaroff, Robert. *Bernard Malamud: A Study of the Short Fiction*. Boston: Twayne, 1989.
Tayyar, Paul Kareem. "'Because We Could Not Walk on Water': Bernard Malamud's 'My Son the Murderer' and the Limits of Magical Realism." *Americana*. Ed. Institute for American Studies and Creative Writing. July 2008. Web. (9 July 2018).

Tyler, Ralph. "A Talk with the Novelist", https://www.nytimes.com/1979/02/18/-archives/a-talk-with-the-novelist-malamud.html. *New York Times Book Review*. 18 February 1979. (8 February 2021).

Watts, Eileen H. "Jewish Self-Hatred in Malamud's 'The Jewbird.'" *MELUS* 21.2 (1996): 157–163.

Wershba, Joseph. "Not Horror but Sadness." 1958. *Conversations with Bernard Malamud*. Ed. Lawrence M. Lasher. Jackson: UP of Mississippi, 1991. 3–7.

Wirth-Nesher, Hana. "Traces of the Past: Multilingual Jewish American Writing." *The Cambridge Companion to Jewish American Literature*. Eds. Hana Wirth-Nesher et al. Cambridge: Cambridge UP, 2003. 110–128.

4.2 Further Reading

Avery, Evelyn, ed. *The Magic Worlds of Bernard Malamud*. Albany: State U of New York P, 2001.

Davis, Philip. *Bernard Malamud: A Writer's Life*. Oxford: Oxford UP, 2007.

Meeter, Glenn. *Bernard Malamud and Philip Roth: A Critical Essay*. Columbia: U of Missouri P, 1973.

Náhliková, Michaela. *Jewishness as Humanism in Bernard Malamud's Fiction*. Olomouc: Palacký U, 2010.

Ochshorn, Kathleen. *The Heart's Essential Landscape: Bernard Malamud's Hero*. New York: Peter Lang, 1990.

Christina Kelley

23 Grace Paley (1922–2007)

Abstract: The legacy and oeuvre of celebrated and respected author and activist Grace Paley remain relevant and almost timeless to this day, over a decade after her death. Constantly trying to reconcile her seemingly conflicting identities of writer, activist, and mother and wife, much of these sentiments show up in her fiction in the connections she draws between the causes to which she was devoted. Paley's interweaving of the themes that occupied her mind mirrors the transition of women's lives from the private (in the home as wives and mothers) to the social sphere (as active members of society), and how these two separate worlds are in perpetual conflict with each other, yet ultimately inextricably linked. Many of her female protagonists deal with similar efforts. This chapter closely examines and compares the two short stories "The Story Hearer" (1982) and "An Interest in Life" (1959) in relation to their female narrators and their struggles to navigate an understanding of womanhood in a patriarchal society.

Keywords: Activism, feminism, patriarchy, women's lives, marriage, motherhood

1 Context

Ten years after her mother's death, in the afterword to *A Grace Paley Reader* (2017), a collection of fiction and non-fiction published after her death, Grace Paley's daughter Nora writes: "My mother's powerful optimism annoyed me as a grouchy teenager. Later, I understood it. In her last illness she wanted to keep that thin silver thread of her life going as long as she could – like she always did in her gardens after the frost with the slim green stem hiding under the frozen ended summer. Her dying was not like her, but she knew everyone else did it" (Paley 2017a, 363). The "powerful optimist" Grace Paley was born in New York City on December 11, 1922 to Jewish Russian immigrants Isaac and Manya Ridnyik Goodside. She "early experienced the hardship as well as the excitement of New York immigrant life," and Paley often "describe[d] how thoroughly a product of the Bronx and of Lower Manhattan she [was]" (Bach and Hall 1997a, xii). Married twice, and the mother of two children, Paley began writing poetry early on. She attended multiple colleges, including the New School, where she studied under W.H. Auden, but never graduated. Apart from her career as a writer, Paley also taught at universities, along with fiction writing classes at Sarah Lawrence. Among several volumes of poems, such as *Leaning Forward* (1985), Grace Paley published the three story collections *The Little Disturbances of Man* (1959), *Enormous Changes at the Last Minute* (1974), and *Later the Same Day* (1985), as well as non-fiction in the form of essay collections and the 1989 War Registers League

https://doi.org/10.1515/9783110587647-024

peace calendar *365 Reasons Not to Have Another War*. Throughout her career, she received multiple awards and honors for her fiction, counting the PEN/Faulkner Prize for fiction in 1986, and, in 1987, the "prestigious Senior Fellowship of the Literature Program of the National Endowment for the Arts given to 'support and honor those who have made a major contribution to American literature over a lifetime of creative endeavor'" (Bach and Hall 1997b, xx). After a lifetime of creative genius, Grace Paley died in Vermont on August 22, 2007 of breast cancer.

To this day, she remains respected as a revered author and activist. She initially wrote as a poet, and then felt a calling to prose, in order to "speak in some inventive way about our female and male lives": "Some knowledge was creating a real physical pressure, probably in the middle of my chest – maybe just to the right of the heart. I was beginning to suffer the storyteller's pain: Listen! I *have* to tell you something!" (Paley 1994c, ix, original emphasis). As part of a generation that lived through a tumultuous time in American and world history, Grace Paley was extremely invested in social justice and political activism. As a self-proclaimed "pacifist," Paley "traveled on political tasks to Vietnam during that war, to Sweden, Russia, Central America, and [saw] China and Chile and reported on these meetings" (Paley 1994c, xi), but was also intensely devoted to the local struggles around her, co-founding the "Greenwich Village Peace Center" in 1960. She was an anti-nuclear activist, took part in many antiwar and peace demonstrations, being arrested on multiple occasions.

Not uncommonly, the time Grace Paley lived in distinctly affected and influenced her art. In her writing, Paley conscientiously interweaves social and political themes with the life stories of her characters, citing the time she lived in as "a big luck" in terms of its inspiration for her craft with its "political movements [and] history that happens to you while you're doing the dishes" (Paley 1994c, xi). Important subjects throughout her oeuvre include feminism, motherhood, marriage, relationships between men and women, questions of Russian and Jewish identity, activism, and antiwar sentiments, which she examines with a unique mixture of sarcastic wit and melancholy (Fox 2007). Paley's feminism especially is a large part of her writing, and as she claimed, an important component of and influence on women's writing of the time, whether they were invested in the movement or not: "Every woman writing in these years has had to swim in that feminist wave. No matter what she thinks of it, even if she bravely swims against it, she has been supported by it – the buoyancy, the noise, the saltiness" (Paley 1994c, xi). Her fiction largely "explored in precise, pungent and tragicomic style the struggles of ordinary women muddling through everyday lives" (Fox 2007). Wanting to move away from the "masculine fiction" of the fifties, Paley realized her own role and contribution as an author in having "[t]wo ears, one for literature, one for home": "I had been sold pretty early on the idea that I might not be writing the important serious stuff. As a grown-up woman, I had no choice. Everyday life, kitchen life, children life had been handed to me, my portion, the beginning of big luck, though I didn't know it" (Paley 1994c, x). As her daughter Nora fittingly states in the afterword to *A Grace Paley Reader*, "[t]his was a time when women's lives were not considered interesting enough to be the subject of stories" (Paley 2017a, 362).

In that sense, very much a writer before her time, Paley knew that her ideological convictions were all connected in intricate and important ways, arguing for example that many contemporary women "[didn't] see the connection between the patriarchy of militarism and the patriarchy of ordinary daily life" (Silesky, Hemley, and Solwitz 1985, 137–138). Paley's own efforts to navigate and forge connections between the causes she was devoted to, and being a writer, as well as a mother and wife, come through distinctly in her stories. The manner in which her writing intertwines the countless themes that occupied her mind strikingly mirrors the transition of women's lives from the inner sphere (in the home as wives and mothers) to the outer sphere (as active members of society), and how these two separate worlds are in perpetual conflict with each other, yet ultimately inextricably linked.

The female protagonists of "The Story Hearer" and "An Interest in Life" are situated at very different stages in their lives, arguably somewhat paralleling Paley's own experience and what moved her – as a mother of young children and later as a political activist – at the respective times the stories were published. While tonally quite different, "The Story Hearer" and "An Interest in Life" not only deal with some of Paley's most notable themes, but arguably also exemplify her craft. Both stories showcase the very real everyday lives of women as Grace Paley encountered them. Dealing with the mundane and quotidian, the selected works are told in a manner that appears to be equally ordinary, and is yet extremely profound in its illustration of the author's perspective on the female experience in both private and public life.

2 Close Readings

Grace Paley's protagonists are constantly navigating what it means to be a woman in a society still dominated by patriarchal structures. They experience various manifestations of sexism on a daily basis, while simultaneously attempting to unify the two seemingly conflicting roles of an inner and an outer life. There is a noteworthy tension between a home life that consists of everyday chores, often complicated relationships with men, and the protagonists' roles as wives and mothers, and a yearning for something larger that often manifests as a strong investment in worldly events and political activism. This speaks to Paley's own experience of having to unify "the often warring interplay of three social factors which have continually shaped her biography – family, writing, and politics" (Bach and Hall 1997a, vii), to which she herself has stated "'I can't live without writing, but I couldn't live without my children. And I couldn't bear not to respond to the awful things that are'" (Bach and Hall 1997a, vii). At first glance, Paley's stories seem to be about an obvious juxtaposition of revolution and the mundane, often mirrored in the simplicity of Paley's language. Yet, in taking a closer look, it becomes clear that revolution is an essential part of the quotidian and the profundity of the inner lives of the women Paley writes about.

"The Story Hearer," originally published in 1982, is no exception, and arguably an exemplary demonstration of this phenomenon. In this regard, Deborah Heller argues that protagonist "Faith sets the two poles – of private life and the wider world – firmly and clearly in opposition at the outset" (Heller 2005, 137). Taken from Paley's 1985 collection *Later the Same Day*, the story centers around the recurring character Faith Darwin, whose journey the reader can follow in Paley's stories through time. She has on multiple occasions been assumed to be semi-autobiographical (Johnson 2017, 38), while Paley herself has often described Faith as someone more akin to a friend of hers (Hulley 1980, 40). As a first-person narrator, Faith recounts telling the events of a day to her current partner Jack, making him the first "story hearer," and the reader the second on a further meta-level. The title speaks to "Paley's conviction that in order to become a storyteller one must first learn to become a 'story-hearer.' This approach not only invites the reader into the text, it also invites the reader to share the complexity of the truth hidden within the text" (Bach and Hall 1997a, xiii). In the story, Faith chronicles her trip to the local grocery store, discussing what and whom she encounters there, focusing largely on the conversations she has. The story ends with Faith's profound musings about life and human existence while her partner Jack drifts into an unburdened sleep next to her.

Immediately at the onset of the story, the reader is informed about Faith's need to concern herself with something larger than herself:

> I am trying to curb my cultivated individualism, which seemed for years so sweet. It was my own song in my own world and, of course, it may not be useful in the hard time to come. So, when Jack said at dinner, What did you do today with your year off? I decided to make an immediate public accounting of the day, not to water my brains with time spent in order to grow smart private thoughts. (Paley 1994b, 337)

Even before leaving her house, Faith discovers the neighbors' newspaper, and promptly criticizes her neighborhood's propensity to ignore the tragedies taking place in the outside world by planting trees that clearly signal wealth, entitlement, and privilege, creating an idealized and idyllic bourgeois microcosm she feels is divorced from actual reality:

> The *Times* was folded on the doormat of 1-A. I could see that it was black with earthquake, war, and private murder. Clearly death had been successful everywhere but not – I saw when I stepped out the front door – on our own block. Here it was springtime, partly because of the time of the year and partly because we have a self-involved block-centered street association which has lined us with sycamores and enhanced us with a mountain ash, two ginkgoes, and here and there (because we are part of the whole) ailanthus, city saver. (338)

Demonstrating her genius craft, Paley often weaves anti-war and anti-capitalist sentiments into her writing in a very matter-of-fact and still subtle fashion. In relation to a household chore, Faith casually yet sharply condemns governmental testing and use of chemical weapons: "I haven't needed to iron in years because of famous American science, which gives us wash-and-wear in one test tube and nerve gas in the other. Its right test tube doesn't know what its left test tube is doing" (338). Furthermore, the conversations Faith has with other characters about her political activism all take place at the neighborhood grocery store, a pinnacle of everyday life. Here, she discusses local issues, such as "guerilla theater demonstrations" and local boycotts, but also international events including the Japanese invasion of Manchuria, and tying them back to her own neighborhood in the reports of a scandal that affected "some young New Yorkers of [her] generation" (338–339). With her old acquaintance Jim, Faith debates Pearl Harbor, the execution of antiwar demonstrations, and the issue of female genital mutilation. She also encounters Treadwell Thomas, who is simultaneously a "fussy gourmet," and a previously commissioned official of the Language Division of the Defense Department who developed euphemisms for questionable government operations (340). All the while, these New York citizens are inspecting meat, fruits and vegetables, and general "comestibles" for their daily use (338). While "Faith affirms [. . .] throughout the stories the value of the quotidian and the personal" (Heller 2003, 89), this constitutes an interesting juxtaposition of the quotidian with revolutionary ideas, and of the microcosm of local life with the macrocosm of international affairs. Paley demonstrates very effectively how these two spheres are and should be inextricably linked, establishing politics as a crucial part of the daily human routine; and rightfully so, as it affects everyone equally and should thus be of everyday interest.

However, as Paley suggests, while these quotidian activities (ironing and grocery shopping) are intrinsically related to the political, they also undoubtedly tie Faith back to the domestic sphere and the implications it carries. Paley's female protagonists are regularly faced with varying degrees of misogyny and sexism from the men in their lives. Faith is very conscious of the fact that her acquaintance Jim only treats her as an authority, because she is older and therefore more experienced than he is: "He was aware that I had witnessing information of events that had occurred when he was in grade school. That respect gave me all the advantage I needed to be aggressive and critical" (Paley 1982, 339). These ageist tendencies continue in her interaction with the butcher, who obliviously calls her "young lady," to which Faith protests (339). She is further prompted to justify her behavior to her partner Jack who is discernibly annoyed by her decision to act on her principles: "It's an insult. You do not say to a woman of my age who looks my age, What'll you have, young lady? I did not answer him. If you say that to someone like me, it really means, What do you want, you pathetic old hag?" (340). Faith is angered by the butcher's audacity and ignorance, yet is not taken seriously by Jack, who asks her if "[she is] getting like that now too," presumably expressing his general disregard for feminist thought (340). When Faith refuses to do business with the butcher if he continues to speak to her in such a way, he replies with: "Whatever you say, honey, but what'll you have?" (340), demonstrating a failure to comprehend the issue at hand. By ignoring Faith's irritation and instead calling her by an alternative pet name, he further displays a lack of interest in altering his behavior.

Apart from her outer life of political and social activism, Faith is equally concerned with her role as a mother and wife, or romantic partner, to the man in her life. From previous stories about Faith, the reader knows that she has been married multiple times. She is now older, her children are grown, and she lives with her current partner Jack. The fact that they have both had several previous partners and children with other people is an accepted normality: "So we slept, his arms around me as sweetly as after the long day he had probably slept beside his former wife (and I as well beside my etc. etc. etc.)" (Paley 1982, 343). While the song Faith's friend Ruthy wrote about them suggests Faith and Jack committed adultery with each other by having respective extramarital affairs, Faith displays a progressive and positive relationship with her own sexuality and that of her partner. She is also not afraid to stand up to Jack when she feels he is treating her unfairly, and the reader frequently witnesses her meet his misogyny with sarcasm. As David Remnick contends, "Faith, like Paley, is now [. . .] [in] a feisty period in which all her experience and thinking has come to an extraordinary maturity" (Remnick 1985, 134). When Faith commences to tell Jack about her day, she suggests starting at the beginning, to which Jack replies: "I've always loved beginnings. Men do, I replied. No one knows if they will ever get over this. Hundreds of thousands of words have been written, some freelance and some commissioned. Still no one knows" (Paley 1982, 337). Here, Faith makes the mocking implication that men are only interested in the beginnings of things, likely referring to the beginning of love or a new romantic relationship, arguably considered to be the most "exciting" time. Jack becomes defensive at Faith's suggestion, proceeding to utilize his default excuse by telling the story of how his father, a "decent man" had "settled into a great appreciation of the middle" when his mother decided to abandon the family, never to be heard from again (338–339). Faith does not enjoy hearing the story, and feels discernibly attacked by Jack's insistence to tell it at every opportunity, presumably as a manipulation tactic, or simply to shut down her argument against him: "Now, I've heard that story maybe thirty times and I still can't bear it. In fact, whenever I've made some strong adversary point in public, Jack tells it to grieve me. Sometimes I begin to cry. Sometimes I just make soup immediately. Once I thought, Oh, I'll iron his underwear. I've heard of that being done, but I couldn't find the cord" (338). Jack's goal seems to be to put Faith back in her place in the domestic sphere, especially when she disagrees with him publicly, which is mirrored in Faith's sarcastic rendering of an inclination to immediately do a chore.

Towards the end of the story, Faith has a dream about her children that prompts her to contemplate motherhood and the concept of childlike innocence. She imagines the literal and metaphorical loss of touch with her two children who both move away, one after "blow[ing] up a very bad bank" after being urged to do so by Faith herself: "The dream continued; no – it circled itself, widening into my very old age. Then his disappearance made one of those typical spiraling descents influenced by film technique. Unreachable at the bottom, their childhood played war and made jokes" (344). In a bout of nostalgia for her own motherhood and a yearning for another baby, Faith reads the biblical story of Abraham and Sarah contemplating the innocence of their son Isaac to a sleeping Jack:

> [B]efore all that popular badness wedged its way into the world, there was first the little baby Isaac. You know what I mean: looking at Sarah just like all our own old babies – remember the way they practiced their five little senses. Oh, Jack, that Isaac, Sarah's boy – before he was old enough to be taken out by his father to get his throat cut, he must have just lain around smiling and making up diphthongs and listening, and the women sang songs to him and wrapped him up in such pretty rugs. Right? (345)

In Faith's dream, her children can be read as split off manifestations of the two conflicting sides of her own identity that she is trying to unify. The child that moves to another neighborhood – thus remaining in the local area – represents Faith's inner emotional life, her role as a mother and wife, and existence in the domestic sphere of the home. Her other son, who is forced to leave the country after committing a presumably politically motivated crime, speaks to Faith's outer life and her existence as a social and political activist. The fact that Faith was the one to prompt her son to blow up the bank arguably results in feelings of guilt on her part. As Deborah Heller argues, "Faith's nightmare at the end of 'The Story Hearer' brings together her fears that her own child may both do harm and be exposed to harm, and that she, the mother, a pacifist but also a social activist, will be responsible for both" (Heller 2003, 83). This stimulates the spontaneous wish to create another life, one not yet corrupted by the "popular badness" of the world (Paley 1982, 345), to have a chance to try again. While Faith is emotionally motivated, Jack, in his slumber, is able to give only surface-level or pragmatic answers to her concerns, such as telling her that her fertile years have passed. "Juxtaposed again here with Jack, Faith is, after all, the one who has faith – in the children, in new beginnings, in better possibilities" (Heller 2003, 83). While Jack is mumbling away in his sleep, Faith has very deep-seated musings about human existence, suggesting that the profundity of the inner lives of the women in Paley's stories often go unnoticed by men or the outside world. Faith is torn by the need to reconcile the conflicting aspects of her identity, of wanting to lead a fulfilling home life but still yearning for something larger than herself: "I am stuck here among my own ripples and tides. Don't you wish you could rise powerfully above your time and name? I'm sure we all try, but here we are, always slipping and falling down into them, speaking their narrow language, though the subject, which is how to save the world – and quickly – is immense" (Paley 1982, 342). Her inner struggle is juxtaposed with Jack's seeming ignorance of her actual feelings on the matter and about the topic in general. Realizing this, Faith resigns to telling a sleeping Jack, "Now all you have to do is be with me" (345). On the one hand, Faith's remark indicates that she only wants Jack's support and is happy to have it. On the other hand, in a further step, it displays the almost melancholy acceptance that he by nature will never completely comprehend her deep-seated conflict, one that she as a woman must deal with on her own.

Apart from the stories about the semi-autobiographical character Faith, Grace Paley's writings for the greater part deal with female protagonists in various social and economic situations, all grappling with what Paley has referred to as the "patriarchy of ordinary daily life" (Silesky, Hemley, and Solwitz 1985, 138). "An Interest in Life," published in Paley's 1959 collection *The Little Disturbances of Man,* is a significant example. Virginia or Ginny is a young mother of four, whose husband abandons her and their children to join the military. Having left her behind in mounds of debt, Ginny is on Welfare, struggling to make ends meet. She begins an extramarital affair with her neighbor's son John Raftery, who ends up lending the emotional and financial support she is not receiving from her husband.

Like Faith, Ginny is also a first-person narrator, providing the reader with in-depth insight into her interior life. Paley also describes Ginny as "not unlike Faith," who also spent an amount of time during her life as a single mother (Hulley 1980, 41). However, at this juncture in Faith's character journey, there is a noteworthy distinction in tone between the two stories, as the protagonists find themselves in different positions at varying stages in their lives. While Faith is old enough to have adult children, twenty six-year-old Ginny is at the prime of her life and peak of her motherhood with four young children to care for. Unlike Faith, Ginny does not have the time or the capacity to occupy herself with worldly affairs or political issues, demonstrating that social activism also presupposes an inherent form of privilege. After being left by her husband "with fourteen dollars, and the rent unpaid, in an emergency state" (Paley 1994b, 53), Ginny is constantly dealing with the Welfare office in an attempt to keep her family above water while simultaneously managing the chaos that comes with being a mom and the sole head of her household:

> But they never fixed my refrigerator. Every time I called I said patiently, 'The milk is sour [. . .] Corn beef went bad.' Sitting in that beer-stinking phone booth in Felan's for the sixth time (sixty cents) with the baby on my lap and Barbie tapping at the glass door with an American flag, I cried into the secretary's hardhearted ear, 'I bought real butter for the holiday, and it's rancid [. . .]' They said, 'You'll have to get a better bid on the repair job.' (54)

While Faith is well-off enough to concern herself with the struggles of others, Virginia is struggling to simply survive. Her situation affords her neither the time nor the possibility to look for work, thus leaving her extremely vulnerable and completely financially dependent on the men in her life. In this regard, her lack of agency in comparison to Faith is strikingly noteworthy, causing Virginia to view John as a desperately needed savior-figure: "it was from observing when I put my lipstick on daily, how my face was just curling up to die, that John Raftery came from Jersey to rescue me" (55). This form of dependency is problematic, yet understandable after years of being publicly and privately mistreated by an abusive husband.

While patriarchal structures treat Faith in an ageist way, Virginia experiences misogyny in a different manner, as she is still young enough to be sexualized – and consequently blamed – by men, her husband and his friends especially: "I couldn't believe that he would turn against me so, while I was still young and even his friends still liked me. But he did, he turned absolutely against me and became no friend of mine" (61). Her husband constantly verbally abuses her and the children. He forces himself on her sexually, and subsequently chastises her for the inevitable consequence – pregnancy: "'Oh, you make me so sick, you're so goddamn big and fat, you look like a goddamn brownstone, the way you're squared off in front [. . .] Your big ass takes up the whole goddamn bed [. . .] All you ever think about is making babies. This place stinks like the men's room in the BMT. It's a fucking *pissoir*'" (61, original emphasis). He goes so far as to blame Virginia and their life together for his own unhappiness, deeming his wife the reason for his ultimate desertion of the family: "'I have come to the end of my rope. I see a black wall ahead of me [. . .] I'll give it to you straight, Virginia, if I stick around, you can't help it, you'll hate me [. . .] This place drives me nuts [. . .] I got to get out of here'" (61–62), ultimately claiming that he is only doing her a favor. Ginny is very aware of the fact that societal norms would have her believe that her husband's unhappiness is in fact her fault: "Still, face facts, if my husband didn't budge on in life, it was my fault. On me, as they say, be it" (60):

> Happiness isn't so bad for a woman. She gets fatter, she gets older, she could lie down, nuzzling a regiment of men and little kids, she could just die of the pleasure. But men are different, they have to own money, or they have to be famous, or everybody on the block has to look up to them from the cellar stairs. A woman counts her children and acts snotty, like she invented life, but men *must* do well in the world. I know that men are not fooled by being happy.
> (60, original emphasis)

In an emotionally abusive and ultimately cowardly manner, Ginny's husband begins to distance himself from her to make her resent him (56). He finally disappears, and in an interview with Cora Kaplan, Paley describes this as "cynical" and the "way men get away from women" (Kaplan 1985, 150). He does not depart without a public and almost violent goodbye, leaving Ginny embarrassed in front of her neighbor: "He took hold of me with his two arms as though in love and pressed his body hard against mine so that I could feel him for the last time and suffer my loss. Then he kissed me in a mean way to nearly split my lip" (Paley 1959, 52). Virginia internalizes the guilt and "every burnt-up day of [her] life smoked in shame" (55) so intensely that she is constantly at pains to remember her last happy memory, describing it as "trying to move back into the dry mouth of a nightmare" (61).

At this juncture, John enters her life to "rescue" her, and Ginny experiences him as glaringly opposite from her husband (55). A married father himself, John comes to visit Virginia and her children every Thursday evening after seeing his mother. He consistently brings thoughtful gifts for the family, and becomes a father-figure of sorts to the children, almost treating them as if they were his own. He is very respectful, romantic, and genuinely sweet to Ginny. Initially, she refuses to have a physical relationship with him out of consideration for his situation, and again knowing that society would deem her culpable as the other woman: "I knew that if he stayed late to love with me, he would not do it lightly but would in the end pay terrible penance and ruin his long life. The responsibility would be on me" (57). However, Virginia and her children come to depend on John financially as well as emotionally, awaiting his visits in happy anticipation on "Johnday" (62). Although John encourages Ginny to talk about her abusive husband, hoping it will help her to heal, she is reluctant to tell him everything: "I think it makes a woman look too bad to tell on how another man has treated her. He begins to see her through the other man's eyes, a sitting duck, a skinful of flaws" (62). Her initial instinct to keep some things to herself is affirmed when, in a drunken state, she decides to make a list of her struggles that "could have brought tears to the eye of God if He had a minute" in the hopes of becoming a contestant on the game show *Strike It Rich* (63). While throughout the story, John shows a genuine investment in Ginny and her well-being, he ultimately displays that, like her husband, he does not fully take her seriously. In simply "giving it a glance," indicating that he denies her the consideration of taking his time to understand, John completely belittles Ginny's legitimate grievances: "'Virginia, for goodness' sakes [. . .] They'd laugh you out of the studio. Those people really suffer [. . .] In addition to all of this – the little disturbances of man' – he waved a scornful hand at my list – 'they *suffer*. They live in the forefront of tornadoes, their lives are washed off by floods – catastrophes of God. Oh Virginia'" (64, original emphasis). He treats her as though she were a child with no comprehension of actual suffering, thus delegitimizing her very real depression. With an allusion to the title of Paley's collection and a discrediting tone, John proves that he, like many of Paley's characters, has also internalized a patriarchal belief system. While his misogyny is more subtle and less outright than that of Ginny's husband, it is nonetheless undeniably relevant, marking a further parallel to Faith and Jack, who while he hears his partner, does not actually listen to what she is telling him.

While Ginny has much less active agency than Faith, arguably also due to the decade she lives in, Grace Paley makes very clear to the reader that she is by no means oblivious to her own situation or maltreatment by the men in her life. She is intelligent, with a quick wit, and not afraid to stand up for herself or speak her mind. Bold enough to tell her husband that she hates him (62), or that "[he] should've been institutionalized years ago" (51), Ginny arguably remains as "fresh," as she is convinced she was when she was a child (56). She displays a sarcastic tone throughout the story and does not shy away from "kidding around" with the Welfare office concerning the whereabouts of her husband (54), or "ask[ing] him if he could wait to join the army in a half hour, as [she] had to get the groceries" (50). In this regard, Victoria Aarons explains that

> Paley's women are strong characters; their strength – to survive, to adapt, to embrace change – lies in the humorously ironic attitudes they take toward themselves and others. Unburdened by pretense, Paley's women use humor to protect themselves from hostility, from indictment, and from unanticipated change. And their own postures in this regard are, characteristically, self-critical and ironic. (Aarons 1996, 152)

While Ginny's narration is often melancholy, and she speaks with self-pitying undertones to John especially, she is self-aware and displays a profound willingness to accept her situation for what it is: "All that is really necessary for the survival of the fittest, it seems, is an interest in life, good, bad, or peculiar" (Paley 1959, 63). In this light, Paley's female protagonists "are not deceived by fanciful illusions about the way the world is but neither are they diminished by seemingly irreconcilable hurdles" (Aarons 1996, 152). When John briefly stops visiting without notice, Virginia's interest in life grants her the composure to know there is absolutely nothing she can do about it, and she thus refuses to lie to her children:

> I had to give him up after two weeks' absence and no word. I didn't know how to tell the children: something about right and wrong, goodness and meanness, men and women. I had it all at my fingertips, ready to hand over. But I didn't think I ought to take mistakes and truth away from them. Who knows? They might make a truer friend in this world somewhere than I have ever made. So I just put them to bed and sat in the kitchen and cried. (Paley 1959, 63)

The tragedy of Virginia's character lies therein that she is cognizant about the gravity of her circumstances, yet she does not possess the means to make an impactful change. Her life appears as a never-ending cycle from which she remains unable to liberate herself. This heartbreaking reality culminates in her final envisioning of her husband's return. He gives her an emotionless and mean greeting, telling her she "look[s] older," and lets her wait on him hand and foot (65). He remains as emotionally and physically abusive as before he left, "rapping her backside," as Ginny "walk[s] around him as though he were a Maypole, kissing as [she goes]" (65), an image epitomizing the way her life revolves around him. She gets down on the floor to help him look for his cuff link that he allegedly lost, all the while "know[ing] he never had a cuff link in his life. Still I would do a lot for him" (65). According to Paley, Virginia "really [is] trying to do everything the man wants, but she is the strong one, all the same" (Marchant and Robertson 1982, 120). This again depicts her explicit awareness yet crippling dependency. The final sentence of "An Interest in Life" aptly summarizes the contradiction that is Ginny's life: "And before I can even make myself half comfortable on that polka-dotted linoleum, he got onto me right where we were, and the truth is, we were so happy, we forgot the precautions" (65). After years of absence, Virginia's husband forces himself onto his wife a final time on the floor of their kitchen. From Ginny's description, the reader can assume that they do not use protection, providing the possibility of another pregnancy, and thus further accusations from her husband. It is telling that Ginny would imagine this scene in such a way. It simultaneously clearly depicts her consciousness of her husband's toxic behavior on the one hand, and the acceptance of her own inability to change it on the other. Ultimately and tragically, she is forced, it seems, to convince herself that she is in fact happy. However, according to Paley, Ginny draws strength from her kids, and "refuses to be a victim": "Having a hard time is not being a victim. It's how you deal with it" (Hulley 1980, 45).

In the end, both stories leave their protagonists with an unresolved inner conflict and a deep-seated lack of understanding from the men in their lives. While Paley's protagonists' outlooks may often seem bleak, initially leaving the reader with a somewhat dissatisfying takeaway, the author remains a master of depicting the sometimes harsh realities of everyday life and human relationships, and especially the simultaneously ordinary and yet profound nature of the inner lives of courageous women. As Bach and Hall contend,

> [T]here always is that unflinching, almost stoic devotion to 'ordinary life,' the fumbling attempts to find a balance between what appear to be irreconcilable opposites. Paley's own multiple lives as a parent, a writer, and an activist are constantly revealed [. . .] as being exemplary of a citizen responsive to social and political demands. Paley herself, not prone to historical, psychological or sociological theorizing, usually has answers so straightforward as to sound simple. (Bach and Hall 1997a, xiii)

While reality is often harsh, and also portrayed that way, Blanche Gelfant views Paley's open endings in a positive light, arguing that their literary unconventionality enables the opportunity for growth: "You never know where a Paley story is heading, though you know the conventions that should rule literary events, especially the convention that stories must have an end. Belief in the possibility of change clashes with the literary convention of closure" (Gelfant 1985, 17).

3 Theoretical Perspectives

Admired by critics and contemporaries alike, Grace Paley's legacy remains influential. Noteworthy Paley scholars, among others, include Neil D. Isaacs, Jacqueline Taylor, Judie Newman, and Judith Arcana, who published an extensive biography of Paley, *Grace Paley's Life Stories: A Literary Biography* (1994). About Arcana's biography Isaacs writes: "[I]ts greatest accomplishment may well lie in drawing increased critical attention to the work of a writer more firmly established among readers, students, fellow activists, and other writers than in academic scholarship" (814). Those critics that have examined her work have often labeled Grace Paley as a realist, but also a postmodern writer. Diane Johnson asserts that "Paley came to represent a category of person only then beginning to be included in the ranks of serious American literary writers – moms" (37), and forges comparisons between Paley and Mary McCarthy, Katherine Anne Porter, and Tillie Olsen. She further points to Grace Paley's Jewish identity to draw parallels to the "new voices of Saul Bellow, Philip Roth, Herbert Gold, Bernard Malamud (↗22 Malamud), J.D. Salinger, and many others who, with their brilliant talent, access to inherited dialects, store of Yiddish folklore and jokes, local references, and the shadow of the Holocaust, had come to have a preeminent place in American literary writing" (37). Paley's significance and influence on the field are thus undeniable. Brooke Allen even goes so far as to call her a "sacred cow" (211).

> Grace Paley's contribution to American Literature, while comparatively small in volume, has been substantial in impact. With a voice very much her own, Paley has been a critical force in post-World War II American culture, particularly at its controversial centers. Reviewers and critics overwhelmingly agree that her short stories are vivid examples of a genre still very much alive in western culture and that one of their distinguishing marks is a language which manages to re-establish the long lost dialogue between writers and their community of readers in America. In this as well as other respects, Paley's writings are judged to be refreshingly, even radically, different from the contemporary American mainstream.
>
> (Bach and Hall 1997a, vii)

While this quote, taken from the 1996 introduction to an extensive volume of interviews conducted with the author herself, is timewise clearly dated, it arguably remains relevant to this day. Ten years after her death in 2007, Ken Bowen and Paley's daughter Nora jointly published *A Grace Paley Reader,* collecting important stories, poems, and essays, proving the timelessness of Paley's oeuvre. Arguably a writer before her time, Paley's concern for politically left-wing causes and the "principled stands she takes may seem self-evident in hindsight but were disputed at the time she wrote about them" (Johnson 2017, 37). As Diane Johnson fittingly claims, "To reread Paley is to re-experience the activism and excitement of the 1960s, their energy and perhaps naïve optimism; her essays quiver with indignation and faith" about "concerns [. . . that] are still present today" (37–38). As Bach and Hall explain, "Paley is more adamant about having different movements and interest groups join forces, particularly the merger of the women's movement, the disarmament movement, and diverse environmental movements" (Bach and Hall 1997a, ix–x). Johnson describes her as a "regional" and an "ethnic" author (37), with a commitment to diminishing both racial and gender inequalities. In this regard, Johnson's review of *A Grace Paley Reader* further cites several of Paley's essays, the concerns of which have undeniably persisted, including "Cop Tales," in which she reflects on police brutality against African Americans especially, and "The Illegal Days" that critically discusses anti-abortion ideology and the attempts to rob women of agency over their own bodies (37). In the current American political climate marred by its divisiveness on environmental issues, militarism, women's rights, and the rights of minority groups, including communities of color and the LGBTQ community, and in the wake of movements, such as #metoo and "Black Lives Matter," Grace Paley's activism is thus pertinent as ever. Bach and Hall mention the Judy Slawson interview for *The Villager*, in which Paley states: "'It's war, whether it's war now or a war being waged against future generations by using up their resources'" (Bach and Hall 1997a, x). However, in this regard, Diane Johnson encourages us to be optimistic, much like Paley herself, and "[t]o look on the bright side, among the many pleasures of Paley's work is the reassurance it offers that, just as we survived the events of those years, by analogy, we may yet survive the things that menace us now" (38). George Saunders ends his introduction to *A Grace Paley Reader* on a similarly positive note, aptly and endearingly framing the importance of the legacy of a cherished author for generations to come: "Grace Paley lives, in the minds of the readers she has moved, and in the minds of those she will yet move. As long as there are human beings wondering who they are, and how they can be better – looking for a more full-hearted way of being in the world – there will be readers for the great, beloved, much-missed Grace Paley" (xxiv).

4 Bibliography

4.1 Works Cited

Aarons, Victoria. "Margins of Hope: Grace Paley's Language of Memory." *A Measure of Memory: Storytelling and Identity in American Jewish Fiction*. Athens: U of Georgia P, 1996. 123–169.

Allen, Brooke. "Cockeyed Optimist: Grace Paley and the Eternal Feminine" (1994). *Twentieth-Century Attitudes: Literary Powers in Uncertain Times*. Chicago: Ivan R. Dee, 2003. 211–215.

Bach, Gerard, and Blaine H. Hall. "Introduction" (1996). *Conversations with Grace Paley*. Eds. Gerard Bach and Blaine H. Hall. Jackson: UP of Mississippi, 1997a. vii–xiv.

Bach, Gerard, and Blaine H. Hall. "Chronology." *Conversations with Grace Paley*. Eds. Gerard Bach and Blaine H. Hall. Jackson: UP of Mississippi, 1997b. xvii–xx.

Fox, Margalit. "Grace Paley, Writer and Activist, Dies." *New York Times*. 23 August 2007. Web. 23 April 2019.

Gelfant, Blanche H. "Grace Paley: Fragments for a Portrait in Collage." *Women Writing in America: Voices in Collage*. Hanover, NH: UP of New England, 1985. 11–29.

Heller, Deborah. "Faith, Optimism, and the Place of the Personal: Grace Paley's *Later the Same Day* and 'Midrash on Happiness.'" *Studies in American Jewish* Literature *(1981–)* 22, The Emerging Landscape: Tradition and Innovation (2003): 79–91.

Heller, Deborah. "The Work of Faith in the Stories of Grace Paley." *Literary Sisterhoods: Imagining Women Artists*. Montreal: MQUP, 2005. 89–146.

Hulley, Kathleen. "Interview with Grace Paley" (1980, from *Delta*). *Conversations with Grace Paley*. Eds. Gerard Bach and Blaine H. Hall. Jackson: UP of Mississippi, 1997. 36–55.

Isaacs, Neil D. Review. *Grace Paley's Life Stories: A Literary Biography*, by Judith Arcana. *American Literature* 65.4 (December 1993): 813–814.

Johnson, Diane. "Postmodern Mom." (Review of *A Grace Paley Reader: Stories, Essays, and Poetry* by Grace Paley). *The New York Review of Books* 64.15 (October 12, 2017): 37–38.

Kaplan, Cora. "Grace Paley Talking with Cora Kaplan" (1985, from *Writing Lives*). *Conversations with Grace Paley*. Eds. Gerard Bach and Blaine H. Hall. Jackson: UP of Mississippi, 1997. 146–152.
Marchant, Peter, and Mary Elsie Robertson. "A Conversation with Grace Paley" (1982, from *Massachusetts Review*). *Conversations with Grace Paley*. Eds. Gerard Bach and Blaine H. Hall. Jackson: UP of Mississippi, 1997. 117–125.
Paley, Grace. "An Interest in Life" (1959). *Grace Paley: The Collected Stories*. New York: Farrar, Straus, and Giroux, 1994a. 50–65.
Paley, Grace. "The Story Hearer" (1982, first published in *Mother Jones*). *Grace Paley: The Collected Stories*. New York: Farrar, Straus, and Giroux, 1994b. 337–345.
Paley, Grace. "Two Ears, Three Lucks" (1994). *Grace Paley: The Collected Stories*. New York: Farrar, Straus, and Giroux, 1994c. ix–xi.
Paley, Nora. "Afterword." *A Grace Paley Reader: Stories, Essays, and Poetry*. Eds. Ken Bowen and Nora Paley. New York: Farrar, Straus, and Giroux, 2017a. 361–363.
Remnick, David. "Grace Paley, Voice from the Village" (1985, from *Washington Post*). *Conversations with Grace Paley*. Eds. Gerard Bach and Blaine H. Hall. Jackson: UP of Mississippi, 1997. 131–134.
Saunders, George. "Introduction: 'The Saint of Seeing.'" *A Grace Paley Reader: Stories, Essays, and Poetry*. Eds. Ken Bowen and Nora Paley. New York: Farrar, Straus, and Giroux, 2017. xv–xxiv.
Silesky, Barry, Robin Hemley, and Sharon Solwitz. "Grace Paley: A Conversation" (1985, from *Another Chicago Magazine*). *Conversations with Grace Paley*. Eds. Gerard Bach and Blaine H. Hall. Jackson: UP of Mississippi, 1997. 135–145.

4.2 Further Reading

Arcana, Judith. *Grace Paley's Life Stories: A Literary Biography*. Champaign: U of Illinois P, 1994.
Isaacs, Neil D. *Grace Paley: A Study of the Short Fiction*. Boston: G.K. Hall/Twayne, 1990.
Lévy, Paule, ed. *Special Issue: Grace Paley. Journal of the Short Story in English* 65 (Autumn 2015).
Paley, Grace. "Cop Tales" (1980, from *Just as I Thought*). *A Grace Paley Reader: Stories, Essays, and Poetry*. Eds. Ken Bowen and Nora Paley. New York: Farrar, Straus, and Giroux, 2017b. 216–219.
Paley, Grace. *Enormous Changes at the Last Minute*. New York: Doubleday, 1974.
Paley, Grace. "The Illegal Days" (1991, from *Just as I Thought*). *A Grace Paley Reader: Stories, Essays, and Poetry*. Eds. Ken Bowen and Nora Paley. New York: Farrar, Straus, and Giroux, 2017. 198–205.
Paley, Grace. *Later the Same Day*. New York: Farrar, Straus, and Giroux, 1985.
Paley, Grace. *Leaning Forward*. Penobscot, Maine: Granite P, 1985.
Paley, Grace. *The Little Disturbances of Man*. New York: Doubleday, 1959.
Taylor, Jacqueline. *Grace Paley: Illuminating the Dark Lives*. Austin: U of Texas P, 1990.

Sascha Pöhlmann

24 Donald Barthelme (1931–1989)

Abstract: This book chapter assesses the vast thematic and formal multiplicity of Donald Barthelme's short story oeuvre by contextualizing it in its particular generic tradition (in contrast to Hemingway and modernism) and in the more general framework of U.S.-American postmodernism (with its elements of metafiction, semiotic playfulness, fragmentation, collage, and a disregard for the binary of high and pop culture). The chapter analyses "The School" as an example of Barthelme's fiction of escalation that combines existentialism with absurd humor, and "The Indian Uprising" as an example of his more abstract, self-reflexive inquiry into meaning-making. It argues for the need to transcend the all too dominant classification of Barthelme as a postmodernist and suggests multimodality as one analytical category among others that seems a particularly fruitful way of taking this step.

Keywords: Postmodernism, humor, collage, metafiction, multimodality

1 Context

Donald Barthelme can best be classified by saying that he is annoyingly unclassifiable. This is especially true of the form of writing he is most well-known for, the short story, since the only thing his short stories all have in common is precisely that which is expressed by their trivial generic marker: they are stories that are short. Beyond that, they contain such a wide array of styles, voices, themes, techniques, and formal features that any attempt to homogenize them under a single label seems utterly misleading, and they integrate themselves into the American tradition of their genre as much as they subvert it. They do conform to the generic characteristics and rules that have accumulated throughout the century after Edgar Allan Poe (↗7 Poe) first theorized the short story in his 1846 essay "The Philosophy of Composition" when he emphasized that such texts need to exhibit a unity of effect and must allow themselves to be read in one sitting. The most recent revolutionary change to the American short story before Barthelme's own was Ernest Hemingway's modernist take on the form (↗16 Hemingway), which took carefully controlled omission rather than explicit presence as its guiding aesthetic principle (summarily known as the "iceberg theory"). If Hemingway may be credited as the main writer (among others) to have updated the short story into a modernist form, then Barthelme is the main writer (among others) to have moved the short story into postmodernism. This is the predominant context in which Barthelme's work – his short stories, but also the novels, essays, and the children's book he is less known for – was received during his career, and this is the context in which he still remains firmly emplaced in contemporary

https://doi.org/10.1515/9783110587647-025

literary histories, although critical interest in Barthelme seems to have waned in the twenty-first century. Regardless, he is of central importance to the innovations and experiments that have come to be known under the label of U.S.-American postmodernist literature, and his postmodernization of the short story is no less relevant than Thomas Pynchon's postmodernization of the novel.

What does this mean, though? Hemingway's modernization meant omission and a further increase of the involvement of the reader in the process of meaning-making that had always been a defining feature of a form based on brevity. Barthelme's postmodernization – which is as much a move away from modernism as its intensification – means a further expansion, if not explosion of this meaning-making process that may best be described as playfulness (a characteristic which, in any given sense of the word, hardly marked Hemingway's aesthetic). If Hemingway's stories show only the tip of the iceberg of meaning but hide most of its significance underneath their surface for the reader to discover, then readers may find just about anything beneath the textual surface of Barthelme's stories: floating bits of meaning that may or may not be related to what is above, a massive block of meaning that seems impenetrable and stretches further than one can see, or why not an arbitrary polar bear who cannot wait to eat the diver. Barthelme may provide his readers with a proper aesthetic response to this openness in the final sentences of "Porcupines at the University," in which car drivers wonder about thousands of porcupines on the road: "What is wonderful? Are these porcupines wonderful? Are they significant? Are they what I need?" (1987, 192)

This proliferation of meaning, in which readers are invited to fill in more gaps than they possibly could, has various consequences. One of them is that Barthelme's short stories often seem to eschew narrative altogether, and, as they dispense with plot, characters, or action, they move closer to poetry than prose. Another consequence is that the troubled subjectivity that dominated Hemingway's short stories has now given way to such a multiplicity of voices that Barthelme's narrators and protagonists cannot be reduced to any such unifying construct, and if there is a tenuous sense of personhood about his wide array of speakers and characters, then it is as much a function of language as anything else that happens in these texts, as if the story were playing at having characters as part of its more general game of language. This is not to say that Barthelme's stories are impersonal or 'merely' (as if that were a simple or negligible feat) aesthetic play, art for art's sake; instead, they are themselves explorations of their postmodern sociocultural climate, and they allow for ethical as much as for political readings precisely for their critique of the concepts of subjectivity and identity in and through language. At the same time, they are often also patently absurd, and their playfulness in this second sense combines the bitterly humorous existentialism of Samuel Beckett (whom Barthelme often acknowledged as his strongest literary 'influence') with the anarchic and often surreal comedy of Monty Python. The third aspect of their playfulness is that they refuse to take anything seriously, such as the boundaries between the fictional and the real, between popular culture and high art, between image and text, or between order and chaos; or the concept of artistic originality and the finer distinctions between collage, appropriation, and plagiarism; or even most of all themselves, for that matter, and this in itself marks a radical break with the often terrible gravity of modernism that managed to make even irony dead serious.

Appropriately, Barthelme's beginnings as a writer coincided with the beginning of the 1960s, a decade that would gradually grow into the most rebellious phase in U.S.-American cultural history since Romanticism. Born in 1931, Barthelme published his first short story ("Me and Miss Mandible") in 1961 and would end up publishing over one hundred in various formats until his death in 1989, most notably in the *New Yorker*, but also in book form. The companion volumes *Sixty Stories* (1981) and *Forty Stories* (1987) provide readers with the best material in this regard, and along with the posthumously published *Flying to America: 45 More Stories* (2007) they form a complete picture of his short story oeuvre. His four novels, *Snow White* (1967), *The Dead Father* (1975), *Paradise* (1986), and the posthumous *The King* (1990), may be classified as postmodernist fiction, and yet their impact on the theorization of this literary aesthetic is relatively minor compared to other works of the 1960s and 70s. His short stories, however, were more than enough to place Barthelme at the center of literary postmodernism as the concept was in the process of being developed in various ways during that time, first as a shorthand to signify any kind of break with the then-dominant modernist aesthetic, then as an aesthetic of its own that is just as multiple and elusive as the one it supposedly replaced. Notably, if Barthelme's texts now tick all the boxes of what makes a short story postmodernist, then it is worth considering that these boxes may have been drawn up to a considerable extent using Barthelme's short stories as a model, and their postmodernist aspects such as semiotic playfulness, fragmentation, collage, absurdity, irony, anarchy, etc. seem as such because these are the stories that helped create that catalog in the first place.

As stated above, it is utterly futile to attempt to categorize Barthelme's whole short story oeuvre in any simplistic way, and the terms that seem to provide an appropriate classification (such as the ones I just listed above) indicate multiplicity and categorial openness rather than provide conclusive generalizations. Barthelme's choice to give the first of his anthologies the unassuming title of *Sixty Stories* is highly telling in this regard. In his introduction to the reprint of this volume in 2003, David Gates speculates about why Barthelme, for whom "title-giving was one of his great knacks," would "settle" for this "modest, no-spin title," and he rules out that he "despaired of coming up with such any one title that could overarch such a various landscape," arguing that "that had never stopped him before" (xi). Yet beyond speculating about his particular psychological motivation, this is precisely the appeal of this ridiculously descriptive title, as it most accurately reflects how much the stories collected in this anthology resist any unifying categorization with regard to any of their aspects one might pick. The same goes for the *Forty Stories* and the *45 More Stories* the subtitle of *Flying to America* promises. These 145 stories – Barthelme's complete short story oeuvre – are classifiable only in terms of family resemblances, of patterns and similarities emerging and overlapping with others, but never in the sense of an overarching, coherent literary aesthetic. If anything, then these stories in their very multiplicity but also in many particular cases challenge and undermine such pretentions to a universal structure, to a stable order and a linear notion of coherence.

As such, they resonate very well with their contemporary poststructuralism, which philosophically and semiotically opposes the notion of stable and universal deeper structures underlying particular cultural artefacts, and instead points out that any such structure is at best provisional, incomplete, and bound to its own historical moment. They share this resonance also with an even more general resistance of postmodernism to modernist ideas of order, and this is indeed one of the distinguishing features that really sets one apart from the other instead of highlighting their continuities. Modernism sought to restore order in and through art in a world that seemed to have become chaotic; as a paradigmatic example, consider T.S. Eliot's review and quasi-manifesto "*Ulysses*, Order, and Myth" and its theorization of an aesthetic "way of controlling, of ordering, of giving a shape and a significance to the immense panorama of futility and anarchy which is contemporary history" (Eliot 1975, 177). Postmodernism, however, decided to play around with chaos by creating provisional orders, perhaps under the fresh impression of what rigid order really meant in both aesthetic and political terms as embodied by the totalitarian regimes and ideologies quite a number of modernists embraced. Thus, the postmodern could be famously described by Jean-François Lyotard as "incredulity toward metanarratives" (Lyotard 1984, xxiv), whereas a new metanarrative seemed precisely what many modernists sought in and through art. In these terms, then, Barthelme's oeuvre resists the metanarrative of a single coherent framework in which to categorize them, but it is rather a loose collection of micronarratives that sometimes resemble each other and form patterns that may or may not overlap, but which are always disrupted somewhere along the way before becoming a universal structure. Stories such as "The Glass Mountain" parody such a desire for order by numbering each of its 100 sections, perhaps in reference to the way Ludwig Wittgenstein structured his *Tractatus Logico-Philosophicus*, which is exemplary of the philosophical desire for order that motivates so much of modernism; the story of climbing the glass mountain ends with the line "100. Nor are eagles plausible, not at all, not for a moment" (Barthelme 1982, 176), as it disavows verisimilitude or 'realism' but highlights and asserts its own fictional ontology.

Along with a playful approach to order, such metafictional strategies may be regarded as the second recurring characteristic in Barthelme's loose set of family resemblances, and again this is one of the few aspects of postmodernism that really breaks with modernism rather than intensify it. Postmodernist fiction routinely highlights its own fictional nature in an antirepresentational and antirealist move, and it exhibits a critical self-reflexivity that is perhaps nascent but far from prominent in modernist texts. Barthelme's short stories often remind their readers – either implicitly or explicitly –

that they are indeed short stories, that readers engage with language first and foremost, and that the reality of the words matters much more than any 'reality' they might convey as fiction. Employing such metafictional strategies, writers such as Vladimir Nabokov, John Barth, Kurt Vonnegut, or Raymond Federman wrote novels that would be classified as postmodernist precisely for this, and the same applies as well to Barthelme's short stories. Their repertoire of metafictional tactics is quite rich, though, and it is hardly ever a simple matter of directly addressing the readers and telling them that they are reading a story (which may well be why Barthelme rejected the label of metafiction for his works). Instead, Barthelme uses mainly formal and structural elements such as the enumeration mentioned above, but also appropriated visual images inserted into the text, or numerous other anti-realist devices that insist on the primacy of language over its representational function. For example, "Concerning the Bodyguard" consists almost entirely of questions; "Sentence" indeed seems like one long sentence until it fails to even end like one with a concluding period; or "Eugénie Grandet" has a paragraph that only consists of the word "butter" 97 times. This last story, drawing on Honoré de Balzac's novel of the same title not directly but through "*The Thesaurus of Book Digests*" (230), indicates that one of the major anti-realist and metafictional devices Barthelme uses is intertextuality, and his short stories often allude to other texts, build on them, rewrite them, or rearrange them in combination with other found and original material. In doing so, they happily disregard the boundary between high and popular literature, mixing erudite intellectual allusions with references to glossy magazines, and the deconstruction of this ideological rather than aesthetic opposition may serve as the final major postmodernist aspect of Barthelme's short stories that can be mentioned here.

2 Close Readings

Given the utter diversity of Barthelme's short stories, picking two out of 145 will always seem somewhat arbitrary, and they may only be presented as somewhat representative and characteristic of his work in the sense mentioned above, in that they indicate a few of the family resemblances and overlapping patterns that may be found across his oeuvre but without any claim to universality. The two stories I will discuss in more detail in the following, "The School" (first published in *Amateurs*, 1976) and "The Indian Uprising" (first published in *Unspeakable Practices, Unnatural Acts*, 1968), exemplify two types of stories whose aesthetic strategies recur in a number of others, and although they are not models of these types in the sense that the other stories are copies, they are particularly well-suited for an exemplary reading in this regard.

"The School" is the more accessible of the two stories, and also the one that may more easily be assigned to a particular narrative mode in Barthelme's tales. It is a monologue by a speaker talking to us as the silent part in a two-way communication, and the precise nature of this situation and their own supposed role in it is one of the things that readers are invited to construct themselves. These monologues are often vindicatory as they justify particular events or acts, as if we as readers were the accusers and the narrator the defendant; more often than not, these defendants only dig themselves deeper into their hole as they ramble on. A good way of describing these texts – other notable and utterly hilarious examples are "Some of Us Had Been Threatening Our Friend Colby" (Barthelme 1987, 167–171) and "The Baby" (244–246) – is to call them stories of escalation. They set out from a beginning that may already be absurd or surreal, and then the situation based on this initial premise quickly goes from bad to worse, usually because the narrator refuses to be anything but consistent in his actions even though they are as weird as the arbitrary rules on which they are based. Thus the father in "The Baby" initially punishes his infant daughter for tearing pages out of books by forcing her to stay in her room alone, where she would tear out more pages so that her sentence quickly accumulates to years on her own; he eventually solves the problem simply "by declaring it was *all right* to tear pages out of books," and the story ends with a most harmonious image of parent and child who "go out on the street and smash a windshield together" (246). "Some of Us Had Been Threatening Our Friend Colby" begins with the premise that Colby, after a long time of unspecified misbehavior, had "gone too far, so we decided to hang him" (167), and they plan the hanging in ever more elaborate ways, and in the end everything "went off very smoothly on the day of the event," and "nobody has ever gone too far again" (171). These stories utilize the same narrative logic of escalation that also drives some of Monty Python's work; just think of the "Dirty Fork" sketch that leaves much of the restaurant staff dead after an initial modest complaint about the cutlery. "The School" is no different, only that it is not so much individual action and arbitrary rule-making that is at stake here, but rather an individual and collective refusal to acknowledge that something more fundamental might be amiss with their situation. These stories derive much of their humorous and philosophical effect from the fact that readers will quickly discern a pattern that the speakers fail or refuse to notice as they seem constrained by a stubborn insistence on an order they believe in but readers will question more with every step they take toward its radical but coherent conclusion.

The only speaker and narrator in "The School" is a teacher, perhaps even the headmaster of a school that is, as he puts it, having "a run of bad luck" (Barthelme 1982, 306) in that animals, plants, and people associated with it keep dying in significant numbers. The story shows Barthelme's masterful precision: as readers will notice only a paragraph or so into the tale, it establishes the narrative situation already with the first word, "Well" (304), as this presents the speaker as somebody verbally addressing an audience, perhaps in response to a question by a journalist or a superior asking him to explain what is going on at the school, and the word combines the casual with the exhausted, the descriptive with the puzzled. The speaker keeps interrupting himself and getting sidetracked, and the style neatly captures in written form the idiosyncrasies of nervous speech groping for an explanation. The use of paragraphs is a major part of this, and Barthelme here again is as exact as the comedian who knows that humor is all about timing, and that this means *not* saying something at just the right moment. Every paragraph escalates the situation further for the reader, and the pauses between them invite us to extrapolate how the speaker moves on from one tragedy to the next while maintaining his all too laconic and apologetic tone as he keeps downplaying each new and progressively more outrageous death with the same casualness. More generally, the story is pushing Hemingway's aesthetic of omission into the realm of the surreal and absurd, as virtually every other sentence hints at a much more complex situation than we are being told about here; for example, the speaker tells us that "the boiler was shut off for four days because of the strike" (304), and the definite article implies that we are supposed to know all about that strike, even if we have no way of deducing anything about it from the text itself, which only adds to the weirdness of the school.

With this absurd poignancy the speaker moves from dead trees to snakes, herb gardens, gerbils, white mice, salamanders, and to tropical fish, and then Barthelme lands a humorous precision strike with a single-sentence paragraph: "We weren't even supposed to have a puppy" (305). The narrative logic of escalation combines the comic and the tragic in the black humor of the absurd, until finally the speaker addresses the elephant in the room that "maybe there was something wrong with the school," and in trying to reject this theory he actually affirms it by saying "we had an extraordinary number of parents passing away, for instance" (306). The speaker is either unaware of the ironic difference between his words and what they mean or is caught in the rhetoric of an official who is trying to put a positive spin on a terrible situation; in any case, the story is built on this discrepancy, and as such it both affirms the power of language to create its own reality for the reader and at the same time denies it the power of representing reality objectively.

Rambling on, the speaker is trying to make sense of what is happening but fails to arrive at any meaningful or definite conclusion. Instead, he admits that "I don't know what's true and what's not. It's been a strange year" (306). This doubt as to truth – not just the truth in a particular case but the concept of truth in general – is characteristic of the postmodernist challenge to objectivity that is part of a larger critique of representation that fundamentally questions the relation between the word and the world. Even more significantly, though, this situation of epistemological uncertainty then suddenly turns into a situation of existential uncertainty, as the story suddenly addresses the absurdity of life in the sense that there is no transcendental meaning to human existence. This indicates the more general impact of the philosophy of existentialism, whose main proponents are Jean-Paul Sartre and Albert Camus, on postmodernism, which also often rejects transcendence for immanence and therefore reveals any human value system as provisional and arbitrary (though notably no less real for it).

Within a single paragraph, then, the narrative changes from a piece of black humor to an existentialist parable. The impact of this shift on the reader is so great because it is marked by a change in tone that may also be read as an anti-realist tactic to undermine any sense of verisimilitude our suspension of disbelief may have left us with: the class discussion on "where did they go" (306) starts out by asking about "the poppas and mommas" but then quickly and surprisingly changes from these childish semantics to a veritable Socratic dialogue:

> And they said, is death that which gives meaning to life? And I said no, life is that which gives meaning to life. Then they said, but isn't death, considered as a fundamental datum, the means by which the taken-for-granted mundanity of the everyday may be transcended in the direction of–
>
> I said, yes, maybe. (306)

The children go on to tell the speaker that they "require an assertion of value, we are frightened," and he tells them that "they shouldn't be frightened (although I am often frightened) and that there was value everywhere" (307), and while this seems like an answer to their common existential dread, it really is not, as he is simply positing the existence of value in their world without giving any evidence of it. Yet the story ends by showing that this feeble strategy of denial seems to work at least momentarily, as if to argue that humans are good at convincing themselves and each other of the meaning and value of life, and notably the children are distracted from thoughts about death by the appearance of life: "Then there was a knock on the door, I opened the door, and the new gerbil walked in. The children cheered wildly" (307). This final image contributes to the surreal tone of the story, as gerbils are not known to knock on doors, and it implies that humans may come to terms with the fundamental absurdity of their existence by both embracing and resisting it through symbolic means (like Camus's Sisyphus), and perhaps most of all by telling themselves and each other stories about it, like this one. We suspect that this gerbil is probably as doomed as the other critters in the school, but at least it is a *new* gerbil so that life, simply and banally, goes on anyway and is celebrated as such in its moment – and even an obviously fictional gerbil can do the job of keeping our existential dread at bay.

The aesthetics of escalation in "The School" is naturally a linear one, but this is rather exceptional in Barthelme's short stories. Instead of a relatively clear and coherent narrative progression that would move plot and characters from A to B and then to C, his stories often exhibit a different kind of coherence that is much more complex. "The Indian Uprising" is a good example of this style, and as such it shares a certain family resemblance with stories such as "Alice," "Daumier," or "Aria" (all collected in *Sixty Stories*), although they are quite different in other regards. In fact, these stories are so disassociated from narrative conventions of plot, character, and action that they test the limits of what a story *is*, and their coherence is not so much narrative but poetic as it comes about through repetition, juxtaposition, and metaphor rather than, say, a protagonist who does something. One might be tempted to describe "The Indian Uprising" as fragmented, but this would seem to imply that the story is told to us in a broken and non-linear way, whereas the more fundamental aesthetic effect of the text is that readers must wonder if they are being told any story at all. The simplest test of this might be to summarize the 'plot,' and while this often results in simplifications that sound beside the point with regard to complex literary works, it seems utterly silly in this case, as none of its aspects will congeal into any logic that could be grasped by narratological means. In other words, the attempt to make sense of the story must be frustrated on the narrative level and thus must shift to the level of language where the very nature of the story as story matters less than how the words on the page relate to each other to evoke patterns of meaning without forming a meaningful whole.

The plot, then, could be summarized by saying that the unnamed first-person narrator is a combatant defending an unnamed city against the Comanches while being too closely involved with their sympathizers Sylvia and Miss R., so that he is eventually captured by them. Yet most readers would note how frustratingly inadequate such a summary is and how little it helps to understand what is going on in this text. On a deeper level, one could interpret this 'story' as the staging of a conflict between plot and text: it pits the coherence of narrative against the incoherence of language, and it frustrates our expectations of order by contextualizing it within another, higher order. As our reading for narrative elements fails, we are forced to make sense beyond this convention and open up to other, more open and elusive ways of language being used, other ways of connecting words, sentences and paragraphs as readers. This is not to say that the story has no 'content,' that it is only about playing with language; in fact, there is no such thing as 'mere play,' as playing is always significant in one way or another. The text addresses numerous issues that range from the philosophical to the political or the personal, and it may be read as a critical commentary on U.S. imperialism in the 1960s, the crisis of white male hegemony and the stereotypical racialized othering it involves, or the irrelevance of so-called high art in a violent world, but none of these readings will ever be more than partial and allusive rather than congeal to become 'the' dominant interpretation that fully resonates with the text as a whole. This is one of the cases where the tip of the iceberg, the text on the page, is not directly connected to something hidden underneath the surface, and the gaps of meaning are so wide that readers cannot fill them so as to connect all the dots to form a complete picture. As Block says in the story: "'The situation is liquid'" (105).

Instead of narrative order, then, readers get a dense and complex web of meaning whose lines cannot be spun into a single thread. One way of navigating this web is to consider each paragraph as a provisional structural unit and establish some kind of coherence on that level before seeing how each level may or may not connect to others. Barthelme's use of paragraphs is as significant here as it is in "The School," though the effect is not that of a humorous progression but rather of a collage-like assembly, a mosaic with linguistic and conceptual patterns that does not form a recognizable picture if viewed from an appropriate distance. One thing these paragraphs have in common is the use of anti-realist elements that disturb rather than satisfy the narrative expectations we bring to the form of the short story. For example, the story is evocatively playing with the concept of setting by telling us that the city has a "Boulevard Mark Clark" (102) and a "Rue Chester Nimitz" (104), combining French synonyms for 'street' with the names of a U.S.-American general and admiral, respectively. Granted, this slightly surreal fusion into a French-American place may still be explained in fantastic narrative terms, just like the involvement of the IRA (104) or other events such as the narrator having "just learned via an International Distress Coupon that Jane had been beaten up by a dwarf in a bar on Tenerife" (103). Yet the text goes further than that as its style becomes less mimetic and more enumerative, and as it comments on this very form in an implicitly metafictional way, so that the sentence in the first paragraph, "People were trying to understand" (102), seems less than a description within the story but rather *of* the story itself.

This begins modestly when we are told that the "table held apples, books, long-playing records" (102) and we might still combine these items into a visual image that is perhaps a still life of aestheticism, but it takes on more massive proportions soon after when we read that the barricades are "made of window dummies, silk, thoughtfully planned job descriptions (including scales for the orderly progress of other colors), wine in demijohns, and robes," and the adjacent ones of

> two ashtrays, ceramic, one dark brown and one dark brown with an orange blur at the lip; a tin frying pan; two-liter bottles of red wine; three-quarter-liter bottles of Black & White, aquavit, cognac, vodka, gin, Fad #6 sherry; a hollow-core door in birch veneer on black wrought-iron legs; a blanket, red-orange with faint blue stripes; a red pillow and a blue pillow; a woven straw wastebasket; two glass jars for flowers; corkscrews and can openers; two plates and two cups, ceramic, dark brown; a yellow-and-purple poster; a Yugoslavian carved flute, wood, dark brown; and other items. (103)

The speaker concludes from this enumeration: "I decided I knew nothing" (103), and he shares with the readers the sense that this is a world of bits and pieces that fail to cohere into a meaningful, graspable whole. Yet this is not merely unsettling in epistemological terms but also in ontological ones, a distinction that according to Brian McHale (cf. 9–10) marks the difference between modernist and postmodernist fiction: This textual world may not be ordered by the narrator or the reader into a sufficiently coherent whole that would make it seem like a world at all. This world is an *assembled* one, and the cracks between its parts are showing clearly to reveal the structure as artifice. The text even reveals itself as a text – a visual artifact made out of words on a page where the signifiers matter as much as their signified – when Miss R. states (of course for no obvious narrative reason) that the "only form of discourse of which I approve [. . .] is the litany" (Barthelme 1982, 106); she exemplifies this by an enumeration that draws on the one above, and then adds that "[v]ertical organization is also possible" and lists the same terms one after the other with line breaks like in a poem. Allegedly she is still speaking, and yet such vertical organization could not occur in oral speech but only in the visual materiality of text.

This may be the most explicit of anti-realist elements that draw attention to the textual nature of this web of meaning that deliberately falls short of cohering into a fictional world, yet the arguably most important one is that of intertextuality, as the 'story' contains a staggering number of allusions. Intertextual references are a favorite element of modernism and postmodernism alike, only that they are employed for very different purposes, and "The Indian Uprising" may be read as a comment on this difference. On a more basic level, they serve yet again as anti-realist and perhaps even anti-narrative elements so that readers will be aware that they are reading a text instead of a story that is at least somewhat mimetic. For example, the tortured Comanche claims that his name is "Gustave Aschenbach" and he was born in Silesia (107); on one level, this ethnic discrepancy may be considered a comment on the value of information obtained by torture, but on another level, it refers to the protagonist of Thomas Mann's *Death in Venice* and thus opens up a connection between one fiction and another instead of a connection between fiction and reality. However, the most significant intertextual connection with regard to the collage-like structure of the story is that to T.S. Eliot's poem "The Waste Land," and it is so important because this poem employs similar means to very different ends. The narrator tells us that they "sent more heroin into the ghetto, and hyacinths, ordering another hundred thousand of the pale, delicate flowers," and Sylvia responds to this highly creative way of dealing with an uprising by saying "'You gave me heroin first a year ago'" (104). This scene echoes the lines from the first part of "The Waste Land," "'You gave me hyacinths first a year ago; / They called me the hyacinth girl'" (Eliot 1952, 38, ll.35–36), and as such they invite a contrasting reading of the poem as a whole in conjunction with the story. Eliot's mythical method, as laid out in his review of *Ulysses*, consists in incorporating and juxtaposing fragments of earlier texts – usually representative of a European high culture – in order to create an aesthetic framework in which to grasp what is perceived as a chaotic contemporary world and perhaps even to restore it. The 'speaker' of the poem's fifth and concluding section, a combination of the seer Tiresias and the Arthurian Fisher King, comments on this technique and its purpose when he first asks "Shall I at least set my lands in order?" (50, ll.426) and then declares "These fragments I have shored against my ruins" (50, ll.431). Barthelme, though, is not shoring fragments against the ruins of (high) culture here, and he does not make these intertextual references in the hope of restoring a destroyed order from them. Some critics read these passages in terms of "dreck" as the communicative noise or linguistic waste of a consumerist society, and yet Barthelme does not simply oppose these negative bits and pieces to a positive and more coherent and valuable high culture (cf. Dini). "The Indian Uprising" is devoid of the nostalgia that makes "The Waste Land" a reactionary poem despite its revolutionary qualities. Instead, the story uses the poetic method of juxtaposition to assemble a collage of high culture and the mundane, and it playfully revels in the multiplicity of an order that it knows is arbitrary, incomplete, and temporary, and which it seems to cherish precisely for these aspects. Here, the erudite allusions to Eliot, Mann, Godard, or Wedekind are not made to indicate a possible high-cultural escape from the chaos of quotidian life, but they are rather revealed to be of no particular value to the project of ordering the world, or at least no more value than the weird assemblage of "zip guns, telegrams, lockets" (Barthelme 1982, 104) others are bringing to the fight. Readers are invited to draw on these sources in their meaning-making, but they are in no way privileged, and there is no high-cultural key that would unlock the story. Instead, its pieces will never all fall into place for the readers, and as such the story aesthetically inquires into the conditions, processes, and consequences of making sense in general, but with no hope of creating a stable order in the end, and no despair either.

3 Theoretical Perspectives

Without a doubt, the potential theoretical perspectives on Barthelme's short stories are as varied as the stories themselves, and one will find in them material for analyses through any of the approaches that have come to be summarily known as capital-T Theory during the very time that Barthelme was active – structuralist, poststructuralist, psychoanalytic, feminist, Marxist, postcolonialist, etc., to name just the most pertinent ones. Furthermore, his short stories will just as easily provide material for analyses from theoretical perspectives that have arrived on the scene later, most notably queer theory and environmental criticism, but also for computer-based 'distant readings' that may offer a very different take on the oeuvre. This is not the place to provide exemplary readings from each of these perspectives, especially as the canon of Barthelme stories is so vast (and vastly diverse) that they would hardly be exemplary at all in a generalizing sense. Critic Wayne B. Stengel has tried to bring order to this multiplicity in *The Shape of Art in the Stories of Donald Barthelme* (1985) by categorizing Barthelme's stories according to "sixteen junctures of theme and process" (5), an undertaking that traces family resemblances but does not succeed in uncovering or imposing a unifying structure on the oeuvre as a whole. Suffice it to say that Barthelme's stories have aged well with regard to such approaches, and they are as much of their particular historical and cultural context of the 1960s through 1980s as they are independent of the narrower confines of this framework.

Given this abundance and relevance, which is not least due to the prominence Barthelme's stories gained even on first publication in the popular and highly visible venue of *The New Yorker*, it may seem all the more surprising that Barthelme's academic reception seems to have hit its peak before the turn of the millennium – no doubt partly due to his untimely death – and that critical attention has turned elsewhere once his complete oeuvre had been published with *Flying to America*, or even before. Instead of becoming a vast resource for various contemporary readings, then, Barthelme's short stories seem to have found their all too secure place in postmodernism, as if their main purpose now was to illustrate what postmodernism was, and as if their importance for this definition was so great that it is for example quite inconceivable that they may transcend the very aesthetics they have come to construct in the first place. In other words, Barthelme's stories are often being read for what they say about postmodernism, but not necessarily for what else they might be saying.

Most of the major readings of Barthelme's short stories occur within this framework of postmodernism, and they constitute the comparatively small canon of critical monographs on his work that all came into being as the theorization of postmodernism was consolidated: Luis Gordon's introductory *Donald Barthelme* (1981), Larry McCaffery's *The Metafictional Muse: The Works of Robert Coover, Donald Barthelme, and William H. Gass* (1982), Timothy Charles Lord's *Postmodernism and Donald Barthelme's Metafictional Commentary on Contemporary Philosophy* (1987), Stanley Trachtenberg's *Understanding Donald Barthelme* (1990), Jerome Klinkowitz's *Donald Barthelme: An Exhibition* (1991), Paul Maltby's *Dissident Postmodernists: Barthelme, Coover, Pynchon* (1991), and Richard Anthony Dooner's *Repetition in Postmodern Fiction: The Works of Kathy Acker, Donald Barthelme and Don DeLillo* (1993). As some of these titles show, Barthelme seems to be an author who lends himself to comparative analyses, and more recent publications such as Barbara Schwerdtfeger's *Ethics in Postmodern Fiction: Donald Barthelme and William Gass* (2005) continue in this vein. A notable recent exception to this pattern of reading Barthelme comparatively in the framework of postmodernism is Daniel Chaskes's dissertation *Beyond Fragmentation: Donald Barthelme and Writing as Political Act* (2012), and it shows just how productive a reading of Barthelme beyond postmodernism can be, as it indicates the larger potential of such analyses that remains mostly untapped.

Recent forays into this potential have taken the form of shorter essays rather than monographs, for example Craig Medvecky's analysis of Barthelme's stories in terms of their construction of masculinity, or Cristina Ionica's interpretation that connects the narrative framing of gender to family violence. Such essays fruitfully indicate other ways of reading Barthelme that either complicate the postmodernist classification or go against its grain entirely. One of the most promising perspectives in this regard is the analysis of Barthelme's work in terms of its material and visual aspects as a textual object and as multimodal fiction (which Alison Gibbons defines as "a body of literary texts that feature a multitude of semiotic modes in the communication and progression of their narratives" (420)). Daniel Punday's essay "Donald Barthelme and the Emergence of the Dynamic Page" (2013) is groundbreaking in this regard, as it concerns itself with "Barthelme's page design" rather than narrative analysis, and sees it as exemplary of an "emerging aesthetics of the page that has only become more important as the dynamic space of the computer screen emerged as a fundamental environment for our everyday reading" (115). This media-centered approach is pursued further by Nicole Sierra in "Surrealist Histories of Language, Image, Media: Donald Barthelme's 'Collage Stories'" (2013) and Granit Zela and Jaup Zenuni in "Donald Barthelme's Visual Approach in the Short Story 'At the Tolstoy Museum'" (2015). Such perspectives move away from the postmodernist obsession with text toward a consideration of its materiality, and they reveal Barthelme's multimodal stories that combine textual and visual elements to be meaningful beyond the metafictional effect of such arrangements, so that they resonate with the visual arts as much as with comics.

Perhaps this is, in summary, the reason why Barthelme's stories have aged so well: they are, whatever their particular content or form, stories about their own mediality, about the coherence and incoherence of symbolic forms, and as such their relevance has only increased in a digital media ecology that is fundamentally multimodal and whose rules of coherence and meaning-making have changed with every new medium. In very abstract terms, Barthelme's stories are about the sheer mass of information that may or may not come together to form messages or noise, either because they explore how characters negotiate this intensity or because they force it upon their readers themselves. This intensity has only increased as the analog aesthetics of postmodernism have given way to the digital era of the Internet age, and Barthelme's stories seem as valid, poignant, and pertinent on this side of that electronic divide as they were on the other.

4 Bibliography

4.1 Works Cited

Barthelme, Donald. *Forty Stories*. 1987. New York: Penguin, 1989.
Barthelme, Donald. *Sixty Stories*. 1982. New York: Penguin, 2003.
Chaskes, Daniel. *Beyond Fragmentation: Donald Barthelme and Writing as Political Act*. Dissertation. University of British Columbia, 2012. Web. 25 January 2021.
Dini, Rachele. "The Writing of "Dreck": Consumerism, Waste and Re-use in Donald Barthelme's *Snow White*," https://journals.openedition.org/ejas/11588. *European Journal of American Studies* 11.2 (Summer 2016): 1–17. (25 January 2021).
Eliot, T.S. "The Waste Land." *The Complete Poems and Plays 1909–1950*. New York: Harcourt, Brace & World, Inc., 1952. 38–55.
Eliot, T.S. "*Ulysses*, Order and Myth." *Selected Prose of T.S. Eliot*. Ed. Frank Kermode. London: Faber & Faber, 1975. 175–178.
Gates, David. "Introduction." Barthelme, *Sixty Stories*. xi–xxii.
Gibbons, Alison. "Multimodal Literature and Experimentation." *The Routledge Companion to Experimental Literature*. Eds. Joe Bray, Alison Gibbons, and Brian McHale. New York: Routledge, 2012. 420–434.

Lyotard, Jean-François. *The Postmodern Condition: A Report on Knowledge*. Minneapolis: U of Minnesota P, 1984.
McHale, Brian. *Postmodernist Fiction*. London: Routledge, 1993.
Medvecky, Craig. "Reconstructing Masculinity: Donald Barthelme's 'Unspeakable Practices, Unnatural Acts.'" *Contemporary Literature* 48.4 (Winter 2007): 554–579.
Ionica, Cristina. "The Nagging Father: Donald Barthelme's Narrative Framing of Gender and Family Violence," https://www.tandfonline.com/doi/abs/10.1080/00111619.2013.854194. *Critique: Studies in Contemporary Fiction* 56.1 (2015): 61–76. (25 January 2021).
Punday, Daniel. "Donald Barthelme and the Emergence of the Dynamic Page." *Mosaic: A Journal for the Interdisciplinary Study of Literature* 46.1 (2013): 113–133. Web. (25 January 2021).
Sierra, Nicole. "Surrealist Histories of Language, Image, Media: Donald Barthelme's 'Collage Stories,'" https://www.ingentaconnect.com/content/intellect/ejac/2013/00000032-/00000002/art00005;jsessionid=y70l0hdp7722.x-ic-live-03. *European Journal of American Culture* 32.2 (2013): 153–171. (25 January 2021).
Stengel, Wayne B. *The Shape of Art in the Stories of Donald Barthelme*. Baton Rouge: Louisiana State UP, 1985.
Zela, Granit, and Jaup Zenuni. "Donald Barthelme's Visual Approach in the Short Story 'At the Tolstoy Museum,'" https://www.mcser.org/journal/index.php/mjss/article/view/8085 *Mediterranean Journal of Social Sciences* 6.6 (2015): 200–202. (25 January 2021).

4.2 Further Reading

Couturier, Maurice, and Regis Durand. *Donald Barthelme*. London: Methuen, 1982.
Gordon, Lois. *Donald Barthelme*. Boston: Twayne, 1981.
McHale, Brian, and Moshe Ron. "On Not-Knowing How to Read Barthelme's 'The Indian Uprising.'" *Review of Contemporary Fiction* 11.2 (Summer 1991): 50–68.
Molesworth, Charles. *Donald Barthelme's Fiction: The Ironist Saved from Drowning*. Columbia: U of Missouri P, 1982.
Nel, Philip. *The Avant-Garde and American Postmodernity: Small Incisive Shocks*. Jackson: U of Mississippi P, 2002.
Patteson, Richard Francis, ed. *Critical Essays on Donald Barthelme*. New York: G.K. Hall, 1992.
Roe, Barbara L. *Donald Barthelme: A Study of the Short Fiction*. New York: Twayne, 1992.
Sacido, Jorge. *Modernism, Postmodernism, and the Short Story in English*. Leiden: Brill, 2012.

Susan Farrell

25 Tim O'Brien (1946–)

Abstract: This essay examines the weightiness of war as depicted by Tim O'Brien, the burdens of guilt and shame that U.S. soldiers fighting in Vietnam carried with them. It explores as well ways that soldiers tried to lighten their burdens: by dreaming of home, by adopting tough poses and language, and by pretending a courage they never actually experienced. O'Brien is a writer particularly interested in what happens when decent human beings are placed in untenable situations and when traditional boundaries are transgressed – borders separating moral from immoral behavior, courage from cowardice, savagery from civilization, war from home, men from women, and above all, truth from fiction. His literary style itself challenges traditional narrative expectations as he blurs the line between fact and fiction, comments metafictionally on his own stories, and asserts the power of storytelling and the imagination as more effective avenues into the truth of wartime atrocity than strict adherence to historical fact.

Keywords: Vietnam War Literature, storytelling, heroism, narrative technique, boundary transgression, metafiction

1 Context

Perhaps the most critically acclaimed and widely read writer of literary fiction about the American war in Vietnam, Tim O'Brien is the author of a war memoir, seven novels, over thirty-five short stories, and numerous essays and reviews. Nevertheless, O'Brien is a war writer who insists that his stories are not about war in the conventional sense. In a lecture at Brown University in April of 1999, O'Brien argued that "war stories aren't always about war, per se. They aren't about bombs and bullets and military maneuvers. They aren't about tactics, they aren't about foxholes and canteens. War stories, like any good story, are finally about the human heart" (O'Brien 1999, n.p.). While O'Brien's service in Vietnam gave him material for his fiction-writing, his novels and stories continue to have a broad appeal as they explore the moral choices human beings make when they find themselves in unbearable situations. Striving for an emotional resonance in his work that "makes the stomach believe" (O'Brien 1990, 78), O'Brien's stories offer, in the words of critic Pico Iyer, "incantatory prose" (Iyer 1994, 74) that often reflects metafictionally on the nature of story-telling itself, on the possibilities and limits of the human imagination.

Growing up in small-town Minnesota in the 1950s, O'Brien, like many Vietnam War writers, lived in the shadow of the Second World War. Both of his parents were World War II veterans, his father a chief yeoman in the U.S. Navy who served in the

https://doi.org/10.1515/9783110587647-026

Pacific theater and his mother a Navy WAVE (Women Accepted for Volunteer Emergency Service). Yet, for O'Brien and others who came of age in the 1960s, his generation's war lacked the clear-cut moral urgency of the so-called good war fought by their parents. O'Brien's anti-war sentiment in high school and college, which included campaigning for protest candidate Eugene McCarthy, did not prevent him from being drafted into the Vietnam War immediately after graduating from Macalester College in St. Paul, Minnesota, despite having a four-year graduate school fellowship to Harvard University's Kennedy School of Government waiting for him. While his year of military service as an infantry soldier in Vietnam put his graduate school plans on hold, it did foster O'Brien's writing. His first published book, *If I Die in a Combat Zone, Box Me Up and Ship Me Home* (1973), is a memoir of boot camp and his service experiences, some chapters and vignettes actually written while still in Vietnam. Although well-received, the memoir did not make O'Brien famous; he remained relatively unknown until his novel *Going After Cacciato*, the story of a soldier who walks away from the war and the platoon that chases him all the way to Paris, unexpectedly won the National Book Award for fiction in 1979.

But the work that most fully captured the reading public's imagination was the 1990 collection *The Things They Carried*. A genre-bending book variously referred to as a novel, a short story cycle, a pseudo-memoir, and a meditation on the art of fiction, *Things* follows several members of the fictional Alpha Company during their time in Vietnam and afterwards. Many of the intertwined stories and vignettes that make up the collection were first published as stand-alone pieces, including the two stories considered in this chapter. "The Things They Carried" initially appeared in *Esquire* magazine in August of 1986, while "Sweetheart of the Song Tra Bong" was originally published in *Esquire* in July of 1989. Both stories explore themes common to all of O'Brien's fiction: the futility and cruelty of war, the shame and guilt soldiers feel, the fine line separating cowardice from bravery, the difficulty at getting at and understanding historical truth, and above all, the way that human experience can be transformed through language and storytelling.

2 Close Readings

"The Things They Carried," which serves as the opening story in the collection of the same name, tells two very different stories in two very different ways. The first is the story of the actual items the soldiers of Alpha Company carry. This story explores the emotional weight of wartime experience as it lists the actual physical objects carried by members of Alpha Company, detailing these items precisely, often down to the exact ounce. Readers learn, for example, that jungle boots weigh 2.1 pounds and that "it was SOP for each man to carry a steel-centered, nylon-covered flak jacket, which weighed 6.7 pounds" (O'Brien 1990, 3). This story, in fact, reads almost like a kind of hyper-realism. It is precise and specialized and makes extensive use of military acronyms such as "RTO" and "PRC-25," which are offered without explanation to the reader. Yet, spiraling around this story is another one – the story of the death of Ted Lavender, a soldier shot by a sniper outside the village of Than Khe after briefly separating from the other men to urinate. Lavender's story is first mentioned very briefly, in an almost off-hand way in the subordinate clause of a sentence that focuses on specific items the men carried: "Ted Lavender, who was scared, carried tranquilizers until he was shot in the head outside the village of Than Khe in mid-April" (2). We then immediately return to the lists: "By necessity, and because it was SOP, they all carried steel helmets that weighed 5 pounds" (2). But Lavender's death will be returned to repeatedly in the story, the narrator providing more explicit details each time we circle back to it. This method of narration has the effect, as do the lists themselves, of piling up weight, as the emotional burden of Ted Lavender's death builds for readers. By the end of the story, Lavender's death and the burden of guilt Jimmy Cross carries in response nudge the initial story out of the way and fully take over the narration.

The narrative technique O'Brien uses in this opening story makes sense in a collection that so fully blurs the boundary between fact and fiction. As many critics have pointed out, the book version of *The Things They Carried* obscures the traditional boundary dividing "truth" from "make-believe," "history" from "fiction," right from the beginning. While the title page bears the phrase "a work of fiction," and the copyright page insists on the "imaginary" nature of the work as well, nevertheless the book is dedicated to "the men of Alpha Company [. . .] in particular to Jimmy Cross, Norman Bowker, Rat Kiley, Mitchell Sanders, Henry Dobbins, and Kiowa," as if these characters were real people. The book's epigraph, from *John Ransom's Andersonville Diary*, a Civil War prison memoir, also suggests that the book should be read as "true": "it is commended as a statement of actual things by one who experienced them to the fullest" (n.p.). And, as readers of the full book will soon realize, the collection is narrated by a 43-year-old Vietnam veteran and writer named Tim O'Brien who has previously published a novel called *Going After Cacciato*. These are certainly identifying features of the real-life Tim O'Brien. Yet, this Tim O'Brien narrator also has life experiences that the biographical Tim O'Brien does not. The real O'Brien, for instance, does not have a daughter named Kathleen and never worked at an Armour meat-processing plant the summer after he was drafted to serve in Vietnam. Even more, those who read the collection in its entirety will experience narrator O'Brien frequently telling stories that he insists are true only to subsequently confess to making up or inventing them out of whole cloth. The book uses these metafictional devices to meditate on the meaning of storytelling, and above all, to point out ways that "story-truth is truer sometimes than happening-truth" (179). While the story version of "The Things They Carried," read apart from the rest of the collection, seems less metafictional, more conventional – it uses a fairly straightforward third-person limited omniscient point-of-view and never comments on its own status as a story – still, we can read the dueling narrative technique used in the story as representing the pull between fact and fiction. The very specific detailed lists represent what seems to be a hard-nosed, "factual" way of talking about war, but these lists compete with and eventually give way to story – the fictional tale of Ted Lavender that emerges slowly, that highlights the emotional burdens of war, and that "makes the stomach believe" (78), as the narrator of the later piece, "How to Tell a True War Story," claims any good war story should.

If we read a war story for simple fact, focusing on the lists of physical objects the men carry, the soldiers seem to become mere pack animals, dehumanized as "legs or grunts" whose job is to "hump" things (3), or carry them mindlessly up the mountains and down into the swamps and rice paddies of Vietnam. But just as the story of Ted Lavender's death winds in and around the hard-nosed lists, the seeming impersonality of the grunts is belied as well when the actual physical objects they carry begin to illuminate the character and concerns of the individual soldiers. Henry Dobbins, for instance, a big man, carries extra rations, while Dave Jensen, obsessed with "field hygiene," carries a toothbrush, dental floss, and small bars of soap, and Kiowa, a Native American from Oklahoma City, carries his "grandfather's old hunting hatchet" (3–4). Soon, readers discover that, alongside the physical objects the men carry, lie emotional burdens even heavier: Jimmy Cross's longing for love and home, Ted Lavender's need to get high, Kiowa's inherited distrust of whites. But above all, the soldiers carry the fear, guilt, and shame of war with them. O'Brien illustrates these burdens with a recurring motif of heaviness versus lightness in the story. The emotional heaviness of war is perhaps best highlighted by descriptions of Lavender's death. Portrayed as being scared several times in the story and as using tranquilizers to keep his emotions in check, Lavender is the soldier who carries the fear of war most heavily. This point is driven home when he is described as going down "under an exceptional burden" when he is shot. He carries "more than 20 pounds of ammunition, plus the flak jacket and helmet and rations and water and toilet paper and tranquilizers and all the rest, plus the unweighed fear" (6). As Kiowa repeatedly describes Lavender's shooting, even to the point of infuriating fellow soldier Norman Bowker, he emphasizes this sense of heaviness: "it was like watching a rock fall, or a big sandbag or something – just boom, then down" (6). Lavender's death concentrates the unbearable emotional weight of war into a single image: a human being, lying on the ground, having become, literally, "dead weight" (6).

But if war weighs O'Brien's soldiers down unbearably, they also try desperately to reduce the heaviness of the things they carry. The men literally shed objects as they march, throwing away rations, blowing grenades and mines; but they work to ease the burdens of war in other ways as well. Many of the men dream of women and home. Henry Dobbins "carries his girlfriend's pantyhose wrapped around his neck as a comforter" (10), while the platoon's leader, Lieutenant Jimmy Cross, cannot stop thinking about Martha, his not-quite-girlfriend back home. He reads her letters over and over, imagines going on camping trips with her in the New Hampshire mountains, and he speculates about whether or not she is a virgin. For Cross, Martha represents the innocence of the home front, a safe space he imagines as unsullied by the death and atrocity he has witnessed in Vietnam. Cross's obsession with Martha's virginity – he both wants and does not want for her to be a virgin – indicates his conflicted desires concerning her. Martha figures both as the object of his strange, violent/tender sexual fantasy in which he ties her to her bed but merely strokes her knee all night long, but also as someone he wishes to stay pure, apart and separate from the war. A college English major, Martha never mentions the war in her letters. She seems to exist in a different world than Cross, a world of literature and the imagination. When Cross reads her letters and dreams about her obsessively in his foxhole, he can seem to escape the war for a time. The pebble from the Jersey Shore that Martha sends Jimmy seems "weightless" (8), and when he imagines himself walking barefoot with her along the beach, he "would feel himself rising [. . .] all love and lightness" (9). A similar feeling of rising above the war visits the men on guard duty at night who dream of "freedom birds" (22), the jumbo jets that will bear them home from Vietnam. They imagine themselves "flying," the "weight" falling off as they soar "naked [. . .] light and free" (22) above the war.

But while these dreams of home must remain fantasies for the time being, the men use more immediate techniques to lighten the burden of war as well. The soldiers adopt tough "poses" because, as afraid as they are of dying, they are "even more afraid to show it" (20). This fear, rather than a desire for honor or glory, is what sends them to war in the first place and what spurs their behavior once in Vietnam: "Men killed, and died, because they were embarrassed not to" (21). In order to hide their fear, the soldiers become "actors" who adopt scripted ways of speaking, tones of "irony mixed with tragedy" (20). They tell jokes; they use a "hard vocabulary" to discuss the "terrible softness" of death, they kick corpses, they "cut off thumbs" (20). O'Brien undoes traditional notions of heroism and cowardice during wartime when he writes that the soldiers "were too frightened to be cowards" (22). This line will be reflected in a later story in the collection, "On the Rainy River," when the narrator, a young O'Brien who cannot make himself flee to Canada, concludes with the words, "I survived, but it's not a happy ending. I was a coward. I went to the war" (61). What traditionally might be viewed as heroism in these stories – young men going off to fight in war – instead becomes a manifestation of fear, the fear of being thought a coward. Like many other American writers of the Vietnam War, O'Brien presents decidedly unheroic narratives, in which his male characters long to behave like the heroes of their past, but for whom old ideas about masculine courage, comradeship, and heroism finally disintegrate.

Readers see this disintegration of moral values when the men of Alpha Company, in an attempt to lighten the burden of Lavender's death, burn the village of Than Khe to the ground, destroying the village well and even shooting the local animals. Lieutenant Jimmy Cross, who blames himself for Lavender's death because he had been in a foxhole dreaming of Martha when the shot was fired, especially feels the heaviness of his shame. Martha's weightless pebble transforms into a "stone" that he will carry "in his stomach for the rest of the war" (16). Cross is not really responsible for Lavender's death. Rather, war deaths seem random and unpredictable in the story. After all, Lavender is shot while going about the ordinary business of living immediately after Lee Strunk had just emerged alive and victorious from searching a Vietcong tunnel, one of the most deadly duties a soldier could draw. But, as Cross thinks to himself in the later story, "In the Field": "When a man died, there had to be blame" (177). Blame brings with it the illusion of control. If someone can be blamed for a death, perhaps something can be done differently next time to avoid death. Cross's name suggests that, in Christ-like fashion, he will take on this burden of guilt and blame for the deaths he and the other soldiers witness. But readers suspect that his resolve to be a better commander, to stop dreaming and to practice "strict field discipline," to distance himself from his men and assert authority and strength (25) – in short, to become the conventional heroic commander he longs to be – will not change anything in this immoral, unheroic war. While Cross burns Martha's picture, the larger burning of the village cannot be undone. When he tells the men to "pack up and head west" the next morning, Jimmy imagines that the soldiers "would do what they had always done" (24). The steady rain will only "add some weight" (24).

O'Brien's interest in traditional expectations of masculine heroism and courage during wartime leads to larger questions about gender behavior – about the relationship between war and home, between the frontlines and the home front. The story "Sweetheart of the Song Tra Bong" addresses these concerns as it describes a small U.S. medical detachment outside of Chu Lai, where a soldier named Mark Fossie schemes to bring his high school girlfriend to Vietnam. The fact that Mary Anne Bell arrives on a helicopter along with the "daily resupply shipment" (93) suggests that the men view her as an object, although, unlike the sexualized Asian "mamasans from Saigon" (93) the men first consider importing, Mary Anne is initially figured as the all-American girl-next-door: seventeen years old, fresh out of high school, sporting blonde hair, blue eyes, and a strawberries and cream complexion. She and Mark, who have been "sweethearts since grammar school," imagine a future for themselves straight out of an American middle-class romance: "someday they would be married, and live in a fine gingerbread house near Lake Erie, and have three healthy, yellow-haired children, and grow old together" (94). And sure enough, the sweethearts initially try to carry this dream of domestic bliss into Vietnam. They "set up house" near the camp's perimeter, "moon over" each other and behave like a "pair of high school steadies" (95).

But this is a story about boundary crossing, about borders transgressed, as the opening line hints when the narrator explains that "the stories that will last forever are those that swirl back and forth across the border between trivia and bedlam, the mad and the mundane" (89). Despite the small camp's best efforts to establish boundaries – the medical unit is positioned on a hill and surrounded with "concertina wire [. . .] bunkers and reinforced firing positions at staggered intervals" (91) – readers soon see that such boundaries are artificial and permeable: "even with decent troops the place was clearly indefensible" (91). The physical borders of the camp come to stand in for other kinds of traditional boundaries. While the line separating officers and enlisted men is often clearly demarcated, the camp at Chu Lai does not have commissioned officers, so the soldiers can grow their hair long, neglect polishing their boots, and ignore the usual required "rear-echelon nonsense" (91). It is precisely this lack of officers that allows the scheme to bring in Mary Anne to come to fruition. And Mary Anne herself soon begins to transgress traditional borders as well. Curious about her new environment, she begins visiting the ARVN soldiers along the perimeter of the camp. While these soldiers of the Army of the Republic of Viet Nam are allies to the Americans, the U.S. soldiers traditionally keep to themselves, despising their allies as "useless" and even "outright dangerous" (91). But Mary Anne spends time with the ARVNs, picking up phrases of Vietnamese and learning to cook rice over a sterno can. She even ventures into the Vietcong-controlled village at the foot of the hill, blurring the line between friend and enemy in wartime.

But, of course, the border that Mary Anne most fully transgresses is the one that traditionally separates the genders. While Western tradition holds that women belong at home, that war is a strictly masculine enterprise in which brave soldiers protect the innocent and pure domestic sphere from external threats, Mary Anne makes herself "entirely at home" in the midst of war (96). Eager to learn and unafraid to get her hands dirty, she assists in surgery and learns how to quickly disassemble an M-16 rifle. As Mary Anne increasingly refuses to behave according to conventional gender expectations, her physical appearance seems to change as well. Mark Fossie feels that "her body seemed foreign somehow – too stiff in places, too firm where the softness used to be" (99). Her nervous giggle disappears and her voice even seems to "reorganize itself at a lower pitch" (99). And Mary Anne's vision of the future alters, too. She begins to talk about herself and Mark living together before getting married, not necessarily having three children or the gingerbread house on Lake Erie. As Mary Anne seems to grow increasingly masculine, Mark Fossie, accordingly, takes on stereotypically feminine qualities. At one point, he is described in terms similar to those originally used for Mary Anne: "Just a boy – eighteen years old. Tall and blond. [. . .] A nice kid, too, polite and good-hearted" (100). Traditional gender boundaries become confused, tangled, and transgressed as the story progresses.

When Mary Anne disappears and is later found inside the tent belonging to six Green Beret soldiers, she seems, in true *Heart of Darkness* fashion, to have crossed the boundary separating so-called civilized behavior from savagery. The Greenies are themselves boundary crossers who specialize in unconventional warfare and counter-terrorism, frequently disappearing into enemy territory for days at a time. Geographically positioned "at the edge of the perimeter" (92) of the medical camp, the Greenies also operate at the edge of morally acceptable behavior, their hootch littered with stacks of bones and emitting a stench "like an animal's den, a mix of blood and scorched hair and excrement and the sweet-sour odor of moldering flesh – the stink of the kill" (110). Mary Anne herself has joined this savagery, seeming to move beyond the human as Mark Fossie, Rat Kiley, and Eddie Diamond hear her half singing, half chanting in an incomprehensible language they cannot understand. Mary Anne's boundary crossing appetite, by this point, has grown monstrous. She has perhaps even crossed the border separating sanity from insanity as she describes her desire to "swallow the whole country – the dirt, the death" and have it "inside" of her (111). She no longer recognizes distinctions between herself and others, between herself and the very landscape. Her ego boundaries have completely dissolved. The necklace of human tongues she wears represents not only the barbarism of this war, but the solipsistic nature of the American imperial project in Vietnam – the country's native inhabitants are silenced by foreigners who, as Mary Anne suggests softly, are "in a place" where they "don't belong" (111).

Just as in the larger collection, O'Brien also works to blur the boundary separating fact and fiction in "Sweetheart of the Song Tra Bong." The story takes a metafictional turn when narrative conventions, or the traditional rules of good storytelling themselves, become one more type of border that is crossed. "Sweetheart" is a story with a complex narrative voice. The narrator, presumably the Tim O'Brien figure of the larger collection, tells readers that he heard the story from Rat Kiley, who claims to have witnessed part of it, but who admits to hearing some things "secondhand" or even "thirdhand" (113). Readers, then, are getting the story at least third or fourth-hand. Despite this, Rat initially swears "up and down" (89) to the truth of the story he will tell about Mary Anne Bell. But the narrator also explains that, "among the men in Alpha Company, Rat had a reputation for exaggeration and overstatement, a compulsion to rev up the facts" (89). His fellow soldiers, in fact, routinely "discount sixty or seventy percent" of anything Rat tells them: "If Rat told you, for example, that he'd slept with four girls one night, you could figure it was about a girl and a half" (89). Of course, the joke is that the "fact" is harder to believe than the "fiction"; in reality, it is impossible – no one can sleep with a girl and a half. We must question our usual notions of what is true and what is invented here, as the narrator further informs us that Rat's point is not to be deceitful, but rather that "he wanted to heat up the truth, to make it burn so hot that you would feel exactly what he felt" (89). As elsewhere in the collection, for instance when Mitchell Sanders in "How to Tell a True War Story" admits to narrator Tim O'Brien that he had to "make up a few things" to make his story "true" (77), truth is not necessarily best achieved by simply relating events that actually occurred in real life – what O'Brien calls "happening truth." Rather, a more metaphorical kind of truth – "story-truth" – might be better at communicating what the experience of war felt like to the men who were there and participated in it.

As Rat tells his story about Mary Anne, readers see that the members of Alpha Company have internalized the traditional rules of good storytelling and that Mitchell Sanders, in particular, insists that these rules be followed. When Mary Anne first disappears, Rat stops, looks at Sanders, and asks where the girl was. Without hesitation, Sanders replies, "The Greenies," adding, "No other option. That stuff about the Special Forces – how they used the place as a base of operations, how they'd glide in and out – all that had to be there for a *reason*. That's how stories work, man" (101–102; original emphasis). Importantly, gender boundaries, like storytelling, also work according to traditional, conventional rules, as Rat Kiley makes explicit when he describes Mark Fossie "setting down some new rules" (103) after Mary Anne returns from ambush. And just as Rat is able to stick to the rules of good storytelling for a time, so Mary Anne returns to her expected gender behavior briefly after her talk with Mark, freshly shampooing her hair, dressing in a demure white blouse and navy skirt, keeping her eyes, down, "subdued to the point of silence" (103). But in this story about boundary transgressions, neither Mary Anne nor Rat follow the conventions for long. When Mark threatens to send his girlfriend home, she acts "as if she had come up on the edge of something, as if she were caught in that no-man's-land between Cleveland Heights and deep jungle" (105). It is an edge she will cross, just as Rat's story, near the end, slips over the boundary of conventional narrative.

Sounding much like the Tim O'Brien narrator of the collection as a whole, Rat refuses to tell his story of Mary Anne simply. Instead, he "had a tendency to stop now and then, interrupting the flow, inserting little clarifications or bits of analysis and personal opinion" in his desire to "bracket the full range of meaning" (106). Mitchell Sanders scolds Rat for this habit, telling him that "all that matters is the raw material, the stuff itself," and that Rat shouldn't "clutter it up with [his] own half-baked commentary" (106). But Rat does not listen, and when he stops his story "almost mid-sentence," it infuriates Sanders: "You can't do that [. . .] Jesus Christ, it's against the *rules* . . . Against human *nature*. This elaborate story, you can't say, Hey, by the way, I don't know the *ending*. I mean, you got certain obligations" (112–113). Sanders' insistence on the rules of proper storytelling is linked to the men's expectations about proper gender conduct. When Sanders says, "it's against the rules [. . .] against human nature" itself, he is referring to gender rules as well as narrative rules. Sanders sees both as natural and innate. But Rat insists on the artificiality of gender boundaries, arguing that Mary Anne was not dumb, just young, that she was like all the male soldiers, "a *girl*, that's the only difference, and [. . .] it didn't amount to jack" (97; original emphasis). O'Brien himself, in discussing this story, has said, "My feeling [. . .] is that what happened to me as a man in Vietnam could happen to a woman as well. And the reasons it didn't were reasons of sociology and demography, not a difference in spirit" (Coffey 1990, 60–61). Gender divisions, like the rules of good storytelling, are shown to be conventional – socially constructed rather than natural or inherent. But just as Sanders has trouble accepting an unconventional narrative style, the men in Alpha Company do not know how to 'read' unconventional gender behavior. When Mary Anne no longer fits into their expectations of what women should be like, she at first slips into the category "masculine" for them – as we have seen, Rat describes her voice deepening and her body hardening – then she disappears from the story altogether. Once Mary Anne fully transgresses established social codes – rules codifying not only gender behavior, but racial and national differences, conduct of war, and ethical mores as well – she moves out of the world that the medics at Chu Lai and the soldiers of Alpha Company know and understand and try to cling to despite their chaotic, confusing, and often horrific experiences in Vietnam.

Lorrie Smith argues that the story peters out at this point because "O'Brien cannot imagine an ending for" it (Smith 1994, 36). In her view, Mary Anne exists only to validate traditional gender expectations: "Having appropriated masculine power within a female body," Smith writes, "she is seen as 'dangerous' to the social order" and is thus banished "beyond the periphery of civilization" (36). What Smith fails to point out, however, is that O'Brien does indeed imagine an end for his story; Rat doesn't simply drop the story when Mary Anne disappears from the camp after last being seen in the Greenies' tent. He turns to his imagination, to "speculation" (O'Brien 1990, 113), to provide an ending. Further, Rat's attempt to imagine a conclusion for Mary Anne serves as a model for how readers must approach the larger collection. As the O'Brien narrator tells readers in "The Lives of the Dead," the final story in the book version of *The Things They Carried*, "stories can save us" (225). While the notion that "you had to be there to understand" has become a cliché associated with the Vietnam War and the literature arising out of it, O'Brien's collection questions the accuracy of experiential truth and instead validates imaginative storytelling as a means of communicating war trauma. Readers who themselves have not experienced war must rely on storytelling and imagination as they try to understand the experiences of soldiers at war, just as Rat relies on imagination to finish Mary Anne Bell's story. As a character who refuses to follow rules, to stay where she "belongs," Mary Anne subverts American imperialistic tendencies as well as traditional gender imperatives. By the end of the story, she is even said to have "crossed to the other side" (116). Has she literally joined the enemy, the Vietcong, who are often described as elusive, ghostlike figures in the collection, just as Mary Anne herself is almost glimpsed "sliding through the shadows" (116) after she disappears? Rumors of just such desertions swirled around Vietnam as the war itself seemed to betray traditional boundaries separating good wars from bad, upsetting myths of American virtue, of masculine courage, and of feminine passivity. When the men of Alpha Company are presented with a woman who transgresses traditional gender roles, they are uncomfortable, unsure how to react. But the story of Mary Anne Bell expands the limits of their imaginations.

3 Theoretical Perspectives

As critic Alex Vernon puts it, the book-length collection *The Things They Carried* "appeared on pretty much every 'best book' list of 1990" (Vernon 2016, 18). It was named a finalist for the Pulitzer Prize, and it was awarded the *Chicago Tribune's* Heartland Prize, the French Prix du Meilleur Livre Étranger, and the Melcher Book Award from the Unitarian Universalist Association. The book received glowing reviews from critics as well. Robert Harris, writing in the *New York Times*, hailed it as being "high up on the list of best fiction about any war" (Harris 1990, 8), and Rosellen Brown wrote in the *Massachusetts Review* that the book was "stunningly original in concept and daring in execution" and that "it ought to become a classic" (Brown 1991, 144). Since its publication, O'Brien's collection has become one of the most taught books by a living author in high school and college classrooms. O'Brien has been highly in demand on the lecture circuit, frequently visiting schools and universities across the country, serving as the featured author for convocations, community Big Reads, and similar events. The two stories specifically discussed in this article, "The Things They Carried" and "Sweetheart of the Song Tra Bong," have been critically well-received as well, "The Things They Carried" winning the National Magazine Award for fiction in 1987 and securing inclusion in the 1999 John Updike-edited volume, *The Best American Short Stories of the Century*. "The Things They Carried," in fact, has been the most frequently reprinted O'Brien story, appearing in numerous anthologies of American literature, in short story collections, and in compilations of war literature. Likewise, "Sweetheart of the Song Tra Bong" continues to be a critical and popular favorite. A finalist for the National Magazine Award in 1989, "Sweetheart" has gained more attention from critics as a stand-alone story than any other O'Brien short piece.

When critics discuss the story version of "The Things They Carried," they most frequently consider it as part of the longer volume of the same title. As such, they are particularly interested in questions of genre and storytelling, especially in how the book blurs the line between truth and fiction. As we have already seen, "The Things They Carried," when read as a stand-alone story, lacks some of the metafictional elements of the collection as a whole. But placing the story in its context within the larger book raises some interesting questions. The second story in the collection, "Love," introduces the first-person Tim O'Brien narrator who does not appear in the opening story. In "Love," an older Jimmy Cross visits his former war buddy O'Brien at the writer's home in Massachusetts. The two men reminisce about their wartime experiences with Cross speaking specifically about Martha. When the narrator O'Brien suggests that he'd like to write a story about some of what Jimmy has been telling him (O'Brien 1990, 29), Cross gives his permission but requests that O'Brien make him out to be "a good guy [. . .] Brave and handsome, all that stuff. Best platoon leader ever" (30). Then he adds, "Don't mention anything about –" and O'Brien assures his friend he will not (30). While it is easy to read the opening story, with its third-person narrative voice and hard-nosed lists of the things the soldiers carried, as simply presenting the "truth" of wartime experience, this second story prompts readers to go back and reconsider "The Things They Carried" not as simply presenting truth, but as a constructed narrative, shaped by a specific writer for specific reasons. Is the portrait of Jimmy Cross presented in that initial story, especially his determined resolve to be a good, by-the-books officer at the end, at least partly a product of O'Brien trying to please his friend? And readers might wonder as well about that blank that ends Cross's request. Is the former soldier asking O'Brien not to mention Lavender's death and Cross's supposed responsibility for it? In this case, the writer breaks the promise he makes to his friend, perhaps showing himself to be an unreliable narrator. Or does Cross's blank suggest something more, some traumatic event that narrator O'Brien does indeed omit from this story? In this case, readers might understand that stories can only be partial and incomplete, that the entire "truth" of wartime experience can never be conveyed fully.

The relationship between trauma and narrative itself – hinted at in the exchange between Cross and narrator O'Brien – has been another frequent window into understanding both the specific stories examined in this chapter and O'Brien's larger body of fiction. Mark Heberle, in *A Trauma Artist: Tim O'Brien and the Fiction of Vietnam* (2001), places O'Brien's work at the center of a more general "post-traumatic culture" (Heberle 2001, xiv) that includes war experiences, but also the psychic wounds caused by, in O'Brien's own words, "the things we do for love" (xx). According to Heberle, O'Brien constantly circles around and revisits trauma in his work, "without linear progression or closure" (33). Critic Tina Chen also focuses on trauma in *The Things They Carried*, arguing that the theme of exile and the impossibility of returning home is expressed through repeated emphasis on dismembered or displaced physical bodies. In the story "The Things They Carried," Jimmy Cross's carrying of Martha's pebble in his mouth and his dreams of home suggest his exiled status, which he must acknowledge after Lavender's death when he burns Martha's pictures. In a similar vein, my article, "Just Listen: Witnessing Trauma in Tim O'Brien's *The Things They Carried*" (Farrell 2015) argues that O'Brien's obsessive emphasis on telling, retelling, and especially on listening to stories illustrates the relationship that medical clinicians have observed between narrativization of trauma and the beginnings of recovery. Images of tongues, mouths, and ears in the collection demonstrate that soldiers like Jimmy Cross in "Things" and Rat Kiley in "Sweetheart" long for a listener who will fully understand their traumatic wartime experiences.

Alongside issues of fact and fiction and how best to communicate wartime trauma, critics have also read O'Brien's fiction in terms of American imperialism and the war in Vietnam. In her book *Friendly Fire: American Images of the Vietnam War* (2000), Katherine Kinney argues that American writers consistently portray a war in which Vietnam becomes merely a staging ground for Americans to fight Americans, for a divided nation to play out its own internal drama in a foreign country, with little or no attention paid to the Vietnamese perspective. While

Kinney focuses more fully on O'Brien's earlier novel, *Going After Cacciato* than on *The Things They Carried*, and while she believes O'Brien presents a more complex, ambiguous understanding of the war than some other writers, she nevertheless represents a school of thought that critiques O'Brien for replicating the effects of American imperialism in his storytelling. One way that we see this imperialistic tendency is through writers' repeated use of American frontier mythology – writing about the Vietnamese as "Indians" and the American soldiers as "cowboys." In "The Things They Carried," for instance, several critics, including Marilyn Wesley, have pointed out that Jimmy Cross's resolve at the end of the story is to be a good officer in the mold of a John Wayne-type hero, and that the story concludes with Cross imagining the soldiers "saddl[ing] up" and moving "west" (Wesley 2002, 26). Critics are divided, however, about whether O'Brien uses this frontier mythology in unwitting support of the American imperialistic project in Vietnam or whether he uses it ironically to undermine just such a project. Issues of American imperialism are raised as well in the numerous articles that read Mary Anne Bell in "Sweetheart" as a female version of Colonel Kurtz in *Heart of Darkness*, Joseph Conrad's seminal 1902 novel about European imperialism in Africa that was subsequently re-imagined in the 1979 Vietnam War film *Apocalypse Now*.

O'Brien's depiction of courage and cowardice in wartime has also been the center of theoretical speculation. Christopher McDonough and Carl Horner, for example, examine how O'Brien defines courage in "The Things They Carried" and in the collection as a whole. McDonough compares the Tim O'Brien narrator in the collection to Hector in Homer's *The Iliad*, arguing that both face similar issues of courage and cowardice "posed by the warrior mentality" (McDonough 2000, 23), while Horner (1999) argues that O'Brien challenges traditional notions of courage we have seen in previous war writing. The issue of courage in wartime, of course, is very much related to cultural expectations about masculinity. Many critics have focused on just such expectations in O'Brien's stories, examining his depiction of both masculinity and femininity. Lorrie Smith draws from a line in "The Things They Carried" to provide the title of her influential article, "'The Things Men Do': The Gendered Subtext in Tim O'Brien's *Esquire* Stories," an essay that turned the feminist spotlight on O'Brien's work. In this article, Smith argues that the short fiction that eventually came together to make up *The Things They Carried* silences women and re-enforces traditional masculine views of war and gender. She points out that Martha, in the story "They Things They Carried," is condemned by the narrator for refusing to listen to or understand Cross's wartime experiences, while Mary Anne Bell in "Sweetheart" is presented as "monstrous" because she "dares to appropriate masculine codes of behavior" (Smith 1994, 32). Later critics, however, have challenged this reading, arguing that paying careful attention to the differences between narrator-O'Brien and author-O'Brien in the collection demonstrates that the novel's gender politics are more complex than Smith allows (see Farrell 2015) or even that O'Brien imagines an idealized female reader for his collection (see Smiley 2002).

Finally, critics have frequently compared O'Brien's stories to other pieces of American fiction. While some readers have looked all the way back to classical Greece for comparisons, Alex Vernon, in *Soldiers Once and Still* (2004), reads O'Brien work in the framework of previous American war writers Ernest Hemingway and James Salter. Stephen Kaplan, in his book *Understanding Tim O'Brien* (1995), pays particular attention to O'Brien's stylistic choices, pointing to Joseph Conrad, Ernest Hemingway (↗16 Hemingway), and William Faulkner (↗18 Faulkner) as particular influences and arguing that the rhythmic intensity of O'Brien's sentences, his strategic use of repetition, and the structural integrity of his chapters as stand-alone pieces give his prose a weight and vividness that is haunting to readers. Many recent critics have argued that the emergent genre of American literature about the wars in Iraq and Afghanistan, especially Phil Klay's 2014 National Book Award-winning collection of stories, *Redeployment,* leans heavily on *The Things They Carried* as a literary influence. Still other critics have compared O'Brien's work to other contemporary American writers who do not focus on war at all. Christopher Donovan (2005), for instance, sees O'Brien as a postmodern ironist alongside of Paul Auster, Don DeLillo, and Charles Johnson, while Brian Jarvis (2008) supplies a reading of *The Things They Carried* that appears in a book examining general American fiction and culture in the 1990s. Chapters on O'Brien also appeared in books that focus on American novelists who confront history (see Franklin 1994), on the American short story (see Harrison 2001; Kaylor 2003); and even on ecocriticism (see Norris 2003).

Critics have explored the line separating history-telling from storytelling in O'Brien's work, the role and function of trauma, and the meaning of courage and cowardice. They have debated his depiction of race and gender and argued about whether he is blind to or critical of U.S. imperialism in Vietnam. They have placed his work in the context of war writers in general, of Vietnam War writers more specifically, and of contemporary American writers who do not address war. They have read his works through the lens of recent theories of postmodernism, feminism, and historiography. What seems not open for debate, however, is O'Brien's status in the canon of American literature as he remains our nation's most acclaimed chronicler of the U.S. War in Vietnam.

4 Bibliography

4.1 Works Cited

Apocalypse Now. Directed by Francis Ford Coppola. United Artists, 1979.

Brown, Rosellen. Review of *The Things They Carried*, by Tim O'Brien. *The Massachusetts Review* 32.1 (1991): 144–146.

Chen, Tina "'Unraveling the Deeper Meaning': Exile and the Embodied Poetics of Displacement in Tim O'Brien's *The Things They Carried.*" *Contemporary Literature* 39.1 (1998): 77–97.

Coffey, Michael. "Tim O'Brien: Inventing a New Form Helps the Author Talk About War, Memory, and Storytelling." *Publishers Weekly*, 16 February (1990): 60–61.

Conrad, Joseph. *Heart of Darkness*. London: Blackwood, 1902.

Donovan, Christopher. *Postmodern Counternarratives: Irony and Audience in the Novels of Paul Auster, Don DeLillo, Charles Johnson and Tim O'Brien*. London: Routledge, 2005.

Farrell, Susan. "'Just Listen': Witnessing Trauma in Tim O'Brien's *The Things They Carried*." *Critical Insights: Tim O'Brien*. Ed. Robert C. Evans. Ipswich, MA: Salem P, 2015. 185–201.

Farrell, Susan. "Tim O'Brien and Gender: A Defense of *The Things They Carried*." *CEA Critic* 66.1 (2003): 1–21.

Franklin, Bruce. "Kicking the Denial Syndrome: Tim O'Brien's *In The Lake of the Woods*." 1994. *Novel History: Historians and Novelists Confront America's Past (and Each Other)*. Ed. Mark C. Carnes. New York: Simon and Schuster, 2001. 332–343.

Harris, Robert. "Too Embarrassed Not to Kill." Review of *The Things They Carried*, by Tim O'Brien. *The New York Times Book Review*, 11 March 1990, 8.

Harrison, Brady. "Tim O'Brien." *A Reader's Companion to the Short Story in English*. Eds. Erin Fallon, et. al. Westport, CT: Greenwood P, 2001. 309–316.

Heberle, Mark. *A Trauma Artist: Tim O'Brien and the Fiction of Vietnam*. Iowa City, IA: U of Iowa P, 2001.

Horner, Carl S. "Challenging the Law of Courage and Heroic Identification in Tim O'Brien's *If I Die in a Combat Zone* and *The Things They Carried*." *War, Literature, and the Arts: An International Journal of the Humanities* 11.1 (1999): 256–267.

Iyer, Pico. "Missing in Contemplation." Review of *In the Lake of the Woods*, by Tim O'Brien. *Time*, 24 October 1994, 74.

Jarvis, Brian. "Skating on a Shit Field: Tim O'Brien and the Topography of Trauma." *American Fiction of the 1990s: Reflections of History and Culture*. Ed. Jay Prosser. London: Routledge, 2008. 134–147.

Kaplan, Stephen. *Understanding Tim O'Brien*. Columbia, SC: U of South Carolina P, 1995.

Kaylor, Noel Harold. "Postmodernism in the American Short Story: Some General Observations and Some Specific Cases." *The Postmodern Short Story: Forms and Issues*. Ed. Farhat Iftekharrudin. Westport, CT: Praeger, 2003. 246–266.

Kinney, Katherine. *Friendly Fire: American Images of the Vietnam War*. Oxford: Oxford UP, 2000.

Klay, Phil. *Redeployment*. Harmondsworth: Penguin, 2014.

McDonough, Christopher Michael. "'Afraid to Admit We Are Not Achilles': Facing Hector's Dilemma in Tim O'Brien's *The Things They Carried*." *Classical and Modern Literature* 20.3 (2000): 23–32.

Norris, Nanette. "Imagination and the Civilized Man: Paris in Tim O'Brien's *Going After Cacciato*." *The Image of the City in Literature, Media, and Society*. Eds. Will Wright and Steven Kaplan. Pueblo, Co: Society for the Interdisciplinary Study of Social Imagery, U of Southern Colorado, 2003. 111–115.

O'Brien, Tim. *Going After Cacciato*. New York: Delacorte, 1978.

O'Brien, Tim. *If I Die in a Combat Zone, Box Me Up and Ship Me Home*. New York: Delacorte, 1973.

O'Brien, Tim. *The Things They Carried*. New York: Houghton Mifflin, 1990.

O'Brien, Tim. Transcript of President's Lecture, http://www.stg.brown.edu/projects/-WritingVietnam/obrien.html. Brown University, Providence, R.I. 21 April 1999. (2 February 2021).

Smiley, Pamela. "The Role of the Ideal (Female) Reader in Tim O'Brien's *The Things They Carried*: Why Should Real Women Play?" *Massachusetts Review* 43.4 (2002): 602–613.

Smith, Lorrie. "'The Things Men Do': The Gendered Subtext in Tim O'Brien's *Esquire* Stories." *Critique: Studies in Contemporary Fiction* 36.1 (1994): 16–40.

Vernon, Alex. "Field Notes on *The Things They Carried.*" *War, Literature, and the Arts: An International Journal of the Humanities* 28 (2016): 1–20.
Vernon, Alex. *Soldiers Once and Still: Ernest Hemingway, James Salter, and Tim O'Brien*. Iowa City, IA: U of Iowa P, 2004.
Wesley, Marilyn. "Truth and Fiction in Tim O'Brien's *If I Die in a Combat Zone* and *The Things They Carried.*" *College Literature* 29.2 (2002): 1–18.

4.2 Further Reading

Calloway, Catherine. "'How to Tell a True War Story': Metafiction in *The Things They Carried.*" *Critique: Studies in Contemporary Fiction* 36.4 (1995): 249–257.
Ciocia, Stefania. "Conradian Echoes in Vietnam War Literature: Tim O'Brien's Rewriting of *Heart of Darkness* in 'Sweetheart of the Song Tra Bong.'" *Symbiosis* 11.1 (2007): 3–30.
Clarke, Michael Tavel. "'I Feel Close to Myself': Solipsism and U.S. Imperialism in Tim O'Brien's *The Things They Carried.*" *College Literature* 40.2 (2013): 130–154.
Herzog, Toby. *Tim O'Brien*. Woodbridge, CT: Twayne, 1997.

Robert C. Clark

26 Raymond Carver (1938–1988)

Abstract: Raymond Carver's short stories generate provocative, impassioned arguments about the interplay of form and content. While he is sometimes thought to be the originator of American literary Minimalism, he is in fact a part of its perpetuation. Following in the stylistic legacy of Ernest Hemingway, William Carlos Williams, and Anton Chekhov, Carver wrote carefully crafted tales in part inspired by his own life experiences. "Cathedral" and "A Small, Good Thing" are among his best. Few edits were made to the original published version of "Cathedral" before its appearance in *Cathedral*; "A Small, Good Thing" underwent numerous transformations over a two-year period, changes often initiated by Carver's long-time editor, Gordon Lish. Carver criticism is bifurcated. Scholars have discussed Carver's themes, politics, use of symbol, and stylistic choices, but in the last twenty years, it has become increasingly important to examine Lish's role in the development of Carver's fiction.

Keywords: First-person narrator, iceberg theory, American literary Minimalism, naturalism, impressionism, imagism

1 Context

Raymond Carver maintains an important position in the American short story tradition, particularly in the development of the Minimalist aesthetic that was relatively dominant in the early-to-mid 1980s. His pathway toward this mode stemmed from his admiration of writers such as Ernest Hemingway (↗16 Hemingway), Anton Chekhov, and William Carlos Williams, but it is equally the product of his life experiences, education, and relationship with editor Gordon Lish. In addition to autobiographical essays such as "My Father's Life" and "Fires," biographical accounts of Carver's life have steadily emerged in the last two decades, among them Carol Sklenicka's *Raymond Carver: A Writer's Life* (2009), Maryann Burk Carver's *What It Used to Be Like* (2006), and the valuable chronology compiled in William L. Stull and Maureen P. Carroll's *Raymond Carver: Collected Stories* (2009). James Carver's recent memoir, *Raymond Carver: Remembered by His Brother, James* (2017), is also of interest given the author's stated desire to correct errors related to various assertions about his brother's childhood experiences.

Born in Clatskanie, Oregon, in 1938, Carver was raised in the Pacific Northwest in a working-class family; his father, Clevie Raymond, was an occasional binge-drinker (see J. Carver 2017, 37), and Raymond himself struggled with alcoholism until he achieved sobriety on June 2, 1977. He said that many of his stories focus

https://doi.org/10.1515/9783110587647-027

upon individuals and families struggling financially because "the things that have made an indelible impression on me are the things I saw in lives I witnessed being lived around me, and in the life I myself have lived" (Gregory and McCaffery 1993, 111). Married as a teenager to Maryann Burk in 1957, he matriculated in Creative Writing 101 with John Gardner at Chico State University in 1959, and Gardner proved to be a mentor who shaped Carver's exacting, scrupulous approach to editing. He ultimately finished a B.A. in English from Humboldt State in 1963, spent a brief time as a student in the Iowa Writer's Workshop, but then moved back west in 1964, working "a series of 'crap jobs'" (Carver, Stull, and Carroll 2009, 963). Several years later, he published his first two books, both poetry collections, titled *Near Klamath* (1968) and *Winter Insomnia* (1970).

While Carver's style is rightly at the center of much of the criticism, recent biographical scholarship more thoroughly explores the manner in which his work was edited. In 1967, Carver was introduced to Lish, who as editor at *Esquire* revised and published Carver's story "Neighbors" in 1970. Six years later, while physically ill and in the depths of alcoholism, Carver saw his first major collection of short stories, *Will You Please Be Quiet, Please?*, go into print. He checked into Duffy's, a voluntary treatment facility in Calistoga, California, but continued to drink until he stopped permanently the following year. Carver's sober years, 1977 until his death in 1988, were quite important to him, and he in part attributed the "generosity" of his later fiction to his sobriety. In recent accounts, however, Carver's view of his own stylistic evolution has been complicated by the role of Lish. During the preparation of Carver's third short story collection, *What We Talk About When We Talk About Love* (1981), Lish's excisions were rather significant, leading Carver to write a letter asking that the book be delayed or canceled (see Max 1998, 40; Carver, Stull, and Carroll 2009, 992–996). However, the project moved forward, and while Carver later attempted to distance himself from its severe terseness, *What We Talk About* is considered among his finest works. In 1983, Carver published *Cathedral*, a short-story cycle that is now thought to more accurately represent his unique authorial vision.

Despite the fact that the stories in *Cathedral* are indeed more generous than those found in *What We Talk About*, Carver is appropriately placed among the American literary Minimalists, a group that includes authors such as Amy Hempel, Susan Minot, Weike Wang, and Cormac McCarthy. Carver did not invent the style for which he is known, but he fundamentally strengthened and sustained it in the decades following the death of Hemingway in 1961. Carver's and Hemingway's fiction at times overlaps in content – "Pastoral" and "Big Two-Hearted River," for example – but their most substantive link is stylistic. Carver admired Hemingway's *In Our Time*, calling the collection "marvelous," and he affirmed the influence of Hemingway's iceberg theory. He said in an interview that "Hemingway is an author whose work I admire greatly. [. . .] You're probably familiar with his comment comparing a literary work to an iceberg: nine-tenths of the iceberg is under water. But as long as the writer knows what he is leaving out, that's okay. If he's just writing and doesn't know that he's leaving out crucial things, then that's not so good" (Pope and McIlhenny 1990, 17). Carver's mentioning of Hemingway's idea says much about his personal aesthetic as well as the mode of which he is an integral part. Their connection illustrates that the moment of Minimalism extends from the late nineteenth century to the present.

The iceberg theory is Hemingway's articulation of the means by which a writer generates implication by the act of omission. He first explained his idea in 1932 in his bullfighting treatise *Death in the Afternoon*. At the end of Chapter Sixteen, Hemingway ponders how an author must interact with tradition, specifically when determining which elements of predecessor novels must be synthesized and which must be discarded, and then says that

> if a writer of prose knows enough about what he is writing about he may omit things that he knows and the reader, if the writer is writing truly enough, will have a feeling of those things as strongly as though the writer has stated them. The dignity of the movement of an ice-berg is due to only one-eighth of it being above water. A writer who omits things because he does not know them only makes hollow places in his writing. (Hemingway 1932, 182)

The passage is an exploration of authorial self-effacement. If executed well, a reader is conscious of what is being implied either through dialogue or description, and, additionally, is not distracted by the presence of a didactic voice. Hemingway's use of the term "feeling" is perhaps imprecise; omission generates an intellectual complex.

The generation of implication by omission, or suggestiveness, is one of three central facets of American literary Minimalism. Similar to poetry, the style requires much from readers. Authors often use allusion to build depth, but this technique places the onus upon the reader to both recognize and perform any necessary research. For example, in his Minimalistic 2007 short story "Good People," David Foster Wallace includes brief quotations and misquotations from biblical passages in Galatians and 1 Timothy as a means to complicate the protagonist's ethical struggle. Lane A. Dean, Jr. and his girlfriend, Sheri Fisher, are evangelical Christians facing an unwanted pregnancy. Distraught over their decision to abort, Lane contemplates scripture in bits and pieces, within and outside of its context, which suggests that his intellectual foundation in his religion is perhaps suspect. In order to notice this, however, the reader needs to either know the texts well or double-check Lane's recitations.

Furthermore, readers are allowed considerable freedom in the construction of theme and meaning due to Minimalist narrators' tendency to use direct reportage. In addition to Hemingway, Carver was heavily influenced by turn-of-the-century Russian short story author Anton Chekhov. While his work is not particularly terse, it is Impressionistic, which means that sensory experience is crystallized in image-based moments in time. In a discussion of Hemingway's use of Impressionist techniques, James Nagel asserts that

> the central premise of this approach was an attempt to render in literature the sensory nature of human life, to present to the reader the sensations of a character so graphically that the reader would experience the scene directly, participating in the action on the same epistemological plane as the character. Impressionistic painting represented a single instant of sensory experience; Impressionistic literature also focused on an intense, abbreviated moment of experience, a Vistazo, and used such moments as either entire works or as units from which to build a larger artistic construct. (Nagel 1987, 19)

"A Small, Good Thing" and "Cathedral" are comprised of a series of moments, reported in direct, sensorial language. The Weisses eat warm bread and drink coffee with an erstwhile antagonist, but the meaning of their visceral reactions is left for readers to assemble. As H. Peter Stowell maintains in his study of Impressionism in the work of Chekhov and Henry James (↗12 James), "the impressionists' belief in the primacy of perception forced the pose of the omniscient, didactic, and discursive author out of the work. He was replaced by an elusive presence who allowed all characters to perceive for themselves the ambiguous and ultimately unknowable surfaces of sensory reality" (Stowell 1980, 17). Protagonists in Carver's tales are often able to clearly describe what is happening, but they implicitly acknowledge their own epistemic limitations.

Narrators focus more upon what the senses convey and generally avoid any form of synthesis. Kronegger stresses that Impressionist authors focused more upon style than conveying messages, asserting that "the writer is neither a sensualist nor a thinker; he is primarily a maker. And to paraphrase the poet, an impressionist piece of prose doesn't mean; it is" (Kronegger 1973, 21). This tendency, adopted by Minimalists, is what has led figures such as Wallace, Wyndham Lewis, and John W. Aldridge, among others, to argue that Minimalist works are empty due to a lack of political, social, or ethical stances. In this regard, the mode is a continuation of the work of the early modernists. Ezra Pound and the Imagists were reacting against voluminous British novels narrated by individuals who not only describe the world in minute detail but then offer didactic interpretations; the works of Charles Dickens and George Eliot often reflect these tendencies. Chekhov's "Lady with the Dog" (1899), like many of Carver's stories, is predicated upon the idea that feelings are powerful yet difficult to synthesize. Gurov and Anna Sergeyevna are directed by their senses, yet they are unable to articulate how the feelings they derive from them should inspire their actions. The ending, marked by a pained conversation between the lovers about how to reconcile their infidelities and continue their affair in the open, harkens back to Gurov's first encounter with Anna's home: "In front of it was a long gray fence with inverted nails hammered into the tops of the pilings." He misinterprets the image, thinking that "a fence like that is enough to make anyone want to run away" (Chekhov 1979, 230). The brutal, unromantic wall represents the dreary reality of their state: it is artless, immovable, and ultimately irreconcilable. Anna and Gurov do not "run away" with one another but rather realize that "the end was still far, far away, and that the hardest, most complicated part was only just beginning" (235). The only thing of which Gurov is certain is that time is inexorably progressing. When he sees his reflection, it "strikes him as strange that he should have aged so much in the last few years" (234). He asks himself why a woman half his age should love him, if 'love' is even the correct term, a question that remains unanswered. Chekhov provides no tidy lesson in the end, and in "Cathedral," Carver proceeds under similar epistemic limitations.

Like many of Chekhov's, Carver's short stories tend to depict the human experience as fundamentally entropic. Relationships disintegrate under the weight of extramarital affairs, fishes are unceremoniously chopped in half, and protagonists construct various types of barriers between themselves and others. Concepts such as 'love' hold no consistent meaning, rendering them unstable and open to unceasing debate; this is the primary theme conveyed in "What We Talk about When We Talk about Love." "Cathedral" and "A Small, Good Thing" are thus unusual within Carver's oeuvre. In their final forms, both are among his later tales, fiction described by the author as emanating from a spirit of generosity. However, this generousness is not readily apparent, largely due to the terseness and subtlety of his prose.

The efficiency of Minimalist fiction emanates from a combination of transpacific influence and the turn-of-the-century reaction against the Maximalist tendencies of Victorian English fiction. Carver, Hemingway, Hempel, and other American Minimalist authors share technical sensibilities with haiku poets, a line of influence that flows from Pound, who was deeply interested in Asian poetics, to the Imagists. In his scholarship on haiku and modernism, Yoshinobu Hakutani asserts that Pound was aware of Yone Noguchi's work "as early as 1911," which includes the first poems written in English by a Japanese author (Hakutani 2009, 2–3). Japanese letters also affected the formulation of the modernist aesthetic of implication, a core tenet in austere fiction. Rebecca Suter asserts that the concept of the "objective correlative," a foundational idea that is related to Hemingway's iceberg theory, was not entirely original to Pound and Eliot but "relied heavily on Morikawa Kyoriku and Matsuo Bashō's notions of *toriawase*" (Suter 2008, 29; original emphasis). Minimalist fiction is often prose-poetic, melding forms that are traditionally placed in opposition. Pound's exhortations towards sparseness were not focused solely on poetry but extended to prose as well, and Noguchi, along with British writer T. E. Hulme, helped direct Pound's program.

Recent evaluations of the moment of Minimalism assert that the program Pound et al. established, and that was continued by Carver and others, now serves as the foundation of post-1945 American creative writing programs. The main proponent of this idea, Mark McGurl, argues that Minimalism is derived from an "aestheticization of shame," causing authors to engage in a form of "self-retraction" (McGurl 2009, 301). In light of the inevitable depoliticization that ensues, McGurl asserts that Minimalism lacks racial and ethnic diversity, and is "despite a few exceptions, quite white" (314). This argument denies the valuable contributions represented by Jamaica Kincaid's *Annie John* (1985), Sandra Cisneros's *The House on Mango*

Street (1984) and *Caramelo* (2001) (↗29 Cisneros), and fully erases the aesthetic tradition of Minimalism inspired by Asian poetics and the South American *crónica*. Additionally, it speaks to why it is essential to establish an accurate chronology: literary Minimalism neither begins nor ends with Carver. Drawing from a range of formal approaches, works within the mode ultimately allow readers interpretive freedom, rewarding in particular those willing to explore the complex implications of what may seem on the surface a rather simple narrative.

2 Close Readings

The composition and publication history of "A Small, Good Thing" is remarkably complex, largely due to Carver's fluctuating relationship with editor Gordon Lish. In total, there are four published versions of "Good Thing" and two iterations titled "The Bath," which is the product of major excisions. According to William L. Stull and Maureen P. Carroll, the original narrative, which Carver submitted to Lish for publication in 1980, was reduced from "11,000 words to fewer than 2,500, the largest cut to any story in the manuscript" (Carver 2009, 983–984). Carver's repeated insistences that he be given editorial "autonomy" during the preparation of *Cathedral*, however, render that iteration the most authoritative. In other words, it is the manuscript that best reflects his aesthetic and editorial vision within the context of his collaboration with Lish.

The basic action in "A Small, Good Thing" is not as mundane as other tales in *Cathedral*, such as "Feathers" and "The Compartment," in that it centers upon a hit-and-run car accident that leads to the death of Scotty Weiss, an eight-year-old boy. The narrative opens on a Saturday afternoon with Ann Weiss, the child's mother, ordering a cake for her son's upcoming birthday celebration. Two days later, Scotty is struck while on the way to school. He turns around and heads home but later demonstrates signs that he has internal injuries. His concerned parents take him to the hospital, and doctors are unable to offer concrete reasons as to why the boy is not recovering. During their son's hospitalization, Ann and Howard take turns returning to their house. While there, they receive increasingly menacing phone calls from the baker, who has been left with a rather expensive cake for reasons that have not been communicated to him. After Scotty dies, the Weisses visit the baker to confront him for his painful, untimely calls, and what begins as a rather intense moment transforms into a communal, perhaps restorative conversation over coffee and fresh-baked bread.

The plot of "A Small, Good Thing" is in part derived from events that took place in Carver's own life. In a discussion of the genesis of "The Bath," Jim Naughton explains that

> when Raymond Carver was a boy he was walking to school with a friend who got hit by a car. The boy wasn't hurt, but years later, Carver wondered what if? Several years ago Carver's

> telephone rang late at night. When he answered, the caller hung up. Carver was not disturbed by the call, but he wondered what if the call had come when he was upset about something else? (Naughton 1990, 28)

Carver perhaps conflated this event with a later accident involving his daughter, Christi. While the girl was riding her bike, a driver hit her, causing an injury that required stitches. Christi recalls that she "had a bleeding gash in my calf, but the mother and son who hit me stopped, looked, and then drove on. Dad couldn't get over that people could be like that. We went over that many times" (Sklenicka 2009, 114). While biographical parallels do not necessarily form a part of the metaphorical iceberg, these events do serve to illuminate the emotional complex that resides below the surface of the tale. During the compositional process, Carver seems to have been considering the role of chance, imaginatively exploring how life might have been different for his friend's family if something more had happened. The author also submerges his own moral response to his daughter's accident; Carol Sklenicka observes that the story "says nothing about the driver's motives and yet disperses huge burdens of guilt and anger among its other characters" (115). The dramatic ending sequence does not include a conversation between the parents and the person who has committed the greatest crime. However, it does afford considerable unity, locating the emotional apogee in the same place that the narrative begins.

The most interesting detail to emerge from the initial action, the first conversation with the baker, is that Ann Weiss orders a sixteen-dollar cake. If the tale is set in or around the year it was submitted to Lish, 1980, that amount, adjusted for inflation, has the same spending power as fifty-one dollars in 2020. This indicates that Ann and Howard Weiss are not a part of the working-class Carver is sometimes said to write about exclusively. Among the many alternative names for Minimalism that have circulated over the decades, some suggest that its protagonists are solely people who are struggling financially: perhaps the most direct examples are "K-mart Realism," "White Trash Fiction," and John Barth's "Post-Alcoholic Blue-Collar Minimalist Hyperrealism" (Herzinger 1985, 8). In writing about a middle- or upper-middle class couple, Carver eschews one of the supposed tenets of American Minimalist fiction. Furthermore, the piece supports the idea that Minimalism is best-defined based on stylistic tendencies rather than content. Despite their affluence, the Weisses suffer from many of the same conflicts as other Carver protagonists.

For example, Scotty's accident is tragic, brought about by forces that his parents are unable to control. In other words, the fact that they have money does not allow them to avoid unwanted phone calls or insulate themselves from others. The narrator says that until the accident, Howard's "life had gone smoothly and to his satisfaction – college, marriage, another year of college for the advanced degree in business, a junior partnership in an investment firm. Fatherhood. He was happy and, so far, lucky – he knew that" (Carver 1989, 62; all subsequent references are to the same edition). He is a man who has avoided the "forces" that could "bring a man down if the luck went bad, if things suddenly turned" (62). At this point in the story, Howard feels confident that Scotty will recover, but within the context of the ending, the boy's father in some manner acknowledges his lack of agency. The entropic elements of Carver's stories have led writers such as Russell Banks (see Banks 1991, 101) and Edward Hoagland (see Hoagland 1987, 745) to include Carver among the literary Naturalists. Literary Naturalism is predicated upon what Donald Pizer termed "pessimistic determinism" (Pizer 1984, 9), or the notion that human life necessarily ends tragically. The movement was programmatic. In the tradition of French author Emile Zola, Naturalist authors sought to identify social problems and offer their fiction as both a diagnosis and corrective. "A Small, Good Thing" is typical of Carver's stories in that while it communicates themes about forgiveness and reconciliation, it is not ultimately didactic.

Despite the assertions of Banks, Hoagland, and others, Carver was not part of the Naturalist tradition. The Impressionistic tendencies he adopted from Chekhov and Hemingway coincide more closely with his epistemic objectives. When asked in an interview if he thought literature "can help readers make sense of their lives," Carver responded that "literature can make us realize that our lives are not being lived to the fullest possible extent. But whether literature can actually change our lives I don't know. I really don't know." Echoing the final line of Hemingway's *The Sun Also Rises*, he concluded that "it would be nice to think so" (Sexton 1990, 129–130). When asked more directly if he tried "to avoid making explicit statements" in his work, Carver firmly maintained that he was "incapable of doing that," explaining that "I wouldn't know the first thing about making statements: the peach pickers aren't getting paid enough for picking a box of peaches, so put a character in there who shows that. No, nothing schematic or programmatic" (O'Connell 1990, 143). Certainly, Carver suggests that it is within community that people can begin to address their pain. Even though sensory experiences are complex and articulating the sources of angst and sadness is fraught, he implies that human beings must try to talk.

"A Small, Good Thing" thematically centers upon epistemic limitation, the idea that filtering and translating sensory experience into concrete thought is challenging if not impossible. Dr. Francis is unable to tell the Weisses what is wrong with their son, why the eight-year-old remains in a coma, and the doctor's lack of knowledge ultimately enervates the couple. The baker's phone calls magnify their exhaustion. They are threatening, primarily because the Weisses are unwilling to decipher who the caller is and why he is saying such things. Howard unwittingly enrages the baker by saying that he knows nothing about the cake, dismissively ending the conversation by saying, "Jesus, what are you talking about?" Howard uses the word "Jesus" as an expletive, but if read as a type of address, his reference to the merchant adumbrates the impromptu communion that occurs at the end. Howard's actions in this moment are driven by the worry he feels for his son. Once Howard returns to the hospital, he attempts to comfort Ann by stroking her hair and insisting that Dr. Francis "knows what's what" (Carver 1989, 64). When the doctor does arrive, his language is rather tentative, and he states that Scotty "could be better, I think. But he's all right. Still, I wish he'd wake up. He should wake up pretty soon" (66). Perhaps the most damning line he utters is his assertion that the boy is "out of any real danger, I'd say that for certain, yes" (66). This of course turns out to be wrong, belying the doctor's certainty and calling into question the very possibility of surety.

After Dr. Francis's departure, the Weisses say to one another that they have been praying, suggesting that they do not fully place their trust in medical science. This is consistent with Howard's earlier feeling that "forces" have overtaken the situation and that something must be done to counter what cannot be accounted for rationally. He weakly states that he too has asked God for help, although his agitation over what day and time his act took place intimates that this might not be true. Ann is reassured, however, and the narrator reveals that "for the first time, she felt they were together in it, this trouble. She realized with a start that, until now, it had only been happening to her and Scotty" (68). Implying that one of the omitted elements is that Ann and Howard have for some time been emotionally distant, the speaker adds that now Ann "felt glad to be his wife" (68). Painful circumstances reshape her perception of him, although the togetherness she senses is immediately placed under tension by the knowledge that the radiologist needs more "pictures" and a "scan."

As Scotty's condition declines and the tests continue, Ann and Howard rely upon the language of the body. Their trust in the doctors eroded and their prayers unanswered, Ann and Howard stare out of the hospital window, which is an image of escape. The narrator explains that "they didn't say anything. But they seemed to feel each other's insides now, as though the worry had made them transparent in a perfectly natural way" (70–71). Despite verbal and scientific attempts to reassure, Carver suggests that the only way for the Weisses to cope is through the relatively intangible and untranslatable communication of feeling. Throughout Scotty's hospitalization, the doctor has hesitated to use the word "coma," but he finally accepts it by saying "we'll call it that for the time being" (71), a seeming admission that language is fluid.

Carver deftly undermines the possibility of certainty by continuously developing the theme that one can achieve emotional comfort through communication, which is admittedly limited in its efficacy, as well as empathy. After Howard convinces Ann to return home, to feed the dog and "check on things" (72), she inadvertently begins talking with a "Negro family" waiting for a young man named Franklin to get out of surgery after being "cut" at a party. The unnamed father says that they are "hoping and praying, that's all we can do now" (74). The implication is that there is awkward tension between Ann and the family due to their racial difference, a difference that the narrator goes out of his way to emphasize by identifying the race of the family. However, Ann "had an urge to ask what" words the group is using to pray, and she "wanted to talk more with these people who were in the same kind of waiting she was in. She was afraid and they were afraid. They had that in common" (74). The speaker conveys Ann's sense of division through the use of the phrase "these people" along with her need to confirm that they all have suffering in common. Carver here is constructing the foundation for the ending, suggesting that meaningful connection often requires one to set aside prejudice.

The power of the ending is also amplified by a second menacing call, an action that causes Ann to feel intense fear and regret. Her anger with the baker is entirely justifiable given the early-morning phone call she answers during her respite at home. The man certainly has no idea what has happened to Scotty, and his anger is motivated purely by money and his perception of class-based condescension. He says that he knows something about Scotty, which then leads Ann to call the hospital in a panic; the fact that she purchased a birthday cake is understandably far from her consciousness. She reaches Howard, and he conjectures that the caller might be the "psychopath" driver of the car. Ann tells him that she is "scared to death," to the point that she regrets having a child (79).

The events leading to Scotty's death reflect some of the central tendencies of Naturalist writing, but "A Small, Good Thing" does not conclude entropically. Medical science ultimately fails, and the Weisses are confronted with the knowledge that their son is lost due to a "one-in-a-million" circumstance (80). Scotty dies defiantly, howling "until he had no more air in his lungs" (80). Dr. Francis apologizes profusely but then says that an autopsy is required. The boy's father says that he understands but then breaks down, saying "Oh, Jesus. No, I don't understand, doctor, I can't, I can't. I just can't" (81). Again, introducing deity and the metaphysical into the circumstance, Dr. Francis puts "his arm around Howard's shoulders" and says, "God, how I'm sorry" (82). Despite all of the pain she has had to endure, Ann detects a "goodness" in the doctor, an indication that she has not given in to cynicism. However, as the couple leaves, she looks back at the hospital, an action that echoes the earlier image of the couple looking out of window in the suppressed hopes of an escape. Now on the outside, the building to her represents death and mourning rather than healing.

Carver communicates the inexorable continuation of life despite tragedy, describing Ann while she sits on the couch, makes a series of calls to family, and listens to the "hissing" of the coffee maker. The phone resumes its function as a conduit to family and the outside world until the baker intrudes yet again, stoking a murderous impulse within the forlorn mother. When he calls "just before midnight," Ann says that she would like to "kill him," "to shoot him and watch him kick," a sad imagistic echo to the convulsions her son makes in the moments before his death. She wants revenge. Ultimately, she remembers the cake, which has transformed into a morose representation of birth and celebration and asks Howard to drive her to the bakery. When the couple arrives at the door of the store, Carver creates an interesting moral dichotomy through the use of light. The narrator describes how Ann and Howard see "a big man in an apron" move "in and out of the white, even light" (84). When Ann first addresses her antagonist, she steps "into the light" (85), suggesting that there is righteousness in her anger. She ultimately tells the man what has happened to Scotty, and this leads to a moment of confession.

The baker tells his own story, and this creates a moment of parallel with Ann's earlier recognition that other people and families are confronting harrowing moments. His apology suggests that this experience will for him prove epiphanic: "'Maybe once, maybe years ago, I was a different kind of human being. I've forgotten, I don't know for sure. But I'm not any longer, if I ever was'" (87). Building upon the image of standing in "the light," the baker delivers lines that the Weisses need to hear from the murderous driver of the car more than their erstwhile harasser: "'Forgive me, if you can,' the baker said. 'I'm not an evil man, I don't think. Not evil, like you said on the phone. You got to understand what it comes down to is I don't know how to act anymore, it would seem'" (88). The man then manifests his need for forgiveness through the act of giving, offering the couple coffee and sweet rolls. They stay, eat, and "listened carefully" as the baker talks about himself. Remarkably, they say nothing back to him, and it is in their bodily act of remaining in the kitchen that the narrator communicates that they accept the apology. Like many Minimalist tales, silence proves to be as important as what is said.

"A Small, Good Thing" thus provides a clear delineation between Carver's Minimalist and Naturalist tendencies. Despite their wealth and success, the Weisses must ultimately negotiate and confront unavoidable destructive forces, but their experiences do not teach an overbearing lesson or suggest a broad corrective for a social ill. Much fuller than the elliptical "The Bath," in this iteration Carver suggests that while bad luck cannot be conquered, there is something therapeutic in communing with others. Forgiveness combats internal corrosion. While Ann and Howard do not ultimately confront the actual culprit, the driver of the car who struck their son, they are given an opportunity to listen, empathize, and forgive.

"Cathedral" is Carver's finest short story. First published in *The Atlantic Monthly* in September of 1981, it changed in small yet significant ways before appearing as the last tale in *Cathedral*. Typical of Minimalist fiction, little happens. Because it is a first-person retrospective, the current action is the telling. The speaker explains how he was initially upset with the notion that a blind man, Robert, whom his wife has known for a longer period, was coming for a visit. Buried beneath his pejorative comments about the visitor are his own inadequacies; he resents that a person who is capable of intimacy was going to be in his home. He talks to his wife about the upcoming visit, indicating through rude questions to her that he is not going to be a receptive host. When Robert ultimately arrives, the speaker resents that he is not included in his wife's process of catching up. The narrator recalls that after eating and smoking marijuana late into the night, he and Robert watched and listened to a television show about cathedrals. Robert could not visualize the buildings and the speaker could not fully describe them, so they joined hands and drew one together.

Although the encounter occurs in the past, the narrator remains unable to articulate or synthesize how he was changed. Even with the benefit of hindsight, he does not elaborate on what he meant when he said, "it's really something" (Carver 1989, 228; all subsequent references are to the same edition). Much like the end of Chekhov's Impressionistic "The Grasshopper" and "Lady with the Dog," the final line of "Cathedral" offers no resolution.

This is not to suggest, however, that the narrative approach does not offer insight into the omitted yet implied elements. Much of the scholarly debate about "Cathedral" has focused upon whether the unnamed narrator experiences an epiphany, but in the courses of their various discussions, critics tend to miss or ignore the fact that the tale is a first-person retrospective narration. The speaker communicates how he has been transformed, but it is subtly conveyed in his use of past tense. In the opening paragraph, he says that he "wasn't enthusiastic" about Robert's prospective arrival, that he "was" unknown, and that his blindness "bothered" the narrator (209). The implication is that these things are no longer true; perhaps the story is being told because Robert is soon going to visit again, and the speaker is looking forward to seeing his friend. While the teller has not completely abandoned his bigotries and prejudicial attitudes, the suggestion is that he has become a more empathic human being.

The opening paragraph also provides details about what the narrator previously valued, further conveying how his present self is different from his past self. He says that "in the movies, the blind moved slowly and never laughed. Sometimes they were led by seeing-eye dogs" (209). His fear was that Robert would prove to be an obstruction, costing him time, testing his impatience. Additionally, he believed that the visitor would be mirthless and serious, perhaps qualities amplified by the fact that, like the Weisses in "A Small, Good Thing," he is in mourning. The need for a seeing-eye dog implies that the speaker assumed that Robert would be dependent, proving to be what he perceived as an awkward burden. However, the teller's admission that he drew information from movies suggests a newfound self-awareness, an acknowledgement that in the past he relied on fantastical or imagined sources of information. In the time since Robert's visit, he has gained something more valuable: truth gleaned through bodily, sensory experience.

The narrator also, in retrospect, affirms the validity of what his wife said to him before and during Robert's visit, suggesting that the evolution of his vision extends to the way he views his partner. He leans upon her authority to give credence to his own narrative. While recounting her time working for Robert in Seattle, the speaker seems to think that his listeners are implicitly questioning the veracity of the story, or that they perhaps suspect he is conjecturing. As if repeating a question that has been posed, he says, "how do I know these things? She told me" (210). He believes her implicitly, and he suggests that his audience should, as well. In his current state, he sees that what she affirmed to him was consistently correct. The tone of the telling suggests that this is a state of mind that the narrator has come to accept, a way of seeing that he has earned through introspection. At the time of Robert's arrival, she was unsure of his love: "'if you love me,' she said, 'you can do this for me. If you don't love me, okay. But if you had a friend, any friend, and the friend came to visit, I'd make him feel comfortable'" (212). This is not playful banter, and it is no doubt painful for the speaker to recollect being told that he was friendless: it indicates that his wife did not consider him a friend, either. His willingness to repeat what she had said about him, details about his life that are perhaps embarrassing, demonstrates a recognition, or admission, that he needed to change.

Also important, however, is that the narrator changes little in respect to the way he views sensory information. Despite the fact that he is unable to ultimately process his experiences into a concrete message, he learns through his body. He contemplates that Robert and his late wife, Beulah, "married, lived and worked together – had sex, sure. [. . .] All this without his ever having seen what the goddamned woman looked like." Recalling the viewpoint of his previous self, he admits that "it was beyond my understanding" (213). Again, the key word is "was," and his use of past tense suggests that he has moved closer to some form of comprehension. He has not abandoned his reliance upon the primacy of the senses, but he has progressed. The speaker is epistemically consistent, and his failure to comprehend Robert's love for a woman he had never seen is echoed by a scene he describes shortly after. After he mockingly asks Robert which side of the train he sat on, the narrator recalls being left to interpret his wife's gaze: "My wife finally took her eyes off the blind man and looked at me. I had the feeling she didn't like what she saw. I shrugged" (215). This marks the beginning of his psychic tension, a state mitigated by the prayer and dinner that follow.

Throughout "Cathedral," Carver communicates the narrator's past and present unease through stylistic shifts. In contrast with Ann Weiss's earnest desire to learn which words to use to ask a higher power for her son's recovery, the speaker recalls using the pre-dinner prayer as another opportunity to make a joke. When he says, "now let us pray," his partner again "looked at" him, "her mouth agape" (217). Despite again being called upon to interpret her expression, he proceeded without thought. His brief, mocking plea that "the phone won't ring and the food doesn't get cold" subtly reflected his precarious mental state: he wants to comfort himself by eating food while simultaneously being excused from having to interact. His prayer is answered rather directly: "We dug in. We ate everything there was to eat on the table. We ate like there was no tomorrow. We didn't talk. We ate. We scarfed. We grazed that table. We were into serious eating" (217). While he was uncomfortable with the rituals required for friendship, like the Weisses, he learned that eating in a time of duress can lead to emotional generosity.

Even though he has progressed in his capacity for intimacy and empathy, the speaker recalls these events with a lingering tension. Like Hemingway does in stories such as "Big Two-Hearted River," Carver implies mental unease through passages written in brief, direct sentences. The diction tends to be repetitious, suggesting that the power of the recalled moment leads to a lexical narrowing. Early in the narrative, the narrator recalls that his wife contacted Robert after she had stopped working for him, saying "she called him up one night from an Air Force base in Alabama. She wanted to talk. They talked. He asked her to send him a tape and tell him about her life. She did this. She sent the tape. On the tape, she told the blind man about her husband and about their life together in the military" (211). The multiple uses of the word "tape" suggests that it is an object viewed and contemplated from multiple perspectives. To the speaker it represents intimacy, a past of which he is not a part, and a personal vulnerability he was unable to undergo.

The most important implication in "Cathedral" is that, through his collaboration with Robert, the protagonist has learned to better understand his wife. His experience teaches him something important about intimacy, that in order to construct a meaningful relationship with another human being he must adopt the same level of openness that Robert possesses. While Carver did not make major changes to "Cathedral" before including it in the eponymous collection, he did omit important details, and these omissions suggest that drawing a cathedral with Robert transformed the narrator's view of his spouse. Early in the narrative, the speaker mocks his wife for attempting to write poetry. He describes how when they first met, she showed him a work that she wrote after Robert touched her face, and the narrator concludes that "maybe I just don't understand poetry. I admit it's not the first thing I reach for when I pick up something to read" (210). On a superficial level, his resistance seems a product of jealousy: another man has touched his wife's face and she has documented her reactions. However, these events are also indicative of the protagonist's unwillingness to engage with his partner in a manner that is uncomfortable. In the version of the story that appeared in *The Atlantic*, however, the speaker attempts and fails to explain what he is seeing on television but frames his description in terms that correlate with the opening. At one point he says that the stone structures "reach way up. Up and up. Toward the sky. They soar. *They're like poetry, that's what they're like*" (Carver 1981, 28; omitted line italicized). Given his earlier resistance to poems, it is interesting that he chooses terms that demonstrate that he is not only attempting to reach out to Robert in a meaningful way, he is also forming a tacit connection to his wife. He has abjured his spirit of rejection.

In addition to his use of past tense, perhaps the most remarkable evidence of an epiphany is the speaker's willingness to share this story in the first place. He is, in effect, creating a "tape" of his own, recording what has happened even if he leaves much of the synthesis to his audience. Following a pattern established by his partner, he adopts a greater openness to vulnerability. The epistemic limitations implicit to the narration, coupled with the generation of implication by omission and sentence-level efficiency, demonstrate that Carver did not abandon the Minimalist aesthetic following *What We Talk About* but rather wrote stories that prove conciseness and brevity do not equate to superficiality.

3 Theoretical Perspectives

Cathedral was received with interest by reviewers, and their comments largely focused upon what they perceived as a gloomy tone and muted style. Several comment directly on "A Small, Good Thing" and "Cathedral." Writing about the collection as a whole, Jeff Clark asserts that "these stories are not happy, but their people find the means to go on when the rules of their small worlds have been changed" (Clark 1983, 1720). In her review for the *Wall Street Journal*, Helen Dudar attributes the stylistic brusqueness not to the tenets of the Minimalist tradition but to communicative failure, maintaining that "Carver country is an emotionally blighted landscape. Nobody connects; everyone is verbally disabled, conversing in commonplaces gruesomely inadequate for expressing pain and anger" (Dudar 1983, 28). James Kauffman, with the benefit of some hindsight, refers to Carver as a "major talent of our time" and "the cartographer of standard lives" (Kauffman 1984, B12).

Other readers were ahead of the scholarly reaction, writing about themes that would become central to the ongoing conversation about the collection. Referring to the ending of "A Small, Good Thing," J. I. M. Stewart calls the conversation between the Weisses and the baker a "strange sacramental conclusion" (Stewart 1983, 17), a description that frames nicely the idea that the story is, in part, about the potential for redemption. Much of the discussion about "Cathedral" explores whether the narrator experiences an epiphanic moment; J. J. Maloney writes that "the husband is transformed when he tries to describe the cathedral to the blind man, finally closing his own eyes to help the man draw one on paper" (Maloney 1983, 61). A number of scholars refer to Carver's increasing level of "generosity" in the stories that comprise *Cathedral*, and it is a term that is thought to signify that these are works about characters who possess the emotional capacity to change.

When the narrator of "Cathedral" states near the end that he did not feel that he was "inside of anything," it suggests that he has perhaps gained a newfound freedom. In *The Stories of Raymond Carver: A Critical Study*, Kirk Nesset articulates one of the central binaries found in *Cathedral*: insularity and freedom. Characters occupy liminal spaces, and Nesset argues that each attempts to gain some form of independence, but in some cases with little success. Positioning the collection within Carver's oeuvre, Nesset concludes that "in *Cathedral* Carver has traversed a great distance; as in no other volume, characters connect with one another and, however briefly, come away changed. Still, such momentary connections do not reflect the tone of the books as a whole" (Nesset 1995, 70). J. Gerald Kennedy discusses this same theme, arguing that in "Cathedral," "A Small, Good Thing," and other stories in the collection, "Carver appears to reaffirm the importance of communal relations and against the alienation and escapism of bourgeois middle-class life – to project unexpected images of understanding, helpfulness, and compassion" (Kennedy 1995, 203). Kennedy's focus largely resides on discussing *Cathedral* within the context of what he terms the "short story sequence," or books comprised of narratives that are independent yet linked by content or style or both.

The intersection of content and style has long been the locus of Carver criticism. His artistic legacy has proven to be rather complicated, but he has for the last two decades maintained extraordinary relevance, in part due to ongoing, impassioned arguments about Minimalism. Adam Meyer (1989) asserts that Carver was a Minimalist in only his second collection, *What We Talk About When We Talk About Love*. In *Technique and Sensibility in the Fiction of Raymond Carver*, Arthur F. Bethea maintains that the individual stories within the collection vary stylistically, and that "while there is impressive evidence in *Cathedral* of his having begun to transcend the haplessness and brittle restraint that commonly besets his characters, Carver by no means forsakes the 'down to the marrow' aesthetic that governed his earlier collections" (Bethea 2001, 125). Manuscript evidence opened over the past decade-and-a-half suggests that Carver was not as prone to cut material as Lish, complicating assessments of Carver's "true aesthetic."

The genesis of the Lish-Carver "controversy" is perhaps traceable to D. T. Max's "The Carver Chronicles," published in the August 9, 1998 edition of *New York Times Magazine*. Max begins his account asserting that Lish had said in private that "he had changed some of the stories so much that they were more his than Carver's" (Max 1998, 34). In a 2015 interview, Lish affirmed Max's claim, saying that he "wrote" Carver's stories (Lish and Winters 2015, 98). While he no doubt made significant cuts to Carver's tales in the preparation of *What We Talk About*, Lish's claims of authorship are rather hyperbolic. Editors at times recommend drastic changes to manuscripts, but this does not lead to shifts in authorial assignation. For example, F. Scott Fitzgerald suggested a number of cuts and changes to Hemingway's *The Sun Also Rises*, but this does not somehow mean that he wrote or co-wrote the book.

Given the wave of articles and book sections that now promulgate the idea that Carver was essentially a victim of Lish – Enrico Monti's "Minimalism, Dirty Realism, and Raymond Carver" and Wells Addington's "Will You Please Be Edited, Please: Gordon Lish and the Development of Literary Minimalism" among them – readers and critics will now have to at least consider "Raymond Carver" as a figurative literary consortium. *What We Talk About*, cuts and all, is an excellent short story collection. Lish's contribution to the development and sustainment of a valid literary mode should not be dismissed or overlooked purely because Carver did not prefer the "Minimalist tag." Carver himself was prone to cutting material. He was by no means a Maximalist, and it is rather difficult to accept Monti's argument that his stories carried weighty, moralistic themes before being subjected to Lish's excisions. Carroll and Stull's publication of the *Beginners* manuscript, however, allows readers to decide for themselves.

Buried in the Max article is a trenchant statement by Carol Polsgrove, who maintained that writing is implicitly collaborative, or "a social act" (Max 1998, 51), intimating that Lish's revelations should have little effect on Carver's legacy. Max then concludes that Polsgrove's "view is becoming more widespread" and that "a new form of literary analysis, genetic criticism" (51) might provide a way forward. Genetic criticism was not "new" in 1998; French scholars had begun to use this approach in the mid-to-late 1970s. Carver scholars are ultimately left with a series of stories and manuscripts perhaps best synthesized under the auspices of *la critique génétique*. The objective is not to decide upon a "definitive text" but to be able to better understand Carver's aesthetic development and creative process. Biography, editorial influence, and the multiple published versions of his tales should be considered within both temporal and atemporal contexts. On the one hand, comparing the *Beginners*, *What We Talk About When We Talk About Love*, and *Cathedral*, iterations of "A Small, Good Thing," along with those that appeared in literary journals, indicates much about Carver's evolution as both a writer and as a human being. However, readers who have worked through the entire arc of the life of the tale must negotiate coexisting, at times contrasting, narratives. "The Bath" shifts all emphasis away from the redemptive and communal elements present in the fuller pieces, yet it amplifies thematic elements common among all the variants. For example, Carver suggests that mundane objects like a phone or doorbell can become tools of menace. "Cathedral," however, seems a much more direct reflection of Carver's late aesthetic sensibilities, indeed a story with a greater spirit of generosity in content but still remarkable for how much it communicates by implication.

4 Bibliography

4.1 Works Cited

Addington, Wells. "Will You Please Be Edited, Please?: Gordon Lish and the Development of Literary Minimalism." *The CEA Critic* 78.1 (March 2016): 1–23.

Banks, Russell. "Raymond Carver: Our Stephen Crane." *The Atlantic* 268.2 (1991): 99–103.

Bethea, Arthur F. *Technique and Sensibility in the Fiction of Raymond Carver*. New York: Routledge, 2001.

Buzbee, Lewis, and Mona Simpson. "Raymond Carver." *Conversations with Raymond Carver*. Eds. Marshall Bruce Gentry and William L. Stull. Jackson: UP of Mississippi, 1990. 31–52.

Carver, James. *Raymond Carver: Remembered by His Brother James*. London: Austin Macauley, 2017.

Carver, Maryann Burk. *What It Used to Be Like: A Portrait of My Marriage to Raymond Carver*. New York: St. Martin's Griffin, 2006.

Carver, Raymond. *Carver: Collected Stories*. Eds. William L. Stull and Maureen P. Carroll. New York: Library of America, 2009.

Carver, Raymond. *Cathedral*. New York: Vintage Contemporaries, 1989.

Carver, Raymond. "Cathedral." *The Atlantic Monthly* 248 (1981): 23–29

Carver, Raymond. "A Small, Good Thing." *Ploughshares* 8.1–2 (1982): 213–240.

Carver, Raymond. "On Writing." *Call If You Need Me*. Ed. William L. Stull. New York: Vintage Books, 2000. 87–92.

Carver, Raymond. “Fires.” *Call If You Need Me*. Ed. William L. Stull. New York: Vintage Books, 2000. 93–106.
Carver, Raymond. “My Father’s Life.” *Call If You Need Me*. Ed. William L. Stull. New York: Vintage Books, 2000. 77–86.
Chekhov, Anton. “The Lady with the Dog.” *Anton Chekhov’s Short Stories*. Ed. Ralph E. Matlaw. New York: Norton, 1979. 221–235.
Clark, Jeff. “Cathedral.” *Library Journal* 108.15 (September 1, 1983): 1719–1720.
Dudar, Helen. “A Storyteller in Touch with the Angels.” *Wall Street Journal* (November 14, 1983): 28.
Gregory, Sinda, and Larry McCaffery. “An Interview with Raymond Carver.” *Raymond Carver: A Study of the Short Fiction*. New York: Twayne Publishers, 1993. 98–114.
Hakutani, Yoshinobu. *Haiku and Modernist Poetics*. New York: Palgrave Macmillan, 2009.
Hemingway, Ernest. *Death in the Afternoon*. Harmondsworth: Penguin Books, 1932.
Herzinger, Kim A. “Introduction: On the New Fiction.” *Mississippi Review* 40–41 (1985): 7–22.
Hoagland, Edward. “A Symposium on Contemporary American Fiction: Edward Hoagland.” Part 1 of 2. *Michigan Quarterly Review* 26.4 (1987): 60–63.
Kauffman, James. “Between Soft Covers.” *The Christian Science Monitor* 76 (October 1984): B12.
Kronegger, Maria Elizabeth. *Literary Impressionism*. New Haven, CT: College and UP, 1973.
Kennedy, J. Gerald, “From Anderson’s *Winesburg* to Carver’s *Cathedral*: The Short Story Sequence and the Semblance of Community.” *Modern American Short Story Sequences: Composite Fiction and Fictive Communities*. Ed. J. Gerald Kennedy. Cambridge: Cambridge UP, 1995. 194–215.
Lish, Gordon and David Winters. “An Interview with Gordon Lish.” *Critical Quarterly* 57.4 (2015): 89–104.
Maloney, J. J. “Cathedral: Stories.” *The Saturday Review* 9 (September/October 1983): 61.
Max, D. T. “The Carver Chronicles.” *The New York Times Magazine* (9 August 1998): 34–40, 51, 56.
McGurl, Mark. *The Program Era*. Boston: Harvard UP, 2009.
Meyer, Adam. “Now You See Him, Now You Don’t, Now You Do Again: The Evolution of Raymond Carver’s Minimalism.” *Critique* 30 (Summer 1989): 239–251.
Monti, Enrico. “Minimalism, Dirty Realism, and Raymond Carver.” *Critical Insights: Raymond Carver*. Ed. James Plath. Ipswich: Salem P, 2013.
Nagel, James. “Literary Impressionism and *In Our Time*.” *Hemingway Review* 6.2 (1987): 17–26.
Naughton, Jim. “As Raymond Carver Muses, His Stature Grows.” *Conversations with Raymond Carver*. Eds. Marshall Bruce Gentry and William L. Stull. Jackson: UP of Mississippi, 1990. 28–30.
Nesset, Kirk. *The Stories of Raymond Carver: A Critical Study*. Athens: Ohio UP, 1995.
O’Connell, Nicholas. “Raymond Carver.” *Conversations with Raymond Carver*. Eds. Marshall Bruce Gentry and William L. Stull. Jackson: UP of Mississippi, 1990. 133–150.
Pizer, Donald. *Realism and Naturalism in Nineteenth-Century American Literature*. Carbondale: Southern Illinois UP, 1984.
Pope, Robert, and Lisa McElhinny. “Raymond Carver Speaking.” *Conversations with Raymond Carver*. Eds. Marshall Bruce Gentry and William L. Stull. Jackson: UP of Mississippi, 1990. 11–23.
Sexton, David. “David Sexton Talks to Raymond Carver.” *Conversations with Raymond Carver*. Eds. Marshall Bruce Gentry and William L. Stull. Jackson: UP of Mississippi, 1990. 120–132.
Sklenicka, Carol. *Raymond Carver: A Writer’s Life*. New York: Scribner, 2009.
Stewart, J. I. M. “Other Things.” *London Review of Books* 30 (November 24, 1983): 16–17.
Stowell, H. Peter. *Literary Impressionism: James and Chekhov*. Athens: U of Georgia P, 1980.

Suter, Rebecca. *The Japanization of Modernity: Murakami Haruki between Japan and the United States*. Cambridge: Harvard UP, 2008.
Wallace, David Foster. "Good People." *The New Yorker* (5 February 2007): 66–69.

4.2 Further Reading

Clark, Robert C. *American Literary Minimalism*. Tuscaloosa: U of Alabama P, 2014.
Facknitz, Mark A. R. "'The Calm,' 'A Small Good Thing,' and 'Cathedral': Raymond Carver and the Rediscovery of Human Worth." *Studies in Short Fiction* 23 (Fall 1986): 287–296.
Johnston, Taylor. "'Inside Anything': The Evacuation of Commodified Space in Raymond Carver's 'Cathedral.'" *The Raymond Carver Review* 5/6 (2017): 13–34.
Leypoldt, Günter. *Casual Silences: The Poetics of Minimal Realism from Raymond Carver and the New Yorker School to Bret Easton Ellis*. Trier: WVT, 2001.
McSweeney, Kerry. *The Realist Short Story of the Powerful Glimpse: Chekhov to Carver*. Columbia: U of South Carolina P, 2007.

Frank Obenland

27 Alice Walker (1944–)

Abstract: Alice (Malsenior) Walker is generally recognized as one of the most influential voices in twentieth-century African American women's literature. Recent criticism has concentrated on Walker's achievements as poet and as the writer of such seminal novels as *Meridian* (1976) and *The Color Purple* (1982). This chapter, though, will focus on two early short stories included in the collection *In Love and Trouble: Stories of Black Women* (1973). Building on earlier criticism that relates these stories to her essayistic and autobiographical reflections on the possibility of a distinctly Black female literary tradition, this essay will discuss the stories "Everyday Use" and "To Hell with Dying" as Walker's exploration of the Black female artist as the determining factor in the production, definition, and reception of African American storytelling.

Keywords: Black feminism, quilting, vernacular tradition, folk culture, blues, storytelling, materiality, scriptive things, womanism

1 Context

Prompted for a summary of her artistic and intellectual "preoccupations" in an interview with John O'Brien in 1993, Alice Walker offered the following explanation:

> I am preoccupied with the spiritual survival, the survival *whole* of my people [sic]. But beyond that, I am committed to exploring the oppressions, the insanities, the loyalties, and the triumphs of black women. [. . .] For me, black women are the most fascinating creations in the world.
>
> Next to them, I place the old people – male and female – who persist in their beauty in spite of everything. (O'Brien 331)

As stated here, Walker's vision not only reflects her personal views and interests, but also exemplifies the objectives of an avant-gardist "renaissance of black women writers" that crested in the 1970s and early 1980s (Washington 1981, 2–3). Part of this diverse and vibrant movement, Walker worked alongside African American novelists, dramatists, poets, and essayists Toni Morrison, Toni Cade Bambara, Sonia Sanchez, bell hooks, Audre Lorde, Gloria T. Hull, Michelle Cliff, Paule Marshall, Gayl Jones, Nikki Giovanni, Ntozake Shange, and Sherley Anne Williams. At the time, Walker helped to uncover a genealogy of Black female writers regularly rendered invisible not only by mainstream white feminist literary criticism, but also by the revolutionary Black Arts Movement of the 1960s and 1970s. In her efforts of reconstructing a line of black women writers Walker reached back to Phillis Wheatley's late

https://doi.org/10.1515/9783110587647-028

eighteenth-century poetry and turned to Zora Neale Hurston's modernist and anthropological writings of the early twentieth century (↗17 Hurston). The recuperative work of Alice Walker and others also involved a re-examination of established critical views about African American literature. By coining the idea of "womanism," Alice Walker joined literary critics Barbara Smith and Mary Helen Washington in thinking through the contours of a Black feminist aesthetic that includes the exploration of new textual strategies and genres, an embrace of Black vernacular tradition and language, as well as the creation of new images of Black womanhood. From the Black feminist perspective of these writers, "queer" Black sexualities promised radical alternatives to heterosexual and racialized social conventions that have shaped Black female identities since the era of slavery.

In a series of autobiographical and critical essays written between 1966 and 1982, Alice Walker provides her readers with the personal, intellectual, and social contexts for her work as writer and activist. Later published in the collection *In Search of Our Mothers' Gardens: Womanist Prose* (1983), these mostly autobiographical essays introduce readers to personal details and formative experiences during her childhood and youth. In her early writings, Walker repeatedly emphasizes the importance of a strong emotional attachment to her mother, as well as the strains of a severe eye injury that blinded her at age eight. Born into a family of sharecroppers in 1944, she attended Spelman College in Atlanta and later graduated from Sarah Lawrence College in Bronxville, New York, where she also became involved in the protests and activism of the civil rights movement. The traumatic experience of an early pregnancy and abortion inspired *Once,* her first volume of poetry published in 1968. While her interracial marriage with Melvyn Leventhal, a white lawyer and civil rights activist, did not last, the years between 1966 and 1972 saw Walker working as a political activist and teacher in Mississippi and Georgia. Walker's formative period as critically acclaimed author of poetry, short fiction, essays, and her first novel in 1970 coincided with her discovery of the work and life of Zora Neale Hurston. At a time, when her own work earned the praise and recognition of Langston Hughes. After briefly teaching at Wellesley College, where she introduced a class on African American women writers, Walker became contributing editor to *Ms.* magazine and published her second novel *Meridian* in 1976. Her career ultimately culminated in the early 1980s with the publication of her most successful novel *The Color Purple* (1982) for which she received the Pulitzer Prize for fiction. With the publication of her essay collection *In Search of Our Mothers' Gardens* in 1983, which won the American Book Award, Walker engages in a form of authorial "self-fashioning" placing before her readers a public persona with a distinctively feminist aesthetic and political agenda (Lauret 2000, 3).

The essays collected in *In Search of Our Mothers' Gardens* have proven particularly important for the subsequent reception and interpretation of her narrative fiction. Readers and critics have become attuned to the thematic connections between Walker's essayistic reflections on her development as a Black female artist and her portrayal of female characters in her novels and short stories. At the very beginning of the volume, Alice Walker famously introduces the idea of an African American "womanist" author and intellectual in the form of a dictionary entry. Lifting the term from a vernacular expression, Walker describes a "womanist" as someone who is characterized by "outrageous, audacious, courageous or *willful* behavior" and a readiness to overcome social restrictions on knowledge, personal growth, and cultural self-determination (Walker 1983, xi; original emphasis). For Walker, the self-determined outlook of a "womanist" also includes assuming responsibility for the well-being of a whole community. In this inclusive and collective sense, a "womanist" is defined, among other things, as follows:

> A woman who loves other women, sexually and/or nonsexually. Appreciates and prefers women's culture, women's emotional flexibility (values tears as natural counterbalance of laughter), and women's strength. Sometimes loves individual men, sexually and/or nonsexually. Committed to survival and wholeness of entire people, male *and* female. Not a separatist, except periodically, for health. (xi; original emphasis)

When she metaphorically explains that "[w]omanist is to feminist as purple is to lavender," Walker presents "womanism" as an extension and response to the feminist movement of the 1960s and 1970s (xii). Obviously alluding to the title of her novel *The Color Purple*, Walker's metaphorical use of colorful and flowery imagery in both instances establishes a transtextual relationship bridging her essays and her narrative fiction and invites readers to consider both texts alongside each other.

These resonances between Walker's essays and narrative fiction underline the importance of her womanist aesthetic as a conceptual framework also for understanding her narrative fiction. Despite the strong autobiographical overtones of her artistic views, Walker also emphasizes the importance of role models and exemplary figures that provide orientation and inspiration, particularly for artists working in isolation. In her essay "Saving the Life That Is Your Own: The Importance of Models in the Artist's Life," Walker takes her cue from Toni Morrison when she argues that the Black female artist "must be her own model as well as the artist attending, creating, learning from, realizing the model, which is to say, herself" (8). Given the paucity of established female role models, Walker maintains that learning about other female Black artists can literally have a life-saving function for gifted but marginalized women of color (14). Next to Kate Chopin (↗11 Chopin), Virginia Woolf, and Flannery O'Connor (↗19 O'Connor), Walker cites Phillis Wheatley and Zora Neale Hurston as the most influential models for Black women who struggle to salvage their "contrary instincts" from the contempt of a hostile social environment (235).

In addition to these literary predecessors, Walker also cherishes generations of "grandmothers and mothers" and their spiritual and artistic sensibilities. In the collection's signature essay "In Search of Our Mothers' Gardens," Walker discovers a vernacular tradition of Black female artists that has flourished in the liminal space proximate yet outside the domestic and agricultural spheres and its respective chores (231–243).

Exemplified by the "creative spark" of her own mother Minnie Lou Walker, this vernacular artistic tradition is characterized, on the one hand, by the importance of oral storytelling. As she recalls, "through years of listening to my mother's stories of her life, I have absorbed not only the stories themselves, but something of the manner in which she spoke, something of the urgency that involves the knowledge that her stories – like her life – must be recorded" (240). On the other hand, the ornamental dimension of art sometimes offers a screen for social and economic hardships – "even my memories of poverty are seen through a screen of blooms," Walker observes (241). Most importantly, however, in her mother's gardening Walker finds a first role model for the self-determined expression of an individual's vision of beauty.

Central to Walkers' conception of the vernacular Black female artist is not only her arduousness, endurance, and individual vision, but also the role she plays for the community. For Walker, bestowing a "heritage" and "legacy of respect" on subsequent generations is the artist's most important task. In this sense, womanist art is not only inherently defiant and "contrary," but also has a decidedly social and communal function. Since a womanist aesthetic includes a sense of "providing" for others, womanist art necessarily has an ethical dimension, in an artistic as well as an intellectual sense. In her essays, Walker suggestively speaks of accidental and intuitive moments that establish emotional as well as intellectual connections between individual voices across generations. Speaking about her re-discovery of Zora Neale Hurston, for example, Walker sees her as a prescient "model" who "had provided [. . .] a nearly complete record of her life" for her to explore and imitate (12). The importance of identifying with an artistic ancestor also explains the importance of (auto-)biographical references and echoes in her narrative fiction.

In addition to her autobiographical essays, Walker's early short fiction depicts female protagonists whose portrayals often echo her personal meditations on the role and function of the Black female artist for the African American community. This is especially true for the two stories "Everyday Use" and "To Hell with Dying," collected in her first volume of short fiction *In Love & Trouble: Stories of Black Women*, published in 1973. As the following discussion of these two early short stories will show, it is here that Walker offers more complex explorations of how forms of female intersubjectivity are established and regulated in and through artistic practices and objects. As these stories also illustrate, for Walker, the female protagonists of these stories participate in traditional African American folk culture that includes such seminal art forms as clothing, cooking, quilting, oral story-telling, and the blues. It is around these artistic practices that interpersonal relationships, communities, and a "womanist" aesthetic are created, nurtured, and sustained.

2 Close Readings

From her 1973 volume of short stories *In Love & Trouble: Stories of Black Women*, Walker emerges as the pre-eminent "apologist and spokeswoman for black women" from the American South (Washington 1979, 133; Davis 1994). Experimenting with different narrative structures and storytelling techniques, Walker's early stories privilege the point of view of her female protagonists. In "Roselily," for example, the female protagonist's interior monologue during a wedding ceremony is contrasted with and framed by the officiating minister's voice. The story "Really, *Doesn't* Crime Pay" (original emphasis) consists of fragments from a heroine's journal, a technique that allusively echoes the narration of Charlotte Perkins Gilman's "The Yellow Wall-Paper." For Barbara Christian, these authorial choices suggest that the award-winning collection of stories is organized around the "contrariness" of these female characters who "strain against their restraints" (Christian 1985, 34). A typical protagonist of these stories is characterized by her "insistence on challenging convention, on being herself, sometimes in spite of herself" (Christian 1985, 33). As for example in the story "Her Sweet Jerome," these women's rebellion against social conventions can also result in considerable self-harm or even increase a protagonist's isolation. Often confronted with mental and physical violence, these female protagonists nevertheless defend their individuality by striving "to be whole rather than a defect of nature or [. . .] nonexistent" (Christian 1985, 45). Particularly in "Everyday Use" and "To Hell with Dying" the individual's aesthetic sensibility and attachment to the artistic traditions of African American folk culture promise a remedy for overcoming the emotional distance, conflict, and alienation that regularly mark the personal relationships of these "contrary women."

Taken together, the stories included in *In Love & Trouble: Stories of Black Women* articulate ideas similar to those of Alice Walker's essays on a womanist aesthetic in African American literature. In her stories, Walker has also repeatedly turned to quilt-making as a central metaphor for describing artistic creation (Christian 1985, 85). For Walker quilt-making describes a vernacular process of artistic creation in which "patterns of clarity, imagination and beauty" are formed "out of seemingly disparate, everyday materials" (85). Understood in terms of quilt-making, a womanist aesthetic allows not only for the expression of an artist's individual voice, but also includes a vision of collective healing and wholeness (85). As a collective form of artistic creation, quilt-making originates from the historical experience of slavery and bondage as well as economic necessity and brings to light a history of "communal bonding that confounds traditional definitions of art and the artist" (Baker and Pierce-Baker 311). This "sorority of quiltmakers" constitutes a collective of Black female artists who preserve a "sacred wisdom" for future generations in the material objects they create (309). Also in the stories "Everyday Use" and "To Hell with Dying," material objects such as quilts or musical instruments are crucial for regulating social and familial relations. In both stories, however, a full and "wholesome" integration of the female protagonists in the African American community always remains at least partially incomplete, and underlying conflicts, contradictions, and antagonisms remain unresolved.

The story "Everyday Use" features three female characters who clash over the quilts made by generations of mothers and grandmothers to preserve the memory of their ancestors. Handed down from one generation to the next, Mrs. ("Mama") Johnson and her daughters Dee and Maggie agree that these quilts constitute the most precious parcels of the family heirloom. In the course of the story, however, the women demonstrate different views on how the family tradition should be maintained and continued. The older daughter, Dee, considers these quilts as objects of art that need to be saved from the "everyday use" in the family household. Instead, Dee wants to use these and other material objects to decorate her home and expose the artistry and history inscribed in these quilts to public view. Her mother, however, had promised the same quilts as a wedding gift to her younger daughter Maggie. As Barbara T. Christian has rightly pointed out, the conflicting outlooks of the three women raises the more fundamental question of how to preserve and honor not only an individual family's history, but also the "heritage" of African American folk art and culture in general (86–87).

Even though the story has been rightly recognized as being centrally concerned with quilting as a trope for Black female art, the artistic process of quilting itself only plays a marginal role in "Everyday Use." While the story underlines the importance of these quilts for mediating and containing the family's history, these material objects also throw into stark relief the conflicting views of Mrs. Johnson and her daughters about the preservation of the family's "heritage." As Mary Helen Washington insightfully remarks, "all three women characters are artists" in that they exemplify different attitudes and outlooks (Washington 1994, 101). "Mama" Johnson uses her power as storyteller to provide the story with narrative and argumentative closure. From her perspective, her daughter Maggie embodies the authentic quilt-making tradition. Maggie is seen as a guarantee that the quilts will retain their relevance and practical function in everyday life. While the reader can recognize Dee as one of Alice Walker's contrary women whose interests and views are excised from the story-world, the mother-narrator's decision to exclude Dee from the family tradition only represents the partial point of view of "Mama" Johnson. Told from the perspective of "Mama" Johnson as the story's homodiegetic narrator and internal focalizer, the story not only shows how she comes into her own rights as storyteller, but also how her storytelling exemplifies and examines the authority and responsibility that she now carries.

In the story Walker uses an unorthodox narrative structure for her narrator's account. Different narrative parts and temporal strands become woven together in the manner of "quilt-making with words" (Cowart 1999, 21). Before the actual family reunion is described, the story opens in the present tense with a paragraph describing the narrator and Maggie sitting in their yard expecting Dee's visit. The narration is then interrupted by a direct address to the reader in which Mrs. Johnson offers a dream fantasy in which she imagines her reunion with Dee in terms of a TV show hosted by Johnny Carson (Walker 1973, 48). Briefly moving back to the present tense, the narrator then offers a flashback to the family's history and her own life story told in the past tense, before the narration continues to describe the events around Dee's visit. During the visit, the narration then suddenly shifts from the mother-narrator's use of the simple present to the simple past. This change occurs at the very moment when Dee denounces her first name and tells her mother about her newly adopted African name "Wangero Leewanika Kemanjo" (53). Here it becomes clear that Dee/Wangero refuses to return on her mother's terms. She defiantly redefines her symbolic place in the family and expresses a general skepticism and distrust of English as the dominant language of an inherently oppressive culture. While Mrs. Johnson initially complies with Dee/Wangero's new name, the temporal shift illustrates the growing distance between mother and daughter.

The temporal change in the narration also indicates that Mrs. Johnson fully emerges as female storyteller. Abandoning the original immediacy in the vernacular tradition of African American storytelling, Mrs. Johnson's voice becomes a more conventional "narrative voice" in a register reminiscent of the American short story tradition. As a consequence, "Mama" Johnson thus overcomes her initial passive behavior which critics have taken to be marked by "acquiescence and defeat" as well as "her own erasure" (Hirsch 1989, 203). In addition, Mrs. Johnson begins to assume full narrative control over the story and emerges as "a wayward artist" who casts her story in the mold of "a narrative frame which creates distance while at the same time [. . .] it makes possible a closeness, a coming home" (Whitsitt 2000, 453). In the act of narrating her story, "Mama" Johnson moreover betrays the search for a distinct and self-determined artistic vision. At the beginning of the story, her artistic sensibility is only partially developed when she describes her front yard as a liminal space just beyond the domestic space of the home where the individual is also exposed to a spiritual and aesthetic dimension beyond everyday experience. Her description of the yard conveys a "meaningful simplicity" and "spiritual serenity" reminiscent of Japanese Zen gardens while its "stark vacuity" also "evokes the minimalist lives of poor people" and "suggestive of spiritual wealth" (Cowart 25). From Mrs. Johnson's description, the yard also shows a remarkable similarity to the artistic and creative accomplishment described by Walker in her account of her mother's gardening skills ("In Search"). In particular, "Mama" Johnson's view of the "elm tree" and her longing for "the breezes that never come inside the house" symbolize the promise of sharing with others an aesthetic experience that has not yet been fully realized.

"Mama" Johnson's emergence as a storyteller in her own right also prepares for the epiphanic moment at the end of the story when the first-person narrator is transformed for a second time when she directly intervenes and prevents Dee/Wangero from taking possession of the family's quilts. This decision follows from a moment of epiphany which Mrs. Johnson describes as a revivalist awakening "just like when I'm in church and the spirit of God touches me and I get happy and shout" (Walker 1973, 58). Mrs. Johnson's epiphany consists of her realizing that Maggie might lack the resistant quality necessary for developing a wholesome identity. When the mother-narrator looks at her daughter, she sees an acquiescent individual who responds "with something like fear" to her overreaching sibling. With her bulging lips and her "dopey hangdog look," her appearance calls to mind the most sinister racial stereotypical images of Black womanhood (58). Confronted with this image of her daughter, "Mama" Johnson distances herself from this passivity when she declares that "[t]his was *Maggie's* portion. This was the way *she* knew God to work" (58; emphasis added). This visual impression of her daughter awakens the narrator to the necessity of assuming responsibility for her. Literally taking matters into her own hands, "Mama" Johnson's yanking away of the quilts substitutes for what she perceives as the lack of a spirited resistance in Maggie. By handing her the quilts, the narrator performs a ritual of initiation in which Maggie is recognized as a female artist of her own within the African American folk tradition. More importantly, it seems that "Mama" Johnson also re-defines Maggie's passivity as a form of wayward resistance and "contrariness." At the end of this initiation ritual, Maggie becomes transfixed by her mother's action who recounts that "Maggie just sat there on my bed with her mouth open" (58).

Many readings of the story have concentrated on "Mama" Johnson's decision as a symbolic gesture to defend the authenticity of traditional quilt-making against an overreaching, urbanized outsider (Winchell 1992, 80–81). However, such readings overlook Alice Walker's own sympathies for Dee (Tuten 1993; Farrell 1998) as well as the mother's efforts to maintain an ironical distance to her daughter without completely rebuffing her (Hirsch 1989, 203). In addition, the story also suggests that the quilts assume the function of narrative agents that influence the characters' actions. In fact, their presence at the pivotal and transformative moment in the story gestures at a form of material agency. Building on Bruno Latour's distinction between mediators and "intermediaries" (Latour 2005, 38–39), Rita Felski has usefully suggested to consider such material actors not as "a self-authorizing subject" but rather as "an independent agent who summons up actions and orchestrates events" (Felski 2011, 582–583). Kindling "Mama" Johnson's spirit of resistance, the quilts in the story are not important because of their representational qualities, but because of their material presence that mediates the interaction of the human agents in the story. The quilts can thus also be understood in terms of what Robin Bernstein has defined as "scriptive things." As a particular form of material agency, a scriptive thing "broadly structures a performance while simultaneously allowing for resistance and unleashing original, live variations that may not be individually predictable" (Bernstein 2009, 69). These scriptive agents mobilize human agents to take action and thus elicit a bodily response or reaction. They also realize unspoken ideas and tacit knowledges. Similar to the Althusserian hailing of an individual, material agents can also draw human agents into a "scripted scenario" so that "the individual is interpellated into ideology and thus into subjecthood" (73). In "Everyday Use" the epiphanic moment that leads "Mama" Johnson to entrust Maggie with the family quilts results in Maggie's initiation into the vernacular tradition of Black female artists. This is made possible by the material objects that have been created and curated for them.

Emerging from the story as an artist figure herself, "Mama" Johnson's decision also appears to be an ethical gesture and an instance of "providing" for Maggie. After Dee/Wangero leaves, Maggie and "Mama" Johnson revel in the intimacy of the moment when they are sitting in the yard "just enjoying, until it was time to go in the house and go to bed" (Walker 1973, 59). Mrs. Johnson, who has previously identified more with her masculine traits than with her role as mother, now accepts her nurturing role when she embraces and tugs Maggie to her. "Mama" Johnson finally enjoys Maggie's company when they chew snuff sitting on the porch and look after Dee/Wangero driving away. A self-conscious storyteller, Mrs. Johnson now arrives at a different form of Black womanhood that allows her to leave behind a more precarious sense of self. Earlier in the story, "Mama" Johnson has expressed her insecurity about her masculine behavior marking her as backward and queer (48; 50–51). Society's normative expectations about female looks and behavior had also seeped into her fantasy of appearing on the Johnny Carson Show. In the fantasy space of the TV show, her mediated body is transformed into a more female appearance with lighter skin color and her intellect matches that of the TV show host. However, the narrator herself interrupts this fantasy by reminding herself of the racialized codes of Southern society when she remarks on the unlikeliness of "looking a stranger white man in the eye" (49). At the end of the story she has finally overcome her initial self-doubts and embraces a more resistant outlook on life. She cherishes the newly-found sorority with her younger daughter and her ancestors, and she has finally become the custodian of the family's artistic heritage, both in the sense that she has assumed responsibility of watching out for Maggie, but also in the sense of protecting the family's heritage.

However, "Everyday Use" also problematizes how Mrs. Johnson's story presents Maggie as an exemplary "role model" for nurturing a legacy of womanist respect and creativity. For one, Maggie's behavior in the story does not always match Mrs. Johnson's perception of her being shy and reclusive (Tuten 1993, 127). In addition, Mrs. Johnson's turning away of Dee/Wangero also illustrates the inherently conflictive and antagonistic character of maintaining a female vernacular tradition. From the very beginning, Mrs. Johnson and Maggie repudiate Dee/Wangero's choice of a male companion. The only male character in the story, he is repeatedly ridiculed because of his name and his Muslim religion (54–55). Him leaving with Dee/Wangero at the end of the story can also be understood as a political renunciation of the Black Arts movement with its cultural nationalism and separatist agenda. More importantly, however, what gets lost in reading Dee/Wangero's expulsion from the family homestead as the defense of an authentic cultural tradition is the importance of nurturing the possibility of and respect for "a ludic play of differences" as a part of that tradition (Whitsitt 2000, 446). Significantly, Dee/Wangero's insistence on a difference between her and the sisterhood formed between her mother and sister is mirrored by "Mama" Johnson's creation of a similar distinction by establishing narratorial control over the story. Dee/Wangero's expulsion symptomatically points at the exclusionary vision and potentially "violent" gesture of "Mama" Johnson's storytelling.

In contrast to "Mama" Johnson's efforts to brand her daughter Dee/Wangero as an outcast, the story provides several instances that gesture towards the possibility that the "wayward" sister indeed shares important qualities with her mother and sister. Before she leaves, she affectionately kisses Maggie which signals her continuing emotional attachment while she insists on a different rationale about the quilts. Also earlier in the story, her mother had readily acknowledged that Dee/Wangeroo was not only a self-determined and "contrary" woman, but also a potential artist who had developed her own "style" to express herself:

> Dee wanted nice things. A yellow organdy dress to wear to her graduation from high school; black pumps to match a green suit she'd made from an old suit somebody gave me. She was determined to stare down any disaster in her efforts. Her eyelids would not flicker for minutes at a time. Often I fought off the temptation to shake her. At sixteen she had a style of her own: and knew what style was. (Walker 1973, 50)

Here it becomes obvious that "Mama" Johnson had been unsuccessful in containing Dee's individual tastes before. Maggie is much more susceptible of Mrs. Johnson's tastes when she subjects herself to her mother's normative gaze, asking "How do I look, Mama?" at the very beginning of the story (49). Even though "Mama" Johnson does not approve of her older daughter's taste, she appreciates the individualism and self-determination that comes along with it. This becomes obvious from her remarks on the flashy clothes Dee/Wangero wears during her visit:

> A dress so loud it hurts my eyes. There are yellows and oranges enough to throw back the light of the sun. I feel my whole face warming from the heat waves it throws out. Earrings gold, too, and hanging down to her shoulders. Bracelets dangling and making noises when she moves her arm up to shake the folds of the dress out of her armpits. The dress is loose and flows, and as she walks closer, I like it. (52)

In "Mama" Johnson's account, this appreciation of Dee/Wangero's defiance and self-assertion is paired with an uneasiness of this "contrariness." The narrator, for example, is still suspicious about her daughter Dee potentially having burned down her older home. An incident which can also explain Maggie's reservation about her sister since she had been scarred in the fire. It is because of her individual, wayward, and self-determined views that Dee/Wangero is finally excluded by her mother and sister.

Her final alienation not only illustrates the dangers of her struggle for intellectual independence, but also indicates that she indeed shares qualities that are considered seminal in the tradition from which she is excluded. When Dee/Wangero grabs the quilts and holds them away from her mother, "Mama" Johnson observes that her daughter treats the quilts as if "[t]hey already belonged to her" (57). The narrator's commentary gives away an ambiguity: while Dee/Wangero appears to be overreaching and possessive, her claim is not completely illegitimate. What Dee/Wangero certainly overlooks, however, is the danger of turning these quilts into lifeless objects of art. Their consecration as "art" hanging in a museum or decorating her apartment denies their original function as "scriptive things" capable of forming and nurturing a genealogy of Black female artists. For this reason, Dee/Wangero in the end is prevented from participating and sharing in the "enjoyment" of Maggie and her mother – who, according to the idiomatic expression, are "up to snuff." Considering the complex interaction of the three female protagonists, the story expands and complicates Alice Walker's discussion of the Black female artist in her essays and interviews. As Mary Helen Washington observes, the story's "opposition between the two daughters [. . .] mirrors her own internal struggles" about making the oral tradition of African American folk culture the *sujet* of her literary texts (Washington 1994, 102). Reading Alice Walker's fictional characters against her discussion of the Black female artists in her essays and interviews, "Everyday Use" elaborates on this autobiographical dimension by focusing on the artistic sensibilities exemplified by the story's protagonists. The story's ultimate lesson would then be that Dee/Wangero's failed attempt of appropriating and objectifying the family's quilts is not illegitimate because it violates the authenticity of African American folk culture. A careful reading shows that Dee/Wangero, who consistently defends her personal integrity and artistic convictions, exemplifies a resistant and resilient quality which is definitely a necessary ingredient of Walker's womanist aesthetic. However, the story also shows that Dee/Wangero cannot be seamlessly integrated into a womanist tradition of Black female art precisely because of her "contrariness."

Alice Walker's first published short story "To Hell with Dying" shares a number of structural and thematic features with "Everyday Use." The story is also told by a homodiegetic female narrator who returns to her Southern home to witness the death of an old friend and blues singer, Mr. Sweet. Similar to Dee/Wangero's visit in "Everyday Use," the narrator returns to her family and community as an intellectual outsider who is called away from completing her dissertation at a university in Massachusetts. Upon her arrival, she visits her dying friend – "a diabetic and an alcoholic and a guitar player" (Walker 1973, 133) – and engages in a ritual of calling him back from the brink of death. During her childhood she had regularly participated in similar ritualistic "revivals" of Mr. Sweet when she and her siblings were called to his deathbed. The children usually succeeded in preventing Mr. Sweet from departing by fondling and tickling the old man. Now visiting her nonagenarian friend for the last time, the narrator comes too late to recall her dying friend from whom she receives a guitar as a farewell gift. At the end of the story, the narrator evolves into a female blues musician singing “Sweet Georgia Brown,” one of her friend’s favorite tunes.

As Alice Walker confided in an essay written several years later, she wrote the story on the day when an actual old friend of her family similar to Mr. Sweet was buried (Walker 1988a). Walker was away at college at the time and could not attend the funeral. In her essay, Walker suggests that the story had an autobiographical background and was written when she recovered from a bout of depression in her early twenties. As Alice Walker also suggests, her friendship with this semi-fictional old blues singer/friend, from whom she receives the “gift of creativity,” also introduces her to the African American folk culture which she also found embodied in Langston Hughes’ work who had included “To Hell with Dying” in his anthology *Best Short Stories by Negro Writers* (1967). Despite the early critical recognition of the autobiographical dimension of Walker’s early short story (Winchell 1992, 12–13), later criticism has been less generous in its evaluations. In comparison to the more experimental narrative forms in other stories, Alice Hall Petry has characterized “To Hell with Dying” as “very conventional in terms of structure, characterization, and action” and has criticized the portrayal of the aging artist friend as “sentimentalized ‘old darky’ character” (Petry 1989, 24). This criticism overlooks, however, one of the central concerns of the story which is – similar to “Everyday Use” – an effort in exploring the relationship between an individual artist and her community. The central theme of the story is the female narrator’s emotional attachment not only to the blues singer, but particularly to the blues and more generally to the African American oral tradition. Singing “in the old, sad, sweet, down-home blues way” (Walker 1973, 136), Mr. Sweet expresses such common blues themes as unrequited love, social isolation, and thwarted economic opportunities. As the narrator’s use of flowery imagery in the story’s concluding passage makes poignantly clear, the blues also has a redemptive function for the artist as well as the community. For Walker, the blues has to be considered alongside the female artistry of quilt-making and the “mothers’ gardens.” When the narrator observes that she starts playing the “old guitar” and “humming” the tune of the love song “Sweet Georgia Brown,” she discovers that “[t]he magic of Mr. Sweet lingered still in the cool steel box” (138). Emphatically declaring her love to the old musician is thus not only a quasi-autobiographical confession on the part of Alice Walker (Winchell 1992, 12–13). It can also be understood as signaling an important change in the oral tradition of the blues when it is now a female blues singer lamenting the loss of a beloved person in her song.

The story’s ending thus signals the conclusion to a development that has begun earlier in the story when the younger narrator describes her experience of Mr. Sweet singing the blues. From early childhood, the narrator has shared her family’s veneration and respect for the old blues singer and also from her earliest years her emotional attachment to the artist has been mediated through his music.

> He liked to play 'Sweet Georgia Brown,' that was what he called me sometimes, and also he liked to play 'Caldonia' and all sorts of sweet, sad, wonderful songs which he sometimes made up. It was from one of these songs that I learned that he had had to marry Miss Mary when he had in fact loved somebody else (now living in Chi-ca-go, or De-stroy, Michigan).
> (Walker 1973, 132).

Later in the story, at one of Mr. Sweet's customary death-bed "revivals," the narrator agrees to her adoption into the blues tradition by returning the affective attention she has been given: "I held his woolly head in my arms and wished I could have been old enough to have been the woman he loved so much and that I had not been lost years and years ago" (132). These mutual exchanges of a nurturing and respectful relationship thus exhibit the "near-tragic, near-comic lyricism" which Ralph Ellison has discovered in the blues as "an autobiographical chronicle of personal catastrophe expressed lyrically" (Ellison 2003, 129).

Another important feature of what Houston A. Baker has called the cultural "matrix" of the blues is its collective and ritualistic nature (Baker 1984, 3). In "To Hell with Dying" this ritualistic dimension of African American folk culture is illustrated by the series of deathbed revivals which all basically follow the same structure. Overcome by depression, Mr. Sweet would regularly lie down to die so that the narrator's family is customarily summoned to the dying artist. While the children's father would repeat his formula of "To hell with dying – these children want Mr. Sweet," the children, usually the youngest, would start to fondle the old man and would revive him by enticing him to smiles and laughs. This suggests that the blues singer, as artist in the vernacular tradition, can only thrive and flourish when he is nourished by community and friends. While these repetitive "revivals" are reminiscent of the biblical figure Lazarus, these re-current and ritualistic enactments also illustrate what Kimberly Benston has called a form of "communal 'helping-out,'" or "methexis," in the sense that the participants of a ritual are invested with the responsibility of bringing about the desired outcome of the ritual, an important feature of African American performative and expressive culture (Benston 2000, 28).

Regarding the ritualistic dimension of the blues, "To Hell with Dying" also traces another revision in the process of group formation. In the course of the story, the female narrator more and more takes over the responsibility of bringing Mr. Sweet back to life. As the narrator observes, "I had been the one chosen to kiss him and tickle him long before I knew the rite of Mr. Sweet's rehabilitation" (Walker 1973, 132). The narrator also declares that "I was very good at bringing him around" (134). Her affectionate physical encounters with the old man also mean that the female narrator more and more replaces her brother who shows himself as not really a suitable follower:

> My brother was very generous in all this, he let me do all the revivaling – [. . .] What he would do while I talked to Mr. Sweet was *pretend* to play the guitar, in fact *pretend* that he was a young version of Mr. Sweet [. . .]. (135; emphasis added)

Instead of her brother's "pretensions," it is the female character who fills the void left behind by her brothers and Mr. Sweet's own son who is described as a kid spoiled by his mother and primarily interested in chasing girls (130). Also, in the final encounter with her old friend, the narrator observes that her brother who had normally accompanied her is now serving "in the war in Asia" (137). In the end, it is the female narrator who picks up the guitar, whose "magic" represents Mr. Sweet's artistic legacy. The story thus gestures at a womanist re-definition of the blues and the vernacular tradition.

As Alice Walker recounts in her essay "The Old Artist," at the time of its writing the story resonated with her on a personal level for several reasons. For one, she was away from her Southern home at Sarah Lawrence College in the state of New York where she grew increasingly fond of remembering Mr. Sweet as "an artist" who "went deep into his own pain and brought out words and music that made us happy, made us feel empathy for anyone in trouble, made us think" (38). Expressing a deep-seated melancholy that resulted from the "racist terrorism of the American South," Mr. Sweet's music offers an example of endurance under hardship: "[. . .] he had continued to share all his troubles and his insights with anyone who would listen, taking special care to craft them for the necessary effect. *He continued to sing*" (39; original emphasis). At the time when she wrote "To Hell with Dying," Alice Walker identified with this resistant quality of Mr. Sweet's blues not only as a response to the pervasive racism in American society. In hindsight, Alice Walker also considers Mr. Sweet an exemplary role model for confronting the issue of poverty and social injustice. Commenting on the milieu of Sarah Lawrence College in New York, Walker observes: "[t]here were only three or four other Black people there, and no poor people at all as far as the eye could see" (38). Making her first steps as a writer of short fiction, Alice Walker bridges the distance she experiences between the college classrooms of her fictional writing classes and the vernacular tradition of African American storytelling by conceiving of her writing as a form of resistance to the social and emotional unfreedom of her poverty: "If in my poverty I had no other freedom – not even to say good-bye to him in death – I still had the freedom to love him and the means to express it, if only to myself" (39). This theme is also present in "To Hell with Dying" in which the narrator also describes how the fictional Mr. Sweet's ambitions for a professional career have been thwarted by society:

> Mr. Sweet had been ambitious as a boy, wanted to be a doctor or lawyer or sailor, only to find that black men fare better if they are not. Since he could become none of these things he turned to fishing as his only earnest career and playing the guitar as his only claim to doing anything extraordinarily well. (Walker 1973, 130)

Because of these early experiences of racial discrimination and social injustice, the fictional Mr. Sweet turns into an artistic counter-model to mainstream American notions of frugality and industry. His abode is described as "a neglected cotton farm" whose inhabitant "never paid the slightest attention to his crops" (129; 131). When the narrator returns for her final visit, she notices that the farmhouse is basically little more than "a shack" and appears "more dilapidated than when I was last there" (136). Ending with the narrator picking up the guitar and song of Mr. Sweet, the story thus suggests that the Black female artist translates her experience of social injustice into the creation of beauty and art in the vernacular tradition, and also promotes a sense of collective belonging.

3 Theoretical Perspectives

Alice Walker's short fiction, like her novels and essays, has not only invited critical readings from different theoretical perspectives, it has also fomented the formation of Black feminist criticism. When Mary Helen Washington included "Everyday Use" in her anthology of Black female writers, *Black-eyed Susans*, in 1975, which became only the first instance of anthologizing the story, Alice Walker's early short fiction also helped to spark a critical debate among Black feminists about the necessity of a distinct branch of literary criticism. As Mary Helen Washington has observed, for most of the 1970s Black female authors have been "searching for a specific language, specific symbols, specific images with which to record the myths of their history" (Washington 1981, 10–11). Barbara Smith also argued for the necessity of establishing a "Black feminist approach to literature that embodies the realization that the politics of sex as well as the politics or race and class are crucially interlocking factors in the works of Black women writers" from the United States as well as from "Third-World" countries (Smith 1979, 185; 183). Smith criticizes a host of male critics and reviewers who have missed the "specifically Black female language" by African American women writers and the "innumerable commonalities" between them (188). Also seminal for establishing Black feminist criticism in the early 1980s are the critical writings of Barbara Christian who pointed at how Black women's writings depict the "many facets of the interrelationship of sexism and racism" which allows them to address "the fact that sexism existed in black communities" and to revise "the prevailing definition of woman in American society" (Christian 1985, 180).

Black feminist criticism has frequently adopted and appropriated Alice Walker's definition of a "womanist" aesthetic (Christian, Washington, et al.) which offers an alternative to white "mainstream" feminism as well as the heterosexual and patriarchal normativities of the Black Arts Movement. While the notion of "womanism" allowed for criticizing often detrimental relationships between Black men and women, Black feminist critics were also intrigued by the term's implication of envisioning alternative forms of female intersubjectivity. While Walker's original definition of a "womanist" allows for sexual as well as non-sexual intimate relationships between women (and sometimes men), her novel *The Color Purple* more pointedly introduces lesbian and queer sexualities as viable forms of resisting the heterosexual and patriarchal normativities that have not only informed Southern "folk culture," but also the cultural nationalism and political separatism of the Black Arts Movement. Because of their womanist critique of African American masculinity, Walker's works have been regularly compared to Toni Morrison imagining alternative women-centered communities in her novels as well as to Ntozake Shange's controversial critique of African American masculinity as harmful for women of color in her choreopoem *for colored girls who have considered suicide when the rainbow is enuf* (1977).

While Walker's focus on lesbianism and the experience of women of color has inspired early Black feminist criticism (Smith 1979), later critics have warned against grounding a womanist criticism in exclusive and potentially essentialist terms along these lines (McDowell 2017; Chay 1993). Instead, critics have argued for the necessity of developing a more comprehensive theoretical foundation of Black feminist practice that moves beyond an author-centered approach to literary criticism and is thus better capable of addressing historical changes in the reading and reception of womanist literature (Chay 1993, 645–646). In addition, a too narrow focus on the mimetic function of literary texts to represent a particular "experience" incurs the danger of overlooking how this experience is ideologically constructed (639). More recently, Deborah E. McDowell has restated her argument for the need of paying closer attention to the details of how Black female writing can be distinguished from other literary traditions as a "gender-specific use of language" (610). As McDowell points out, this raises intricate questions not only about the content ("experience") of womanist literature, but also about the position of Black female writing *vis-à-vis* the vernacular tradition in African American literature more generally (607).

Critical assessments of Alice Walker's treatment of the vernacular tradition as a distinct literary form and practice have not arrived at an overall consensus. Walker's use of mostly standard English in her early short fiction has been understood as reflecting the general double-voicedness of African American literature that has always existed in tension with standard English and European literary forms. In this sense, Walker's "Everyday Use" and "To Hell with Dying" can be understood as offering a textualization of the vernacular traditions of singing the blues, quilt-making, and storytelling. While Walker wrote her early fiction when she was either enrolled at college or employed as editor of the *Ms.* magazine, her major novels demonstrate a greater liberty with regard to the use of African American vernacular English. Particularly in *The Color Purple,* Walker introduces Black dialect as a form of literary discourse by modelling her novel on the European epistolary novel as well as the *Bildungsroman* (Gates 1989, 243–255; Boesenberg 1999, 275–276). For Henry Louis Gates, Jr., Walker's exclusive use of black vernacular English in *The Color Purple* clearly echoes Zora Neale Hurston's "speakerly" novel *Their Eyes Were Watching God* which allows him to place Walker firmly in the African American literary tradition of repetition and revision (Gates 1989, 251).

While Henry Louis Gates's exploration of the rhetorical and literary structure of Walker's fiction mostly concentrates on her signifying on literary ancestors, other critics have maintained that the "process" of Walker's narrations reveals the underlying structure of a "quest for personal and communal wholeness" (King 2004, 239). While this points at the ethical responsibility of the storyteller to offer a role model for others to follow, it also has important implications for Walker's characterization of her female protagonists and narrators in terms of their distinct and individual use of language. Giving literary treatment to her narrator's female "voice" also aligns her fiction more broadly with feminist criticism. For the most part, critics have suggested to read Walker's handling of an individual "voice" as the struggle for "self-expression" and "self-empowerment" if not the redemption of an "attempt to express the totality of the self" that has characterized Black female writing since the nineteenth century (Christian 1985, 172). A further instance that connects Walker's writings to forebears such as Frances Ellen Watkins Harper, Pauline Hopkins, Jessie Fauset, Nella Larsen, and Zora Neale Hurston is Alice Walker's rescinding of an objectifying and judgmental heterodiegetic form of narration that privileges a third person point of view. Walker's style of narration emphasizes the importance of an individual narrator's voice which becomes "a vehicle of self-expression and a synecdoche of identity for womanist storytellers" allowing for "alternative definitions of identity and a recasting of relationships among writer, narrator, and audience" (Jablon 1997, 49).

In discussing the literary form of Alice Walker's narrative fiction, critics have also often followed the leads in her own writings that suggest quilt-making as a metaphor for understanding her craft of storytelling. As Sam Whitsitt has pointed out, however, this analogy also includes a number of limitations (Whitsitt 2000, 445). Instead of exploring how quilting actually features as a metaphor for storytelling in Alice Walker's writings, critics have mostly discussed the cultural and contextual implications of quilting. The metaphorical use of "quilting" can thus be expanded into imagining "the promise of creating unity among disparate elements, of establishing connections and in the midst of fragmentation" (Kelley 1994; 176). This offers the cultural practice of quilting not only as a heuristic category for understanding a feminist aesthetic, but also lends itself to envisioning a more multicultural and diverse society (Showalter 1991, 169).

As the above readings of "Everyday Use" and "To Hell with Dying" have shown, critical readings can come dangerously close to projecting a critical point of view onto these texts instead of developing a critical reading out of a close textual analysis. As the story "Everyday Use" suggests, it is the material presence of the ancestral quilts that allows the individual consciousness of the mother-narrator to embrace a different form of intersubjective relationship. Paradoxically, it could be suggested, then, that the materiality of the family quilts in "Everyday Use," similar to the guitar in "To Hell with Dying," offers a material form of resistance that subverts the historical script of American slavery and racism that has insisted on reducing the humanity of the slave to the thingness of chattel property (Brown 2006, 199; Bernstein 2009, 88). What matters more in these stories by Alice Walker, however, is the creative, and individual response of the black female artist that relates her individual experience to a collective aesthetic practice.

4 Bibliography

4.1 Works Cited

Baker, Houston A. *Blues, Ideology, and Afro-American Literature: A Vernacular Theory*. Chicago: U of Chicago P, 1984.

Baker, Houston A. and Charlotte Pierce-Baker, "Patches: Quilts and Community in Alice Walker's 'Everyday Use.'" *Alice Walker: Critical Perspectives Past and Present*. Ed. Henry Louis Gates Jr. and Kwame Anthony Appiah. New York: Amistad P, 1985. 309–316.

Benston, Kimberly W. *Performing Blackness: Enactments of African-American Modernism*. London: Routledge, 2000.

Bernstein, Robin. "Dances with Things: Material Culture and the Performance of Race." *Social Text* 27.4 (2009): 67–94.

Boesenberg, Eva. *Gender, Voice, Vernacular: The Formation of Female Subjectivity in Zora Neale Hurston, Toni Morrison and Alice Walker*. Heidelberg: Winter, 1999.

Borgmeier, Raimund. "Alice Walker: Everyday Use (1973)." *The African American Short Story 1970 to 1990: A Collection of Critical Essays*. Eds. Wolfgang Karrer and Barbara Puschmann-Nalenz. Trier: WVT, 1993. 59–73.

Brown, Bill. "Reification, Reanimation, and the American Uncanny." *Critical Inquiry* 32 (Winter 2006), 175–207.

Chay, Deborah G. "Rereading Barbara Smith: Black Feminist Criticism and the Category of Experience." *New Literary History* 24.3 (1993): 635–652.

Christian, Barbara T. "Alice Walker: The Black Woman Artist as Wayward." *Black Feminist Criticism: Perspectives on Black Women Writers*. New York: Pergamon P, 1985. 81–101.

Cowart, David. "Heritage and Deracination in Walker's 'Everyday Use.'" *Critical Essays on Alice Walker*. Ed. Ikenna Dieke. Westport: Greenwood P, 1999. 21–32.

Davis, Thadious M. "Alice Walker's Celebration of Self in Southern Generations." *Alice Walker: "Everyday Use."* Ed. Barbara T. Christian. New Brunswick: Rutgers UP, 1994. 105–121.

Ellison, Ralph. "Richard Wright's Blues" (1945). *The Collected Essays of Ralph Ellison*. Ed. John F. Callahan. New York: Modern Library, 2003. 128–144.

Farrell, Susan. "Fight vs. Flight: A Re-Evaluation of Dee in Alice Walker's 'Everyday Use.'" *Studies in Short Fiction* 35.2 (1998): 179–186.

Felski, Rita. "'Context Stinks!'" *New Literary History* 42.4 (2011): 573–591.

Gates, Henry Louis. *The Signifying Monkey: A Theory of African-American Literary Criticism*. Oxford: Oxford UP, 1989.

Hirsch, Marianne. "Clytemnestra's Children: Writing (Out) the Mother's Anger." *Modern Critical Views: Alice Walker*. Ed. Harold Bloom. New York: Chelsea House, 1989. 195–213.

Hubbard, Dolan. "Society and Self in Alice Walker's 'In Love and Trouble.'" *Obsidian II* 6.2 (1991): 50–75.

Hurston, Zora Neale. *Dust Tracks on a Road*. 1942. New York: Harper Perennial, 1996.

Jablon, Madelyn. “Womanist Storytelling: The Voice of the Vernacular.” *Ethnicity and the American Short Story*. Ed. Julie Brown. New York: Garland, 1997. 47–62.
Kelley, Margot Anne. “Sisters’ Choices: Quilting Aesthetics in Contemporary African-American Women’s Fiction.” *Alice Walker: Everyday Use*. Ed. Barbara T. Christian. New Brunswick: Rutgers UP, 1994, 167–194.
King, Lovalerie. “African American Womanism: From Zora Neale Hurston to Alice Walker.” *The Cambridge Companion to the African American Novel*. Ed. Maryemma Graham. Cambridge: Cambridge UP, 2004. 233–252.
Latour, Bruno. *Reassembling the Social: An Introduction to Actor-Network-Theory*. Oxford: Oxford UP, 2005.
Lauret, Maria. *Alice Walker*. Houndmills: Macmillan, 2000.
McDowell, Deborah E. “New Directions for Black Feminist Criticism.” *The African American Review* 50.4 (2017): 606–612.
O’Brien, John. “Alice Walker: Interview with John O’Brien,” *Alice Walker: Critical Perspectives Past and Present*. Eds. Henry Louis Gates, Jr. and Kwame Anthony Appiah. New York: Amistad, 1993. 326–346.
Petry, Alice Hall. “Alice Walker: The Achievement of the Short Fiction.” *Modern Language Studies* 19.1 (1989): 12–27.
Showalter, Elaine. “Common Threads.” *Sister’s Choice: Tradition and Change in American Women’s Writing. The Clarendon Lectures 1989*. Oxford: Clarendon P, 1991. 145–175; 192–195.
Smith, Barbara. “Toward a Black Feminist Criticism.” *Women’s Studies International Quarterly* 2.2 (1979): 183–194.
Tuten, Nancy. “Alice Walker’s ‘Everyday Use.’” *The Explicator* 51.2 (Jan. 1993): 125–28.
Walker, Alice. “The Old Artist: Notes on Mr. Sweet.” *Living by the Word: Selected Writings 1973–1987*. San Diego: Harcourt Brace Jovanovich, 1988a. 37–40.
Walker, Alice. *Everyday Use*. Ed. Barbara T. Christian. New Brunswick, N.J.: Rutgers UP, 1994.
Walker, Alice. *In Love and Trouble: Stories of Black Women*. New York: Harcourt Brace Jovanovich, 1973.
Walker, Alice. *In Search of Our Mothers’ Gardens*. New York: Harcourt Brace Jovanovich, 1983.
Walker, Alice. *Meridian*. New York: Harcourt Brace Jovanovich, 1976.
Walker, Alice. *The Color Purple*. New York: Harcourt Brace Jovanovich, 1982.
Walker, Alice. *To Hell with Dying*. San Diego: Harcourt Brace Jovanovich, 1988b.
Washington, Mary Helen. “A Postscript to My 1979 Essay on Alice Walker.” *Everyday Use*. Ed. Barbara T. Christian. New Brunswick, N.J.: Rutgers UP, 1994. 100–104.
Washington, Mary Helen. “An Essay on Alice Walker (1979).” *Alice Walker: Critical Perspectives Past and Present*. Eds. Henry Louis Gates, Jr. and Kwame Anthony Appiah. New York: Amistad, 1993. 35–49.
Washington, Mary Helen. “New Lives and New Letters: Black Women Writers at the End of the Seventies.” *College English* 43.1 (1981): 1–11.
Whitsitt, Sam. “In Spite of It All: A Reading of Alice Walker’s ‘Everyday Use.’” *African American Review* 34.3 (2000): 443–459.
Winchell, Donna Haisty. *Alice Walker*. London: Twayne, 1992.

4.2 Further Reading

Baraka, Amiri. *Blues People: The Negro Experience in White America and the Music That Developed from It*. New York: Morrow, 1963.
Bobo, Jacqueline., ed. *Black Feminist Cultural Criticism*. Malden, MA: Blackwell, 2001.

Brown, Julie, ed. *Ethnicity and the American Short Story*. New York: Garland, 1997.
Collins, Patricia Hill. *Black Feminist Thought: Knowledge, Consciousness, and the Politics of Empowerment*. 2nd ed. New York: Routledge, 2000.
Graham, Maryemma and Jerry J. Ward, Jr., eds. *The Cambridge History of African American Literature*. Cambridge: Cambridge UP, 2011.
Guy-Sheftall, Beverly, ed. *Words of Fire: An Anthology of African-American Feminist Thought*. 2nd. ed. New York: The New P, 1996.
hooks, bell. *Ain't I A Woman: Black Woman and Feminism*. Boston: South End P, 1993.
hooks, bell. *Feminism is for Everybody: Passionate Politics*. Boston: South End P, 2000.
Hughes, Langston, and Arna Bontemps, eds. *The Book of Negro Folklore*. New York: Dodd and Mead, 1958.
Hull, Gloria T. et al., eds. *All the Women Are White, All the Blacks Are Men, but Some of Us Are Brave: Black Women's Studies*. New York: Feminist P, 1982.
Hurston, Zora Neale. "Characteristics of Negro Expression." *Norton Anthology of African American Literature*. Eds. Henry Louis Gates Jr. and Nellie Y McKay. New York: W.W. Norton, 1997. 1019–1032.
James, Joy, ed. *The Black Feminist Reader*. Malden, MA: Blackwell, 2000.
Jarrett, Gene Andrew, ed. *A Companion to African American Literature*. Chichester: Wiley-Blackwell, 2010.
Levine, Lawrence W. *Black Culture and Black Consciousness: Afro-American Folk Thought from Slavery to Freedom*. Oxford: Oxford UP, 1977.
Lorde, Audre. *Sister Outsider: Essays and Speeches*. Freedom: The Crossing P, 1984.
Morrison, Toni. *Playing in the Dark: Whiteness and the Literary Imagination*. Cambridge: Harvard UP, 1992.
Patterson, Robert J. "African American Feminist Theories and Literary Criticism." *The Cambridge Companion to African American Women's Literature*. Eds. Angelyn Mitchell and Danille K. Taylor. Cambridge: Cambridge UP, 2009. 87–106.
Smethurst, James Edward. *The Black Arts Movement: Literary Nationalism in the 1960s and 1970s*. Chapel Hill: The U of North Carolina P, 2005.
Smith, Barbara, ed. *Home Girls: A Black Feminist Anthology*. New York: Kitchen Table, 1983.
Spillers, Hortense J. "Mama's Baby, Papa's Maybe: An American Grammar Book." *Diacritics* 17.2 (1987): 64–81.
Traylor, Eleanor W. "Women Writers of the Black Arts Movement." *The Cambridge Companion to African American Women's Literature*. Eds. Angelyn Mitchell and Eleanor W. Traylor. Cambridge: Cambridge UP, 2009. 50–70.
Wall, Cheryl. *Worrying the Line: Black Women Writers, Lineage, and Literary Tradition*. Chapel Hill: The U of North Carolina P, 2005.

Kerstin Knopf

28 Leslie Marmon Silko (1948–)

Abstract: This article engages with two seminal short stories by Leslie Marmon Silko, "Yellow Woman" and "Storyteller." It introduces Silko as a writer deeply embedded in the oral traditions, cultures, and landscapes of the Laguna Pueblo people. After briefly discussing the two editions of the book *Storyteller* that contains the stories, the article provides an overview over the Kochininako stories about Yellow Woman being abducted by a male spiritual being and sacrificing herself for the good of her people, or returning to the community. The analysis brings to the fore the beauty and richness of the story that is steeped in cultural lore, the land and its stories. Silko's short stay in Alaska has inspired her to write "Storyteller," a story narrating on three different levels, bringing alive oral traditions and connecting them with Western literary tradition. This text tells of a young Yupik woman who takes revenge upon the store clerk who killed her parents, ingeniously protecting herself against settler colonial law through Yupik knowledge and studying the land and river as the white intruders will never do. This story, on the realistic and symbolic level, engages with the onslaught of settler colonialism and violence perpetuated against Indigenous life, land, and knowledge. Finally, the article summarizes and evaluates theoretical approaches to Silko's influential and distinguished work.

Keywords: Native American Literature, Indigenous Literature, "Yellow Woman", "Story Teller", Laguna Pueblo, Yupik

1 Context

"[W]e [Pueblos] have always been able to stay with the land. Our stories cannot be separated from their geographical locations, from actual physical places on the land. [. . .] And our stories are so much a part of these places that it is almost impossible for future generations to lose them – there is a story connected with every place, every object in the landscape," writes Leslie Marmon Silko (Silko 2001, 165). Silko is a storyteller, whose texts (poetry, short stories, novels, autobiographical, and non-fiction writing) are born within an epistemological and ontological realm where land, water, flora and fauna, human beings, their knowledges, narratives, beliefs, and lives are integrated equally and holistically, where none takes precedence and where recognition and respect for all is imperative. "In the view of the old-time people," writes Silko, "we are all sisters and brothers [. . .], clanspeople of all the living beings around us. The plants, the birds, fish, clouds, water, even the clay – they are all related to us. The old-time people believe that all things, even rocks and water, have spirit and being" (Silko 1997, 64). This philosophical concept

https://doi.org/10.1515/9783110587647-029

of the Pueblo people reflects many others throughout Indigenous North America; similar views of land and people as integrated in complex structures of relationships have governed people's understanding of being in the world before colonial contact and often continue to do so.

Scholar and writer Jeannette Armstrong, for example, explains that the Okanagan (Syilx) people in British Columbia see human beings as "intricately woven into the very fabric of the life force of the land" (Armstrong 2007, 31). This view is well demonstrated with the Syilx word for 'land,' *tmxwulaxw*, translated as "from nothing, the life force spreading outward," "in many individual strands," "is here in continuous cycles" (30). Syilx people see themselves as one of those strands, "which are continuously being bound with others to form one strong thread coiling year after year into the future as the life force of the land" (31). She says further, "language arises as an expression of the 'land' and [. . .] 'Indigeneity,' therefore, generates literatures so deeply enmeshed in nature as to be the 'metasource' of the environmental ethic of the people" (31). Silko's as well as Armstrong's writings and those of many other Indigenous authors reflect such views that people, language, knowledge, and stories are tied to the land that produces them. Silko says: "The Pueblo people and the land and the stories are inseparable. [. . .] Human identity, imagination and storytelling were inextricably linked to the land, to Mother Earth" (Silko 1997, 14, 21). Despite print as a medium pressing their writing and telling in certain forms, and English imposing Western epistemes onto the texts as language implicitly contains its culture's views and concepts, both writers integrate oral traditions and narratives, original language words and concepts, oral telling styles, traditional and modern cultural knowledges, community and family "stories into their texts, which Armstrong terms 'oraliture'" (Armstrong 2009; ↗4 Textual Materiality).

Leslie Marmon Silko was born into a prominent mixed-race Laguna family in March 1948 and raised with her two sisters on the southeast edge of Old Laguna village (Silko 1997, 196; Nelson 2005, 245). Her upbringing included much training in cultural history and oral tradition; Silko's "Grandma A'mooh" (Maria Anaya) and her husband Robert (he was the source of two storytelling events collected in Franz Boas's *Keresan Texts*) were both conversant in English and Laguna oral tradition and were sources of oral history Silko grew up with. Such rich sources were also her Aunt Susie (Susie Reyes Marmon), who was a schoolteacher, a Keresan cultural historian and a storyteller, and Aunt Alice (Alice Marmon Little) (Silko 2012, xxv). While "Grandma A'mooh," because of her Presbyterian conviction acquired at Carlisle Indian School, told her family stories and community history, she refrained from telling creation mythology – the wealth of *hummah-hah* stories that Silko heard from Aunts Susie and Alice (Silko 2012, xxii; Silko 1997, 62–63). Aunt Susie "was of a generation, / the last generation here at Laguna, / that passed down an entire culture / by word of mouth / an entire history / an entire vision of the world / which depended upon memory / and retelling of subsequent generations" (Silko 2012, 4). Her influence is made most explicit in the narrative poem on Waithea, an Acoma girl committing suicide, as the poem begins with "This is the way Aunt Susie told the story. / [. . .] I write when I still hear / her voice as she tells the story" (7). This metanarrative beginning demonstrates the idea that words and stories, oral and in print, are means that link the past and present, the dead and living (Pellérin 1997, 120).

Silko's father, working for the tribal council and as an independent photographer, headed a project that in 1977 republished the collection of Laguna-Acoma Keres stories *Schat Chen* (Nelson 2005, 246), which was published by Paula Gunn Allen's mother's uncle John M. Gunn in 1917 (Gunn Allen 1993a, 84). This Laguna-Acoma oral heritage was complimented by many books in the Silko household, also Western children's books and the Bible from which "Grandma A'mooh" read to her, making her an artist and writer steeped in both Indigenous and Western knowledges and traditions (Nelson 2005, 246). Silko attended the Bureau of Indian Affairs (BIA) school at Laguna, parochial schools in Albuquerque, the University of Mexico, graduating with a B.A. in English, and the American Indian Law Program at the University of New Mexico (UNM) for three semesters to finally join the M.A. Program in Creative Writing at UNM (Nelson 2005, 247; Silko 1997, 197).

She published her first short story "The Man to Send Rain Clouds" in 1969 (*New Mexico Quarterly*) and by 1974 had launched her career as a writer with the poetry chapbook *Laguna Woman* and seven stories in Kenneth Rosen's *The Man to Send Rain Clouds*, a collection of Native American short fiction. "Lullaby," widely anthologized, was published in the same year (*Chicago Review*), gaining her a National Endowment for the Arts (NEA) writing fellowship (Nelson 2005, 247–248), which should be followed by others, such as the MacArthur Fellowship that allowed her time and funds to write. She lived for two years in Ketchikan, Alaska, with her then husband John Silko, who worked there as the supervising attorney for the Alaska Legal Services, and their two young sons Robert and Cazimir. Despite missing Pueblo land and culture enormously and suffering "a terrible sense of isolation" (Silko 1997, 193), it was in Alaska where she wrote most of *Ceremony* with a Rosewater Foundation grant (Silko 2012, 197; Nelson 2005, 248). She returned to the Southwest, moved to Arizona in 1978, and continued her writing life on a ranch outside Tucson (Silko 1997, 138; Nelson 2005, 248), an area that should become central to her novel *Almanac of the Dead*. Besides being a very accomplished writer, Silko also featured as a storyteller in the 1978 documentary film *Running on the Edge of the Rainbow* (dir. Larry Evers), and then forayed into filmmaking herself. She founded the Laguna Film Project that was to capture Laguna stories together with their narrative locations, and filmed *Arrowboy and the Witches* on the basis of the Laguna story "Estoy-eh-muut and the Kunideeyahs," also filming in the mesa country around Old Laguna (253). Likewise, she experimented with photography, early on influenced by her father's photography, and tried to establish a meaningful relationship between landscape and narrative, between photography and printed text. She explains:

> I am interested in photographic images that obscure rather than reveal; I am intrigued with photographs that don't tell you what you are supposed to notice, that don't illustrate the text, that don't serve the text, but that form a part of the field of vision for the reading of the text and thereby become part of the reader's experience of the text. (Silko 1997, 169)

In film and books, she repeated her interest in showing the intricate connection between land, people, knowledge, and narratives; and she has been creating storied land, visually and in print. The Southwestern landscape and its oral tradition have informed most of her writing: her non-fiction works *Sacred Water* (1993), *Yellow Woman and a Beauty of the Spirit* (1996), and *The Turquoise Ledge: A Memoir* (2011) mix autobiographical and political essay, memoir, and metatextual notes on writing and photography, often departing and returning to the Laguna/Acoma cultural and geographic center. Her three novels *Ceremony* (1977), *Almanac of the Dead* (1991), and *Garden in the Dunes* (1999) brim with the rich Pueblo lore, have storytelling characters, protagonists that must seek healers and knowledge keepers to return to home and culture or restore equilibrium; these novels have mythical narratives and ritual poetry included in the larger plots and have plots that enliven between book pages the Southwestern landscapes from which the mythic and community stories spring. Epistemological and ontological perceptions of the land create meaning for characters and the plots in which they act. Her characters are often transgressor beings in liminal zones at the boundaries between the natural and supernatural worlds, between reality and myth, between different cultures, between human beings and animals, between mind and body. Silko makes also clear that in contrast to anthropological practice, which tended to attach most importance to the old, sacred, and traditional stories, also the family and community stories enjoy similar recognition in Pueblo tradition: "we make no distinctions between types of story – historical, sacred, plain gossip" (Silko 2001, 161). She says further: "Accounts of the appearance of the first Europeans (Spanish) in Pueblo country or of the tragic encounters between Pueblo people and Apache raiders were no more or no less important than stories about the biggest mule deer ever taken or adulterous couples surprised in cornfields and chicken coops" (Silko 1997, 31).

Her book *Storyteller* is a collage of Laguna myth, family history, and community stories in poetry and prose form, letters, and short stories. Like in her book of nonfiction *Yellow Woman and a Beauty of the Spirit*, Silko has inserted many photographs of herself, family members, Laguna people, and landscape (most taken by her father Lee H. Marmon) in order for the reader to get a sense of the storytellers and context of the Pueblo village that her writing emerges from. "The photographs," Silko says, "are here because they are part of many of the stories and because many of the stories can be traced in the photographs" (Silko 2012, 1). They are interwoven with the multigeneric texts that in themselves appear as a weblike structure of stories (Rainwater 2016, 5; Peterson 2016, 40). The texts and the photographs write Indigenous history from a Laguna-centric perspective. Both the 1981 and 2012 editions use a wider book format that creates more space for the poetry and allows Silko to experiment with "using space on the page, with indentations and various line spacings to convey time and distance and the feeling of the story as it was told aloud" (Silko 2012, xxvi). While the earlier edition was published in a horizontal format (9 inch wide and 6⅞ inch tall) and gave ample space for the stories in poetry form and the photographs, the newer edition comes in a more vertical format, but is still wider than regular book formats (7⅜ inch wide and 9¼ inch tall). Silko also replaced more than half of the original 26 photographs with family photographs, omitting some of the imposing landscape images that came across so beautifully in the horizontal, enlarged format (Petersen 2016, 40). The two short stories discussed in the following are part of the web-like text structure that should be understood as trans-Indigenous history that includes perspectives from the Laguna, Acoma, Hopi, Navajo, Apache, Yupik, and Euro-American worlds. As her novels, the short stories show the beauty of Silko's words, the depths of her writing.

2 Close Readings

"My thigh clung to his with dampness, and I watched the sun rising up through the tamaracks and willows. The small brown water birds came to the river and hopped across the mud, leaving brown scratches in the alkali-white crust. They bathed in the river silently" (Silko 2012, 52). Thus begins the short story "Yellow Woman" based upon one version of the Laguna-Acoma Keres stories of Kochininako, in which Yellow Woman has romantic encounters with a Kat'sina, Whirlwind, Sun, Buffalo Man, or other supernatural beings (50), with Yellow Woman usually abducted and brought to live with the male being, staying, being killed, or returning home in order to achieve survival for her people. Paula Gunn Allen explains that Yellow Woman stories are always female-centered, told from Yellow Woman's point of view: "Kochinnenako, Yellow Woman, is in some sense a name that means Woman-Woman because among the Keres, yellow is the color for women (as pink and red are among Anglo-European Americans), and it is the color ascribed to the Northwest. Keres women paint their faces yellow on certain occasions and are so painted at death" (Gunn Allen 1993a, 88). Silko has several versions in her book *Storyteller* (mostly in the form of poetry to reflect the orality of the tale) that embed the short story analyzed here firmly in the Keresan oral tradition where many Kochininako versions exist (cf. Gunn Allen 1993a, 87–89). The versions emphasize that Yellow Woman is a being from the rich Pueblo cultural lore that has certain responsibilities to ensure the survival of the people through heroic acts (Silko 1997, 71). They also explain the emergence of certain geographical formations, forms of sustenance, and connected rituals. Gunn Allen makes clear that Yellow Woman is usually a bit different from the rest of the people if not alienated: She may live on the edge of the village, may have some special talent, may be quick-witted and resourceful (Gunn Allen 1993a, 88). She says: "The stories do not necessarily imply that difference is punishable; on the contrary, it is often her very difference that makes her special adventures possible, and these adventures often have happy outcomes for Kochinnenako and for her people" (89).

In the first version in Silko's book, Whirlwind travels with Kochininako "swiftly / this whole world / with dust and with windstorms" (52). In the second version, "Cottonwood Part One: Story of Sun House," she leaves her husband, three children, and clan to go with Sun in order to prevent an apocalypse at winter solstice (Lorenz 1991, 65) as a sacrifice "for the world / to continue / Out of love for this earth / cottonwood / sandstone / and sky" (62–63). In the third version, "Cottonwood Part Two: Buffalo Story," Kochininako walks far away from the village suffering from drought and hunger and is abducted at a water hole by Buffalo Man; her husband Estoy-eh-muut, Arrowboy, comes to rescue her and upon flight she betrays their hiding place in a cottonwood tree; he shoots all the Buffalo people and eventually her, since she was in love with Buffalo Man and wanted to stay. He thus sacrifices Yellow Woman for the starving people to have buffalo meat "and they would bring home / all that good meat. / Nobody would go hungry then" (72); the story explains how the Pueblo people have built a relation with the buffalo to the East that they hunted at certain times: "It was all because / one time long ago / our daughter, our sister Kochininako / went away with them" (72). It is Yellow Woman's task to forge a connection to the buffalo people, who sacrifice themselves because she is their sister, and thus to ensure the survival of her people. She is the transgressor, whose actions are possible because of her readiness to give in to the "overpowering sexual attraction" she feels, as Silko said in an interview (Barnes 1993, 57; see Thompson 1989, 23; see also Ruoff 1993, 73).

The fourth version is prefaced by a short narrative text informing readers that Aunt Alice told the Marmon girls this story when Silko, at the age of seven, was sad not to be able to accompany the men deer-hunting in the mountains. Silko authenticates this version, names the originator as in an Indigenous oral copyright system based on trust (Wolfe 1988, xiv; see Lutz 2002b, 113), and relates the circumstances of the storytelling situation. One day Kochininako (a great rabbit hunter) is cornered by the giant *Estrucuyu* who demands her game, her bow and arrows, and her clothing and ravenously swallows all; Kochininako realizes he will devour her as well, ingeniously flees into a cave he is too large to enter and calls the Hero Twins for help, who kill *Estrucuyu* with their *hadti*, their flint knives. They take out the heart, throw it far away, "and it landed right over here / near the river / between Laguna and Paguate / where the road turns to go / by the railroad tracks / [. . .] that big rock there / looks just like a heart" (Silko 2012, 83; cf. Silko 1997, 33). This version is rendered in print with Silko inserting in italics metanarrative commentaries her aunt had given while telling the story, which explain the context and provide information for her nieces to better comprehend what is told: e.g. that women back then could be good hunters, and the meaning of *hadti* and *Estrucuyu* (see Thompson 1989, 24). The last version, "Storytelling," is a modernized version also in poetry; it begins with a metanarrative comment by the storyteller: "You should understand / the way it was back then, / because it is the same even now," and continues with a typical beginning of a *hummah-hah* story: "long ago it happened / that her husband left / to hunt deer" (Silko 2012, 90). The kidnapping part about Buffalo Man abducting Yellow Woman from a water hole is very brief; more emphasis is placed on the mother/father who defends the actions of the woman and on the reaction of her husband: "'You better have a damn good story,' [. . .] 'about where you have been for the past / ten months and how you explain these / twin baby boys'" (91). The middle part of this version humorously tells of four Laguna women being abducted by three Navajo men – in the end it turns out that the women abducted the men, leaving a trail of wine bottles and size 42 panties. The last part again talks about a modern woman being off with a tall Navajo man and coming home with a rather lame excuse derived from myth claiming that he abducted her threatening her life and that she was not able to come home because the roads have been bad for a long time. Her husband leaves her with the narrating voice proclaiming: "It was my fault and / I don't blame him either. / I could have told / the story / better than I did" (93). Silko thus links the realm of myth with modern women's experiences, the realm of storytelling with every-day life, and she emphasizes the quality of telling over the content of a story or explanation: "I could have told / the story / better than I did" (93).

Silko's short story "Yellow Woman" is another modernized version that connects myth and reality, the profane and the sacred, tradition and modernity, women's roles in old times, in modernity, and in myth. Like the traditional models, this version tells the story from Yellow Woman's perspective and stresses her agency (see Barnett 2005, 22). Although the woman protagonist is abducted, the unnamed personal narrator is in control of her actions throughout the story. She meets the stranger Silva at the river, succumbs to the lure of an intimate romantic relationship and follows him, being enamored by his secretive behavior. He tells her that he is a mountain ka'tsina taking her to his home on the top of a mesa. Throughout the story, the narrator is increasingly drawn into the realm of myth, believing she is Yellow Woman following a path foretold in the oral lore she grew up with. And yet, the story rather neglects details of the myth, such as abduction by threat, impossible tasks Yellow Woman has to perform, and her mythical task to sacrifice for her people. Rather, it focuses on the narrator's relationship with Silva, her wavering thoughts about leaving for home or staying with him in a limbo world, and her perception of home when she returns. Silko's version thus does not only focus on Yellow Woman from an outside perspective (as the other versions do) but also from an inside one, offering her thoughts throughout and, in this way, stressing the female principle and female agency.

Accordingly, the story begins in medias res with the narrator's perception of the morning after the intimate encounter and emphasizes the sexual relation, the lovers' rootedness in nature and relation to the sun, trees, earth, river, and birds. The river appears and reappears as geographical marker: the couple sleeps at the river, travels along the river, and the narrator feels embedded in the river landscape. The narrator's feelings toward the stranger are focused on several times, again stressing her sensual perception that often overpowers her rational thoughts: "All I could know was the way he felt, warm, damp, his body beside me. This is the way it happens in the stories, I was thinking, with no thought beyond the moment she meets the ka'tsina spirit and they go" (Silko 2012, 54). The narrator tries to find out more about Silva, but her questions are met with evasive remarks. Time and again, she tries to see a glimpse of home but cannot: "I tried to look beyond the pale red mesas to the pueblo. I knew it was there, even if I could not see it" (52). Likewise, time and again the narrator attempts to remove herself from the mythic realm and enter back into reality: "But I only said that you were him and that I was Yellow Woman – I'm not really her – I have my own name and I come from a pueblo on the other side of the mesa. Your name is Silva and you are a stranger I met by the river yesterday afternoon" (Silko 2012, 53). She attempts to conjure up the physical reality she knows: "I will see someone [. . .] and I will be sure that I am not Yellow Woman. Because she is from out of time past and I live now and I've been to school and there are highways and pickup trucks that Yellow Woman never saw" (54). Her family is introduced rather late in the story, shortly before they arrive at Silva's hut, when she thinks of her own home; this is a narrative device to stress the woman's thoughts and sensual perceptions and not her role as mother and wife.

Silva's hut on top of the mesa seems unearthly, made of black lava rock and red mud, with a smell of "pitch and buck brush" (54), a bare interior, and "one small window facing east" (54), towards sunrise. At the top of the mesa it seems as if they can see the world: the Navajo reservation, the Pueblo boundaries, Texan ranches, and Mexican cattle herds (55); it is as if they are in some godly all-seeing realm being in the center of the world they know, and Silva being a spiritual figure ruling over the world they see. When the narrator again tries to find out if he is a Navajo, he insists he is a spirit; he "shook his head gently. 'Little Yellow Woman,' he said, 'you never give up, do you? I have told you who I am. The Navajo people know me, too'" (56). The narrator is completely enchanted by the eerie atmosphere up in the sky: "I was standing in the sky with nothing around me but the wind that came down from the blue mountain peak behind me" (55), which keeps her from returning although she had already decided to go home: "There was no wind and the sun warmed me. I was surrounded by silence. I drowsed with apricots in my mouth, and I didn't believe that there were highways or railroads or cattle to steel. [. . .] I remembered that I had meant to go home. But that didn't seem important any more" (56). There are small hints that Silva has the potential to use force and even to be violent: "I was afraid lying there on the red blanket" (54); "his hand around my wrist. I had stopped trying to pull away from him" (54); and "I lay underneath him and I knew that he could destroy me" (56). But each of these text passages is followed directly by the narrator being overpowered by her sexual desire or by feelings of love for Silva: "All I could know was the way he felt, warm damp, his body beside me" (54); and: "But later, while he slept beside me, I touched his face and I had [. . .]

the kind of feeling for him that overcame me that morning along the river" (56). Possibly these hints of violent force are inserted because they are elements in the traditional Kochininako stories, or because they add to the sexual attraction of Silva; what is important is that the narrator is not forced and acts out of her own accord – she assumes female agency and power that her ancestor Kochininako did not enjoy in all versions of the story.

Throughout her stay with Silva, the narrator is kept in a liminal state, the boundaries between the spiritual and natural world blurred, unsure even of her own thoughts and feelings about home and the man with her. This liminal state allows her to live out her sexual desires. She might have wandered off her home to the river aimlessly and have intimate contact upon unconscious desire; she might have left her home consciously because of the irresistible promise of a sexual experience with a stranger; she might have left her home because her married life had turned bland and disenchanted. Her knowledge about Yellow Woman she gained from her grandfather has bolstered her thoughts and actions. Asked if Yellow Woman flees from social and sexual domination, Silko says:

> The kind of things that cause white upper-middle-class women to flee the home for a while to escape or get away from domination and powerlessness and inferior status, vis-à-vis the husband, and the male, those kinds of forces are not operating at all. What's operating in those stories of Kochininako is this attraction, this passion, this connection between the human world and the animal and spirit worlds. (Barnes 1993, 57; see Ruoff 1993, 72, 75, 76)

Into "Yellow Woman" Silko included a story the narrator's grandfather has told her about Badger and Coyote being invited to sleep with Yellow Woman. This story within a story continues the old lore and makes it meaningful and important through the analogy to the narrator's story. Also, the narrator muses: "I was wondering if Yellow Woman had known who she was – if she knew that she would become part of the stories. Maybe she'd had another name that her husband and relatives called her so that only the ka'tsina from the north and the storytellers would know her as yellow woman" (53). She thinks again if there is a possibility of her being like Yellow Woman, maybe this could serve as an excuse for her enjoying sexual freedom similar to her ancestor. Moreover, in the metanarrative the narrator ponders the origin of Yellow Woman in myth or in reality. Hers will be a new Yellow Woman story told by the people (see Ruoff 1993, 77). This example illustrates the metanarrative quality of the book *Storyteller*: "*Storyteller* epitomizes a metacritical text; every piece can be read for the story it tells and for its story about storytelling and the role of stories" (Krumholz 1999, 64). The narrator imagines her husband calling the tribal police for help and her family's life going on without her; moreover, she imagines that there will be a new story about her disappearance, a story the storytellers like her grandfather will include in their lore. Through these metanarrative devices, Silko has her story function as continuation of the oral tradition, here in printed form.

Only when Silva returns to the hut with a rustled cow and asks her to accompany him to the market to sell the meat, she starts to see him as a man. Also, on their way down the mountain, they are embedded in nature, the narrator seeing the wildflowers and "every color that a cactus blossom can be" (58). They meet an unarmed rancher who exposes Silva as the rustler they had been looking for a long time since. In contrast to the couple, he seems out of place and not to belong, being fat, sweating in the heat, and smelling rancid. Now there was "something ancient and dark" (59) around Silva with his finger on the trigger of his rifle; he sends her back to the hut to save herself and the meat. Only when the narrator hears four shots in the distance Silva appears human again, he probably has shot the farmer. Either as man or spirit Silva has removed the intruding force from the land, has killed the alien, evil principle like evil witches had to be killed in Navajo and Pueblo lore (cf. Silko's short story "Tony's Story" and her novel *Ceremony*). She returns home "because [she] thought it was safer" (59) and sends the horse uphill to the hut; it is clear that now she is sure Silva is a man and alive. When she comes to the place where they met, she is overcome by desire again "to kiss him and to touch him – but the mountains were too far away now" (60); meaning she is back in the real world, left with the hope that he will appear again. She will tell that a Navajo had kidnapped her, as the Pueblo oral lore is full of such stories of Navajo raids, stealing food and abducting women (see Ruoff 1993, 77). But she knows it will not be believed, she needs her grandfather to authenticate her story "because it was the Yellow Woman stories he liked to tell best" (Silko 2012, 60; see Ruoff 1993, 78). Now, with her grandfather gone, it is upon her to become the next storyteller and add her story to the rich lore of Yellow Woman stories.

Silko's stint in Alaskan lands and her contact with its Indigenous people, their history, and traditions has left a deep impression upon her; so much so that she set one of her few short stories in an Alaskan Yupik community. Her story "Storyteller" is strongly aligned with Western postmodern narrative practices and employs a frame story and several non-chronological flashbacks. It tells a story about a young unnamed woman, having grown up with her grandmother and an old man, steeped in the cultural and oral traditions, taking revenge for the death of her parents when she was a little girl and thus establishing "justice (or harmony) in her world" (Lorenz 1991, 69). Also this story, written in a figural narrative situation with the protagonist as the reflector figure/focalizer, centers on the thoughts, perceptions, and memories of the protagonist; only once the text changes its focalization for a few lines to the man she kills. The text is organized as a frame plot, the protagonist being in jail and awaiting trial, and a revenge plot, the protagonist killing a white man, which becomes part of the story the grandmother tells about the death of the protagonist's parents, her actions continuing the grandmother's story. On the metanarrative level, the text explains storytelling traditions: appearing in the protagonist's memories, the old man tells a story about a lone hunter being stalked by a giant polar bear, which functions as an analogy to the revenge plot. In the end all three plots merge and connect the narrative and metanarrative levels. While the short story appears to lean towards Western postmodern literary traditions, a closer look shows its indebtedness to Indigenous narrative traditions: what appears as a frame story is in fact also aligned with circular storytelling in Indigenous traditions as the tale returns to where it began (see Pellérin 1997, 122; see also Lutz 1989, 2002a); likewise the telling in non-chronological flashbacks manifests the digressive loops Indigenous tales often take, where various temporal levels exist simultaneously as such simultaneity of past, present, and future is one key understanding in Indigenous philosophies. Thus, "Storyteller" is an exercise in storytelling, grown in both traditional and modern Indigenous and postmodern Western narrative traditions.

Right at the beginning the story drops readers into the realm of the supernatural, because with the protagonist we witness the sun stopping in its track and the sky appearing frozen like ice, trapping the sun:

> Every day the sun came up a little lower on the horizon, moving more slowly until one day she got excited and started calling the jailer. She realized she had been sitting there for many hours, yet the sun had not moved from the center of the sky. The color of the sky had not been good lately; it had been pale blue, almost white, even when there were no clouds. She told herself it wasn't a good sign for the sky to be indistinguishable from the river ice, frozen solid and white against the earth. The tundra rose up behind the river but all the boundaries between the river and hills and sky were lost in the density of the pale ice. (Silko 2012, 17)

The old man's prophesy, the protagonist realizes, that the sun, sky, and earth, and thus life and time itself, will come to be frozen and stop, has come true. The eerie atmosphere is supported by the borders between sky, river, tundra, and hills being blurred in the protagonist's perception, only the words 'jailer' and 'every day' firmly root her in empirical reality as they indicate that she has been jailed for quite some time now. The jailer, an assimilated Inuk, who refuses to speak Yupik, does not seem to understand her warnings. He, neglecting language and culture, cannot connect to the traditional lore, to the land, and to the knowledge necessary to realize what is going on. The protagonist, however, is rooted ontologically and epistemologically in Yupik culture and is ever curious and open to all other knowledges and experiences.

This curiosity has made her leave the old people to go to the BIA school, "because she was tired of being alone with the old woman whose body had been stiffening for as long as the girl could remember. Her knees and knuckles were swollen grotesquely, and the pain had squeezed the brown skin of her face tight against the bones" (18). The grandmother suffers from somatic reactions to her pain and anger over the death of her children, "the joints are swollen with anger" (18). When she tells the girl about her parents' death she breaks into a heavy sweat, "telling the story was like laboring to walk through deep snow" (24). As the grandmother talks less and less, the old man talks more: "in a soft deliberate voice, he told stories, moving his smooth brown hands above the blankets. He had not fished or hunted with the other men for many years although he was not crippled or sick. He stayed in his bed, smelling like dried fish and urine, telling stories all winter; and when warm weather came, he went to his place on the river bank" (18). It is as if he refuses to participate in life, out of protest against the intruding Euro-American culture that brings death and upturns balance both in nature and the community; or because he immerses himself in the role as knowledge keeper and storyteller, a role respected by the community as people repeatedly bring him dried fish and venison. He confines himself to immobility except to indulge in his pedophilic advances that transform into sexual relations as the protagonist turns older. These relationships are not judged by the text as inacceptable, rather they are presented as the maturing girl knowingly entering into intimate contact with the old man as much as she consciously goes with the oil workers since she "wondered what they looked like under their quilted goose-down trousers; she wanted to know how they moved" (20). She is a woman who actively seeks sexual relations, regardless of gossip, and controls her body and actions. She even uses men as part of her plan to kill the trading post clerk, who had sold her parents fuel as drinking alcohol, consciously risking their excruciating deaths; their bodies are found by the grandmother "twisted on the sand" in the summer shelter that seemed to have been abandoned after that.

The intruders bring not only physical death but also cultural, spiritual, linguistic, and psychological death, as the children are assimilated in the boarding schools and many do not return to the village: "They were used to the town where the boarding school was located, with electric lights and running water. After all those years away at school, they had forgotten how to set nets in the river and where to hunt seals in the fall. Those who left did not say it, but their confidence had been destroyed" (21). Also the landscape will be destroyed once oil extraction is developed. The old man rants about the 'Gussocks' now stealing the oil after fur animal, seal, and fish populations are decimated. In this regard, Patrick Wolfe speaks of the elimination of the Native at the hands of the settler state (Wolfe 2006). The apocalypse is approaching because of the death of the Native: "'It is approaching. As it comes, ice will push across the sky'" (Silko 2012, 21).

The old man begins to tell the bear story right that day: "It began with a giant bear which he described muscle by muscle, from the curve of the ivory claws to the whorls of hair at the top of the massive skull. And for eight days he did not sleep, but talked continuously of the giant bear whose color was pale blue glacier ice" (21) – pale blue, the same color the sky assumes when the prophesy comes true. The giant polar bear stalking the lone hunter symbolizes the settler state, aggressive, devouring, and lethal, pale blue standing for giantism and coldness as much as yellow stands for the things related to the white men and their aggression against the land and humans, their reckless behavior, their noise, and their smell: the heavy yellow machines that freeze to immobility or sink into the summer mud and bogs, nature defending itself against attempts at exploitation – futile attempts indeed, as "they would come back again" the old man says (22); the smelly yellowish dog that is allowed into their houses, perceived by the Yupik as dirty and inappropriate while the Euro-Americans insist that the Yupik are dirty and smell of dried fish (21); the flat yellow holes the feeble light bulbs leave in the darkness (22); the stringy yellow wadding that starts coming apart in the trading store, increasingly unable to insulate the Euro-Americans against the cold (17, 27); and the yellow stains the oil drillers leave after urinating in the snow (21).

Red stands for Euro-American violence as the fuel that killed the protagonist's parents comes in red barrels. While the fuel is lethal to her parents, the red wine the red-haired man pours for her is enlivening; she is not made drunk and raped as so many Native women that face extreme sexual violence at the hands of Euro-American and Euro-Canadian men. She consciously approaches the back room of the store, where Inuit are not welcome, to pick up one of the oil workers, the red-haired one, who makes attempts at creating a romantic atmosphere with candles and music. Despite his efforts he cannot connect to the young woman, he tapes a pornographic image over the bed to get himself going and he is repeatedly cold, shivers, or has rattling teeth; he and the other oil workers are out of place, alien to the land and climate that reject them. They do not realize that the ice is coming, the wind tearing at their yellow insulation; neither do they have the knowledge to live on the land and off its sources. The text highlights the abundance of fish and venison – with the grandmother preparing a tub full of fish for drying, the people repeatedly bringing dried fish and meat to the old man and the protagonist, even a king salmon when she becomes the storyteller – while the workers eat canned food that is presented as rather unsavory: "The long wooden shelves were jammed with cans of beans and potted meats. On the bottom shelf a jar of mayonnaise was broken open, leaking oily white clots on the floor," the smelly dog lying right beside the shelves (21). The young woman turns the tables on power relations and utilizes the oil workers for her plan: she makes the store clerk so jealous and horny with her repeated pick-ups of men that he easily falls for her when she comes to the store for a last time.

The woman and the Yupik people are presented as embedded in the land, easily move on it, live off it by hunting and fishing, wearing clothing made of animal skin, and knowing the land and its stories. The grandmother's voice "flowed with the wind blowing off the river," the river connecting both women after the old woman's death as "she listened to the wind off the river; its sound was like the voice that day long ago" (24). The protagonist wears mukluks and her grandmother's wolfskin parka, creamy colored, silver, and almost white making her and the old woman "invisible in the snow" (20). The white and dark green colors of the tundra, snow, ice, animal skin, whitefish, and the winter landscape stand for the village Yupik, who are ontologically and epistemologically connected to the land. This is supported by repeated descriptions of the Indigenous space and land and the limited descriptions of the enclaves of colonized Euro-American space on Yupik land. While the village people use the red tin from the barrels to mend walls and roofs of their houses, red penetrates the white and green also in a symbolic way: "But the red tin penetrated the thick white color of earth and sky; it defined the boundaries like a wound revealing the ribs and heart of a great caribou about to bolt and be lost to the hunter forever" (26).

Thoughts appear in high frequency that the protagonist saw something red in the grass the morning of her parents' death, the red tin standing for their death as much as the red tin she nails to her house looks like a wound. But she disturbs the established power relations, and deliberately plans her use of the red wine (when picking up men) and moreover the red tin. Time and again, she walks on the river ice, studies the colors of stable and unstable ice, studies the patterns of currents underneath, studies how to navigate on the ice in a landscape with blurred boundaries between river ice, white tundra, and pale sky: "in the distance she could see the red tin nailed to the log house, something not swallowed up by the heavy white belly of the sky or caught in the folds of the frozen earth. It was time" (27). She prepares for her move while the old man tells the story of the bear and hunter in every little detail, describing each motion of the bear and hunter, each sound, each smell, for days on end, in his sleep, and even when he starts suffering from old age dementia. He loves the story; he treats it like a lover: "The giant bear was creeping across the new snow on its belly, close enough now that the man could hear the rasp of its breathing. On and on in a soft singing voice, the old man caressed the story, repeating the words again and again like gentle strokes" (26). Once the polar bear has caught the hunter, the world will freeze; the story of settler colonialism must be told without lies, as much as the grandmother's story must be told to the end. The young woman enacts the end of the story so that the conclusion of the grandmother's story becomes the young woman's story (Hirsch 1988, 5).

When she playfully escapes his embrace, the store clerk follows her half mad with desire – he without jacket or mittens, blind with horny rage, she in the wolfskin parka, breathing through her mittens to protect her lungs. She lures him onto the river, carefully navigating, he running after her so sure he will catch her, taking straight short cuts. After he has fallen through the ice to his death, the woman carefully probes the ice with her boots searching for "sinews of ice to hold her" (Silko 2012, 28) and for the red beacon on her house to return. Her knowledge of the river and her ingenious plan protect her life and her freedom as well, because the settler state cannot try her for murder. As the big yellow machines are driven into the bogs by men ignoring local knowledge, so does the storekeeper ignore local knowledge of the river and weather, and runs to his death. Silko highlights Indigenous knowledge of the land and Indigenous assessment of Euro-American male arrogance as helping overpower Euro-American political and material superiority.

The protagonist is in control of all situations throughout the text, supported by the often repeated phrase "she laughs out loud," which displays her self-confidence. She is in control even when facing trial and defends herself against imposition of Euro-American authority: the attorney insists that it was an accident, that she could not have planned the death of a white man, and that he will plead her innocence on the grounds of her suffering from mental confusion. This attitude betrays his Euro-American arrogance and underrates Indigenous intellectual capacity that had cost the store clerk his life. But the young woman insists that she killed the man and that she will not lie and instead tell the story – in the jail cell to the attorney, in court, and in the community, "year after year as the old man had done, without lapse or silence" (30). Only then does the sun break free from the ice, the apocalypse is prevented because she takes over from the old people and tells the stories of her parents' death and of the giant polar bear and the lone hunter. She tells two stories of colonization, one of committed crime and elimination of the Native as Patrick Wolfe would have it, and one in the mythical realm where lies "could not stop what was coming" (Wolfe 2006, 30). Mishuana Goeman argues that repetitive colonial practices (here, violating Indigenous life through selling fuel as alcohol, violating Indigenous lands through oil drilling, and violating Indigenous epistemologies through forced assimilative education) sediment "normative modes of settler colonialism" (Goeman 2014, 236–37). "Yet space is fluid," she explains, "it is only in the constant retelling and reformulating of colonial narratives that space becomes place as it is given structure and meaning" – what she calls "settler grammar" (237). Likewise, she insists that "Native stories speak to a storied land and storied peoples, connecting generations to particular locales and in a web of relationships" (Goeman 2013, 37). In this light, I argue that Silko's text continues oral tradition and storytelling in print, both on the narrative and the metanarrative level, simultaneously recreates storied people and storied land that resists 'settler grammar' and "make[s] impossible a sense of complete U.S. or Canadian conquest" (2014, 237).

Like the young woman, who, with knowledge of the land and without weapons, has turned the devouring greed and sexual desire of white men against themselves, killing one man, the hunter (standing for Indigenous people), weak, with no weapons, forever stalked and hunted, and reduced to bare life, in the end must face the giant polar bear (the devouring settler state) with ingenuity and cultural knowledge alone: "The hunter had been on the ice for many hours. The freezing winds on the ice knoll had numbed his hands in the mittens, and the cold had exhausted him. He felt a single muscle tremor in his hand that he could not stop, and the jade knife fell; it shattered on the ice, and the blue glacier bear turned slowly to face him" (Silko 2012, 30).

3 Theoretical Perspectives

Critical readings of Silko's work and *Storyteller* come foremost from the field of Indigenous Studies, where we can discern somewhat differing tendencies. Bernhard Hirsch (1988) and Linda Danielson (1993) were among the first to look at the text through a holistic lens and intensely discuss the structure and design of the array of texts as a complete whole rather than collection of assorted text types. While Hirsch arranges the texts in four thematic clusters around cultural motifs, among them a Survival and a Yellow Woman section, Danielson discusses *Storyteller* as a coherent work about tribal survival with a spider web structure, storytellers in the book and Silko herself "making stories [and continuing] the traditions of Thought Woman and Grandmother Spider" (Danielson 1993, 201). Hirsch also outlines how *Storyteller* draws from and supports the oral tradition and how the photographs interact with the texts. Likewise, Joan Thompson (1989) analyzes *Storyteller* as oral storytelling and sees it as continuing the oral tradition, albeit in a different medium. "By writing from memory and imagination, rather than from written sources and by including variations as signifiers of change," she holds that "Silko moves her stories and poems closer to the intent of oral tradition" (Thompson 1989, 23). LaVonne Ruoff (1993) embeds her rich reading of "Yellow Woman" in Keres traditions and other Yellow Woman stories. Lynn Domina (2007), Linda Krumholz (1999), Arnold Krupat (1993), and Hertha Wong (1992) read *Storyteller* as autobiographic writing; Krupat discusses the book in Bakhtinian terms as polyphonic text, the texts in the book in dialogue with each other, and the book's storytellers and Silko also in dialogue with each other the surrounding, also Euro-American, world of storytellers (Krupat 1993, 190). Likewise, Krumholz analyzes the text with a dialogistic lens, looking at how it creates a dialogue between the Indigenous and Western spheres of discourse in order to "'read' the Other across cultural differences" (Krumholz 1999, 66). Domina discusses Silko as a shape-shifting storyteller/trickster moving easily between worlds mirrored "in the generic indeterminacy of the text" (Domina 2007, 46).

A strong feminist trend in discussing Silko's work is initiated by herself when she insists that traditional Pueblo cultures did not know inequality and labor division according to different genders as much as "marriage" did not keep people from having extramatrimonial sexual relationships (Silko 1997, 66–67). Louise Barnett centers her analysis of Yellow Woman stories on the "female creative principle that actively contributed to the well-being of the people through such female deities as Thought Woman, Spider Woman, Corn Mother, and others" (Barnett 2005, 19). Barnett holds that Kochininako protects the Pueblos affirming the life force of nature through her heroic and sexual quests, assuming roles traditionally associated with men (19–20). She thus generates life sustenance for example by establishing a relationship with the buffalo people or with the sun. Patricia Jones similarly employs a women-centered approach and traces the "absent mother" in Silko's writings, matrilineality and ensuing female power in Pueblo culture, and discusses Yellow Woman as mother figure and creator: "She is simultaneously the 'good' mother who fulfills the traditional role of wife and nurturer and the 'bad' mother whose sexuality is a powerful force, capable of both creation and destruction" (Jones 1993, 216). Paula Gunn Allen approaches feminist issues from a tribal perspective, she terms the study of Indigenous literature, history, and culture consequently from a feminist perspective "feminist-tribalism" (Gunn Allen 1993a, 83). This approach is necessary because, as she says, the study of Indigenous life and thought was dominated by male-centered modes of consciousness right from its beginnings in the fifteenth century, distorting understandings of Indigenous life, philosophy, gender roles, and sexual practices into ideas incongruent with Indigenous understandings and practices (83–84). "A feminist approach reveals not only the exploitation and oppression of the tribes by whites and white government but also areas of oppression within the tribes and the sources and nature of that oppression" (84). Such a critical analysis, she says, can generate strategies to work against conditions and practices caused by patriarchal colonialism as much as patriarchal-colonial influences on oral traditions and people's consciousness (84–86). She voices a critical perspective on Kochininako stories and highlights the fact that she is often abducted by force, the abductor threatening to kill her if she does not follow and that she might possibly be raped (Gunn Allen 1993b, 113–114). Both ideas usually do not emerge in the various versions as Yellow Woman is said to go with the abductor on her own terms. With her critical stance Gunn Allen possibly does not so much critique the stories themselves but patriarchal attitudes, dominance, and violence against Indigenous women at the hands of both Indigenous and non-Indigenous men.

Lee Schweninger makes specific ecocritical and ecofeminst approaches to Silko's texts productive, specifically analyzing *Ceremony*. He contends that Silko's texts render opposing Euro-American and traditional Indigenous attitudes towards land and nature visible – a concept of ownership and domination derived from the Old Testament and a holistic concept of relationality that understands the world as an interconnected web of relations between land, water, sky, plants, animals, and human beings and that knows no distinction between wilderness and civilization, or humans and non-humans (Schweninger 1993a, 48). Western attitudes towards land were (more or less successfully) imposed on Indigenous thought and practices under colonialism, and together with settler colonial geopolitics of appropriation and displacement, have rendered many Indigenous people physically dislocated and/or spiritually alienated from traditional lands and connected beliefs and practices (e.g. Tayo). An ecofeminist approach helps to see that the "same patriarchal world view motivating the oppression of women and minorities motivated human oppression of nonhuman nature as well" (Schweninger 1993b, 38). While one must be careful not to truncate the conspiracy of Euro-American women in settler colonialism, this approach can focus sharply on dominating, controlling, and exploitive practices of white patriarchal regimes. Ecofeminism, so Schweninger, attempts to challenge patriarchal hierarchal norms and beliefs and substitute them with "different patterns of belief, de-homocentric patterns without oppression, exploitation, and domination" (40).

René Dietrich applies a rigorous settler colonial studies approach to Silko's *Storyteller*, and here specifically to the often-anthologized short story "Lullaby." With Mark Rifkin he argues that Indigenous people are pressed into the paradoxical relation of "inclusive exclusion" with the settler state through the state's colonial geo- and biopolitics. Giorgio Agamben's *homo sacer*, who is included in society as body under control of the sovereign but has no right to meaningful life, serves Rifkin and Dietrich as concept applicable to Indigenous people who are similarly included and excluded: The settler state exerts a particular form of biopower, controlling certain parts of the population, and turning problematic and unwanted groups into a state of "bare life" and a "state of exception." The settler state can "include them in the sovereign sphere of the U.S.," and exert control and the rule of law, "while at the same time excluding them from the possibility of claiming rights and enjoying the same living conditions usually available to those existing within the same sovereign sphere" (e.g. the characters Chato, Ayah, and Jimmie; Dietrich 2012, 305–306).

In her edition of new perspectives on *Storyteller*, Catherine Rainwater confirms Brewster Fitz's statement that Silko is indeed a "writing medicine woman" that seeks ways that work, modifying, adapting, and breaking, as she goes, "many rules of both Indigenous oral and Western written storytelling" (Rainwater 2016, 2). She has, so Rainwater, through "metatextual instruction," cultivated her audience to perceive various dimensions of her creative process (2). New perspectives on *Storyteller* include, for example, Susan L. Dunston's essay, arguing that Silko's writings emerge from a distinct Indigenous/Pueblo epistemology that needs to be considered lest one comes up with constrained readings that obscure Native ways of knowing and the science and teaching contained in oral and written stories, e.g. environmental ethics or land knowledge, among others. This can be knowledge and science directly displayed by characters as for example the protagonist of "Storyteller," or people's awareness of "intricately interrelated systems that form the self-weaving, self-renewing fabric of life" (Dunston 2016, 22). Nancy J. Peterson applies to Silko's writing LeAnne Howe's concept of "tribalography" – Indigenous storytellers interweaving all major elements of life, people, and land, grounding their work in the web of their relations, and easily moving between genres established in Western realms of knowledge. Comparing the 1981 and 2012 editions of *Storyteller*, she insists that the book contains an interconnected web of stories – a tribalography – "a rich and multifold story of Laguna past and present, told in order to make possible an empowering future" (Peterson 2016, 41). Lee Schweninger discusses the inserted photographs of the early edition, arguing that they are arranged to engage in dialogue with the texts and captions as well as amongst each other. They thus claim "visual sovereignty," a concept originally employed by Michelle Raheja in view of Indigenous filmmaking, and they answer to ethnographic visual simulations of Indigeneity in nineteenth-century "Indian photography" (Schweninger 2016, 59, 62). Finally, Elizabeth Archuleta looks at *Storyteller* through a spatial lens, i.e. arguing that beyond being firmly rooted in Pueblo land and landscape, the texts hold a complex way of thinking about and creating trans-Indigenous space, of relations between race and space, and of connecting "race to physical space or territory, thereby racializing space and spatializing race" (Achuleta 2016, 98). Partly discussing captions in

Storyteller, Archuleta exposes how practices of the settler state enacted violence against Indigenous epistemology through renaming Indigenous geography with English terminology, as for example Deetseyamah into McCartys, or Kaweshtima into Mt Taylor, against Indigenous bodies, forcing them onto reservations and prohibiting them to enter fenced-off farmland, and against Indigenous lands, for example ripping parts open and setting its nuclear toxicity in motion when creating the Laguna uranium mine, in all cases racializing spaces (102–107). These newer perspectives according to new theoretical developments will soon be complimented by even newer approaches – what remains are Silko's texts, ever so stable and dynamic.

4 Bibliography

4.1 Works Cited

Allen, Paula Gunn. "Kochinnenako in Academe: Three Approaches to Interpreting a Keres Indian Tale." *"Yellow Woman" Leslie Marmon Silko*. Ed. Melody Graulich. New Brunswick: Rutgers UP, 1993a. 83–111.

Allen, Paula Gunn. "Whirlwind Man Steals Yellow Woman." *"Yellow Woman" Leslie Marmon Silko*. Ed. Melody Graulich. New Brunswick: Rutgers UP, 1993b. 83–111.

Archuleta, Elizabeth. "Space, Place, and Violence in Storyteller." *Leslie Marmon Silko's Storyteller: New Perspectives*. Ed. Catherine Rainwater. Albuquerque: U of New Mexico P, 2016. 97–114.

Armstrong, Jeannette. "Kwtlakin? What is Your Place?" *What is Your Place? Indigeneity and Immigration in Canada*. Eds. Hartmut Lutz and Thomas Rafico Ruiz. Augsburg: Wissner, 2007. 29–33.

Armstrong, Jeannette. "Constructing Indigeneity: Syilx Okanagan Oraliture and tmix^{w}centrism," http://ub-ed.ub.uni-greifswald.de/opus/volltexte/2012/1322/. Diss. U of Greifswald, 2009. (06 January 2019).

Barnes, Kim. "A Leslie Marmon Silko Interview." *"Yellow Woman" Leslie Marmon Silko*. Ed. Melody Graulich. New Brunswick: Rutgers UP, 1993. 47–65.

Barnett, Louise. "Yellow Woman and Leslie Marmon Silko's Feminism." *SAIL* 17 (2005): 18–31.

Danielson, Linda. "The Storytellers in *Storyteller*." *"Yellow Woman" Leslie Marmon Silko*. Ed. Melody Graulich. New Brunswick: Rutgers UP, 1993. 201–212.

Dietrich, René. "The Inclusive Exclusion of Native Americans: Indigenous Life Writing and the Threat to U.S. Nationhood in Leslie Marmon Silko's *Storyteller*." *Transnational American Studies*. Ed. Udo Hebel. Heidelberg: Winter, 2012. 305–322.

Domina, Lynn. "'The Way I Heard It': Autobiography, Tricksters, and Leslie Marmon Silko's Storyteller." *SAIL* 19.3 (2007): 45–67.

Dunston, Susan L. "Storytelling Science." *Leslie Marmon Silko's Storyteller: New Perspectives*. Ed. Catherine Rainwater. Albuquerque: U of New Mexico P, 2016. 21–38.

Goeman, Mishuana. *Mark My Words: Native Women Mapping Our Nations*. Minneapolis: U of Minnesota P, 2013.

Goeman, Mishuana. "Disrupting a Settler-Colonial Grammar of Place: The Visual Memoir of Hulleah Tsinhnahjinnie." *Theorizing Native Studies*. Eds. Audra Simpson and Andrea Smith. Durham and London: Duke UP, 2014. 235–265.

Gunn, John M. *Schat Chen. History, Traditions, and Narratives of the Queres Indians of Laguna and Acoma*. Albuquerque: Albright and Anderson, 1917; reprint, New York: AMS, 1977.
Hirsch, Bernard A. "'The Telling Which Continues': Oral Tradition and the Written Word in Leslie Marmon Silko's *Storyteller*." *American Indian Quarterly* 12.1 (Winter 1988): 1–26.
Jones, Patricia. "The Web of Meaning: Naming the Absent Mother in *Storyteller*." *"Yellow Woman" Leslie Marmon Silko*. Ed. Melody Graulich. New Brunswick: Rutgers UP, 1993. 213–232.
Krumholz, Linda. "Native Designs: Silko's *Storyteller* and the Reader's Initiation." *Leslie Marmon Silko: A Collection of Critical Essays*. Eds. Louise K. Barnett and James L. Thorson. Albuquerque: U of New Mexico P, 1999. 63–86.
Krupat, Arnold. "The Dialogic of Silko's *Storyteller*." *"Yellow Woman" Leslie Marmon Silko*. Ed. Melody Graulich. New Brunswick: Rutgers UP, 1993. 185–200.
Lorenz, Paul. H. "The Other Story of Leslie Marmon Silko's 'Storyteller.'" *South Central Review* 8.4 (Winter 1991): 59–75.
Lutz, Hartmut. *Approaches: Essays in Native North American Studies and Literatures*. Augsburg: Wissner, 2002.
Lutz, Hartmut. "The Circle as Philosophical and Structural Concept in Native American Fiction Today." *Native American Literatures*. Ed. Laura Cotelli. Pisa: Servicio Editoriale Universitario, 1989. 85–100; repr. in Harmut Lutz, *Approaches: Essays in Native North American Studies and Literatures*. Augsburg: Wissner, 2002a. 195–208.
Lutz, Hartmut. "Native Literatures in Canada Today: An Introduction." *Zeitschrift für Kanada-Studien* 17.2 (1990): 27–47; repr. as "Native Literatures in Canada Before Oka: An Introduction" in Hartmut Lutz, *Approaches: Essays in Native North American Studies and Literatures*. Augsburg: Wissner, 2002b. 109–125.
Nelson, Robert M. "Leslie Marmon Silko: Storyteller." *The Cambridge Companion to Native American Literature*. Eds. Joy Porter and Kenneth M. Roemer. Cambridge: Cambridge UP, 2005. 245–256.
Pellérin, Simone. "Death and the Power of Words in Leslie Marmon Silko's *Storyteller*." *Native American Women in Literature and Culture*. Eds. Susan Castillo and Victor M. P. Da Rosa. Porto: Fernando Pessoa UP, 1997. 119–126.
Peterson, Nancy J. "*Storyteller* as Tribalography." *Leslie Marmon Silko's Storyteller: New Perspectives*. Ed. Catherine Rainwater. Albuquerque: U of New Mexico P, 2016. 39–55.
Rainwater, Catherine. "Introduction: New Walks on Old Trails." *Leslie Marmon Silko's Storyteller: New Perspectives*. Ed. Catherine Rainwater. Albuquerque: U of New Mexico P, 2016. 1–19.
Ruoff, A. LaVonne. "Ritual and Renewal: Keres Traditions in Leslie Marmon Silko's 'Yellow Woman.'" *"Yellow Woman" Leslie Marmon Silko*. Ed. Melody Graulich. New Brunswick, NY: Rutgers UP, 1993. 69–81.
Schweninger, Lee. "Writing Nature: Silko and Native Americans as Nature Writers." *Melus* 18.2 (1993a): 47–60.
Schweninger, Lee. "A Skin of Lakeweed: An Ecofeminist Approach to Erdrich and Silko." *Multicultural Literatures through Feminist/Poststructuralist Lenses*. Ed. Barbara Frey Waxman. Knoxville: U of Tennessee P, 1993b. 37–56.
Schweninger, Lee. "'Serious Business': The Visual and the Verbal in Silko's Storyteller." *Leslie Marmon Silko's Storyteller: New Perspectives*. Ed. Catherine Rainwater. Albuquerque: U of New Mexico P, 2016. 57–77.
Silko, Leslie Marmon. "Language and Literature from a Pueblo Indian Perspective." *Nothing But the Truth. An Anthology of Native American Literature*. Eds. John L. Purdy and James Ruppert. New Jersey: Prentice-Hall, 2001. 159–165.
Silko, Leslie Marmon. *Yellow Woman and a Beauty of the Spirit. Essays on Native American Life Today*. 1996. New York: Touchstone, 1997.

Silko, Leslie Marmon. *Storyteller*. 1981. London: Penguin Books, 2012.
Thompson, Joan. "Yellow Woman, Old and New: Oral Tradition and Leslie Marmon Silko's *Storyteller*." *The Wicazo Sa Review* 5.2 (Fall 1989): 22–25.
Wolfe, Alexander. *Earth Elder Stories: The Pinayzitt Path*. Saskatoon: Fifth House, 1988.
Wolfe, Patrick. "Settler Colonialism and the Elimination of the Native." *Journal of Genocide Research* 8.4 (2006): 387–409.
Wong, Hertha Down. *Sending my Heart Back Across the Years: Tradition and Innovation in Native American Autobiography*. New York: Oxford UP, 1992.

4.2 Further Reading

Arnold, Ellen N. *Conversations with Leslie Marmon Silko*. Jackson: UP of Mississippi, 2000.
Barnett, Louise K., and James L. Thorson, eds. *Leslie Marmon Silko: A Collection of Critical Essays*. Albuquerque: U of New Mexico P, 1999.
Graulich, Melody, ed. *"Yellow Woman" Leslie Marmon Silko*. New Brunswick, NY: Rutgers UP, 1993.
Krumholz, Linda. "The Web of Stories: Reading and Change in Leslie Marmon Silko's Storyteller." *Leslie Marmon Silko's Storyteller: New Perspectives*. Ed. Catherine Rainwater. Albuquerque: U of New Mexico P, 2016. 115–131.
Nelson, Elizabeth Hoffman, and Malcolm A. Nelson. "Shifting Patterns, Changing Stories: Leslie Marmon Silko's Yellow Women." *Leslie Marmon Silko: A Collection of Critical Essays*. Eds. Louise K. Barnett and James L. Thorson. Albuquerque: U of New Mexico P, 1999. 121–133.

Jochen Achilles

29 Sandra Cisneros (1954–)

Abstract: This chapter discusses Sandra Cisneros's short fiction as a literary expression of Chicana feminism, as it manifests itself in Gloria Anzaldúa's concept of the borderland existence of the new *mestiza*. In "Woman Hollering Creek" and "*Bien* Pretty," major contributions to her story cycle *Woman Hollering Creek* (1991), Cisneros experiments with hybrid redefinitions of Chicana womanhood, which rely on a complex intercultural reinterpretation of Aztec and Spanish religious myths. Mythopoeic femininity counteracts the oppressive effects of both Mexican *machismo* and the US patriarchy, which are reinforced in popular culture by the submissive female role models of *telenovelas* and romance novels. As in *The House on Mango Street* (1984), Cisneros's first composite novel or story cycle, artistic activity as a writer or painter emerges as the *via regia* towards women's emancipation in these liminal narratives of borderland communities.

Keywords: Chicana feminism, new *mestiza*, intercultural mythopoiesis, critical regionalism, liminal narratives, borderland community

1 Context

Together with Denise Chavéz, Ana Castillo, Helena Maria Viramontes, and others Sandra Cisneros belongs to the most famous representatives of Chicana literature. Chicanas/os are hybrid people "of mixed Spanish, Indian, and Anglo descent" who live in the United States (Madsen 2000, 6). Whereas the early Chicano movement as part of the civil rights movement of the 1960s and 1970s was male-dominated and chiefly protested against U.S. supremacy in general, Chicana feminism mainly strives for liberation from *machismo*, a term popularized by Chicanas (3), and "the transformation of the patriarchal structure of Mexican social and gender relations" (28). The search for viable concepts of identity in Chicana literature takes the shape of a fictional inquiry into the multiple factors that determine the lives of Chicanas and the liminal forms of existence that result from negotiations of such multiplicity (Fitts 2002, 8; see also Ketz 2010, 221).

In *Borderlands/La Frontera: The New Mestiza* (1987), her seminal collection of essays and poetry, Gloria Anzaldúa bases the identity of Chicana women on the border. Borderland existence has psychological consequences, as it triggers "preoccupations with the inner life of the Self, and with the struggle of that Self amidst adversity and violation" (Anzaldúa 2012, 19). This mentality creates a space of its own, "a shock culture, a border culture, a third country, a closed country" (33), based on a visionary mindset derived from Indian culture and called *la facultad*.

https://doi.org/10.1515/9783110587647-030

This capacity manifests itself in marginalized individuals and tries to reconcile the empirical and the spiritual. The most conspicuous feature of this interstitial and liminal borderland mentality, which Anzaldúa calls “new *mestiza* consciousness, *una conciencia de mujer*” (99; original emphasis), is its multidimensionality. In her autobiographical essay collection *A House of My Own* (2015), Cisneros shares this view: “Our work as bicultural citizens is to help others to become visionary, to help us all to examine our dilemmas in multiple ways and arrive at creative solutions” (Cisneros 2015, 196). This productive awareness of diversity, expressing itself in the works of Anzaldúa, Cisneros, and other Chicana writers, is part of what Julia Kristeva explains in *Strangers to Ourselves* (1991) as a Freudian dialectic of repressed otherness: “The foreigner is within me, hence we are all foreigners. If I am a foreigner, there are no foreigners” (Kristeva 1991, 192). The transnational decentering processes of the self, which Kristeva aptly describes, are also at the heart of the culture-specific new mythology developed in Chicana literature.

In the case of the new *mestiza* consciousness, the mediation between foreignness and familiarity takes the shape of negotiations between Indigenous traditions and emancipatory intentions. The liberation of the new *mestiza* is based on a revision of both the territorial and identitarian implications of Indigenous myths. These myths are chiefly tied to three female figures: La Virgen de Guadalupe, La Malinche, and La Llorona. In seminal works of Chicana cultural theory and literature, such as Gloria Anzaldúa’s *Borderlands/La Frontera* and Sandra Cisneros’s *Woman Hollering Creek*, new directions for Chicanas are based on innovative recombinations of the ambiguities inscribed in the myths surrounding La Virgen de Guadalupe, Coatlicue/Tonantsin, La Malinche/La Chingada, and La Llorona (Madsen 2000, 9).

The Mexican veneration of La Virgen de Guadalupe goes back to the apparition of the Holy Virgin to the native Mexican peasant Juan Diego on a hill named Tepeyac in the north of modern Mexico City shortly after the Spanish conquest in December 1531. The basilica, erected near this place, supersedes a site of veneration for the Aztec goddess Coatlicue/Tonantsin previously destroyed by the Spanish conquistadors. The basilica can thus be considered a manifestation of the colonialist superimposition of Spanish Catholicism on Aztec religion (Lafaye 1987; see also Miller/Taube 1993, 26–35; Townsend 2009, 220–241). La Malinche also bridges the intercultural divide between Indians and Spanish colonizers:

> Among those associated with Cortés, his Indian courtesan La Malinche has entered into Chicano/a mythology. Some doubt attaches to her willingness to assist the conquistadors – whether she was raped or seduced or gave herself willingly – and her motives are also ambiguous – whether she acted to preserve her people by forging an alliance with the invaders or acted more purely from self-interest. Her status as the mother of the Chicano is unquestioned; from her there issued the mestizo, the people of mixed Indian and Spanish blood whose descendants in the United States call themselves Chicano. (Madsen 2000, 7)

In this context Gloria Anzaldúa argues that Chicanas have three symbolic mothers: "All three are mediators: *Guadalupe*, the virgin mother who has not abandoned us, *la Chingada* (*Malinche*), the raped mother whom we have abandoned, and *la Llorona*, the mother who seeks her lost children and is a combination of the other two" (Anzaldúa 2012, 52; original emphasis). As all three figures are surrounded by ambiguity, writers such as Sandra Cisneros or Ana Castillo imaginatively reshuffle the predominant qualities of these precolonial, legendary or Christian representations of Mexican womanhood and use them for redefinitions of contemporary Chicana femininity.

In *Woman Hollering Creek*, Cisneros's second volume of fiction, she bases several stories on these mythical figures. "Little Miracles, Kept Promises" reconciles the humble submissiveness of La Virgen de Guadalupe with the fierceness of Tonantsin and her other incarnations. "Woman Hollering Creek," the titular story, reinterprets La Llorona and the either resigned or furious qualities of her wailing. And both "Never Marry a Mexican" and "*Bien* Pretty" revise the role of La Malinche. These redefinitions are tentatively forged into new emancipatory myths that replace their repressive earlier versions. They are not new in the sense of complete reinvention but arise from the interstices of precolonial and colonial traditions as a new synthesis of permanent liminality (Castro-Dopacio 2010; see also Harvey 2010; Rebolledo 1995, 49–81). In this regard Chicana writing is moving in the direction of a redefinition of the postcolonial situation as such, also noticeable in other cultures and in postcolonial theory since Homi Bhabha's *The Location of Culture* (1994).

Cisneros's main works are story cycles or novels of initiation like *The House on Mango Street* and *Caramelo* (2002), which delineate the emancipatory development of their respective narrators, protagonists, and initiands. They describe the economic, interethnic, and gender repression that determines the lives of Chicanas as well as the solidarity of families and communities, which also function as stabilizing forces. Cisneros's linguistic liminality, sandwiched between Spanish and English, is immediately tied up with a more profound identitarian liminality that can only be resolved by a transnational imagination: "She wants to write stories that ignore borders between genres, between written and spoken, between highbrow literature and children's nursery rhymes, between New York and the imaginary village of Macondo, between the U.S. and Mexico" (Cisneros 2009, xvi–xvii; see also Achilles 2017). Cisneros's writing emerges as a process of both distancing and belonging.

Woman Hollering Creek and Other Stories is loosely held together by the common theme of female emancipation but does not share the composite novel-structure of *The House on Mango Street*, its predecessor. There is no central intelligence providing cohesion like Esperanza Cordero in *The House on Mango Street*. Different girls and women tell their own stories in *Woman Hollering Creek*. In addition, a few stories, such as "The Marlboro Man," "Remember the Alamo," and "*Los* Boxers" (original emphasis), deal chiefly with forms of masculinity. *Woman Hollering Creek* falls into three parts. The first part, titled "My Lucy Friend Who Smells Like Corn," continues where *The House on Mango Street* left off. It contains stories about stages of childhood, about children's problems at home and in school, and about the care for family members. The stories in the second part, titled "One Holy Night," depict macabre predicaments of adolescents. "There Was a Man, There Was a Woman," the third section of *Woman Hollering Creek*, opens up new terrain, as it contains elaborate stories about love relationships and gender problems, largely from the perspective of middle-aged women in mythical contexts. These tales deal with more advanced stages in the lives of Chicanas, who are nevertheless still groping for the meaning of their lives. The criticism of the Hispanic *machismo* tradition and the successive repressiveness of fathers and husbands, a dominant theme in *The House on Mango Street*, is continued in these stories.

Different from *The House on Mango Street*, *Woman Hollering Creek* experiments with cultural, religious, and media-driven frameworks, such as Indigenous myths, Aztec and Christian religious beliefs as well as the tremendously influential contemporary everyday myths disseminated by *telenovelas* and popular romance novels. This syncretic integration of the diversity of the cultural imaginary in order to forge a liminal new *mestiza* identity counteracts traditional forms of submissive womanhood as well as TV character stereotypes and is fuelled by the feminist reinterpretation of mythical and religious lore noticeable in a substantial number of stories in *Woman Hollering Creek*. Cisneros's technique of interreligious and intercultural contextualization, developed in *Woman Hollering Creek*, leads to added layers of semanticization. But it also diminishes the immediacy of episodic as well as epiphanic observation, which is so striking in *The House on Mango Street* and in some vignettes in *Woman Hollering Creek*, such as "My Lucy Friend Who Smells Like Corn," "Eleven," or "*Los* Boxers." By contrast, individual experience tends to become part of an overriding pattern of intercultural mediation in major stories of *Woman Hollering Creek*.

2 Close Readings

In the following, "Woman Hollering Creek" and "*Bien* Pretty," the first and the last story in the third and most innovative section of *Woman Hollering Creek*, will be discussed in detail. "Woman Hollering Creek," presented in time leaps, is about a liminal form of emancipation in the intercultural realm between Mexico and Texas, mapped out in Anzaldúa's *Borderlands/La Frontera*. The story is rather about the protagonist's awareness of, and potential for, a more self-determined life than about the actual living of such a freer existence. Cleófilas Enriqueta DeLeón Hernández, the central character, remains on the threshold of freedom without passing it (Harvey 2010, 177–184; see also Ketz 2010, 214). She is in an intermediate position between her timid Seguin, Texas, neighbors Soledad and Dolores, who passively accept their fate, and the bold nurses Graciela and Felice, who assist her, when she is about to give birth to her second child. Graciela and Felice also demonstrate to her that

Chicana women can live uninhibited and free lives. Reminiscent of stories such as "Sally," "What Sally Said," and "Linoleum Roses" in *The House on Mango Street*, "Woman Hollering Creek" maps out a circular movement for Cleófilas. She moves across the border from caring for her father and her "six good-for-nothing brothers" (Cisneros 1991, 43) in her hometown Monclova, Caohuila, Mexico, to married life with her increasingly violent husband Juan Pedro Martínez Sánchez in their new home in Seguin, Guadalupe County, Texas. During her second pregnancy she leaves her husband with their son Juan Pedrito and makes it by the Greyhound from San Antonio back to her Mexican hometown with the help of Graciela and Felice. Cleófilas perceives and mentally adopts their different, self-determined attitude but only manages to flee back to her father's home and "to the chores that never ended" there (43).

Cisneros excellently describes Cleófilas's disappointment with the discrepancy between her romantic expectations of married life on the one hand, fueled as they are by the *telenovelas* she used to watch in her Mexican home throughout her adolescence and the romance novels she reads in her Texas home devoid of a TV set, and, on the other, the realities of living with her husband. It turns out that Cleófilas's husband Juan Pedro does not hold "the very important position" (45) Cleófilas optimistically credits him with but works with a beer company. She is reduced to the boredom of the Texas town where she has followed him, which begins to resemble more and more the uneventful ennui she hoped to leave behind in her Mexican hometown.

Cleófilas's disappointment with crossing the border is matched by her disappointment with "the man I have waited my whole life for," "this man, this father, this rival, this keeper, this lord, this master, this husband till kingdom come" (49). The disillusionment with their marital situation is mutual, as Cleófilas realizes. Not only does she wax less than enthusiastic about Juan Pedro's looks and ways, he also comes to detest his own yoke of marital duties, when he "kicks the refrigerator and says he hates this shitty house and is going out where he won't be bothered with the baby's howling and her suspicious questions, and her requests to fix this and this and this" (49).

At the beginning of their marriage Cleófilas is invited to "the ice house" (48) where her husband and his cronies meet in the evening to drink and talk. She is introduced to the pastimes of the men, talking and drinking beer. Soon she sees through their repetitive ways "and finally becomes good at predicting where the talk will lead" (48). While Cleófilas is vainly in search of the hazy sensual love fulfilment the *telenovelas* promise, the men also seem to be after some evasive, ungraspable dream: "Cleófilas concludes each is nightly trying to find the truth lying at the bottom of the bottle like a gold doubloon on the sea floor" (48). Intoxicated as they are, the men's yearnings evaporate in "a belch" (48), however, as Cleófilas's grind to a halt in the violent eruptions of her husband. In different, gender-specific ways men and women are similarly dissatisfied with their lives in both Mexican and Texan small towns.

While Cleófilas's and Juan Pedro's disillusionment with each other and the monotony of their married life may be inevitable, the degree of male violence that pervades Cisneros's story is certainly not. Different from her life at home in Mexico, Cleófilas, who was never beaten by her parents, is surprised when her husband starts beating her (47–48). Like Sally's father in "What Sally Said" (Cisneros 2009, 93), Juan Pedro repents and weeps after each of his bouts of violence, which continue all the same, as Cleófilas does not fight back, although "the lip split and bled an orchid of blood" (Cisneros 1991, 47). Juan Pedro's ostentatious repentance for his violence does not lead to change. His violence fits into a communal and perhaps national pattern of masculinity. Maximiliano, one of Juan Pedro's drinking friends from the ice house, even shoots his wife when she attacks him with a mop (51). Both Maximiliano's lethal brutality and Juan Pedro's violence stand in for a more general phenomenon in Mexican society, the many stories of marital abuse and extreme violence which fill the newspapers (52).

The only antidote to the pervasive threat of male violence seems the mental flight to a both printed and televised dream world, which Cleófilas has been conditioned by since early youth (Almería 2010, 69–75; see also Harvey 2010, 173–176; Saldívar-Hull 2000, 108–117). The marriages of women like Cleófilas suffer not only from the *machismo* of their husbands but also from the illusionary expectations, unrealistic standards of beauty and fashion as well as a feminine passivity deemed noble, which are all transmitted by popular culture and cement *machismo*. More and more Cleófilas fails to bridge the distance between her life and her media-induced hopes of a fairy tale come true. Following Cleófilas's inner as well as outer borderland existence between Mexico and the United States – between the rosy dream world of *telenovelas* and the real as well as emotional chill of the ice houses of Texan breweries –, Cisneros's story demonstrates both the omnipresence of media entertainment, projecting submissive images of femininity which function like an ideological drug, and the need of more viable self-projections for women like Cleófilas.

The story provides such an option, as it maps out a heterotopic territory of Mexican legend and myth and thereby goes back to the historical roots of colonial subjugation and its offshoot of women's oppression. What is special about the location where Juan Pedro and Cleófilas live is the titular Woman's Hollering Creek behind their house. It is named after La Gritona, the Screaming Woman (Cisneros 1991, 46). This place name refers to the Mexican tradition of Texas before the annexation by the United States and to the Mexican legend of La Llorona, the beautiful woman who drowned her children in a river as a revenge on her unfaithful husband, who had left her for another woman. After she also drowns herself in the river, she is not allowed to enter the gates of heaven until she has found her children again. A ghostly presence wandering in a liminal realm between heaven and earth, she is considered searching for her lost children, constantly weeping mournfully.

La Llorona, the weeping woman, is sometimes conflated with La Malinche, the Nahua woman who became Hernan Cortés's interpreter and mistress. She bore him a child believed to be the beginning of *mestizaje*, Mexican mixed-race identity. According to legend she was abandoned by Cortés, when he wanted to marry a Spanish woman. Both characters taken together symbolize the oppression and abandonment of women by both Mexican and white men, which becomes one of Cisneros's central topics in *Woman Hollering Creek* (Doyle 1996; see also Fitts 2002, 7; Saldívar-Hull 2000, 117–123; Wilson 2010, 48–50; Wyatt 1995). Cleófilas keeps wondering whether the hollering of the woman that gives the creek its name is caused by pain or rage. She exposes herself to the humiliation, sorrow, and rage, embedded in the La Llorona and La Malinche myths, and wants to explore the emotional and existential potential of the screaming and hollering that, as she senses, is more promising than both the submissiveness and unreachably elevated elegance *telenovelas* and sentimental novels exude.

More and more Cleófilas begins to think of alternatives to, and ways out of, her oppressive married life. She thinks of her father's offer of perennial support but does not see how she could overcome the communal prejudices against women leaving their husbands, especially when children are involved (Cisneros 1991, 50). Neither her present situation of married life in Seguin nor a return to her Mexican hometown appear as viable solutions. Both communities appear equally patriarchal, repressive, boring, and hopeless. Their cultural differences and thereby the border between them are negligible by comparison: "There is no place to go" (51), she concludes.

For a viable identity Cleófilas is thrown back on the mythical story of La Llorona that is deeply embedded in her cultural heritage and seems to emanate from the creek in the heterotopic terrain behind her house in Seguin. The precolonial myth of female humiliation and pride cannot be silenced and reverberates in the postcolonial Southwest:

> The stream sometimes only a muddy puddle in the summer, though now in the springtime, because of the rains, a good-size alive thing, a thing with a voice all its own, all day and all night calling in its high, silver voice. Is it La Llorona, the weeping woman? La Llorona, who drowned her own children. Perhaps La Llorona is the one they named the creek after, she thinks, remembering all the stories she learned as a child. La Llorona calling to her. She is sure of it. (51)

What follows is a postmodern reenactment and actualization of what La Llorona's hollering may signify. In this and other stories, Cisneros is developing a mythopoeic dimension as a platform for women's emancipation. This platform cannot be found on either side of the border between Mexico and the United States. It is to be built on the attitudinal development inspired by the resistance to women's oppression.

When Cleófilas is about to have a second baby, it becomes difficult to cover the medical expenses, as Juan Pedro bought a new truck. It is also difficult for Cleófilas to hide the physical traces of Juan Pedro's violence from the doctor in the hospital (53). In a switch of perspective, Graciela, the nurse who examines Cleófilas in hospital, has a telephone conversation with her friend Felice. Graciela asks Felice to give pregnant Cleófilas a ride to the Greyhound station in San Antonio, as she has "black-and-blue marks all over" (54) and should travel home to Mexico. Graciela and Felice are Chicanas, as they intersperse their English conversation with Spanish words, while Cleófilas – like Mamacita in "No Speak English," a vignette in *The House on Mango Street* – "does not even speak English" (54). Graciela describes the case of Cleófilas and her likes as a "regular soap opera sometimes" (55). Whereas women like Cleófilas drug themselves both mentally and emotionally by the sentimental falsities of *telenovelas*, Graciela and Felice have adapted to life in the United States by learning English and getting jobs, whereby they become independent.

Driving pregnant Cleófilas and her son Juan Pedrito to the Greyhound Station in San Antonio, Felice crosses the creek behind Cleófilas's house and lets out "a yell as loud as any mariachi" (55). She wonders why "nothing around here is named after a woman? Really. Unless she's the Virgin. [. . .] That's why I like the name of that *arroyo*. Makes you want to holler like Tarzan, right?" (55). In a sense, Felice answers Cleófilas's mute question about whether women holler out of pain or rage at the creek (47). She alludes to the docile and passive femininity of La Virgen de Guadalupe, Mexico's patron saint, and replaces it by her shout of emancipatory self-affirmation and joy about her own Tarzan-like strength. Felice is an independent woman, who incorporates qualities considered male by Cleófilas. Unmarried, she herself owns the pickup truck she drives, which she has exchanged for her previous Pontiac Sunbird, "a *real* car" for what she calls "[p]ussy cars" (55; original emphasis). Felice thus represents both emancipation and self-propulsion by the adoption of attributes commonly assigned to males. Her hollering and laughter are addictive. Cleófilas will bring the spirit of her new and different, adopted womanhood to her father and brothers in Mexico: "Then Felice began laughing again, but it wasn't Felice laughing. It was gurgling out of her own throat, a long ribbon of laughter, like water" (56).

Felice's behavior both reinterprets and transforms the sad and depressing legend of La Llorona, who sacrificed her children as a reaction to being maltreated by their father. While Cleófilas's neighbors Soledad and Dolores share La Llorona's tearful depressiveness, Graciela and Felice reenact La Llorona's rage in much more constructive ways. They help Cleófilas save both her son and unborn child from marital violence. By her hollering Felice teaches Cleófilas a self-confidence which triggers joy rather than pain or despair. In the process, Felice's Tarzan-style independence and self-determination also eliminate the prepackaged submissive femininity of the *telenovela* heroines from Cleófilas's mind. Women's hollering at the creek deconstructs such commodified patriarchal self-images many Latina women still adhere to in consequence of the detrimental influence of popular visual and print media. Although Cleófilas returns to the patriarchy of her father and brothers and "the chores that never ended" (43), she does so with a radically altered mindset, with Gloria Anzaldúa's new *mestiza* mentality.

If a girl, Cleófilas's second child to be born in Mexico will in all likelihood be named after Graciela and Felice, who function as her *comadres* ["godmothers," "female friends"], as Graciela suggests to her friend on the phone (55). Cleófilas's all-male family will be faced with a new and differently calibrated femininity. Cleófilas represents a liminal stage in this process of female self-determination. She succeeds in dissociating herself from the world of and oppression by her married life as well as the sugarcoated female submission recommended by *telenovelas*. But she does not quite succeed in reaching Felice's fearless self-determination, sparked by an Indigenous spirit of female battle-readiness long buried unrecognized and untapped in communally shared mythological lore. Cleófilas's daughter yet to be born promises to become such a new *mestiza*, however.

In "*Bien* Pretty," the last story in *Woman Hollering Creek*, Lupe Arredondo moves to Texas from northern California and accepts "an art director's job at a community cultural center in San Antonio" (142) because she split with her friend Eduardo in San Francisco. As his contribution to women's emancipation, left-wing Eduardo, called Eddie, betrays Lupe with a blonde "financial consultant for Merrill Lynch" (142). Notwithstanding his humanitarian ideology, Eddie, like most Chicanos in Cisneros's stories, apparently has a predilection for white women (142). Lupe has been invited to live for a year in the San Antonio, Texas, home of "Irasema Izaura Coronado, a famous Texas poet who carries herself as if she is directly descended from Ixtaccíhuatl or something. Her husband is an honest-to-God Huichol *curandero* [healer], and she's no slouch either, with a Ph.D. from the Sorbonne" (139; original emphasis). The intercultural couple, syncretically associated with European erudition, poetic production, Indigenous healing practices, and the Aztec goddess Ixtaccíhuatl, has left on a Fulbright scholarship for a year and Lupe can house-sit. In several respects Lupe's situation is close to aspects of Cisneros's own life (Brackett 2005, 36–39, 91–96).

In near-Whitmanesque fashion, Lupe makes a whole list of the furnishings of the house, which is as intercultural as its owners. It is decked out with paraphernalia from different Mexican, Texan, and other regional traditions from "Oaxacan black pottery pieces," a "punched tin Virgen de Guadalupe chandelier" (139), a "seventeenth-century Spanish *retablo* [altarpiece]" to an "eye-of-God crucifix," a "death mask of Pancho Villa with mouth slightly opened" and a "Texana chair upholstered in cowskin with longhorn horns for the arms and legs" (140). To this mixture Lupe adds a "stainless-steel wok," her "grandmother's *molcajete* [mortar]," "flamenco shoes," "*huipiles* [tunics]," "*rebozos* [scarves]," and "all my Latin tapes" (141; original emphasis). The house of Lupe's hosts brings together a broad range of objects from even broader precolonial and colonial, Mesoamerican and Texan traditions. Lupe adds her own possessions to this hybrid mix. Her new San Antonio home thus provides Lupe with "a rich matrix in which to practice her evolving amphibious ways" (Carroll and Maher 2010, 10) by entering "a liminal period of growth and healing" (11; see also Tontiplaphol 2010, 20, 22). As the story's bilingual title already indicates, "*Bien* Pretty" is obviously about the reconciliation of diverse cultural, linguistic, religious, and mythical influences and trends.

Like "Woman Hollering Creek," "Never Marry a Mexican," and "Eyes of Zapata," "*Bien* Pretty" is also the story of a doomed love relationship. Lupe Arredondo falls in love with Flavio Munguía, a Mexican father of seven sons by two women, as it later transpires (Cisneros 1991, 138, 156). Flavio is a kindred spirit of faithless Eddie, his predecessor. Flavio signs the poetry he writes in his free time as "Rogelio Velasco" (138) and is obviously identical with the protagonist of the same name in "Tin Tan Tan," the previous vignette, which reproduces his love letters and poems to a woman with whom he claims to be madly but vainly in love – a prefiguration of the plot of "*Bien* Pretty."

Their relationship begins, when Lupe, her house infested with vermin, calls La Cucaracha Apachurrada Pest Control and Flavio appears at her door for the first time. Admiring his perfect build, Lupe begins to think that, beyond roach extermination, she might also use him artistically as a model for a painting based on the legend of Popocatépetl and Iztaccíhuatl she has been thinking about for some time (143). In her painting she intends to reactivate the authenticity and emotional power of the so-called volcano myth, which dissolved in its numerous commodified renditions:

> I'd always wanted to do an updated version of the Prince Popocatépetl/Princess Iztaccíhuatl volcano myth, that tragic love story metamorphosized from classic to kitsch calendar art, like the ones you get at Carnicería Ximénez or Tortillería la Guadalupanita. Prince Popo, half-naked Indian warrior built like Johnny Weissmuller, crouched in grief beside his sleeping princess Ixtaccíhuatl, buxom as an Indian Jayne Mansfield. And behind them, echoing their silhouettes, their namesake volcanoes. (144)

Lupe tries to free the volcano myth from its commercial vulgarization, suggestive of Hollywood icons of strength and beauty and available in butcher shops or taco joints. Integrating Flavio, Lupe wants to tease out this mythical love story's existential implications, immortalized by their projection into the shape of the adjacent volcanoes Popocatépetl and Iztaccíhuatl and reminiscent of Romeo and Juliet. In Lupe's version of the volcano myth, Prince Popocatépetl and Princess Iztaccíhuatl will be made to represent a fresh view of the relationship between men and women that pays tribute to Lupe's actual experiences with Eddie and Flavio (Carroll and Maher 2010, 13–14; see also Rader 2010, 147).

When she begins to transform Flavio artistically into a mythical being, he also becomes Lupe's lover. In her painting Lupe tries to capture both the hybridity of her life in San Antonio and the intensity of her new passion. She wants to integrate and synthesize the combined utopian potential of her Mexican roach exterminator, her artist hostess, and the palatial home she has temporarily inherited (Cisneros 1991, 139). But Lupe's art will also register the ways in which realities fall short of their sheen and promise. As the roaches threaten to disfigure her home, the moral deficiencies of her lovers tend to disfigure her life. Lupe's painting of the volcano myth echoes what Cisneros strives for in the major stories of *Woman Hollering Creek*. She wants to embed individual characters and their relationships in larger mythical structures and (inter)cultural ensembles to highlight their significance. And she wants to reinterpret and recontextualize these myths in ways which unleash their contemporary emancipatory power for the formation of a *mestiza* identity (Ketz 2010, 201; see also Lewis 1994, 71–73; Rader 2010, 148–149).

Flavio comes from a poor Mexican family and has a checkered career behind him as a dishwasher in Corpus Christi, shrimper in Port Isabel, field worker, and, finally, exterminator (Cisneros 1991, 146–147). An artist manqué, he works as "a bug assassin" (145), as he cannot live from his poetry. Lupe and Flavio discuss and act out their cultural differences, as Flavio is genuinely Mexican, whereas Lupe was born in the United States. This constellation mirrors the respective ethnic affiliations of Cisneros's parents (Brackett 2005, 12–18). Flavio considers Lupe's American sense of time peculiar. His own sense of time proves peculiar, too, when he says: "I believe love is always eternal. Even if eternity is only five minutes" (Cisneros 1991, 150). Lupe invites Flavio to dinner, dresses up for him "like Carmen in that film by Carlos Saura" (150) and he gives her dance lessons in several Latin American dances (151). By implication, Lupe and Flavio negotiate the hybridity of their identities. Pointing at "the alligator on his shirt," Lupe calls Flavio "a product of American imperialism" (151). In response, he claims not to define himself by such superficial markers but to know in his heart where he belongs: "I don't have to dress in a sarape [poncho] and sombrero to be Mexican [. . .] I *know* who I am" (151; original emphasis). She wants to be Mexican like him but is also aware of her ineradicable Chicana identity: "I was not Mexican" (152). In any case, she thinks that, like Prince Popocatépetl and Princess Iztaccíhuatl, Flavio and herself belong to a golden-brown race, neither black nor white and privileged by God. Lupe admires Flavio's skin, "sweet as burnt-milk candy, smooth as river water. He made you *bien* pretty even if I didn't always know it. Yes, he did" (152; original emphasis). But when Lupe examines Flavio's naked body, she also finds tattoos celebrating other relationships, one for a Romelia on his arm and another one for an Elsa on his upper body (153).

In the negotiations of their cultural identities Flavio increasingly appears as a perfect embodiment of Mexican specificity, created by a Godhead who, like Lupe, considers Mexicanness the climax of humanity. This also holds true linguistically. Lupe and her former lover Eddie "were products of our American education" and produce love talk in Spanish that "always came off sounding like the subtitles to a Buñuel film" (153). By contrast Flavio speaks genuine Mexican Spanish: "*That* language. That sweep of palm leaves and fringed shawls. That startled fluttering, like the heart of a goldfinch or a fan. Nothing sounded dirty or hateful or corny. How could I think of making love in English again?" (153; original emphasis). Lupe's love for Flavio more and more reveals itself as the love for an undiluted Mexicanness which she idolizes, since it remains unreachable for herself. By comparison, her own Mexicanness appears contaminated by an American upbringing. Whereas the story in its entirety pursues the formation of hybrid cultural and personal identities as liberating and meaningful, in Lupe's infatuation with Flavio it also celebrates Mexicanness as an unattainable miracle from across the border (see Carroll and Maher 2010, 11; see also Ketz 2010, 202).

But Flavio's Mexicanness does not only include marvelously brown skin and impeccable Spanish but also an ethics that accepts promiscuity, as long as it is that of men. During one of their Sunday breakfasts at Taco Heaven Flavio suddenly announces that he has to leave. He hesitantly explains his "family obligations" (Cisneros 1991, 156), which consist of four sons from his wife and another three from his lover. He explains: "Loving one person doesn't take away from loving another. It's that way with me with love. One has nothing to do with the other. In all seriousness and with all my heart I tell you this, Lupe" (156). In one version of the volcano myth, Prince Popo seems to have died in battle, therefore the princess kills herself or dies from sorrow. Likewise Lupe suffers heavily from Flavio's silence and traceless disappearance (157–158). In her isolation Lupe is completely thrown back on her art. Different from the relationship to Flavio, which celebrates unitary Mexicanness, Lupe's painting turns into a highly syncretist activity, in which diversity spells empowerment and not deficiency. Joining Aztec and Christian myth and legend in her collages, she combines "Church-sanctioned powers" (158) and "folk powers" (159), mixing Christian and voodoo traditions.

Telescoping into each other the qualities of the Aztec goddesses Coatlicue and Tonantsin on the one hand and La Virgen de Guadalupe on the other, Lupe tries to combine and unite the romantic and passive femininity of Christian saints and *telenovela* heroines with the fierceness of pagan goddesses, a union which Cisneros fully elaborates in "Little Miracles, Kept Promises" (126–129). In this vignette, Chayo, the narrator, suggests that a syncretist female spirit, composed of both Aztec and Christian elements, pervades Mexican culture and is responsible for its emancipation: "Chayo heads toward salvation and liberation once she is able to understand that Guadalupe contains Indigenous deities within herself who have been fully endowed with power, something that marks a difference from the Virgin Mary" (Castro Dopacio 2010, 163–164; see Carbonell 1999; Doyle 2004; Veauthier 2011, 263–281; Wilson 2010, 36–42). Chayo finally embeds her synthesis of the belligerent femininity of Aztec Coatlicue/Tonantsin and the forgiving mildness of La Virgen de Guadalupe in a transnational and universalist religious horizon of gender politics: "When I could see in all your facets, all at once the Buddha, the Tao, the true Messiah, Yahweh, Allah, the Heart of the Sky, the Heart of the Earth, the Lord of the Near and Far, the Spirit, the Light, the Universe, I could love you, and, finally, learn to love me" (Cisneros 1991, 128).

In "*Bien* Pretty" Lupe, named after La Virgen de Guadalupe, pursues a similarly comprehensive program of reconciling diverse notions of femininity. Lupe's revolutionary new *mestiza* spirit of hybrid feminism gives additional impetus to her painting. She reaccentuates "the twin volcano painting. Got a good idea and redid the whole thing. Prince Popo and Princess Ixta trade places. After all, who's to say the sleeping mountain isn't the prince, and the voyeur the princess, right? [. . .] I think I'm going to call it *El Pipi del Popo*. I kind of like it" (163; original emphasis). In Lupe's new version, the princess watches over the sleeping or dead prince, not the other way round as in traditional renderings. In a sense, Lupe activates the Princess. The princess is wide awake and so is Lupe who goes on to intensely scrutinize her life. While Lupe cannot change Flavio in real life to make him a more acceptable partner, she can transform both his masculinity and her own femininity artistically.

Nature seems to applaud and to promise betterment. The story ends in January when thousands of magpies, *urracas*, arrive. The arrival of the magpies signals the continuation of ambivalence, of illusion and deception, but also of joyful hope for spiritual renewal:

> And every bird in the universe chittering, jabbering, clucking, chirruping, squawking, gurgling, going crazy because God-bless-it another day has ended, as if it never had yesterday and never will again tomorrow. Just because it's today, today. With no thought of the future or past. Today. Hurray. Hurray! (165)

Cisneros is determined not to "clip her heroines' wings [. . .] The *urracas* at the collection's end bespeak the hope of a new cultural order" (Carroll and Maher 2010, 15; original emphasis; see also Griffin 1997, 93–94; Ketz 2010, 202; Kevane 2003, 66–68; Payant 1999, 104–107; Veauthier 2011, 377–426).

In several of the stories in *Woman Hollering Creek* Cisneros is trying to redefine traditional femininity marked by subservience and acquiescence to men, most notably in "Woman Hollering Creek" and "*Bien* Pretty." She attributes this traditional meekness to the influence of the notions of womanhood propagated by Catholicism and by forms of popular culture. She believes that both an infusion and restitution of the qualities of the Aztec tradition, the fierceness of goddesses like Tonantsin and mythical figures like La Llorona, help change the Latina concept of femininity in an emancipatory direction. This discourse of proper forms of femininity and their cultural derivation supersedes the relation of individual fate. At the end of "Woman Hollering Creek" Cleófilas returns to her father and brothers, an environment in which it will not be possible to live an emancipated life in the style of Felice, her mentor. Like Esperanza Cordero in *The House on Mango Street* and like Sandra Cisneros herself, Lupe Arredondo has the chance to reorganize her life as an independent artist, who gives expression to revisions of cultural myths and religious beliefs. Like Esperanza and Cleófilas, Lupe remains in a liminal state betwixt and between male oppression and independent self-determination, but she is certainly the most developed embodiment of women's emancipation in Cisneros's short fiction.

3 Theoretical Perspectives

In their attempts to reshuffle conceptions of cultural and gender identities – Aztec and Christian, Mexican and Anglo, traditional and emancipatory – "Woman Hollering Creek" and "*Bien* Pretty" contribute substantially to the development of what may be defined as liminal narratives of borderland communities. *Woman Hollering Creek* adds a new dimension of intercultural mythopoesis to Cisneros's work, which both shares and influences the related concerns of other Chicana writers, such as Denise Chavez, Bernice Zamora, or Ana Castillo. The fiction of these writers helps develop the liminal identity of what Gloria Anzaldúa calls the new *mestiza*, a woman whose self-determination plays itself out in ambivalent renegotiations of her borderland existence (Anzaldúa 2012, 101). The concrete qualities of these borderlands are specific but they also represent a broader, transcultural phenomenon.

In a wider anthropological framework, intercultural positionings and the acceptance of ambivalence are also at the core of ethnologist Victor Turner's concept of liminality. Developed in the context of African initiation rituals, Turner expands his theory of liminality to a transcultural expression of both existential insecurity and a dynamics of transition and change: "Liminality may perhaps be regarded as the Nay to all positive structural assertions, but as in some sense the source of them all, and, more than that, as a realm of pure possibility whence novel configurations of ideas and relations may arise" (Turner 1973, 97). As Anzaldúa describes the new *mestiza* and Cisneros characterizes the protagonists of her stories as members of communities, Turner defines liminality as a communal sphere, "where there is a certain freedom to juggle with the factors of existence" (106). In this sense Chicana/o communities and the narratives of initiation and emancipation they yield are based on liminalities.

Community is also at the background of the attempts of Cisneros's protagonists to liberate themselves. Writing about meetings with other ethnic writers, Cisneros foregrounds the communal spirit of their art: "We are men and we are women. What we have in common is our sense that art should serve our communities" (Cisneros 2009, xvii). This fits Sandra Zagarell's definition of narratives of community, which take as "their subject the life of a community (life in its *everyday* aspects) and portray the minute and quite ordinary processes through which the community maintains itself as an entity. The self exists here as part of the interdependent network of the community rather than as an individualistic unit" (Zagarell 1988, 499; original emphasis). But these communities are far from homogeneous. Echoing Victor Turner, they rather represent a "state of being betwixt and between" (Alvarez 2012, 234).

The communal aspect of liminal Chicana identities, as theorized by Gloria Anzaldúa and fictionalized by Sandra Cisneros, is also a fluid form of regionalism manifesting itself in the border existence of the new *mestiza*. In its liminalities the movable South West where Cisneros's and other Chicana writers' texts are located, is associated with the recent resurgence of critical regionalism as defined by Kenneth Frampton among others. In his seminal essay "Towards a Critical Regionalism" (1983) Frampton negotiates between the highly mobile technological universalism of a commodified global civilization on the one hand and a staid, nature-wary regionalism on the other. Frampton criticizes the abstractions of universalist civilization, which threaten to drown out specific cultural ensembles: "Ever since the beginning of the Enlightenment, *civilization* has been primarily concerned with instrumental reason, while *culture* has addressed itself to the specifics of expression" (Frampton, 1983, 17; original emphasis). As Cisneros's fiction mediates between precolonial, colonial and postcolonial traditions, ancient animistic belief systems and contemporary materialist commercialization, Mexican lifestyles and a globalization that spells Americanization, critical regionalism emerges as a liminal meandering between universalizing civilization and cultural specificities in Frampton's influential argumentation, which became a model also for other forms of art: "Critical Regionalism is contingent upon a process of double mediation. In the first place, it has to 'deconstruct' the overall spectrum of world culture which it inevitably inherits; in the second place, it has to achieve, through synthetic contradiction, a manifest critique of universal civilization" (21).

In their deepest layers of meaning, the liminal positionings in Cisneros's and other Chicana writers' stories touch upon the necessity of both rootedness and transformation and the desire for both stability and mobility, built into human existence and especially into contemporary global culture. On a theoretical plane, Klaus Benesch (2012 and 2016) has explored conflicts between adherents of acceleration, mobility, and development on the one hand, and, on the other, thinkers who emphasize rootedness and the specificity of places. Whereas Emmanuel Levinas argues for the embrace of technology-driven mobility and change as an antidote to the discrimination of otherness (Benesch 2016, 103–105), Henry David Thoreau and Martin Heidegger emerge as "representatives of a counter-model to the modern paradigm of spatial-horizontal mobility" (Benesch 2012, 406). In his essay "Bauen Wohnen Denken" / "Building Dwelling Thinking" (1951) Heidegger is amazingly close to Anzaldúa's and Cisneros's emphasis on the constitutive centrality of the border for human existence: "die Grenze ist jenes, von woher *etwas sein Wesen beginnt*. [. . .] Raum ist wesenhaft das Eingeräumte, in seine Grenze Eingelassene" (Heidegger 2000, 156; original emphasis) / "the boundary is that from which something *begins its presencing*. [. . .] Space is in essence that for which room has been made, that which is let into its bounds (Heidegger 2001, 152; original emphasis). By the same token, Cisneros's perennial desire for a house of her own gains a transcultural, ontological dimension in light of Heidegger's definition of humans as dwellers: "Das Wohnen aber ist *der Grundzug des Seins*, demgemäß die Sterblichen sind" (Heidegger 2000, 163; original emphasis) / "Dwelling, however, is *the basic character of Being* in keeping with which mortals exist" (Heidegger 2001, 158; original emphasis). In this light, the oscillation of Chicanas between different options offered in their lives relates to both the fear of unhoused indeterminacy and the yearning for reliable bearings of us all.

4 Bibliography

4.1 Works Cited

Achilles, Jochen. "Changing Cultures of Solitude: Reclusiveness in Sandra Cisneros's *The House on Mango Street*." *Cultures of Solitude: Loneliness, Limitation, Liberation*. Eds. Ina Bergmann and Stefan Hippler. Frankfurt am Main: Peter Lang, 2017. 217–228.

Almería, Ana María. "Male and Female Roles in Mexican-American Society: Issues of Domestic Violence in 'Woman Hollering Creek.'" *Sandra Cisneros's Woman Hollering Creek*. Ed. Cecilia S. Donohue. Amsterdam: Rodopi, 2010. 61–77.

Alvarez, Julia. "Gloria Anzaldúa, *que en paz descanse*." Gloria Anzaldúa. *Borderlands/La Frontera: The New Mestiza*. 1987. San Francisco: Aunt Lute Books, 2012. 232–234.

Anzaldúa, Gloria. *Borderlands/La Frontera: The New Mestiza*. 1987. San Francisco: Aunt Lute Books, 2012.

Benesch, Klaus. "Cultural Immobility: Thoreau, Heidegger, and the Modern Politics of Place." *Amerikastudien/American Studies* 57.3 (2012): 403–418.

Benesch, Klaus. "Space, Place, Narrative: Critical Regionalism and the Idea of Home in a Global Age." *ZAA* 64.1 (2016): 93–108.

Bhabha, Homi. *The Location of Culture*. London: Routledge, 1994.

Brackett, Virginia. *A Home in the Heart: The Story of Sandra Cisneros*. Greensboro: Morgan Reynolds, 2005.

Carbonell, Ana Maia. "From Llorona to Gritona: Coatlicue in Feminist Tales by Viramontes and Cisneros." *Melus* 24.2 (Summer 1999): 53–74.

Carroll, Michael, and Susan Naramore Maher. "Amphibious Women: The Complexity of Class in Sandra Cisneros's *Woman Hollering Creek and Other Stories*." *Sandra Cisneros's Woman Hollering Creek*. Ed. Cecilia S. Donohue. Amsterdam: Rodopi, 2010. 1–16.

Castro Dopacio, María Jesús. "Resemanticization of Chicana Motherhood and Sexuality Through the Virgin of Guadalupe." *Sandra Cisneros's Woman Hollering Creek*. Ed. Cecilia S. Donohue. Amsterdam: Rodopi, 2010. 155–168.

Cisneros, Sandra. *The House on Mango Street*. 1984. New York: Vintage Books, 2009.

Cisneros, Sandra. *Woman Hollering Creek and Other Stories*. New York: Random House, 1991.

Cisneros, Sandra. *Caramelo*. New York: Alfred A. Knopf, 2002.

Cisneros, Sandra. *A House of My Own: Stories of My Life*. New York: Alfred A. Knopf, 2015.

Doyle, Jacqueline. "Haunting the Borderlands: La Llorona in Sandra Cisneros's 'Woman Hollering Creek.'" *Frontiers: A Journal of Women Studies* 16.1 (1996): 53–70.

Doyle, Jacqueline. "Faces of the Virgin in Sandra Cisneros's *Woman Hollering Creek*." *Things of the Spirit: Women Writers Constructing Spirituality*. Ed. Kristina L. Groover. South Bend: U of Notre Dame P, 2004. 256–283.

Fitts, Alexandra. "Sandra Cisneros's Modern Malinche: A Reconsideration of Feminine Archetypes in *Woman Hollering Creek*." https://journals.lib.unb.ca/index.php/IFR/article-/view/7712. *International Fiction Review* 29.1–2 (2002): 11–22. (3 August 2018).

Frampton, Kenneth. "Towards a Critical Regionalism: Six Points for an Architecture of Resistance." *The Anti-Aesthetic: Essays on Postmodern Culture*. Ed. Hal Foster. Seattle, WA: Bay P. 1983. 16–30.

Griffin, Susan E. "Resistance and Reinvention in Sandra Cisneros's *Woman Hollering Creek.*" *Ethnicity and the American Short Story*. Ed. Julie Brown. New York: Garland, 1997. 85–95.

Harvey, Brandy A. "The Cries of *La Llorona*: Maternal Agency in 'Woman Hollering Creek.'" *Sandra Cisneros's Woman Hollering Creek*. Ed. Cecilia S. Donohue. Amsterdam: Rodopi, 2010. 169–188.

Heidegger, Martin. "Bauen Wohnen Denken." 1951. Martin Heidegger. *Vorträge und Aufsätze*. Gesamtausgabe Bd. 7. Frankfurt am Main: Vittorio Klostermann, 2000. 145–164.

Heidegger, Martin. "Building Dwelling Thinking." *Poetry, Language, Thought*. Transl. and intr. Albert Hofstadter. New York: Harper Perennial Classics, 2001. 141–159.

Ketz, Victoria. "Voicing Taboos in Sandra Cisneros's *Woman Hollering Creek and Other Stories*." *Sandra Cisneros's Woman Hollering Creek*. Ed. Cecilia S. Donohue. Amsterdam: Rodopi, 2010. 189–225.

Kevane, Bridget A. "The Fiction of Sandra Cisneros: *The House on Mango Street* (1984) and *Woman Hollering Creek* (1991)." *Latino Literature in America*. Westport: Greenwood P, 2003. 47–69.

Kristeva, Julia. *Strangers to Ourselves*. Transl. Leon S. Roudiez. New York: Columbia UP, 1991.

Lafaye, Jacques. *Quetzalcoatl and Guadaloupe: The Formation of Mexican National Consciousness, 1531–1813*. Chicago: U of Chicago P, 1987.

Lewis, L.M. "Ethnic and Gender Identity: Parallel Growth in Sandra Cisneros's 'Woman Hollering Creek.'" *Short Story* 2.2 (Fall 1994): 69–78.

Madsen, Deborah L. *Understanding Contemporary Chicana Literature*. Columbia: U of South Carolina P, 2000.

Miller, Mary, and Karl Taube. *An Illustrated Dictionary of the Gods and Symbols of Ancient Mexico and the Maya*. London: Thames & Hudson, 1993.

Payant, Katherine B. "Borderland Themes in Sandra Cisneros's *Woman Hollering Creek*." *The Immigrant Experience in North American Literature: Carving Out a Niche*. Eds. Katherine B. Payant and Toby Rose. Westport: Greenwood P, 1999. 95–108.

Rader, Pamela J. "Boys to Men: Redefining Masculinities in *Woman Hollering Creek and Other Stories*." *Sandra Cisneros's Woman Hollering Creek*. Ed. Cecilia S. Donohue. Amsterdam: Rodopi, 2010. 131–149.

Rebolledo, Tey Diana. *Women Singing in the Snow: A Cultural Analysis of Chicana Literature*. Tucson: U of Arizona P, 1995.

Saldívar-Hull, Sonia. *Feminism on the Border: Chicana Gender Politics and Literature*. Berkeley: U of California P, 2000.

Tontiplaphol, Betsy Winakur. "So You'll Know Who I Am: Inventory and Identity in *Woman Hollering Creek and Other Stories*." *Sandra Cisneros's Woman Hollering Creek*. Ed. Cecilia S. Donohue. Amsterdam: Rodopi, 2010. 17–29.

Townsend, Richard F. *The Aztecs*. Third Ed. London: Thames & Hudson, 2009.

Turner, Victor. *The Forest of Symbols: Aspects of Ndembu Ritual*. New York: Cornell UP, 1973.

Veauthier, Ines E. *Perra, Padre, Promise: Identitätskonstruktion im Erzählwerk von Sandra Cisneros*. Frankfurt am Main: Peter Lang, 2011.

Wilson, Shannon. "The Chicana Trinity: Maternal Mestiza Consciousness in *Woman Hollering Creek and Other Stories*." *Sandra Cisneros's Woman Hollering Creek*. Ed. Cecilia S. Donohue, Amsterdam: Rodopi, 2010. 31–52.

Wyatt, Jean. "On Not Being La Malinche: Border Negotiations of Gender in Sandra Cisneros's 'Never Marry a Mexican' and 'Woman Hollering Creek.'" *Tulsa Studies in Women's Literature* 14.2 (1995): 243–271.

Zagarell, Sandra A. "Narrative of Community: The Identification of a Genre." *Signs* 13.3 (Spring 1988): 498–527.

4.2 Further Reading

Augenbraum, Harold, and Margarite Fernández Olmos, eds. *U.S. Latino Literature. A Critical Guide for Students and Teachers*. Westport: Greenwood P, 2000.
Bolaki, Stella. *Unsettling the Bildungsroman: Reading Contemporary Ethnic American Women's Fiction*. Amsterdam: Rodopi, 2011.
Brady, Mary Pat. "The Contrapuntal Geographies of *Woman Hollering Creek and Other Stories*." *American Literature* 71.1 (1999): 117–150.
Fiore, Teresa. "Crossing and Recrossing: 'Woman Hollering Creek' by Sandra Cisneros." *Prospero* 1 (1994): 61–75.
Heide, Markus. *Grenzüberschreibungen: Chicano-Erzählliteratur und die Inszenierung von Kulturkontakt*. Heidelberg: Winter, 2004.
Ikas, Karin. *Die zeitgenössische Chicana-Literatur: Eine interkulturelle Untersuchung*. Heidelberg: Winter, 2000.
Jacobs, Elizabeth. *Mexican American Literature: The Politics of Identity*. London: Routledge, 2006.
Kelley, Margot. "A Minor Revolution: Chicano/a Composite Novels and the Limits of Genre." *Ethnicity and the American Short Story*. Ed. Julie Brown. New York: Garland, 1997. 63–84.
Kevane, Bridget A., and Juanita Heredia. *Latina Self-Portraits: Interviews with Contemporary Women Writers*. Albuquerque: U of New Mexico P, 2000.
Martínez, Julio A., and Francisco A. Lomelí. *Chicano Literature: A Reference Guide*. Westport: Greenwood P, 1985.
Socolovsky, Maya. *Troubling Nationhood in U.S. Latina Literature: Explorations of Place and Belonging*. New Brunswick: Rutgers UP, 2013.

Frank Newton

30 Louise Erdrich (1954–)

Abstract: This chapter analyzes two of Louise Erdrich's short stories, "Fleur" and "The Butcher's Wife," both published in *The Red Convertible: Selected and New Stories, 1978–2008* (2009). The short stories exemplify Erdrich's exploration of her mixed Native American and European American heritage alongside their respective oral and written traditions. They were published both as independent short stories and as chapters or parts of her novels, which highlights the cyclical nature of Erdrich's writing. When read together, these two short stories show how Erdrich successfully explores and navigates relatable Native and non-Native American experiences within their respective "re-publications." This chapter discusses the two short stories in- and outside of Erdrich's broader literary works and argues that the shift in focus from Native American to European American stories within the fictional setting of her works defines Erdrich not as a mainly female Native American author, but as a distinct American storyteller who connects communities outside geographical and ethnic or racial boundaries.

Keywords: Authenticity, orality, storytelling, community, Native Americans

1 Context

Louise Erdrich is, first and foremost, a storyteller. "In a tribal view of the world, where one place has been inhabited for generations, the landscape becomes enlivened by a sense of group and family history. Unlike most contemporary writers, a traditional storyteller fixes listeners in an unchanging landscape combined of myth and reality. People and place are inseparable" (Erdrich 1985, 1). Louise Erdrich's writings are as deep and rich as they are manifold: a total number of 78 publications to date, including fiction, non-fiction, poetry, children's literature, and a memoir of early motherhood attest to her tremendous literary oeuvre. Most famous for her novels, Erdrich evokes in them a strong sense of community that combines her European American and Native American heritage with a distinct focus on her Chippewa (Ojibwa) roots. It is this combination, the European American written tradition together with the Native American oral tradition characterizes Erdrich as a distinct American storyteller. Consequently, categorizing Louise Erdrich mainly as a *Native* American writer does not take into account the diversity and breadth of her background. Similarly, her novels are a hybrid form which are best situated in the genre of the short story cycle (see Nagel 2001, 19; Smith 2018, 79). What is characteristic of her stories is not her identity in terms of race and gender, but Erdrich's ability to connect her readers to her fictional world(s) in a manner that makes them believe

https://doi.org/10.1515/9783110587647-031

her fiction "could be true" (Erdrich 1986, 35). In other words, she reflects upon and develops her lived experiences in her writing, transmuting them into stories her readers can believe in and (to a certain extent) identify with. This process is both static and fluid. As for Erdrich's prose, this somewhat paradoxical process and narrative mode derives itself from the intersection of the European American written tradition and the Native American oral storytelling tradition. For her writings are static, in a sense that her written word has been put into print and published. Fluid, in a sense that she has revised her manuscripts (e.g., *Love Medicine* was published in 1984, with revised and expanded editions in 1993 and 2009). In terms of form and genre, her writing is very much in between the novel, the short story, and poetry; oftentimes one emerges from the other, developing the stories further and attesting to the cyclical nature of her storytelling (see Nagel 2001, 24). Or, as Nowick has put it, Erdrich is a writer who "specializes in hybrid genre, fiction with the intensity and lyricism of poetry, short story sequences that transcend themselves to become novels" (Nowick 1986, 74). This blend of narrative traditions into a "hybrid genre" is best articulated in Erdrich's understanding of the way stories are communicated within her writings: the characters' voices are those of storytellers, and these storytellers transmute her readers into listeners (see Chavkin and Chaikin 1994, 231; Stookey 1999, 13). As for categorizing Erdrich according to race and gender, her choice to identify as Native American (see Jacobs 2001, 179), along with ongoing debates on what exactly constitute authentic and legitimate literary contributions of a given marginalized group (in this case, Native American) have shown that this is an issue impossible to solve in a satisfying manner (176–178). However, what these discussions reveal is an ever-increasing bifurcation of marginalized categories which will split into ever smaller categories until they reach the level of the individual. Ultimately, literary output will be viewed with "writing [a]s the denominator" (177).

Born on June 7, 1954, in Little Falls, Minnesota, Karen Louise Erdrich was the oldest of seven children. The daughter of Ralph Louis and Rita Joanne (Gourneau) Erdrich, she grew up in a closely knit family which encouraged her to write stories at an early age (see Stookey 1999, 1). Her family instilled in her a strong sense of community, and Erdrich's understanding of what a story is and how to communicate it has been highly influenced by them: "[w]hen I think what's a story, I can hear somebody in my family, my Dad or my Mom or my Grandma, telling it. There's something particularly strong about a *told story*. You know your listener's right there, you've got to keep him hooked – or her" (Bruchac [1987] 1994, 103–104). The emphasis Erdrich puts on orality here is a reflection of her way of absorbing stories, and "listen[ing] in on everything else" (Wong [1987] 1994, 46) was as important in shaping her narrative style as pouring over books (see Jacobs 2001, 54). It also exemplifies the nature (if not dating the origin) of Erdrich's "hybrid genre" (Nowack 1986, 74); a blend of German American (paternal) and Chippewa (Ojibwa) French (maternal) heritage which brings together written and oral narrative traditions in a manner that gracefully sidesteps attempts to place her works in either one of those camps. The active choice to *be* a Native American stretches the boundaries of her inherited identity even further, to an extent that the seemingly non-overlapping magisteria of ethnic authenticity and legitimacy become heavily intertwined. Her maternal grandparents – with whom she spent so much time on the North Dakota Turtle Mountain Reservation that it practically became a "second home to her" (Jacobs 2001, 54) – are the reason she qualifies (legally speaking) for membership on a tribal roll. However, as has been pointed out by scholars, "being an Indian is often a matter of identification with a worldview and a way of life rather than a classification based on actual amount of Indian blood" (53). Erdrich's parents were teachers at a Bureau of Indian Affairs school in Wahpeton, North Dakota; and it is here that she spent most of her youth. When she decided to enroll at Dartmouth College and left for New Hampshire in 1972, it was the first time she had travelled so far from her home and community (Stookey 1999, 2).

Famously established in 1769 by Congregational minister Eleazar Wheelock in Hanover, New Hampshire, Dartmouth College was initially founded to assimilate Native Americans into Christian-European culture by means of education (see Childs 1957, 16). While this endeavor proved to be unsuccessful and the college mainly trained Congregational ministers during its early years, when Louise Erdrich enrolled at Dartmouth in 1972, it was under the auspices of two distinct amendments: first, 1972 marked the year that allowed women to be admitted to the college and, second, the inclusion of a Native American Studies program (see Stookey 1999, 3). Both relevant to Erdrich's own identity and meaningful in terms of making education accessible for women, as well as enabling research in Native American Studies as a program, she also met her future husband Michael Dorris, a man she would marry in October 1981 (3–4). Dorris played an important role in Erdrich's writing process, and the way they interacted with each other to produce some of Erdrich's writings highlights the community-driven aspect which is often placed at the center of her writing. The encouragement Erdrich received from Dorris certainly adds to her "breakthrough into the established literary world" (Jacobs 2001, 55). "The World's Greatest Fishermen," later featuring as the opening chapter of Erdrich's first novel *Love Medicine* (1984; with subsequent and expanded editions in 1993 and 2009), was written in only two weeks and then submitted for the 1982 Nelson Algren Prize (see Stirrup 2010, 165). Out of the 2,000 stories submitted to the newly established fiction contests, "The World's Greatest Fishermen" won and "propelled the couple into literary acclaim" (Jacobs 2001, 56). "Scales," another short story to be featured in her first novel, was printed in *The Best American Short Stories of 1983*. In the early stages of their marriage Dorris and Erdrich wrote together. They published co-authored pieces under the pseudonym Milou North; "*Mi*chael plus *Lou*ise plus where we live" (Grantham 1985, 46). Because the nature of their collaboration had been something of a secret, the revelation that the two had individually published writings born from a collaborative process "captured the imagination of the literati" (Stirrup 2010, 166), further heightening Erdrich's literary acclaim. It also exemplifies the blend of European American written tradition and the Native American oral storytelling tradition: "[t]he basis of the collaboration, says Dorris, is talk – 'weeks of conversation' about a character's looks, his clothes, her jewelry. 'The details of clothes and action are invented and thought about between the two of us,' says Erdrich" (Grantham 1985, 44). Community, too, as a central part of Erdrich's writings, can thus be found not only within her works, but outside of her literary productions as an integral part of her everyday life. In a literal sense, Erdrich's stories *store* her narratives. This idea of narratives storing a story also implies that retelling a story is not done word for word but that the idea itself is being communicated. This "temporary storage" (Chavkin and Chavkin 1994, 234), as Erdrich describes publication, further substantiates the claim that fluidity and mutability are integral and defining characteristics of her narrative style.

While Louise Erdrich has continued to publish a wide variety of short stories and novels (most notably and recently *The Night Watchman*, 2020), this article focuses on two of her short stories: "Fleur" and "The Butcher's Wife." These two short stories, as they appeared in *The Red Convertible: Selected and New Stories, 1978–2008* (2009), give an insight into the development of Erdrich's short prose. They do this not only as individual short stories, but within the context of the novels they have been published in – *Tracks* and *The Master Butchers Singing Club*. What a close reading of the two short stories reveals, is, firstly, how character development and story progression shifts from a geographically bound area to a more global setting, rooted in the articulation of Erdrich's diverse ethnic background and, secondly, how retelling the characters' experiences within short stories (i.e., Pauline in "Fleur" and Delphine in "The Butcher's Wife") moves Erdrich's writing from a community-oriented narrative to an exploration of the characters' nature. It is both the cyclical nature of Erdrich's writing and the different framing of the stories in their respective iterations, particularly as short stories, that reveals how the characters situate themselves within the broader narrative of Erdrich's writings.

2 Close Readings

The short story "Fleur" ([1986] 2009) gives a loose outline of the life of Fleur Pillager, a Chippewa woman, and focuses on her brief stay in Argus, North Dakota, during the summer of 1920. Originally published in *Esquire*, August 1986 and then as the third-place entry of *O. Henry Prize Stories, 1987*, the short story also appeared as part of the second chapter of *Tracks* (1988). Republished in an altered and abbreviated form, it is part of Erdrich's *The Red Convertible: Selected and New Stories, 1978–2008* (2009, 176–189). The story is narrated from the first-person perspective of Pauline Puyat and begins with an account of "[t]he first time" Fleur Pillager drowned (Erdrich 2009, 176). Her boat had tipped and two men, witnessing her struggle in the water, rushed to save her. This proved to be their downfall, and "it wasn't long afterward that those two men disappeared" (176). The strange occurrence is amplified by Fleur's second drowning. This time, "no one touched her" (176). However, upon closer inspection by George Many Woman, Fleur opens her eyes and prophesizes – maybe a curse, even – he will take her place (see 176). The earlier drowning had made people wary, and George Many Woman became paranoid to the point where he would neither leave his house nor go near water. Yet, the first time he took a bath in his "new tin bathtub," George Many Woman slips and drowns in it, the macabre scene intensified by his wife obliviously preparing breakfast in the room next to him (176). These circumstances make it clear to the narrator (yet to be named as Pauline Puyat) and the people within the community that "Misshepeshu, the water man, the monster, wanted her for himself" (177). The trope revolving around Native American spirituality is emphasized by the proclamation that drowning is "the death a Chippewa cannot survive" (177). It is further developed by Fleur's ability to transform her body: upon following her tracks, people from the community – made clear by the collective "we" throughout the paragraph – realized that their shape shifted from human footprints to claw tracks (177). At the end of this introductory account of Fleur's early years (the second drowning occurred when she was 20 years old (176) and her subsequent mischievous and in part frightening behavior, the narrator points out that the near-destruction of Argus is "what this story is about" (178).

Fleur arrives at the small town, possibly drawn to it by the town's catholic church, the "sign of pride, that slim prayer, that marker," and begins to work at the town's butcher's shop. It is here that the narrator reveals herself to be the stepdaughter of Dutch James, one of three men working at the butcher's (178). Dutch "took me [Pauline] out of school to take her [Pauline's mother's] place" (178), making it clear that the narrator was forced by her stepfather to perform menial tasks. Additionally, the narrator emphasizes that she was invisible to the people in the butcher's shop (unless she was needed for a particular task), and that it was her invisibility that gave her an insight into "everything, what the men said when no one was around, and what they did to Fleur" (179). The outright statement that the three men will harm Fleur, ominous as it is on its own, also foreshadows the men's impending doom, seeing how the three men who had previously come too close to her had all died. Next to a (self-)description of Pauline, the butcher's meat locker is described in all its "marvel" (179). It is "where Fleur worked," whom the butcher, Pete Kozka, had "hired for her strength" (179), and Fleur works together with Kozka's wife, Fritzie Kozka, who teaches Fleur how to cut meat.

The three men – Dutch James, Lily Veddar, and Tor Grunewald – routinely play poker and are occasionally joined by Pete Kozka. They converse while they play, Pauline listening in the shadows; although she points out that "there wasn't much to hear since almost nothing ever happened in Argus" (179). When Fleur arrives at Argus and starts working at the butcher's, this changes as she "finally gave them a subject" (180). Since "[w]omen didn't usually play with men, [sic] the evening that Fleur drew a chair up to the men's table without being asked [sic] was a shock of surprise" (180). Lily is the first to comment, and his dog, the "mean little bull," tries to bite Fleur "but cringe[s] back" (180). The narrator, who in this section is addressed by name for the first time, hides in the shadows and watches Fleur play cards with the three men. Pauline is asked to "stake" Fleur when the three men simultaneously leave the room for a break (181). Fleur looks at Pauline with a "white wolf grin a Pillager turn on its victims, except that she wasn't after me" (181). This grin is a direct expression of Fleur's dual spirit heritage, part human and part animal. The specific role it takes within the short story is both to emphasize Fleur's connection to the spirit world and to highlight Fleur's hunter-like characteristics. The powerful, wolf-like grin has an immediate effect on those it is aimed at; Pauline seems intimidated yet realizes she is not the victim of Fleur but an assistant to her. Pauline compares herself to Fleur and finds herself lacking, i.e., neither as attractive nor youthful as Fleur, despite the fact that Pauline is younger than Fleur. In this comparison she sees the reason she is not noticed by anyone: "I wasn't anything like Fleur. My dress hung loose and my back was already curved, an old woman's. Work had roughened me, reading made my eyes sore, caring for my mother before she died had hardened my face. I was not much to look at, so they never saw me" (181). Fleur plays with the men, and after Tor drops a non-chalant, sexist-racist remark in regards to Fleur's gender and ethnicity, Pauline falls asleep and is carried and put to bed by her in the closet in Fleur's room (181).

Fleur wins a dollar during the poker game and repays Pauline "five times over" (181). She continues to win exactly one dollar every night playing cards with the men, making them "lit with suspense" to the point they make Pete join the poker nights (181). The three men's animosity towards Fleur continues to rise, while Lily, in his ignorant, utter disbelief cannot fathom "that a woman could be smart enough to play cards, but even if she was, that she would then be stupid enough to cheat for a dollar a night" (182). He consequently wishes to break her run of luck more than anything, by however small a margin. Here, we find out that "Fleur bathed in the slaughtering tub" (183), a somewhat macabre and paradoxical scene: she sleeps in the very same tub the animals are not put to rest but slaughtered. For Fleur, who is said to be able to shape shift into an animal, to bathe in the very spot animals are slaughtered brings to the fore the connection between Fleur's prior drownings, the death of the three men who had tried to save her or came too close to her, and the water man, Misshepeshu, within the sterilized setting of a butcher's slaughtering area. Additionally, Pauline confesses a certain attachment to Fleur, describing herself as becoming "her [Fleur's] moving shadow that the men never noticed," yet also as "the shadow that could have saved her" (183), further consolidating the earlier foreshadowing of what is to happen.

After a month and once Pete and Fritzie Kozka "left for Minnesota to escape the heat" (183), Lily sees his chance to break Fleur's streak. One particularly hot evening, while Fleur is boiling heads in the room next door, the three men eagerly wait to play cards, Lily urging her to finish work for the day and join them for the game (183). After Fleur "stood in steam, turning skulls through a vat with a wooden paddle" (183), the four start playing cards. Lily raises the stakes, and after several rounds of Fleur winning and loosing, while "[t]he pot grew, round by round, until it held all the money" (184). When Fleur wins the pot, Lily – who had been caressing and talking to his dog while they were playing – "squeezed the dog up like a fist of dough and slammed it on the table" (184) once he realizes Fleur has won the game. She refuses to keep on playing and leaves. The three men are left behind; first dumbfounded, then (incited by the whiskey from Pete Kozka's private stock they had helped themselves to) they silently plan what to do with Fleur, stand up and go outside (185). Pauline follows them, staying hidden in the dark, and observes the scene that is to unfold: Lily assaults Fleur who is on her way to tend to a sow. After a brief struggle, Fleur gets knocked over and the money she won from the poker game is scattered on the ground, while Lily stumbles over the sow and gets bitten multiple times (185–186). Fleur runs to the smokehouse trying to escape, but the men catch her and have their way with her. All the while, Pauline is rigid with fear and unable to move. She knows that she "should have got to Fleur, to save her, throw [herself] on Dutch" (186), but she is only able to close her eyes and stays hidden, not seeing what had occurred but clearly hearing "Fleur's hoarse breath, so loud it filled me [Pauline], her cry in the old language, and my name repeated over and over among the words" (186). The next day, the three men sit in the butcher's shop, still hungover but drinking before noon (186). When Pauline steps outside to clear her mind and catch a breath, she notices that the hot, sultry weather was heavily clouded. As soon as one of the clouds "flicked out and became a delicate probing thumb" (186–187), she ran inside looking for the three men. They had left, and Pauline ran through the shop looking for them. When she "stopped at the heavy doors of the lockers" (187), she realizes the three men must have taken shelter in it. Suddenly, Pauline "heard a cry building in the wind, faint at first, a whistle and then a shrill scream that tore through the walls [sic], [it] spoke plain so I understood that I should move, put my arms out, and slam down the great iron bar that fit across the hasp and lock" (187). Pauline had – finally – managed to take action against the three men, her pent up terror and her timid, shadow-like existence channeled into decisive, vengeful action by the tornado. The tornado, a force of nature, is described to have agency and shares similarities with the sow from the earlier part of the story; looking for its target and unleashing all its power onto those it sought to punish. It seems to have been conjured up by Fleur, who (by way of the wind) instructed Pauline to close the locker from the outside, trapping the three men inside it. Pauline "understood" (187) what to do, and in conjunction with the wind ordering her to close the lock, it further seems as if Fleur took control of Pauline through the medium of wind; making sure that the three men who had callously assaulted her found their certain end in the place they thought themselves to be safest. Meanwhile, Pauline went outside and was picked up by the tornado, sending her flying but landing her safely "somewhere in a ball" (187). She witnesses stranger things still: "[a] herd of cattle flew through the air like giant birds, [sic] [a] candle, still lighted, blew past, and tables, napkins, garden tools, a whole school of drifting eyeglasses, jackets on hangers, hams, a checkerboard, a lampshade, and at last the sow from behind the lockers, [sic] as everything in Argus fell apart and got turned upside down, smashed, and thoroughly wrecked" (187). Especially the still lit candle and the sow are noteworthy. A candle staying lit while it is being blown around in a tornado, while physically impossible, emphasizes the selective nature of what the tornado seeks to destroy. It also highlights the Native American spirituality and powerful nature of Fleur's ability; harnessing and manifesting the might of nature to smite those who cross her. With this, "what this story is about" (178) becomes clear: Fleur had almost destroyed the town by unleashing a tornado on the three men in Kozka's butcher shop in vengeful punishment.

In the aftermath of the disaster, the people in the town do not realize that Dutch, Lily, and Tor are missing until the Kozkas come back from Minnesota. The damage dealt to the buildings in the town is described as extensive, and the steeple of the Catholic church that had possibly drawn Fleur to Argus "had been torn off like a peaked cap and sent across five field," while "the iron bathtub had been flung – a hundred feet" (187). Other objects, even entire rooms, remained unscathed: "[t]he glass candy went fifty [feet], and landed without so much as a crack," and "the back rooms where Fritzie and Pete lived were undisturbed" (188). Not even a speck of dust inside the back rooms had been moved by the tornado, and when Fritzie gazed at the ruins of the smokehouse, where Fleur had slept, and she inquired about her and the three men, "the town understood that three men were missing" (188). Fleur's whereabouts, however, seem to be irrelevant to the people in Argus, and in "a rally of help" (188) they remove the debris from the locker until they are able to enter the freezer without – as Fritzie insisted – damaging it and spoiling the meat (188). How the people in Argus knew to look for the men in the meat locker is not explicitly stated, although one may assume that, since everyone knew that the three men worked at Kozka's Meats, they might have sought shelter in the sturdy freezer. Pauline is, together with Fritzie and Paul, the first to enter the freezer "since it was my stepfather lost" (188). In it, they discover the three men; "hunched around the barrel where the game was still laid out, and a dead lantern and an empty bottle too," the dog "perched among them, heavy as a doorstop" (189). The three had tried to cover themselves in bearskins to escape the cold, but without avail. That the three had specifically chosen bearskins is a reference to Fleur's ability to shape shift into one (177). Neither alcohol nor the bearskins provided the three men with enough warmth to survive, and that the thick bearskins gave no shelter from the cold inside the freezer can be read as an additional spiritual intervention on the part of Fleur to make sure they received the punishment she intended.

The short story ends with Pauline being drawn back to the reservation by blood, “as if it runs through a vein of earth. I’ve come home and, except for talking to my cousins, live a quiet life” (189). Although implied earlier, it is now clearly stated that Fleur, too, returned to the reservation and lives “down on Lake Turcot with her boat” (189). People keep avoiding her but gossip extensively; they think she could have married Misshepeshu, lives “in shame with white men or windigos, or that she’s killed them all” (189). She has almost no visitors, except Pauline, who helps her deliver a baby during a winter that is specified merely as the “last” (189). The baby girl, “whose green eyes and skin the color of an old penny made more talk” (189) and whose father none but Fleur knows for certain, smiles while she sleeps, “as if she knows what people wonder, as if she hears the old men talk, turning the story over” (189). The baby’s inherited smile is a reference to Fleur’s smile, which is the hallmark expression identifying her as a member of the bear clan.

“The Butcher’s Wife” was originally published in *The New Yorker*, April 17, 2000 and it appeared in *O. Henry’s Prize Stories, 2002*. Similar to “Fleur,” the short story came in third and it is also a chapter in one of her novels, *The Master Butchers Singing Club* as well as *The Red Convertible* (399–420). It tells the story of Eva and Fidelis Waldvogel, two German immigrants, from the third-person perspective of Delphine Watzka, daughter of Polish Roy Watzka, the town drunk. Roy is described as a “dangerous romantic” (399); one who fell to drink after the death of his wife, Minnie Watzka. What remains of her are photographs and their daughter Delphine. Roy chooses to “worship” the photographs instead of his daughter, and he occasionally speaks to Minnie during the night while drinking “until, from deep in his cups, she answered” (399). Growing up, Delphine came to know her father as a “pickled wreck,” one who “destroyed the organ he’d mistaken for his heart” (399). Roy cannot sustain himself on his own, and in early 1936, the adult Delphine “quit secretarial school and moved back from the Twin Cities to their Minnesota farm to care for him” (399). Delphine cares deeply for her father, yet his alcoholism instills in her a sense of helplessness and resignation, in part fueled by the lack of a supporting parental figure in her life. It is in the context of “longing to see her mother’s face” (400) that Delphine meets Eva Waldvogel. The “mentally worn-out” Delphine, who comes into Waldvogel’s Meats to buy half a pound of lard, is still preoccupied by Roy’s “insistence that since he was dying anyway he might as well kill himself more pleasurably with schnapps” (400). After Eva advises Delphine on different types of lard, advertizing that her “husband was trained back in Germany as a master butcher [and] uses a secret process to render his fat” (400), she inquires about Delphine’s father. The “direct but polite manners” (400) of Eva have a likable effect on Delphine, who notices “an odd golden streak that would turn black when the life finally left her body, like a light going out behind the crack in a door” (400) – a concrete foreshadowing of an event that will not come to pass within the frame of the short story: Eva’s death. The choice of words here also suggests a prolonged period of suffering that is to take place in the future, with the loss of life being more of a blessing than a curse. Upon finding out that Delphine's father is Roy Watzka, Eva takes her behind the counter and into the kitchen to have a cup of coffee and "teach [Delphine] to make a mincemeat pie better than [she had] ever eaten" (400). When the two of them enter the brightly lit kitchen, "all time stopped. As [Delphine] took in the room, she experienced a profound and fabulous expansion of being" (401). Delphine is "[s]uddenly extremely happy" (401), marking the kitchen as a place of respite, the solemn and serene atmosphere giving the scene an almost supernatural touch. She is rendered speechless, but "that was fine, for Eva had already whipped one of the pencils out of her hair and taken up a pad of paper to set down the mincemeat recipe" (401). Here, Delphine was "grateful" to notice the only flaw she sees in Eva: "she wasn't a good speller" (401). It makes Eva approachable, and Delphine, "who had never really had a mother, much less a sister, who cleaned up shameful things in her father's house, who had been toughened by cold and hunger and was regarded as beneath notice by the town's best society, and yet could spell – stole confidence from the misspelled recipe" (401). It is quite clear that Delphine is drawn to Eva from their very first encounter onwards, and the brief stay in the kitchen beatifies Eva almost to the point she is beyond approach. Yet, in her description of Eva, one can already discern how Delphine's longing for her mother Minnie might be exchanged by a wish to be accepted and liked by Eva. Recalling the "golden streak that would turn black when the life finally le[aves]" (400) Eva's body and the emphasis on all things German in Waldvogel's Meats, one can see how Erdrich's writing moves from a community-oriented narrative to an exploration of her characters' heritage and nature.

The second time Delphine visits Waldvogel's Meats, the descriptions of her surroundings are a continuation of the religious imagery evoked earlier: the "cheerful shop bell," Eva's hair "bound back in a singular knot, in the shape of the figure eight, which Delphine knew was the ancient sign for eternity," and the sun meeting Eva's hair in a halo-like fashion (402). Delphine had not come to simply place the same order as before, but to ask Eva to help her provide a "cleanser [sic] that would be strong enough for the work Delphine had to do to make Roy's place habitable again" (402). The notion of purification can be read as one of Eva's almost supernatural abilities, especially since "Eva wanted to concoct something of her own" (402). The calm, serene atmosphere in the kitchen is suddenly disturbed when Delphine "sensed" Fidelis Waldvogel before the master butcher enters the room, "like a surge of electric power in the air when the clouds are low" (403). The preoccupation with Roy Watzka and his alcoholism is abandoned in light of the "heaviness" Delphine experiences (403). When Fidelis "suddenly filled the doorway," it is not helplessness and longing filling her, but a profound sense of awe at the man "shed[ding] power, as though there were a bigger man crammed into him" (403). Fidelis looks at Delphine and the "shock of his gaze caused her to pick up one of Eva's cigarettes. To light it" (403). Similar to how Delphine was rendered speechless the first time she was in the kitchen, the sudden presence of Fidelis seems to make her deaf: the first words Fidelis utters in Delphine's presence are not recounted and Delphine merely notices that he "conversed with his wife" (403). Delphine is intimidated by Fidelis and "hoped that she could avoid him [sic] so long as she could still be friends with Eva, and hold the job that she soon was offered, waiting on customers" (403).

Delphine, who from then on "used the back door" and "heard the customer bell ring from the other side of the counter," "had met most of the regular customers" after her first week (403). It is here that she meets Tante Marie-Christine, Fidelis's sister. Tante – the German word for "aunt" – is described as moving with "elegance," and Delphine thinks her "feet [are] slim and pretty" (403). Other than that, Delphine is intimidated by her, noticing that Tante "resembled her brother, replicated his powerful neck and too stern chin, and the eyes that on him were commanding on her were a ghostly blue that gave Delphine the shivers" (404). When Tante asks Delphine who she is, "without deigning to accept an answer" (404), Delphine is cast into a self-deprecating existential crisis as she ponders the questions's meaning. She imagines the deeper meaning of Tante's question to be a judgement of Delphine's being and her past; "Delphine Watzka, you drunkard's child, you dropout secretary, you creature with a belly of steel and a heart that longs for a mother? Who are you, *what* are you – born of a dirty Pole in a Polack's dirt? [sic] What makes you think you belong anywhere near this house, this shop, and especially my brother Fidelis, who is the master of all that he does?" (404). It becomes clear that Delphine does not view herself as a member of the community, be it the town of Argus as a whole or the German immigrant community, i.e., the Waldvogels. Furthermore, she thinks she is unfit to be a member of a community; her self-described dysfunctionality and that of her father have no place within a nurturing, wholesome and productive environment. Given that Delphine's goal is to stay "friends with Eva, and hold [a] job" (403) within the context of her saint-like view of Eva, one can assume Delphine tries to redeem herself through both her friendship with Eva and the work she does in the butcher's shop.

During the "brutal heat" of the summer of 1936, Delphine notices "how her friend suffered" (405). An analogy is drawn between Eva's decreasing stamina and the meat in the meat lockers, which "had to be checked constantly for rot" (405). Yet the connection is not explicitly stated and Eva's failing health, while alarming to Delphine, is not viewed as life-threatening. One of the reasons for this may be Delphine's elevation of Eva, whose saint-like character also exhibits a tremendous strength and resilience; Delphine simply cannot imagine Eva being toppled by anything. Fidelis, however, has a more distressing impact on Delphine. One morning, while he is sharpening his knives with his back to Delphine, she hears him sing for the first time. She "realized, with a strange shock, that his voice was very beautiful [sic] and his voice dismayed her, so pure in a room that was slippery with blood" (406). Fidelis's singing disrupts and distracts Delphine, but the real shock is to occur the next morning. Eva, who had been busy baking during the night, admits to

Delphine that she is not feeling well (407). After a brief conversation between the two in which they hint at the possibility that Eva's discomfort is due to menstrual cramps (407), Eva seems to regain some of her strength and continues to work. Delphine is at first reassured by this, but when Eva "took one of Delphine's hands in hers [sic], she said the words that even in the heat chilled her friend. 'Take me to the doctor'" (407). Delphine rushes Eva to the local doctor, Dr. Heech, who arranges an operation at the clinic to the south. Eva is drugged with morphine to counter the pain, and it becomes clear to Delphine that Eva's "illness was remarkably advanced" (408). The illness Eva suffers from is cancer, and although she is operated on at the clinic, the subsequent treatment is not successful: "[t]he cancer spread" (409).

Delphine stays at Eva's side throughout the treatments, "even [sharing] a sympathetic morphine sweat" (409) with Eva. Delphine is powerless and curses God, taking little comfort in the fact that she is not a hypocrite (410). The religious come knocking at Eva's door in the form of Tante, with "a host of pious Lutheran ladies" (410). They try to convert the Catholic Eva on her deathbed, "crowding around the bed like a flock of turkey vultures in a gloating prayer circle" (410). Delphine, as opposed to Tante, is making herself useful around the Waldvogel's house, washing the laundry and cleaning the house now that Eva is incapacitated. Delphine considers doing these chores a "good-bye gift" (410) and plans to leave after Eva dies. She knows that Fidelis has people who can look after him and the children, for example Tante, who will "find stepping in to care for the boys and her brother a perfect showcase for her pieties" (410).

One of the people taking care of Eva and making himself useful is Roy Watzka. The "pickled wreck" (399), as Delphine had come to know him, suddenly "came to the shop almost every afternoon, sometimes stinking of schnapps" (411). In contrast to Tante and her friends, who "gorged on Eva's pain and her signature Linzer torte" (410), Roy takes care of the menial labor and entertains Eva with miraculous stories of his adventurous youth. Delphine, who never experienced any sort of interaction like this with her father, "was both glad of Roy's adept distractions and resentful" (411). She can quite clearly see the potential familial community Roy and Delphine could have shared, yet she had "cleaned up after him all her life and never had he talked to her like this" (411). Delphine wanted a father figure to look up to and rely on, but was deprived of this relationship all her life due to Roy's alcoholism. The fact that Roy shows a capacity to help and entertain, to make himself useful and be reliable but does not extend this caring nature to Delphine makes it clear that she is not – maybe even cannot – be part of a familial (i.e., Delphine and Roy) or geographical (i.e., the people of Argus) community. This is further substantiated when Eva and Delphine spend the evening in Eva's garden and talk about death and their thoughts on the afterlife. The two women spend a "very good day" (412), one of the last that Eva would have, sitting in the garden and sipping on a few bottles of beer while Delphine thinks about the upcoming events: "[she] had always had a tendency to think about fate, but she did so more often now that Eva's sickness put her constantly in mind of mortality" (412). Delphine is attached to Eva, the only person she seems to share a real connection with. The fact that Eva is close to death can be read as an ultimately unsuccessful effort by Delphine to become part of a community, further emphasized by the first articulation of her earlier thought to leave (412). Yet, the connection Delphine and Eva have is dear to both of them, and they jokingly extend it to the afterlife: "[s]ave me a cloud and I'll play a tune with you" (413).

During the first weekend in September, Labor Day, the town comes together at the Waldvogels's yard to celebrate. While they are outside, playing games and getting drunk, Eva and Delphine stay inside – Eva unable to move from the bed and Delphine keeping her company, staying close to her for when she requires her medicine or a dose of opium to relieve her pain (413). Eva's bed had been placed in front of the window so she could see the celebrations unfold, and the two of them watch the townsfolk guiltily drinking liquor hidden away in the gooseberry fronds (413). Delphine can sense it is time for Eva's dose, for her "sense of time passing had to do only with the duration of a dose of opium wine" (414), and she sets out to prepare the needles for the injection. Delphine rummages through the icebox and notices the "morphine that Fidelis had labored with vicious self-disregard to pay for and which Delphine had guarded jealously was gone" (416). Delphine acts immediately. She runs outside to inform Fidelis, spoon-feeds Eva the rest of the opium wine that is left (Eva cannot keep it down), and then – following a sudden thought that comes to her mind – drives to Tante's house (417). When Tante opens the door and asks Delphine what she wants (in German, to highlight the "great pretense of having trouble understanding [Delphine]" (417)), Tante's face "had all the knowledge in it and Delphine knew she'd guessed right" (417). Delphine is visibly upset by Tante's lack of compliance, and she tries to get Tante to tell her where the opium is, at first politely, then angrily screaming, and finally begging Tante to relieve Eva of the horrible pain she is in (417–418). Tante, who has no idea of the pain she is inflicting on Eva, stubbornly refuses to tell Delphine what has happened with the morphine. Her "cold twist of a smile" (417) is a reflection of the superiority Tante feels, and "with the upper hand, she enjoy[s] her moment" (418). While Eva is suffering from excruciating pain, Tante tells Delphine that Eva's dependency on morphine is seen as an addiction by Tante's Lutheran friends and tarnishes the reputation of her brother: "[t]his cannot be. The wife of my brother? It is a great shame on us" (418). Tante is under the impression that she knows more than the doctor, and when Delphine almost falls to her knees begging Tante to give her the morphine, "Tante's frozen little mouth twitched" (419). When Tante finally reveals that she has "thrown [the morphine] down the sinkhole" (419), Delphine drags her out of the house and into the car, dropping Tante off at the Waldvogels's house in a fit of rage. "With a smug grimness of a woman who has at last been allowed to take charge" (419), Tante enters the house while Delphine drives off to find a way to fetch the much needed medicine for Eva. Tante seemingly thinks that this situation is "a perfect showcase for her pieties" (410), yet in her actions clearly reveals herself to be the devil of the story. Her understanding of the situation is grossly erroneous, and her preoccupation with the perceived failing reputation of her brother in combination with the attempted death-bed conversion of Eva marks Tante not as the saint-like figure she thinks herself to be, but as a devil in disguise, a disruptor of the final stage of Eva's life.

Roy Watzka, however, who caused Delphine so much grief in her life, acts as a life saver. Delphine is unable to find either Dr. Heech or the pharmacist and is forced to return "empty-handed, driving back to the house, weeping angrily, when she saw her father stumbling along the road" (419). At this point, Delphine is filled with a helpless rage and contemplates running her father over, when she notices something different about her father. He is indeed holding a bottle, as she suspects, yet it is not the usual schnapps but "a bottle labeled 'Sulphate of Morphia'" (420), the very substance Delphine had so desperately tried to obtain. Roy had acquired the bottle by breaking into the pharmacy and Delphine can finally prepare the medicine to relieve Eva of her pain. When Delphine enters the kitchen, she finds "Tante, white and sick with shock, slumped useless in the corner of the kitchen, on the floor" (420). Tante has completely lost control of the situation (if she was able to manage Eva in her pained state at all is highly doubtful), showing how unprepared she is when confronted with the harsh reality of a condition she had believed to be harmful only to the Waldvogels's reputation. Eva tries to get hold of a knife – presumably to end her pain by ending her life – when Delphine takes charge of the situation. Delphine is used to "scenes of mayhem" because she has been taking care of her alcoholic father, and "now a cold flood of competence descended on her" (420). She takes the knife away from Eva and instructs the frightened children to make Eva more comfortable and to tell her "[Delphine is] making the medicine now. It will be soon. It will be soon" (420) while the medicine is prepared.

The short story ends on these instructions. Delphine's predisposition to help others, especially Eva to whom she has formed such a strong bond, is not diminished in the slightest by her friend's agony and misery. Rather, she is able to calmly muster the mental fortitude to deal with pain and suffering due to her experience with horrible situations and misfortunes. Her actions are in stark contrast to Tante, who in her self-aggrandizement thinks she can present herself as a savior, yet does nothing but make the situation worse for all who are involved in it. Whether Delphine successfully relieves Eva of her pain or not is not recounted in the short story, and when viewed in the Native American oral tradition (retelling stories from different perspectives as part of a larger communication within a community), this is no dissatisfying ending to the story. "The Butcher's Wife," when read outside of the context of *The Master Butcher's Singing Club* (i.e., as a short story), explores Delphine's nature as a character with an almost explicit exclusion of the community which she hopes to be a part of. It focuses on her as a character and her actions. The consequences of her actions are further developed in the novel, but the short story itself – and this is the point of the narrative as a short story – emphasizes Delphine as an individual character, her strengths, weaknesses, and ultimately her resolve. And while the notion of community (so relevant as an overarching theme in Erdrich's writing as a whole) occupies an important function within the short story, it is the presentation and exploration of Delphine as an individual character that is the driving factor in "The Butcher's Wife."

3 Theoretical Perspectives

Erdrich's success as a writer only began in 1982. In 1984, with the publication of *Love Medicine*, her first novel instantly found widespread recognition and fame. Among other awards, *Love Medicine* received "the *Los Angeles Times* Award for Best Novel of the Year, the Janet Kaufman Award for Best First Novel, and the Virginia McCormack Scully Prize for Best Book featuring Indians or Chicanos," became a bestseller, and "rapidly made its way onto the syllabus in many literature classrooms" (Stookey 1999, 29). That same year, Erdrich also published her first collection of poems, *Jacklight*. Most of the poems were written in 1977/1978, and they address many of the major themes that would appear in her later fiction and exemplify the cyclical nature of her writing as a whole. For example, "Love Medicine" features as the second poem of *Jacklight*, and the entire section "The Butcher's Wife" foreshadows her 2003 novel *The Master Butchers Singing Club*, which picks up on the German American part of Erdrich's heritage most directly. With *The Master Butchers Singing Club*, Louise Erdrich published a novel telling her stories not mainly from the perspective of Native Americans, but rather focuses on the lives of European immigrants in the fictional town of Argus, North Dakota. It is often considered her first novel to explicitly deal with her German heritage, with "all the German material that was previously latent in Erdrich's fiction finally com[ing] to the fore" (Austenfeld 2006, 3–4, 6).

Far from being 'stand alone' publications, her novels are also heavily related to one another. Her first four novels, referred to as a tetralogy (*Love Medicine*, *The Beet Queen*, both 1986; *Tracks*, 1988; *The Bingo Palace*, 1994), do so quite clearly through recurring characters telling and re-telling stories of people living on and near the North Dakota Turtle Mountain Reservation (see Stirrup 2010, 67). *Tales of Burning Love* (1996) shares many of the same characters as the four earlier novels, although it was never intended as a part of the series (106). Since they do share common characters and geographical location, the five novels have been referred to as the North Dakota novels (see Jacobs 2001, 56). In addition to this list, *The Last Report on the Miracles at Little No Horse* (2001) and *Four Souls* (2004) also fit into this series. With the publication of *The Blue Jay's Dance* in 1995, Erdrich released a memoir of motherhood, which is complemented by *Books and Islands in Ojibwe Country* (2003).

Both works specifically address motherhood, *The Blue Jay's Dance* dealing with Erdrich's experiences during pregnancy and the birth of her child, while in *Books and Islands in Ojibwe Country* she "recounts a 'border-incident' in which she is quizzed over the maternity of her infant child" (Stirrup 2010, 14). Since these two works are part of her non-fiction literary output, Erdrich's lived experiences play a much more concrete role in her non-fiction writing than, say, the fictional reservation in North Dakota whose community, not its individual inhabitants, is the central figure.

Reviews of Erdrich's *The Red Convertible: Selected and New Stories, 1978–2008* highlight both Erdrich's talent as an engaging writer; her ability to authentically depict the "moods" (Schillinger 2009) of a variety of complex characters over the course of generations; as well as the different types of genres represented in the collection (see Seligmann 2009 and Freeman 2009). These reviews partly echo the critical scholarly reception of Erdrich's writing as a whole, which is focused on the joint categories of community, genre, and identity (see Hessler 1995, Ferguson 1996, Ward 1997, Cox 1998, Reid 2000, Nagel 2001, Martinez-Falquina 2006, Stirrup 2010, Smith 2018). Community, understood to represent the "relationship between people, place, and politics" (Stirrup 2010), does have its focal points: "[Erdrich's] works are primarily about Native American characters who live on or near the reservation, who think of themselves in the context of tribal history, and who collectively tell their stories as part of a community of speakers" (Nagel 2001, 22). The shifting perspectives and emphases on different narrators, in addition to the "limited locality" (Smith 2018, 38) of her writings, complete intricately woven relationships between the characters, stories, settings, and Erdrich as a writer. That this process articulates itself in the publication of short stories as independent works and as parts of a larger work (e.g., "Fleur" as an independent short story and as the second chapter of *Tracks*) is picked up by Ferguson in her discussion of "Erdrich's 'new' kind of story-sequence novel" (Ferguson 1996, 542). Unreliable as the narrator in "Fleur" (and the second chapter of *Tracks*) is, the short story emphasizes the narrator (singular), while the novel highlights the function of the narrators (plural). As Reid pointed out: "Erdrich's novels, in fact, demonstrate how the use of multiple narrators helps alleviate the alienation of individual characters; how stories which are half-told, re-told, and left un-told suggest a common base of knowledge that ties characters together" (Reid 2000, 70). The short stories "Fleur" and "The Butcher's Wife" are told from the perspective of a single narrator, yet the "common base of knowledge" is apparent in the novels as well as – to a lesser extent – in the short stories. The most obvious commonality is a geographical one. Both short stories take place in the fictional town of Argus, and beyond the shared location both female narrators are somewhat forced to return to their (step-)fathers to assist them. Moreover, the limits of a small geographical region are extended to a more global setting when the major character's heritage is taken into account: both Fleur and Delphine share names of French origin, and this is noteworthy inasmuch as it is a reflection of mixed ethnic and national identities on (U.S) American soil. And even though "Fleur" highlights how the narrator has almost no agency of her own ("Dutch took me out of school to take [my mother's] place"; "[t]he blood draws us back," Erdrich 2009, 178, 189), while Delphine in "The Butcher's Wife" seems to make a conscious decision to leave ("[f]or once Eva left, Delphine would be leaving too," 410), the shared geographical location in both cases has an episodic character to it and the focus is quite clearly placed on the exploration of the characters' nature. This enables, even necessitates, a shift from a community-driven narrative to a narrative of character exploration. Within the context of the novels, the characters' relationships are explored in more detail and community is highlighted as one of the major binding factors in the characters' lives. The short stories, however, omit certain characters (e.g., Pauline's cousin Russel in "Fleur" and Delphine's would-be lover Cyprian in "The Butcher's Wife") in order to draw special attention to the main characters and narrators. This exemplifies how the stories themselves, as Erdrich notes in the preface to *The Red Convertible* "are rarely finished with me" (Erdrich 2009, ix). But it is not only the stories themselves that stir a desire to re-tell them. It is the individual characters who – in the Native American oral tradition – have yet to tell each other's stories.

Joseph Ward, who analyzes how folklore is oftentimes presented in a humorous fashion in Native American literature, offers an insightful assessment of Erdrich's characters: "[a]lthough most of the characters are Indian and appear always conscious of the fact, they lead lives that seem little different from other poor Americans" (Ward 1997, 275). This sentiment is echoed by Martinez-Falquina, who draws attention to the fact that "Erdrich has repeatedly expressed her claim to be considered not just a Native writer but an American writer capable of articulating the complexity and richness of the American experience" (Martinez-Falquina 2006, 13–14) in her essay on the construction of American ethnicity in *The Master Butchers Singing Club*. This becomes apparent in "The Butcher's Wife," whose characters of European descent are not presented as a homogenous block but as distinct individuals with distinct national roots (e.g., French, German, Polish). Similar to how Erdrich presents the complexity of her indigenous North American characters and their relationships to each other in her earlier writings, the focus on her German heritage encapsulates national European identities and differences, brings them together in a small geographical location, and focuses on character development within the context of "a dynamic process where the local and the transnational are both emphasized, and where borders are not fixed but unstable, and constantly reconstructed" (22). The depth of a specific German heritage and culture in this novel is discussed in detail by Austenfeld. He also points out that Erdrich "suspends her Native American identity temporarily in an inclusive, unflinching, yet celebratory gesture toward German culture" (Austenfeld 2006, 10). This is partly true for "The Butcher's Wife," although Tante, as a character, takes on a more dominant role in the short story and the "celebratory gesture" (10) points to a more nuanced view of the immigrant characters' attitudes and lives. It again echoes Ward's point, and the "little" (Ward 1997, 275) differences – while certainly not negligible – are ultimately an expression of

Erdrich's attempt to bring the various characters, their backgrounds, identities, and relationships together "to consider what it means to understand Erdrich, in her own words, as an *American* author" (Stirrup 2010, 4; original emphasis).

4 Bibliography

4.1 Works Cited

Austenfeld, Thomas. "German Heritage and Culture in Louise Erdrich's *The Master Butchers Singing Club*." *Great Plains Quarterly* 26.1 (Winter 2006): 3–11.

Bruchac, Joseph. "Whatever is Really Yours: An Interview with Louise Erdrich." *Survival This Way: Interviews with American Indian Poets*. Tucson: Sun Tracks and U of Arizona P, 1987. 73–86. Reprinted in *Conversations with Louise Erdrich and Michael Dorris*. Eds. Allan Chavkin and Nancy Feyl.

Chavkin, Nancy Feyl, and Chavkin, Allen. "An Interview with Louise Erdrich." *Conversations with Louise Erdrich and Michael Dorris*. Eds. Allan Chavkin and Nancy Feyl Chavkin. Jackson: UP of Mississippi, 1994. 220–253.

Childs, Francis Lane. "A Dartmouth History Lesson for Freshman." *Dartmouth Alumni Magazine* 50.3 (December 1957): 16–21.

Erdrich, Louise. "Louise Erdrich Interview." *CutBank* 1.27 (1986): 30–35.

Erdrich, Louise. *The Red Convertible: Selected and New Stories, 1978–2008*. New York et al.: Harper Perennial, 2009.

Erdrich, Louise. "Where I Ought to Be. A Writer's Sense of Place." https://www.nytimes.com/1985/07/28/books/where-i-ought-to-be-a-writer-s-sense-of-place.html New York Times, July 28, 1985. (11 May 2020). 1.

Ferguson, Suzanne. "The Short Stories of Louise Erdrich's Novels." *Studies in Short Fiction* 33 (1996): 541–555.

Freeman, John. "Raw Power. Thirty Years of Short Fiction by Minnesota Author Louise Erdrich Show Her to be a Master of the Form," https://www.newspapers.com/-newspage/250608607/. *Minneapolis Star Tribune*, January 9, 2009. (11 May 2020).

Grantham, Shelby. "Intimate Collaboration, or, What *The Wall Street Journal* calls 'a Novel Partnership.'" *Dartmouth Alumni* Magazine 77.6 (March 1985): 43–47.

Jacobs, Connie A. *The Novels of Louise Erdrich: Stories of Her People*. New York: Peter Lang, 2001.

Nagel, James. *The Contemporary American Short-Story Cycle: The Ethnic Resonance of Genre*. Baton Rogue: Louisiana State UP, 2001.

Nowick, Nan. "Belles Lettres Interview: Louise Erdrich." Belles Lettres 2.2 (Nov./Dec. 1986): 70–74.

Martinez-Falquina, Silvia. "Louise Erdrich's *The Master Butchers Singing Club*, or the Construction of American Ethnicity." *Revista de Studios Norteamericanos* 11 (2006): 13–24.

Reid, Shelley E. "The Stories We Tell: Louise Erdrich's Identity Narratives." *MELUS* 25.3/4 (Fall/Winter 2000): 65–86.

Schillinger, Liesl. "All American," https://www.nytimes.com/2009/01/04/books/review/-Schillinger-t.html *New York Times*, January 2, 2009. (11 May 2020).

Seligman, Craig. "Expect the Unexpected in the Short Stories from Louise Erdrich," https://www.eastbaytimes.com/2009/01/01/review-expect-the-unexpected-in-short-stories-from-louise-erdrich/. *East Bay Times*, January 1, 2009. (11 May 2020).

Smith, Jennifer J. *The American Short Story Cycle*. Edinburgh: Edinburgh UP, 2018.

Stirrup, David. *Louise Erdrich*. Manchester: Manchester UP, 2010.
Stookey, Lorena L. *Louise Erdrich. A Critical Companion*. Westport, Connecticut: Greenwood P, 1999.
Wong, Hertha D. "An Interview with Louise Erdrich and Michael Dorris." *North Dakota Quarterly* (Winter 1987): 196–218. Reprinted in *Conversations with Louise Erdrich and Michael Dorris*. Eds. Allan Chavkin and Nancy Feyl Chavkin. Jackson: UP of Mississippi, 1994. 30–53.
Ward, Joseph. "Prayers Shrieked to Heaven. Humor and Folklore in Contemporary American Indian Literature." *Western Folklore* 56.3/4 (Summer/Autumn 1997): 267–280.

4.2 Further Reading

Castelluci Cox, Karen. "Magic and Memory in the Contemporary Story Cycle. Gloria Naylor and Louise Erdrich." *College English* 60.2 (1998): 150–172.
Erdrich, Louise. *The Master Butchers Singing Club*. New York et al.: Harper Perennial, 2003.
Erdrich, Louise. *Tracks*. New York: Henry Holt and Company, 1988.
Hessler, Michelle R. "Catholic Nuns and Ojibwa Shamans. Pauline and Fleur in Louise Erdrich's 'Tracks.'" *Wicazo Sa Review* 11.1 (Spring 1995): 40–45.
Lee, A. Robert. "Native American Short Story: Authorships, Styles." *The Routledge Companion to Native American Literature*. Ed. Deborah L. Madsen. New York: Routledge, 2016. 412–422.
Nagel, James. *The American Short Story Handbook*. Chichester: Wiley Blackwell, 2015.
Scheiding, Oliver. "The American Short Story." *Handbook of Transatlantic North American Studies*. Vol. 3. *Handbooks of English and American Studies*. Ed. Julia Straub. Berlin/Boston: De Gruyter, 2016. 234–250.

Lynn Blin

31 Lydia Davis (1947–)

Abstract: This chapter is a paired reading of Lydia Davis's "Negative Emotions" and "A Story of Stolen Salamis" from her collection *Can't and Won't.* The stories have been chosen because they amply illustrate how Davis's pieces are reading lessons in themselves. By their very brevity, they invite us to pay attention to language in a very privileged way. They enable us, in the words of Roland Barthes to "see language" (Barthes 2002, 735). The chapter will seek to explain how these short-short pieces invite us to pay attention to details down to the very comma. My analysis will try and show how this invitation to pay attention to the minutest details of language gives readers indications on how it can be a useful tool for efficient close reading. As we examine Davis the translator, Davis the adaptor, and Davis the short story writer, we will discover how her very dexterous use of language conveys poignancy, humor, violence, and ultimately life.

Keywords: Short-short fiction; translation; adaptation; grammar; tone; politeness; face

1 Context

When Lydia Davis was seven years old, her father brought the family to Vienna, where he was invited to teach for a year and she was enrolled in an Ursuline convent school. In an interview with Eleanor Wachtel, Davis tells of finding herself in a Vienna still reeling from the war, surrounded by a language she did not understand. She speaks of how difficult that year was for her emotionally, because her mother was hospitalized shortly after their arrival. Though she managed to learn German in a month, the fact that she went from not understanding anything to things becoming 'transparent,' acquainted her with the notion of 'not feeling at home in a language.' She concedes that this experience was probably an important influence on her decision to become a translator. Later, when the family moved to New York City, she once again found herself in a strange environment in a new school much larger than the small-town school she had known in her hometown of North Hampton in Massachusetts. She had to have private tutoring in French to catch up with her classmates who had been learning it since kindergarten. These French lessons, she confides, with a teacher whom she liked very much, and a French grammar book which she refers to as her "Rosebud," enabled her to "come to a comfortable place" in a strange city (Wachtel 2008).

These notions of "not feeling at home in a language" and finding "a comfortable place" are primordial to us as Lydia Davis readers. This chapter shows that for

https://doi.org/10.1515/9783110587647-032

us to feel at home in her art we too have to develop at least some of the reflexes of a translator, and an eye and an ear for grammatical details with a regularity that is perhaps not habitually part of our reading practice.

Today, Davis is considered one of the most important voices in American fiction, but it was as a translator, notably of Maurice Blanchot and Michel Leiris that she first gained renown. Virtually ignored at the beginning of her career as a writer, it was only in 2001, when Dave Eggers published her collection, *Samuel Johnson is Indignant*, that she started to gain attention – attention that grew to acclaim when in 2013 she was awarded the Man Booker International prize for the ensemble of her works. To date she is the author of one novel, *The End of the Story* (2004 [1995]) and seven short story collections, including *Break it Down* (1986), *Almost No Memory* (1997), *Samuel Johnson is Indignant* (2001), *Varieties of Disturbance* (2007), all to be found in her *Collected Stories* (2013 [2009]), and *Can't and Won't* (2015 [2014]). Her translation of Marcel Proust's *Swann's Way* won the 2003 French-American Foundation Translation Prize – an honor that was repeated in 2011 for her translation of Gustave Flaubert's *Madame Bovary*. In 2019, she published the first volume of a collection of her essays, *Essays One* – a compilation of pieces from the 1970s to the present day – on writing about literature and on writing itself, but also covering other topics such as the visual arts, history, and religion. The second volume, due in 2021, is to be devoted to her writings on and about translation.

Her short-short prose pieces, for which she is the most known, "miniatures" as they are sometimes called, take the reader by surprise. It is perfectly understandable that a reader will be taken aback by a piece such as "Collaboration with Fly," which reads:

> "Collaboration with Fly"
> I put that word on the page, but he added the apostrophe. (Davis 2013, 508)

> *Can't and Won't*
> I was recently denied a writing prize, because, they said, I was too *lazy*. What they meant by *lazy* was that I used too many contractions: for instance, I would not write out in full the words *cannot* and *will not*, but contracted them to *can't* and *won't*. (2015, 46)

> "Circular Story"
> On Wednesday mornings early there is always a racket out there on the road. It wakes me up and I always wonder what it is. It is always the trash collection truck picking up the trash. The truck comes every Wednesday morning early. It always wakes me up. I always wonder what it is. (2015, 3)

We might simply think of "Collaboration with Fly" as an intertextual reference to Emily Dickinson's "I Heard A Fly Buzz When I Died," as does Shelley Jackson writing in the *Los Angeles Times* (2007); "*Can't* and *Won't*" a tongue-in-cheek comment on the arbitrary choice of literary awards; and "Circular Story," as John Winters writes, "typical of the dead ends that occupy most of [*Can't* and *Won't*'s] pages" (Winters 2014, n.p.).

Davis leaves so much space for the reader, there can be as many interpretations as there are readers. However, if we consider them as an invitation to see language, we discover aspects that help explain these explanations. Conjuring up the fly on the page, we might consider what type of words need apostrophes – possessives like *John's* or contracted forms like *can't* and *won't*. Substitutes for *cannot and will not*, as the story recounts. But then we perhaps wonder why *cannot* is written as one word and *will not* as two. And this takes us to our grammar books, and we see that according to Huddleston and Pullum these forms are not completely interchangeable with the short forms. Not only are they considered more formal, they do not express negation in the same way. And if "will not" is written in two words and not one, it is because negation does not work the same way with each. When "not" follows "will," it negates the clause that follows. When "not" follows "can," it negates the auxiliary. Thus "*She cannot write well*" means "*It is not possible for her to write well*". Whereas, "*She can not write well*" means "*On occasion it is possible for her to not write well*" (Huddleston and Pullum 2002, 799–803). Since the theme of negation is a recurring one in Davis's works, we might want to look at what it entails – literally entails, as we will see below. We also notice that these words are all in italics, as is the title. And words in italics, the equivalent of emphasis in speech, are used to draw attention of the reader. And if they are emphasized, we are meant to pay attention. This piece might further be a reflection on fiction. Is the narrator speaking of Lydia Davis herself? Surely not. Lydia Davis readers know that she rarely uses short forms. It is one of the reasons her stories read so much as though they are 'over-translated' in English. A closer reading of "Circular Story" reveals itself to be about the pronoun "it" and the slipperiness of language. I will return to this at the end of the chapter.

We might also think that in leaving us so much space for interpretation, she is putting her readers through the same experience she had as a child, making us not feel perfectly at home in language, albeit our mother tongue. We understand the words, but 'we don't get it.' Her careful attention to syntax, to grammar, to form, to rhythm may give the impression of an 'over-correct' text.

Davis dates her earliest inspiration to becoming a writer to her discovery of Kafka and Beckett – Kafka for "the sparseness, the humility (whether assumed or real) or a combination, the imagination" (qtd. in Gunn 2014, n.p.) and Beckett for teaching her that "you could write about the self watching the self-write, and also that you could write about anything" (Wachtel 2008, n.p.). She tells of her excitement upon discovering *Malone Dies*: "There was so little material, such plain language, no attempt at literalism whatsoever, or flowery language, that he would spend a page or two about how he dropped his pencil was so utterly strange and wonderful" (Wachtel 2008, n.p.). Our impression as Davis readers of "seeing language" as Barthes once expressed it (Barthes 2002, 735) may come partly from her teenage pastime of cutting out her favorite Beckett sentences to study the sheer beauty of them (Manguso 2008, n.p.). This invitation to "see language" is a solicitation that is constantly renewed in our reading of Davis. For indeed, if, in her stories, there is a dearth of character, of place, of plot, it is amply made up for in the profusion of ways Davis summons us to examine language. The following close readings will show that Davis is first and foremost a reader's reader.

2 Close Readings

In Davis's preoccupation with the poetics of the normal, there is, as Eliza Haughton-Shaw points out, "an invitation and the permission to become unfamiliar to ourselves in our most ordinary moments" (Haughton-Shaw 2020, n.p.). The two stories "Negative Emotions" and "A Story of Stolen Salamis" are from her last collection *Can't and Won't*. The two ordinary moments rendered unfamiliar are an email, and a news brief. It is in this collection that the trademark characteristics of her writing are the most salient: (1) the short-short pieces – out of the 122 stories 97 fall into this category, 70 are less than a page long – the other 7, a page and a bit. (2) Found material: extracts from Flaubert's letters to his mistress Louise Colet, translated and turned into stories; dreams – her own and those of her friends turned into stories; complaint letters she wrote herself and sent – one to a frozen peas manufacturer complaining that the color of the peas in the can does not correspond to the color on the label.

Lydia Davis comments on "Negative Emotions" in *Essays One*. She speaks of the influence and her great admiration for the stories of Thomas Bernhard, notably "their tight structure, their completeness, and negativity of attitude. But also the hypercomplex syntax of some of the sentences" (Davis 2020, n.p.). More specifically in regard "Negative Emotions" she specifies the direct influence Bernhard's story "Near Sulden" exerted on her:

> (1) [. . .] my analytical readings of Bernhard's short pieces, all the closer since I was working on them in class; and more immediately, (2) the raw material – the email I encountered. The emotional impetus for the piece – since there is always, for me, strong feeling behind a piece of writing – was at least twofold: (1) amusement at what Bernhard does in his brief stories, and at the content of the email; and (2) admiration for Bernhard's writing and a desire to do something similar, although I did not see the influence of Bernhard (obvious to me now) until after I had written it. (n.p.)

The following close reading is an attempt to explain how "emotion," "amusement," and "negativity of attitude" – all characteristics of Davis's works – are carried out grammatically and thus stylistically. We will notably be looking at the grammatical notions of direct and indirect reported speech, free indirect speech, pronouns, and negation. These notions are admirably woven into the story and explain how politeness theory and the notion of face intertwine in how we interpret what we hear or read. We will discover how these notions enable us to discover unsuspected clues concerning narrative voice.

"Negative Emotions" (282 words) tells the story of a well-meaning teacher who sends an article that he found in a book written by a Vietnamese Buddhist monk giving advice on what to do about emotions to all the teachers in his school. The message is badly received by the colleagues resulting in an angry backlash with them writing back an irate message with an ironic twist: "They said, they in fact liked having negative emotions, particularly about him and his message" (Davis 2015, 103). It is a story that could be included in a textbook on professional writing entitled, "How Not to Write an Email." Though this will be part of my analysis, the essential of my task will be devoted to how we as readers interpret what is not said – a task required by every reader of great literature, but even more so with a Lydia Davis story, where the reader might conclude to there being a little too much silence.

My resume above of "Negative Emotions" is a combination of indirect reported speech, and a direct quotation from the story. The direct quotation is in inverted commas and is a faithful rendering of how the story ends. The reported speech, on the other hand, in being a resume is much less faithful. Not only have I left out details, I have also added my own interpretation – there is no mention of an email in the story, only a message.

Communicatively speaking, a message conveyed via indirect reported speech can be considered a regular culprit in failed communication. This is because it is, like the above, often a summing up of what the other person said – words are exchanged for others, things are left out, and a certain amount of subjectivity can enter into the transformed message. Here is the actual opening paragraph:

> A well-meaning teacher, inspired by a text he had been reading, once sent all the other teachers in his school a message about negative emotions. The message consisted entirely of advice quoted from a Vietnamese Buddhist monk. (103)

Though the message is not explicitly referred to as being an email, we infer that this is how he sent the message to "all the other teachers." And we imagine the email sent via the professional mailing "reply-all" list to all his colleagues in the "to" section, with his name in the "from" section will be primarily received as a communication of professional tenor. However, there is no direct message from the colleague himself, just "advice quoted from a Vietnamese Buddhist monk about negative emotions." Though the narrator uses the term "quoted," we notice in the paragraph that follows that there are no quotation marks, and though there is the reporting frame "said the monk," and the absence of the inverted commas in direct reported speech is grammatically accepted (Huddleston & Pullum 2002, 1023), the effect of direct speech and the constraint upon this structure "to use the actual wording" of the speaker (1023) can be held in doubt. We might question the reliability of the term "quoted," because, if taken literally this could also infer that the message would be in Vietnamese. Since the quote is in English, we must further infer that the monk in question was either fluent enough to write in English or, if not, that the text was translated. The formal, rather restrained tone of the message (for instance, the use of the formal "one") can be interpreted by the fact that English is not his mother tongue. But this can also be interpreted by a perception of Buddhist spirituality as being itself formal; or it can be interpreted as a general cultural abeyance to the oriental code of politeness. We might further ponder the idea of "negative" emotion, because the monk only speaks of "emotion":

> Emotion, said the monk, is like a storm: it stays for a while and then it goes. Upon perceiving the emotion (like a coming storm), one should put oneself in a stable position. One should sit or lie down. One should focus on one's abdomen. One should focus, specifically on the area just below one's navel and practice mindful breathing. If one can identify the emotion as an emotion. It may then be easier to handle. (103)

Emailing, Business Letters, Context, and Politeness Theory

There is nothing immediately offensive about this email. "Emotion" is in itself not necessarily negative, but as the simile "emotion is like a storm" infers, it can be strong, powerful, and destructive. And, since, as the definition of the word "emotion" specifies, it is: (1) a "strong feeling deriving from one's circumstances, mood, or relationship with others; (2) an instinctive or intuitive feeling as distinguished from reasoning of knowledge," the introduction of such notions in a professional milieu can be understood as being inappropriate. If it had been sent to a few close friends, close enough for them to interpret it as a catechism for fellow Buddhists, for example, or as something to laugh about, either because they were all sharing negative emotions about a shared experience, or because they would all have in mind a friend or colleague Buddhist whose image of him/herself as the epitome Buddhist detachment and serenity is so far off the mark that it is laughable, the outcome would probably have been different. But assuming "all the teachers in his school" were not personal close friends, which is probable because being "close friends" with the entire staff is not possible ("can't" be friends with everyone and "won't" even try), the vicious harangue that followed is natural.

To better grasp what is at stake here, we are going to use the linguistics field of pragmatics or, as George Yule has put it, "the study of 'invisible' meaning" (Yule 2014, 126) as our guide. Invisible meaning often means the gap between the intended message and what is understood and what is meant (129). Inference in pragmatics is additional information used by the listener to logically create a connection between what is said and what is meant. Gaps between what is said and what is meant are not always a source of polemic. If someone says, "Could you please pass the salt?," even though it is in interrogative form, we know that it is a polite way of asking for the salt, and not a question about our physical capacity to pass the salt, in which case we would answer, "yes" or "no." This is called an indirect speech act. And we use indirect speech acts to make requests because it is more polite than an imperative “Pass the salt” and “removes the assumption of social power” (132). Our adeptness in using indirect speech acts is tied to our sensitivity to the other person’s ‘face,’ and in politeness theory is called a “face-saving act.” We all have a ‘negative face’ and a ‘positive face.’ Negative face is the need to be independent and free from imposition; positive face the need to feel connected, to have a sense of belonging, to be part of a group. As the colleagues understood this message not only as an imposition, but as an accusation, the well-meaning teacher has managed not only to disregard their negative face, in threatening their negative face, he has unintentionally managed to address their positive face – creating a group of colleagues whose sense of belonging to a group has taken on a whole new dimension. And has excluded the sender.

Different cultures respond to these notions of face in various ways. Oriental cultures being more deferential than Western cultures is one possible explanation for the choice of the pronoun “one” instead of the more direct “you” in the monk’s advice on emotion. But more specific to the story in regards to face and culture are those required of the professional world in regards to politeness. The inference of the remark made by a friend at a party will not be the same as the same comment made by a colleague or a superior in a professional context. Thus, simply the name of the sender would immediately condition the reaction of his colleagues. Factors such as their relationship to him and his and their place in the hierarchy of the school would all be determinant in their reception of his advice:

> The other teachers were puzzled. They did not understand why their colleague had sent them a message about negative emotions. They resented the message, and they resented their colleague. They thought he was accusing them of having negative emotions and needing advice in how to handle them. Some of them were, in fact, angry. (103)

Face and Culture; Context and Relationships; Negative Inference and Added Meaning

We have already noted that the term “negative” appears in the title of the story as well as in the summing up by the narrator, “sent them a text about negative emotions.” I have suggested that it is this interpretation by the sender that led the colleagues to interpret it this way. But what if the title of his email were not “negative emotions” but “advice from a Vietnamese Buddhist monk”? Would the negative inference still be present? It would for the reader because of the title of the story, “Negative Emotions.” But even without this title the adjective “well-meaning” in the opening noun phrase is also a clue. “Well-meaning” only means “to have good intentions.” But there is an added meaning that is “implicated” – the added meaning being something like “gone about in the wrong way thus doomed for failure.” Note the term inference refers to the act or process of inferring by deduction, while implicature is an implied meaning. In the term “well-meaning” there is the additional conveyed negative meaning. We detect further implicature in the modal auxiliary "should" as advice is going to be understood by the listener as a suggestion to correct a behavior. Even though the pronoun "one" could be paraphrased by the general "everyone," thus less threatening than "you," colleagues might not appreciate being lumped into this "everyone" and being given the unsolicited advice to regard the area below their navel. They do not wish their positive face to be addressed in such a way.

Vocabulary

A term in this email that might take on additional meaning and could be considered as offensive is "Buddhist monk." Though Buddhism, in being nontheistic, is not strictly speaking a religion but a philosophy, the interpretation of the message as an inflicting of spiritual views upon others through the bias of a professional mailing list is a possible construal. And that the term 'emotion,' let alone those of 'abdomen' and 'navel' might cause offense, is linked to the fact that these terms belong not only to the personal domain, but to the intimate. We are beginning to have a clearer picture of a very authoritative narrative voice deftly guiding our way to the dramatic, albeit ironically funny denouement.

Narrative Voice

We note so far that there have been no outright negative sentences – only negative inference and implicature. This negative implicature continues with the adjective "puzzled," which implies "did not understand." However, it is not until "They did not understand why their colleague had sent them a message about negative emotions," that we encounter a negative sentence constructed via "analytic primary negation." That is, a sentence formed by a "primary clause containing a primary form of an auxiliary verb" negated by adding "not" after the verb (for instance, "did not"). It is opposed to secondary synthetic negation where the verb is inflected (for instance, "didn't") (Huddleston & Pullum 2002, 799). We further note that throughout the rest of the story, each time there is an outright negative sentence, it is systematically an analytic negative. We thus encounter a narrative voice that is omnipresent in Lydia Davis's stories. The somewhat stiff, overly pedantic, working a topic to the bone, non-conversational voice that is rather distant. The use of the full forms in the negative construction is partially responsible for this. Huddleston and Pullum explain that analytic and synthetic negation are not fully interchangeable, the main difference being one of style level, the synthetic forms, being more informal, are considered the default form. Analytic negation is preferred in formal writing and solemn contexts. Synthetic negation is preferred in ordinary conversation and is considered to increase the sense of familiarity, intimacy, and accessibility (800). Huddleston and

Pullum go on to specify that analytic forms sound unnatural in many conversational contexts unless there is some clear reason for their use, such as emphasis on the word "Not." This is because emphasis, which is not possible on the contracted form N'T, is possible on the full form NOT. This use is thus understandable in the contrastive sentences:

> The teachers did **not** choose to regard their anger as a coming storm. They did **not** focus on their abdomens. They did **not** focus on the area just below their navels.
> (Davis 2015, 103; emphasis added)

But it is not necessary in the explanatory, "They did not understand why their colleague had sent them a message [. . .]," as there is no contrast to the preceding sentence expressing the idea that the teachers were "puzzled." This sentence simply explains why they were puzzled.

We further notice that the narrative voice does not make use of grammatical ellipsis. For example, the above sentence could have been more concise. In coordinate sentences with "and" the repetition of the noun and verb in the compound clause is not compulsory. Thus, in preferring the more formal: "**They resented** the message, and they **resented their** colleague" (emphasis added) to the more concise and spontaneous "**They resented the message and their colleague**" (emphasis added), the narrator can thus literally pound home the point.

The verb phrases "resent," "accusing," "needing advice," "handle," "get over," "caused," along with the adjective "other" in "The **other** teachers were puzzled" (emphasis added) all combine in the mounting crescendo of negation, building up to the exclusion of the well-meaning colleague. The adjective "other," excludes the sender of the email. Idem for "**Some** of them were, in fact, angry" (emphasis added). "Some" implies "not all of the teachers were angry" and excludes the teachers who wrote back:

> Instead, they wrote back immediately, declaring that because they did not understand why he had sent it, his message had filled them with negative emotions. They told him that it would take a lot of practice for them to get over the negative emotions caused by his message. **But**, they went on, they did not intend to do this practice. Far from being troubled by their negative emotions, they said, they in fact liked having negative emotions, particularly about him and his message. (103; emphasis added)

The coordinating conjunction "but" sets up an opposition to something expressed or inferred in the previous clause, "it would take a lot of practice for them to get over the negative emotions caused by his message." This opposition can be argumentative or explanatory. Argumentative "but" does not oppose the preceding clause, but something inferred. "He is rich but happy," infers the rich are not happy. Explanatory "but," which we have here, introduces a clause that develops or explains the preceding clause. The negative inferring verbs, "get over" meaning to recover or survive a difficulty, and the verb "cause" implying to make something (usually bad) happen, clearly enable the sentence introduced by "but" to be an explanatory refutation of even attempting the practice. But who in fact is explaining?

Reported Speech, Parentheticals, Supplements, and a Duo of Narrative Voices

On a first reading, the story seems to be a stable third person narration, as all pronouns in the text are all third person pronouns: "he," "they," "them," and "their." It is in the narrative past tense, and when there is reported speech, the necessary grammatical adjustments are in place. In direct speech the angry message of the angry teachers would read: "Because **we do not understand** why **you have sent** it, **your** message **has filled us** with negative emotions [. . .]" (emphasis added).

Did each angry teacher individually send this message – this exact message? In which case were they authorized by the others to use the collective "we," or did they use the first person singular "I"? Or was one teacher chosen to send it with all the angry teachers in copy? And what about the adverb "immediately." If taken literally this paragraph would lead us to believe that all the angry teachers were at the same place at the same time and speaking in one harmonious voice dashed off the angry email. Of course, this is not how we interpret the story, because we know that indirect reported speech is a retelling of what was actually said. The reporter in the story is the narrator and as narrator she has direct access to the content because the author has made her the spokesperson. The original speakers are the monk, the other teachers, and then some of the other teachers – the angry ones. The only information the story gives us about the original sender is that there was no message from him, just the quote.

The major difference between reported speech and direct speech is the question of deixis, that is the personal pronouns, time and place adverbs, the demonstratives, and tense. Reported direct speech, with quotation marks reports the words of an "I" talking to a "you," about some topic. In a normal conversation there is turn taking, where the "I" becomes the "you," when it is the other's turn to speak. In indirect reported speech, "I" usually becomes a third person – in our story "they" – and the present tense of the moment of utterance becomes the past tense, "here" becomes "there." In speech utterance linguistics, the third person pronouns are considered *non-persons*:

> The third person pronoun does not refer to a person because it refers to an object located outside direct address. But it exists and is characterized only by its opposition to the person "I" of the speaker who is uttering it, situates it as a non-person. Here is its status. The form "he' takes its value from the fact that is necessarily part of a discourse uttered by "I."
>
> (Benveniste 1971, 229)

Since the pronouns "they," "their," and "them" appear 21 times and "he," "his," and "him" 9 times, this repetition (nearly 12% of the text) merits a closer look into how they participate in the orchestration of narrative voice.

Framing Structures

The framing structures, in bold print in my examples below are the words of the narrator.

1. Emotion, **said the monk**, is like a storm;
2. Instead they wrote back immediately **declaring that**;
3. They **told him that**
4. But, **they went on**, they did not intend;
5. Far from being troubled by their negative emotions, **they said,** they in fact liked [. . .].

In indirect reported speech, there is a choice between an "embedded subordinate framing structure," found in 2 and 3 above, which are grammatically like any subordinate content clause introduced by "that," or a "non-embedded framing structure" which either follows the content clause, or like in our story, is in median position as in 1, 4, and 5. These non-embedded framing structures are considered less common than their embedded alternatives (Huddleston & Pullum 2002, 1025). We notice that, once again, as for the case of the analytic negation and coordinate sentences, Davis has chosen the less common of two possible grammatical structures.

Third Person Personal Pronoun and Content Clause

When the third person pronoun used in indirect reported speech is singular, it is easier to accept that the content corresponds to what the person said. However, when it is the plural "they," corresponding to a group, it is a little more complicated, as seen above. What is of interest in these non-embedded framing structures in the indirect reported speech without the inverted commas, is that in splitting the reported speech in two, we cannot entirely ascertain whether the beginning belongs to the original speakers or to the narrator. In 4 is "But, they went on," and in 5 "Far from being troubled by their negative emotions, they said" the words of the angry teachers, or the comment of the narrator?

There are several types of blurring going on in this story. In choosing not to use the inverted commas to introduce the reported speech, there is a constant blurring between the direct speech of the narrator and the indirect speech of the teachers. The non-embedded framing structures are termed parentheticals and act as supplements (952–953, 1024). Thus, like a passage between parentheses in a text, that is, a word or phrase inserted as an explanation or afterthought into a passage which is grammatically complete without it, in writing usually marked off by brackets, dashes, or commas (1748), we are invited to pay attention to why Davis chose this structure and not the more common one.

Free Indirect Speech

Free indirect speech is a grammatical and stylistic means of allowing an author to express the thoughts of a character. There are no inverted commas, and the third person pronoun of the narrative voice remains. Free indirect speech seamlessly allows the author to surreptitiously slide from the third person narrative voice into the minds of the characters. The reader can sometimes detect free indirect speech through the use of deixis – the passage remains in the past tense, but 'now' is used instead of 'then,' 'here' instead of 'there'. The reader understands a "s/he thought/felt" etc. that is not present in the text. Our story, on the surface at least, seems to be entirely about the thoughts of the characters so free indirect speech may not seem to be immediately pertinent. However, upon examining these parentheticals as well as the other grammatical choices involving negation and coordination, the choice not to use inverted commas to introduce the monk's text, and the insistent use of third person pronouns we detect another very subtle narrative voice that stealthily makes its way into the story. Rather than the detached narrative voice of the third person narration, this voice becomes a double voice and takes on the voice of the community, one of the voices Rimmon-Kenan identifies as the unreliable narrator (Rimmon-Kenan 1989,100–103), here ironically recounting the hilarious but violent reaction of the teachers. We hear not only an echo of their anger, but also gain understanding of the first rule of professional writing: "Never write an email when you are angry."

We also better understand how this story echoes the three dimensions Davis admired in Bernhard's story: (1) Davis's particular use of negative structures has captured the negative attitude admired in Bernhard's story; (2) her ironic twist on positive face, where sense of belonging to a group concludes in the hilarious ending of the sheer jubilation ignited by the sharing of negative emotion, captures the amusement Bernhard's story procures for her; (3) in the response to a colleague, who has in fact said nothing, we detect an interesting twist on a theme that is very dear to another inspiration to Davis – Maurice Blanchot who wrote about how "the need to write is linked to the approach toward this point at which nothing can be done with words" (Blanchot [1955] 1992, 52). The "well-meaning" colleague, in not including a personal message has literally "done nothing with words," and thus captures the emotional impetus that makes this more than a humorous anecdote. In inviting us to examine reported speech, it is in fact a story about how we interpret text – the added meaning. Email is like conversation without the intonation, and without the immediate turn taking possible when we speak. The unsuspected violence of the colleague's message, and the violence of the reaction is language at work. Davis's over-careful narrator joining her voice with that of the community reminds us that it is only in taking great care with language that we can begin to feel at home with it and in it.

Reported speech and the undetected violence of language is present also in our second story, which is the opening story in the collection *Can't and Won't*. Davis's stories are better understood when perceived in their cohesive whole. The second close reading will demonstrate how it prepares us for the reading of much that goes on in her very short prose pieces, as she herself prefers to call them.

> "A Story of Stolen Salamis"
> My son's Italian landlord in Brooklyn kept a shed out back in which he cured and smoked salamis. One night, in the midst of a wave of petty vandalism and theft, the shed was broken into and the salamis were taken. My son talked to his landlord about it the next day, commiserating over the vanished sausages. The landlord was resigned and philosophical, but corrected him. "They were not sausages. They were salamis". Then the incident was written up in one of the city's more prominent magazines as an amusing and colorful urban incident. In the article, the reporter called the stolen goods "sausages". My son showed the article to his landlord, who hadn't known about it. The landlord was interested and pleased that the magazine had seen fit to report the incident, but he added: They weren't sausages. They were salamis. (Davis 2015, 3)

To analyze this story, I will be using the theoretical tools which are those of M.A.K. Halliday and Ruquia Hasan in their work *Cohesion in English*. Their work explains how a text functions as a semantic edifice, how one element connects to another via the following 4 tools: (1) Reference: that is, Pronouns: A noun is replaced by a pronoun; (2) Substitution: one word is replaced by another, that is, **A dish** on the table was cracked. **The plate** had been in the family for centuries; (3) Conjunction: A term is not only substituted, it is redefined; for instance: **The plant her mother gave her** before she died was losing its flowers. **That blooming symbol of her love** was disappearing also. (4) Ellipsis, which is substitution by zero. The easiest example is coordination ellipsis, which was explained in the previous close reading. These four tools will be applied to two ideas: One, by the linguist André Joly who has demonstrated how the title and first sentence of a literary text is the conditioning framework for the whole novel. I am extending this to apply to the first story in the collection. The first sentence is a suggestion of the text that is to follow – and what is to follow is a transformed version of this initial suggestion (Joly 2001, 151). The second idea is Lydia Davis's. In her chapter on writing in *Essays One* she writes of the importance of ending: "Do work hard on the very last words – they can sometimes make all the difference as to whether or not a story seems finished" (Davis 2020, n.p.).

Reference: There are several incidences of reference: "My son's Italian landlord" referred to as "he" twice; "my son" becomes "him"; "shed was broken into and the salamis were taken" beomes "it"; "sausages" become "they" four times; "the article" becomes "it." "The landlord" is also replaced by "who" in a relative clause, and the "shed" by "which" in an adverbial clause introduced by the preposition "in."

Substitution: The term "stolen," as specified in the title, is replaced by "taken" and "vanished." "Salamis" is substituted by "goods" and "sausages" twice, once by the boy, the second time in article.

Conjunction: Conjunction is used in the most interesting manner here by Davis and in so doing she calls to mind what we have just discovered about indirect reported content. For when salamis is substituted by the word "sausages," it is not only substituted it is redefined. And this redefinition is at the heart of two challenges of translation. The first concerns what Walter Benjamin termed "the manner of meaning" (qtd. in Jacobs 1975, 761). Benjamin remarks that though the English word "bread" and the German word "Brot" refer to the same foodstuff made with water and flour and yeast – the manner of meaning is not the same. If a salami connotes a sausage, a sausage does not connote a salami. A sausage is what we in the English-speaking world eat with mashed potatoes, or for breakfast with eggs. A salami is an antipasto eaten perhaps with a glass of wine. The journalist who wrote the article has translated the term, and, in so doing, has ignored the landlord's cultural heritage.

Reported speech: The title "A Story of Stolen Salamis" is not exact. It is a story of a story of stolen salamis. Written up, not as "an incident in a wave of petty vandalism and theft," but as "an amusing and colorful incident." There is indeed here what Lecercle has termed a "violence of language" (Lecercle 1990, 141) – perhaps even an ever-so-slight mocking of the expected appreciation of the "woke" readers of "one of the city's more prominent magazines." We begin to touch here on a poignancy that can go undetected in Davis's fiction, if we are not conscious of how language can trip us up – a theme that is central to all of Davis's writing.

Ellipses: The direct speech of the landlord at the end of the story, "They weren't sausages. They were salamis," which at first seems to be a repetition of "They were not sausages. They were salamis" is not a repetition. The long form analytic negation has been replaced by the default, more spontaneous, more conversational synthetic negation. The landlord in his repeated complaint uses the contracted form of "were not". This, of course, is not an ellipsis of a word but a simple grapheme. But we can take this one step further and interpret it metaphorically. The substitution of NOT by a zero and an apostrophe could be metaphorically interpreted as a restitution of the landlord's identity. Davis, in substituting one form of negation for another uses this zero to indicate the integration of the immigrant landlord – and this through his use of a contracted form of negation. When looking through Lydia Davis's translations and adaptations of Flaubert's letters, we notice that contrary to most of her fiction, she does in fact use the contracted forms. She has explained that when she was translating *Madame Bovary* she went to Flaubert's letters to find a more spontaneous Flaubert (Davis 2020, n.p.). The narrative tells both the story of an unrecognized other but ends with a transformation. A more spontaneous landlord expressing himself via synthetic negation has reminded us of his desire to be an integrated American.

3 Theoretical Perspectives

Any theoretical approach to Lydia Davis's work will bloom and flourish by an examination of what she herself has written, not only about translation and writing, but about the visual arts as well. Because of the very tight control she wields over the English language and her own particular manner of deploying it in the disarming brevity of most of her pieces, the ideal place to begin is a study of her *Essays One* (2020). Davis's concern for form reveals itself to be a necessary prerequisite to emotion. It is in her essay on artist Joan Mitchell that we find one of the best ways of apprehending her fiction. She speaks of a visit she made to Mitchell in the French village of Vétheuil and wandering out to her studio to look at the painting, "Les Bluets," which she liked very much. And yet, she admits to naively seeing it only as "shapes and colors, white and blue." It was only when someone told her that it was in fact a painting of the Vétheuil landscape, more precisely the cornflowers, and this came as a shock to her:

> Apparently I had not known before that an abstract painting could contain references to concrete objective, identifiable subject matter. Two things happened at once: the painting went beyond itself, lost its isolation, acquired a relationship to fields, to flowers; and it changed from something I understood, into something I did not understand, a mystery, a problem.
>
> (n.p.)

Even though Davis's stories in their seemingly obsessive manner of presenting us with the ordinary, the quotidian, appear to be about things we know, she, like Mitchell, throws new light on the concrete, the objective, the identifiable matter. "She is our Vermeer, patiently observing and chronicling daily life, but from angles odd and askew," writes Parul Seghal (2019, n.p.). And it is first and foremost through her use of the English language that she takes us unawares. This is because, as Paul De Man points out in his Introduction to Walter Benjamin's *The Task of the Translator*:

> We think we are at ease in our language, we feel a coziness, a familiarity, a shelter in the language we call our own in which we think we are not alienated. What the translation reveals is that this alienation is at its strongest in our own original language, that the original language within which we are engaged is disarticulated in a way which imposes upon us a particular alienation – a particular suffering. (De Man 1986, 84)

Davis exposes the capacity of language to go wrong, but she does not write against language. On the contrary, in making language her leading character, the main action, and the principal setting, she gives its study a whole new role in the understanding of a literary text. Since Davis's stories are sometimes seen as a defamiliarization of English, or even, as someone once remarked, a translation of English into English, looking to the works on translation helps to better grasp this dimension of her work. It is via the concept of the defamiliarization of language brought to light by Walter Benjamin in *The Task the Translator* that we can better apprehend Davis. This is because, like for Benjamin, Davis firmly believes, what we take for granted in our language is up for questioning, that is the question of the compatibility between grammar and meaning (87).

It is Jonathan Evans's *The Many Voices of Lydia Davis* (2016) that gives us the most in-depth insight on how her literary works can be regarded as being in dialogue with the works she has translated. From Blanchot, whom Evans cites as being Davis's most enduring influence as a translator, she has learned and continualy sought to remain as faithful as possible in regards to syntax, word order and style of the source author's text (Evans 2016, 26. In his detailed analysis of her translation of *Madame Bovary* as well as in her reworking into short stories of various short extracts of Flaubert's letters to Louise Colet, we better discover not only Davis the translator, but also Davis the adaptor and Davis the appropriator. Moreover, Evans further contends that not only is Davis's novel *The End of the Story* a rewriting of Proust's *Swann's Way,* but in her extending of the practices employed in her translations of Leiris, he also contends that her translation of *Swann's Way* is a veritable rewriting of the canonical translation by Scott Moncrieff and the lesser known translation by James Grieve (71–88).

Our analysis in both the stories studied above exposed how a harmless email and an amusing news brief dissimulated a violence of language linked to the characters' disregard for how language means and does not mean. For insight into the defamiliarization and the violence of language, we can profitably look to Lecercle's *The Violence of Language* (1990) where he introduces the notion of the "remainder" – "that dimension of language that is there beneath the surface and resists the prescriptive rules of grammar" (141). The "remainder" is also studied in Venuti's *The Translator's Invisibility* ([1985] 2008) where the Davis reader can gain more insight into the concept of the foreignization of language with the "remainder" allowing for the cultural "other" to emerge from the text. As Davis's fiction often reveals ourselves as 'other,' these works in their delving beyond the surface are useful guides. Moreover, a look at how nonsense texts work is yet another theoretical context that is of help. In their study *The Force of Language*, Lecercle and Riley (2005) investigate how in refusing meaning, nonsense texts provoke interpretation. They are "forceful attempts to force meaning out of them or into them. Such forceful attempts develop intuitions about the workings of language, even as nonsense itself does" (81).

Davis does not like the term flash fiction, nor does she like her work to be considered experimental:

> To me, *experimental* implies that the writer had a plan to test some preconceived writing strategy and see if it would work, that what resulted might or might not prove anything, and might or might not be successful. It seems to me both preplanned, deliberate, conceptual, and at the same time rather tentative. Since I generally prefer to start a piece of writing without much of a plan, and not to be sure exactly what I'm doing, I do not consider the stories that result, in any way experimental. (Davis 2020, n.p.; original emphasis)

However, to get a view of how her fiction fits into the general picture of the short story, and more precisely short-short fiction, Kasia Boddy on Davis in her textbook *The American Short Story from 1950* (2010) and Julie Tanner's chapter in Cocchiarale and Emmert's *Flash Fiction* (2017) each in their own way demonstrate how, if Davis is to be classified in the above genres, she takes them to a new level.

To conclude, as this chapter's close readings have striven to show, the very best place to start analyzing Davis's work is with a grammar textbook in hand – my favorite being the *The Cambridge Companion to the English Language*. In saying this, I am simply repeating the advice Davis herself gives to young writers. The book she advises is Virginia Tufte's *Artful Sentences: Syntax and Style* (2006). It is because the short story cannot afford the luxury of the novel to include passages that are present simply for the sheer pleasure of the skillful writing, we are called upon to take into account every word. In Lydia Davis's work, not only is it every word, it is every sign down to the tiniest phoneme or unobtrusive lexeme. The reader who, at first might believe that there is little to work on can discover, such as we did rapidly above, what the modest apostrophe can bring to the work; what a little verb phrase, like, "said the monk," can tell us about narrative voice, and how, when the grammar of a language gives us a choice between two seemingly synonymous structures, or words that may mean the same, we discover that what is expressed is different. But this difference often reveals itself to ring a bell with a theme, or an idea expressed further back in the text, or announces something we will discover further down.

Repetition in the short-short story is another source of investigation. Let us return briefly to "Circular Story," cited at the beginning of this chapter and examine the repetition. The pronoun "it" is used four times. Three of those times, the referent for "it" is clearly the noun anaphorically referred to in the preceding sentence; but in the last sentence we cannot replace "it" by the "the truck." The referent can only be "the racket out there on the road" located at the beginning of the text. And that is why it is a "circular" story. It brings us back to the beginning. And in bringing us back to the beginning it becomes what Roland Barthes considers to be the ideal text – the reversible text open to the greatest variety of independent interpretations and not restrictive in meaning (Barthes 1970, 11). Just like the pronoun "it."

While the study of grammar is an entry point to all her stories, grammar itself as title and/or theme predominates some of her most intriguing stories. The haunting "Grammar Questions" (Davis 2013, 527), the story of the narrator's dying father recounted through grammatical questionings: "In the phrase 'he is dying,' the words '*he is*' with the present participle suggest that he is actively doing something. But he is not actively dying. [. . .] 'He is not eating' sounds active, too. But it is not his choice. He is not conscious that he is not eating. He is not conscious at all" (528; original emphasis). Or "Example of the Continuing Past Tense in a Hotel Room" which reads, "Your housekeeper has been Shelley" (715). Or again, the use of the subjunctive in her story "Honoring the Subjunctive": "It invariably precedes, even if it **do** not altogether supersede, the determination of what is absolutely desirable and just" (377; emphasis added). The choice between the five aspects of the present indicative tense in English (compared to only one in French) cannot be carried out adequately if the translator is only applying the grammar rules. The subjunctive mode has disappeared in British English but is still used in formal writing in American English, but as Davis said herself in her interview with Wachtel, "It's so wrong to the ear" (Wachtel 2008, n.p.). The study of grammar is at the heart of the very long story "We Miss You: A Study of Get Well Letters from a Class of Fourth-Graders" (534–560), which is a socio-anthropological study of 27 children via the letters they sent to her brother who was hospitalized after being hit by a car. Their use of verb phrases, their command of complex sentences, their choice of coordinate conjunctions and formulaic expressions are described in the minutest detail in the dry, analytic voice of a sociologist, which leaves the reader not only with a sentiment of deep poignancy at these letters, but also a fine lesson on how language means and does not mean.

Like Lydia Davis looking at Mitchell's "Cornflowers," we might naively think, "Oh, it's only an apostrophe, only a contracted negation. We understand that." But then we realize, we do not. Davis concludes: "I became willing to allow aspects of the painting to remain mysterious, and I became willing to allow aspects of other problems to remain unsolved as well" (Davis 2020, n.p.). And so it is with her stories – these ever-yielding miniatures. Lydia Davis has been said to be a writer's writer. But she is first and foremost, a reader's reader. And each of her stories contain a lesson in close reading.

4 Bibliography

4.1 Works Cited

Barthes, Roland. *S/Z*. Paris: Seuil Points – Essais, 1970.

Barthes, Roland. *Oeuvres Complètes IV*. Ed. Eric Marty. Paris: Seuil, 2002.

Benjamin, Walter. *The Task of the Translator*. Ed. Hannah Arendt. Transl. Harry Zohrn. New York: Schocken Books, 1969.

Benveniste, Émile. *Problems in General Linguistics*. Transl. M. E. Meek. Miami: U of Miami P, 1971.

Blanchot, Maurice. *The Space of Literature*. Transl. A. Smock. Lincoln, Nebraska: Nebraska UP, 1992 [1955].

Boddy, Kasia. *The American Short Story since 1950*. Edinburgh: Edinburgh UP, 2010.

Cocchiarale, Michael, and Scott D. Emmert, eds. *Flash Fiction*. Ipswich, MA: Salem P, 2017.

Davis, Lydia. *The End of the Story*. London: Picador, 2004 [1995].
Davis, Lydia. *The Collected Stories of Lydia Davis*. London: Penguin, 2013 [2009].
Davis, Lydia. *Can't and Won't*. London: Penguin Random House, 2015 [2014].
Davis, Lydia. *Essays One*. New York: Picador eBook, 2020 [2019].
De Man, Paul. *The Resistance to Theory*. Minneapolis: U of Minnesota P, 1986.
Evans, Jonathan. *The Many Voices of Lydia Davis*. Edinburgh: Edinburgh UP, 2016.
Gunn, D. "The Lydia Davis Interview," http://quarterlyconversation.com-/the-lydia-davis-interview. The Quarterly Conversation. March 10, 2014. (4 July 2018).
Halliday, M. A. K., and Ruqaiya Hasan. *Cohesion in English*. London and New York: Routledge, 1976.
Haughton-Shaw, Eliza. "Essay – The Eccentricity of Lydia Davis's Essays," https://www.thelondon magazine.org/essay-the-eccentricity-of-lydia-davis-s-essays-by-eliza-haughton-shaw/. *The London Magazine*. August 18, 2020. (20 May 2021).
Huddleston, Rodney, and Geoffrey Pullum. *The Cambridge Grammar of the English Language*. Cambridge: Cambridge UP, 2002.
Jackson, Shelley. *But Is It a Story?* https://www.latimes.com/archives/la-xpm-2007-may-27-bk-jackson27-story.html. *Los Angeles Times*. May 27, 2007. (27 April 2021).
Jacobs, Carols. "The Monstrosity of Translation." *MLN* 90.6 (1975): 755–766.
Joly, André. "Lucy was Beautiful." *De la Syntaxe à la Narratologie*. Eds. Monique De Mattia and André Joly. Paris: Ophry's, 2001. 147–170.
Lecercle, Jean-Jacques. *The Violence of Language*. London: Routledge, 1990.
Lecercle, Jean-Jacques, and Denise Riley. *The Force of Language*. Basingstroke: Palgrave, 2005.
Manguso, Sarah. "An Interview with Lydia Davia," https://believermag.com/an-interview-with-lydia -davis. *The Believer's*. January 1, 2008. (2 May 2021).
Rimmon-Kenan, Shlomith. *Narrative Fiction: Contemporary Poetics*. London: Routledge, 1989.
Seghal, Paruhl. "Lydia Davis Loved Learning the Word 'Look.' These Essays Show Why," https://www.nytimes.com/2019/11/05/books/review-essays-one-lydia-davis.html. *New York Times*. November 5, 2019. (4 April 2021).
Tanner, Julie. "'Art is not in some far-off place': Lydia Davis's 'One-Paragraph Freedom,'" *Flash Fiction*. Eds. Michael Cocchiarale and Scott D. Emmert. Ipswich, MA: Salem P, 2017. 130–145.
Tufte, Virginia. *Artful Sentences: Syntax and Style*. Cheshire, CN: Graphics P, 2006.
Venuti, Lawrence. *The Translator's Invisibility*. London: Routledge, 2008 [1995].
Wachtel, Eleanor. "An Interview with Lydia Davis," https://brickmag.com/an-interview-with-lydia-davis. *Brick Magazine*. June 1, 2008. (24 April 2021).
Winters, John. "Lydia Davis's New Collection Can't and Won't . . . Doesn't," https://www.wbur.-org /artery/2014/05/13/lydia-davis-story-collection. *The ARTery*. May 13, 2014. (2 May 2021).
Yule, George. *The Study of Language*. Cambridge: Cambridge UP, 2014 [2006].

4.2 Further Reading

Cohen, Josh. "Reflexive Incomprehension – On Lydia Davis." *Textual Practice* 24.3 (2010): 501–516.
Culbert, John. "Lydia Davis in 2000 Words," https://johnculbert.wordpress.com/.../lydia-davis-in -2000-words/. *The Purgatory Press*. June 3, 2015. (11 April 2018).
Davis, Lydia. "Eleven Pleasures of Translating," https://www.nybooks.com/articles-/2016/12/08/ eleven-pleasures-of-translating/. *The New York Review*. December 8, 2016. (16 April 2021).
Fludernik, Monika. "Second Person Fiction: Narrative You as Addressee And/Or Protagonist." *Arbeiten aus Anglistik und Amerikanistik* 18.2 (1993): 217–247.

Gardelle, Laure, and Sandrine Sorlin. *The Pragmatics of Personal Pronouns*. Amsterdam: John Benjamins, 2015.
Goodyear, Dana. “Long Story Short: Lydia Davis's Radical Fiction.” *New Yorker*. March 17, 2014. 24–31.
Grice, Paul H. “Logic and Conversation.” *Syntax and Semantics: Speech Acts*. Vol. 3. Eds. Peter Cole and Jerry L. Morgan. New York, San Francisco, London: Academic P, 1975. 41–58.
Jakobson, Roman. *Language in Literature*. Cambridge, Massachusetts: Harvard UP, 1987.
Klinedinst, Nathan, and Paul Egre. *Vagueness and Language Use*. Basingstoke: Palgrave Macmillan, 2011.
Knight, Christopher J. “An Interview with Lydia Davis.” *Contemporary Literature* 40.4 (Winter 1999): 525–551.
Mars-Jones, Adam. “Reality is Worse,” https://www.lrb.co.uk/the-paper/v36/n08/adam-mars-jones/reality-is-worse. *The London Review of Books* 36.8, April 17, 2014. (20 May 2021).
Matthews, Brendan. “Lydia Davis: ‘I kind of like the fact that my work isn't for everyone,’” https://www.salon.com/2014/04/28/lydia_davis_i_kind_of_like_the_fact_that_my_work_isn%%80%99t_for_everybody/. *Salon*. April 29, 2014. (20 May 2021).
McCaffery, Larry. *Some Other Frequency: Some Interviews with Innovative American Authors*. Philidelphia: U of Pennsylvania P, 1996.
Parkes, Malcom B. *Pause and Effect: An Introduction to the History of Punctuation in the West*. New York: Routledge, 2016 [1992].
Pennebaker, James W. *The Secret Life of Pronouns: What Our Words Say About Us*. New York: Bloomsbury Press, 2011.
Ricks, Christopher. “The Art of Lydia Davis.” *The Collected Stories of Lydia Davis*. London: Penguin, 2013. xii–xxii.
Sorlin, Sandrine. *Inventive Linguistics*. Montpellier: Presses Universitaires de la Méditerranée, 2009.
The Harlequin. “An Interview with Lydia Davis,” http://www.theharlequin.org-/davis5/. *The Harlequin*. 2014. (22 April 2021).

Philip Coleman

32 George Saunders (1958–)

Abstract: Widely regarded as one of the most important and influential writers of short fiction working in the United States today, George Saunders's reputation to date has been built on the critical and popular success of four collections of short stories published between 1996 and 2013: *CivilWarLand in Bad Decline* (1996), *Pastoralia* (2000), *In Persuasion Nation* (2006), and *Tenth of December* (2013). In this chapter Saunders's development is contextualized in relation to "post-postmodern" writers such as David Foster Wallace, with whom he shared a desire to create a new kind of fiction that would be technically accomplished but, at the same time, would engage meaningfully with readers. Focusing on Saunders's stories "Sea Oak" (from *Pastoralia*) and "The Semplica Girl Diaries" (from *Tenth of December*), the essay will examine the technical achievement of Saunders's work while also recognizing some of the ethical and political challenges it presents to readers of short fiction in the early twenty-first century.

Keywords: Contemporary short story, post-modernism, post-postmodernism, New Sincerity

1 Context

Until his first novel, *Lincoln in the Bardo*, was published in 2018, George Saunders was known almost exclusively as a writer of short fiction. While he has also published four collections of nonfiction – and a poem ("Black Widows") in *The New Yorker* in 2005 – Saunders's commitment to the art of short fiction, in particular, is noteworthy among contemporary writers. This is evident not just in his own creative output as a writer of stories but also in his detailed consideration of the short story form in relation to texts by Anton Chekhov, Ivan Turgenev, Leo Tolstoy, and Nikolai Gogol in *A Swim in the Pond in the Rain* (2021). There is a gap of a decade between the publication of his first short story, "A Lack of Order in the Floating Object Room" (1986), and his first collection, *CivilWarLand in Bad Decline* (1996), but in the period since then he has published three further volumes of short stories that have established Saunders, along with writers such as Sherman Alexie and Lydia Davis (↗31 Davis), as one of the leading figures in the development of the American short story in the early decades of the twenty-first century. As one might expect of a writer who came of age in the 1980s, Saunders's work shares a number of similarities with others who came to prominence during that decade, most notably David Foster Wallace. To put Saunders's work in context, then, one needs not only to consider the extent to which his work engages with problems and issues in the world at

https://doi.org/10.1515/9783110587647-033

the present time – from people trafficking to climate change – but it is also important to appreciate a number of biographical details that help to explain both the thematic focus of many of his best stories as well as the creative impetus behind their particular approach to language and form.

George Saunders was born on 2 December 1958 in Amarillo, Texas, but grew up in Chicago. In part because of the boom (and subsequent glut) in the oil market in the United States and globally during the 1980s, he studied geophysical engineering at Colorado School of Mines. He graduated in 1981 with a qualification that allowed him to "prospect for oil," as he put it in an online interview in 2000 (Childers 2000, n.p.), and he travelled to Sumatra and Indonesia, where he worked for two years as a scientist. During the same time, however, he was writing fiction, even though he was "awash in technical homework" (n.p.). "In the end," he says in the same interview, "the two interests dovetailed":

> the tech writing I was doing influenced both style and subject. And any claim I might make to "originality" in my fiction is really just the result of this odd background: basically, just me working inefficiently, with flawed tools, in a mode I don't have sufficient background to really understand. Like if you put a welder to designing dresses. (n.p.)

The dovetailing of his scientific background and training with his interest in literature is then a crucial detail in understanding Saunders's early development as a writer, and he often reminds readers of these beginnings in his nonfiction: he does so, for example, in the opening pages of *A Swim in a Pond in the Rain* (Saunders 2021, 2–3). It relates in an important way to the pervasiveness of the theme of work in his stories – a topic that has been explored in detail by Kasia Boddy (2017) – but it also justifies the comparison between Saunders and Wallace.

Wallace – born four years after Saunders – is also a writer whose work is, perhaps to a greater degree, pervaded by what Saunders terms "tech writing." However, while this is evident in many of Wallace's works – most notably in the long mathematical digression in the "Eschaton" section of *Infinite Jest* (Wallace 1996, 321–342) – it often has less to do with the use of terms or concepts derived from the hard sciences than it has with a certain approach to form and language that is related to these writers' engagements with earlier post-modern writers such as John Barth and Donald Barthelme (↗24 Barthelme). In an interview with Larry McCaffery in 1993, Wallace explored his great admiration for, but also his acute sense of distance from these writers, and the "built-in meta-games and intertextualities" that often characterize their work (McCaffery). The challenge, as Wallace saw it, was to create a new kind of fiction that would not succumb to the recursive trap of metafictional irony that one finds, for example, throughout the stories in Barth's *Lost in the Funhouse* (1968). Instead, he wished to create fiction that would allow readers to access to what he termed "the art's heart's purpose":

> [. . .] it seems like the big distinction between good art and so-so art lies somewhere in the art's heart's purpose, the agenda of the consciousness behind the text. It's got something to do with love. [. . .] [I]t seems like one of the things really great fiction-writers do – from Carver to

> Chekhov to Flannery O'Connor [. . .] – is "give" the reader something. The reader walks away from the real art heavier than she came into it. Fuller. All the attention and engagement and work you need to get from the reader can't be for your benefit; it's got to be for hers. What's poisonous about the cultural environment today is that it makes this so scary to try to carry out. Really good work probably comes out of a willingness to disclose yourself, open yourself up in spiritual and emotional ways that risk making you really feel something. (McCaffery)

There was a sense, for Wallace, that earlier postmodern fiction often did not allow readers to "feel something": the focus, rather, was on trying to work out what the writer was saying as a kind of technical problem. In Wallace's later short fiction, in particular, there is a discernible shift towards the question of empathy and what it means to achieve intersubjective connection between individuals that builds on his stated desire to "open up [his work] in spiritual and emotional ways" (McCaffery). For Wallace, then, the writer had a responsibility to "'give' the reader something" (McCaffery) in return for whatever the reader was willing to invest in terms of intellectual, interpretive, and critical labor. Critics such as Adam Kelly have described this in terms of a movement towards what has been characterized as a "New Sincerity" (Kelly 2017, 41–58) in contemporary fiction that may be found, also, in the works of Jennifer Egan, Dave Eggers, and Jonathan Franzen.

Reflecting on his early development as a writer, Saunders has identified with exactly the same issues raised by Wallace, as he explained in an essay-interview with Joel Lovell in *The New York Times Magazine* in 2013 (reprinted as an introduction to the 2013 Bloomsbury paperback edition of *Tenth of December*). In the course of their conversation, Lovell asks Saunders about his "relationship to Wallace" and he records that:

> For all the ways in which their fiction might seem to be working similar themes, they were [. . .] "like two teams of miners, digging at the same spot but from different directions." He [Saunders] described making trips to New York in the early days and having "three or four really intense afternoons and evenings" with, on separate occasions, Wallace and Franzen and Ben Marcus, talking to each of them about what "the ultimate aspiration for fiction was." Saunders added: "The thing on the table was emotional fiction. How do we make it? How do we get there? Is there something yet to be discovered? These were about the possibly contrasting desire to: (1) write stories that had some sort of moral heft and/or were not just technical exercises or cerebral games; while (2) not being cheesy or sentimental or reactionary."
>
> "Those guys came from a much better trained place," he said. "They had a very strong and passionate involvement with postmodernism when it was still hot off the griddle." Whereas, for him the question wasn't how to move beyond the postmodern fathers who shaped current American literary sensibility; it was how "to mimic the emotional conditions of my actual working life" – how to, as he later put it, arrive at a voice that was informed by "the mild ass-kickings" he suffered or witnessed in his adult life "that had the effect of politicizing and tenderizing me." (Lovell 2013, xxii–xxiii)

It is important to note Saunders's sense of closeness to Wallace, Franzen, and Ben Marcus here – notwithstanding their different intellectual and educational backgrounds – but also the fact that he shared their desire to create works of "emotional fiction" that were neither "cheesy or sentimental or reactionary" or "just technical exercises or cerebral games." In short, Saunders recognized the need to create a new kind of "post-postmodern" fiction that would have "spiritual" and "emotional" value, to use Wallace's terms. However, his work would be different from theirs because it would engage more directly with the world of "working life" that is so central to his short stories.

This helps to explain why Saunders's stories, at their best, are rewarding both in terms of their formal and technical achievement but also for what they have to say about the ethical and political challenges of living in the world today. For all of its "fantastical elements," as Kasia Boddy has argued, Saunders's fiction "depicts a recognizable America" (Boddy 2017, 4) – replete with many of the social, political and cultural markers that make the world of his stories so uncomfortably proximate and real. At the same time, however, as the close readings of two of his stories in the following sections of this essay will show, his formal skill as a writer is evident in every sentence of his carefully crafted prose.

2 Close Readings

First published in *The New Yorker* in December 1998 – like many other contemporary short fiction writers, Saunders's stories often appear in magazines before they are collected in book-length volumes – "Sea Oak" is a darkly comic tale that brings together several of the author's major themes and concerns: work, family, responsibility, money, the environment, language, and the American Dream; the story's treatment of the relationship between the living and the dead also connects it with *Lincoln in the Bardo*, albeit in very different ways. Structured in unnumbered sections and delivered in a clipped, pared-back prose style, the title of the story refers to a community of "a hundred subsidized apartments" where "there's no sea and no oak [. . .] and a rear view of FedEx" (Saunders 2000, 93) somewhere in the United States in the late twentieth century.

As with much of Saunders's work, then, readers have a sense that they are being told a story about the contemporary moment in a cultural and social context that seems recognizably real, but there are elements that seem to belong to a dystopian near-future. At one point, for example, a character refers to "a light from the last remaining farm," suggesting that the story is set against a background of environmental collapse in a world where farms and the idea of living off the land is becoming, or has already become, a thing of the past (99). Elsewhere, a raccoon is seen "nibbling at a rusty bike" (96). "Last time I checked this was still America" (101) a character announces early on in the story, however, and of course this mirrors what often happens in Saunders's writing: the readers have to keep reminding themselves that the world is *not* really as Saunders describes it, but it *could* be. His stories are terrifyingly prescient in their representations of contemporary reality, especially in relation to real contemporary concerns such as climate change. Like Flannery O'Connor (↗19 O'Connor), in a sense, Saunders is a writer who frequently draws "large and startling figures" in his stories because he wants his readers to see and hear the world in a radically new way (O'Connor 2000, 34). His fictions strain the reader's credibility but, at the same time, they present urgent ethical and political questions that go to the heart of contemporary experience. "Sea Oak" is set in a world that is very much on the threshold of the anthropocene, but it is one, too, where human beings also struggle for survival.

Emphasizing Saunders's preoccupation with "actual working life" – to borrow a phrase from his interview with Lovell (2013, xxiii) – "Sea Oak" begins in the unnamed protagonist's workplace, a male stripper club called "Joysticks":

> At six Mr. Frendt comes on the P.A. and shouts, "Welcome to Joysticks!" Then he announces Shirts Off. We take off our flight jackets and fold them up. We take off our shirts and fold them up. Our scarves we leave on. Thomas Kirster's our beautiful boy. He's got long muscles and bright-blue eyes. The minute his shirt comes off two fat ladies hustle up the aisle and stick some money in his pants and ask will he be their Pilot. He says sure. He brings their salads. He brings their soups. My phone rings and the caller tells me to come see her in the Spitfire mock-up. Does she want me to be her Pilot? I'm hoping. Inside the Spitfire is Margie, who says she's been diagnosed with Chronic Shyness Syndrome, then hands me an Instamatic and offers me ten bucks for a close-up of Thomas's tush.
>
> Do I do it? Yes I do. (Saunders 2000, 91)

Saunders's attention to certain details – the specificity of time ("At six"), place ("Welcome to Joysticks!") and the naming of characters and their physical characteristics ("long muscles and bright-blue eyes") – helps the reader to apprehend the world of the story, for all of its weirdness. Very quickly, however, it is clear that there is not much *joy* in "Joysticks." The employees labor under a system where they are rated according to their popularity with patrons (referred to as "guests") who rank them as "Knockout, Honeypie, Adequate, or Stinker" (92). The first section of the story ends with one employee, Lloyd Betts, losing his job because "[l]ately he's put on weight and his hair's gone thin" (91). As the narrator puts it: "What a stressful workplace. The minute your Cute Rating drops you're a goner. [. . .] Not that I'm complaining. At least I'm working. [. . .] I'm a solid Honeypie/Adequate, heading home with forty bucks cash" (92).

What the narrator does, then, in the world of "Sea Oak" is a kind of work that most people would find profoundly demeaning. He takes off his clothes for money and, if he is lucky, he gets to earn a little more by going further with the customers to either perform sexual acts or take pornographic photographs of himself or his colleagues (such as the "close-up of Thomas's tush" requested by Margie in the opening scene). His success at work depends on the physical attractiveness and well-being of his body, but if the work he does in itself seems "stressful" – not to mention the "workplace" in which he must do it – his "home" environment is not much better. There the narrator lives with his sister, Min, their cousin, Jade, their babies Troy and Mac, and an aunt, Bernie, who is presented as "a peacemaker" (94). The six of them live in an apartment in Sea Oak, a home for which everyone "should be thankful" (97) according to Aunt Bernie but which, in the narrator's view, is "a pit" (95). Aunt Bernie belongs to an earlier generation that seems to still believe in the American Dream: if you work hard enough you can eventually succeed. The narrator has given up on this idea, however, and if he could he would "move everybody up to Canada" (97). Min and Jade do not seem to have a view either way, obsessed as they are with so-called Reality TV, spending their days watching shows with titles such as *How My Child Died Violently* and wondering "how many sides a triangle has" (93–94). Interestingly, all of these characters are named in the story, but the narrator is not. In a sense he is merely there to serve others, whether he is working in Joysticks or at home in Sea Oak.

Between the sad debasement and stress he experiences daily as an employee at Joysticks and his deeply dysfunctional and demanding homelife in Sea Oak, things could hardly get any worse for the narrator but in the fourth section of the story Aunt Bernie – the only other member of the family who has a job – dies of fright after an intruder breaks into their home. "All her life she worked hard. She never hurt anybody. And now this" (100): Saunders's prose expresses the sense of exhausted dismay with which so much of what happens in "Sea Oak" is narrated, a feeling of disbelief coupled with an inability to really express what is going on in this strange end-time world. This is further compounded by Aunt Bernie's transformation into a zombie, which represents an absurd escalation of the action in Saunders's story, but it is also a surreal re-writing of the Christian resurrection tale: following her violent death she returns to instruct the living on how they should live their lives so that they can improve their general state of affairs. Towards the end of the story, the narrator wonders if "other than Jesus, this has ever happened before" and surmises: "Maybe it happens all the time. Maybe there's angry dead all over, hiding in rooms, covered with blankets, bossing around their scared, embarrassed relatives" (123–124). What the narrator imagines does not happen all the time, of course, but the idea of one individual in a community dying before their time so that those who live on might learn from their example is an old one, going back at least to the figure of Christ in the New Testament. In a sense, then, "Sea Oak" is a post-postmodern re-writing of the Christian message in a late capitalist U.S. American context. Aunt Bernie's "resurrection" serves to awaken in her family a positive sense of purpose and possibility that they had forgotten, or possibly never known. The narrator acknowledges that change will be gradual and many things will remain the same – he continues to work in Joysticks and Min and Troy still spend much of their time watching TV – but their experience dealing with Bernie as a zombie brings them closer together as a family and they agree to take small steps towards creating a better home for them all to live in going forward. The story ends with the narrator admitting that he does not know why "[s]ome people get everything and I [get] nothing" (125) – a question put to him in a dream by Aunt Bernie – but he seems to have learned that dwelling on that question alone will not change anything.

"What fun is life when you don't know nothing?" Bernie demands of her nieces and nephew shortly after she comes back from the dead (119). In a sense, however, Saunders is less concerned with affirming the American Dream in "Sea Oak" than he is with promoting ideas of self-knowledge and awareness as essential aspects of selfhood that are often undermined and erased in contemporary culture, and not just in the United States. "Sea Oak" is a story that scrutinizes the way that selfhood is constructed and performed in the United States today, but everything from the Joysticks' employees' use of "Penile Simulators" to Min and Jade's arguments about language suggests that this is a world in which it is no longer possible to achieve authenticity. The narrator's closing acknowledgement that he does not know why some people are better off than others is a way of saying that there may be more important questions worth considering – questions that he can actually address, such as the possibility of moving out of Sea Oak into a place like "Swan's Glen" where "[t]hey've never had a shooting or a knifing and the public school is great and every Saturday they have a nature walk for kids behind the clubhouse" (125). Bringing together a keen sense of the important interconnection between education, community, and responsibility, then, Saunders offers a pragmatic vision of the future for his characters that frees them from the idealistic delusions that have often thwarted those who have pursued the American Dream. Like his story "Brad Carrigan, American" (see Saunders 2006), as Michael Trussler has argued, "Sea Oak" is also a text that "presents zombie figures in order to critique the nihilist dimensions in present America" (Trussler 2017, 211). However, the final section of "Sea Oak" suggests a movement from nihilism towards a new-found optimism based on principles of patience and kindness that have more to do with spiritual than national self-recovery. For all of its grim and occasionally dark comedy, this allows "Sea Oak" to be read, finally, as a fable of hope. As Gillian Elizabeth Moore has put it, this is a text that does "not point toward an exceptional utopia, but instead toward the need not to make America *the best*, but to make many of its many disempowered citizens' lives better" (Moore 2017, 73; original emphasis). By directing the reader towards the question of what might make an individual life go better rather than thinking of what it might mean to be the best in a narrow individualistic sense, Saunders creates a work of fiction that raises important ethical questions about the relationship between self and world in the contemporary era.

Like "Sea Oak," "The Semplica Girl Diaries" is also set in a world that is immediately recognizable to the reader – a reference to U.S.-American office supplies retailer OfficeMax in the story's opening sentence accomplishes this – but it also contains elements that are disturbing in their visceral surrealism. Martin Ansin's image that accompanied the story when it was first published in *The New Yorker* in October 2012 reinforces the text's social and cultural immediacy with its intense red and blue coloring of a suburban clapboard home, but the feet of four women who appear to be floating in the air above a pool suggests something more sinister in this otherwise very ordinary representation of contemporary American life. In a sense, "The Semplica Girl Diaries" is a story about ordinariness and what happens when one becomes obsessed with achieving or possessing what is thought of as extraordinary or exceptional. The central character – the narrator around whose seventeen dated diary entries the story is told – desires, on the one hand, to create "a picture of [. . .] how life really was/is now" for the benefit of future generations, but it quickly becomes apparent that he is deeply unhappy with how life is for him and his family at the present time (Saunders 2013, 109). While he claims that he wishes to record his "philosophical speculations" for the benefit of "future people" – "Will future people be aware of concept of 'demons'? Will they find our belief in 'demons' quaint? Will 'windows' even exist?" – his chief preoccupation is with the here-and-now and his sense of dissatisfaction with how things actually are (110). This is bound up with the narrator's misguided sense that "In America of [his] time [. . .] anything [is] possible" (129) but over the course of the story he comes to realize that while "Happiness" (131) *is* attainable it does not always follow from greater material wealth. "The Semplica Girl Diaries" is then a story that also interrogates the American Dream and, like many of Saunders's stories, it does this through an examination of questions to do with the nature of family, community, work, and responsibility. One of the things that makes it particularly compelling in Saunders's *œuvre*, however, is its treatment of larger, global issues including people-trafficking and the status of refugees.

The narrator of "The Semplica Girl Diaries" believes that if he becomes wealthier, he will have more time to "reflect deeply on meaning of life, etc." (112–113) but his desire for wealth is quickly revealed to have more to do with the acquisition of material objects than any kind of deep philosophical understanding. Indeed, after a visit to the home of one of his children's schoolfriends, the Torrinis – with his wife, Pam, and their two daughters, Lilly and Eva – he becomes more obsessed with comparing himself with what other families appear to have, leading him to write the following prayer in his diary: "Lord, give us more. Give us enough. Help us not fall behind peers. Help us not, that is, fall further behind peers. For kids' sake. Do not want them scarred by how far behind we are" (121). This is troublingly ironic given what happens in the rest of the story, not least in terms of the physical scars left on the bodies of the so-called Semplica Girls after they have been prepared for "instalment" (116) in the gardens of those who can afford to buy them. Fitted with "brain inserts" (116) through which "microlines" (116) are run, the narrator sees them "hanging silently, white smocks in moonlight" (121) in the gardens of his neighbors. Seeing them there "[m] akes [him] think of ancient times and affluent men of those times building great gardens, roaming through while holding forth on philosophy, bounty of earth having been lassoed for the pleasure of etc., etc." (121). The reader is not told precisely what "the pleasure of etc., etc." is, but the narrator believes firmly that the acquisition of these and other material things will make his life better. As it happens, a brief spell of financial good fortune only makes him more miserable, not least because of his decision to spend the bulk of his winnings – "approx. $7400" – on four Semplica Girls for his garden (153).

The Semplica Girls are effectively slaves, bought by homeowners to serve as garden furniture. To those who own and promote the business of buying and selling them, however, they came to the United States of their own volition. As a police officer tells the narrator at one point in the story, after his four Semplica Girls have gone missing, they were there because they were "Smelling that American dream, baby" (152). The barbaric logic of neo-imperialism in the world of the story does not allow for sympathy, insisting instead that the missing "SGs" would become "illegal immigrants" and "take jobs away from 'legit Americans'" if they were not found (162). Coming from Moldova, the Philippines, Somalia, and Laos, it transpires that they were freed by the narrator's own daughter, Lilly, because she could see the violent racism that underpinned the whole Semplica Girl enterprise. When the narrator discovers that the young women have gone missing – even their names are not their own, as he reveals in one diary entry (166) – he writes down the following questions:

> No money, no papers. Who will remove microline? Who will give her job? When going for job, must fix hair so as to hide scars at Insertion Points. When will she ever see home + family again? Why would she do? Why would she ruin it for all, leave our yard? Could have had nice long run w/us. What in the world was she seeking? What would she want so much, that would make her pull such a desperate stunt? (167)

Conflating the identities and motivations of the escaped Semplica Girls with his daughter – the questions are addressed to both at the same time in a rush of anger and disappointment – the narrator exposes his own deep-felt racism and misogyny. At the end of the story, he makes a note to call the company from which he bought the "SGs" to come and "take [the] ugly thing away" (167), but one has the sense that he has not learned or changed very much from the experience. The whole experience causes him more embarrassment than shame. "Empty rack in yard," he notes, "looking strange in moonlight" (167). The haunting image of the "rack" – a common metonym for cruelty and torture – is "ugly" to him because it is no longer fulfilling its terrible function. To his daughter, however, and the enlightened reader, it has become a powerful symbol of contemporary society's failure to address the persistence of racism and forms of slavery in the twenty-first century, despite all of the so-called progress that has been made in science, for example, and medicine. The narrator describes the creator of the Semplica Girls to his daughter Eva at one point in the narrative as a "doctor + smart cookie" and stresses that the process of inserting microlines in the SGs' heads is "very gentle, does not hurt" (142). His ignorance and naivety are reflective of a cultural and political moment in which personal advancement is prized over communal well-being, when there is a general acceptance of the idea that the happiness of the few depends to a large degree on the misery of the many.

At one point in "The Semplica Girl Diaries" one of the narrator's workmates, Todd Grassberger, dies. Following his funeral, the narrator writes in his diary that he must "try harder, in all things, to be better person" (149). In the course of the story the reader does not perceive a convincing shift in the narrator's attitude towards others, however, but he does record many ethical challenges that are a direct consequence of his experience of the contemporary world. In a way, Saunders's story serves as a kind of thought experiment, prompting the reader to ask what they might do – and what one ought to do – in situations such as those experienced by all of the characters in the world of the Semplica Girls. Can we understand why the narrator feels a need to possess the same material objects as his neighbor and, more fundamentally, is his sense of the connection between material wealth and personal well-being justified? Notwithstanding the justifications of medical science presented in the story, is it right that human beings should be used in the way that the Semplica Girls are, even if the work they do as SGs allows them to escape greater poverty in their home countries? What is poverty and to what extent, if at all, should someone from another country be responsible for the experience of citizens of other states? Saunders's story raises these and many other pressing ethical questions that are relevant not only to our sense of what happens in his fiction but to how we conduct ourselves in our own daily lives today, in the United States and beyond. While the basic idea of "The Semplica Girl Diaries" may have come to the author in a dream, as Saunders has explained in an interview with Deborah Treisman (2013, n.p.), it is a text that reflects in direct and powerful ways on what it means to live in the world in the early twenty-first century.

3 Theoretical Perspectives

In an essay on Donald Barthelme's short story "The School" collected in his volume of non-fiction *The Braindead Megaphone* (Saunders 2007, 175–185), Saunders begins with a brilliant discussion of what he calls "the hardest thing in storytelling: getting one's action to rise" (175). His discussion of Barthelme is useful for what it reveals about Saunders's approach as a teacher – the piece is framed in terms of his work as a professor of creative writing – and for its analysis of a major text in the development of twentieth-century short fiction. However, it also signals Saunders's appreciation of plot in his own stories and the strategic importance of narrative pacing and design. Saunders's stories are rewarding, then, to close readers who are interested in questions of form and language, and it is not surprising that several critics have used formalist accounts of his work as a basis for deeper philosophical and ideological engagements with his fiction. Clare Hayes-Brady, for example, has examined the relationship between language, subordination, and freedom in Saunders's stories, beginning with what she sees as his "constant attention to how American identity is formed and determined by American speech patterns" (Hayes-Brady 2017, 24). Robert Cameron Wilson has also paid close attention to Saunders's language from a formalist perspective, using the theoretical perspectives of Mikhail Bakhtin and Caryl Emerson to explore the writer's use of microdialogues and polyphony. Both Hayes-Brady and Wilson emphasize the ways in which Saunders engages with contemporary political, cultural, and social discourse in the United States, which in turn they see filtered into the language spoken by Saunders's characters. By analyzing the ways his characters speak to each other, these critics provide important perspectives on the connection between Saunders's use of language and the way people speak to each other in the world today. As Wilson puts it: Saunders's "stories reveal how much of thought is born in response to the voice of another human being, real or imagined, and how the voices one channels in everyday life can be the resources for survival or the gadflies that nag or denigrate" (Wilson 2017, 232).

Saunders's fiction has generated a wide variety of critical responses, especially since the publication of *Tenth of December* (2013). Taking seriously Saunders's admission that the idea for "The Semplica Girl Diaries" came from a dream, Steve Gronert Ellerhoff has offered a reading based on post-Jungian psychoanalysis. In a spirit of critical empathy that is aligned with the same principles of "kindness" Saunders promoted – perhaps most famously in his Syracuse University convocation address in May 2014 which was subsequently published as *Congratulations, by the way: Some Thoughts on Kindness* (Saunders 2014) – Ellerhoff contends that the reader is encouraged to feel concern for the narrator of "The Semplica Girl Diaries" who he says is "sleepwalking through life [and] desperately needs to wake up" (Ellerhoff 2017a, 202). The humane generosity of Ellerhoff's reading is prompted by the promotion of empathy as a fundamental and necessary human trait in Saunders's fiction, which has also been analyzed by Michael Basseler in the light of recent critical and theoretical work within the field of affect studies. As Basseler has put it:

> Despite all [the] satire, surreal settings, outlandish protagonists, stylistic coolness, and deadpan humor, there is always Saunders's deep humanism serving as a counterbalance and, perhaps, constituting something like the aesthetic *and* ethical core of his writing.
> (Basseler 2017, 153)

These ideas are contextualized in relation to the emergence of the New Sincerity by critics such as Adam Kelly, for whom the political and ethical questions raised by literary texts are "not fully articulated but waiting to be taken up" (Kelly 2017, 54). Kelly's reading stresses the crucial importance of readers engaging not just in the critical interpretation of Saunders's work – with matters of language, structure, and style – but with the larger ethical and political questions it raises about what it means to live in the world today. Indeed, David Huebert has gone further to suggest that Saunders takes us beyond the human and is also receptive to a Foucauldian / biopolitical reading that "presents a consistently unsettling vision of the human body and its persistently pathological relationship to the synthetic environments in which it is housed" (Huebert 2017, 105).

Saunders's work is then amenable to traditional formalist and historicist readings, but it also yields useful and challenging questions when considered from within a range of critical and theoretical perspectives that have evolved in the early decades of the twentieth century, from post-Jungian analysis to biopolitics and theories of affect. In some ways the critical work on Saunders is still in its very early stages, as one might expect from a writer who is still in full flight. Apart from Jurrit Daalder's analysis of the trope of cruelty in Saunders's fiction and his critique of a general "fixation on the author's moral vision" (Daalder 2017, 169) that has dominated the critical response to date – and, indeed, the readings offered in this essay – most critics have been very positive about his work so far, especially his short fiction. However, Saunders's openness to various forms of cultural production – he was involved in an adaptation of "Sea Oak" for television serialization in 2017 starring Glenn Close, for example – also suggests that this is a writer whose work may yet develop in unexpected and exciting directions. Certainly, the publication of his first novel, *Lincoln in the Bardo*, came as a great and welcome surprise to many readers. In his interview with Steve Gronert Ellerhoff, Saunders said of his stories, and stories in general, that: "The story is talking back to you and then you have to sort of start serving it" (Ellerhoff 2017b, 243). Readers have been serving and, indeed, "talking back" to Saunders's stories in increasingly complex and revealing ways in recent years and there is every reason to believe they will continue to do so for some time to come. However, Saunders's detailed engagement with the stories of Chekhov, Gogol, Tolstoy and Turgenev in *A Swim in a Pond in the Rain* confirms his critical commitment to the short story too and his desire, in all of his work, to promote new ways of reading and understanding the possibilities of the genre.

4 Bibliography

4.1 Works Cited

Ansin, Martin. Illustration for George Saunders, "The Semplica-Girl Diaries" [sic], https://www.newyorker.com/magazine/2012/10/15/the-semplica-girl-diaries. *The New Yorker* 8 October 2012 (30 May 2019).

Basseler, Michael. "Narrative Empathy in George Saunders's Short Fiction." *George Saunders: Critical Essays*. Eds. Philip Coleman and Steve Gronert Ellerhoff. New York: Palgrave Macmillan, 2017. 153–171.

Boddy, Kasia. "'A Job to Do': George Saunders on, and at, Work." *George Saunders: Critical Essays*. Eds. Philip Coleman and Steve Gronert Ellerhoff. New York: Palgrave Macmillan, 2017. 1–22.

Childers, Doug. "*The Wag* Chats with George Saunders", http://www.thewag.net/interviews/saunders.html. *The Wag: A Magazine for Decadent Readers* 1 July 2000 (30 May 2019).

Coleman, Philip, and Steve Gronert Ellerhoff, eds. *George Saunders: Critical Essays*. New York: Palgrave Macmillan, 2017.

Daalder, Jurrit. "Cruel Inventions: George Saunders's Literary Darkenfloxx™." *George Saunders: Critical Essays*. Eds. Philip Coleman and Steve Gronert Ellerhoff. New York: Palgrave Macmillan, 2017. 173–188.

Ellerhoff, Steve Gronert. "Dreaming and Realizing 'The Semplica Girl Diaries': A Post-Jungian Reading." *George Saunders: Critical Essays.* Eds. Philip Coleman and Steve Gronert Ellerhoff. New York: Palgrave Macmillan, 2017a. 189–204.
Ellerhoff, Steve Gronert. "'A Little at a Time. And Iteratively': A Conversation with George Saunders." *George Saunders: Critical Essays.* Eds. Philip Coleman and Steve Gronert Ellerhoff. New York: Palgrave Macmillan, 2017b. 237–243.
Hayes-Brady, Clare. "Horning In: Language, Subordination and Freedom in the Short Fiction of George Saunders." *George Saunders: Critical Essays.* Eds. Philip Coleman and Steve Gronert Ellerhoff. New York: Palgrave Macmillan, 2017. 23–39.
Huebert, David. "Biopolitical Dystopias, Bureaucratic Carnivores, Synthetic Primitives: 'Pastoralia' as Human Zoo." *George Saunders: Critical Essays.* Eds. Philip Coleman and Steve Gronert Ellerhoff. New York: Palgrave Macmillan, 2017. 105–120.
Kelly, Adam. "Language Between Lyricism and Corporatism: George Saunders's New Sincerity." *George Saunders: Critical Essays.* Eds. Philip Coleman and Steve Gronert Ellerhoff. New York: Palgrave Macmillan, 2017. 41–58.
Lovell, Joel. "Introduction: George Saunders Has Written the Best Book You'll Read this Year." *Tenth of December: Stories.* London: Bloomsbury Paperbacks, 2013. xi–xxxii.
McCaffery, Larry. "A Conversation with David Foster Wallace" (1993), https://www.dalkeyarchive.com/a-conversation-with-david-foster-wallace-by-larry-mccaffery. (30 May 2019).
Moore, Gillian Elizabeth. "'Hope that, in Future, All Is Well': American Exceptionalism and Hopes for Resistance in Two Stories by George Saunders." *George Saunders: Critical Essays.* Eds. Philip Coleman and Steve Gronert Ellerhoff. New York: Palgrave Macmillan, 2017. 59–75.
O'Connor, Flannery. *Mystery and Manners: Occasional Prose.* Eds. Sally and Robert Fitzgerald. 1957. New York: Farrar, Straus and Giroux, 2000.
Saunders, George. *CivilWarLand in Bad Decline.* New York: Random House, 1996.
Saunders, George. *Pastoralia.* New York: Riverhead Books, 2000.
Saunders, George. *In Persuasion Nation.* New York: Riverhead Books, 2006.
Saunders, George. *The Braindead Megaphone: Essays.* New York, Riverhead Books, 2007.
Saunders, George. *Tenth of December: Stories.* New York: Random House, 2013.
Saunders, George. *Congratulations, by the way: Some Thoughts on Kindness.* New York: Random House, 2014.
Saunders, George. *Lincoln in the Bardo.* New York: Random House, 2017.
Saunders, George. *A Swim in a Pond in the Rain: In Which Four Russians Give a Master Class on Writing, Reading and Life.* London: Bloomsbury, 2021.
Treisman, Deborah. "This Week in Fiction: George Saunders on Authorial Bullying, Parenting, and Consciousness After Death", https://www.newyorker.com/books/page-turner/this-week-in-fiction-george-saunders. *The New Yorker* 1 February 2016 (30 May 2019).
Trussler, Michael. "Everyday Zombies: Ethics and the Contemporary in 'Sea Oak' and 'Brad Carrigan, American.'" *George Saunders: Critical Essays.* Eds. Philip Coleman and Steve Gronert Ellerhoff. New York: Palgrave Macmillan, 2017. 59–75.
Wallace, David Foster. *Infinite Jest.* 1996. New York, Boston and London: Little, Brown and Company, 2006.
Wilson, Robert Cameron. "'Third-person Ventriloquism': Microdialogues and Polyphony in George Saunders's 'Victory Lap.'" *George Saunders: Critical Essays.* Eds. Philip Coleman and Steve Gronert Ellerhoff. New York: Palgrave Macmillan, 2017. 221–235.

4.2 Further Reading

Doyle, Jon. "The Changing Face of Post-postmodern Fiction: Irony, Sincerity, and Populism." *Critique: Studies in Contemporary Fiction* 59.3 (2018): 259–270.

Kelly, Adam. "The New Sincerity." *Postmodern/Postwar – And After: Rethinking American Literature*. Eds. Jason Gladstone, Andrew Hoberek, and Daniel Worden. Iowa City: U of Iowa P, 2016. 197–208.

McLaughlin, Robert L. "Post-Postmodern Discontent: Contemporary Fiction and the Social World." *symplokē* 12.1/2 (2004): 53–69.

Nele Rodiek

33 Junot Díaz (1968–)

Abstract: The short stories analyzed in this chapter are taken from Junot Díaz's two short story collections, *Drown* (1996) and *This Is How You Lose Her* (2012), respectively. These collections constitute part of a story cycle, together with Díaz's debut novel *The Brief Wondrous Life of Oscar Wao* (2007), which examines the Dominican American experience focusing on suppressed trauma and toxic masculinity. The protagonist and narrator Yunior serves as the main linking element of the otherwise fragmented cycle and is overtly present in most of Díaz's short fiction. This chapter discusses the interconnection of trauma, ethnicity, and toxic masculinity in "Drown" and "The Cheater's Guide to Love" and offers both an independent reading of each story as well as an enriching contextualization within the larger cycle.

Keywords: Short story cycle, toxic masculinity

1 Context

James Nagel contends that "the conflicts inherent in the cultural duality of immigrant groups is one of the things that unites writers from a broad spectrum of countries of origin, since virtually all of them deal with the isolation and identity crisis faced by recent arrivals and yet the desire to sustain the traditions, language, and religions of their birth" and highlights the United States as "the only country to present a rich literary legacy of short stories on this theme, one that begins with European settlers in the 1850s and continues into the twenty-first century" (Nagel 2015, 54), yet his latest *American Short Story Handbook* only includes a single Latinx short story giving creative insight into the lives of the U.S.' largest and fastest growing minority group.

Junot Díaz is one of the most recognized contemporary Latinx authors and widely considered the most renowned and commercially successful Dominican American writer to date (Morán González 2016, 82). He arrived on the literary scene in 1996 with his first collection of short stories, entitled *Drown*, which received general praise by critics though Díaz later stated it "sold nothing but got me more press than any young writer should ever have" (Díaz 2018). Díaz's debut novel, *The Brief Wondrous Life of Oscar Wao*, was published in 2007 to great critical acclaim, culminating in winning the Pulitzer Prize for fiction in the following year. His second short story collection, *This Is How You Lose Her*, was released in 2012. He is also the author of a children's picture book, *Islandborn* (2018). Notably, Díaz is not among the most productive authors in terms of quantity. He spends years, sometimes closer to decades, revising and editing his works (Díaz 2012c). It is believed that Díaz is currently developing a second novel

https://doi.org/10.1515/9783110587647-034

based on a sci-fi Afrofuturist short story, "Monstro," which was initially published in 2012 (Díaz 2012b).

With Díaz's focus on Dominican immigration to the US, his oeuvre discusses the identity struggles arising from this immigration background. Díaz's protagonists and narrators are usually males of color from working-class immigrant families, grappling with their machista upbringing and unable to sustain healthy relationships. In *Oscar Wao* the machismo evident in all of Díaz's adult fiction is linked to the Dominican Republic's history of political instability and dictatorships, specifically that of Rafael Leónidas Trujillo Molina, whose authoritarian rule lasted from 1930 till his assassination in 1961. Maja Horn argues in her work *Masculinity After Trujillo: The Politics of Gender in Dominican Literature* that Díaz's fiction does not exhibit 'traditional' or 'standard' Latino machismo initially formed during the patriarchal colonial era of Latin America but rather a very particular set of hegemonic notions of Dominican masculinity which she traces to Trujillo's regime and propaganda to nationalize a superior male identity (Horn 2014, 1). Horn shows that gender in Díaz's works cannot be read independently of ethnicity or, in this case, national background (131). Hence, Díaz's fiction centers on both immigration background and how that ties into his male characters' masculine and sexual identities. The inescapability of trauma, born of (sexual) violence, colonialism, and diaspora, and passed along to the next generation as a certain inheritance, is the central topic of Díaz's fiction.

Díaz's works do not only share topics and similar themes but are also unified by a recurring protagonist and narrator: Ramón de las Casas, mostly referred to as Yunior (cf. Díaz 2012d), who is the narrator and second protagonist in *Oscar Wao* and also appears in five of ten short stories in *Drown*. The stories of the 1996 publication revolve around Yunior's fatherless childhood in the Dominican Republic, the reunion of the family in New Jersey, and Yunior's experiences as an immigrant adolescent of color. The five stories not directly associated with him have nameless tellers that could or could not be Yunior, but share very similar features with him regardless (Loss 2004, 804). *This is How You Lose Her*, Díaz's second short story collection, again centers on Yunior and his family. Seven of nine stories directly mention Yunior, while the other two can be connected to him via his featured family members, Ramón, his father, and Rafa, his brother, who eventually dies of cancer. In an interview with the magazine *Complex*, Díaz explains that "my idea, ever since *Drown*, was to write six or seven books about him [Yunior] that would form one big novel" (Díaz 2012d). Fragmentation is the single most dominant feature of Díaz's narrative prose, which requires an active reader to piece together not only who the (narrative) characters are, how they are involved with one another, which time frame the individual story adheres to and how it chronologically fits into the larger narrative, but also to realize how much of each story is left untold. As such, Díaz's short story collections more specifically constitute a short story cycle and each 'individual' publication ties into the larger narrative enabling the reader to discover a broader and more collective story. The following close-reading in part two will showcase an independent reading of the interconnection of trauma, ethnicity, and gender identity within Díaz's stories "Drown" and "The Cheater's Guide to Love," which have been selected from the larger cycle and are contrasted only with each other. The third part of this article will focus on theoretical perspectives of the short story cycle in general and Díaz's own particular rendering of the genre.

Díaz is not only considered Dominican American but also as part of the Dominican diaspora. His parents left the Dominican Republic with their five children in 1974 – Díaz was six years old at the time – and relocated to New Jersey, where Díaz's father had already been working for some time (Loss 2004, 803). The Afro-Dominican immigrant population that settled along the east coast of the United States at the time was characterized by its working-class orientation and formed a large labor force for factories and public services; herein Díaz's family did not pose an exception (Heredia 2010, 208). Díaz went on to receive his Bachelor of Arts degree from Rutgers University, his Master of Arts from Cornell University, and has since taught creative writing at Syracuse University and, currently, at MIT (Loss 2004, 803). Still, Díaz would rather not be held up as an example of the American Dream, as he states in an interview with the Global Times in 2010: "I may be a success story as an individual. But [. . .] my family tells a much more complicated story. It tells the story of two kids in prison. It tells the story of enormous poverty, of tremendous difficulty" (Díaz 2010). In an interview ten years earlier, he had already stressed that he believes success in the United States to be arbitrary and a matter of luck rather than hard work and dedication (Díaz 2000, 893). This reflects in his frequent portrayal of the American Dream as an illusion in his fiction. In the same interview, he explains that he aims to write "politics without letting the reader think it is political" and refers to this goal as his main agenda as a writer (901).

Díaz's works contain large autobiographical components and many journalists, critics, and readers have therefore been too quick in focusing on either only the documentary or only the literary nature of his writing as this ignores the fact that Díaz's works heavily point at the category of authenticity itself already being scripted (Loss 2004, 804). The fictional character Yunior immigrates to the U.S. from the Dominican Republic at the age of nine after his father comes to retrieve him; apart from the age gap, this exactly applies to Díaz as well. Yunior attends Rutgers, as Díaz did, and writes fiction, as Díaz does. Both *Drown* and *This Is How You Lose Her* contain short stories on Yunior's experience of losing his brother to cancer, a traumatic experience that also occurred in Díaz's family. However, Díaz continues to stress that "Yunior is a fictional character crafted by a professional writer" (804) while at the same time referring to Yunior as his "alter-ego" (Díaz 2012c).

Various literary traditions are associated with Díaz's fiction: U.S. Latinx, Caribbean, Dominican diasporan, and African diasporan. As such, Díaz's works are not only multilingual with persistent codeswitching between English and Spanish but also multidialectal; African American dialects and contemporary youth modes of speech reflect in both his Spanish and English usage. In "Fiction Is the Poor Man's

Cinema," Díaz voices the opinion that employing italics or quotation marks to demarcate Spanish code switches is to 'other' a major language of the Americas. In his fiction he strives to keep Spanish as "normative in a predominantly English text," and he thus challenges the majority position of the English language (Díaz 2000, 805). Díaz neither provides translations nor a glossary for his readership – although his debut *Drown* is a notable exception – to ensure that they gain better insight into the immigrant experience of struggling with a world in an incomprehensible language (805).

2 Close Readings

The short story collection *Drown* opens with a poem by Gustavo Pérez Firmat on the inadequacy of the English language to express immigrant concerns and firmly situates the ensuing stories within the topics of displacement and alienation even before having read their first words. Likewise, language identity and minority representation are thus established as concerns in the reader's mind. Unsurprisingly therefore, the collection as a whole is strewn with code switches to the characters' native or second language, Spanish, thereby marking the story as an immigrant narrative in form as well as in content. In "Drown," the fifth story in the collection of the same title, Spanish switches are mostly employed for swear words. This is often a strategy to obscure offensive language or to root characters in their native tongue and thus establish a language community. The neighborhood "abuelas" [grandmothers] call out rowdy behavior with "sinvergüenzas" [scoundrels] (Díaz 1996, 72) while the unnamed protagonist's mother watches TV only in Spanish (71, 74, 84). Additionally, some of her direct speech is characterized by overt translation, a literary strategy employed to make an act of translation more visible to the reader (see 74). This 'inbetweenness' of language underlines the issue of 'inbetweenness' in identity for the immigrant characters. Noticeable, too, is the typical postmodern element of unmarked dialogue and direct speech, leaving the readers to decipher for themselves who is speaking when (see 75).

"Drown" has an unnamed protagonist who is also the autodiegetic first-person narrator. The story's focalization is fixed on the perspective of the young Latino male growing up with a single working-class mother in a run-down New Jersey neighborhood. The setting may be referred to as an immigrant ghetto since it is mainly inhabited by Latinxs and people of color who live in an uninviting area close to a garbage dump in crumbling houses. The protagonist's mother constantly checks door and window locks for fear of being assaulted at home (75). A crack dealer is mentioned as the only "full time commuter in the area" (73) pointing to a lack of job opportunities for the community. Space is seen as territorialized by color in the narrative; a vandalized sign at the community pool reads "No Whites" (73) and "redneck territory" is an obstacle to be overcome on the way to the mall (75).

Political criticism of this ghettoization of a colored community is most apparent when a "teacher, whose family had two grammar schools named after it" abuses his position of authority and power to erode his underprivileged students' self-worth by telling them, including the protagonist, that most of them are "going nowhere" (83–84). Further political criticism is implied in the protagonist's encounter with a U.S. Army recruiter on his regular rounds in the neighborhood. Well aware of many local young men's lack of perspective, he tries to entice adolescents to enlist by showing them his gun and photos of "skinny Filipino girls sucking dick" (79). The protagonist acknowledges himself so tempted by this ticket to escape – although he does not consider himself "Army material" – that he hides from the recruiter to avoid any further temptation (79). The empty promise of the American Dream looms in these passages.

The story starts out as an achronological interpolated narrative, weaving together past and present episodes, of the protagonist hearing that his (former) best friend Beto is home from college and him going out to look for Beto despite them having had a fallout that is immediately connected to Beto now being a "pato," a derogative term for a homosexual man (71). Seeking Beto, the protagonist pays the community pool a forbidden nighttime visit, a habit with the neighborhood teens, and reminisces their time spent there together. A mall trip with his mother a few days later again prompts memories of Beto and his shoplifting days there. Throughout these switches between subsequent and simultaneous narration, the narrator compares and contrasts himself to his former best friend. Although the protagonist left school a year ago, he continues to live at home with his mother while earning money by selling marihuana in his neighborhood. Beto on the other hand was "delirious" (71) to leave as soon as possible and saw his admission to college as his chance "to walk the world" (81). That the narrator too yearns to break free from the confines of the neighborhood is made obvious in the Army recruiter episode.

In his narrative of their friendship, the protagonist portrays Beto as the more adventurous, stronger, and braver of the two. Their masculinity is constructed in relation to each other. "Beto was the best" at shoplifting, bravely small-talking to mall security with a bag full of contraband while the narrator was "shitting his pants" for fear of getting caught (76). When they are inevitably apprehended, the protagonist cries whilst gray-faced Beto squeezes his hand trying to console his friend (78). Where the narrative thus exposes Beto's function as a role model for his friend, the protagonist also comments how, even though Beto is "the one leaving for college," he has a more expansive vocabulary – though he is reluctant to admit that this is a result of liking to read (73). This allusion to an unresolved tension between the narrator's interest in learning and knowledge and his masculine 'street' identity is most obvious when he relates playing truant to stay at home and watch TV but then loses interest and visits the local library instead (80). Meanwhile Beto uses this 'spare' time to visit other neighborhoods and later refuses to tell his friend where he went or what he did.

After having confessed to their juvenile criminal behavior, the shoplifting, the truancy, a new paragraph starts with the narrator stating "Twice. That's it." (82) and goes on to describe two homosexual encounters that took place between Beto and himself. He pointedly describes Beto's initiation of this intercourse and his own passivity. While watching porn together Beto reaches into his friend's shorts and does not react to the protagonist exclaiming "what the fuck are you doing?" (82). The protagonist remembers being scared and leaving silently with shaking legs after orgasming (82). He fears turning out "abnormal, a fucking pato" but states that he went to meet Beto at the pool the next day anyway because their friendship mattered more to him than anything (82). When Beto suggests going back to his place after swimming, the protagonist agrees and again remains silent when Beto tells him in a position suggestive of fellatio that he will stop if his friend wants him to (83). Afterwards, agitation is once more only portrayed when a door crashes and the protagonist fears being caught with his pants down and Beto's head in his lap (84). It is very conspicuous that the narrator gives little detail of these sexual encounters and why he submitted to or took part in them. The narrative does not allow a definite interpretation of whether the interaction was consensual or sexual abuse. However, it is indicative that the protagonist's voiced fears surrounding these encounters center on being found out and suffering social stigma.

Sexuality is connected to silence in "Drown." When Beto watches porn with his father "neither of them would say a word" (82). Both boys part in silence after their first sexual encounter. When Beto voices concern with the protagonist's evasive behavior the next day he is silenced by the phrase "nothing to worry about" (83). Later, when Beto asks him whether he would like him to stop before sexual intercourse, the protagonist does not respond. In this light, Beto's refusal to speak of his truant hours spent away from their neighborhood suggest a connection to (homo) sexual activity. The narrators's own internalized prejudice against homosexuality is made very clear from the first page with the derogative language he uses and his given assumption that Beto and he are no longer friends simply because Beto is now openly gay (71). In one of the simultaneous episodes, the protagonist and a group of male friends pass a queer club on their usual way home from a night out partying. Their demeaning behavior and vicious tricks against the queer partygoers further illustrate the homophobic attitude manifested in the protagonist's homosocial group (81).

Both father figures included in "Drown," the protagonist's own father and Beto's, are deeply machista characters and therefore contributors to the sexist and homophobic environment of their homes and neighborhood. As destined role models to their sons both treat their wives disdainfully and disrespectfully and do not productively participate in raising their children. In the simultaneous narration, the protagonist's father is physically absent and living with an affair in Florida. While still at home during his son's school years, he exhibits unsupportive and disinterested behavior when confronted with his son's shoplifting and is characterized as "a real asshole" (77).

Beto's father collects and watches porn with his son "in the middle of the day, not caring a lick about his moms" who exiles herself to the kitchen to cook (82). Apart from neglect, emotional abuse is also a regular occurrence in the protagonist's family. The protagonist's father calls his wife from Florida to beg for money with promises of renewing their relationship should she comply (80). Because his mother too initiates these calls, her son feels compelled to intervene to spare her "wrecking her sleep for days" (80). Their child-parent role is reversed with the protagonist looking out for her mental well-being and trying to put a stop to her self-destructive dependent behavior. He has an acute sense of needing to take care of his mother both emotionally – to the extent of his (limited) socio-emotional abilities – and financially. The narrator at times describes his mother more like a piece of furniture within the apartment than a human being: "I'll enter a room, and she'll stir, detaching herself from the crackling plaster wall, from the strained cabinets" (74). Here the narrator's masculinity is constructed in relation to a woman. His mother is the fragile and vulnerable female whom he as a son feels bound to protect and take care of. On the other hand, there is also a striking parallel established between the mother and son: Both have been left behind by a beloved person and are unable to cope with this loss. The mother continues to actively seek contact with the father while the protagonist's passive but continuously resurfacing memories indicate his ongoing emotional involvement with Beto and their relationship.

The concluding episode of the story shows the protagonist giving up his (half-hearted) efforts to find Beto and settling into domestic routine with his mother. He reminisces seeing Beto off to college and then throwing away his fare-well gift, a book. While stating that he never even opened it, he contradictorily claims not having bothered to read what Beto had written inside. Again, he chooses silence on their relationship. The reader is not afforded insight whether this silence is born from sexual trauma or from the narrator's reluctance to reveal himself homosexually interested. Meanwhile, Beto's gift shows the depth of his understanding of his friend's character and interests. The tossing of this object of knowledge can also be read as the protagonist's decision to give up any academic interest or ambitions. The story's tone is confessional and defeatist. Ultimately, the young protagonist finds himself unable to deviate from the stereotypes and expectations levelled at him due to his gender, ethnicity, and class background.

The attentive reader of Díaz's oeuvre can identify sufficient similarities of the unnamed narrator in "Drown" with the cycle-linking protagonist Yunior to assume that they are in all likelihood the same person. This adds to the suspicion that "Drown" may lack a truthful and unbiased narrator since Yunior proves himself an unreliable narrator, focused on his ego and attempts to make himself seem a better person than his actions betray him to be, in many short stories, and above all in *Oscar Wao*. "The Cheater's Guide to Love," the final story included in Díaz's second collection, *This Is How You Lose Her*, culminates in the insight that he, Yunior, has been both unable to prevent himself internalizing the machista role his gender and ethnicity 'assigned' him as well as having failed to admit this to himself, thus making internally motivated change impossible.

The first story of the collection *This Is How You Lose Her* opens with Yunior's words: "I am not a bad guy. I know how that sounds – defensive, unscrupulous – but it's true. I'm like everybody else: weak, full of mistakes, but basically good" (Díaz 2012a, 3). All further details provided in the story of him cheating on his girlfriend, like him calling her to say he misses her while he is in bed with another woman, sow the first seeds of doubt regarding his reliable self-perception and thus the trustworthiness of his narrative. These opening lines of the collection and the slow realization – and even slower acceptance – of Yunior in "The Cheater's Guide" that he is, in fact, not a good guy, summarize his development as a character.

"The Cheater's Guide" is divided into episodes, each with a subtitle of "Year 0" up to "Year 5." These indicate the unusually expansive story time of the narrative, which also constitutes the longest story – page wise – included in the collection. Díaz habitually employs the second person point-of-view; in "The Cheater's Guide" he merges Yunior and the implied reader as addressee. While such openings as "You, Yunior, have a girlfriend named Alma" (49) announce the primary addressee to be the protagonist himself, second person narratives by definition minimize the distance between the narrator and the implied reader, pulling 'you' into the story world. The collection presents itself as 'The Cheater's Guide to Love' that Yunior, the narrator, writes in a self-therapeutic attempt to come to terms with and rise above the toxic masculinity he has internalized. This metafictional technique of blurring authorship serves as an invitation to the reader to ponder the author-narrator relationship. "The Cheater's Guide" is narrated retroperspectively whereby Yunior evaluates his machista attitude in romantic relationships and includes moralizing comments on his former self's behavior from a perspective of supposed improvement. "Year 0," the starting point of the story, begins with the self-depreciative comment that "your fiancée [. . .] could have caught you with one sucia, she could have caught you with two, but as you're a total batshit cuero who didn't even empty his e-mail trash can, she caught you with fifty" (185). This quote is also an example of the continuation of Díaz's codeswitching strategy in his second story collection.

In his initial attempts to save their relationship after the disclosure of his infidelity, Yunior exposes the insincerity of his regret and remorseful behavior as well as his deeply ingrained conviction that love is connected to deception. While he claims to "*try every trick in the book to keep her*. You write her letters. [. . .] You quote Neruda. [. . .] You claim you're a sex addict and start attending meetings" (186; emphasis added), he immediately drops all pretense when she leaves him: "Of course you go back to smoking, to drinking, you drop the therapist and the sex addict groups and you run around with the sluts like it's the good old days" (188). His lack of internal motivation to address the underlying issues of his inability to uphold loving and stable relationships, also illustrated by the fact that all of the couple's formerly mutual friends "went to her" and Yunior's mother refusing to speak to her son because "she liked the fiancée better than she liked you" (189), keep him locked in a cycle of addiction – to sex, to drinking and partying, and later to sports and physical exertion – to avoid depression and to numb the pain of losing the woman he truly loved.

The interconnection of gender identity and race or ethnicity, most importantly toxic masculinity and latinidad, is the dominant topic of the story and is crafted more obviously than in "Drown." Yunior and his friend Elvis, who is also Dominican American and an Iraq veteran, both exhibit typical machista behavior in their relationships with women. Both men cheat on and betray their long-term partners, whom they consider "good Dominican" girls, and as such 'relationship material' (192), with women they demean "sucias," dirty women (see 185), for no other apparent reason than for consenting to have casual sex with them. This dichotomy does not seem to apply to them as men in their worldview. Later, Elvis comes to serve Yunior as a mirror to observe with at least some reflective distance his own (former) machista behavior (215).

Yunior is disgusted with the racism he faces as a person of color in Boston and on Harvard campus. He narrates various instances of white people insulting him in passing by and of racial profiling by police or campus security (189). On a date with a Dominican visiting scholar, he portrays her as ignorant of racial discrimination yet he, and Elvis too, have pervasive biases against women of a certain race or ethnicity. In their worldview, non-Dominican women, or at least Latinas, are not eligible for a serious relationship. During a yoga class, a "blanquita," a diminutive form of white woman, approaches Yunior but he "skitter[s] away from her cornpoke grin" asking himself "what the hell are you going to do with a blanquita?" (200). On the opening page, Yunior already states that he believes white women to be "super open-minded" with respect to fidelity (185) – apparently not a characteristic desirable in a long-term partner.

When Yunior does meet a 'good' Dominican American woman, Noemi, he is interested in, her reluctance to consent to sexual intercourse in the first weeks of their dating, leaves him resentful and feeling entitled to have sex with her. When he tries to assert 'his right' in an aggressive tone, she terminates contact with him (196). Yunior further exhibits racist assumptions of women of his own ethnicity with a misogynistic journal entry he composes in the belief of having impregnated an affair, a student from Harvard law school: "*Only a bitch of color comes to Harvard to get pregnant. White women don't do that. Asian women don't do that. Only fucking black and Latina women. Why go to all the trouble to get into Harvard just to get knocked up? You could have stayed on the block and done that*" (210; original emphasis). Speaking of their former affairs, Elvis asks whether it is "written on the genes" that Asian girls marry white men (205).

Their convictions show both men to hold preconceived notions of women's characteristics determined by the race or ethnicity they belong to. Yet when Yunior experiences rejection from "one girl [who] when you tell her you are Dominican actually says, Hell no and runs full-tilt toward the door" he is incredulous (193). Likewise, he cannot understand that Noemi is guarded in their burgeoning relationship because of a past experiences with a Dominican man, who abandoned her and their child, and is angered by Noemi drawing conclusions about him, a professor at Harvard, on the basis of this past relationship with a man his 'inferior' concerning education and income (195).

Dominican identity in "The Cheater's Guide" is mainly constructed through a lens of the 'non-Dominican' as can be seen in the treatment of women and race/ethnicity discussed above. This strategy turns against Yunior when the Dominican visiting scholar he dates excludes him from Dominicanness for not speaking Spanish fluently (205). The hypocrisy of the protagonist's own sexism and racism and simultaneous condemnation of discrimination is aptly commented on in the story: "Your only student of color for that semester writes: he claims that we don't know anything but doesn't show us any way to address these deficiencies" (214). Díaz's fiction does focus on the toxicity of machismo in Latinx society and sexism in society at large, and it does highlight the traumatic effect toxic masculinity has on both men and women. Yet criticism has been voiced time and time again that his writing does not offer ideas on how this may be overcome. "Drown" settles for passivity and learned helplessness, but "The Cheater's Guide" does take the first step towards change: self-reflection and realization of the need to reform toxic behavior.

Initially, both male characters are caught between two scripts of internalized masculine behavior that prove themselves incompatible. They both desire a long-term relationship with a 'good' Dominican woman and the stability and intimacy of raising a family together, yet at the same time they feel the need to validate their virility with casual sexual encounters. Yunior acknowledges having been aware that cheating on his fiancée was "the one thing [. . .] she would never forgive" (185) and yet he also admits that he could not be bothered to cover up his promiscuity. Shortly after the breakup that marks ground zero of "The Cheater's Guide," Yunior professes "feeling terribly guilty and terribly alone. You keep writing letters to her [. . .]. You also keep fucking everything that moves" (189) and thus shows a continued state of denial regarding his responsibility for the behavior that led to his misery.

Realization starts to creep into the narrative in "Year 2" where Yunior declares: "You clean up your act. You cut it out with all the old sucias [. . .] you finally break clear and when you do you feel lighter" (193). Though Yunior wants to "focus on me for a while" (197), he ends up with various injuries sustained from overindulging in his new addiction, physical exertion. He also begins a brief affair with a young African American law student even though he suspects "she feels sorry for you" most of the time (201). In this episode, the advice he gives after learning of Elvis's most recent extra-marital affair reveals his increasing awareness of his loneliness, frustration, and yearning for true human connection: "Just don't end up like me" (202).

This comment foreshadows the events of "Year 4." Yunior's former affair, the Harvard law student, claims to be expecting his child and Yunior plays along as 'responsible' father, taking her in, accompanying her to doctors' appointments, and picking up all the bills, although he acknowledges having ambivalent feelings towards this alleged paternity. Elvis shows enthusiasm for the situation for one reason only: He expects Yunior to have a son. Elvis thereupon reveals that he fathered a son outside of his marriage during a trip to the Dominican Republic. Both men switch into a mode of perceived familial responsibility and both are misled by their own desires. Yunior desperately longs for a stable and loving relationship in his life and therefore pushes aside his own doubts concerning his paternity of the child. This is best illustrated by him keeping the law student's pregnancy from his friend Arlenny because "Arlenny you know would march right in and boot her ass out on the street" (211). Meanwhile Elvis, who has a wife and daughter, values the sex of a boy higher than his existing family. When Yunior is thrown out of the law student's birthing room and learns that he is not the father of the baby, Elvis claims the betrayal to have been Yunior's own mistake (214). However, Elvis's yearning for a son led him to fall for paternity fraud just as Yunior did.

"Year 4" shows both male characters the hurt and humiliation – the trauma – of being on the victim's side of betrayal. Even after all the pain, physical and mental, that Yunior experiences in the narrative, he still comments how he "didn't think anything could hurt so bad" (213) as his anticipated fatherhood turning out a lie. When Elvis and Yunior fly to the Dominican Republic to visit Elvis's supposed son, Yunior is appalled by Elvis's "cool as fuck" goodbyes to his wife and daughter and remorsefully comments that "this used to be me [. . .] me me me" (215). Indicative of his maturation as a character, he even goes so far as to criticize his best friend's deception of his wife and family and refuses to let Elvis remain in denial of his son's biological origins. In consequence, Elvis casts off his presumed baby mother and her son because "there is some shit that can't be forgotten" (220). Afterwards, Yunior experiences remorse for dooming the mother and child to abject poverty and envies his friend his "phlegmatic" dismissal of the topic (221). Elvis, too, comes to suffer the consequences of continuing to mask his true feelings when he tries to reconnect with the mother and her son – to no avail.

As in "Drown," "The Cheater's Guide" aesthetically constructs masculinity by contrasting the two main male characters against one another. Initially, both lie and cheat in an attempt to uphold their identity and self-worth of both internalized masculine scripts, as loving and providing husbands and fathers and of sexually insatiable libertines. Yunior eventually realizes the incompatibility of these scripts and their damaging effect on his love life, mental and physical health. Hereby Elvis increasingly serves as the foil of toxic masculinity against which Yunior's character development is constructed.

Finally, in "Year 5," Yunior reads "The Doomsday Book" for the first time, which is revealed to the reader – for the first time – to be a compendium compiled by his ex-fiancée with all evidence of his betrayal, including emails and photos (225). Attached is a note "*for your next book*" (225; original emphasis). While reading, Yunior ultimately comes to realize his own cowardliness and baseness and admits to himself the justification of the break-up. Metafiction climaxes when Yunior begins to write "the cheater's guide to love" (225) because it "feels like hope, like grace" (226). Writing as therapy is a motif in Díaz's fiction, especially in "The Cheater's Guide" and *Oscar Wao*. Effectively, the metafictional element in the story asks – or lures – the readers to believe that they are holding the product of the fictional story writing in their physical hands. Again, the question of narrative authority and authenticity is brought to the reader's attention.

David A. Colón writes that Yunior eventually accepts his inability to overcome compulsive self-sabotage in "The Cheater's Guide" and that his honesty to himself and the implied reader in the face of this flaw make him a "pitiable, vulnerable hero" (147). I, however, disagree with this premature conclusion. "The Cheater's Guide" ends with a confession and with "hope": "you know in your lying cheater's heart that sometimes a start is all we ever get" (226). *This Is How You Lose Her* concludes with a protagonist who has seen the error of his ways and thus stands at the brink of change. Yet, actual change is not the topic of Díaz's second story collection. Where *Drown* focuses on an adolescent and witnessing perspective on toxic masculinity, machismo, and trauma, *This Is How You Lose Her* exposes the wreckage left behind where suppressed trauma and internalized toxic masculinity lead to reckless, egoistic, and aggressive behavior. However 'pitiable' and 'vulnerable' Yunior portrays himself to be, he is an aggressor, who repeatedly exhibits racist and sexist convictions, and he can therefore by definition not serve as a heroic figure. Yunior is not a hero. He is a deeply conflicted character struggling with his identity on the intersections of being a man and a Latino. This is his foremost appeal. It remains to be seen how – or if – Díaz will further develop his cycle-linking character.

3 Theoretical Perspectives

The story cycle as a genre has its roots in ancient narrative tradition, where it historically centered on a memorable event or an outstanding character and was usually written either in prose or in verse. Meanwhile, the modern concept of the short story evolved in the nineteenth century as a distinct form positioning itself against the tale and the sketch (↗1 Sketches and Tales; see also Fash 2020). Many acclaimed local color writers, such as George Washington Cable and Kate Chopin (↗11 in this handbook), linked stories to depict communities with regional customs and characters. In the course of the twentieth century, the short story cycle came to prove itself a distinctly multicultural genre and a favorite format for ethnic minority authors to explore and depict the complexity of ethnicity, gender, and individuality in their communities (Nagel 2001, 2–5). Among the canonical Latinxs contributions from this time are Denice Chavez's *The Last of the Menu Girls* (1986), where seven stories depict the maturation of a unifying character, and Sandra Cisneros's *The House on*

Mango Street (1983; ↗29 Cisneros), which was marketed as a novel in vignettes but is unarguably part of the story cycle tradition as the vignettes are simultaneously self-contained and unified by setting and characters (Nagel 2001, 8–9; see Scheiding 2016, 240–246; Smith 2018, 1–11).

Literary scholars are largely in accord that at its most basic "there is only one essential characteristic of the short story cycle: the stories are both self-sufficient and interrelated" (Mann 1989, 15). These short stories will usually involve an internal or external conflict of one or more main character(s), whereby a resolution of the conflict is commonly achieved in a concluding story. This resolution will often lead to an obvious or more implied change in the main character. Here, "The Cheater's Guide" is a case in point. The story constitutes the concluding narrative (up to now) of Yunior's cycle-stretching struggle with unaddressed trauma and internalized toxic masculinity. His newly found self-awareness and honesty in the face of his shortcomings are implied to bring change for his long-term happiness. These principles apply to "Drown," too. The protagonist's conflict centers on Beto and their (sexual) relationship and concludes with the protagonist choosing to suppress either his trauma of sexual abuse or his homosexuality by severing all ties with his former best friend. Implied is a change for the worse, since the protagonist does not realize his desire for learning nor to escape from the confines of his neighborhood. Each story can thus stand on its own.

Subordinate units of a novel, such as chapters or episodes, are incomplete by and of themselves. Artistic unity can only be achieved when all units are regarded as a whole. Within a short story cycle internarrative unity is not necessary, but reading all interrelated stories does enrich the narrative. Linking elements are most commonly (a) recurring protagonist(s) – Yunior for Díaz's cycle, (b) evolution of a theme – such as adolescent witnessing of toxic masculinity in *Drown*, deepening conflicts in Yunior's own life due to internalized machismo, and finally insight into toxicity, and (c) recurring people, places, and situations, such as Yunior's family members, New Jersey settings, and narratives of emotional and sexual abuse. Taken together, *Drown* and *This Is How You Lose Her* paint a very nuanced picture of a young Latino protagonist traumatized by child molestation ("Ysrael"), his family's immigration experience ("Invierno"), and his brother's and father's emotional and physical brutality ("Fiesta, 1980," "The Pura Principle"). Where a 14-year-old Yunior is still unabashed about his intellect and love of reading in "Nilda," he has given up all academic interest in "Drown" in favor of a more 'street' masculine identity.

Short story cycles lend themselves to diegetical discontinuities, to fragmentation of chronology, character development, narrative perspective, space, and time (see Nagel 2001, 16). In Díaz's rendering of the genre, fragmentation of chronology, and therefore also character development, is most conspicuous. There are, however, also some switches in narrative perspective or regarding the protagonist, such as "Negocios" in *Drown*, which portrays Yunior's father's immigration experience to the U.S., or "Otravida, Otravez" in *This Is How You Lose Her*, which adds a woman's perspective to the "Negocios" narrative. The relationships of some characters and incidents do present interpretive complexities, which is best illustrated by the ambivalence with which the homosexual interaction in "Drown" is treated. The silence Yunior chooses on his sexual engagement with Beto could be born of trauma or shame, or simply indicate his own lasting confusion.

In April 2018, amidst the influential #MeToo movement and a promotional phase for his children's book *Islandborn, The New Yorker* published an article by Díaz in which he made public for the first time that he himself is a survivor of rape and child abuse. Entitled "The Silence," Díaz details the consequences the sustained trauma had on his childhood and adolescence, resulting in mental health issues and an inability to uphold loving relationships. "[The trauma] leaked into my writing, too, but you'd be amazed how easy it is to rewrite the truth anyway" (Díaz 2018). This quote ties into Díaz's recurring literary engagement with the dictatorial element of authorship. Yunior narrates *Oscar Wao* and comes to openly address his own unreliability – after having attempted to mask himself as an omniscient narrator for half of the novel that is – and hereby likens himself to a dictator in his power and authority to construct reality. Naturally, the narrator is created by the author and thus it is the author who rivals the dictator in creating reality. Since *Oscar Wao* is penultimately narrated in first-person perspective, this also plays with the readers' tendency to confuse author and narrator in this point of view. The metafictional strategy employed in "Year 5" of "The Cheater's Guide" serves a similar purpose. Asking the reader to believe that they are holding and reading a story created within a fictional narrative by a fictional protagonist is only possible if the reader assumes the actual author to be the same person as the narrator. Instead of focusing on the supposed autobiographical nature of Díaz's writing and assuming that Díaz is Yunior and Yunior is Díaz, the postmodern narrative invites readers to question the boundaries of non-fiction and fiction, authenticity and unreliability.

Because Yunior is blatantly and, for many, revoltingly misogynistic and because Díaz's fiction mostly includes sexist and rigid female roles, Díaz has often been accused of supporting toxic masculinity and also of embodying these values himself. Díaz's fiction has been subject to accusations of sexism since the publication of *Drown* in 1996. The debate peaked shortly after the publication of "The Silence" in 2018 when several women, among them authors and former students of Díaz's, publicly accused him via Twitter of sexual harassment and misogynistic behavior. His employer, the MIT, subsequently investigated the charge and officially cleared him of any accusations of misconduct.

Following the publication of *This Is How You Lose Her* in 2012, Díaz told *The Atlantic* that "there's plenty of people out there who are like, 'Fuck you. You are endorsing this shit. Your portrayal of women is fucked up'" and he ensuingly quotes a line from James Baldwin (↗20 Baldwin): "Not everything that is faced can be changed, but nothing can be changed until it is faced" (Fassler 2012). Díaz claims to be more interested "in the kind of sophisticated art" where "the larger structural rebuke has to be so subtle that is has to be distributed at an almost sub-atomical level" to avoid falling "into the kind of preachy, moralistic fable that I don't think makes for good literature" (Fassler 2012). In the interview, he also indicates aiming at crafting "better women characters" in the long run "because I know that as a dude, my privilege, my long-term deficiencies work against me in writing women" (Fassler 2012).

4 Bibliography

4.1 Works Cited

Colón, David A. "Heroic Insecurity in Junot Díaz's 'Drown' and 'This Is How You Lose Her.'" *Critical Insights: Contemporary Immigrant Short Fiction*. Ed. Robert C. Evans. Ipswich: Salem P, 2015. 134–147.

Díaz, Junot. *Drown*. London: Faber and Faber, 1996.

Díaz, Junot. "Fiction is the Poor Man's Cinema: An Interview with Junot Díaz." By Diógenes Céspedes & Silvio Torres-Saillan. *Callaloo* 23.3 (2000): 892–907.

Díaz, Junot. *The Brief Wondrous Life of Oscar Wao*. London: Faber and Faber, 2008.

Díaz, Junot. "Writing Wrongs," https://www.globaltimes.cn/content/522054.shtml. Interview by Hao Ying. *Global Times*, 14 April 2010. (28 November 2019).

Díaz, Junot. *This Is How You Lose Her*. New York: Riverhead, 2012a.

Díaz, Junot. "Junot Díaz Aims to Fulfill His Dream of Publishing Sci-Fi novel with Monstro." https://www.wired.com/2012/10/geeks-guide-junot-diaz/. Interview. *Wired*, 10 March 2012b. (18 Jan 2021).

Díaz, Junot. "The Sunday Rumpus Interview: Junot Díaz," https://therumpus.net/2012/09/the-sunday-rumpus-interview-junot-diaz/. By Gina Frangello. *The Rumpus*, 30 Sep 2012c. (28 November 2019).

Díaz, Junot. "Interview: Junot Díaz Talks Dying Art, the Line Between Fact and Fiction, and What Scares Him Most." By Ross Scarano. *Complex*, 17 Dec 2012d. Web. 28 Nov 2019.

Díaz, Junot. "The Silence: The Legacy of Childhood Trauma," https://www.newyorker.com/-magazine/2018/04/16/the-silence-the-legacy-of-childhood-trauma. *The New Yorker*, 16 April 2018. (28 November 2019).

Fash, Lydia G. *The Sketch, the Tale, and the Beginnings of American Literature*. Charlottesville, VA: U of Virginia P, 2020.

Fassler, Joe. "How Junot Diaz [sic] Wrote a Sexist Character, but Not a Sexist Book," https://www.theatlantic.com/entertainment/archive/2012/09/how-junot-diaz-wrote-a-sexist-character-but-not-a-sexist-book/262169/. *The Atlantic*, 11 September 2012. (28 November 2019).

Heredia, Juanita. "The Dominican Diaspora Strikes Back: Cultural Archive and Race in Junot Díaz's 'The Brief Wondrous Life of Oscar Wao.'" *Hispanic Caribbean Literature of Migration: Narratives of Displacement*. Ed. Vanessa Peréz Rosario. New York: Palgrave Macmillan, 2010. 207–221.

Horn, Maja. *Masculinity After Trujillo: The Politics of Gender in Dominican Literature*. Gainesville: U of Florida P, 2014.

Loss, Jacqueline. "Junot Díaz: (1968–)." *Latino and Latina Writers*. Ed. Alan West Durán. New York: Thomson Gale, 2004.

Mann, Susan. *The Short Story Cycle: A Genre Companion and Reference Guide*. New York: Greenwood, 1989.
Morán González, John, ed. *The Cambridge Companion to Latino/a American Literature*. Cambridge: Cambridge UP, 2016.
Nagel, James. *The Contemporary American Short-Story Cycle: The Ethnic Resonance of Genre*. Baton Rouge: Louisiana State UP, 2001.
Nagel, James. *American Short Story Handbook*. Chichester: Wiley Blackwell, 2015.
Scheiding, Oliver. "The American Short Story." *Handbook of Transatlantic North American Studies*. Vol. 3. *Handbooks of English and American Studies*. Ed. Julia Straub. Berlin/Boston: De Gruyter. 2016. 234–250.
Smith, Jennifer J. *The American Short Story Cycle*. Edinburgh: Edinburgh UP, 2018.

4.2 Further Reading

Boddy, Kasia. *American Short Story since 1950*. Edinburgh: Edinburgh UP, 2010.
Delaney, Paul, and Adrian Hunter, eds. *The Edinburgh Companion to the Short Story in English*. Edinburgh: Edinburgh UP, 2019.
González, Christopher. *Reading Junot Díaz*. Pittsburgh: U of Pittsburgh P, 2015.
Hanna, Monica, Jennifer Harford Vargas, and José David Saldívar, eds. *Junot Díaz and the Decolonial Imagination*. Durham: Duke UP, 2016.
Ostman, Heather. *The Fiction of Junot Díaz: Reframing the Lens*. Lanham: Rowman & Littlefield, 2017.

Johanna Hartmann

34 Yiyun Li (1972–)

Abstract: This chapter is a paired reading of the Chinese American writer Yiyun Li's short stories "Immortality" and "A Sheltered Woman." "Immortality" is a village community's mythologizing historiography of the twentieth century. In "A Sheltered Woman," the protagonist resolves to detach herself from the world and others. The resulting liberty comes however with an existential homelessness. Both stories illustrate the psychological, emotional, and physiological impact of growing up and living in a totalitarian country, and the effects on both individual human beings and whole communities. The stories portray survival strategies that lead to dissociations between public and private, friends and enemies, and self and other, and result in cultural as well as individual states of dissociation and paralysis. Both short stories represent how Yiyun Li's writings – without being didactic – constitute a contemporary *littérature engagée* and can be evaluated as counternarratives to China's dominant historiography.

Keywords: Chinese American Literature, collective narrator, fictional historiography, *littérature engagée*

1 Context

Yiyun Li is a much acclaimed but not yet well-studied contemporary Chinese American writer of non-fiction and prose literature. Her work consists of the two short story collections *A Thousand Years of Good Prayers* (2005), *Gold Boy, Emerald Girl* (2011); the four novels *The Vagrants* (2010), *Kinder than Solitude* (2015), *Where Reason Ends* (2019), and *Must I Go* (2020); the memoir *Dear Friend, from My Life I Write to You in Your Life* (2017), and a number of non-fictional essays. Together with Marco Lorenzetti she adapted *The Story of Gilgamesh* for children in 2014. Some of her short stories have appeared in *The New Yorker*, *The Paris Review*, or *A Public Space*, a magazine of fiction where she is a contributing editor (↗4 Textual Materiality). The title story of her first short story collection was turned into a movie for which she also wrote the screen play. She has won numerous prizes, among them the *Guardian* First Book Award, the Frank O'Connor International Short Story Award, the PEN/Hemingway Award, and the California Book Award for first fiction – all for *A Thousand Years of Good Prayers* (Brada-Williams 2017, n.p.). Recently she won the Windham-Campbell Literature Prize and was awarded a Guggenheim Fellowship.

Yiyun Li rejects labels like Chinese American writer, political writer, or autobiographical writer and would rather see herself as an author of "Chinese stories" (Chin 2008, 319) "will[ing] [her]self into a nonentity" through writing (Li 2017, 46). However,

https://doi.org/10.1515/9783110587647-035

she acknowledges that her fictional as well as non-fictional works are shaped by her personal experiences, her childhood and youth in a rather dysfunctional family in China and her adulthood in the U.S. (Laity 2017, n.p.), and her struggle with mental health issues. Yiyun Li was born in China in 1972, came to the U.S. in 1996 as a student, and gained American citizenship in 2012. The characters in her works are either Chinese or Chinese Americans who struggle with their lives. Stories set in China portray the psychological and economic hardships under the changing undemocratic regimes throughout the twentieth century. Stories which feature Chinese American protagonists thematize their diasporic existence and their disconnect from Chinese culture, as becomes apparent in the depiction of intergenerational relationships ("Thousand," "Son"). The focus on intergenerational conflicts with the tension between maintaining Chinese traditions and assimilating to the American society is a main source of conflict between parents and their children and a common focus within Chinese American literature in general and Yiyun Li's texts in particular. The description of dysfunctional families, and more specifically, dysfunctional mother-daughter relationships are prominent themes in Yiyun Li's texts. This focus connects her to other Asian American authors like Maxine Hong Kingston, Amy Tan, or Fae Myenne Ng (Ho 1999, 21).

Suffering from depression or suicidal ideation is a prominent theme in many of Yiyun Li's stories (Khakpour 2017, n.p.). For Yiyun Li, writing constitutes a therapeutic means which allows her to grapple with her own mental illnesses but also to work through disturbing experiences in her life. As a child, she, for example, witnessed a public denunciation and was aware of public executions (Gathman 2008, n.p.) which becomes central in the novel *The Vagrants*. After losing her son to suicide, she wrote the novel *Where Reasons End*, an imaginary conversation between a mother and her dead son. Yiyun Li's corpus of non-fictional essays contemplates and gives witness to the experience of having been socialized in China, serving in the military, emigrating to the U.S. as a young adult, and provides an interculturally informed view on American culture. She uses the genre of the personal essay to philosophically reflect on seemingly mundane observations which she utilizes for deeper insights into both American and Chinese culture. In her essay "A Clean, Well-Lighted Place," she, for example, reflects on the commonplace American phrase "to have fun" and the overabundance of resources and commodities which stands in contrast to her essays "The Man Who Eats" or "A Soldier Home" where she highlights the frugality necessary to sustain oneself in China during her childhood (the title of this short story pays homage to the influence of Ernest Hemingway on her writing; ↗16 Hemingway). Also, American consumer culture and the myth of the American Dream as the promise of social mobility stand in stark contrast to the highly hierarchical societal system in China.

Li was academically socialized both in China and the U.S., holding a B.S. from Beijing University and an M.S. and M.F.A. from the University of Iowa (Creative Nonfiction). At the moment she is a professor of Creative Writing at the Lewis Center for the Arts at Princeton University. Her native language is Mandarin but her fiction is exclusively written in English, a language in which – unlike in Mandarin – she feels at liberty to express herself freely without "censor[ing] myself" (Li qtd. in Thompson 2005, n.p.). Self-expression and the use of language are also recurring themes in her works, for instance, when it pertains to (self-)censorship in China, freedom of speech in the U.S., writing and expressing one's thoughts in English as a foreign language instead of Mandarin, or the speechlessness that results from language being a communication barrier (e.g., "Thousand"). Her writing is marked by her awareness of being an immigrant and her diasporic existence (see Lam 2011, 311–312), and the experience of (internal) exile is common in her characters.

In particular the distance to her former home country – and of course the right of freedom of speech and artistic expression – allow her a pronounced view on life in China before her emigration. This bears political implications. Writing from a position of not political but emotional exile, her texts are sensitive portrayals of life in China and the Chinese American diaspora. Specifically the description of China under Communism cannot be in the interest of the Chinese leadership: The most obvious example is that she writes about the Tiananmen massacre of 1989, which is omitted from China's official historiography, implemented through a comprehensive system of material and digital censorship. But it is not only the inclusion of "the June Fourth incident" – the euphemism used in China, but also the purges throughout the Cultural Revolution, or the death toll of the Great Famine that renders Li's work problematic for the Chinese regime. She also undermines China's official narrative by portraying the emotional and psychological toll that life under a totalitarian regime has on individuals, families, and whole communities. Despite or rather because her texts are not didactic, they constitute both sensitive psychological portrayals of the people and a revisionist historiography, and as such are a form of resistance, a *littérature engagée* (cf. Sartre 1965), in which China is described as a prison-house of culture, marked by the impact of totalitarianism, omitted processes of working through the traumas of the twentieth century, and an ingrained misogyny.

Yiyun Li's stories topicalize the tensional fields between private and public, self and other, between the individual, her family, the immediate community and, on a larger scale, society as a whole. By describing an all-encompassing surveillance system and the remuneration of denunciation, her stories portray the politicization of the private and the subsequent psychological toll. Her characters constantly monitor their behavior, being cautious of what they can say to whom. Li portrays the psychological schizophrenia and dissociation that pertains to individual human beings but also to society as a whole, as social cohesion is undermined by an erosion of trust which makes intersubjectivity, let alone intimacy, difficult for the characters in her stories. They struggle to decide between conformity and self-realization, isolation and intersubjectivity, but mostly – like in the story "Sheltered" – reject intimacy and attachment in favor of isolation. The political impetus of Yiyun Li's works then results from their perspective on concrete individuals who are embedded in

Chinese and American history of the twentieth and twenty-first centuries, their striving for love, happiness, political and personal freedom, and economic stability.

Her work not only sheds light on the psychological toll of being socialized in a totalitarian regime but also describes the effect the politicization of life has on the bodies of both men and women. Throughout her works, the human body is shown as deeply politicized, be it in the depiction of public executions (*The Vagrants*, "Immortality"), the tradition of castration in Imperial China ("Immortality"), motherhood and the one-child policy ("Sheltered"), or abortion ("Princess from Nebraska"). Female protagonists are shown to be suffering from a strictly hierarchical and patriarchal society in which the value of individual women depends on their marital status and the number of sons they have given birth to ("Extra"). In Li's stories men and women suffer under each other but also under a patriarchal system in which women are suppressed ("Immortality") and men are under extreme pressure to perform a certain form of masculinity ("After a Life").

Because her stories are always focused on individuals and their communities, Li's works shed enough light on the circumstances of life in Chinese society of the past and the present to be interpreted as a critique of the political regime. Lu Wang's article on ethical dimensions in Li's work written in Mandarin is one trace of the awareness of her work in China (Wang 2017). However, so far, her works have not been translated into Mandarin, as "the country is 'not ready' for what she has to say" (Rustin 2012, n.p.).

2 Close Readings

"Immortality" is a fictional village historiography, told from the perspective of a collective first-person narrator and spans the turbulent Chinese history during the twentieth century: The story chronicles the fall of the last emperor, the Cultural Revolution, the establishment of the Communist regime, both the Great Leap Forward (1958–1962) and the Great Chinese Famine (1959–1961), the economic opening to the West, the student protests in 1989, and the 1990s. Against this historical backdrop, the story contrasts and parallels the personal fates of two significant sons of one rural village: the protagonist in the first and shorter part of the story is one of the last eunuchs in Imperial China; the protagonist in the second and longer part is a young boy who is to become a Mao impersonator. Both, through the narrative perspective and the juxtaposition of two members of the village community who are embedded in a supra-individual genealogy and continuity, the story stages the fields of tension between past and present, private and public, self and other, self-realization of the individual and the survival of the larger community, tradition and change, and processes of inclusion and exclusion and as such undermines the myth of an egalitarian society. The short story was first published in *The Paris Review* and subsequently in

Yiyun Li's first short story collection *A Thousand Years of Good Prayers* (2005). It won the Plimpton Prize for New Writers and the Pushcart Prize.

"Then we have a short decade of republic, the warlords, two world wars, in both of which we fought on the winning side yet winning nothing, the civil war, and finally we see the dawn of communism" (Li 2006b, 47). This laconic summary connects the tumultuous decades between Imperial China and the Communist regime. It is a history of continuities that prevail discontinuities of various changes of the political system. A young carpenter, the father of the Mao-impersonator to-be is convinced that "new [regime, JH] or old, life is the same" (48) pointing to the prevalent distribution of power within society both between men and women, and the ruling class and the poor population. However, by joking about the regime's population policy, he violates the rule forbidding any public criticism of the regime and is publicly hanged in the wake of the counterrevolutionary purges during the first year of the Cultural Revolution. Life indeed is "the same." Also under communism, both the economic oppression of the rural population and the retaliation for "the smallest mistakes" (45) is continued.

The narrative perspective is integral for the analysis of the story. Not unlike the collective narrator in William Faulkner's "A Rose for Emily" (Faulkner 1950, 119–130; ↗18 Faulkner), the rhetoric in "Immortality" can be characterized as one of guilt, complicity, and survival. Early in the story, the narrator introduces to their audience the societal functions of eunuchs in Imperial China. After the boys chosen to become eunuchs are castrated (the villagers' euphemism is "*cleaned*" (Li 2006b, 44) which points to a more prevalent somatophobia), they serve at the emperor's court in order to earn enough money for their siblings to start families of their own. The collective narrator time and again highlights the socially superior position of these eunuchs – their being endowed with the honorary titles "Great Papas" (44), cared for by their nephews during old age, and honored with lavish funerals. Despite being introduced as a source of pride, the story unmasks this social institution as a necessary survival strategy resulting from the economic precarity and vulnerability of the villagers: "[...] uncles, brothers, and cousins [...] gave up their maleness so that our names would not vanish in history" (44). The individual's bodily integrity and happiness is sacrificed for the survival of the family and village community. The word "eunuch" etymologically meaning "bedroom guard" thus symbolizes the social and political control of both men's and women's bodies and sexuality, subjugating both to the power of the government.

Even more so, under communism the strength of the regime depends on the uniformity and homogeneity of society, manifested in the adherence to a set of strict rules which is enforced against alleged dissenters who are subject to social exclusion and ostracism, and even death – as in the case of the young carpenter. That everybody can become a "witness" (49) at any time effectively establishes an all-encompassing surveillance society (cf. Foucault 1995). In this way, conforming behavior is ensured and the regime appears strong albeit at the price of the erosion of trust between its citizens.

While the collective narrator insists on the authenticity, trustworthiness, historiographical authority, and accuracy of their narrative – "Trust us" (Li 2006b, 66) – which is supposed to be a textual monument of the town's exceptionality, it at the same time discloses its own mechanisms of construction. It shows itself to be the result of processes of narrativization, interpretation, moral and cultural evaluation, selection (e.g., in terms of the sources interpreted), highlighting and omission (e.g., by focusing on the history of eunuchs instead of the handmaids), remembering and enforced forgetting.

However, over the course of the story, the attitude towards the past and the emotional involvement undergo significant transformations. The more distant historical past is evaluated much more positively than the more immediate past or the lived present. As such, the nostalgic romanticization of Imperial China has a fundamental role in creating a relevant group identity and community narrative. Imperial China with the tradition of eunuchs – which was in fact the enslavement and mutilation of young boys – is reinterpreted as a period of economic stability, "pride" and "glory" (46), and mythologized as "the brightest page in our history" (47). The long timespan between this period in Chinese history and the time of narration and the subsequent lack of contemporary witnesses turn public monuments into the only sources that lend themselves to a historicizing interpretation – albeit them being purposely constructed for a glorifying future reading: "What we believe is what we have seen—the exquisitely carved tombstones in our cemetery, the elegantly embroidered portraits in our family books. [...] If not for them, who were we, the small people born into this no-name town?" (46). Stories which counter this dominant narrative – accounts of sexual abuse, loss, trauma, social injustice, torture, and death as punishment for the "smallest mistakes" – are discounted by the townspeople as "rumors" (45) which were "made up to attack the good name of our town" (45–46). This is just one example of how, in this short story, the attempt to constitute a narrative of exceptionality and a textual monument is deconstructed by pieces of information that turn the narrative instance into an unreliable narrator.

Even more so, the politicization of the private sphere and the concomitant separation between authentic private sentiment and public performance manifest themselves in a schizophrenic and fragmented narrative perspective. Life under the communist regime shows more sites of fracture and is less coherent than the more distant past. Whereas the view on Imperial China is predominantly nostalgic, the narrative focusing on Communist China shows the succession and simultaneity of various sentiments: The blind and reckless euphoria during the early days of the regime, the ensuing disillusionment with the regime and regrets about their complicity, fear of becoming casualties in the regime's authoritarian regiment, and a growing apathy. Already at the "dawn of communism" (47) the public display of enthusiasm is overshadowed by reflections on the ethical dimension of the purges during the Communist revolution.

The town members find "it [...] difficult [...] to look into one another's eyes" (49) and wonder "[w]asn't the dead young man one of our brothers?" (49) after the young carpenter is executed. By denouncing him, the community becomes complicit with the regime, effectively suspending the boundaries between public and private spheres and establishing a dichotomy between "friends" and "enemies," characteristic of totalitarian states. The deaths of "counterrevolutionar[ies]" (50) and "enem[ies] of our newborn communist nation" (49) are subsequently downplayed, justified, or repressed. At the same time, the prevailing moral value system forbids mourning for its victims and, additionally, prescribes to ostracize the relatives of alleged counter-revolutionaries, for example, the young man's widow and her son.

However, the narrators' questioning sentiment intensifies over the course of the story, for instance when they allege their "hidden hatred of the dictator" (54) or when they wonder in light of the propagandistic mainstream entertainment on TV – at least "those of us who think a little more than others" – whether "our people are growing down, instead of growing up" (59). At the same time, the public display of subservience is an integral strategy of survival: When learning that the regime would happily sacrifice half the population during an atomic war if it would save the communist regime, the villagers reinterpret this as an opportunity "to prove to the dictator our courage, and our loyalty" (50). However, the public display of loyalty is just a thin veneer over the underlying terror. The atrocities of the regime eventually lead to a collective apathy and feeling "[n]umbed" (61). The violent suppression of the student revolts in 1989 then "slips out of our memory as soon as the dead are burned to ashes in the state-supervised crematoriums" (61) but is nevertheless registered – another instance in the story where the narrative includes subversive elements which are omitted in the dominant discourse.

The second part of the story focuses on the young carpenter's son. For stealing a sparrow during the Great Chinese Famine, a mob is "ready to knock him out. Some of us bare our teeth, ready to eat him alive" (52), effectively scapegoating him and using him as a release valve for "our nameless rage" (52). His mother saves his life by claiming – what until then went unnoticed – that the young boy has "the face of the dictator" (52). On this day, the boy becomes an index and an icon, pointing to the leader and, at the same time, becoming "the emperor himself" (66). As an uncanny doppelgänger, an *eidolon*, he is both himself and someone else. This simultaneity of identity and non-identity, and the superstition that "[t]here are things that are not allowed to exist in duplicates" (53) posits a conundrum for the townspeople who struggle to find the right conduct around him: "We treat the boy and his mother as the most precious and fragile treasure we have, never breathing one word about them to an outsider" (53) fearing that the town as a whole will be retaliated against: "We worry about not paying enough respect to the face, an indication of our hidden hatred of the dictator. We worry about respecting that face too much, which could be interpreted as our inability to tell the false from the true, worshipping the wrong idol" (53–54). As Paul points out, Mao's image was ubiquitous in communist China, a quasi-transcendent entity that symbolizes the ubiquitous presence of the leader and the surveillance society as such (326). Saving her son's life by calling attention to his similarity to the dictator, his mother has transformed the village into a sphere where the leader is constantly present and also eliminated his options of marriage and leading a private life.

Once the dictator is dead, the young man is no longer a "duplicate" who is "not allowed to exist" (53). With the complicity of the villagers who had "concealed the history of our young man being the son of an executed counterrevolutionary" (57) he is chosen as the dictator's impersonator, quoting the dead leader with the sentence "*A man cannot conceal his reactionary nature forever, just as a widow cannot hide her desire to be fucked*" (58; original emphasis). By doing so he simultaneously betrays both his father and his mother, effectively discarding his role as a son. He becomes part of the death cult around the dead leader whose corpse, exhibited in a glass coffin, is turned into a place of pilgrimage. Installed with the "aura" (60) of the emperor, he is idolized, revered, and even called by the dictator's name. In fact, the young man in his role as dictator allows for a collective "[n]ostalgi[a]" (59), arresting time before China is radically changed by opening to the West: the arrival of a new consumer culture (symbolized by the names of big corporations) and the critical evaluation of life under the dictator which is however quickly suppressed and criticism once more relegated to rumors which travel "mouth to mouth" (60). Eventually, these societal changes turn the young man into an anachronism of an outdated version of communism, a symbol of an earlier stage of history.

Ironically, the young man's life as an impersonator of the dictator comes to an end when his perusal of a "yellow book" (pornography) and a "red book" (*The Private Life of Chairman Mao* by Li Zhisui, 1994) convinces him to being entitled to "an endless supply of women" (Li 2006b, 63–64). When he calls on a prostitute, she takes incriminating nude pictures of him and uses them to blackmail him. Like his father, the young carpenter misjudges the situation, believing his political, symbolic, and cultural capital make paying the ransom unnecessary. However, once the pictures circulate in the public realm his face becomes a prison that makes him effectively unemployable.

After having returned to his village, the young man – like Emily in Faulkner's story – becomes "a tradition, a duty, and a care, a sort of hereditary obligation upon the town" (Faulkner 1950, 119), an uncanny revenant of the past. Having lost his status, he castrates himself at his mother's grave (the story suggests that his mother commits suicide in the aftermath of his public scandal). The town reacts with incomprehension, wondering "[w]hat the meaning of such an act is" (Li 2006b, 66). From the perspective of the young man, who grew up with legends of highly respected eunuchs and the story of this one particular boy, who, after castrating himself, "walked across the town, his male root dripping blood in his hand" (46), this act might have seemed like a way to reconnect to a glorious past and to inscribe himself into the town's history. At the same time, his self-castration can be interpreted as a futile attempt to destroy the relationship between his own embodied reality and the nude photographs of himself. Also, his act is coherent with practices of the regime to erase material evidence, not realizing that even though "[n]one of us in town has seen the pictures" (65), the village community has constructed imaginary versions which are at least as powerful as the material images. However, reality again conflicts with the mythologized past. The villagers are woken up to a "long howl" (66) and find the young man "roll[ing] over on the ground, his face smeared with tears and blood" (66–67). The discrepancy between past legend and contemporary reality thus again casts doubt on the authenticity of the collective self-narrative that pertains to the more distant past.

The story presents itself as a history of exceptionality but deconstructs itself by pointing out the generic experience of suppression, exploitation, and dissociation between private and public: For instance, to the shock of the townspeople, the young man is only one of several contestants who compete to become Mao's impersonator. The collective narrator claims that "[f]or dynasties, our town provided the imperial families their most reliable servants" (44), however, we learn that at the Imperial court there were thousands of eunuchs and even more handmaids who are, as are other female protagonists in the story, constantly subjected to misogyny, exploitation, and discrimination.

Already the story's first sentence – "His story, as the story of every one of us, started long before we were born" (44) – epitomizes the tensions brought forth in "Immortality": between individual and community, a historicized past and a lived present. The story thus interconnects individual and collective experiences, establishes a continuity, and epitomizes the shared collective experience of rural communities in China throughout the twentieth century. The story portrays that under both regimes – Imperial China and the communist regime – the state claims sovereignty over the "social body" (Foucault 1995, 90) and the embodied existence of individuals as inherently politicized. Despite the "epic" dimensions of "Immortality," the selective focus on two unnamed individual characters embedded in a single community which represents society as a whole and the narrative strategies of the collective narrator turn this text into a highly effective short story.

"A Sheltered Woman" was published in *The New Yorker* in 2014 and won the Sunday Times Short Story Prize of the same year. Whereas "Immortality" is a collective historiographical narrative spanning the twentieth century in China, "A Sheltered Woman" is set in San Francisco and covers only a couple of days in October, presumably during the 1990s. The story is focalized through the protagonist of the story, Auntie Mei, a "first-month nanny" (Li 2014a, 63) who grew up in China and who now helps affluent Chinese American parents of newborn babies to settle into a breast feeding routine. Her own life too is organized by a strict routine and rhythm: She lives with a new family for exactly one month before moving on to live with the next family. In her current assignment, Chanel, a mother of a newborn son, refuses to take care of him only six days after giving birth. Forced to assume the role of the mother, Auntie Mei faces an existential crisis, being torn between the desire to settle down and the resolution to stick to her self-imposed lifestyle. In the short story, the ambiguity of relationships and intersubjectivity is intricately connected to the significance of place. Already the title of the story encapsulates the tensional field Auntie Mei is caught up in: on the one hand the desire to form meaningful relationships, the wish to be recognized by others, and to settle down and, on the other hand, her fear of losing her autonomy which comes at the price of an existential homelessness, moving from shelter to shelter.

Auntie Mei belongs to the Chinese American diaspora, having moved from China to the U.S. after the deaths of her mother and grandmother. As such, she is caught up between past and present, isolation and communality, self-identity and intersubjectivity, proximity and distance. Speaking with Wong, these poles constitute "two contrasting modes of existence and operation": "*Necessity*" (*force, demand*, or *constraint*) as the default mode, "[*e*]*xtravagance*" referring to "*urge*[*s*], *impulse*[*s*], or *desire*[*s*]" (Wong 1993, 13; original emphasis). Auntie Mei has built a system of necessities in the form of her job profile and the fulfilling of her professional duties which help her control her desires and emotions. Her striving to adhere to the mode of necessity corresponds with the *Tao Te Ching*-inspired aphorism "Be content with what you have. Desire nothing and nothing will be lacking" – a motif in Li's essay "Listening is Believing." Desire for extravagance then only exists at the expense of its other – extravagance (Li 2014b). In "A Sheltered Woman" – in contrast to Li's autobiographical essay – these "modes of existence" (Wong 1993, 13) do not pertain to material goods but to human relationships.

Liberty to Auntie Mei means not "knowing" anybody and not "being known" (Li 2014a, 68). Growing up in China has primed Auntie Mei's attitude towards intersubjectivity. To her, relationships are deeply ambiguous, illustrated in the semantic spectrum of the word "attachment," semantically encompassing on the one hand "to fasten" or "to fix" but on the other hand also encapsulating the emotional connection between human beings. Being close to others requires "knowing" the other person and being known in turn comes at the price of losing her freedom – a price, Auntie Mei is unwilling to pay: "Auntie Mei wondered if knowing someone – a friend, an enemy – was like never letting that person out of one's sight. Being known, then, must not be far from being imprisoned by someone else's thought" (68). This feeling of being trapped, incarcerated, and under constant surveillance is not restricted to the present but also holds for relationships in her past – in Auntie Mei's case her maternal ancestors and her deceased husband. She was raised by her mother and grandmother "who could not understand themselves, and in not knowing themselves they had derailed their men and orphaned their children" (69). As a result, she "had the good fortune of not wanting to know anyone who came after" (68). During her marriage, she wishes it would end only to be guilt-ridden when her husband is killed in a robbery, "a senseless wish granted senselessly" (65). In the act of "knowing," epistemological and ethical dimensions intersect. Knowing herself and being aware of the history of mental illness in her family, and of the destructive force of herself on others, she has "talked herself into being a woman with an ordinary fate" (69) and to adopt a life philosophy of detachment and isolation. Being unable to find a middle ground within the continuum between over-identification and isolation she allows herself only so much human contact as necessary.

Her grandmother had left her husband and Auntie Mei's mother shortly after her birth, thus depriving her of a mother. Auntie Mei's mother then deprived her of a father by threatening to commit suicide if her husband would not leave. Her being orphaned as "a condition of her birth" (Li 2014a, 67) has had a lasting and damaging impact on Auntie Mei's ability to connect with others. From early childhood on, to Auntie Mei human relationships have been inherently connotated with loss and deprivation. Being an orphan and a widow (both words share the semantic meaning bereft, deprived), having left China behind for a new life in the U.S. after her mother's and grandmother's deaths however does not free her of the "ghosts" of her ancestors "enter[ing] and le[aving] the minds freely" (65). Quite to the contrary: "death does not take the dead away; it only makes them grow more deeply into you" (69). Alluding also to a history of mental illness in her family, she remains a prisoner caught up in a relationship with her maternal ancestors.

Apart from her experiences as a child which primed her view on intersubjectivity, observing the relationships in her immediate surrounding reinforces her insistence on her isolated existence. In her current assignment, she works for Chanel, who is trapped in a marriage to a man who does not love her and overwhelmed with taking care of a child she does not want. In particular nursing him – "this thing" (65) like a "cow" (63) – feels especially degrading and as a violation of her own corporeal integrity. After six days she "wean[s]" (65) and thereby "orphans" him. Chanel's adopted name – she "surely had a Chinese name" (63) – reflects her aspirations for a life of luxury and elegance. She insists on the exceptionality of her experience and claims that "[n]obody" (63) could understand how it feels to be in her situation. Auntie Mei however, massaging her breasts, trying to turn Chanel's body into a source of nurture, maintains that "a mother is a mother" (63), classifying her feelings in a communal way and insisting on her fulfilling the duties that come with her role as a mother.

Auntie Mei is also wooed by Paul, an elderly Vietnamese handyman, who deems her way of life improper and suggests she "should find someone" (66–67). He too has formed negative attachments throughout his life. He compares his life with others, e.g. his siblings who already have grandchildren, and a former rival who his late wife once mentioned favorably. He even tries to use Auntie Mei and the baby to pose as his wife and grandson in order to impress this man. Observing the proof for her belief that "[a]n enemy could be as eternally close as a friend; a feud could make two men brothers for life" (69) reinforces her to eventually maintain her solitary existence. These examples in her proximity – "Chanel's depraved; Paul's unbending" (69) – in addition to her self-knowledge reassert her in her monadic existence.

Her liberty and autonomy result from resilience, self-reliance, and, most importantly, a lack of intersubjectivity. This is manifested in her communication with others, which she keeps to a minimum. She answers questions only vaguely and refuses to listen to "stories" (67) – wishes Chanel and Paul disregard. Instead, she has taken to talking to herself. In the text, this is represented in a doubled discourse in which her sparse and diplomatic utterances to others are accompanied with a sarcastic but mute commentary. The only human contact she allows herself is with newborn babies who will not remember her, whom she cannot harm, and who cannot harm her. "Because it was simple that way" – in fact as a strategy to avoid attachment – she calls the babies she cares for "Baby" and their mothers "Baby's Ma" (63). Only in her "palm-size notebook" (63) does she note down the names and contact information of the families she worked for.

In "A Sheltered Woman" the ambiguity of relationships is intricately related to notions of place. Intersubjectivity (both in its positive form as mutual recognition and as imprisonment) is metaphorized by words pertaining to places, dwelling, and home: "She had been brought up in two worlds: the world of her grandmother and her mother, and that of everyone else; each world had sheltered her from the other, and to lose one was to be turned, against her wish, into a permanent resident of the other" (69). Both, people and places, are portrayed to be "habitat[s]" (67), living spaces. With the death of her female ancestors, "two mythic women" (67), she has been orphaned, simultaneously losing her "habitat." Losing her husband too but still not being free of the relationships of the past, she resolves to not "accumulate" (66) more people in her life, claiming that "[f]ortunate are those for whom everyone can be turned into a stranger" (69). As a consequence, she finds herself displaced and in a permanent state of emotional exile and alienation, with the "impossibility of returning home" (Madsen 2009, 66).

As intersubjectivity is intricately connected to the theme of space, her belief that knowing her is harmful to others, her wish to refuse any attachments requires a nomadic existence which results in an existential homelessness, her being a fugitive and a nomad, leading the self-imposed life-style of a "vagrant" – to borrow the title of Yiyun Li's first novel. The transitions from one family to the next however have a more existential undertone. She does not merely move on, but loses her shelter, she is "pluck[ed] out" (Li 2014a, 66) in order to care for someone else, and finds herself in places where she is not wanted, but "needed" (64) and treated like "poor relations" (66). Working as a doula has set up an automatic mechanism that removes her forcefully from her shelters, a word which hints at the lack of a home, habitat, or dwelling but also implies the need for protection from a perceived threat or harm. The passive construction in the story's title hints at the lack of agency and the precarity and vulnerability of the protagonist's existence.

Auntie Mei's belief that relationships are a source of harm, bereavement, and loss (instead of security, joy, love, and trust), turns her into a static character who believes that there is an unalterable "correct" way and a "different" way (64). Her migratory routine ensures detachment and prevents her from harming others and being harmed but also evolving and growing as a human being. Her lifestyle is symbolized by an egret Auntie Mei observes in the garden with an artificial pond and goldfish. "[E]njoying, as she always did, a view that did not belong to her" (64) she sees what she believes is the statue of an egret, "its neck curved into a question mark" (64). All of a sudden, what she thinks is a statue comes to life and flies away. The egret symbolizes her own migratory existence, her not paying for rent nor for food, providing herself with what she needs but not owning anything herself. Its curved neck symbolizes the ethical questions "What is the meaning of life?" or "What is a good life?" but, having been mistaken for a statue instead of a real bird, it also signifies her own static personality and the lack of interpersonal recognition she receives in her life. As soon as she is seen and recognized she moves on. On the other hand, the fleetingness of life and in particular her existence are symbolized by the fish in the pond. They too, are "pluck[ed] out" of their environment, living in constant danger of threats to their freedom. In this respect, her profession as a doula fulfills her own and her clients' most basic needs: food and shelter. It is a non-verbal, corporeal form of care both for herself and the people she cares for. When Chanel assumes that Auntie Mei must love the children she cares for, she vehemently protests, comparing herself to a bricklayer and his bricks. However, while performing her professional duty, she herself is provided with a basic form of human contact in which the roles between caregiver and caretaker are blurred. This is symbolized in the act of rocking. At one point, Auntie Mei wonders who enjoys the rocking more, the rocking chair or herself. The rocking is also a reminder that she too must have been the recipient of maternal care once. Her work then provides her with a space between the opposite poles of total isolation on the one hand and being consumed by human relationships on the other hand.

Auntie Mei's actions are characterized by what can be called a negative ethics of responsibility that point at containment, not at therapy, nor working through nor healing, believing that "living is a most disappointing business" (65). She states that her job as a doula has "ch[o]se[n]" (64) her after her journey from the East Coast to the West Coast, a journey of self-reckoning and self-realization after the death of her husband. However, her westward journey is a self-realization *ex negativo*, and, in a way, a reversal of the American Dream. She does not strive for greatness but ordinariness, not to make a difference in the world but in fact to not leave a mark. Her form of self-realization is based on an invisible work for the community which co-aligns with her view of women fulfilling their duty. In her notebook, she chronicles her professional assignments, but apart from that, she claims that "there's not enough life to fill one [book, JH]" (63). Her decision to not have children of her own consequently constitutes an intervention against the cycle of destructive relationships that have kept her ancestors – both male and female –, her own husband, and eventually herself deeply damaged.

Nevertheless, her life is shown as a constant succession of temptations – "urge[s]" that come "in waves" – which tempt her to actually break her habits and routines and abandon her self-imposed restrictive lifestyle (69). When Chanel "orphans" her son and when Paul suggests she settle down with him, this culminates in a crisis hallmarked by various conflicting emotions, desires, and wishes. On the one hand, she seems annoyed that her routines are disturbed, on the other hand she pities the boy and desires to see him grow up – "a few months, a year, two years" (68). Despite her desires, the ending of the story suggests that Auntie Mei will resist her impulses and stick to her identity construction and continue her migration:

> She was getting older, more forgetful, yet she was also closer to comprehending the danger of being herself. She had, unlike her mother and her grandmother, talked herself into being a woman with an ordinary fate. When she moved on to the next place, she would leave no mystery or damage behind; no one in this world would be disturbed by having known her. (69)

In "A Sheltered Woman," we see a protagonist who denies herself any deeper human contact as a result of her upbringing, her self-recognition, and the experiences she witnesses in her surroundings. Being free from others however also means isolation and an existential homelessness. Self-realization to Auntie Mei does not mean the reinvention of her life but the striving for invisibility and a lack of recognition. As such, the default mode of her life is the necessity of survival, not the fulfillment of her desires. Auntie Mei's answer to the question of how to live responsibly in the face of emotional and psychological impairment is to live according to a self-defined ethics of isolation as the only form of humanity that she can envision.

"Immortality" and "A Sheltered Woman" are just two examples of how Yiyun Li interweaves the personal and the political and embeds them in sociohistorical and geopolitical contexts. She sensitively portrays characters who struggle to find a place, a "habitat" in the world. Often, they suffer the psychological toll of mental illness, having been socialized in a totalitarian regime. Self-realization and surviving in a hostile environment, individuality and one's place in the community are often irreconcilable poles in her texts.

3 Theoretical Perspectives

The definition of Chinese American literature and its hypernym Asian American literature underwent significant changes and redefinitions since their inception as both the terms "Asian" and "American" pose definition-related problems (Zhou 2009, 42–45; Xu 2010, 436). According to the three largest immigrant groups to the U.S., Asian American literature first included Chinese, Japanese, and Filipino literature but now encompasses the works by authors from South East Asia, as well as authors with an Indian, Korean, Bangladeshi, or with Middle Eastern backgrounds. Asian American literature is thus an umbrella term which encompasses a widely diverse spectrum of literatures.

For Chinese American literature, "there exists no consensus on what properly falls within [its] purview [as, JH] its boundaries [...] have fluctuated with changing historical conditions" (Wong 1997, 39). Yiyun Li's background is a case in point in order to probe the changing definitions of Chinese American literature. During the institutionalization of Asian American Literary Studies following the *Aiiieeeee*-anthologies (*Aiiieeeee!* from 1974 and *The Big Aiiieeeee!* from 1991 for which Frank Chin wrote the "Preface") being born in the United States, and a certain "sensibility" were prerequisites to be called a Chinese American author. This subjective and vague term allowed for the exclusion of authors who did not write in a certain spirit, authors who lacked a certain agenda and who were seen as puppets who themselves had internalized anti-Chinese stereotypes and who were perpetuating and reinforcing an Orientalist agenda in works that catered to a predominantly white audience (see Roh 2009, 14). Chin's anger was directed particularly against female writers who, in his view, perpetuated the stereotype of Chinese men as being effeminate and thereby reinforced the "emasculation of Asian men" in general and Chinese men in particular. This "misogynistic, chauvinistic, and, again, narrow-minded" stance (Roh 2009, 14) led to a wider discussion on how to define Asian American literature in a more inclusive way. Cheung's definition of Asian American literature for instance comprises "works by people of Asian descent who were either born in or who have migrated to North America" (Cheung 1997, 1). Meanwhile the term encompasses texts that are written in English but also in Chinese languages (see Hillenbrandt 2013). Measured against the early standards formulated by Chin, Yiyun Li's work would not fulfill the requirements of "adequate" Chinese American literature, as she was born in China, immigrated to the U.S. in 2003 and became a citizen only in 2012. Furthermore, her work explores a variety of themes that were seen as anti-Chinese.

Short stories by Chinese American authors have become a popular genre with Ha Jin, Russell Leong, David Wong Louie or Maxine Hong Kingston, Fae Myenne Ng, and Amy Tan as famous proponents. However, research on the Chinese American short story in general and works by Yiyun Li in particular is scarce (see Xu 2010). Except for a handful of articles, Yiyun Li's work is almost untouched by academic attention, which is surprising as it lends itself to a variety of analytical perspectives. Her rich oeuvre, for instance, lends itself to the analyses of the relation between the private and the political, mental illness, suicide, parent-child/mother-daughter relations (cf. Ho *In Her Mother's House*), Chinese philosophy, intertextuality, motherhood, or queerness ("Nebraska") – themes that pervade her fictional work but also her non-fictional essays.

The last five decades have seen a succession but also a continuity of research paradigms towards Asian and Chinese American literature: for instance transnational studies as a result of the "transnational turn" (Amy Kaplan in Gomaa 2016, 1) focused their attention on two "set[s] of dynamics"; first, "the nation-formation themes, often intrinsically tied to language strategies and formal features," and second, "the diasporic, mobile, transmigratory nature of Asian American experience" (Lim et. al 2006, 1), a focus on interconnections and the diaspora experience rather than literature as a national construct (see Dong 2010). Further research angles have focused on Chinese American literature as ethnic or minority literature, or e.g. from the angles of hemispheric studies and diaspora studies (see Kong 110–112).

Thinking along the lines of ethnicity, race, and gender was a logical consequence in the wake of identity politics and activist "cultural nationalist discourses" (Ho 1999, 24) and important for the discursive establishing and canonization of Asian American literature and the institutionalization of Asian American literary and culture studies as an academic field. Evaluating the works by ethnic authors only as a reflex of their identity however masks important dimensions in their work. Yiyun Li's oeuvre is a panorama of Chinese and Chinese American experience in which marginalized outsiders are the central characters. To only see her as a minority writer with a focus on identity politics misses the depth of her texts that shed light on the complex and conflicting psychological and social make-ups that her characters are embedded in. Her oeuvre of "Chinese stories" – both as "a brief moment in time" as well as "epic" (Ellis 2009, 206) – chronicle the experience of Chinese and Chinese Americans in the past and the present and as such (as a fictional historiography) constitutes a particular form of knowledge not limited to "ethnocentric conceptions of identity" (Matthews 2017, 506), or Chinese and American experiences, but a dedication to humanity.

4 Bibliography

4.1 Works Cited

Brada-Williams, Noelle. "An Unfinished Conversation: An Interview with Yiyun Li," https://scholarworks.sjsu.edu/aaldp/vol8/iss1/3/. *Asian American Literature: Discourses & Pedagogies*. Vol. 8, Article 3. (2017). (15 March 2019).

Cheung, King-Kok. "Re-viewing Asian American Literary Studies." *An Interethnic Companion to Asian American Literature*. Ed. King-Kok Cheung. Cambridge: Cambridge UP, 1997. 1–36.

Chin, Frank et al., eds. *Aiiieeeee! An Anthology of Asian-American Writers*. Washington, D.C.: Howard UP, 1983.

Chin, Frank, ed. *The Big Aiiieeeee! An Anthology of Chinese American and Japanese American Literature*. New York, NY: Meridian, 1991.

Chin, Vivian. "I Still Have So Many Chinese Stories to Tell: An Interview with Yiyun Li." *Michigan Quarterly Review* 47.2 (2008): 317–322.

Dong, Lan, ed. *Transnationalism and the Asian American Heroine: Essays on Literature, Film, Myth and Media*. Jefferson, NC: McFarland & Co. Publishers, 2010.

Ellis, Shelly. "Yiyun Li: Tell, Don't Show." *Illuminating Fiction: A Collection of Author Interviews with Today's Best Writers of Fiction*. Ed. Shelly Ellis. Los Angeles: Red Hen P, 2009. 201–210.

Faulkner, William. "A Rose for Emily." *Collected Stories of William Faulkner*. New York: Random House, 1950. 119–130.
Foucault, Michel. *Discipline and Punish: The Birth of the Prison*. New York: Vintage Books, 1995.
Gathman, Roger. "*PW* Talks with Yiyun Li: Chinese Gothic." *Publishers Weekly*, 8 December 2008, 49.
Gomaa, Dalia M.A. *The Non-National in Contemporary American Literature: Ethnic Women Writers and Problematic Belongings*. New York: Palgrave Macmillan, 2016.
Hillenbrand, Margaret. "Letters of Penance: Writing America in Chinese and the Location of Chinese American Literature." *MELUS* 38.3 (2013): 44–66.
Paul, Gerhard. "Mao: Das Portät als Reliquie und Pop-Ikone." *Das Jahrhundert der Bilder*. Ed. Gerhard Paul. Sonderausg. Bonn: Bundeszentrale für Politische Bildung, 2008. 322–329.
Ho, Wendy. *In Her Mother's House: The Politics of Asian American Mother-Daughter Writing*. Walnut Creek: AltaMira P, 1999.
Khakpour, Porochista. "Unsettled: On Yiyun Li, Mental Illness, and Writing in the West." *Virginia Quarterly Review* 93.2 (2017): 182–184.
Kong, Belinda. "In and Out of the Spectacle: The Beijing Olympics and Yiyun Li's *The Vagrants*." *Ariel: A Review of International English Literature* 42.1 (2011): 107–128.
Laity, Paul. "Interview – Yiyun Li: 'I Used to Say that I Was not an Autobiographical Writer – that Was a Lie,'" https://www.theguardian.com/books/2017/feb/24/yiyun-li-interview-dear-friend-from-my-life-i-write-to-you-in-your-life. *The Guardian*, 24 Feb. 2017. (1 January 2019).
Lam, Melissa. "Diasporic Literature: The Politics of Identity and Language." *Journal of Asian Pacific Communication* 21.2 (2011): 309–318.
Li, Yiyun. "A Clean, Well-Lighted Place." *The New Yorker*, 11 April 2011, Web. (26 January 2021).
Li, Yiyun. "After a Life." *A Thousand Years of Good Prayers*. London: Harper Perennial, 2006a. 23–43.
Li, Yiyun. "A Sheltered Woman." *The New Yorker*, 10 March 2014a. 62–69. Web. (10 July 2018).
Li, Yiyun. "A Soldier Home." *The New Yorker*, 8 June 2009. Web. (10 July 2018).
Li, Yiyun. *Dear Friend, from my Life I Write to You in Your Life*. New York: Random House, 2017.
Li, Yiyun. "Immortality." *A Thousand Years of Good Prayers*. By Yiyun Li. London: Harper Perennial, 2006b. 44–67.
Li, Yiyun. "Inner Worlds: Listening is Believing." *The New Yorker*, 22 December 2014b. Web. (10 July 2018).
Li, Yiyun. "Son." *A Thousand Years of Good Prayers*. London: Harper Perennial, 2006c.
Li, Yiyun. "The Man Who Eats." *The New Yorker*, 6 September 2004, 25: 106–112.
Li, Yiyun. *The Vagrants*. New York: Random House, 2009.
Li, Yiyun, and Marco Lorenzetti. *The Story of Gilgamesh*. London: Pushkin Children's Books, 2014c.
Lim, Shirley Geok-lin, et al. Introduction. *Transnational Asian American Literature: Sites and Transits*. Eds. Shirley Geok-lin Lim et al. Philadelphia: Temple UP, 2006, 1–26.
Lo, Yi-Jou. "From What We Eat to How We Are: Food and the Father-Daughter Relationship in Yiyun Li's A Thousand Years of Good Prayers." *The Explicator* 73.1 (2015): 65–68.
Madsen, Deborah L. "Asian Diasporas." *The Greenwood Encyclopedia of Asian American Literature*. Vol. 1. Ed. Guiyou Huang. Westport: Greenwood P, 2009. 61–67.
Matthews, Graham J. "Framing Risk in China: Precarity and Instability in the Stories of Li Yiyun." *TexP* 31.3 (2017): 505–521.
Pu, Ruoqian. "Chinese American Short Story." *The Greenwood Encyclopedia of Asian American Literature*. Vol. 1. Ed. Guiyou Huang. Westport: Greenwood P, 2009. 198–201.
Roh, David. "Aiiieeeee!". *The Greenwood Encyclopedia of Asian American Literature*. Vol. 1. Ed. Guiyou Huang. Westport: Greenwood P, 2009. 13–15.
Rustin, Susanna. "Interview – Yiyun Li: A Life in Writing," https://www.theguardian.com/culture/-2012/apr/13/yiyun-li-life-writing-interview. *The Guardian*, 13 Apr. 2012. (12 April 2018).

Sartre, Jean-Paul. *What Is Literature?* New York: Harper & Row, 1965.
Thompson, Bob. “Will Words Fail Her?,” https://www.washingtonpost.com/archive/lifestyle-/2005/12/21/will-words-fail-her/4b15c36f-fa53-4897-b64b-1fcc53598f7b/. *Washington Post*, 21 December 2005. (15 April 2018).
王璐/Wang, Lu. “转型期社会中的“存在之思”–论李翊云短篇小说中小人物的伦理困境与伦理选择/“Thinking of Being” about the Transitional Society: The Ethical Predicaments and Ethical Choices of Insignificant People in Yiyun Li’s Short Stories.” *Interdisciplinary Studies of Literature* 1.2 (2017): 148–160.
Wong, Sau-Ling C. “Chinese American Literature.” *An Interethnic Companion to Asian American Literature*. Ed. King-Kok Cheung. Cambridge: Cambridge UP, 1997. 39–60.
Wong, Sau-Ling C. *Reading Asian American Literature: From Necessity to Extravagance*. Princeton: Princeton UP, 1993.
Xu, Wenying. “The Asian American Short Story.” *A Companion to the American Short Story*. Eds. Alfred Bendixen and James Nagel. Malden, MA: Wiley-Blackwell, 2010. 436–449.
Zhou, Xueping. “Asian American.” *The Greenwood Encyclopedia of Asian American Literature*. Vol. 1. Ed. Guiyou Huang. Westport: Greenwood P, 2009. 42–45.

4.2 Further Reading

Feng, Pin-chia. *Diasporic Representations: Reading Chinese American Women’s Fiction*. Münster: Lit-Verlag, 2010.
Kim, Elaine H. *Asian American Literature: An Introduction to the Writings and Their Social Context*. Philadelphia: Temple UP, 1982.
Lee, Rachel C., ed. *The Routledge Companion to Asian American and Pacific Islander Literature*. London, New York: Routledge Taylor & Francis Group, 2016.
Lee, A. Robert, ed. *Multicultural American Literature: Comparative Black, Native, Latino/a and Asian American Fictions*. Edinburgh: Edinburgh UP, 2003.
Morgan, Nina, Alfred Hornung, and Takayuki Tatsumi, eds. *The Routledge Companion to Transnational American Studies*. New York: Routledge, 2019.
Thoma, Pamela S. *Asian American Women’s Popular Literature: Feminizing Genres and Neoliberal Belonging*. Philadelphia: Temple UP, 2014.
Yin, Xiao-huang. *Chinese American Literature Since the 1850s: Foreword by Roger Daniels*. Urbana: U of Illinois P, 2000.

Damien B. Schlarb

35 N.K. Jemisin (1972–)

Abstract: N.K. Jemisin's stories illustrate her political commitment to social justice and diversity. She cannily plays with American literary traditions, from romanticist poetry to the African American and the New Orleans short story, balancing between neo-romanticism, steampunk, and Afrofuturism. In doing so, she explores themes such as the transformative potential of lesbian sexuality, Black consciousness, and the prospects of social and economic justice in a globalized late-capitalist world.

Keywords: Afrofuturism, Afrodiaspora, speculative fiction, science fiction, lgbtq+

1 Contexts

Few speculative fiction writers have garnered the transmedia appeal of Nora Keita "N.K." Jemisin. Like Octavia Butler, the trailblazer of African American science fiction (sf), Jemisin has won the Hugo Award, the Sputnik Award, as well as the prestigious MacArthur Grant, and was nominated for several other sf accolades. Her "Broken Earth" trilogy of novels, *The Fifth Season* (2016), *The Obelisk Gate* (2017), and *The Stone Sky* (2018), each received the Hugo award, making her the first and only author to win the prize thrice and in a row. Film adaptations of the novels are on the horizon, with Jemisin personally overseeing development (Fleming 2021, n. pag.). In 2019 she penned the mini-series *Far Sector* for D.C. Comics, set in the publisher's popular *Green Lantern* line of superhero comic books about an intergalactic, interspecies police force. A resident New Yorker and Brooklynite with a master's degree in education and a career in counseling, she also maintains an active online and social media presence, running her own website, where she bloggs and engages with commenters. Given this variety, Jemisin may be said to walk in the footsteps of sf icons such as Butler and Ursula K. Le Guin, but I will argue here that she is also one of the most versatile writers of contemporary short fiction, with deep literary historical roots to American romanticism and a strong commitment to social activism and justice.

Jemisin considers herself first and foremost a sf writer in the feminist tradition of Le Guin, Butler, and Nalo Hopkinson. Contrary to mainstream fiction, which focuses on the "intricacies of inter-human relationships, the discourse of sf is about our relationship to the world and the universe" (Mendelsohn 2003, 9). Still, Jemisin seems interested in how her characters navigate their sexual, ethnic, and class identities and how that navigation constitutes a kind of resistance to repressive hegemonies. Many hold sf to be "the last real bastion of Romantic fiction" as it stretches romance beyond character dynamics to suggest that romance exists "*out there* rather than internal," a conundrum that Jemisin explores as well (Mendelson 2003, 9). Jemisin's fictions play out in

https://doi.org/10.1515/9783110587647-036

anti-utopian and dystopian worlds and deal with ecological catastrophe, community, and cultural memory and contemplate technology, racism, capitalism, and even food culture. In the past decades, sf has sprouted a thicket of specialized subgenres – e.g. lesbian Steampunk – as it transformed from a domain of white male fanboys into a more inclusive genre. She also helped diversify and decolonize a genre in which she, like Octavia Butler, saw too little of herself and her experiences reflected. And so, she engages in specifically American sf themes, like the concepts of the frontier and the adverse effects of late global capitalism (Canavan and Link 2015, 9).

Jemisin is part of a generation of "postracial" and "post-postmodern" speculative realists, that is, writers of color whose outlook has been shaped by systems of American racism yet for whom "postmodernism has proven to be simply too distantly removed from the real world of justice and injustice and too pessimistic about the possibility of freedom and right to make it the basis for an attractive form of imaginative creativity" (Saldívar 2015, 519). Jemisin asserts the social function of literature, arguing that it can provide explanations and even survival strategies to those who have been ostracized by U.S. socio-economic, political, and media regimes: "American society is dedicated to weaving the illusion that what you see is not actually what is happening. The endless excuse-making is part of our mythologizing these days. But fixing that is part of the job that science fiction and fantasy can do" (Bereola 2018). Unlike American political discourse, speculative fiction can make clear moral interventions and may (re-)create coherence between the social world and language. For people of color and LGBTQ+ persons, fiction, at its best, may juxtapose feelings of invisibility and victimization with illustrations of what prolific public intellectual and activist Cornel West calls "prophetic fightback – a moral and spiritual awakening that puts a premium on courageous truth telling and exemplary action by individuals and communities" and that responds to "spiritual blackout" (West 2017, xv).

Many writers of ethnic futurisms (e.g. Afrofuturism, Indigenous Futurism) have come to see their work on a continuum with political activism and scholarship. Jemisin stands shoulder-to-shoulder with leftist activists, writers, and scholars like adrienne maree brown, Walidah Imarisha, and Grace Dillon, who consider social justice inextricably linked with economic, technological, and political justice and see racism and inequality as symptoms of capitalism. These commitments distinguish these futurisms from the more conservative political attitudes conveyed in mainstream sf. This "visionary fiction," as Imarisha calls it, "encompasses all of the fantastic, with the arc always bending toward justice" (Imarisha 2015, 3). Such fiction wants to advance the "decolonization of the imagination" from which, she believes, other forms of decolonization follow (3). Artist-activist scholar Yatisha Womack similarly sees Afrofuturism as combining "imagination, hope, and the expectation for transformative change" (Womack 2013, 42).

Afrofuturism, which Jemisin often is grouped with, is a transgenerational arts movement that combines literature, music (esp. jazz), and visual art. Its roots reach back to nineteenth-century abolitionist utopian literature, the early twentieth century

Black Renaissance in Harlem and Chicago, and the Black Arts Movement (1960s). Other luminaries include music and video artist Janélle Monae, of whom Jemisin is a big fan, and Ryan Coogler, director of the recent blockbuster superhero movie *Black Panther* (2018). The term "Afrofuturism" originally referred to "[s]peculative fiction that treats African-American themes and addresses African-American concerns in the context of twentieth-century technoculture" (Dery 1994, 180). Afrofuturism imagines possible futures and the uplifting of people of color through technology but also critically reads those futures back into the present. Literary Afrofuturism considers "race as a central factor informing the relations of science and society" (Yaszek 2015, 61). Beyond the recovery of African history and the participation in Afrodiaspora, Afrofuturist literary texts "locate the possibility of these new futures in the figure of the black genius," a, usually male, scientist who uses his intellect to save his community (Yaszek 2015, 59).

Afrofuturism today builds on the concept of Afrodiaspora, the idea that Black experiences are diasporic – dispersed and fractured – and can therefore not be explained within historical paradigms that focus on nation states (Gilroy 2002, 3). Some attribute the connection between Afrofuturism and Afrodiaspora to the notion that "Afrodiasporic subjects live the estrangement that science-fiction writers envision. Black existence and science fiction are one and the same" (Eshun 2003, 298). Under the Afrofuturist paradigm, scholars have also begun conceptualizing the interstices between Blackness, digital technology, and the humanities in the "digital black Atlantic" to continue shifting focus away from Eurocentrism in the digital humanities (Risam and Josephs 2021).

Meanwhile, Jemisin began her literary career as an author of short stories. She wanted to turn her "side hustle" writing hobby into a profession and attended a writing workshop, where she was advised to write stories to help her think about novel ideas (Jemisin 2018, vii). She retains some of these utilitarian attitudes. On the one hand, she sees short fiction as a protean version of long-form writing. She writes "'proof of concept' stories in order to test-drive potential novel worlds" (xi). "The City Born Great," for instance, serves as the prologue to her 2020 novel *The City We Became*, which in turn represents the first part of a trilogy of books. Jemisin here follows a wider cotemporary media trend of creating interconnected, and even transmedial, storyworlds (Ryan 2014, 41). However, transgeneric writing is not new in African American women's writing, e.g. Sherley Anne Williams' "Meditations on History" (1976) and her 1986 neo-slave novel *Dessa Rose* (Lucky 2009, 254). "Writing short stories," Jemisin notes on the other hand, "taught me about the quick hook and the deep character. Shorts gave me space to experiment with unusual plots and story forms – future tense, epistolic format, black characters – which otherwise I would've considered too risky for the lengthy investment of a novel" (Jemisin 2018, ix). Accelerated media consumption; the ubiquity of cheap, small, and portable screens; and the wider availability of software have arguably created an ideal media ecosystem for short textual formats that can be consumed in one sitting (↗0

Introduction). The emergence of micro and flash fiction, as exemplified for instance in John Edgar Wideman's collection *Briefs* (2010; ↗4 Textual Materiality), as well as by hybrid forms, such as Jennifer Egan's episodic novel *A Visit from the Goon Squad* (2011), illustrate that "short" does not mean "inchoate" and that the story is not merely the handmaiden to the novel and that short fiction allows for greater freedom.

Her collection, *How Long 'Til Black Future Month?* (2018), from which I cite both stories, intersects with various projects of literary production, political activism, and historical consciousness, but at times also seems skeptical about the prospect of social justice. The title facetiously suggests the commodification of Black history in American culture. It connotes Black History Month (BHM), which comes around every February with its TV specials and soundbites of MLK's "I have Dream Speech." In his memoir, the journalist Ta-Nehisi Coates wonders about the valorization of non-violence during BHM: "Why are they showing this to us? Why were only our heroes nonviolent?"; he is skeptical, not about the ethics of nonviolence but about the insinuation that "blacks are in especial need of this morality" (Coates 2017, 32). Read in this way, BHM figures as part of a repressive hegemonic superstructure, which swaths histories of African slavery as well as Black suffering, liberation, and ingenuity, and curtails deeper conversations about these events, in a bid to prevent them from spilling over into the regularly-scheduled programming. Jemisin's title echoes the Afro-Pessimist notion that blackness is not just one political cause among others but constitutive of white power regimes, and that "the politics of coalition," aligning African American interests with commercial causes, always comes at the price of justice (Sexton 2016, sec. 6). As per her title, the project of envisioning Black futures will end when it, too, receives its own commemorative holiday. As a collection of stories, *How Long 'Til Black Future Month?* foregrounds the conundrum of contemporary African American artists to speak without feeling the need to be political, yet to articulate visions of Black futures that respect Black history without kowtowing to concerns about institutional respectability.

The stories I discuss in this chapter outline a view of identities (race, gender, class) for the age that comes after the American cultural project of postmodernity and after the postracial hopes of the Obama presidency. Jemisin sees these projects not as dashed but as superseded by an onrushing materiality as well as a renewed spirit of solidarity and a need for intersectional analysis. These issues become deepened by their contacts with each other: Protagonists make their future because they are Black *and* gay, Creole *and* lesbian. Navigating those identities is bound up with global historical materialism. Both stories, then, demark a new, broader cultural spectrum, not just for African American lives, but for Black lives.

2 Close Readings

"The City Born Great" presents us with a cosmopolitan protagonist-narrator, a gay, unhoused, Black, male graffiti artist (New York), who roams the city, creating his art, living hand-to-mouth. He is mentored by Paulo (São Paolo), the eponymous elder avatar of another great city. New York must discover his instinctual and psychic connection to the city and its denizens to help midwife it into existence as a "great" city. But he finds himself opposed in this task by agents of "The Enemy," an eldritch entity which seeks to destroy great cities before they can be fully born. New York battles the Enemy throughout the city, eventually harnessing its collective spirit of place to vanquish the creature, and assumes the mantle of NYC's avatar.

Jemisin uses an unnamed protagonist, an everyman who allegorizes not just African Americans but arguably millennial urban professionals, who constitute her main audience. This crowd holds down temporary and even precarious jobs, owns little property, and, most importantly, leads an itinerant lifestyle. NYC scholar Ross Wilson asserts that "New York's literature is an account of movement, a transition from one state to another," Jemisin here writes a typical New York story (Wilson 2020, 2). The various physical and psychic transformations, accelerations, and groupings she shows us connote an anxiety of connectedness and a simultaneous appreciation for the exhilarating independence urban life provides. New York must reconcile his transience with his obligation to care for others. This is also part and parcel of what Paul Gilroy identifies as "the restlessness of spirit which makes that [transnational black Atlantic creativity] diaspora culture vital" (Gilroy 2002, 16).

In this close reading, I focus on the city as a *genius loci*, New York's role as an African American nomadic man, the manner in which he instrumentalizes his body and desire, and his threefold role as an artist, caretaker, and avatar of the city. New York speaks with charisma and ambivalence about the city. He loves and hates and paints it. His story is simultaneously part of a global narrative about urban modernity. His identity seems fluid. His gayness, blackness, and unhousedness render him both more outside the crowd of white, middle-class spectators that he encounters and makes him more of an everyman, an avatar, of cosmopolitan urbanity. At the same time, Jemisin's is not a utopian vision: Black and gay urban dangers, such as the ubiquitous specter of police brutality; the colonizing white gaze; and the implicit fear of AIDS, always linger in the background in this world. The story contemplates community, connectivity, and life in small spaces within larger contexts.

Jemisin roots the story in her version of the concept *genius loci*, that is, "places with minds of their own" (Jemisin 2018, xi). In its original online version, a short expository abstract precedes the story explaining its premise: "Like all great metropolises before it, when a city gets big enough, old enough, it must be born; but there are ancient enemies who cannot tolerate new life. Thus New York will live or die by the efforts of a reluctant midwife [. . .] and how well he can learn to sing the city's mighty song" (Jemisin 2016). Jemisin ponders the ambivalent notion of *genius loci* as both spirit of place and protective spirit.

Greatness seems to mean something like "cultural density" here. Great cities, New York recounts at one point, "make a weight on the world, a tear in the fabric of reality, like [. . .] like black holes maybe. [. . .] As more and more people come in and deposit their strangeness and leave and get replaced by others, the tear widens. Eventually it gets so deep that it forms a pocket, connected only be the thinnest thread of [. . .] something to [. . .] something. Whatever cities are made of" (Jemisin 2018, 20–21). The mechanics of this process remain vague, also because she cleverly relegates explaining them to a narrator whose frantic, artistic nature makes him unreliable. Jemisin's premise playfully invokes the age-old fiction adage of making the setting a character in its own right. "The City Born Great" thus is not just a story about community and identity but also a meta-narrative about how a literary setting may – narratologically as well as mimetically – become a character over the course of a short story.

The idea of a place that develops independent volition and personal traits, which it then articulates via a human avatar, invokes the tenets of speculative realism. A philosophical and artistic movement, speculative realism contemplates "a world independent of the human mind" and arrives at explanations that seem "counterintuitive and even downright strange" (Harman 2018, 2). In that sense, "speculative philosophy rather resembles speculative fiction, for it cannot do without extrapolation" (Shaviro 2014, 10). It postulates the "interrelatedness of all things and the ways that their interactions and variations continually lead to consequences that are new and unforeseen" (5). This philosophical paradigm understands existence not as a static state but as a process of continuous becoming, as it is with New York and the city.

New York begins his narration by commenting, "I sing the city. Fucking city. I stand on the rooftop of a building I don't live in and spread my arms and tighten my middle and yell nonsense ululations at the construction site that blocks my view. I'm really singing to the cityscape beyond. The city'll figure it out" (Jemisin 2018, 14). In singing the city, the narrator invokes romanticist notions of poetic authorship, specifically that of Ralph Waldo Emerson and famous NYC poet Walt Whitman. A Brooklynite like Jemisin, Whitman has said that his time in NYC (1838–1853) inspired the first draft of his collection *Leaves of Grass* (Karbiener 2018, 15). The gay, dandyish, school dropout may even have served as a template for New York (20). Jemisin insinuates the connection between the two artists via intertextual allusion. As the sun rises over the NYC skyline, New York muses,

> It's dawn. The damp of it makes my jeans feel slimy, or maybe that's 'cause they haven't been washed in weeks. Got change for a wash-and-dry, just not another pair of pants to wear till they're done. Maybe I'll spend it on more pants at the Goodwill down the street instead . . . but not yet. Not till I've finished going *AAAAaaaaAAAAaaaa* (breath) *aaaaAAAAaaaaaaa* and listening to the syllable echo back at me from every nearby building face. In my head, there's an

> orchestra playing "Ode to Joy" with a Busta Rhymes backbeat. My voice is just tying it all together.
>
> *Shut your fucking mouth!* someone yells, so I take a bow and exit the stage.
>
> (2018, 14; original emphasis)

As a prospective bard of the city, New York picks up where Whitman leaves off. At the end of his famous epic poem "Song of Myself" (1855, 1888), Whitman's speaker stands "not a bit tamed," sounding his own "barbaric yawp over the roofs of the world," having traversed space and time (Whitman 2002, sec. 52, lines 1333–1334). New York sounds his own voice, which becomes the city's voice as it reverberates through the clefts of the concrete jungle, as the scene emblemizes the story's central theme – avatar and city becoming one. Jemisin comically offsets this transcendence by disrupting it with stereotypical New York irreverence. As references to rap, connoting African American musicological practices of sampling and cultural memory, and poetry indicate a virtuoso aesthetic, the crude explicative, yelled from offstage, cements the image of a city of iconoclasms, where European and Afro-American art, Beethoven and Busta Rhymes, romanticism and rambunctiousness clash.

New York is a graffiti artist, painting his illicit works all over the city. The fact that he paints mouths – symbolic orifices through which his city can breathe – reflects his role as caretaker and eventually midwife of the city. The idea that an unhoused, gay, black man should be the avatar of the capital of American empire subverts expectations and cultural stereotypes. And yet Jemisin's casting choice for the role corresponds to some of the longer arcs in Western cultural tradition. In his *Republic*, Plato envisions the caretaker of the "healthy city," Caliopolis, as a philosopher-king, a person of exceptional experience and reason, who uses these traits to rule the city state and shape the moral character of its citizens. Yet beyond rationalism, this "constitution painter" must weigh considerations of security, stability, social harmony to guide his brush strokes (Plato 2000, 206). New York, too, is a constitution painter, even if he does not consider himself an artist (Jemisin 2018, 17). His art is part of his role as caretaker. While he does not "rule" the city in any formal, legislative sense, he has special revealed knowledge of its needs that are simultaneously more profound and diffuse than conventional notions of governance. Like Plato, Jemisin considers the city a living creature, a hybrid of materiality (buildings, parks) and people. It is made alive by the lives its denizens make in it. His signature virtue being "[f]ortitude, not [p]atience," New York excels through his commitment to the cacophonic multifariousness of his city (24).

But Jemisin's NYC is far from a multicultural melting pot. Racism and repressive social practices and regimes of power determine how New York navigates the city. Traversal, in fact, depends on his ability to appear and disappear at will, to evade the white gaze of the authorities. Sitting in a diner, he gets startled:

> Then a cop comes in, fat florid guy buying hipster joe for himself and his partner in the car, and his flat eyes skim the shop. *I imagine mirrors around my head, a rotating cylinder of them*

> *that causes his gaze to bounce away. There's no real power in this – it's just something I do to try to make myself less afraid when the monsters are near.* For the first time, though, it sort of works: The cop looks around, but doesn't ping on the lone black face. Lucky. I escape.
>
> (Jemisin 2018, 16; emphasis added)

In a sequel to this moment, he explains how he resorts to imagining himself as "nothing, beneath notice, not even worth beating for pleasure" to evade the gaze of the law (Jemisin 2018, 18). Coates agrees that "racism is a visceral experience, that it dislodges brains, blocks airways, rips muscle, extracts organs, cracks bones, breaks teeth" (Coates 2017, 10). New York's apprehensiveness suggests that he has experienced such victimization in the past. His body is precariously situated in a society in which arbitrary violence may be exacted on it at any moment. Racism is a belief system that constitutes reality, argue historians Karen and Barbara Fields (Fields and Fields 2012, 19). Analogously to witchcraft in the early-modern world, "racecraft" posits race as "objective reality" and turns racism from a social practice, "an action and a rationale for action" and "something an aggressor *does*, into *race*, something a target *is*" (17; original emphasis). In this world, violence lies coiled beneath sumptuary rules and "everyday routines that organize racism" and may always explode (37). The force of these explosions land on Black bodies, and so African Americans must spend their lives protecting their bodies from "the terror of disembodiment" (Coates 2017, 12). For New York to evade such abuse, he must reckon with "the profound sense of psychological depression, personal worthlessness, and social despair" that marks many Black American lives (West 2017, 12). Becoming invisible is a survival tactic that shields him from physical violence and spiritual breakdown. For "being black in America means being minimally subject to white supremacist abuse *and* being part of a rich culture and community that has struggled against such abuse" (25; emphasis added). Maintaining the integrity of his body, for New York, is a constant negotiation of avoiding the gaze of power. Were it to fall on him, that gaze would define and perhaps destroy it, rendering him a victim, an easy target, a blank canvas on which to manifest power. Yet his imaginary mirror helmet is a momentary measure with no assured outcomes, without "real power" to create perpetual safety. Eventually, he must run to dodge the advances of power.

New York's itinerancy ensures his survival. Unlike Whitman's speaker, he is not a transcendental loafer but a man on the run. He subverts the system as a matter of course through his mobility. Like a fish, he has to stay in motion to keep breathing. Analogously, early twentieth-century NYC literature often painted African Americans as "edgy and urban existential heroes who lived on the margins of an increasingly commodified postwar society" (Field 2020, 242). And if NYC stories are typically about movement and transformation, New York is the avatar of nomadic movement. Nomadic blackness is "[t]he art of spatial literacy – the ability to read one's environs in order to decode and disrupt the center of discourse – is the lingua franca of blackness" (Jones 2016, sec. 9). Creating graffiti art and moving a Black body through the city disrupts its power structures by exposing their inner workings. The omnipresent threat of violence against Black bodies necessitates that such bodies be itinerant if they want to be safe from harm, but in doing so, they decrypt the coded logic of the modern city: "[A] black hole on the white face of modernity, my body," Jones says, "is the demystification of the mystification of value" (sec. 9).

But what of those who would destroy Black bodies? What animus mobilizes and drives them? In Jemisin's story, NYC's birth is threatened by an entity called "The Enemy," a tentacled, dragon-like, eldritch creature, who dwells within the bowels of the earth, bend on devouring young cities before they can become full born. Jemisin here draws on sf history. The novel version of the story reveals the creature to be R'lyeh, the sunken city where H.P. Lovecraft's chaos monster Chthulhu resides. In horror stories such as "The Call of Cthulhu" (1928), Lovecraft describes how Chtulhu holds sway over the instinctual, unconscious part of the human psyche, as a – latently racist – Other. Jemisin inverts Lovecraft's symbolism and aligns the creature and its prison with the state apparatus, which nominally represents the very Enlightenment rationalism that Lovecraft's heroes seek to defend. Jemisin sees the Other not on the margins of society but at the center of Western civilization, where it dehumanizes its agents while using law and order as window dressing for its destructive purpose. This is evidenced by the appearance of the "Mega Cop," a grotesque amalgamation of two police officers, who melt into a blob of human flesh and chase New York through the city:

> I don't have words for it. Too many arms, too many legs, too many eyes, and all of them fixed on me. Somewhere in the mass I glimpse curls of dark hair and a scalp of pale blond, and I understand suddenly that these are – this is – my two cops. One real monstrosity [. . .]. I cross the street and risk a look back and see something roll onto the sidewalk on at least eight legs, using three or four arms to push itself off a building as it careens a little. (Jemisin 2018, 26)

As metaphor, the Mega Cop asserts that "[t]he destroyers [police officers] are merely men enforcing the whims of our country, correctly interpreting its heritage and legacy" (Coates 2017, 10). Jemisin suggests that the police are mobilized by a collective consciousness. As the Fieldses show, structural racism indeed seems to spellbind its agents, and the horror of the chase scene cited above, similarly, arises from the eradication of individuality and reason. However, this is a story of becoming and emancipation. In running, New York acquires the means to hit back, to inflict the very violence dished out on the monstrous system that produces it.

Eventually, open battle with the Enemy unfolds and coincides with NYC birth, as well as with New York's embracing his role as the city's avatar. New York utilizes the city's cultural idiosyncrasies to battle the creature. The Enemy "drag[s] itself up from the dark toward the world, toward the warmth and light, toward the defiance that is *me*, toward the burgeoning wholeness that is *my city*" (Jemisin 2018, 29; original emphasis). The imagery and vocabulary here suggest the roundness of pregnancy, indicating a shift in gender markers between New York and NYC:

> On Second, Sixth, and Eighth avenues, my water breaks. Mains, I mean. Water mains. Terrible mess, gonna fuck up the evening commute. I shut my eyes and I am seeing what no one else sees. I am feeling the flex and rhythm of reality, the contractions of possibility. I reach out and grip the railing of the bridge before me and feel the steady, strong pulse that runs through it. *You're doing good, baby. Doing great.* (29; original emphasis)

As he midwifes the city's birth, New York inhabits it somatically. Jemisin unsettles categories of male and female corporeality by assigning the vocabulary of birthing to a male body. New York transcends these binary categories and comes to inhabit a dilated body and experiences the city's sense of place as corporeality:

> Something begins to shift. I grow bigger, encompassing. I feel myself upon the firmament, heavy as the foundations of a city. There are others here with me, looming, watching – my ancestors' bones under Wall Street, my predecessors' blood ground into the benches of Christopher Park. No, *new* others, of my new people, heavy imprints upon the fabric of time and space. (29; original emphasis)

Where before he had to oscillate between different points on the city map, he now feels and encompasses it as a unified whole. The scene calls back the advice Paulo gave him earlier, "you need not be them [the city's denizens], but know that they are part of you" (22). Jemisin's rendering of New York's dilation invokes that of Whitman's in "Song." It recalls the latter's reflections on the metaphysics of intergenerational sympathies in "Crossing Brooklyn Ferry" (1856, 1885): "It avails not, time nor place – distance avails not, / I am with you, you men and women of a generation, or ever so / many generations hence" (Whitman 2002, sec. 3, lines 20–23). Midwifing NYC allows New York to experience a similar compression of history. He simultaneously beholds a history of repression – of indigenous and homosexual people – and the promise of a new, multicultural community. Jemisin thus presents NYC as a quasi-utopian space, where, despite past cruelties, new connections may be formed in defense of external threats. Yet the Enemy should not be read in xenophobic terms. For one thing, Jemisin is blunt about the internal social strife that exists in the city. For another, the description of the aftermath of the battle suggests that the story is a self-reflective allegory about American society and human nature.

The outcome of the battle remains contingent and preliminary: "The Enemy," New York explains, "is as quintessential to nature as any city. We cannot be stopped from becoming, and the Enemy cannot be made to end" (Jemisin 2018, 32). Such historical determinism suggests an allegorical reading of the story, by which the battle with the Enemy, that is, between community and destruction, may be waged within human consciousness.

In *Beyond the Pleasure Principle* (1920) Sigmund Freud conjectures that this dichotomy is the effect of two competing yet intertwined drives within all organic life: one bend towards love, community, and multiplication (Eros), the other towards death, isolation, and entropy (Thanatos). Freud identifies an "*an urge inherent in organic life to restore an earlier state of things*" (Freud 1989, 612; original emphasis). The Enemy's "old, old evil" (Jemisin 2018, 29) echoes the "*old* state of things," Freud finds as "the initial state from which the living entity has at one time or other departed and to which it is striving to return" (Freud 1989, 613).

In the story's epilogue, New York has become the elder statesman avatar, a role previously filled by Paulo. Fifty years later, he has come to mentor L.A. as the next great city. As Wilson notes, New York literature "not only addresses the experiences of those living within the city but the city's relationship with the rest of the world" (Wilson 2020, 1). Read from this angle, the story figures as an alternative history of globalization, told from the vantage point not of capital but of people, who band together in local communities that become embedded in a global, transhistorical, and postracial project of civilization.

The second story I will discuss proceeds similarly from the idea that ethnically and sexually diverse places hold revolutionary potential. With "The Effluent Engine" Jemisin writes a multilayered story that is equal parts alternative history, steampunk fantasy, spy thriller, and lesbian romance. Both, nineteenth-century New Orleans, the story's actual setting, and Haiti, its aspirational setting, are postcolonial spaces that enable different types of hybrid identities. The story follows the protagonist-spymaster Jessaline Dumond, who has been dispatched to the United States by the post-revolutionary Haitian government to procure designs for the eponymous effluent engine that can turn the waste products of Haiti's rum production into clean and abundant energy. Assembling the engine would provide the young nation with a technological edge and a means to defend itself militarily against its rivals. To build the machine, Jessaline must engage the services of Eugenie and Norbert Rillieux, an engineer Creole sibling couple. In a take-off on the traditional, heterosexual romance plot, the narrative revolves around Jessaline and Eugenie's fledgling love. Eugenie's mind ultimately holds the key to completing the engine, and her commitment to Jessaline and her political cause will determine Haiti's fate. Jessaline's counterpart is Raymond Forstall, an agent of a clandestine organization, The Order of the White Carmelia, who represent those political interests that would see the newly formed nation of former slaves fail. Jessaline and Eugenie eventually overcome Forstall and decide to escape to Haiti where Jessaline promises they may openly display their homosexual affections. Jemisin juxtaposes New Orleans as a site of colonial repression and Haiti as a socially progressive, ecologically sustainable, as well as technologically and militarily formidable Black utopia.

First published in the magazine *Steam-Powered: Lesbian Steampunk Stories* (2011), the story combines the sf genres of steampunk and alternative history. Steampunk implies "a historical setting in the nineteenth century characterized by technologies extrapolated from the science of that era, but which were not invented at that time" (Prucher 2007). Meanwhile, alternative histories trace the lives of fictitious characters in historical settings, dramatizing some "divergence" from the normative historical record to make a larger ideological point; such stories assume "equal interest" on the part of readers in changes in characters and their world (Duncan 2003, 210). Considering genre prompts us to connect historical setting and character relations. Expounding on these connections may explain how, for instance, lesbian sexuality can transform socio-political relations.

The setting calls up a key moment in European American colonial history. Within the last decade, several historians have stressed the significance of the Haitian Revolution for the histories of Europe and the Americas. As a series of related, world-historical events, the slave revolt in what was then St. Domingue against France (1791–1804), Haitian Independence (1802), the Louisiana Purchase of the Jefferson Administration from Napoleon (1803), and the U.S. embargo policies that followed (1806) reshaped the Atlantic world. Jemisin actually inserts herself into current historicist debates when she examines the "related origins and geopolitical similarities" between Haiti and the United State in her story (Dillon and Drexler 2016, 3). "At the moment of revolution," Elizabeth Maddock Dillon and Michael Drexler write, "white colonials and free and enslaved blacks in St. Domingue and British North America confronted precisely the same situation: how to reconcile the development of modern doctrines of universal political liberty and equality with the economic engine of modernity – namely, a colonial system of labor and production premised upon dehumanizing practices of race slavery" (3). The Haitian Revolution effected changes in "the *geography* of slavery, in the *economics and labor practices* of slavery, and in the *ideological justification* of slavery" in the United States (9). The Revolution transformed the "geographical imaginary" of the United States, effectively turning it from a littoral nation that saw itself as part of the Atlantic world to a continental confederation of states (10).

Haiti's independence from France under its most famous leader L'Overture Toussaint – who in the story turns out to be Jessaline's father – initially played into the hands of then President John Adams, who also quarreled with the French. But the political landscape quickly shifted. The rise of Napoleon (1799) and the presidency of Thomas Jefferson (1801), a slave-owning planter from Virginia, augured tensions between the United States and Haiti. When Napoleon, having been bested by the Haitian insurrectionists, abruptly abandoned American colonialism in 1803, Jefferson procured France's continental territories in the Louisiana Purchase, adding several slavery states to the Union. At the time, U.S. military and economic hegemony were by no means ensured. The Purchase put the new nation's long-term prospects on firmer ground, as it grew the cotton trade, but it also cemented its dependence on slave labor. And so, anxieties over the spread of slave revolt began to fester in the United States. These anxieties manifested as economic skirmishes and trade embargoes, designed to prevent the Haitian example from spreading to continental America. Jemisin's narrator sardonically caricatures the political climate of the time as "a graceful waltz," which "only occasionally devolv[ed] into a knife fight" (Jemisin 2018, 78).

The Haitian Revolution did eventually spark the imaginations of generations of African American thinkers and artists, such as Langston Hughes and C.L.R. James, but its cultural reach extends across the Atlantic world. "[R]adical" Black writers saw the Revolution as a "transformational historical moment," even as its political legacy remains contested to this day (Kaisary 2014, 3). Jemisin seems to sympathize with the minority position of framing Haiti as a "radically universalist vision of freedom" (Kaisary 2014, 10). Yet she also qualifies her utopian vision, for instance by having Jessaline declare to Eugenie: "men and women of all shades are free. I will not pretend that this makes us perfect; I have gone hungry many times in my life. Yet there, a woman such as yourself may be more than the coddled sister of a prominent scientist, or the mistress of a white man" (Jemisin 2018, 94–95). This places her in a tradition of Black utopianists, that is, speculative writers who do not provide escapism, engage questions of political theory, and construct social visions based on "pragmatic experimentation and critical reexamination" (Zamalin 2019, 12–14).

The story also has non-speculative roots, for instance, in its New Orleans setting. Jemisin draws on a tradition of African American short stories about the city, like the realist fiction of Alice Dunbar-Nelson, whose New Orleans story cycles explore "the culture of creoles of color in New Orleans, depicting their fundamental humanity, social conflicts, and dilemmas" (Nagel 2014, 91). Romance, issues of ethnic mixture, and Creole values also figure prominently in New Orleans stories, and so Jemisin explores these themes through Jessaline and Eugenie's blossoming relationship (see Nagel 2014, 117). When the spy informs the scientist that she would have to take the engine prototype back to Haiti without her, Eugenie, seeing Jessaline previous romantic advances rescinded and her hopes at escape dashed, launches into a fiery indictment of Jessaline's arrogance and exploitative pragmatism:

> In your perfect, wonderful land, everyone is free to live as they please. It is the rest of us, then, the poor wretched folk you scorn and pity, who have no choice but to endure it. Perhaps we should never bother to love at all, then! That way, at least, when we are used and cast aside, it will at least be for material gain. (Jemisin 2018, 104)

The charges here are severe. Chattel slavery exploits Black bodies without the pretense of fellow feeling, whereas Jessaline's manipulation of her heart and intellect register as invasive. Although the daughter of Toussaint plays the sardonic revolutionary, it is arguably the bookish Eugenie who truly understands the stakes of freedom. Jessaline's nomadism and perpetual masquerade as a spy render her blind to New Orleans' material and social structures. These forms fetter racially ambivalent individuals within the anxiety-inducing shadow of Southern slavery. Here women of color face long-term prospects that range from physical and sexual abuse to a morally corrosive life as concubines. Jessaline demonstrated her arrogance early on, when she labeled Norbert naïve for being angry after becoming the victim of racism. Norbert, having devised "ingenious means of draining its [New Orleans'] endless, pestilent swamps [. . .] – had been turned down. Six months later, a coalition of city engineers had submitted virtually the same plan and had been heaped with praise and funds to bring it about. The men of the coalition were white, of course. Jessaline marveled that Rillieux even bothered being upset about it" (80). Jemisin here follows the tenets of Afro-Pessimism by having Jessaline dismiss the prospects of either reform or cooperation with a racist state apparatus. Meanwhile, Norbert seems driven by chauvinist narcissism, patronizing his sister and wanting to see his name recognized, all while internalizing the very racism that keeps him locked out of social respectability. An example of this is his disregard for any science without "European roots" (81).

Ultimately, Eugenie creates the effluent engine to aid Jessaline in her country's struggle. The Rilleux siblings both carry aspects of the type of the "black genius" (Yaszek 2015, 64). Jemisin imagines the magnetic pull of Haiti as a colored steampunk utopia and allegorizes what it would mean for people of color to abandon the American experiment and pool their faculties in the pursuit of technological and economic independence and universal equality. The story retains elements of the Slave narrative, insofar as it addresses violent racial persecution and racial passing and casts spaces like the Caribbean as promised lands of salvation. Yet the story's utopian vision remains notably intertwined with the romance plot, arguably staking the future of Black nationhood on homosexual love.

Eugenie is more than she seems. While her wardrobe makes her initially look "rather plain," Jessaline makes her out to be more perceptive than her brother. For starters, she immediately detects a pistol in Jessaline's travel bag. The stares they exchange indicate a moment of mutual recognition: "as the young woman [Eugenie] lifted her gaze [. . .] Jessaline thought she saw a hint of challenge lurking behind those little round glasses, and above that perfectly pleasant smile" (Jemisin 2018, 83). On the one hand, the scene conveys sexual and violent tension. Both women are aware of the real stakes of the interview, while Norbert remains oblivious of it. Eugenie actually saves Norbert's life, for Jessaline had already decided to dispense of him enjoying the exercise "to an unseemly degree" (83). The scene shows how women of color are used to read beyond surfaces because they face more dire threats the safety of their bodies. Eugenie's glasses synecdochally represent her diminutive physical frame and seemingly timid nature. On the other hand, Jemisin uses free-indirect speech to reveal how Jessaline becomes smitten with Eugenie. Over the course of this and subsequent interviews, Jessaline's thoughts regarding the young scientist shift from dismissal, to intrigue, to infatuation.

During their meetings, Eugenie inspires in Jessaline a series of "curious epiphanies" that change her perspective on love and her mission (91). With the meeting in the nuns' garden, Jemisin conjures a scene seemingly taken from the pages of a nineteenth-century romance novel. For her mission to succeed, Jessaline must convince Eugenie to build the engine, but doing so requires manipulating the scientist's tender feelings for her. Jessaline remains hesitant. Acting "on impulse," she pulls Eugenie through the curtain-like branches of a willow tree (91). This "willow fall" forms a place apart from New Orleans' normative social universe, where Jessaline and Eugenie may be true to their feelings. The scene invokes the concept of wilderness as a transformative space in romanticist literary history. Nature, in this tradition, is a liminal, transgressive space in which human beings may escape the restrains of social convention and may rediscover their universal connectedness. The most famous example for this type of depiction is Nathaniel Hawthorne's *The Scarlet Letter* (1851; ↗6 Hawthorne), in which the minister Arthur Dimmesdale secretly meets his lover Hester Prynne, who stands accused of adultery, and their illegitimate daughter Pearl to confess his love for them. Similarly, Jessaline can give in to her attraction to Eugenie and even propose them running away together:

> 'And in my land, it is not uncommon for a woman to head a family with another woman, even to raise children if they so wish' [. . .]. Eugenie started – and to Jessaline's delight, her blush deepened. She darted a half-scandalized, half entranced look at Jessaline, then away, which Jessaline found deliciously fetching [. . .]. Eugenie looked up then, and Jessaline was surprised to see a peculiar daring enter her expression, though her flush lingered. 'And [. . .]' She licked her lips, swallowed. 'Do such women [. . .] ah [. . .] behave as a family in [. . .] *all* matters?' [. . .].
>
> Jessaline got tired of teasing. It was not proper, she knew; it was not within the bounds of her mission. But – just this once – perhaps –
>
> She shifted just a little, from brushing shoulders to pressing rather more suggestively close, and leaned near, her eyes fixed on Eugenie's lips. 'And conjugal,' she added. Eugenie stared at her, eyes huge behind the spectacles, 'C-conjugal?' she asked, rather breathlessly.
>
> (Jemisin 2018, 96)

The scene crackles with sexual innuendo. Initially, Eugenie's scandalized blushing calls up the heroines of nineteenth-century sentimental novels who must fend off the advances of unscrupulous suitors. But her moistening her lips adds a suggestive somatic symbolism to their flirting. Within the parallel universe of the willow fall, Jessaline effectively proposes marriage to Eugenie – although the offer is insincere, at this point – with the promise of familial and sexual union. To wit, lesbian critical theory thinks sexuality beyond genital sex and in terms of companionship, an idea I explore below. Suffice it to say for now that Haiti becomes the projection screen for Eugenie's desires and thus a kind of promised land. Before Jessaline can offer "a demonstration" of this freedom, though, the two are interrupted by the voices of the nearby nuns, Haitian agents and symbols of a repressive, heteronormative social order intervening in their courtship. With that the spell of the willow tree's magic circle is broken, and both return to the task of completing the engine.

Jemisin identifies a major juncture in the world-historical record, a moment at which the fortunes of both Haiti and the United States are intertwined and undecided. Through her story she opens up a speculative space: What if, in this crucial moment, the fates of these countries were in the hands of two utterly ostracized persons, whose transgressive love (by the standards of the time) may help birth a new nation? The romanticist, character-driven "The Effluent Engine" thus foreshadows the process-oriented "The City Born Great" in its contemplation of transformative love and community.

3 Theoretical Perspectives

The two stories provide grounds for a variety of critical approaches. And while I will not be able to sketch out all of them, I will focus here on several obvious and not so obvious ways of reading that deepen the close readings I have provided above. These include Afrofuturism and Afro-Pessimism, as well as Black Diaspora Studies. These related approaches can help us understand Jemisin's characters as mobile subjects in search of a place in the world. Postcolonial theory as well as gay and lesbian criticisms can help us make sense of their resistances and subversions. Meanwhile, cultural game theory may aid us in conceiving of the settings and places these characters inhabit and the conceptual role movement plays in her stories.

From the perspective of African American Studies, both stories illustrate a turning point in Black cultural history, chronicling a deepening of African American Literature and indicating that it has outgrown its former exclusively "indexical and instrumental imperative" that had been foisted on it by Jim Crow (Warren 2010, 14). African American Literature, by some accounts, has only now begun to *be* rather than to portend, by reflecting critically on its own past while articulating its own aesthetic (Warren 2010, 42). This "post era" marks a moment in which Blackness becomes marked by "resistance to claims of authenticity; assertions of individual agency over the meaning of Black identity and Black art; rejection of the idea that Black art is primarily concerned with responding to the ongoing history of American white supremacy; and often parodic relationship to key figures and historical moments of Black history" (Leader-Picone 2019, 4–5). Afrofuturism, in its original form, anticipated this paradoxical position, registering a "troubling antinomy" in its own project when asking "Can a community whose past has been deliberately rubbed out, and whose energies have subsequently been consumed by the search for legible traces of its history, imagine possible futures?" (Dery 1994, 180).

Under the critical register of Afro-Pessimism, being Black necessarily means being hunted and considered inhuman (Weier 2015, 429). New York's breathless chase through the city emblemizes this feeling; he is "the black subject, never grounded, always on the run, is accustomed to navigating the enigmatic line between text and subtext, between the terrain and the subterranean" (Jones 2016, sec. 5). This movement occurs in American urban spaces marked by slippery surfaces that grant no purchase, "no sign of sovereign space for me to call my home, no autonomous zone left for me to suture myself on to, no under-the-commons to use as an escape hatch and no over-the commons" (Jones 2016, sec. 5).

In response to this transience, New York adopts a lusory attitude towards the world and even his body, a stance according to which, "a game player accepts the limitation of means which the rules demand 'just because such acceptance makes possible such activity'" (Suits 2014, 142). This strategic position allows him to see the world in terms of the claim that "Utopian existence is fundamentally concerned with game-playing" (Suits 2014, ix). Avoiding nihilism demands adopting this autotelic attitude, in which things become commutable and actions are their own ends.

Sexuality, desire, and aspirations all seem to exist on a flattened-out plane in New York, that is, before he embraces his role as avatar. For instance, when Paulo invites him to lunch and leaves a twenty-dollar bill on the table, New York contemplates the transaction in terms of sexual barter: "I pick up the twenty, finger it, pocket it. Would've done him for the sandwich, or because I like his eyes, but whatever" (Jemisin 2018, 16). He is as likely to offer his body up for sustenance (the sandwich) as he would for aesthetic pleasure (Paulo's eyes). The scene underscores how emotions, aesthetics, bodies, and goods become mutually commutable in the modern urban economy. More, the playfully sexual connotation of New York's "fingering" the twenty illustrates what Karl Marx has called commodity fetish, that is, the charging of objects with sexual energy. New York reveals these hidden logics simply by being in this space as a function of his nomadic Blackness. And it is arguably for this reason that the other patrons consider his presence an "existential crisis" (16). The "crisis" here is that New York performs openly the clandestine, lusory logic that underpins the edict of universal commutability of art, food, sex, money in urban upper middle-class society. His performance disrupts the pantomime of dressing this logic up as natural and moral.

New York and Jessaline both stand in the African American tradition of the trickster or "Signifyn' Monkey," who manipulates surfaces to evade repressive power structures (Gates 2014, 51). Jemisin thus iterates on the type of the colored trickster and thus echoes twentieth-century writers like Charles W. Chesnutt (↗10 Chesnutt), who use African folklore for social criticism. This is nowhere more clearly displayed than in Jessaline's meetings with her adversary, Raymond Forstall, who is also an emissary of colonial power.

In their initial meeting, Forstall, playing the bumpkin, accosts Jessaline, complimenting her on her appearance and apologizing for not knowing the proper address. Still he routinely goes to work in fixing Jessaline in racist language. He sheepishly comments on the incessant heat, speculating about Jessaline's constitution on the basis of snap judgment: "A fine Negress like yourself would naturally deal better with such things [. . .]. Oh dear, my sister warned me about this. You aren't Creole, are you? I understand they take it as an insult to be called, er [. . .] by certain terms" (Jemisin 2018, 76). He assigns her several labels, thus defining her and making her the subject of a colonizing discourse: "Negress" and "Creole," while qualitatively different, construct "the colonial subject in discourse, and the exercise of colonial power through discourse, demands an articulation of forms of difference – racial and sexual" (Bhabha 1996, 38). Forstall tries to fix Jessaline in discourse to render her fully visible and thus controllable and predictable. Jessaline, on her part, feels insulted by the term – "*do I look like one of them?*" – her inner monolog illustrates her internalized racial apprehension, further proof that Jemisin explores the linguistic thicket of colonial language (Jemisin 2018, 76). Deconstructionists would argue that we can find the sum total of colonial practices manifested in such stereotyping, as an "ambivalent text of projection and introjection, metaphoric and metonymic strategies, displacement, overdetermination guilt, aggressivity; the masking and splitting of 'official' and phantasmatic knowledges to construct the positionalities and oppositionalities of racist discourse" (Bhabha 1996, 51).

This seemingly polite conversation marks an instance of a colonial power struggle over Jessaline's identity. To understand the stakes of this confrontation, we must recall that postcolonial theory, drawing on language philosophy and deconstructive analysis, considers the subject, or self "always production rather than ground" (Spivak 2006, 292). Colonialism demonstrates that human beings are not individuals – free, self-sufficient entities – but products of "ideology, discourse and language" (Ashcroft et al. 2009, 202).

Jessaline strategically adopts the role of the colonial subaltern, aiming for the "correct mix of reserve and deference" (Jemisin 2018, 75). Both attributes and her concern illustrate the overdetermination of women of color in colonial contexts. Etiquette dictates that she let herself be objectified. She must also defer to Forstall's male and white authority to quell any suspicion of aggression, even if New Orleans' class structure allows her to patronize him, as obviously being "not from around here" (75). Forstall's name gives away his role in the story. We learn later on that this scene is an elaborate charade, constructed by him to prod Jessaline and foil her mission. For if "[c]olonial fantasy is the continual dramatisation of emergence – of difference, freedom – as the beginning of a history which is repetitively denied," then Forstall must prevent such emergence (Bhabha 1996, 52). Jemisin envisions a beginning of a postcolonial emergence, both in the state of Haiti and in the lesbian love of Jessaline and Eugenie.

Tracing the Queer and Lesbian strand in Jemisin's texts suggests not only that sexuality is socially constructed but also, mutates mutandis, that societies are built on specific forms of sexual attachment. Both stories feature gay and lesbian characters respectively and thus avail themselves to Gay and Lesbian critical readings. I already performed some of this work in my reading of "The City Born Great." New York's awareness of being different and his "coming out" as the city's avatar and caretaker make the story a gay text (Tyson 2014, 315). Meanwhile, "The Effluent Engine" explores lesbian sexuality and companionship.

Haiti itself has been marked a queer utopian space in its political and literary history. Stereotypical renderings of the island (post-revolution) label it "bizarre, unnatural, odd, queer, freakish, or grotesque [. . .] erratic, and therefore unexplainable" (qtd. in Dillon and Drexler 2016, 1). Forstall follows suit, noting about Jessaline's sexuality, "I heard you Haitians were unnatural [. . .] but this?" (Jemisin 2018, 104). Haiti itself thus opens up a queer creative horizon. Jessaline and Eugenie's love safeguards Haiti as a political entity and simultaneously constitutes it as a utopian space. Jemisin makes this an explicitly lesbian utopia by tying Haiti's political fortunes to their companionship and ability to mutually support each other (Lothian, 2015, 71).

Jessaline's sexuality morphs from a narcissistic search for doppelgangers to a reverence for companionship. Initially, her sexuality is steeped in the language of conquest and aggression, a vocabulary that normally connotes patriarchal sexuality. Jessaline eventually overcomes these chauvinist gender patterns, which initially cause her to objectify Eugenie and judge her entirely on her looks. The free indirect discourse foregrounds Jessaline's masculine gaze: When Eugenie's "eyes sparkle" talking of chemistry, she appears "handsome" (Jemisin 2018, 84). Moments later, Eugenie looks "not handsome, pretty" and a "new angle [that] would be *most enjoyable* to pursue" (86). From a lesbian point of view, the story represents Jessaline's emancipation from these aggressive and ultimately destructive patterns.

The sentimental subplot reveals how Jessaline strings the inexperienced Eugenie along to acquire the engine, while the young scientist changes Jessaline's heart, or, in psychoanalytical terms, her object choice. Jessaline's stated sexuality involves strong-willed, independent partners. But finding the sleeping Eugenie reveals to her an inner longing for parity:

> *Beautiful*, Jessaline thought, and marveled at herself. Her tastes had never leaned toward women like Eugenie, pampered and sheltered and shy. She generally preferred women like herself, who could be counted upon to know what they wanted and take decisive steps to get it. Yet in that moment, gazing upon this awkward, brilliant creature, Jessaline wanted nothing more than to be holding flowers instead of a fake package, and to have come for courting rather than her own selfish motives. (102; original emphasis)

Nurturing is specific to lesbian companionship and exists in contradistinction to patriarchal logics of conquest and hierarchy. More, lesbian criticism conceives of companionship as "redemptive" and notes that "[i]nterdependency between women is the way to a freedom which allows the I to be, not in order to be used, but in order to be creative. This is a difference between the passive *be* and the active *being*" (Lorde 2007, 111). Having decided to flee to Haiti together to start a family, the ending offers Jessaline the "mad fortune" of emotional redemption from a life of perpetual masquerade (Jemisin 2018, 111). Jessaline becomes a "woman-identified woman," in the sense that "directing the bulk of one's attention and emotional energy to other women and having other women as one's primary source of emotional sustenance and psychological support" (Tyson 2014, 310).

Jessaline and Eugenie at different times in the narrative can only subsist through their mutual nurturing. In a chiasmic moment, Eugenie shows to have held the character trades Jessaline desired all along. She demands Jessaline keep her promise to marry her and settle into a domestic life, staking their future together on Jessaline giving up her spy games. "I am not fond of you keeping up this dangerous line of work," she notes, "[m]y inventions should certainly earn enough for the both of us [. . .]. So there is no reason for you to work when I can keep you in comfort for the rest of your days" (Jemisin 2018, 111). This final scene arguably qualifies the supposed radicalness of the story. For Haiti becomes a Black lesbian utopia which rests not on technology or revolution but interpersonal relations, domesticity and the traditional family unit.

Much of the work of situating Jemisin in literary history still awaits. While some have already focused on her substantial novel oeuvre, I have undertaken a foray into the rich archive of her short fiction in which many themes beckon to be explored. I conclude by sketching out some of them. "Too Many Yesterdays, Not Enough Tomorrows" (2004), for instance, is a dystopian allegory about computer-mediated communication and online culture. Her first story, "L'Alchimista" (2004) comments on food culture and invokes magical realism in the vein of Zora Neal Hurston (↗17 Hurston) and Octavia Wynbush (Lucky 2009, 251–252). In the story, Franca, an Italian master chef down on her luck, who works as a cook in a local kitchen, one day encounters a mysterious traveller, who offers her a satchel of strange items. The contents turn out to be ingredients for what we may call *cibus vita* – a culinary fountain of youth. The dishes Franca creates rejuvenate both her and the Faustian stranger, and they live happily ever after. Likewise, New York's ingestion, or lack thereof, in "The City Born Great" casts him as a late modern Bartleby (↗8 Melville), who survives without eating. He may likewise be examined from the vantage point of Disability Studies, seeing that his unhousedness may well result from an undiagnosed mental condition (Jemisin 2018, 18). The story may thus allegorize the current mental health crisis in the U.S. Ecocritical concerns permeate Jemisin's texts, and "The Effluent Engine" partially stakes Haiti's existence on the invention of a clean abundant energy source, made from the effluent of rum (80).

4 Bibliography

4.1 Works Cited

Ashcroft, Bill, et al. *Post-Colonial Studies: The Key Concepts*. 2nd ed. London and New York: Routledge, 2009.

Bereola, Abigail. "A True Utopia: An Interview With N. K. Jemisin," https://www.theparisreview.org/blog/2018/12/03/a-true-utopia-an-interview-with-n-k-jemisin/. *The Paris Review*, 3 Dec. 2018. (19 June 2021).

Bhabha, Homi. "The Other Question." *Contemporary Postcolonial Theory*. Ed. Padmini Mongia. London: Arnold, 1996. 37–54.

Canavan, Gerry, and Eric Carl Link. Introduction. *The Cambridge Companion to American Science Fiction*. Eds. Eric Carl Link and Gerry Canavan. New York: Cambridge UP, 2015. 1–16.

Coates, Ta-Nehisi. *Between the World and Me*. New York: One World, 2017.

Dery, Mark. *Flame Wars: The Discourse of Cyberculture*. Durham: Duke UP, 1994.

Duncan, Andy. "Alternate History." *The Cambridge Companion to Science Fiction*. Eds. Edward James and Farah Mendelsohn. Cambridge: Cambridge UP, 2003. 209–218.

Ellison, Ralph. *Invisible Man*. London: Penguin, 2001.
Eshun, Kodwo. "Further Considerations of Afrofuturism." *CR: The New Centennial Review* 3.2 (2003): 287–302.
Field, Douglas. "Beats, Black Culture and Bohemianism in Mid-Twentieth-Century New York City." *New York: A Literary History*. Ed. Ross Wilson. Cambridge: Cambridge UP, 2020. 240–251.
Fields, Karen E., and Barbara J. Fields. *Racecraft: The Soul of Inequality in American Life*. London: Verso, 2012.
Fleming Jr., Mike. "N.K. Jemisin Book Series 'The Broken Earth' Lands At Sony's TriStar In 7-Figure Deal; Author To Adapt," https://deadline.com/2021/06/the-broken-earth-nk-jemisin-sony-tristar-7-figure-deal-author-adapts-1234769656/. *Deadline*, 4 June 2021. (19 June 2021).
Freud, Sigmund. *The Freud Reader*. Ed. Peter Gay. New York: Norton, 1989.
Gates Jr., Henry Louis. *The Signifying Monkey: A Theory of African-American Literary Criticism*. Anniversary Edition. New York: Oxford UP, 2014.
Gilroy, Paul. *The Black Atlantic: Modernity and Double-Consciousness*. 1993. London: Verso, 2002.
Harman, Graham. *Speculative Realism: An Introduction*. London: Polity, 2018.
Imarisha, Walidah. Introduction. *Octavia's Brood: Science Fiction Stories from Social Justice Movements*. Eds. Adrienne Maree Brown and Walidah Imarisha. Oakland: AK, 2015. 3–5.
Jemisin, N. K. *How Long 'Til Black Future Month?* London: Orbit, 2018.
Jemisin, N. K. "The City Born Great," https://www.tor.com/2016/09/28/the-city-born-great/. *Tor. com*, 28 Sept. 2016. (19 June 2021).
Jemisin, N. K. *The City We Became*. London: Orbit, 2020.
Jemisin, N. K. *Press Kit*, https://nkjemisin.com/about/press-kit/. Dec. 2020. (19 June 2021).
Jones, Dalton Anthony. "Northern Hieroglyphics: Nomadic Blackness and Spatial Literacy." http://www.rhizomes.net/issue29/jones/index.html. *Rhizomes: Cultural Studies in Emerging Knowledges* 29 (2016). (19 June 2021).
Kaisary, Philip. *The Haitian Revolution in the Literary Imagination: Radical Horizons, Conservative Constraints*. Charlotte: U of Virginia P, 2014.
Karbiener, Karen. "Brooklyn and Manhattan." *Walt Whitman in Context*. New York: Cambridge UP, 2018. 15–26.
Leader-Picone, Cameron. *Black and More Than Black: African American Fiction in the Post Era*. Jackson: U of Mississippi P, 2019.
Lorde, Audre. *Sister Outsider: Essays and Speeches*. Berkley: Crossing P, 2007.
Lothian, Alexis. "Feminist and Queer Science Fiction in America." *The Cambridge Companion to American Science Fiction*. Eds. Gerry Canavan and Eric Carl Link. New York: Cambridge UP, 2015. 70–82.
Lucky, Crystal J. "African American Women Writers and the Short Story." *The Cambridge Companion to African American Women's Literature*. Eds. Angelyn Mitchell and Danielle K. Taylor. Cambridge: Cambridge UP, 2009. 245–261.
Maddock Dillon, Elizabeth, and Michael Drexler. *The Haitian Revolution and the Early United States: Histories, Textualities, Geographies*. Philadelphia: U of Pennsylvania P, 2016.
Mendelsohn, Farah. Introduction: Reading Science Fiction. *The Cambridge Companion to Science Fiction*. Eds. Edward James and Farah Mendelsohn. Cambridge: Cambridge UP, 2003. 1–14.
Nagel, James. *Race and Culture in New Orleans Stories: Kate Chopin, Grace King, Alice Dunbar-Nelson, and George Washington Cable*. Tuscaloosa: U of Alabama P, 2014.
Plato. *The Republic*. Ed. G.R.F. Ferrari. Transl. Tom Griffith. Cambridge: Cambridge UP, 2000.
Prucher, Jeff. "Steampunk," www.oxfordreference.com,https://www.oxfordreference.com/view/10.1093/acref/9780195305678.001.0001/acref-9780195305678-e-713. *The Oxford Dictionary of Science Fiction*. Ed. Jeff Prucher. Oxford UP, 2007. (19 June 2021).

Risam, Roopika, and Kelly Baker Josephs, eds. *The Digital Black Atlantic*. Minneapolis: U of Minnesota P, 2021.
Ryan, Marie-Laure. "Story/Worlds/Media: Tuning the Instruments of a Media-Conscious Narratology." *Storyworlds Across Media*. Eds. Marie-Laure Ryan and Jan-Noël Thon. Lincoln: U of Nebraska P, 2014. 25–49.
Saldívar, Ramón. "Speculative Realism and the Postrace Aesthetic in Contemporary American Fiction." *A Companion to American Literary Studies*. Eds. Caroline F. Levander and Robert S. Levine. Malden: Wiley Blackwell, 2015. 517–531.
Scofield, Martin. "Charles Chesnutt, Richard Wright, James Baldwin and the African American Short Story to 1965." *The Cambridge Introduction to the American Short Story*. Cambridge: Cambridge UP, 2006. 183–194.
Sexton, Jared. "Afro-Pessimism: The Unclear Word." http://www.rhizomes.net/issue29/sexton.html. *Rhizomes: Cultural Studies in Emerging Knowledges* 29 (2016). (19 June 2021).
Shaviro, Steven. *The Universe of Things: On Speculative Realism*. Minneapolis: U of Minnesota P, 2014.
Spivak, Gayatri Chakravorty. *In Other Worlds: Essays in Cultural Politics*. New York: Routledge, 2006.
Suits, Bernard. *Grasshopper: Games, Life and Utopia*. 1978. 3rd ed. Peterborough: Broadview, 2014.
Tyson, Lois. *Critical Theory Today: A User-Friendly Guide*. 3rd ed. New York: Routledge, 2014.
Warren, Kenneth W. *What Was African American Literature?* Cambridge: Harvard University P, 2010.
Weier, Sebastian. "Consider Afro-Pessimism / Responses by Christopher M. Tinson, Rinaldo Walcott and Elizabeth J. West." *Amerikastudien/American Studies* 59.3 (2015): 419–445.
West, Cornel. *Race Matters*. 25th Anniversary Edition. Boston: Beacon, 2017.
Whitman, Walt. *Leaves of Grass and Other Writings*. Ed. Michael Moon. New York: Norton & Company, 2002.
Wilson, Ross. Introduction: The History of New York Literature. *New York: A Literary History*. Ed. Ross Wilson. Cambridge: Cambridge UP, 2020. 1–16.
Womack, Ytasha. *Afrofuturism: The World of Black Sci-Fi and Fantasy Culture*. Chicago: Chicago Review P, 2013.
Yaszek, Lisa. "Afrofuturism in American Science Fiction." *The Cambridge Companion to American Science Fiction*. Eds. Eric Carl Link and Gerry Canavan. New York: Cambridge UP, 2015. 58–69.
Zamalin, Alex. Black *Utopia: The History of an Idea from Black Nationalism to Afrofuturism*. New York: Columbia UP, 2019.

4.2 Further Reading

Anderson, Reynaldo, and Charles E. Jones, eds. *Afrofuturism 2.0: The Rise of Astro-Blackness*. Lanham: Rowman and Littlefield, 2017.
Bould, Mark, et al. *Africa SF*. Vashon Island: Paradoxa, 2013.
Dillon, Grace L. *Walking the Clouds: An Anthology of Indigenous Science Fiction*. 2nd ed. Tucson: U of Arizona P, 2012.
Keeling, Kara. *Queer Times, Black Futures*. New York: New York UP, 2019.
Lavender, Isiah, III. *Afrofuturism Rising: The Literary Prehistory of a Movement*. Columbus: Ohio State UP, 2019.

Index of Names

https://doi.org/10.1515/9783110587647-037

Index of Subjects

https://doi.org/10.1515/9783110587647-038

List of Contributors

Jochen Achilles is Emeritus Professor of American Studies at the University of Würzburg and Adjunct Professor at the University of Mainz, Germany.

Nassim W. Balestrini is Professor of American Literature and Intermediality, and Director of the Centre for Intermediality Studies in Graz (CIMIG), at the University of Graz, Austria.

Ina Bergmann is Associate Professor of American Studies and co-head of Irish Studies Würzburg (ISWÜ) at the University of Würzburg, Germany.

Michael Basseler is Academic Manager in the International Graduate Centre for the Study of Culture (GCSC) at the Justus Liebig University Giessen, Germany.

Lynn Blin is Honorary Professor of Translation and English Linguistics (Stylistics) at Paul Valéry University in Montpellier, France. Until her retirement in 2015 she was co-coordinator of the Masters in Specialized Translation.

Mirosława Buchholtz is Professor and Chair of American Literature and Literary Translation at Nicolaus Copernicus University in Toruń, Poland.

Robert C. Clark is Assistant Professor of English at the College of Coastal Georgia, USA, and he serves as the Associate Editor for *Studies in the American Short Story*.

Philip Coleman is an Associate Professor in the School of English and a Fellow of Trinity College Dublin, Ireland.

Birgit Däwes is Professor of American Studies at Europa-Universität Flensburg, Germany, and currently serves as General Editor of the journal *Amerikastudien / American Studies*.

Susan Farrell is Professor of English at the College of Charleston, South Carolina, USA.

Lydia G. Fash is Assistant Professor of English at Simmons University, Boston, Massachusetts, USA.

Mark S. Graybill who previously served as President of the Flannery O'Connor Society, is Professor of English at Widener University in suburban Philadelphia, Pennsylvania, USA.

Michael Grimwood is Alumni Distinguished Undergraduate Professor of English at North Carolina State University, in Raleigh, North Carolina, USA, and co-founder and former co-director of the World Literature Program there.

Gero Guttzeit is Assistant Professor of English Literature at the University of Munich, Germany.

Johanna Hartmann is Assistant Professor (Wiss. Mitarbeiterin) of American Literature at the University of Halle-Wittenberg, Germany.

https://doi.org/10.1515/9783110587647-039

Alfred Hornung is Research Professor and Speaker of the Obama Institute for Transnational American Studies at the University of Mainz, Germany, and Honorary Chair Professor at Shandong University, Jinan, China.

Christina Maria Kelley received her master's degree from the Obama Institute for Transnational American Studies at the University of Mainz, Germany. She recently taught an English literature and writing course at the University of California, Davis, USA.

Holger Kersten is Professor of American Literature and Culture at the University of Halle-Wittenberg, Germany.

Kerstin Knopf is Professor of North American and Postcolonial Literary and Cultural Studies, and Director of the Institute for Postcolonial and Transcultural Studies (INPUTS) and the Bremen Institute of Canada and Québec Studies (BICQS) at the University of Bremen, Germany.

Timo Müller is Professor of American Studies at the University of Konstanz, Germany.

James Nagel is the Emeritus Eidson Distinguished Professor of American Literature at the University of Georgia, USA, and the President of the Society for the Study of the American Short Story.

Frank Newton is a PhD candidate researching periodicals and Indigenous modernities between 1890 and 1920s in the Obama Institute of Transnational American Studies at the University of Mainz, Germany.

Frank Obenland is Senior Lecturer for American Studies in the Obama Institute for Transnational American Studies at the University of Mainz, Germany.

Sascha Pöhlmann is Professor of North American Literature and Culture at the University of Innsbruck, Austria.

Erik Redling is Professor of American Literature and Managing Director of the Muhlenberg Center of American Studies at the University of Halle-Wittenberg, Germany.

Nele Rodiek is Research Associate and PhD candidate at the University of Halle-Wittenberg, Germany.

Stefanie Schäfer holds a venia legendi in North American Literature and Culture and is currently Marie Curie Fellow at the University of Vienna's Department of English and American Studies.

Oliver Scheiding is Professor of North American Literature and Early American Studies in the Obama Institute for Transnational American Studies at the University of Mainz, Germany.

Damien B. Schlarb is Assistant Professor (limited term) of American Studies in the Obama Institute for Transnational American Studies at the University of Mainz, Germany.

Philipp Schweighauser is Professor of North American and General Literature at the University of Basel, Switzerland.

Julia Straub is Professor of Modern English Literature at the University of Fribourg, Switzerland.

Ross K. Tangedal is Assistant Professor of English and Director of the Cornerstone Press at the University of Wisconsin–Stevens Point, USA.

Steven C. Tracy is Distinguished Professor of Afro-American Studies at the University of Massachusetts at Amherst, and Distinguished Overseas Professor at Central China Normal University, Wuhan, China.

Sabrina Völz is Senior Lecturer at Leuphana University Lüneburg, Germany, a post-doctoral researcher, and co-editor of the American Studies Blog (http://blog.asjournal.org/).